Animal Social Behavior

Animal Social Behavior

James F. Wittenberger

University of Washington

DUXBURY PRESS · *Boston*

Animal Social Behavior was prepared for publication by the following people:
Production Editor: Ellie Connolly
Copy Editor: Carol Beal
Interior Designer: Joanna Prudden Drummond
Cover Designer: Clifford Stoltze
Cover photograph by: Gunter Konrad
Line drawings and graphs by: F. W. Taylor Associates

Chapter opening photo credits
Chapter one: Photo courtesy of the American Museum of Natural History, L'Anthropologie, Masson Publishers, Paris.
Chapter two: Photo by Eric Hosking.
Chapter three: Photo by Douglas Faulkner.
Chapter four: Photo by Toni Angermayer.
Chapter five: Photo by Finn Allan/Frank W. Lane.
Chapter six: Photo by Lincoln P. Brower.
Chapter seven: Photo by M. H. MacRoberts.
Chapter eight: Photo by Richard R. Tenaza.
Chapter nine: Photo by Franz Sieber.
Chapter ten: Photo by W. James Erckmann.
Chapter eleven: Photo by Gunter Konrad.
Chapter twelve: Photo by John B. Free.
Chapter thirteen: Photo by Ron Tilson.
Chapter fourteen: Photo by Marjorie Shostak/Anthro-Photo.

Library of Congress Cataloging in Publication Data

Wittenberger, James F
Animal social behavior.

Bibliography: p.
Includes index.
1. Social behavior in animals. I. Title.
QL775.W57 591.51 80–27980
ISBN 0–87872–295–5

Printed in the United States of America
1 2 3 4 5 6 7 8 9 — 85 84 83 82 81

Contents

Preface ix

Chapter one
Introduction 2

Questions of why and two kinds of answers 4
Sociobiology: A revolution in behavioral biology 5
Criticisms of sociobiology 9
Tenets of sociobiology 15
Logic of sociobiology 15
General methodology 19
Conclusion 23

Chapter two
Selection theory 24

Theory of natural selection 25
Some basic molecular genetics 26
A molecular view of natural selection 31
Relation between genotype and phenotype 35
Nature-nurture dichotomy 44
Fitness 45
Analyzing adaptive value 47
Adaptive landscapes 49
Types of individual selection 53
Kin selection 60
Group selection 63
Conclusion 72

Chapter three
Altruism: Does it exist? 74

Categories of behavioral interactions 75
Reciprocal altruism 76

Alarm signals 80
Underutilization of resources and reproductive curtailment 92
Helping others reproduce 95
Conclusion 108

Chapter four
Cooperation 110

Cooperative defenses against predation 111
Cooperative enhancement of foraging efficiency 122
Reducing the costs of grouping behavior 129
Conclusion 133

Chapter five
Aggression 134

What is aggression? 135
A controversy: Opposing motivational models 139
Developmental origins of aggressiveness 142
Proximate causation 147
An evolutionary model of aggression 163
Benefits of aggression 165
Costs of aggression 178
Who wins? 179
Tactics 180
Conclusion 191

Chapter six
Environmental bases of behavior 194

An overview of ecological problems faced by animals 195
Competition and niche space 197
Resource characteristics 206
Plant-herbivore interactions 213
Predator-prey interactions 221
Conclusion 245

Chapter seven
Territoriality 246

The concepts of home range and territory 247
History of the territory concept 251
Why exclusive territories evolve 252
What is defended? 254
Effect of territoriality on population density 276

Defensibility theory 285
Who should be excluded? 295
Conclusion 299

Chapter eight
Coloniality 302

Types of clumped spacing patterns 303
Costs of coloniality 305
Some extremes in colony sizes 310
Benefits of coloniality 311
Effects of predation on colony organization 328
Effects of competition on colony organization 337
Conclusion 342

Chapter nine
Life history patterns and parental care 344

Life history patterns 345
Parental behavior 362
Brood parasitism 383
Parent-offspring conflict 390
Conclusion 394

Chapter ten
Sex and sexual selection 396

Why sex? 397
Why two sexes? 404
Sex ratio theory 405
Why male competition and female choice? 415
Female competition and male choice 417
Evolution of secondary sexual characteristics 420
Intrasexual selection 422
Intersexual selection 431
Conclusion 444

Chapter eleven
Mating systems 448

Some definitions 449
Territorial polygyny 452
Promiscuity 478
The evolution of leks 494
Polyandry 498

✓Monogamy 504
Conclusion 512

Chapter twelve
Insect sociality 514

Degrees of sociality 515
Caste differentiation in eusocial insects 517
Theories for the evolution of sterile castes 524
Predictions and evidence: The indirect (kin) selection hypothesis 527
Queen control of caste differentiation: Evidence for parental manipulation 542
Genetic models 549
Conclusion 549

Chapter thirteen
Mammalian sociality 552

Costs and benefits of group living 553
Determinants of group size 574
Dynamics of group size 579
Social dominance 587
Conclusion 596

Chapter fourteen
Human sociality 598

Processes of cultural change 599
Conclusion 611

Glossary 612

References 623

Index 705

Preface

An evolutionary approach to animal social behavior has been with us since Charles Darwin first published *On the Origin of Species*, but scientific interest in the approach has mushroomed during the past two decades. An increasing number of university courses cover this topic, often under the rubric of *sociobiology*, and yet no adequate textbook exists to accompany such courses. In writing the present book, I hope to fulfill the need for such a book by providing a balanced account of current knowledge and theoretical issues. The book is intended for advanced undergraduate and graduate students in behavioral biology, but I hope students with less background and professional scientists will also find it useful. With those goals in mind, I have included considerable discussion of basic concepts in ecology and evolution that are necessary for fully understanding current sociobiological thought. At the same time, I have included extensive bibliographic citations to help both students and professionals gain easier access to the primary scientific literature.

My general organizational approach has been to present every major hypothesis advanced to explain a given behavioral phenomenon and then discuss the evidence and arguments supporting or refuting each hypothesis. Not all pertinent evidence could be discussed fully or even mentioned, since the relevant scientific literature is enormous, but I have attempted to provide a balanced and representative discussion of each issue.

The order in which topics are presented is a matter of personal preference as well as logical progression. Sociobiology is a multidimensional web of interconnected ideas and facts that is not easily flattened into a linear sequence. Hence compromises and value judgments were necessary in ordering topics of discussion. I have included some cross-referencing to aid in connecting related topics that are not given in linear sequence, but many more cross-references could have been included.

No book of such broad scope is possible without the aid of numerous individuals and organizations. For many, a short "thank you" is wholly inadequate for expressing my sense of obligation for their contributions, and their true reward can only be the availability of a text that aids them in teaching and research. I hope the end product has justified their willingness to help.

This book began as a joint project by William J. Hamilton III and myself. Without Bill Hamilton's instigation, it would never have been written. After Hamilton withdrew from the project during its early planning stages, I would not have continued the project had it not been for his encouragement and the encouragement of Gordon H. Orians and Jerry Lyons (then managing editor of Duxbury Press). To them I owe an eternal debt of gratitude.

Many persons read part or all of the manuscript and made numerous helpful comments that greatly improved the book's quality. Although I did not always follow their suggestions, the value of their comments was immeasurable. I thank the following persons for commenting on various drafts of one or more chapters (indicated in parentheses): Jerram L. Brown (1–4, 9), Paul W. Ewald (7), Mart R. Gross (9), I. Lorraine Heisler (2, 10, 14), John L. Hoogland (8), George L. Hunt, Jr. (3, 4, 7, 8, 11, 13), Donald A. Jenni (11), Kenneth E. Moyer (5), Gordon H. Orians (3–8, 11, 13, 14), Wanda K. Pleszczynska (11), Sievert Rohwer (1, 10), Paul W. Sherman (3, 4), David Sloan Wilson (2), and Edward O. Wilson (12). I also thank those who reviewed the entire manuscript: Terry Christiensen, Tulane University; Martin Day, McMaster University; H. B. Graves, Pennsylvania State University; Richard Howard, Purdue University; F. E. Wasserman, Boston University; and Randy Thornhill, University of New Mexico.

Numerous persons provided unpublished or difficult-to-obtain materials and manuscripts that helped me to remain more abreast of current developments in sociobiology. For this help I thank Jack W. Bradbury, Jerram L. Brown, Robert K. Colwell, Paul W. Ewald, John C. Fentress, Mart R. Gross, William J. Hamilton III, I. Lorraine Heisler, John L. Hoogland, George L. Hunt, Jr., Devra G. Kleiman, Gerd Knerer, Adriaan Kortlandt, Burney J. Le Boeuf, Edward H. Miller, J. Peter Myers, Gordon H. Orians, Wanda K. Pleszczynska, David F. Rhoades, Peter J. Richerson, Matthew P. Rowe, William A. Searcy, Paul W. Sherman, Randy Thornhill, Ronald L. Tilson, Kenneth Yasukawa, and Thomas M. Zaret.

I am grateful to the following individuals and organizations for granting permission to reproduce copyrighted materials and, in many cases, for supplying glossy prints of photographs: Academic Press, Inc., New York and London; Aldine Publishing Co.; Finn Allan; Stuart A. Altmann; American Association for the Advancement of Science; American Institute of Biological Sciences; American Museum of Natural History; American Ornithological Union; Toni Angermayer; Annual Reviews, Inc.; Ballière Tindall, Ltd.; Frank V. Blackburn; Blackwell Scientific Publications, Ltd.; P. Dee Boersma; K. T. Briggs; E. J. Brill Co.; Lincoln P. Brower; Jerram L. Brown; Donald Bruning; Jane Burton; California Department of Fish and Game; Cambridge Entomological Club; Cambridge University Press; F. Lynn Carpenter; A. Christiansen; Martin L. Cody; Bruce Coleman, Ltd., New York and London; William Collins Sons and Co., Ltd.; Columbia University Press; Cooper Ornithological Society; Edward S. Deevey, Jr.; Nancy S. DeVore; Robin I. M. Dunbar; Duxbury Press; G. Gray Eaton; Ecological Society of America; Malcolm Edmunds; Thomas Eisner; W. James Erckmann; Douglas Faulkner; Sally Faulkner; Christopher J. Feare;

John B. Free; F.R.E.E., Ltd.; W. H. Freeman and Co., Publishers; Gautier-Villars; Valerius Geist; Frances Hamerstrom; William D. Hamilton; William J. Hamilton III; Harper and Row, Publishers, Inc.; Harvard University Press; Berthold K. Hölldobler; John L. Hoogland; Eric Hosking; Houghton Mifflin Co.; Richard C. Howard; Sarah Blaffer Hrdy; Japan Monkey Center; Robert L. Jeanne; S. Karger, A. G.; C. B. Keil; Alan C. Kemp; Gerd Knerer; Günter Konrad; Adriaan Kortlandt; John R. Krebs; Hans Kruuk; Hans Kummer; Laboratory of Ornithology, Cornell University; Frank W. Lane Agency; Baron Hugo van Lawick; Burney J. Le Boeuf; Walter Leuthold; Hans Löhrl; Terry Mace; MacMillan Journals, Ltd., Michael H. MacRoberts; John Maynard Smith; Methuen & Co., Inc.; Donald E. Mullins; Museum National d'Histoire Naturelle; Norman Myers; National Audubon Society; National Film Board of Canada; A. v.d. Nieuwenhuizen; W. W. Norton and Co.; Hideyuki Ohsawa; Oxford University Press; Jaspar Parsons; Hans Pfletschinger; Nicholas Picozzi; Plenum Publishing Corp.; Axel Poignant; Roslyn Poignant; George V. N. Powell; Neil L. Rettig; Richard Rosolek; Edward S. Ross; Royal Society of London; Ivan Sazima; W. B. Saunders; George B. Schaller; Paul W. Sherman; Franz Sieber; Sinauer Associates, Inc.; Neal Smith; Smithsonian Institution Press; Society for the Study of Evolution; Springer-Verlag; Stony Brook Foundation, Inc.; Richard R. Tenaza; Ronald L. Tilson; Robert L. Trivers; University of Chicago Press; Verlag Paul Parey; Kenneth E. F. Watt; Wolfgang Wickler; R. Haven Wiley; George C. Williams; David Sloan Wilson; Wilson Ornithological Society; Wisconsin Primate Center; Wistar Institute Press; Glen E. Woolfenden; Richard I. Yeaton; Zoological Society of London.

I am indebted to the many persons at Duxbury Press who helped the book see the light of day. I especially thank Jerry Lyons for encouragement and advice; Jean-François Vilain for providing encouragement and good humor during a tight and sometimes tense production schedule and for supervising production; Sylvia Dovner for shepherding the book through production and keeping publication on schedule; Carol Beal for editing the manuscript; Jane Lovinger for helping put the artwork manuscript together; Jane MacQuarrie and Victoria Keirnan for assisting with the art manuscript and permissions; Ellie Connolly for finalizing the artwork and text manuscripts; and the many other staff persons whose names I never learned for their contributions during all phases of production and promotion. I would also like to thank my friend Sally White for helping organize my permissions file and Sally Beane for heroically typing the entire manuscript despite the obscurities of my handwriting.

Finally, I owe a special debt of gratitude to my family and friends, most of whom have nothing to do with biology. Their encouragement, friendship, and good company provided support and diversion through the long and often unrewarding process of writing the manuscript. To them I gladly dedicate this book.

James F. Wittenberger
Seattle, Washington

Animal Social Behavior

Introduction

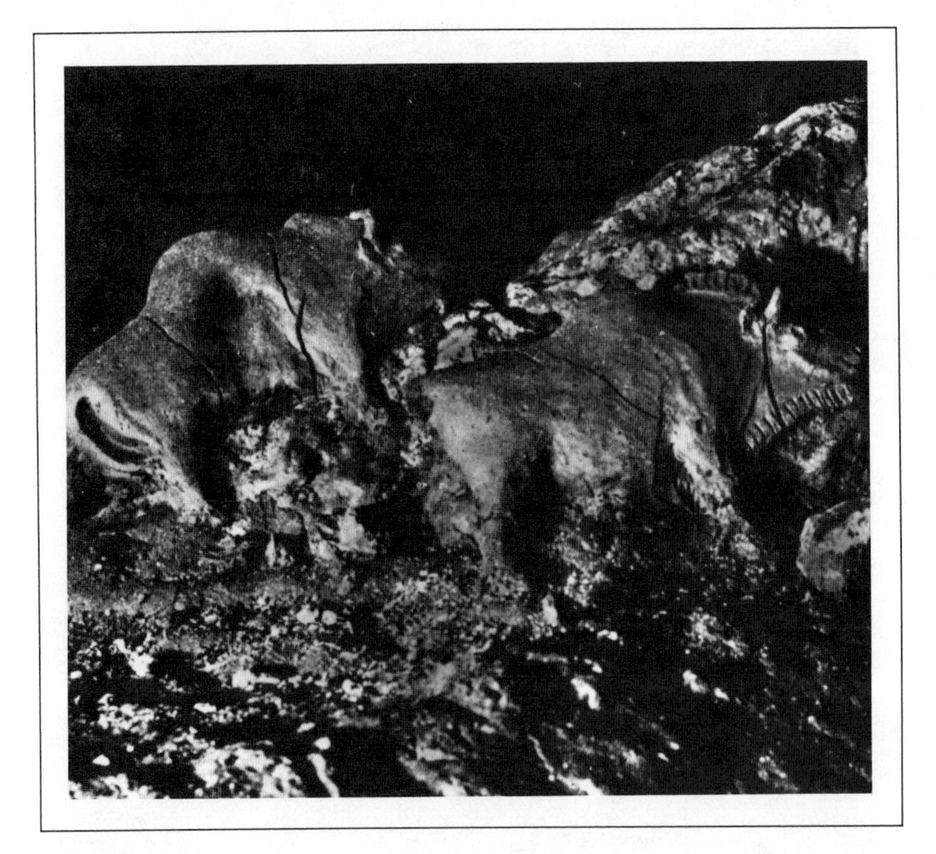

Animals have always been intimately involved in human affairs, an involvement long embodied in humanity's art and religion. Game animals were once an important source of food and raw materials, predatory and venomous animals were a threat to human survival, and scavenging animals were a threat to food stores and human habitations. Other animals were parasitic on the human body, competed with humans for food, or ravaged human crops or flocks. Thus animals were an important part of early human environments, and people depended on knowledge of them to survive and reproduce. This knowledge often became embodied in magical and religious rites, in which art played an important role (Photo 1–1).

Today the majority of people are insulated from direct contact with wild animals. We now rely on domestic animals for food and raw materials. In many parts of the world we have decimated or eliminated the populations of many animals capable of inflicting harm upon us. Technology and institutions take care of our basic needs in exchange for labor or special skills, making knowledge of animal behavior unnecessary for most of us to survive and reproduce. Why, then, do we still seek a better understanding of animal behavior?

Part of the reason is simply to satisfy our curiosity, a curiosity instilled in us because it was once important to our survival. Spending time outdoors and seeing behavioral events happening around us sparks curiosity and raises questions that demand answers. People derive pleasure from understanding the world around them, and animals are still an important part of that world. If we could not satisfy our curiosity about animals, many of us would find life less enjoyable and would feel that the quality of our lives had depreciated substantially.

There is also a more practical reason for studying animal behavior. Despite our walls and pavement, we are not as insulated from the world of animals as we might think. Animals still provide our industries with raw materials. They have a major impact on our agricultural prosperity. They are the source of much medical knowledge and the inspiration for some technological innovations. They are vectors of disease and parasites. They are plant pollinators, nutrient recyclers, sewage decomposers, and pollutant disposers. In short, they are an essential and integral part of natural ecosystems, without which we could not long survive. Our future prosperity is closely tied to the continued maintenance and well-being of our surrounding environments, and we cannot hope to perpetuate the quality of those environments indefinitely without a clear understanding of how

Chapter one

Photo 1–1

Clay relief of bisons mating.

In earlier times human survival and reproduction were greatly affected by animal behavior, and, consequently, animals were a frequent subject of early art. This photograph shows a famous relief that was modeled in clay by a Magdalenian-era artist at Le Tuc d'Audoubert, France, about 10,000 B.C. It was probably used in magical rites to encourage the procreation of game.

Photo courtesy of The American Museum of Natural History and L'Anthropologie, Masson Publishers, Paris.

they are organized and why they change. We can no longer afford the attitude that animals exist solely to fulfill human needs. We must recognize that animals have needs of their own, and we must allow animals to fulfill those needs. This will only be possible if people gain a better awareness of what animals are doing in their natural environments, and why.

Questions of why and two kinds of answers

Why *do* animals do what they do? Behavioral biologists have traditionally offered two fundamentally different kinds of answers. One involves the physiological and biochemical mechanisms underlying behavior, while the other involves its evolutionary origins.

The most direct answer is concerned with immediate or **proximate causation.** An animal's behavior results from internal motivational states or physiological requirements and is elicited by external stimuli appropriate for satisfying its needs (Marler and Hamilton 1966; Hinde 1970). Early models of proximate causation ignored internal states and focused on stimulus-response interactions. Animals were viewed by many scientists as black boxes, lacking spontaneity and capable of giving only automatic responses to external stimulus situations. Today animals are given credit for initiating and modifying their behavior spontaneously. They are no longer viewed as automatons, and their internal states are now recognized as important components of proximate causation.

Models of proximate causation explain behavior in terms of physiological or motivational mechanisms, but they fall short of explaining why animals do what they do because they fail to explain why the internal mechanisms are organized the way they are in the first place. This shortcoming has led to explanations based on early development, or **ontogeny.** Animals are organized as they are and behave as they do because of previous developmental processes. They learned to behave in certain ways or were born with instincts that are automatically evoked by specific stimuli. In short, the internal mechanisms arose from the carefully orchestrated physiological and biochemical processes of development and maturation.

Developmental explanations are really just an indirect way of describing proximate causation. They are still based on physiological mechanisms, and they cannot explain why the developmental processes are organized and channelized the way they are. These processes are ultimately under the influence of each animal's genetic material, or **genome.** The directions and timing of developmental processes are often affected by external events and experiences as well as by the genome, but the *susceptibility* of developmental processes to such influences is programmed genetically. Why development proceeds as it does—and hence why an animal's internal physiology is organized as it is—therefore depends to a considerable extent on the animal's genetic makeup, which in turn is a result of evolutionary processes. Developmental processes, physiological mechanisms, and motivational states have evolved through natural selection to increase each animal's lifetime reproductive success. Whatever processes or mechanisms best enhance that success are favored by selection and consequently possess greater **adaptive value** than alternative processes or mechanisms that might have evolved in their place. Animals ultimately behave as they do because of evolutionary processes, implying that the **ultimate causation** of behavior must be analyzed in evolutionary terms.

Thus studies of animal behavior are conducted at two different levels of causation. Studies of proximate causation are concerned with the physiological and motivational mechanisms underlying behavior and sometimes with the ontogeny of behavior. Studies of ultimate causation, on the other hand, are concerned with the evolutionary origins of behavior. This distinction is important, since data from studies of proximate causation usually have only limited value for understanding ultimate causation, and vice versa.

Sociobiology: A revolution in behavioral biology

A revolution happened in biology during the 1960s and 1970s. Behavioral biologists began to realize how remarkably well animal social behavior is attuned to environmental and social circumstances and began to develop a clearer understanding of why this is so. That led to an explosive surge of theoretical and empirical investigations devoted to the adaptive significance of complex animal social systems, and new information and new ideas are now proliferating so rapidly that books and articles are often out of date before they are printed. A conceptual revolution is truly happening, and that revolution is sociobiology.

Scope of sociobiology

Sociobiology is an amalgamation of behavioral biology, modern population ecology, and evolutionary theory (Figure 1–1). Its central concern is to understand how and why animal social behavior has evolved. It includes the spacing, mating, and parental behavior of so-called solitary animals as well as that of social animals, since even solitary animals interact with each other to some degree. As diagrammed in Figure 1–1, sociobiology does not encompass the study of human behavior, a point of view that deviates from current usage of the term. Indeed, to many social scientists sociobiology is virtually synonymous with *human sociobiology*, or the sociobiology of human behavior. To avoid implications that sociobiology will eventually take over the social sciences, an erroneous assessment in my opinion, I suggest that **humanology** is a better term for referring to the scientific study of human behavior. Sociobiology can make important contributions to humanology, as can all the social sciences and even the humanities, but it does not encompass humanology.

Although sociobiology focuses primarily on ultimate causation, proximate mechanisms do not lie entirely beyond its purview. Physiological mechanisms and developmental processes could be profitably viewed from an evolutionary perspective, even though the organizing principles underlying them have barely been viewed from that perspective up to now. Further, sociobiology is not an island in a scientific ocean; it is part of the fabric that interconnects all of science, and as such the interconnections need to be kept clear. Thus sociobiology does not have discrete boundaries. It is interwoven with many disciplines, including physiology, embryology, and ultimately anthropology, psychology, and even history.

Figure 1–1 *The conceptual organization of sociobiology and its place in contributing to studies of human behavior.*

Theories of sociobiology arise from the interrelated disciplines of behavioral biology, population ecology, and evolutionary biology. Topics of particular relevance to sociobiology are indicated along connecting arrows in the diagram. Sociobiology may make important contributions to the study of human behavior, which I refer to as humanology, but so also will the social sciences and humanities.

Source: Based on E. O. Wilson, *Sociobiology: The New Synthesis* (Cambridge, Mass.: Belknap Press of Harvard University Press, 1975), p. 5. Reprinted by permission.

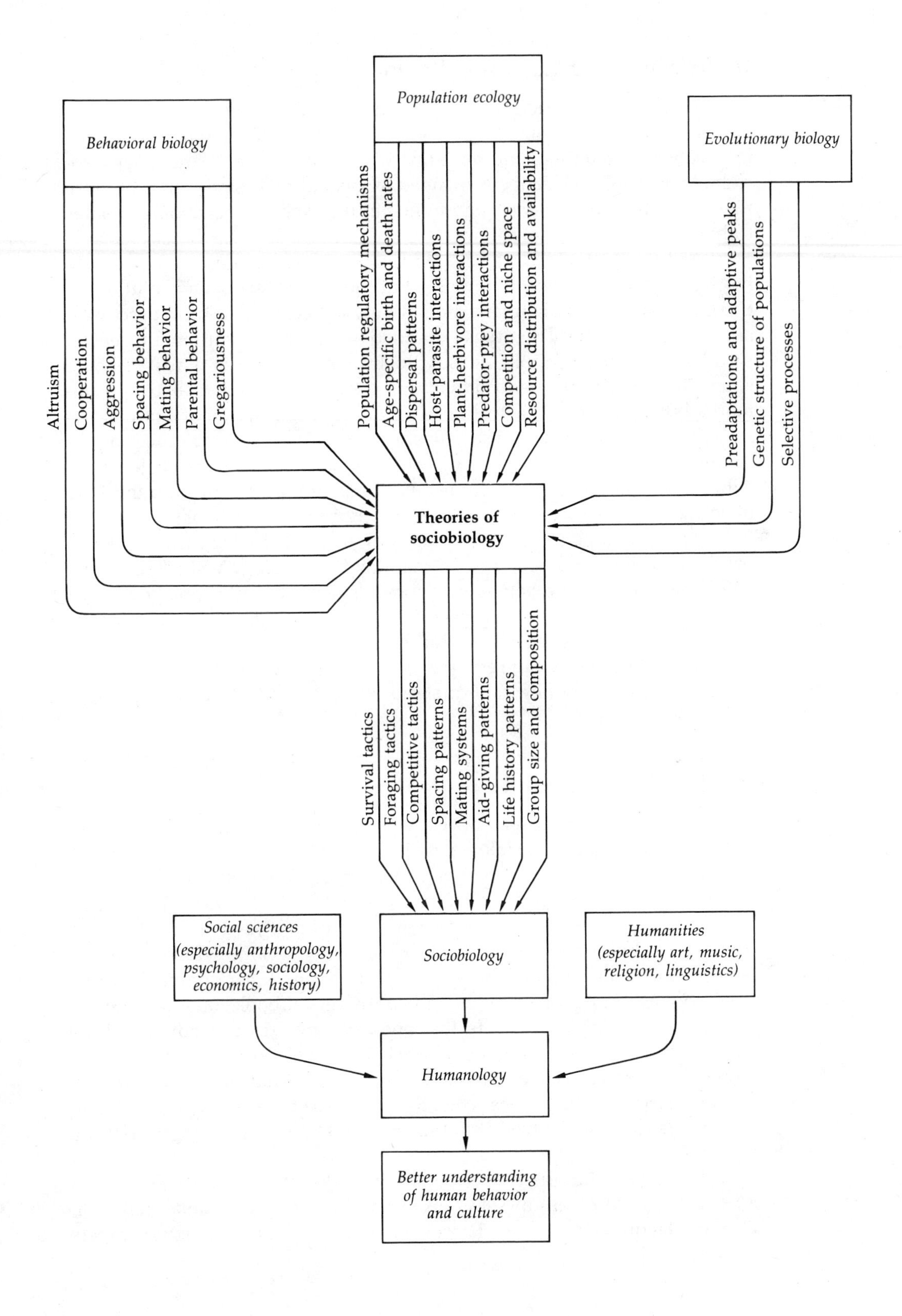
Behavioral biology
Altruism
Cooperation
Aggression
Spacing behavior
Mating behavior
Parental behavior
Gregariousness
Population ecology
Population regulatory mechanisms
Age-specific birth and death rates
Dispersal patterns
Host-parasite interactions
Plant-herbivore interactions
Predator-prey interactions
Competition and niche space
Resource distribution and availability
Evolutionary biology
Preadaptations and adaptive peaks
Genetic structure of populations
Selective processes
Theories of sociobiology
Survival tactics
Foraging tactics
Competitive tactics
Spacing patterns
Mating systems
Aid-giving patterns
Life history patterns
Group size and composition
Social sciences (especially anthropology, psychology, sociology, economics, history)
Sociobiology
Humanities (especially art, music, religion, linguistics)
Humanology
Better understanding of human behavior and culture

Implications for human behavior

This book is principally concerned with the evolution of social behavior in nonhuman animals, but the theories discussed in it have potential implications for understanding human behavior as well. Since many sociobiologists are now actively engaged in evaluating the relevance of these theories to human behavior, an understanding of how sociobiology relates to human behavior is important.

As I see it, the ideas developed from research on animal behavior do have potential relevance to humans, but the present scope of evolutionary theory is too narrow. Humans are strongly affected by cultural processes, and until evolutionary theory is expanded to encompass cultural processes, sociobiology cannot apply directly to humans. There is a real need to elucidate the relevance and limitations of natural selection theory to humans, but there is also a real need to develop a valid theory of cultural selection as it affects human societies. Anthropologists are destined to play an important role in formulating such a theory, and when a theory of cultural selection does emerge, it must be incorporated into the thinking of sociobiologists who are actively studying human behavior.

It is to be hoped that one day a true synthesis of biological and cultural processes will be attained, and on that day humanology will come of age. Until then, a dialogue between sociobiologists and anthropologists must be maintained. An attempt at wedding the two fields of thought will be presented in the last chapter. Until such a merger is achieved, theories developed for animals can represent a source of ideas for studying human social behavior, but they cannot stand alone as a valid body of thought applicable to humans.

Historical background

Sociobiology has at its roots Darwin's (1859) contention that behavior, like morphology and physiology, evolved through natural selection. However, interest in the evolution of behavior emerged only slowly, remaining dormant, for the most part, until David Lack (1954, 1966, 1968) championed the view that many aspects of behavior could be analyzed and understood from an evolutionary perspective. Lack's views became widely accepted in the 1950s and were not seriously challenged until V. C. Wynne-Edwards (1962) published his provocative and controversial book *Animal Dispersion in Relation to Social Behaviour.* In that book Wynne-Edwards advocated the view that virtually all animal social behavior evolved to prevent overexploitation of food resources. His argument required that animals voluntarily curtail reproductive activities when food is scarce to benefit the population as a whole, a notion directly contrary to the Darwinian theory of natural selection.

To explain the origin of such restraint, Wynne-Edwards postulated a pervasive role of group selection in the evolution of animal behavior. **Group selection** is a process whereby whole groups or populations compete with and replace one another through evolutionary time. It favors behavioral traits beneficial to entire groups or populations, unlike Darwinian natural selection, which can only favor traits beneficial to individuals.

The challenge of Wynne-Edward's ideas spurred an immediate and extensive reappraisal of both evolutionary theory itself and the adaptive significance of all aspects of animal behavior. The result has been an overwhelming reaffirmation of natural selection theory as initially conceived by Darwin and espoused by Lack (Lack 1966; Wiens 1966; Williams 1966a; Lewontin 1970).

In the 1960s a renewed impetus for evolutionary analyses of animal behavior developed from the theoretical perspectives initially set forth by Lack (1954) and the earlier work of R. A. Fisher (1930) (e.g., Brown 1964, 1969a, 1974; Crook 1964, 1965; Hamilton 1964; Verner 1964; Lack 1966, 1968; Orians 1969). This impetus recently culminated in E. O. Wilson's (1975a) encyclopedic compendium *Sociobiology: The New Synthesis,* which laid the cornerstone for an interdisciplinary approach to both animal and human behavior. Wilson's views on human behavior were quickly attacked on both political and scientific grounds (Sociobiology Study Group of Science for the People 1975, 1976), and his lectures were disrupted and picketed by demonstrators. The sociobiological approach has instigated a whole new way of analyzing human behavior (see Barash 1979a; Chagnon and Irons 1979; Freedman 1979; Devore 1980) while at the same time becoming the centerpiece of vigorous debate about how human behavior evolves.

What is this phenomenon and why has it sparked so much antagonism? Is sociobiology a new and more subtle justification of discriminatory practices, as its critics maintain, or is it a new and exciting form of scientific enlightenment? This book explains what sociobiology is about, and I hope it will provide a sound basis upon which readers can answer this question for themselves.

Criticisms of sociobiology

Sociobiologists do not agree among themselves about many theoretical issues regarding how evolutionary processes affect behavior, though they do agree that studying the evolution of behavior is a legitimate area of inquiry. Attempts to apply evolutionary theory to human behavior have met with severe criticisms on both political and scientific levels. The principal cause for controversy is not the synthesis of sociobiology itself but rather its extrapolation to human behavior and social organization. The central issue is whether sociobiological principles should be placed at the foundation of the social sciences and humanities or be kept confined to the realm of nonhuman animals.

The most outspoken critics of sociobiology have been members of the Sociobiology Study Group of Science for the People at Harvard University. Membership in the group is heterogeneous, consisting in 1976 of eight professors from several scientific disciplines, at least one psychiatrist, a secondary school teacher, students, and graduate research assistants (Wilson 1976). They are a self-avowed political enclave, espousing the view that human behavior is highly malleable and completely under the control of environmental influences. They view sociobiology as a threat to their Marxist political philosophy because sociobiologists ascribe human behavior partially to genetic influences. But the issue is not entirely a political one. Sociobiology is not a political philosophy; it is a science, subject to the same rules of verification required of any other science. For this reason the Sociobiology Study Group has also attempted to discredit sociobiology as a legitimate science.

In the following sections the main objections raised by the Sociobiology Study Group (1975, 1976) are discussed, as well as some of the replies given to them by two of sociobiology's more outspoken proponents, Edward O. Wilson (1975b, 1976a) and David P. Barash (1979a). The discussions in the sections below only summarize the principal arguments, and the serious student is urged to read the original texts, including the relevant parts of Wilson's (1975a) book, to gain a better understanding of the issues.

Members of the Sociobiology Study Group are not the only critics of sociobiology. Criticisms have also been leveled at it by many social scientists, particularly anthropologists, who question the relevance of natural selection theory to human societies. The interested reader should consult Sahlins (1976), Caplan (1978), Gregory, Silvers, and Sutch (1978), and Washburn (1978).

Political criticisms

Sociobiology is a new, more subtle doctrine of biological determinism. Biological determinism is basically the idea that human behavior is so strongly governed by genetics that it is largely inflexible and cannot be changed to create a better personality or society. Biological determinism is a denial of free will and progressive social change.

According to sociobiologists, this objection rests on a naive view of how genetics influences behavior. It resurrects the now defunct nature-nurture controversy and makes the claim that behavioral biologists still separate behaviors into learned and innate components. Sociobiology is not built on the premise that behavior is genetically determined or inflexible. It depends only on the premise that genetics *influences* behavior to some degree. Behaviors develop through complex processes that are influenced at every stage by both genetics and environment, and the interaction between genetic and environmental factors is impossible to separate into component parts at any stage of development. A behavior need not be entirely determined or even strongly influenced genetically before natural selection can affect it. Natural selection operates on any genes that affect behavior, however small their effects might be. It favors a *propensity* for animals to respond in adaptive ways in any given situation. The propensity can be either strong or weak, depending on the species and behavior involved. That is much different than saying that natural selection rigidly molds behavior into fixed patterns. Sociobiologists recognize the undeniable importance of environmental factors in promoting behavioral flexibility, especially among humans, though they do argue that behavioral flexibility is itself an adaptive attribute.

Sociobiology is racist and sexist. The members of the Sociobiology Study Group claim that sociobiology is a justification for the status quo and that it represents an insidious return to Social Darwinism. They fear that it will be used to justify social inequities based on race or sex. They claim that sociobiology gives these inequities a biological basis and hence makes them seem justifiable and inevitable. They liken sociobiology to earlier evolutionary doctrines that were used to justify poverty, starvation, and Nazi extermination camps as appropriate agents for cleansing society of biologically inferior or unfit individuals (e.g., Spencer 1851; Lorenz 1940; see also Miller 1976).

Sociobiology is not a political philosophy, and sociobiologists as a group do not advocate such views. Any science can be distorted and misused to promote undesirable ends, but this does not make the science less valid. Current attempts by sociobiologists to understand human behavior have sought to identify universal human attributes, traits that are shared by all people of all cultures. Since racism and prejudice are based on perceived differences among categories or classes of people (Allport 1954), the identification of commonalities should, if anything, act as an antidote to racism (Barash 1979a). Even if some of these universals turn out to be socially or politically undesirable, sociobiology does not justify their existence or deny opportunities for social change. It simply offers a better understanding that could facilitate rather than impede social progress.

Most sociobiologists accept the notion that sexual differences exist in behavior as well as in morphology of both animals and people. They ascribe some, though certainly not all, of these differences to the actions of natural selection. They do not maintain that one sex is biologically superior to the other; nor do they condone social inequities based on sexual discrimination. On the contrary, sociobiologists consider the sexes of equal and complementary importance in the roles they play within social systems.

Overview. The political dialogue has been a heated one. The Sociobiology Study Group's crusade has taken on aspects of academic vigilantism, and Wilson (1976a) has accused them of launching a malicious personal vendetta against him and his ideas. At the very least, the political arguments have been made on an emotional as well as a rational level. This turn of events is unfortunate, as the prospect for a science of human sociobiology does raise important ethical questions deserving of serious discussion (Caplan 1976). For example, what significance, if any, does scientific theory and empirical fact have for ethical, religious, and political issues? What obligations, if any, do scientists have in guarding against the abuse or misuse of their speculations about human behavior? These are not easy questions to resolve, and they will certainly not be resolved by emotional exchanges of accusations and counteraccusations.

It is also unfortunate that Wilson's (1976a) disclaimer of political motives has not prevented sociobiology from having political and sociological repercussions (Alper et al. 1978). The potential implications of sociobiology for understanding human behavior have received extensive publicity, usually without any mention of their tentative nature. Harvard University Press promoted Wilson's (1975a) *Sociobiology: The New Synthesis* and its revolutionary implications with a massive public relations campaign, rare for a technical book, even before the book was published. Wilson was interviewed extensively on television and in the press. Favorable stories have appeared in numerous popular magazines, including *Time, People, The National Observer, Mother Jones, Psychology Today,* and *Playboy.* Two eminent sociobiologists, Irven Devore and Robert Trivers, have developed a high school curriculum entitled *Exploring Human Nature,* which essentially presents sociobiological interpretations of human behavior as fact rather than hypothesis. The curriculum is used in over a hundred high schools in 26 states and teaches students biological explanations of sex role differences, male competition and female choice of mating partners, and other behavioral phenomena in human societies while downplaying or ignoring cultural factors that may be equally or more important in their evolution. Relatively few sociobiologists are involved in such endeavors, but their potential impact is enormous.

These few individuals have gained widespread public attention, and their opinions and interpretations are usually treated as representative of sociobiologists in general. The fact is, however, that many behavioral biologists involved in studying the sociobiology of nonhuman animals do not agree with such applications (or misapplications) of sociobiological theory to humans. Many sociobiological interpretations of behavior are still highly controversial, even when applied to animals, and their applicability to humans is far from generally accepted among biologists (e.g., Parker 1978a). Cultural processes clearly have important effects on human behavior, and it is unfortunate that the vogue for interpreting human behavior in sociobiological terms has developed almost entirely in ignorance of how cultural processes work.

Scientific criticisms

The scientific criticisms of sociobiology are less laden with emotion and hence more amenable to objective discussion. The principle criticisms are as follows:

There is no evidence that specific genes influence specific behaviors in humans. Critics of sociobiology claim there is no direct evidence for a genetic basis to human behavior because no identifiable gene has been shown to specifically control any particular behavior (see Feldman and

Lewontin 1975). The critics further point out that one cannot attribute a particular behavior to a particular gene, as is often done by sociobiologists. Assuming the existence of entities such as homosexuality genes, altruism genes, spitefulness genes, and so on is completely out of keeping with our modern understanding of molecular genetics. Such genes do not exist, just as genes for toes, fingers, and noses do not exist. The critics therefore assert that arguments attributing specific behaviors to specific genes are naive and erroneous.

While this objection is technically valid, it does not necessarily negate evolutionary explanations of behavior. True, sociobiologists sometimes refer to altruism genes and similar entities as a convenient kind of shorthand, but they are not so naive as to believe that such entities actually exist. Most sociobiologists are well aware that genes affecting behavior act only through complex developmental processes (e.g., Barash 1979b). They know that no single gene is responsible for any given behavior. However, the existence of such complexities does not deny the potential for evolutionary change. Genetics can affect behavior despite the complex interactions involved, just as it can affect morphology and physiology, and there is plenty of evidence to show that it does (see Brown 1975a; Ehrman and Parsons 1976).

More to the point is the question of whether genes ever influence specific human behaviors. The evidence here is largely inferential because genetic breeding experiments cannot be performed on humans. Abundant evidence demonstrates the importance of genetic effects on the development and expression of behavior in nonhuman animals. Abundant evidence also demonstrates that human morphology, physiology, and neurology all develop along pathways *homologous* to those observed in nonhuman primates and, to a lesser extent, more distantly related vertebrates. These processes are known to be affected genetically (Bonner 1974), and it is eminently reasonable to assume that behavior is based upon them. The same argument used by the Sociobiology Study Group to deny genetic influences on human behavior could be applied equally well to human morphology and physiology. Yet it is clear that genetics does influence morphology and physiology. Unless we are willing to accept a qualitative discontinuity between humans and other animals, the notion that genetics influences human behavior cannot be denied. What remains to be determined is the nature of those influences, not their existence.

Here is where many sociobiologists and their critics disagree. Some sociobiologists claim that genetic influences extend to specific human behaviors, making the behaviors susceptible to natural selection processes. But most critics claim that genetic influences only place very general physiological and morphological constraints on the extent to which human behavior can be modified in response to cultural circumstances. Which point of view is most correct will remain debatable until more information becomes available. The Sociobiology Study Group has argued for an end to sociobiological studies of humans on the grounds that incorrect assumptions about human nature have led to injustices in the past and present and will lead to further injustices in the future, but it seems that continued research is the only way to determine which assumptions are correct and which are not.

The theory of sociobiology is untestable. According to the critics, sociobiological theory explains everything and hence explains nothing. The theory is untestable because no hypotheses lying outside the theory are possible. Critics of sociobiology assert that a tortured logic can be erected to explain away every exception one might conceivably use to reject sociobiological theory. The theory is therefore not subject to falsification. As such, it is not a scientific theory at all; it is instead merely an elaborate rationalization of behavior.

This criticism stems from the mistaken notion that sociobiology is a theory. It is not. Sociobiology is a body of knowledge and interpretations, just as are genetics, anthropology, and psychology. One does not test *the* theory of sociobiology any more than one tests *the* theory of genetics, of anthropology, or of psychology. On the other hand, sociobiology, as a discipline, is based on the major premise that behavior is shaped by natural selection, and this premise *is* subject to verification, in the same way that the premises underlying other disciplines are subject to verification. Many hypotheses have been offered to explain the adaptive value of each specific behavior exhibited by animals. Most are based on the Darwinian theory of natural selection, though some are not. Each hypothesis generates specific predictions that are amenable to empirical verification. Some hypotheses are supported by the evidence, while others are rejected. Refuting any one hypothesis does not refute the entire discipline of sociobiology, though the strengths and limitations of sociobiology will ultimately be determined by the range of hypotheses that gain empirical support.

The fact that many predictions based on evolutionary theory have gained empirical support is a strong endorsement for the assumption that animal behavior is, by and large, adaptive. Nevertheless, there is a very real danger lurking here. Theoretical work is progressing far ahead of the empirical evidence, as is true in any actively growing science, and some specific theories proposed by sociobiologists are very likely to prove erroneous. These will certainly collapse under the eventual weight of empirical evidence, but unfortunately they may lead to social repercussions before this collapse occurs. Attempts to apply sociobiological concepts to human social problems must therefore be made with particular caution and restraint. Premature application of unsupported sociobiological theories to human behavior may have serious repercussions unless the public is informed of the limitations and potential dangers.

Natural selection is not the only possible explanation of cultural change. Cultures change too rapidly for the changes to be ascribed to natural selection. Organic evolution is a rather slow and tedious process, but cultural attributes sometimes change practically overnight. Social and economic forces, along with unique historical events, are major causes of cultural change. These are far removed from the forces of natural selection, and they do not necessarily promote increased biological fitness.

This claim is certainly a valid one. Sociobiologists do not deny the importance of cultural selection in shaping human societies. They view natural selection and cultural selection as complementary processes whose specific effects are open to empirical analysis (e.g., Wilson 1975a). Unfortunately, some sociobiologists have overemphasized natural selection, to the exclusion of cultural processes, in their zeal to identify the biological bases of human behavior. This bias has caused many critics to disregard the work of sociobiologists altogether.

Coherent theories of cultural selection are still lacking, and both biologists and anthropologists are busy seeking a theory that will prove both valid and comprehensive. In the absence of such a theory, the social sciences are left without a unifying theme for understanding how cultures change. Sociobiologists submit that an evolutionary approach to culture is vital for providing a unified theoretical structure. They suggest that both natural selection and cultural selection will prove to be important in such a structure, though they disagree as to the specific role each will play. An adequate theory will require both components, and this is where a melding of biology and the social sciences could prove most productive.

Tenets of sociobiology

Although the controversy surrounding sociobiology is a volatile one, the science itself is not. Sociobiology is a new and useful way of organizing and seeking knowledge, and for that reason it is rapidly expanding. It is primarily concerned with the adaptive significance of social behavior in nonhuman animals, and identifying the adaptive significance of behavior depends on an understanding of evolutionary theory and ecological principles. This is why the domain of sociobiology lies at the interface where behavioral biology, population ecology, and evolutionary biology meet.

Sociobiology rests on three main tenets. First, it assumes that all behavior, both human and nonhuman, results from complex interactions between heredity and environment. As such, behavior is subject to the effects of natural selection. Second, it assumes that natural selection acts upon the genetic underpinnings of behavior in such a way that animals are predisposed to behave adaptively in their natural environments. Third, it assumes that cultural selection acts upon environmentally influenced components of behavior, at least in humans, to cause directional changes according to as yet poorly understood principles. These changes may or may not proceed in directions that increase biological fitness.

Logic of sociobiology

Science consists of both theory and data. Students sometimes have a tendency to discount one or the other as unimportant. However, when theory is neglected, science reverts to simple natural history; when data are neglected, it reverts to mathematical speculation. Sociobiology, along with all other sciences, is created by the interaction of theory and data, and it is this interaction that is most difficult to understand.

Role of theory

The role of theory is to enhance our understanding of observed phenomena. Theory is a systematic and verifiable set of interpretations. Good theory does more than just organize observations into convenient categories. It generates falsifiable predictions or hypotheses about causation that can be tested with obtainable sets of data. The most useful theories are built upon well-established precepts, since science consists of an *integrated* set of verifiable theories, and they are most interesting when they generate surprising or even counterintuitive predictions (Forrester 1971).

Models are explicitly formulated statements of theory. A good model is an accurate portrayal of reality. It usually emphasizes certain key factors and relationships, while de-emphasizing details that are unimportant for the question or problem at hand.

Models represent our interpretations of observed phenomena, and they are often presented in mathematical form. Everyone interprets their sensory experiences, and perceptions are simply mental models of reality. When we must communicate our mental models (i.e., interpretations, opinions, or conclusions) to others, we usually wish to convey them as clearly and accurately as possible. Clarity and accuracy become doubly important when we wish to test the validity of our models, and then mathematical or graphical representations of them are helpful. As our models become increasingly complex, computer analyses gain ascendancy in efficiently and accurately making and testing predictions. Then we must be concerned with how well the computer models represent our mental models, in addition to our initial concern about how well our mental models correspond to reality.

Theoretical models are not always used to interpret data. They are sometimes entirely abstract and not yet supported by empirical evidence. Such models are useful for limiting hypotheses to the realm of possibility and for raising questions that would not otherwise be asked, but they do not constitute fact.

Role of data

The role of data is to falsify or verify predictions or hypotheses. Good data preclude methodological interpretations, thus allowing them to be evaluated solely on the basis of theoretical considerations. The quality of actual data sets varies considerably in this regard. Data are easy to collect, but not all data are useful for testing hypotheses. Descriptive data are important for identifying and posing interesting questions, but they cannot answer those questions. They place limits on the hypotheses that might plausibly explain a phenomenon, but they cannot distinguish among alternative hypotheses that remain plausible. Good data are collected with specific hypotheses already in mind, and they are collected in a way that makes them capable of refuting every hypothesis under consideration. Hypotheses not disproven remain credible, while the others are rejected.

Three kinds of data are especially pertinent in sociobiological studies. They pertain to the effects a particular behavior has on (1) an individual's survival, (2) the number of progeny an individual or its relatives can produce, and (3) the survival and reproductive success of those progeny. These three effects reflect the adaptive advantages of the behavior and hence the probable reasons why it has evolved.

Hypothetic-deductive logic

Good science operates through a process of hypothetic-deductive logic. The method is conceptually very simple, though in practice it may be difficult to achieve. Platt (1964) has outlined the steps necessary for making strong scientific inferences:

1. Explicitly state just what the question or problem is.
2. Devise and explicitly state all possible alternative hypotheses for answering the question or solving the problem.
3. Devise crucial experiments or comparative analyses, with alternative possible outcomes, designed to *be capable of disproving* one or more of the hypotheses unambiguously.
4. Carry out the experiment or comparative analysis so as to avoid methodological ambiguities in the data.
5. Repeat the procedure, refining and testing the hypotheses that remain, until only one hypothesis remains.

The essential ingredient here is the *disproving* of hypotheses. Science progresses by eliminating as many possible explanations of a phenomenon as possible. Hypotheses are accepted only because they have been confirmed by supporting data in spite of all attempts to disprove them. They are never proven valid; some later explanation may turn out to be as good as or better than the previously accepted one (Popper 1959). On hearing a scientific explanation of some phenomenon, one should always ask, "What evidence could *disprove* the explanation?" On hearing a scientific experiment described, one should always ask, "What hypothesis was the experiment intended to *disprove?*" (Platt 1964). If no answer can be given, the theory is not scientific or the experiment has little scientific value.

Systematic evaluation of multiple alternative hypotheses is crucial. Hypotheses may be generated from already existing theory or from prior observations of the phenomenon under study. In either case the hypotheses should be posed as alternatives. When hypotheses are compatible instead of competitive, more than one of them may be true. Then the relative importance of each must somehow be assessed, and the problem becomes a difficult quantitative question. In sociobiology compatible hypotheses are often unavoidable because behavior is not the result of unitary selective pressures. Each potential advantage or disadvantage of a behavior corresponds to a separate hypothesis, and many behaviors have several advantages and disadvantages associated with them.

Dispassionate consideration of all possible hypotheses is essential to good science. All too often scientists adhere to a single favorite theory and only consider hypotheses consistent with that theory. It is easy for this to happen when a scientist ties his or her prestige to a particular theory, and then science becomes a conflict between scientists rather than a conflict between ideas (Platt 1964). Schools of thought arise, each with its outspoken proponents, and science degenerates into polemics. Good researchers do not become enamored with any particular hypothesis and do not grieve over its demise. They do not link their prestige to any one hypothesis or theory, and therefore they have nothing to lose should the hypothesis or theory be rejected. The essence of objectivity is to consider all possible alternatives without prior favoritism, rejecting those that are disproven by empirical evidence and tentatively accepting those that have withstood all attempts to disprove them. This ideal is rarely achieved in any one study, making independent confirmation by other investigators essential.

Scientific advocacy

Scientists frequently do not follow the method of strong inference outlined by Platt (1964). They often advocate their own preferred interpretations without evaluating all possible alternatives, and they often design experiments or collect data with the intent of proving that their preferred theory is correct. E. O. Wilson (1975a, p. 28) has provided a cogent description of this process in action:

> *Author X proposes a hypothesis to account for a certain phenomenon, selecting and arranging his evidence in the most persuasive manner possible. Author Y then rebuts X in part or in whole, raising a second hypothesis and arguing his case with equal conviction. Verbal skill now becomes a significant factor. Perhaps at this stage author Z appears as an* amicus curiae, *siding with one side or the other or concluding that both have pieces of the truth that can be put together to form a third hypothesis—and so forth seriatum through many journals and over years of time. Often the advocacy method muddles through to the answer. But at its worst it leads to "schools" of thought that encapsulate logic for a full generation.*

Two major flaws in the advocacy method make it poor science. First, only one hypothesis is evaluated by each advocate, and since the advocate is attempting to support the hypothesis, data are often selectively chosen to support it. Attempts to test the hypothesis experimentally may unintentionally contain hidden biases that lead to the preferred result. Selective choosing of data and hidden biases in experimental or comparative studies are best avoided by trying to disprove rather than prove hypotheses.

Second, the favored hypothesis is often formulated in a way that makes it unfalsifiable. A hypothesis has no explanatory power when every conceivable form of evidence presented against it can be explained away. The problem here is a subtle one. A hypothesis does not lose its explanatory power simply because it is capable of explaining both a general trend and all known exceptions to that trend. Nor does it lose its explanatory power because it is consistent with every available piece of evidence. The theory of natural selection, for example, has great explanatory power and cannot be considered unfalsifiable simply because all observations are consistent with it. The important issue concerns whether contrary evidence is conceptually possible. A hypothesis has no value if it can be twisted to fit the results of every conceivable test, no matter what the results turn out to be. It does have explanatory power if it can fit only one of several possible outcomes of each test, even though the outcome of every test is the one expected according to the hypothesis.

Only a fine line separates hypothetic-deductive science from scientific advocacy. A common procedure is to generate several hypotheses or predictions and then present data that were collected with the intent of falsifying hypotheses. The scientist must then interpret the data and decide which hypotheses have been disproven and which have not. The need to justify these decisions automatically places the scientist in an advocacy position. This kind of advocacy is justifiable and, indeed, indispensable. Advocacy is poor science when it is used to promote unfalsifiable or favorite hypotheses, but not when it is used to justify rejection or acceptance of alternative hypotheses after they have been subjected to testing.

General methodology

Two general approaches are used in sociobiology to gather data: the comparative method and the experimental method. Each has its advantages and its drawbacks.

Comparative method

The comparative method exploits natural variations in environmental settings by correlating them with observed changes in behavior. Field studies of behavior often rely on the comparative method. The behavior of individuals at different times, in different locations, or within different populations is related to the environmental or social contexts within which it occurs, and observed differences in behavior are attributed to observed differences among contexts. For example, herons in one area were observed nesting in colonies, while in another nearby area they were observed nesting as dispersed pairs (Krebs 1974). Since this behavioral difference was correlated with an inferred difference in the spatial and temporal distribution of food resources in the two areas, nesting dispersion was interpreted to be a response to food distribution.

The most unambiguous comparisons are those made between individuals or populations of the same species because observed differences in behavior can then be attributed, with considerable confidence, to differing environmental conditions (Altmann 1974). When comparisons involve different species, behavioral differences may result from differing phylogenetic heritage instead of differing environmental settings. That is, species may differ because they evolved from different ancestral forms, not because they evolved in different environments. Distinguishing between these two possibilities becomes more difficult when the species being compared are more distantly related to each other.

The comparative method is especially useful for identifying environmental variables affecting a behavior or type of social structure. It is also useful for evaluating the generality of a hypothesis across a broad spectrum of species. However, a major drawback limits its usefulness for demonstrating causation. Because the comparative method relies on correlative analyses, it has difficulty establishing causal relationships. Two variables may be correlated because one causes the other or because they are both caused by some unidentified third variable. The latter possibility is almost impossible to eliminate with comparative analyses, though with perfect controls it could conceivably be done. The problem here is that perfect controls generally cannot be achieved without experimental manipulation. Moreover, comparative studies may give few clues as to which of two variables causes the other. Causal relationships are inferred from the temporal relationship between cause and effect, and correlational studies provide no information on temporal relationships between variables. For example, a common observation in vertebrates is that extensive male parental care is correlated with monogamous mating relationships. Some authors attribute monogamy to extensive male parental care, while others consider extensive male parental care a consequence of monogamy. Both hypotheses are equally plausible in the absence of additional evidence. This problem mainly arises when relationships between two behaviors or two social structures are being studied. When relationships between behavior and environmental conditions are studied, it is usually safe to assume that behavior is a response to the environment rather than vice versa.

Experimental method

The experimental method is well known and hardly needs describing. A control group and an experimental group are set up such that they differ with respect to only one or a few known variables. The behavior of individuals in experimental and control groups is then compared, and any observed differences are attributed to the variable being manipulated. Cause-and-effect relationships can be inferred because the temporal relationship between experimental manipulation and behavioral response is known and because with good controls no third variable could be correlated with both manipulation and response.

Two kinds of experiments are important in sociobiology: field experiments and laboratory experiments. In field experimentation the investigator deliberately manipulates the variable of interest within an otherwise natural setting, and he or she relies on unmanipulated social systems in similar settings for controls. However, since manipulated and unmanipulated settings may differ with respect to variables other than those manipulated by the investigator, only partial controls are ever achieved. Hence complete confidence that observed results were caused by the manipulated variable is not possible in any one study, as the results could always have been caused by some unnoticed difference between experimental and control settings. Confidence in interpretations drawn from the data is increased if the observed results were predicted in advance by theoretical considerations, if all noticeable differences between experimental and control settings were monitored and shown to be incapable of producing the observed results, and if repeated experiments in different environments or with different species consistently yield the same results. Despite the problem of incomplete controls, field experiments allow inferences about causal relationships present in natural environments that cannot be obtained in any other way.

Laboratory experimentation allows more complete control over extraneous variables and enables the use of more sophisticated instrumentation and techniques. Many variables cannot be manipulated in the field and must be studied in the laboratory. The problem is that laboratory settings differ drastically from natural environments. Observed behaviors may therefore be aberrant or pathological, making the results difficult to generalize to natural environments. This problem is especially acute when studying the adaptive significance of behavior. The only way around the problem is to compare the behavior of laboratory control animals to that of free-ranging animals in undisturbed natural environments. Then deviant behaviors caused by the laboratory setting can be distinguished from naturally occurring behaviors. The necessity for making such comparisons makes continued maintenance of undisturbed natural environments and continuation of quantitative field studies in those environments imperative. Indeed, this is one major justification for preserving wilderness areas. Comparison with free-ranging animals is too often neglected in laboratory studies, and in such cases conclusions about the adaptive value of behavior must be viewed with caution.

Learning to be critical and degrees of certainty in accepting hypotheses

Hypotheses should be evaluated at two levels. Evaluations should consider how well an hypothesis explains the specific phenomenon at hand, and they should consider whether the hypothesis is compatible or consistent with previously accepted theory. Hypotheses should not be rejected simply because they do not conform to established theory, but the inconsistencies must be reconciled before the hypothesis can be accepted.

Either the hypothesis is wrong despite its supporting evidence, or the established theory requires modification.

The aim of scientific investigations is to eliminate as many plausible but erroneous hypotheses as possible and confirm those nearest to truth. The purpose of scientific critique is to evaluate how well an investigation has achieved that aim. The credibility of hypotheses supported by a study is only as strong as the attempt made to discredit them. The following guidelines are useful for evaluating the strength of a study:

1. Were sample sizes large enough to allow reliable interpretation of the results? Small sample sizes can more easily give biased or chance support for a hypothesis. Inflated sample sizes based on many observations of a few individuals are not as reliable as smaller sample sizes based on many individuals. Ideally a large sample size for each of many individuals should be achieved. Tests of statistical significance provide the sole criterion for judging the adequacy of sample size, but statistical testing can give misleading results if done improperly.

2. Can the results be attributed to biases in the particular methodology used, or were the results likely to be the same regardless of the particular methodology? Were the data manipulated or transformed in any way? If so, did the manipulations or transformations affect the statistical outcome?

3. Were multiple hypotheses proposed and evaluated, or was the study intended to test only a single hypothesis?

4. Were any plausible hypotheses overlooked? These can often be identified from similar studies or from theoretical considerations. Here is a place where creativity in thinking up alternative interpretations plays an important role.

5. Does the author advocate a favored hypothesis, or does he or she objectively discuss the evidence relevant to each hypothesis?

6. What assumptions were made in the theoretical arguments, and were they reasonable? Does the validity of the conclusions depend on the validity of the assumptions?

7. Do the results justify or substantiate the conclusions, or are the conclusions too general and sweeping for the scope of the evidence presented?

The degree of certainty in accepting hypotheses depends on two considerations: the strength of the evidence used to refute alternative hypotheses and the strength of the evidence used in an attempt to refute the hypothesis accepted. Of course, this assertion assumes that all alternative hypotheses have been identified and evaluated. The strength of the evidence can be assessed by using the guidelines given above and by considering the number of independently conducted studies that have yielded the same results.

The validity of general theories is strengthened if numerous studies of the same or different species, conducted by different investigators, consistently support the theory and if differing predictions drawn from the theory all gain support from independent lines of evidence. The general credence of a theory grows as the number of species and the kinds of evidence supporting it increase. It does not gain credence simply because it attracts many advocates among the scientific community. After all, scientists are human and can be misled by current theoretical fads. Likewise, the limitations of a theory are revealed by the species or kinds of evidence that fail to support it, not by its failure to attract adherents. One of the great strengths of science is that a minority view can eventually prevail when it is supported by irrefutable logic and facts.

Learning to be critical can be bothersome, but it is a vital part of good science. Critical abilities cannot be learned from a textbook or a lecture series alone. Firsthand knowledge of particular methodologies or subjects of study are indispensable to good critical ability. Nevertheless, good critical ability also depends upon knowledge of published theory and data, and this is one important justification for continuing to learn about all matters affecting one's life and interests.

Conclusion

Animal behavior is a continuing source of awe and mystery that sparks the imagination of scientist and public alike. The reasons animals do what they do demand elucidation, both to satisfy our curiosity and to better understand how our activities affect natural ecosystems.

Sociobiology is a new and exciting field of study that promises to go far in answering the question of why animals behave as they do. To this end, it is a legitimate, as well as a rapidly expanding, field of inquiry. Sociobiology may eventually provide insights into human behavior as well, but theories derived solely from research on animals cannot provide an adequate basis for understanding people. There is a real need to develop a theory of cultural evolution to explain how human societies change. Once a general theory of evolution, based on both natural and cultural processes, is validated, the wedding of sociobiology with the social sciences should provide many new insights into human behavior. The major task of sociobiology is to clarify why animals behave as they do, but combined with cultural anthropology and other social sciences, it could also go far in explaining why people do what they do.

Selection theory

The great diversity of living organisms is remarkable. More than a million and a half species of plants and animals are known to science, with perhaps as many as 3 to 4 million more remaining to be described (Dobzhansky 1970), and each possesses its own unique set of behaviors. To be sure, the behavioral repertoire of plants is rather limited, but that of animals is generally quite complex. One of the great challenges in biology has been to explain how this diversity came about.

Two kinds of explanations have been offered to explain the diversity of life. One is based on notions of God and divine creation, the other on processes of organic evolution. Although these two explanations are often treated as mutually exclusive alternatives, they are not. Acceptance of one explanation does not necessarily preclude acceptance of the other.

Theories of organic evolution are testable and falsifiable. Several have been proposed, and one, the modern synthetic theory, has been overwhelmingly affirmed. The occurrence of organic evolution is now accepted as fact, even though the processes involved are still the subject of active research. Much is known about how evolution occurs, but much also remains to be learned. In contrast, concepts of God and divine creation are not testable or falsifiable by any known scientific procedure, because there is no conceivable kind of evidence that cannot be explained as the work of a Creator. Hence the whole idea of divine creation lies beyond the purview of science. Beliefs in a Divine Creator can be superimposed upon scientific knowledge, but they cannot be offered as legitimate alternatives to it.

The central problem of sociobiology is to develop a scientific explanation of how the great diversity of behavioral adaptations has come about. To do this requires some understanding of evolutionary processes. In particular, sociobiological theories are built on the basic premise that behavior has evolved through a process of natural selection. Understanding how natural selection works is therefore essential. The present chapter explains how natural selection works and explores some of the conceptual problems encountered when one tries to deduce the effects of natural selection on behavior.

Theory of natural selection

What is natural selection? In a nutshell, it is the process whereby some genes are propagated to subsequent generations faster than others. At least

Chapter two

three levels of organization must be distinguished when studying natural selection: the level of naked genes or DNA, the level of individual organisms, and the level of populations. Selection at each level is given a different name. **Genic selection** occurs at the genetic level; **individual selection** occurs at the individual organism level; and **group selection** occurs at the group or population level. Most sociobiologists currently interpret behavior in terms of individual selection, and that will be the process emphasized here.

Like most good ideas, that of natural selection is both simple and insightful. Natural selection is a consequence of two universal properties of natural populations: All populations contain genetic variability, and all produce more progeny than can survive to reproductive age in the next generation. The individuals comprising every population vary in their ability to survive and reproduce, and this variability results in part from their differing genetic makeup. Some individuals produce more progeny during their lifetimes than others, and the progeny of some individuals are better able to survive to reproductive age than those of others. Thus individuals propagate a variable number of genetic replicas to future generations. Genes that increase the number of replicas an individual can propagate will become increasingly common each generation, simply because they are propagated more often than alternative genes. Those are the genes that finally come to prevail in a population.

Natural selection occurs whenever some types of genes are replicated and transmitted to the next generation consistently more often than others. The differential propagation of certain genes may occur because they enable individuals to survive longer, reproduce at a faster rate, or produce offspring with greater prospects for survival and reproduction. When evaluating the selective value of a trait, these are the three main consequences of the trait requiring examination.

The above description of natural selection is, of course, an oversimplification. For one thing, it only explains how selection occurs at the level of individuals; it completely ignores the ways that selection might operate at the level of groups. For another, individual selection and group selection affect organisms, not genes, and yet organic evolution consists of genetic changes occurring between generations. More will be said later about the difficult problem of translating differential propagation of progeny into differential propagation of genes. This issue will be skirted for the moment by explaining how natural selection works at the genetic level.

Some basic molecular genetics

Molecular genetics is not really a part of sociobiology, but some knowledge of it helps to grapple with several basic issues. Natural selection is perhaps easiest to understand at the genetic level, and genic selection cannot be discussed without some knowledge about the structure and nature of genetic materials. Understanding how selection at the level of individuals or groups translates into changes in the genetic makeup of populations is a basic issue, and this issue cannot be resolved unless the relationship between genetics and behavior is understood. Finally, the age-old problem of whether behaviors are instinctual or learned still represents a source of confusion, and a little understanding of molecular genetics goes a long way toward reconciling that issue. Thus a brief synopsis of molecular genetics will be presented before these topics are addressed.

Structure of hereditary materials

In most organisms genetic information is encoded in the chemical structure of deoxyribonucleic acid, or DNA, although in a few viruses it is encoded in a related compound, ribonucleic acid, or RNA. Each DNA molecule and its associated proteins comprise a single chromosome. Most species of animals are **diploid,** meaning that every cell possesses two copies of each chromosome. Some species are **haploid,** meaning that every cell possesses just one copy of each chromosome. Finally, a few species are **haplodiploid,** with females being diploid and males being haploid.

DNA consists of two long molecular strands twisted around each other to form a spiraling double helix (Figure 2–1). The two long strands are polynucleotide chains, made up of many nucleotides connected together. Each nucleotide consists of a pentose sugar (S), called deoxyribose, a phosphate group (P), and a nitrogen base. Deoxyribose sugars and phosphate groups of adjacent nucleotides form the outer backbone of each DNA strand, with strong covalent chemical bonds holding the alternating sugars and phosphate groups together. The nitrogen base of each nucleotide is covalently bonded to the deoxyribose sugar and projects toward the center of the molecule. The two polynucleotide chains of a DNA molecule parallel each other along a double spiral and are held together by weak hydrogen bonds between complementary nitrogen bases of the two chains.

Complementarity refers to the property that only certain nitrogen bases can pair with each other. DNA contains four kinds of nucleotides, each with a different nitrogen base. The nitrogen bases are either purines or pyrimidines. The most common purines in DNA are adenine (A) and guanine (G); the most common pyrimidines are thymine (T) and cytosine (C). The chemical structure of these bases is such that adenine and thymine can only form hydrogen bonds with each other. Similarly, guanine and cytosine can form hydrogen bonds only with each other. Thus adenine and thymine are complementary base pairs, as are guanine and cytosine.

Genetic information is encoded by the sequence of nitrogen bases along each DNA strand. A **codon** consists of three consecutive nucleotides along a strand, and it codes for a specific amino acid. Since there are four kinds of nucleotides, there are 4^3, or 64, different codons. They code for 20 amino acids, plus termination signals that indicate the end of a functional unit or gene. The genetic code, which relates the codons to their respective amino acids, has now been fully worked out (Table 2–1). It is identical for all organisms, suggesting that all living organisms have evolved from a

Figure 2–1

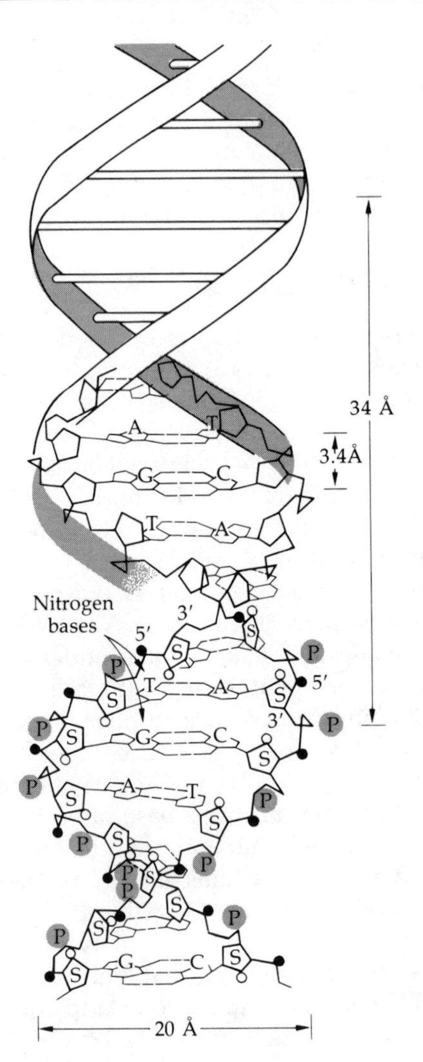

Double-stranded helical structure of the DNA molecule.

The spiraling, outward backbone of the model consists of alternating deoxyribose sugars (S) and phosphate groups (P). The phosphate group is covalently bonded to the fifth (5′) carbon atom of its associated sugar and to the third (3′) carbon atom of the next adjacent sugar along the backbone. Nitrogen bases attached to the sugars project inward. Each sugar–phosphate–nitrogen–base group is a single nucleotide, and hence the backbone of the molecule consists of two spiraling polynucleotide chains. These chains are held together by hydrogen bonds between complementary nitrogen bases (see text). The complementary base pairs are stacked flat, one above the other, at intervals of 3.4 Å (1 angstrom Å = 10^{-7} millimeters), with each pair rotated 36°. Thus each complete rotation along the spiral contains ten base pairs and spans 34 Å along the length of the molecule.

Source: Modified from *Evolution* by Theodosius Dobzhansky, Francisco J. Ayala, G. Ledyard Stebbins and James W. Valentine, p. 22. W. H. Freeman and Company. Copyright © 1977. Reprinted by permission.

single ancestral form (Crick 1968). Note that several different codons can specify a particular amino acid, making the code partially redundant. This means that some changes or **mutations** in a nucleotide sequence can occur without affecting the type of amino acid specified by a codon.

The number of different messages that can be encoded in even a short DNA strand is enormous. For example, a polynucleotide chain consisting of 600 nucleotides contains 200 codons. Each codon could contain any of 21 messages (i.e., it could code for any of 20 amino acids or for termination of the amino acid sequence). Hence the number of possible messages in a chain of 200 codons is $21^{200} = 2.8 \times 10^{264}$, a number far larger than the number of atoms in the known universe (Dobzhansky et al. 1977). More-

Table
2–1 *The genetic code.*

First letter	Second letter: T		Second letter: C		Second letter: A		Second letter: G		Third letter
T	TTT	Phe	TCT	Ser	TAT	Tyr	TGT	Cys	T
	TTC		TCC		TAC		TGC		C
	TTA	Leu	TCA		TAA	End	TGA	End	A
	TTG		TCG		TAG		TGG	Trp	G
C	CTT	Leu	CCT	Pro	CAT	His	CGT	Arg	T
	CTC		CCC				CGC		C
	CTA		CCA		CAA	Gln	CGA		A
	CTG		CCG		CAG		CGG		G
A	ATT	Ile	ACT	Thr	AAT	Asn	AGT	Ser	T
	ATC		ACC		AAC		AGC		C
	ATA		ACA		AAA	Lys	AGA	Arg	A
	ATG	Met	ACG		AAG		AGG		G
G	GTT	Val	GCT	Ala	GAT	Asp	GGT	Gly	T
	GTC		GCC		GAC		GGC		C
	GTA		GCA		GAA	Glu	GGA		A
	GTG		GCG		GAG		GGG		G

Note: Tabulations show the amino acids or termination signals specified by each codon (nucleotide triplet) in DNA. Abbreviations for nucleotide bases are as follows: A = adenine, C = cytosine, G = guanine, T = thymine. Abbreviations for amino acids are as follows: Ala = alanine, Arg = arginine, Asn = asparagine, Asp = aspartic acid, Cys = cysteine, Gln = glutamine, Glu = glutamic acid, Gly = glycine, His = histidine, Ile = isoleucine, Leu = leucine, Lys = lysine, Met = methionine, Phe = phenylalanine, Pro = proline, Ser = serine, Thr = threonine, Trp = tryptophan, Tyr = tyrosine, Val = valine. End = termination signal.

over, an average DNA strand is much longer than 600 nucleotides, with some containing as many as 100,000 of them. A strand of that size could code for $21^{33,333}$ or $3.6 \times 10^{44,073}$ different messages. Add together the codons on all of an animal's chromosomes and the total number of possible messages is inconceivably large. Thus the possibility for genetic variation is virtually infinite.

A **gene** is a sequence of codons that functions as a single unit. An average gene contains about 1500 nucleotide pairs or 500 codons, which closely corresponds to the average size of known proteins (300–500 amino acids) (Watson 1965). The end of each gene is marked by a termination codon. Genes come in two functional types. *Structural genes* encode amino acid sequences of proteins, while *regulatory genes* control the activity of structural genes and other regulatory genes. A few genes are both structural and regulatory in their actions.

At any given position, or **gene locus,** on a chromosome, different individuals in a population may possess different nucleotide sequences.

These different sequences are referred to as alternative **alleles** for that particular gene locus. Natural selection occurs when one allele is more likely to be propagated to future generations than any of its alternative alleles.

Replication and accompanying genetic changes

DNA replication involves an unwinding of the two strands in the double helix. The sequence of nitrogen bases along each strand unambiguously specifies the sequence along the other strand, because base pairs along the two strands must be complementary. For example, if the sequence along one strand is ATAGCAGT, the sequence along the complementary strand must be TATCGTCA. As the strands unwind, new base pairs are formed along each strand to produce two daughter double helices. The point where the parent double helix is unwinding is called the *replication fork,* and microphotographs of the replication fork have confirmed that the unwinding process occurs (Cairns 1963).

Each daughter double helix contains one strand from the parent double helix and one new strand synthesized along the parent strand. Hence both daughter DNA molecules are virtually identical to the parent molecule. On rare occasions errors do occur during the replication process. Such errors are called *mutations.* Some mutations consist of base-pair substitutions (for example, from A–T to C–G, G–C, or T–A). Single base-pair substitutions are generally reversible, giving rise to *back mutations.* Other mutations involve the loss *(deletions)* or gain *(insertions)* of nucleotides. Sometimes hundreds or thousands of nucleotides may be involved, and in rare cases whole genes are lost. Deletions and insertions are nearly always irreversible, except when they involve just one nucleotide pair.

Mutations are not the only way DNA strands change in structure during replication. Reordering of nucleotide sequences can occur by crossovers, inversions, and translocations. **Crossovers** occur when two paired chromosomes break during meiosis and exchange parts. **Inversions** occur when a chromosome breaks in two places and reunites with the middle piece inverted. An inversion changes an original gene sequence of *abcdefgh* to *abfedcgh* following breaks between the *b* and *c* genes and between the *f* and *g* genes. **Translocations** occur when two chromosomes from different chromosome pairs break and exchange parts, resulting in a transfer of some genes from one pair to the other.

Crossovers, inversions, and translocations are important because they affect **linkage** patterns between genes. Genes are linked if they are located on the same chromosome, and the closeness of linkage is determined by how close they are to each other. Genes often function together as coordinated units, and it is advantageous for them to be propagated to progeny as single units. If genes are closely linked, they will probably not be separated by sexual recombination or chromosome breakage during meiosis. If they are more distantly linked, they may become separated by a crossover, an inversion, or a translocation. If they are not linked at all, they will be propagated independently of each other.

New linkage patterns arise from crossovers, inversions, or translocations. When the new pattern is more advantageous than the previous one, progeny possessing it will propagate the new pattern faster than progeny possessing the old one will propagate their pattern. At equilibrium most individuals should theoretically possess the best possible linkage pattern, and virtually all new patterns should be deleterious. Inversions are often exceptional in this regard, since selection to preserve linkage groups can maintain inversions at intermediate frequencies in a population. Crossovers within an inversion lead to immediate genetic death of new linkage patterns if one paired chromosome contains the inversion while the other does not (Figure 2–2). Any linkage group contained within such an inversion is therefore less likely to be broken up by crossovers. Even so, inversion polymorphisms do not prevent linkage changes entirely. Linkage patterns can still change when individuals possess either two inverted or two standard chromosomes. Such individuals will always exist in populations with inversion polymorphisms. Better preservation of linkage patterns is achieved when inversions within inversions exist in a population.

A molecular view of natural selection

In higher organisms DNA is far removed from most environmental influences. Natural selection operates on external **phenotypes,** the outward morphology, physiology, and behavior of organisms, not on genetic materials, or **genotypes.** Selection on phenotypes ultimately leads to changes in the genetic makeup of subsequent generations, but translating selective effects at the phenotypic level to evolutionary consequences at the genotypic level is complicated. Natural selection is easier to analyze in a situation where this translation can be ignored. An illustrative situation of this sort is the evolutionary origin of life. The following description is based on Miller and Orgel (1974), Chai (1976), Dobzhansky and colleagues (1977), and Fox and Dose (1977).

Primal conditions

Numerous experiments have simulated the atmospheric and geological conditions that probably prevailed at the earth's surface before life began (Fox and Dose 1977). Results of these experiments show that amino acids, purines, pyrimidines, simple sugars, and various other organic compounds arise spontaneously under prebiotic conditions with surprising ease. These compounds are precursors to more complex nucleotide chains and proteins, but they probably did not accumulate in a free state. They are reactive materials susceptible to rapid breakdown when steadily irradiated by ultraviolet light, and they are particularly unstable in aqueous solutions (see Fox and Dose 1977). Thus they must have become concentrated by additional mechanisms.

How organic precursors became concentrated is a major unsolved (and perhaps unsolvable) problem. Several concentrating mechanisms have

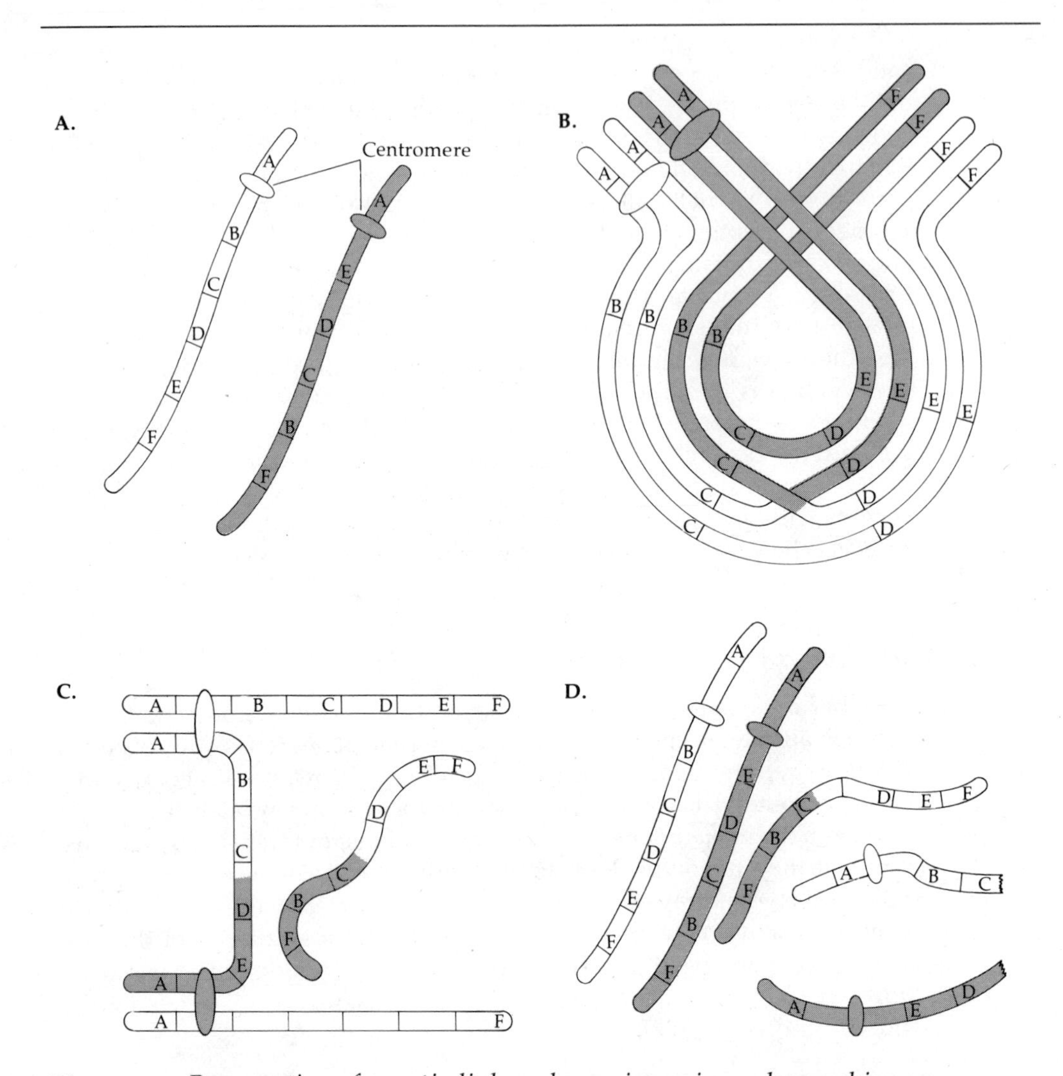

Figure 2–2 **Preservation of genetic linkage by an inversion polymorphism.**
A. When the sequence of alleles (indicated by the letters A–F) along a chromosome is inverted about half the time, many individuals will possess one standard chromosome and one inverted chromosome. B. In such individuals a crossover following chromosome breakage requires a looping configuration of the chromosome pairs during meiosis in order for each allele to align with its opposite on the other chromosome. C. When the chromosomes separate during the first meiotic division, one segment has two centromeres and will break, while another has no centromere and will be lost. D. At the end of meiosis only two chromosomes have complete sets of genes, and those are identical to the two original chromosomes. The remaining genetic material is not incorporated into gametes and hence cannot be passed to offspring. Thus an inversion polymorphism reduces the likelihood that changed linkage patterns resulting from a crossover will be propagated to future generations.

been proposed, including evaporation from shallow pools, freezing, formation of organic colloidal particles, and accumulation of organic materials within crystalline latticeworks of inorganic minerals. Regardless of which concentrating mechanisms were involved, a variety of complex organic polymers, or *protobionts,* arose. These varied in chemical stability, with some breaking apart faster than others, so that a process of chemical selection gradually led to the accumulation of increasingly stable organic molecules.

Eventually chemical associations between nucleotides and proteins formed in combinations capable of self-replication, thereby creating the first primitive organisms. Some kind of chemical selection process must have led up to this event. Miller and Orgel (1974) estimate that a primitive self-replicating system would require about 30 specific molecular constituents. The combination is so unlikely to occur by chance alone that no one seriously defends the possibility that it arose in the absence of chemical selection (Quispel 1968). The nature of this selection process has been the subject of considerable speculation (see Chai 1976), but it is still poorly understood.

Natural selection within gene pools

Once self-replicating systems arose, natural selection began to operate. The process can be understood by considering a hypothetical population of numerous self-replicating systems, or *replicators* (after Dawkins 1976, 1978). Each replicator consisted of some genetic material such as DNA, which acted as a template for protein synthesis, and various structural molecules that probably included proteinoids, proteins, sugars, polysaccharides, and polyphosphates. The various replicators may have arisen through several independent origins of life or as descendants of a single ancestral molecule. In either event mutations, deletions, and insertions eventually produced considerable genetic variability within the replicator population.

Each replicator contained one or more genes, and the entire collection of all genes carried by all replicators in the population comprised a **gene pool.** The frequency of genes present in a gene pool and the linkage patterns of those genes were both likely to change through time. At any point in time the commonness of a gene in the gene pool depends on three properties: copying fidelity of the gene, chemical and physical stability of the replicator, and rapidity of replicator duplication.

Low copying fidelity results in gradual loss of a gene from the gene pool through its transmutation into other types of genes. In short, it results in mutation. Mutations usually reduce the commonness of the original gene in the gene pool. Through a simple statistical process, genes with high copying fidelity become increasingly common relative to genes with low copying fidelity. Although this process leads to the eventual predominance of genes with high copying fidelity, it can never eliminate mutations altogether because perfect replication is a chemical impossibility. New genetic variability is therefore always being introduced into gene pools.

If a gene's effect is to increase a replicator's chemical stability, it will remain in the gene pool longer and be duplicated more times than alternative genes whose replicators do not survive as long. Genes that prolong a replicator's existence therefore become increasingly common in the gene pool relative to genes that do not.

Similarly, genes that enable a replicator to duplicate itself more often will be copied at a faster rate and will also become increasingly common relative to genes lacking that effect. Such genes may enable a replicator to obtain energy and nutrients necessary for duplication at a faster rate. Or in situations where replicators compete for a limited supply of energy or nutrients, they may enable a replicator to outcompete rival replicators.

Some complexities of the selection process

The three processes above all cause certain alleles in a gene pool to become increasingly common relative to alternative alleles. This is the essence of natural selection, although natural selection is not quite that straightforward a process. Genes are likely to affect a replicator's stability and replication rate simultaneously, and genes at some loci are likely to modify the effect of genes at other loci.

When a gene influences both replication rate and replicator stability, its fate depends upon the net effect, as measured by the number of copies propagated per unit of time. If a particular allele causes a replicator to duplicate itself faster, at the expense of reduced survival, it may or may not leave more descendants than alternative alleles. The same is true for an allele that enables a replicator to survive longer at the expense of reduced replication rate. The important variable determining whether the allele will increase or decrease in frequency is the number of descendants produced per unit time, not the fate of the replicator itself.

Since replicators may consist of several interacting genes, a particular allele may improve a replicator's duplicating ability better when linked with some genes than when linked with others. Then some linkage patterns are more likely to persist than others, and any mechanisms that preserve favorable linkage groups will be favored by selection. In higher organisms linkage is often preserved by inversion polymorphisms, though this only works in sexually reproducing species (see Figure 2–2).

External environments vary geographically. As a result, natural selection favors different genes in different geographical areas. This situation can be true even at a very local level. Gene pools are not homogeneous mixtures of replicators. They are heterogeneous assemblages containing many local variations in gene frequencies and linkage patterns. Speciation is one consequence of this genetic heterogeneity. New species are thought to originate primarily after gene pools become geographically isolated and are subsequently subjected to different selective regimes (Mayr 1963; Dobzhansky et al. 1977), but they may also arise without geographical isolation (White 1978; Futuyama 1979).

Relation between genotype and phenotype

Organisms today are the ultimate descendants of replicators that existed several billion years ago, and their *raison d'être* remains the propagation of genes. Individuals are, from an evolutionary point of view, merely genes dressed up with elaborate exterior phenotypes. The same selective processes that once determined the commonness or rarity of genes among replicators still determine the commonness or rarity of genes in contemporary gene pools. However, the effect of natural selection on gene frequencies is much more indirect today than it was earlier because natural selection now acts upon phenotypes far removed from their underlying genotypes. This situation creates a conceptual gap in understanding how selection on phenotypes ultimately affects genotypes.

The conceptual gap

Environmental conditions directly affect the survival and reproductive success of phenotypes, not the frequency of genes or linkage groups in gene pools. To be sure, lifetime reproductive success of phenotypes affects the rate at which genes and linkage groups are propagated to future gene pools, but the nature of the connection is by no means obvious. There are two levels of selection operating here. *Individual selection* results from variability in the reproductive success of individual phenotypes, while *genic selection* arises as an indirect consequence of individual selection. The problem is to understand how individual selection translates into genic selection.

Richard Lewontin (1974) has graphically illustrated this problem (Figure 2–3). At the beginning of each new generation, a population can be thought of as a large collection of fertilized eggs. As the eggs develop, the genes in their genomes interact with each other and with the external environment to produce a variety of phenotypes. Such interactions continue to affect phenotypes, though to a lessening extent, throughout life. As individuals mature, they disperse, select mates, reproduce, and die. Their success or failure in these activities determines the types and linkage patterns of genes inherited by the next generation. During maturation some phenotypes survive better than others, thereby altering the frequency of phenotypes present in the adult population. During reproduction some phenotypes survive longer or produce progeny at a faster rate than others, thereby leaving more mature progeny for the next generation. Such environmentally induced changes in phenotype frequencies result from individual (and sometimes, perhaps, group) selection. With a change in phenotype frequencies, gene frequencies also change. However, gene frequencies usually cannot be measured directly. They must be inferred from

Figure 2–3

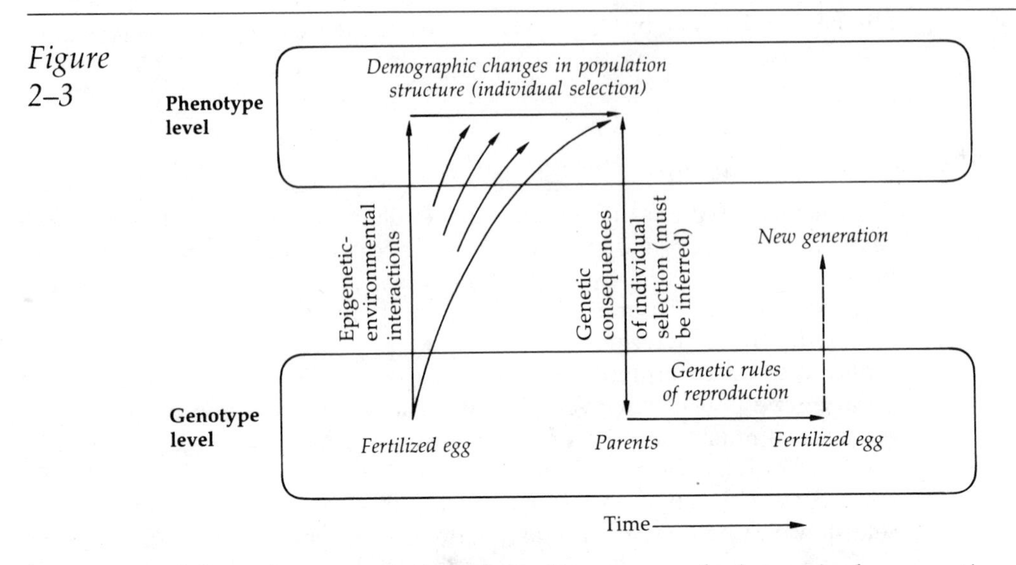

Schematic representation of selection process during a single generation. *Oblong surfaces represent phenotypic and genotypic levels of organization, and the solid line indicates the trajectory a population follows from one generation to another. Beginning at the lower left, a fertilized egg develops from a phenotypic embryo (upper left) into a mature adult (upper right). Demographic changes and individual selection occur during development and determine which individuals survive to reproduce. The genetic makeup underlying the behavior of reproducing adults is generally unknown, and hence the genetic consequences of demographic and selective processes can only be inferred.*

Source: Modified from R. C. Lewontin, *The Genetic Basis of Evolutionary Change* (New York: Columbia University Press, 1974), p.14. Reprinted by permission.

phenotype frequencies. Therein lies the problem. How do changes in phenotype frequencies translate into altered gene frequencies? The answer to this question is a crucial issue, as it will ultimately determine the proper approach to analyzing the adaptive significance of phenotypic traits.

Bridging the gap

The translation from phenotype to genotype can be clarified by understanding how genotypes produce phenotypes in the first place. The process is exceedingly complex, and only a broad outline can be given here. It involves the transcription and translation of DNA molecules, the biochemical differentiation of cells, and the formation of organs and organ systems (see Watson 1965; Markert and Ursprung 1971; Bonner 1974).

Transcription and translation of DNA. **Transcription** is the synthesis of an RNA molecule with a nucleotide sequence complementary to a DNA segment. It occurs in essentially the same way as DNA replication, except that the nitrogen base uracil (U) replaces thymine in base pairs containing adenine (Figure 2–4A). Since RNA is a single-stranded helical molecule, only one DNA strand is transcribed at a time (Marmur et al. 1963). Either strand may be involved during any given transcription.

Translation is the synthesis of proteins from an RNA template, called messenger RNA (mRNA), which was originally transcribed from a DNA strand in the cell nucleus. Each mRNA molecule contains the coded information from a single gene. Newly synthesized mRNA moves from the cell nucleus into the cytoplasm, where it becomes associated with small organelles called ribosomes. That is where translation takes place (Figure 2–4B). In the ribosomes transfer RNA (tRNA) molecules line up along the mRNA strand in the specified sequence. Each tRNA has a chemical structure that can attach to only one type of amino acid, and each possesses three nucleotides complementary to the nucleotides on mRNA that code for the particular amino acid carried by the tRNA molecule. Thus when tRNA molecules line up along an mRNA strand, they align amino acids in the sequence originally coded in DNA. Enzymes then facilitate chemical bonding of adjacent amino acids to produce proteins. Synthesis of the protein is terminated by a termination codon on the mRNA. Thus the biochemical role of the genome is to produce proteins, which act as building blocks, enzymes, and regulators of cell activity.

Control of cell differentiation. The way a cell or tissue differentiates is determined by which of their genes actively synthesize proteins during development. Every cell contains a full complement of genes, but not all genes function in any given cell or at any given time. Thus the course of development is controlled by the mechanisms regulating gene activity. These mechanisms are only now becoming understood.

The simplest control system is the operon found in viruses and bacteria. The operon model of gene regulation was first proposed by Jacob and Monod (1961) and is now generally accepted with a few modifications. An **operon** is a DNA segment consisting of three functional components: one or more promoter regions, one to nine structural genes, and one or more regulatory sites (Figure 2–5) (Campbell 1979; Goldberger 1979). The promoter regions act like on-off switches. They are responsible for initiating transcription along the structural genes. The regulatory sites act like rheostats or valves. They control the rate at which transcription occurs by acting on the promoter regions.

A clearer picture can be gained with a little more detail. Each promoter site is a segment of DNA where mRNA polymerase, an enzyme that increases the rate at which transcription occurs, binds to the operon. If the enzyme is present at the promoter site, transcription occurs rapidly; if not, transcription proceeds very slowly.

Binding of mRNA polymerase to the promoter regions is in turn controlled by the regulatory sites. Two kinds of regulatory sites are known (Goldberger 1979). One exerts negative control (i.e., blocks transcription) and the other exerts positive control (i.e., activates transcription).

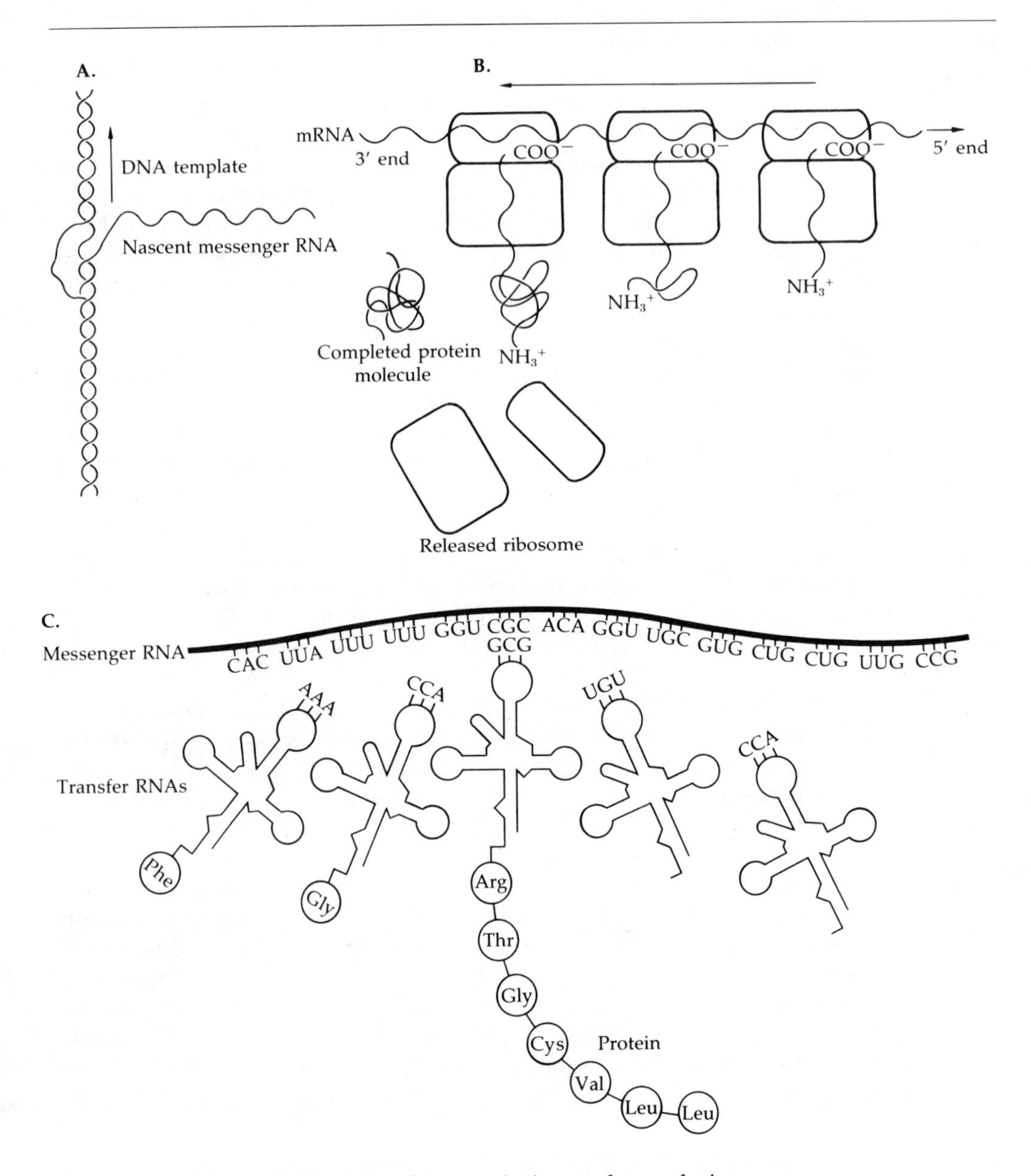

Figure 2–4 *Schematic diagram of transcription and translation.*

A. During transcription one strand of the DNA molecule serves as a template for synthesis of a complementary strand of nascent (i.e., newly synthesized) messenger RNA (mRNA). B. Before translation can occur, mRNA must move from the nucleus into the cytoplasm. There, several ribosomes attach to the mRNA molecule in progression and proceed along the molecule until they encounter a termination signal. Each ribosome synthesizes a single protein molecule in stepwise sequence as it moves along the mRNA molecule. C. A more detailed look at the translation process shows how translation occurs. Within a given ribosome a series of transfer RNA (tRNA) molecules attach along the mRNA molecule in succession. Each tRNA molecule contains a sequence of three nucleotides

Negative control is exerted by a regulatory site called an *operator*, where a repressor substance binds to the operon (Chai 1976; Maniatus and Ptashne 1976; Goldberger 1979). An active repressor substance bound to the operator blocks the binding of mRNA polymerase at the promoter site and hence diminishes the transcription rate. Repressor substances are produced by regulatory genes, which are not necessarily linked closely to the operon. They are active when they are not combined with a smaller inducer substance, which is usually an intermediate or end product of the metabolic pathway controlled by the structural genes.

Positive control is exerted by a regulatory site called an *initiator*, where an activator substance binds to the operon (Chai 1976; Goldberger 1979). An active activator bound to the initiator facilitates binding of mRNA polymerase at the promoter site and hence speeds up the transcription rate. Activator substances, like repressor substances, are produced by regulatory genes and are active when not combined with an inducer substance.

The genes responsible for producing the various enzymes involved in a complex metabolic process are not necessarily clustered together on a single DNA segment (Goldberger 1979). When several operons are involved, they are controlled as a unit by having the same operator or initiator. This characteristic allows a single regulatory gene to control several operons at once. Such a unit is called a *regulon* (Maas et al. 1964).

External control of regulons is exerted through regulatory genes. The activity of regulatory genes is influenced by the biochemical environment of the cell. When external substances enter the cell, new processes, such as breakdown of nutrients or initiation of cell fission, are called for. Particular substances presumably act as triggers to initiate or suppress activity of appropriate regulatory genes, but the exact mechanisms are still unclear. Thus at every level the genome interacts with the biochemical makeup of the cell to control cellular activities.

Operons are unknown in multicellular organisms, and the details of gene regulation in those organisms have not yet been worked out. Several models have been proposed, based on current knowledge of bacterial genetics (e.g., Georgiev 1969, 1972; Crick 1971; MacGillivray et al. 1972; Britten and Davidson 1969, 1971; Davidson and Britten 1973). One plausible model is the one proposed by Britten and Davidson. Their model is con-

---

complementary to a particular codon on mRNA, and each carries a specific type of amino acid. That is how the genetic code specifies particular amino acids. The sequence of codons along the mRNA molecule determines the alignment of tRNA molecules within the ribosome and hence the sequence of amino acids in the newly synthesized protein. The configuration of the tRNA molecules depicted is an outline of the known structure of alanine tRNA. Undoubtedly other types of tRNA have different shapes. NH_3^+ *represents the amine group at one end of a polypeptide (protein) chain, and* COO^- *represents the carboxyl group at the other end.*

Source: From *Evolution* by Theodosius Dobzhansky, Francisco J. Ayala, G. Ledyard Stebbins and James W. Valentine, p. 26. W. H. Freeman and Company.

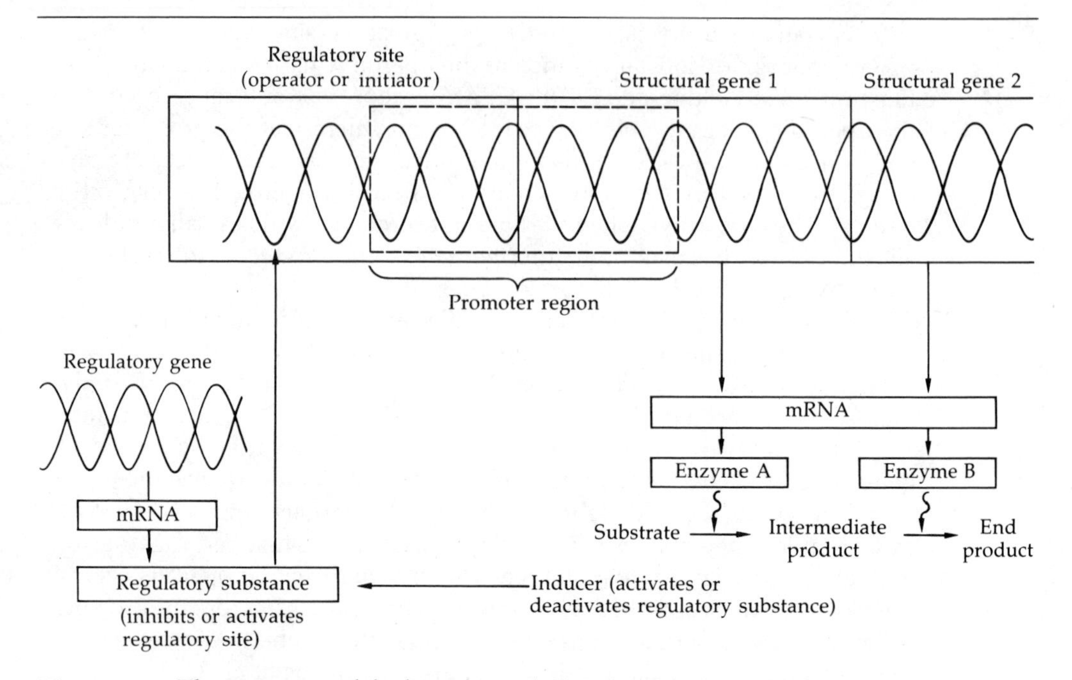

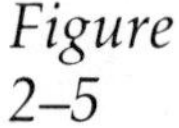

Figure 2–5 *The operon model of gene regulation in viruses and bacteria.*
The operon consists of a regulatory site and several structural genes (two shown here) on a chromosome. A regulatory gene at some other gene locus produces a regulatory substance, which binds to the regulatory site of the operon. When activated, this substance either inhibits or activates transcription of mRNA by blocking or facilitating attachment of mRNA polymerase, an enzyme that speeds up the rate of transcription, to the promoter region. The structural genes act as templates for synthesizing mRNA, which in turn acts as a template for synthesizing a particular set of proteins (enzymes A *and* B *in the present example). The resulting proteins affect a particular biochemical reaction in the cell. Either an intermediate or an end product of that reaction acts as an inducer substance. The inducer activates the regulatory substance bound to the regulatory site. Presence of an inducer can either block (negative control) or facilitate (positive control) binding of mRNA polymerase to the promoter region. Hence an inducer can either speed up or slow down further synthesis of mRNA.*

sistent with current knowledge of genetic systems but may require future modification. According to the model, each structural gene (S) is closely linked to a receptor gene (R) (Figure 2–6). When the receptor gene complexes with a specific *activator RNA* molecule, transcription of the structural gene begins. Activator RNA is produced by a regulatory gene, called an *integrator* (I), somewhere in the genome. The integrator gene lies adjacent to a *sensor* gene (SEN), which initiates transcription of activator RNA at the integrator locus. The sensor gene is activated in turn by some outside stimulus such as a hormone or a metabolic end product. Here is where external influences interact with the system. Whole suites of structural genes may be controlled by a single sensor gene if they all have identical receptor genes associated with them. This may be how the many structural

Figure 2–6

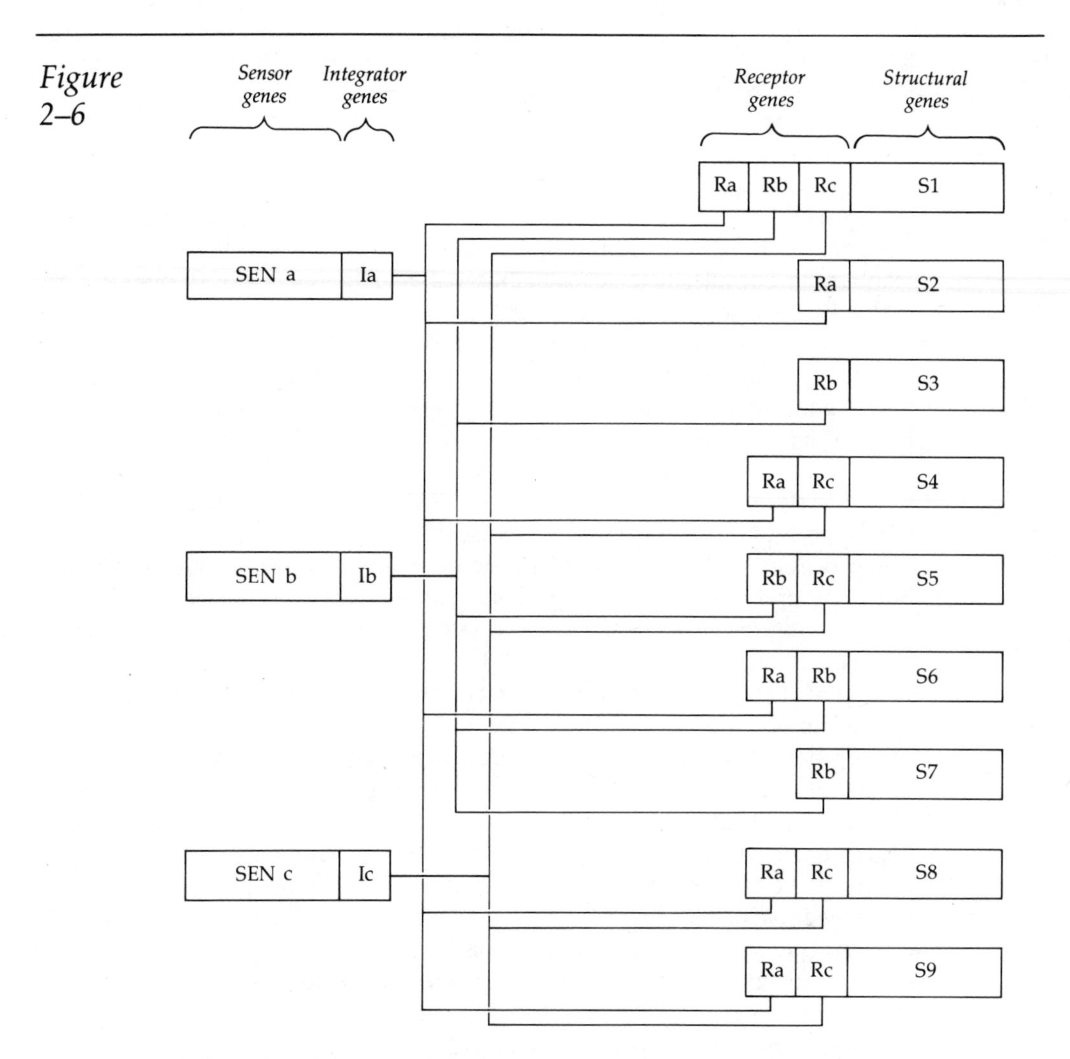

Britten-Davidson model of gene regulation for multicellular organisms.

The presence of particular inducing substances (such as hormones) causes sensor genes (SEN) to activate adjoining regulatory genes called integrators (I). The integrators then produce specific types of RNA molecules, which diffuse through the genome and activate their corresponding receptor genes (R). The receptor genes are closely linked to structural genes (S), and when a receptor is activated, transcription along the adjoining structural gene is initiated. Sometimes more than one receptor gene may be associated with a particular set of structural genes. Many structural genes, located at different gene loci, may be activated by a single sensor, and functionally related suites of sensor genes may be controlled by a single, master sensor gene.

Source: From *Evolution* by Theodosius Dobzhansky, Francisco J. Ayala, G. Ledyard Stebbins and James W. Valentine, p. 257. W. H. Freeman and Company. Copyright © 1977. Reprinted by permission.

genes involved in complex biochemical processes operate in the integrated fashion necessary for cell differentiation to take place.

By now, one might be asking what all this has to do with animal behavior. The answer is twofold. First, genotypes develop into phenotypes through a complicated series of developmental steps. Each step involves synthesis of enzymes, which are encoded in DNA. By understanding how the genome functions to produce these enzymes, one can begin to understand how complex the relationship between genotype and phenotype really is. Second, many people still have the notion that behaviors are either genetically determined or learned. By understanding the interactions occurring between genome and environment at every stage of development, one can see the naivete of such a view. More will be said about this issue in a later section.

Development of organ systems. Returning to the relation between genotype and phenotype, one must consider how development occurs. The processes certainly cannot be covered in any comprehensive way here, but a general overview is helpful. Organ systems develop as a result of three simultaneously occurring processes: cell growth, cell differentiation, and cell movement (Bonner 1974). Control over these processes can be exercised at transcription, at translation, or during enzymatic activity. Control at transcription or translation involves close interactions between regulatory genes and the cellular environment, as described above. Control of enzymatic activity may involve mainly biochemical feedback loops, without any direct intervention from the genome. The controlling agents may be produced either within the cell or in other cells or tissues. The factors controlling tissue differentiation, for example, are mainly produced outside the cells in which they act.

Although control over developmental events may quickly become divorced from direct genetic intervention, the controlling substances are still produced primarily within the self-contained internal environment of the organism, where structural and biochemical makeup were largely orchestrated by the genome. Environmental factors external to the organism play an important role in development, especially in timing developmental events, but the genome channels the course of events and limits the range of phenotypes that can possibly develop (Waddington 1957). As a result, quite different genomes can lead to rather similar phenotypes. On the other hand, environmental and experiential factors generate considerable phenotypic variability, and hence similar genomes can also lead to very different phenotypes. This situation is what makes the relationship between genotype and phenotype difficult to infer.

Relating phenotypes to genotypes. Despite the difficulties, some attempt at relating phenotypes to genotypes should be made. One place to start is with two major generalizations that can be made about genetic influences on phenotypes. First, most genes affect many traits, a phenomenon called **pleiotropy.** Structural genes code for enzymes or other proteins that affect the development of many different phenotypic traits. Regulatory genes control the activity of many structural genes and hence indirectly affect all the traits associated with those structural genes. Second, phenotypic traits are generally affected by many different genes, a phenomenon called **polygenic inheritance.** This term is usually restricted to quantitative characters such as height, weight, body conformation, or intelligence, but even qualitative characters like eye and hair color are affected by many genes. The latter characters undergo conspicuous qualitative changes as a result of single mutations, but their development still rests on the integrated activity of numerous structural and regulatory genes. The genome consists of many coordinated, interacting units. It is not simply a collection of independently operating genes.

Now consider a structural gene with many pleiotropic effects, including effects on a phenotypic trait we are interested in studying. Some effects of the gene may be advantageous, but many others are likely to be deleterious. Hence the gene will be favored by selection acting on some traits it affects, while being selected against by selection acting on other traits it affects. The net effect of these selective pressures initially determines how the gene's frequency will change. However, the expression of structural genes is controlled by regulatory genes, which can activate structural genes where they promote favorable phenotypic traits and suppress them where they do not. Thus selection for a particular phenotypic trait should favor a combination of structural and regulatory genes promoting the appropriate expression of that trait, while at the same time favoring regulatory genes that suppress deleterious pleiotropic effects of the structural genes. Exactly how this effect comes about is unknown, but selection for most traits should cause extensive reorganization of that portion of the genome involved in their expression.

Since regulatory genes can suppress pleiotropic effects, one might argue that each phenotypic trait can be considered separately. If all pleiotropic effects can be ignored, one could treat each phenotypic trait *as if* it resulted from a single allele or linkage group. This is indeed the usual approach taken in sociobiology today.

Unfortunately, there is a serious problem with this approach. The approach rests on the argument that all structural and regulatory genes associated with a phenotypic trait are propagated as a closely linked unit. If regulatory genes are not propagated together with structural genes, they cannot consistently suppress deleterious pleiotropic effects and such effects could then not be ignored. Certainly selection should favor closer linkage of functional units in the genome, and in some cases such units are known to be closely linked (Giles 1978). However, in many other cases regulatory genes are not linked to structural genes. Hence the role pleiotropic effects plays in selective processes may be important, and it needs more study (see Gould and Lewontin 1979 for further discussion).

At the present time no one really understands how genotypes translate into phenotypes or how selection on phenotypes translates into changes in gene pools. The conceptual gap has not yet been bridged with any real assurance. The main procedure used at present is to consider each trait as if it were coded by a single nonpleiotropic gene or linkage group, but one should realize that this approach may result in wrong assessments being made about the adaptive value of particular traits.

Nature-nurture dichotomy

Instincts and learned behaviors

Many complex behaviors are repeatedly performed in the same stereotyped manner by all like-sexed individuals of a species. They require little or no practice, show little variability between performances, and are consistently elicited by specific stimuli. Such behaviors have proven extremely useful when analyzing proximate causation and phylogenetic affinities, and they have traditionally been labeled as **instincts** (Lorenz 1950; Tinbergen 1951). Because they are extremely stereotyped and require little or no prior practice, instincts have generally been attributed largely or entirely to genetic heritage.

Many other behaviors are extremely variable or flexible in their expression and are not performed by every like-sexed individual of a species. They arise from each individual's prior experiences, generally require practice before they are perfected, and can be promoted by appropriate training methods. These are **learned** behaviors, and because they result from individual experiences, they have been attributed largely or entirely to environmental influences rather than genetic heritage.

Problems arising from dichotomizing behavior

The problem of dichotomizing behavior arises because it implies simplistic notions about its developmental origins. Instincts are *attributed* to genetic factors, and learned behaviors are *attributed* to experiential factors. In each case an inference has been made in light of only partial evidence, often with little understanding of developmental processes. A great semantic morass has grown out of such inferences, and only recently has a semblance of agreement been reached about the roles that genetics and experience play in behavioral development. Key papers resolving the debate were those of Schneirla (1952), Hebb (1953), Lehrman (1953), Beach (1950), Hinde (1959, 1968), and Tinbergen (1963). The debate identified three major reasons why dichotomizing behavior into instincts and learned habits is a gross oversimplification of behavioral development (see Hinde 1970).

First, the innate-learned dichotomy implies a separation of environmental and hereditary influences during development when in fact no such separation exists. The discussion in the preceding section makes this point clear. Genetic and environmental factors interact at every stage of development to create an organism and its behavioral manifestations. To simply label a behavior as innate or learned completely ignores such interactions and precludes any real understanding of how development proceeds.

Second, the innateness of a behavior is often documented in practice solely by means of deprivation experiments, in which animals reared in an environment lacking particular stimuli are compared to ones reared more normally. When no environmental factors can be found that affect development of the behavior, the behavior is said to be innate. However, since no series of experiments can exclude all conceivable environmental influences on behavior, such influences cannot be disregarded.

The best that can be achieved by deprivation experiments is a demonstration that development of a particular behavior pattern is or is not affected by the specific stimuli or experiences eliminated by each experiment. Deprivation experiments can show that individuals of a single genetic strain differ in their behavior because they have had different prior experiences. They can also show that individuals are similar in their behavior because they share a common genetic and embryonic heritage and not because they were exposed to different postnatal experiences. Behavioral differences can be ascribed to differing genetic heritage when individuals from different genetic strains or with differing genotypes are reared in environments identical with respect to all environmental variables known to affect the behavior being studied. The behavior of any one individual cannot be attributed to either genetics or experience. Both are involved. Only *differences* or *similarities* between two or more individuals can be interpreted in those terms (see Marler and Hamilton 1966; Hinde 1970; Brown 1975a). One can say, for example, that the stereotyped pecking responses given by newly hatched baby chickens results from similarities in genetic heritage and embryonic environments, along with the inability of environmental factors to modify the behavior after hatching. One cannot say that the behavior results entirely from genetically controlled processes.

Finally, dichotomizing behavior into innate and learned components diverts attention away from some important questions. It is all too easy to use instinct as an explanatory device. A common answer to the question "Why do animals behave that way?" is "The behavior is instinctive." This is not much of an answer. It does no more than give the behavior a name and attribute it to genetic heritage. A better answer is obtained by focusing on the factors and processes involved in its development or on the selective factors favoring its evolution. Labeling a behavior as instinctive often closes the mind to more careful investigations into the developmental processes and selective factors involved in its expression. In the same way, labeling a behavior as learned often closes the mind to possible genetic and selective influences affecting the behavior. This sort of thinking is particularly prevalent in discussions of human behavior, and it represents a real obstacle in analyses of how human behavior evolves.

Fitness

Fitness is one of the most basic concepts in evolutionary biology. It can also be a major source of confusion because it is used without any qualification in several different contexts. In general, **fitness** refers to the tendency for a trait or allele to increase or decrease in a population. The meaning of the term *fitness* depends on the level of selection being discussed. Sometimes the terms **selective value** or **adaptive value** are used instead of *fitness,* and their meanings also depend on the level of selection being discussed.

Individual fitness

At the phenotypic level an individual is said to be more fit if it produces more progeny during its lifetime than a competing individual in the same population. **Individual fitness** is a measure of an individual's ability to produce mature descendants in comparison with other individuals in the same population at the same time. High fitness implies that the individual leaves more descendants in future generations *relative to* individuals with lower fitness. The term is most commonly embodied in the phrase *survival of the fittest,* but this phrase is misleading because individual fitness reflects output of mature offspring, not longevity per se.

Another type of fitness at the phenotypic level is **population fitness,** which refers to the tendency for one population to replace another within a group of populations. Population fitness is not the same as *average population fitness,* which refers to the average individual fitnesses of all individuals within a population. The term *population* is used here in a very general sense to refer to any group of individuals upon which group selection acts. More will be said about this in a later section.

Darwinian fitness

At the genetic level fitness has a more precise meaning. **Darwinian fitness (**W**)** measures the tendency for a particular allele at a gene locus to change in frequency relative to other alleles at that locus. The Darwinian fitness of an allele may vary, depending on which environmental context is being considered. Hence it can be given a value only with respect to a specified context.

Population genetics gives Darwinian fitness a precise mathematical meaning. If the rate at which individuals propagate one type of allele is arbitrarily set equal to 1 (i.e., $W_1 = 1$), then the rate at which a second allele is propagated may be higher (i.e., $W_2 = 1 + s$) or lower (i.e., $W_2 = 1 - s$) than that of the first allele. The value s is the *selection coefficient,* which measures the strength of genic selection. Darwinian fitness specifies how fast and in what direction one allele will change in frequency *relative to* alternative alleles. Thus Darwinian fitness is a relative, not absolute, measure of adaptiveness.

Distinctions

The difference between Darwinian fitness and individual fitness cannot be overemphasized. Darwinian fitness measures the rate at which gene frequencies change, while individual fitness measures the reproductive success of phenotypes. Two different levels of selection are involved here. Individual fitness and Darwinian fitness are not the same thing, even though the former often translates into the latter. Some phenotypic attributes of individuals may affect reproductive success without affecting gene frequencies in subsequent generations because they do not have any genetic basis. Such attributes affect individual fitness but not Darwinian fitness. Also, there is no quantitatively precise relationship between reproductive success and rates of change in gene frequencies because of the complex developmental processes intervening between phenotype and genotype. The distinction between individual and Darwinian fitness is an important one to keep in mind.

Theoretical models are generally based on Darwinian fitness, while empirical tests of those models are generally based on measures of individual fitness. Many discussions of fitness do not specify which type is meant, and one must always be careful about which meaning is intended when only the unqualified term *fitness* is used.

Analyzing adaptive value

General approach

To analyze the adaptive value of a behavioral trait, sociobiologists begin with the assumption that some underlying set of genes is associated with it. They do not know how or, in most cases, to what extent those genes affect the behavior, but they assume that the genome is indirectly involved in its expression. In other words, they assume that the trait has been favored by natural selection. Only rarely is it possible to test this assumption quantitatively in natural populations, but many experimental studies show that behavioral traits are affected by artificial selection analogous to natural selection.

When studying the adaptive value of a behavior, sociobiologists evaluate how the behavior affects lifetime reproductive success or some correlate of lifetime reproductive success. Such correlates may include annual reproductive rate, annual survival rate, energy intake, or energy expenditure. Any behavior that increases lifetime reproductive success should increase the rate at which its underlying alleles are propagated to subsequent generations. If the increase is greater than would have been possible had some other behavior, based on alternative alleles, been expressed in its stead, the alleles associated with that behavior should increase relative to alternative alleles. This result should, in turn, make the behavior more prevalent in the population, though considerable variability may continue to exist because experiential factors also affect behavioral expression.

The reproductive consequences of a particular behavior often cannot be measured. Reproductive success results from the combined effects of many different behaviors, and the effects of any one behavior often cannot be disentangled from the effects of other behaviors. In most cases comparing the success of individuals who differ solely with respect to a single behavior simply cannot be done. Sociobiologists therefore resort to cost-benefit analyses of the behavior. They first list the potential costs and benefits that might be associated with the behavior and then seek evidence that indicates which costs and benefits are important.

Cost-benefit analyses

A **cost** is any consequence of a behavior that is likely to *decrease* the rate at which alleles underlying the behavior are propagated to future generations. It may involve predation risk, energy expenditure, or any other consequence that affects an individual's capacity to survive and reproduce or the capacity of its offspring to survive and reproduce.

A **benefit** is just the opposite. It is any consequence of a behavior that is likely to *increase* the rate at which alleles underlying the behavior are propagated to future generations. It involves the same sorts of consequences as are involved with a cost. The only difference is that a benefit increases individual fitness, while a cost decreases individual fitness.

Costs and benefits are ordinarily not automatic consequences of a behavior. When an animal behaves in a certain way, it has some probability of suffering the costs and some probability of gaining the benefits. In other words, each behavior entails an **expected cost** or **risk** and engenders an **expected benefit.** The behavior is advantageous if (1) the expected benefit exceeds the expected cost or risk and (2) the **net expected benefit** (the difference between expected benefit and expected cost or risk) is greater than the net expected benefit of every alternative behavior that could have been performed instead.

Cost-benefit analysis has a serious limitation. Since costs and benefits can rarely be quantified, analyses based on them are rarely capable of demonstrating that a behavior really is adaptive. One usually cannot determine whether expected benefits exceed expected costs. Similarly, one usually cannot determine whether the net expected benefit of one behavior exceeds that of all other behaviors possible within a given context. About all cost-benefit analyses can do is show that specific benefits are gained by performing a particular behavior and that specific costs are incurred at the same time. The adaptiveness of behavior is best demonstrated by observing how it changes under a selective regime, which is not always possible. Hence for many behaviors adaptiveness is an article of faith rather than established fact.

Goal-directedness

Sociobiologists often discuss cost-benefit analyses *as if* behavior results from a conscious, decision-making process. They talk about animals choosing from among behavioral options in light of the relevant costs and benefits so as to increase fitness, and they ask what animals should choose to do in each behavioral context. This procedure is just a shorthand logic used for convenience. We cannot assume that animals make conscious decisions because we cannot monitor what goes on inside their heads. Nevertheless, it really does not matter what the proximate bases of those decisions are when the evolutionary reasons underlying behavior are our principal concern. Animals do make choices between behavioral options, and those choices reflect the past actions of natural selection. The question of whether those choices are conscious or unconscious need not concern us, as long as we remember that tacit assumptions about purposiveness are just that.

Two questions commonly asked are "How does an animal know what to do?" and "How does it know whether or not a behavior is adaptive?" The answers involve proximate mechanisms. When animals choose between alternative behaviors, their choices are governed by underlying physiological and biochemical mechanisms that were ultimately shaped by natural selection. Particular stimuli or contexts elicit particular behaviors. An animal need not know why those stimulus-response relationships exist. It need only know what the relationships are. This knowing need not involve conscious awareness, though in many cases animals are undoubtedly conscious of what they are doing; it need only involve the appropriate neurological connections. Thus animal behavior is goal-directed. Animals make choices that lead to short-term or proximate goals. They need not be aware of how the behavior leads to a particular goal or why it should be directed toward that goal in the first place. Of course, the reasons involve natural selection, but animals do not have to know that in order to behave appropriately. Animals can be goal-directed without being purposeful, and they can behave appropriately without knowing why.

Adaptive landscapes

An adaptive landscape is a model used to visualize the adaptive features of a gene pool. It is like a topographical map, with the altitude of terrain at each location representing the Darwinian fitness of a particular allele at a given frequency. A separate dimension is required for each allele, requiring a multidimensional landscape to represent all the possible genetic combinations a gene pool might contain. In terms of the landscape analogy, two questions are especially important in studies of adaptation (Wallace 1968). First, where on an adaptive landscape does a population lie? Second, why does it lie there instead of somewhere else?

Progressively more complex landscapes

The simplest adaptive landscape involves a single gene locus. It is simply a two-dimensional graph, with the frequency of a single allele plotted along the *X* axis and Darwinian fitness of that allele plotted along the *Y* axis. Since selection always moves toward increased fitness, it continually acts to increase the allele's frequency if the slope of the curve is positive and to decrease the allele's frequency if the slope of the curve is negative.

When two gene loci are considered simultaneously, a three-dimensional landscape results. The landscape is a surface above the horizontal plane, and it maps the fitnesses of the two alleles as a function of all possible combinations of allele frequencies. The first thorough analysis of such a system involved the Australian grasshopper *Moraba scurra* (Lewontin and White 1960). Relative fitnesses were estimated for each of nine genotypes based on a polymorphism in each of two chromosomes, and these could be used to compute fitness at every location on the landscape (Figure 2–7) (see Wallace 1968). The resulting landscape is a saddle-shaped surface. Surprisingly, the actual genotype frequencies found in the study population lie on the ridge of the saddle rather than on either peak. This result was apparently caused by inbreeding in the population. *Moraba scurra* is a sluggish, sedentary grasshopper that rarely disperses far to find mates. Mating patterns are therefore nonrandom, and individuals mate with rel-

Figure 2–7

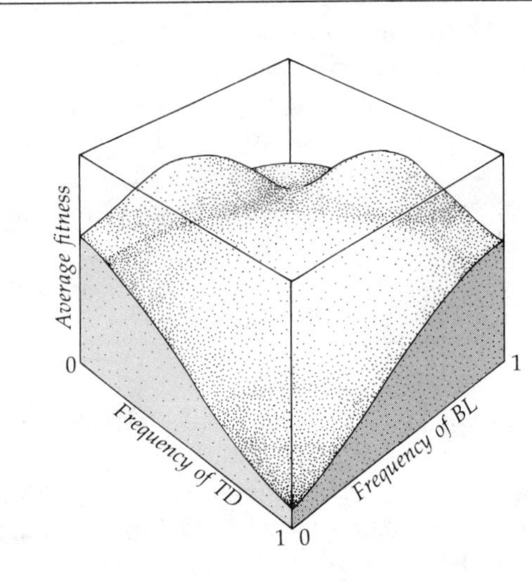

A three-dimensional adaptive landscape for the Australian grasshopper Moraba scurra.

The landscape was constructed by analyzing the joint effect of two independent chromosomal polymorphisms on average fitness of a population. The **CD** *and* **EF** *chromosomes of* Moraba scurra *occur in two cytologically distinguishable forms, a standard V-shaped form (ST) and an atypical rod-shaped form. The rod-shaped form of the* **CD** *chromosome is designated Blundell (BL), and the rod-shaped form of the* **EF** *chromosome is designated Tidbinbilla (TD). The four chromosomal types can occur in nine different combinations. [The* **CD** *chromosome can occur as ST/ST, ST/BL, or BL/BL. The* **EF** *chromosome can occur as ST/ST, ST/TD, or TD/TD. This allows for nine possible combinations of* **CD** *and* **EF** *chromosome pairs.] After an estimation of the relative fitness of each combination is obtained, the average fitness of a population is easily calculated for every possible frequency of BL and TD chromosomes.*

Source: Reproduced from *Topics in Population Genetics* by Bruce Wallace, p. 287, by permission of W. W. Norton & Company.

atives more often than was assumed in the initial computations of fitness. The shape of the adaptive landscape changes when the data are corrected for inbreeding effects. After this correction is made, the landscape has only one peak, and the population lies virtually at the top (Wallace 1968). Thus for this particular example selection has indeed led to maximum fitness, as expected according to natural selection theory.

In real populations average fitness of allele combinations results from interactions among many gene loci. Plotting each locus along a separate axis results in a hopelessly complicated, multidimensional landscape. It is difficult to visualize even a three-locus system. One way might be to plot the frequency of an allele at each locus to form a three-dimensional surface and then indicate the relative fitness at each location on the surface by the darkness of shading. To expand the picture to even more loci becomes a formidable task in abstract geometry. One way to conceptualize a landscape involving many gene loci was devised by Sewall Wright (1932). Instead of single loci being plotted along each axis, combinations of loci are plotted. The result is a landscape with many peaks and valleys (Figure 2–8). Although such a landscape is a gross oversimplification, it does help conceptualize the adaptive features of an entire gene pool. Selection on such a landscape still moves populations uphill toward adaptive peaks, but it is now clear that populations may not reach the highest peak. When one speaks of maximizing fitness, as is often done in sociobiology, maximization refers to fitness changes on a particular peak and not on the landscape as a whole.

Some implications of adaptive landscapes

Adaptive landscapes illustrate several important points about evolution. First of all, the starting point of the population is important in determining which adaptive peak is reached. A population starting at point 1 in Figure 2–8 will automatically move uphill until it reaches peak A, while a population starting at point 2 will move uphill until it reaches peak B. The starting point is determined by the population's previous phylogenetic history. When a population is pushed into an adaptive valley by environmental changes and begins evolving uphill, its starting point or genetic heritage is referred to as its **preadaptations.** The preadaptations of a population may greatly affect which adaptive peak is reached. Two populations exposed to identical selective regimes could theoretically reach different adaptive peaks simply because they started with different preadaptations.

Secondly, a population may become stranded on a minor adaptive peak because selection cannot move populations across valleys from one peak to another. Although a higher adaptive peak may lie nearby, the population cannot reach it because populations do not possess foresight or any prior notions about where they are going. In other words, evolution is not purposeful.

Finally, the existence of many adaptive peaks helps to explain in very simple terms why there are so many species. Simply stated, there are many

Figure 2–8

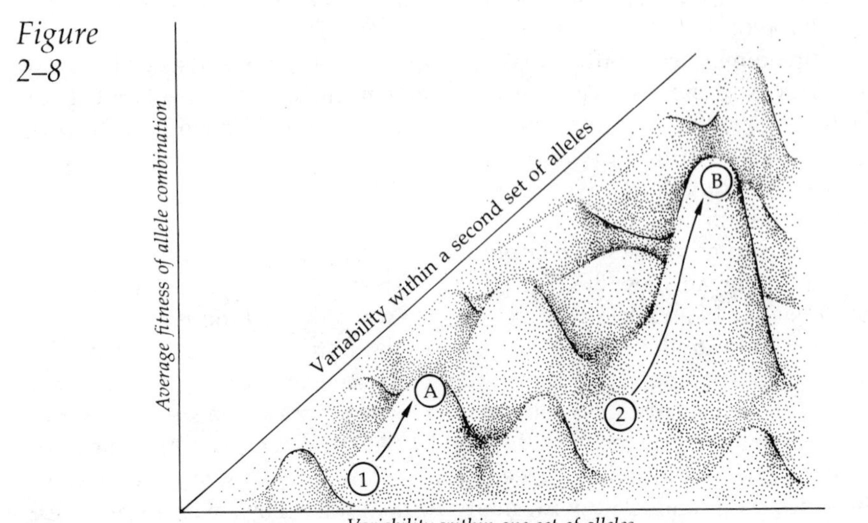

A multiple-locus adaptive landscape.
Each point on the horizontal plane represents a particular combination of alleles within a genotype, and the vertical height of the surface above that point represents the average fitness of a population whose members possess that combination of alleles. Natural selection causes a population to move uphill until it reaches a nearby peak. For example, a population starting at point 1 might move to peak A, and a population starting at point 2 might move to peak B.

species because there are many adaptive peaks (Dobzhansky et al. 1977). Speciation occurs when for some reason a population separates, with each portion moving toward a different peak. The division may be caused by geographical isolation or by disruptive selection (defined below) (Dobzhansky et al. 1977; White 1978; Futuyama 1979). Not all adaptive peaks are occupied by real populations, and some populations occupy more than one peak. The latter are said to be **polymorphic** because population members occur in more than one discrete form. The average fitness of polymorphic populations lies between two or more adaptive peaks, but the average fitness of each morph lies on an adaptive peak.

Dynamics of landscapes

The average fitness of gene combinations never remains fixed. It varies continually as environments fluctuate or change. A better picture of an adaptive landscape is obtained by thinking of it as a constantly shimmering and shifting jellylike globule. Major peaks and valleys persist for long periods of time, but minor peaks bounce up and down, continually subsiding into valleys or becoming elevated into new peaks. This process occurs simultaneously but asynchronously in all geographical areas occupied by a population. The adaptive landscape will be slightly different in each region, especially with respect to minor peaks and valleys. Selection can move a population in a consistent direction only if adaptive peaks are high enough to persist over much of the population's geographical range. Some peaks may differ between regions, and when they do, they can lead to spatial heterogeneity in the gene pool.

Types of individual selection

Adaptive landscapes portray the fitness of alleles as a function of gene frequencies, but they do not depict the extent that individuals vary within a population. Phenotypic variability based on genetic variability is necessary for individual selection to occur. No selective process is possible unless there is underlying variability upon which it can act.

The variability of individual phenotypes commonly follows a normal or bell-shaped frequency distribution, with a majority of individuals resembling the population norm and a few individuals deviating considerably from it. This variability can be affected by individual selection in three ways: selection can change the population norm, maintain the existing norm in the face of mutations and other random processes, or change the shape of the distribution around the norm. Each change is referred to as a specific form of individual selection.

Directional selection

Most people think of directional selection when they think of evolution. **Directional selection** produces a change in the population norm. It occurs when phenotypes near one end of a population distribution reproduce more successfully than individuals at the norm (Figure 2–9). Directional selection begins operating whenever a population no longer rests on an adaptive peak. The adaptive peak may have subsided as a result of changes in the abiotic environment or the biotic community. In either event, the population norm moves toward a new adaptive peak in response to directional selection.

The evolution of sexual dimorphism in the ruff (*Philomachus pugnax*), a promiscuous European shorebird that breeds in grassy fields, provides an example of how directional selection operates. Male ruffs gather together each spring on traditional display grounds, called **leks,** where they strut and display toward one another and toward visiting females (Photo 2–1) (Hogan-Warburg 1966). Females visit leks for only brief periods, primarily during early morning hours, to meet males and eventually to copulate with them. Since females show a marked preference for males on centrally located display courts, male competition for those courts is severe. Males are larger than females, and they possess an unusual ruff of feathers around their necks, which females lack. While competing for display courts, males

Figure
2–9

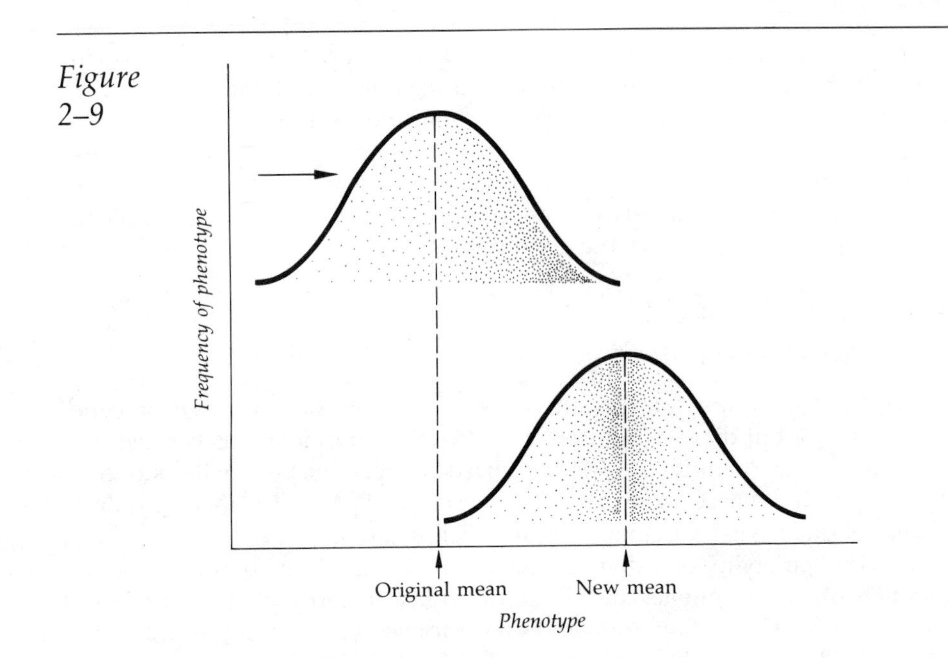

How directional selection leads to a change in phenotypic frequencies within a population.

When phenotypes at one end of the frequency distribution have higher average fitness than any other phenotype (upper curve), the distribution changes until the most fit phenotypes lie at the population mean (lower curve). Density of stippling indicates relative fitness of each phenotype, and arrow indicates direction of selection pressure.

rely on large body size and erection of ruff feathers to intimidate their rivals. In an ancestral population consisting largely of female-like males, any mutant males who were larger in size or appeared larger in size (i.e., by erecting somewhat elongated neck feathers) would have had an advantage in acquiring central display courts. This advantage would have allowed them to fertilize more females and hence leave more descendants in future generations. Thus the advantage of acquiring central display courts led to directional selection toward larger male body size and more elaborate ruffs of neck feathers on males. The selective process was initiated by a female preference for centrally located males, which originally evolved for other reasons. Larger size and elongated neck feathers are not the only ways males could intimidate rivals, but they lie on the adaptive peak reached by ruffs.

Stabilizing selection

Stabilizing selection is much more prevalent than directional selection, but it is less conspicuous because it simply maintains the status quo. As

Photo 2–1

A group of male ruffs (Philomachus pugnax) *displaying on a mating ground or lek.*

Note the dimorphism in male plumage. The male with white neck ruff and ear tufts is a nonterritorial satellite male, while the remaining males with dark neck ruffs and ear tufts are territorial resident males.

Photo by Eric Hosking.

a general rule, the more an individual phenotype deviates from the population norm, the less its genetic representation in future generations will be, because extreme phenotypes usually suffer higher mortality or reproduce less successfully than intermediate phenotypes. Thus **stabilizing selection** discriminates against phenotypes diverging from the population norm, thereby maintaining the existing norm in the face of new mutations arising in the population.

Stabilizing selection can be followed by measuring changes in phenotypic variability as each new generation grows older (Figure 2–10). Genetic recombination and mutation increase phenotypic variability within each new age cohort of offspring. As the cohort grows older, the more extreme phenotypes die at a faster rate than do intermediate phenotypes, with the result that phenotypic variability gradually decreases. In an unchanging environment this process eventually reduces phenotypic variability within the cohort to that found in the parent generation.

Stabilizing selection is easiest to detect during periods of cataclysmic mortality. For example, songbirds are often energetically drained by long migrations or reproductive activities and possess few reserves to cope with

Figure 2–10

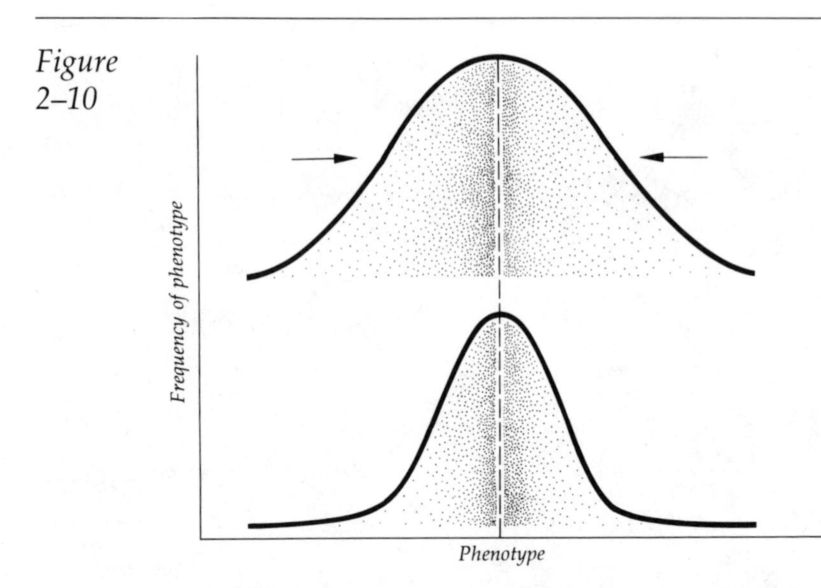

How stabilizing selection maintains the same mean phenotype in a population.

Genetic recombination and new mutations lead to increased phenotypic variation among offspring (upper curve), and selection against extreme phenotypes reduces phenotypic variation among surviving adults to a level similar to that found in the previous generation of adults (lower curve). Density of stippling indicates relative fitness of each phenotype, and arrows indicate directions of selection pressure.

unusually severe weather. The result is sometimes substantial mortality. In one study house sparrows (*Passer domesticus*) with longer or shorter wings than average were more likely to succumb in bad weather than individuals with average wing lengths (Bumpus 1899). Even slight deviations from the norm can be significant. In Bumpus's study the differences in wing length were a matter of only a few millimeters, and yet they significantly decreased the birds' ability to cope with storms.

Returning to the ruff example, stabilizing selection is responsible for maintaining male body size and length of ruff feathers at an equilibrium level. Overly large males probably survive less well than average males (not proven), because they should have greater difficulty satisfying their food requirements when food is scarce and they may be more vulnerable to predators or stormy weather. Males with overly long ruff feathers probably survive less well (not proven), because they should be less able to maneuver during flight, making them more vulnerable to predators or stormy weather. At equilibrium such costs should just balance the benefits gained while competing for females. Males above and below the norm should therefore achieve lower lifetime reproductive success than males at the norm, though the difference may be too small to detect.

An important assumption in evolutionary biology is that the costs and benefits associated with stabilizing selection are similar in kind to the costs and benefits associated with directional selection. Sociobiologists often assume that they can infer the selective origins of a trait by studying the costs and benefits present at equilibrium. They argue that behavioral traits have evolved in response to advantages still present at equilibrium and that the costs responsible for maintaining the equilibrium were present to a lesser degree throughout much of the earlier evolutionary process. This assumption is often necessary, because in undisturbed environments most populations are probably near equilibrium at any given time, making directional selection rare and difficult to study.

Disruptive and frequency-dependent selection

Disruptive selection occurs when phenotypes at each end of a population distribution reproduce more successfully than intermediate phenotypes. This process often occurs because natural environments vary in space and time. Then slightly different phenotypes have the advantage in each part of the environmental mosaic, creating a crosscurrent of selective pressures that maintains spatial or temporal heterogeneity in the gene pool.

In the more extreme cases disruptive selection may result in a bimodal or multimodal distribution of phenotypes throughout a population's range (Figure 2–11). Such polymorphisms are often a result of **frequency-dependent selection**, a form of selection in which rare phenotypes gain an advantage because of their rarity. The ruff again provides a good example.

Disruptive selection has led to two types of male ruffs: resident males and satellite males (see Photo 2–1) (Hogan-Warburg 1966; Rhijn 1973). The two types of males differ with respect to both plumage and behavior, and these differences persist in the population because each male type becomes more fit relative to the other when its commonness decreases.

Resident males typically possess either black or dark neck ruffs and head tufts or white neck ruffs with black head tufts. They are highly aggressive and defend small display territories on the lek, each of which is separated from the others by undefended neutral zones. Resident males initially establish display territories by threatening or attacking any males possessing resident male plumage. After territories are well established, they show little overt aggression toward neighboring resident males, but they continue to attack any nonterritorial intruders with resident male plumage.

Satellite males typically possess white neck ruffs and either white or colored head tufts. Instead of establishing territories of their own, they usually gain entry to territories of resident males by behaving submissively whenever they are threatened. Resident males usually respond to satellite males in much the same way that they respond to females, until many females are attracted to the territory. They are initially hostile but quickly become tolerant following submissive displays of the satellite male.

Each satellite male associates with only a few resident males. Two

Figure 2–11

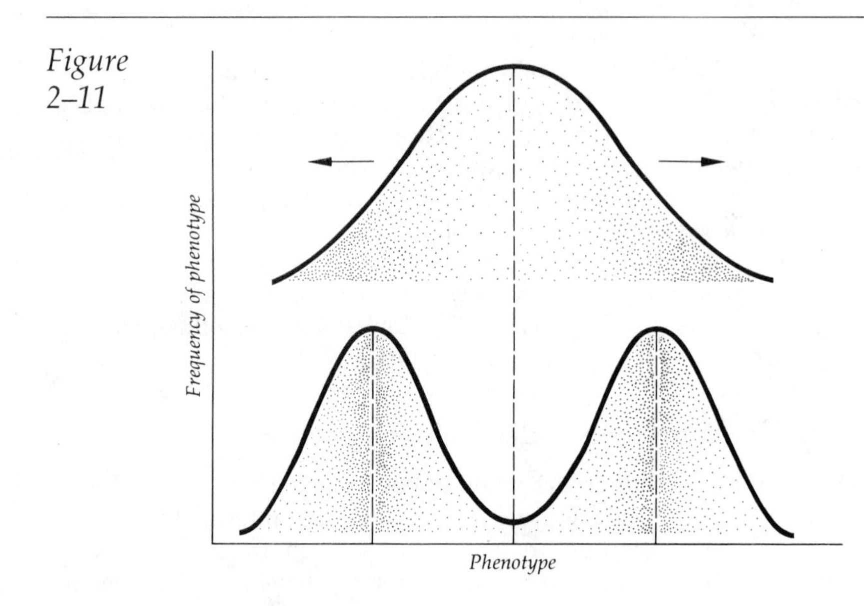

How disruptive selection can lead to a phenotypic polymorphism.
When phenotypes at both ends of a frequency distribution are more fit than average phenotypes (upper curve), selection leads to a bimodal distribution, with the two most fit phenotypes lying at the two modes (lower curve). Density of stippling indicates relative fitness of each phenotype, and arrows indicate directions of selection pressure.

satellite males may associate with a single resident male, but they never visit the resident male at the same time. Centrally located satellite males are not aggressive toward each other, but they attack any peripheral satellite males that try to enter the center of the lek. These centrally located satellite males are tolerated by resident males except when many females are on their territories. (Resident males are intolerant of satellite males only when the frequency of female visits, copulations, or both becomes high.) Consequently, satellite males can copulate successfully with some visiting females.

The evolution of this behavioral and morphological polymorphism can be understood by examining the reproductive advantages gained by each type of male. Resident males gain access to females by excluding other resident males from the center of the lek. By being territorial, they can copulate with more females and hence leave more descendants in future generations. Centrally located satellite males gain this same advantage without paying the high cost of competing for display courts. However, they can gain this advantage only because resident males tolerate them in the lek center. Why resident males tolerate satellite males on their territories is therefore the key question requiring analysis.

The evolved objective of resident males is to fertilize as many females as possible. Hence resident males should not tolerate satellite males unless such tolerance allows them to attract more females. For unknown reasons the presence of satellite males does allow a resident male to attract more females (Rhijn 1973). Females are more likely to enter a central territory if it contains both a resident male and a satellite male. However, once there, females rarely copulate with the resident male while the satellite male is still present, although they do sometimes copulate with the satellite male. The resident male is therefore faced with a choice. He can either chase off the satellite male and try to copulate with the females already present, or he can continue to tolerate the satellite male and try to attract even more females. Not surprisingly, resident males become increasingly intolerant of satellite males as female density increases on their territories, until finally they aggressively chase the satellite males off. Then they are often able to copulate with the females still remaining on their territories.

A simulation model of resident male, satellite male, and female behavior shows that copulatory success of resident males depends on the proportion of time that female visits are high and satellite males are absent (Rhijn 1973). This proportion is high if the resident male has a low tolerance threshold for satellite males once females are attracted to the territory. The most successful males alternate between tolerance and intolerance of satellite males, tolerating satellite males during periods of infrequent female visits to the territory but not tolerating them once many females are on the territory.

The advantages gained by each type of male are frequency-dependent. That is, resident males reproduce more successfully than satellite males if resident males are relatively uncommon, while satellite males reproduce more successfully than resident males if satellite males are relatively uncommon. If resident males become uncommon, competition for central territories is relaxed, and more resident males can acquire good territories. Hence their average reproductive success increases. Meanwhile, satellite males are too common compared to the number of central locations available to them on leks, and, consequently, their average reproductive success decreases. As long as resident males reproduce more successfully than satellite males, they will leave more progeny than satellite males in the next generation. This trend continues until resident males and satellite males reproduce equally well. Similarly, if resident males are too common, competition for central territories increases and average reproductive success of resident males decreases, competition among satellite males for central locations is relaxed, and average reproductive success of satellite males increases. Then satellite males leave more progeny than resident males in the next generation. Again, this trend continues until resident males and satellite males reproduce equally well. Thus individuals of both phenotypes have a reproductive advantage when their phenotype is relatively uncommon but are at a disadvantage when their phenotype is too common. At equilibrium both phenotypes exist at intermediate (though not necessarily equal) frequencies in the population.

Males with intermediate phenotypes are probably less successful than either of the two extreme phenotypes. They are less able to defend a territory because their plumage is less threatening and their behavior is less aggressive. They are also less able to appease resident males and hence cannot compete with satellite males for favorable positions on resident male territories. However, the situation is complicated, as the relative success of males with atypical plumages depends on how common they are on a lek (Rhijn 1973). When they are more common than typical males, they have higher success than typical males. When they are less common, they have lower success. Probably for this reason males show a distinct tendency to leave leks where they are an uncommon plumage type and remain on leks where they are a common plumage type.

As with directional selection, the equilibrium condition reached by disruptive selection is maintained by stabilizing selection. Phenotypes more extreme than those found at each mode of the population distribution suffer the same costs discussed earlier, while intermediate phenotypes cannot compete as well for mates.

Kin selection

Two modes of gene transmission

The classical theory of individual selection was based on the notion that adults attempt to maximize fitness by leaving as many mature progeny as possible in the next generation. Viewed from this perspective, some behaviors appear altruistic in the sense that individuals seem to sacrifice their own reproductive potential to enhance the reproductive success of others. This apparent altruism poses a difficult problem for the classical theory of individual selection.

From a genetic perspective such behaviors usually no longer appear altruistic. Genic selection favors alleles that are transmitted most frequently to future generations, but this transmission process can occur along two different reproductive pathways. Individuals can propagate genes by producing direct progeny or by helping nondescendant relatives produce progeny (Hamilton 1964). Hence altruism directed toward helping nondescendant relatives increases Darwinian fitness in the same manner as individual selection. The only difference lies at the phenotypic level. At that level, helping relatives is an interesting phenomenon, and the concept of individual selection has been expanded to encompass it. The expanded concept of individual selection is termed **kin selection,** and it includes selective processes operating through production of direct progeny, progeny of relatives, or both (after Maynard Smith 1964 and Brown 1966).

Sociobiologists often find it convenient to separate the two modes of gene transmission, but the terminology they use is not applied consistently. A widespread practice is to use *kin selection* when referring to selection resulting from production of progeny by relatives and to use *individual selection* when referring to selection resulting from production of direct progeny. A movement is now afoot to use *kin selection* in the original sense proposed by Maynard Smith (1964) and to adopt a different terminology to separate the effects of producing direct progeny from the effects of helping relatives produce progeny. Brown (1981) proposes that selection resulting from production of direct progeny be referred to as **direct selection,** while selection resulting from production of relatives' progeny be referred to as **indirect selection.** Later in this chapter the term *indirect selection* will be extended to include progeny of individuals who share genes in common but are *not* relatives (see p. 71). In an attempt to establish a more consistent terminology, I will adopt Brown's terminology here but give the older terms in parentheses, where appropriate, to avoid confusion.

When is helping relatives advantageous?

The conditions determining which mode of gene transmission is best in a given situation can be calculated with some simple algebra. Suppose that individuals in a population are regularly faced with the choice of either helping a relative or producing progeny of their own. The best choice will be the one that propagates genes fastest. Hence the problem can be solved by calculating the number of genes propagated by individuals of each phenotype.

Consider two kinds of phenotypes. One type, which I will refer to as ego, behaves selfishly by producing progeny of its own, while the other type, which I will refer to as alter ego, behaves altruistically by helping a relative. By behaving selfishly, ego will produce, on average, say N mature progeny during its lifetime. For a sexually reproducing diploid animal, half the genes carried by each progeny will be identical to those carried by ego. Hence a typical ego will propagate (1/2) N genes to the next generation via its own progeny. Meanwhile, ego's relative, whom ego could have helped but did not, also reproduces, and it produces M mature progeny during its lifetime. Each of the relative's offspring will carry some of the same genes carried by ego, since they are genetically related. The probability that particular alleles are shared because of common ancestry is measured by the coefficient of relatedness r. The rules for computing r are explained in most introductory genetics texts. In the absence of inbreeding, r is an exact measure of the fraction of genes shared, on average, by ego and ego's relative, but when inbreeding is important, r becomes an inexact estimate (Michod and Anderson 1979). For an outbred population the number of ego's genes propagated by ego's relative is (1/2) rM, and the total number of genes propagated by ego and ego's relative combined is

$$S = (1/2)N + (1/2)rM \qquad (2\text{–}1)$$

Now consider the number of genes propagated by alter ego. Because alter ego helps its relative, it produces C fewer direct progeny than ego. Hence alter ego propagates only (1/2) $(N - C)$ genes through direct progeny. On the other hand, alter ego's relative can now raise B more offspring than would otherwise have been possible, thanks to alter ego's help. The relative's progeny will therefore carry (1/2)r (M + B) replicates of alter ego's genes, and the total number of genes propagated by alter ego and its relative combined is

$$A = (1/2)\ (N - C) + (1/2)r(M + B) \qquad (2\text{–}2)$$

Alter ego will propagate genes faster than ego whenever A is greater than S (i.e., $A > S$). This outcome occurs when

$$(1/2)(N - C) + (1/2)r(M + B) > (1/2)N + (1/2)rM \qquad (2\text{–}3)$$

Simplifying, A is greater than S when

$$\frac{B}{C} > \frac{1}{r} \qquad (2\text{–}4)$$

For convenience, the ratio B/C is sometimes referred to as k.

Thus altruism toward a relative should evolve when the ratio of benefits (B) to costs (C) exceeds the reciprocal of genetic relatedness ($1/r$) between altruist and relative. If, for example, an individual propagates only half as many genes per offspring by helping a relative (i.e., $r = 1/2$), its help must increase its relative's reproductive output by double the amount that its own reproductive output is reduced (i.e., $B/C = 2$) before altruism will pay. When relatedness is less than 1/2, the ratio of benefits to costs must be proportionately larger. In many social animals average relatedness among group members is quite low. Nevertheless, altruism can still evolve through indirect (kin) selection if the associated benefits are high enough or if the costs are low enough (see West Eberhard 1975).

Many behaviors are not directed at helping a single other individual. When a behavior affects several relatives, Equation 2–4 must be modified. The mean relatedness of all recipients ($\bar{r}$) must then be substituted for the relatedness of a single recipient (r), and the benefits (B) and costs (C) of the altruistic act must be summed across all relatives.

Inclusive fitness

Kin selection theory requires an extension of the concept of individual fitness. Each individual's fitness depends not only on its own reproductive success but also on its contribution to the reproductive success of relatives. Thus **inclusive fitness** refers to the sum of an individual's own individual fitness plus all effects the individual's activities have on the individual fitness of relatives other than direct descendants (Hamilton 1964). There are two components to inclusive fitness. *Direct fitness* refers to fitness resulting from production of direct progeny, while *indirect fitness* results from production of progeny through relatives (Brown and Brown 1980; Brown 1981). The sum of direct fitness and indirect fitness equals inclusive fitness.

Kin selection theory and the concept of inclusive fitness are relevant to all aspects of sociobiology, but they have proven most useful in analyses of alleged altruistic behaviors. These behaviors will be discussed more fully in the next chapter and in Chapter 12.

Group selection

A legacy from early misconceptions about how natural selection works is the belief that evolution proceeds for the good of the species. This notion is not true of individual selection (Fisher 1930). Individual selection can be either beneficial or detrimental to a species' continued existence. Genes can change in frequency without regard to what is happening to total population size. Individuals may gain a net benefit from the behavior no matter how the behavior affects the population as a whole. That is, individual selection is indifferent to the perpetuation of a species.

Selective processes that could promote the welfare of entire species or even whole ecosystems have been postulated. These processes all involve some form of **group selection,** which brings about genetic change by the differential extinction or proliferation of populations or other groups of organisms.

The two principal kinds of group selection are interdemic selection and trait-group selection (Wade 1978). **Interdemic selection** involves group selection among local populations, or *demes,* while **trait-group selection** involves group selection among subgroups within demes. Note that the more general term *intrademic selection* encompasses all selection processes occurring within demes, including both individual selection and trait-group selection (Brown 1981). Kin selection is a special case of trait-group selection to the extent that it operates by favoring some lineage groups over others (Charnov 1977; Wilson 1977). Some authors do not consider any intrademic selection processes as forms of group selection (e.g., Maynard Smith 1976a; Brown 1981). Instead, they view intrademic selection as synonymous with direct and indirect selection as defined by Brown (1981) (see p. 71).

Selection processes operating at higher levels of organization, such as among species or ecosystems, can occur (e.g., Dunbar 1960, 1972; Van Valen 1975), but they are not likely to affect the behavior of individuals within populations. These processes will therefore not be discussed here.

Group selection is a controversial topic. The controversy revolves primarily around the issue of whether group selection can ever act in opposition to individual selection to give rise to altruistic behaviors among nonrelatives. The impetus for this controversy stems largely from V. C. Wynne-Edwards's (1962) book *Animal Dispersion in Relation to Social Behaviour,* which postulated that interdemic selection was responsible for the evolution of most social behaviors.

The Wynne-Edwards view of behavior

Wynne-Edwards (1962) argued that food availability could not act as a proximate mechanism regulating population density because, if it did, populations would often overeat their food supplies and suffer catastrophic declines. Since such declines are atypical of most populations, he theorized that individuals prevent overpopulation by voluntarily curtailing reproductive output whenever population densities become too high. This kind of homeostatic mechanism would require behavioral conventions designed to mediate reproductive curtailments, along with behavioral means for regularly assessing population density. Wynne-Edwards termed the latter behaviors *epideictic displays* to distinguish them from displays that evolved primarily for courtship purposes, which he referred to as *epigamic displays*.

Wynne-Edwards's basic thesis was that most social behaviors evolved as conventions for curtailing reproductive output or as epideictic displays for assessing population density. His examples included such diverse behaviors as territory defense, coloniality, preroosting aggregations of birds, insect mating swarms, cannibalism, fratricide, schooling behavior of fish, and vertical migrations of zooplankton. Since voluntary curtailment of reproductive effort reduces individual fitness for the benefit of whole populations, behavioral mechanisms mediating such curtailments could only evolve by group selection.

Wynne-Edwards's view of social behavior as an evolved mechanism for regulating population density has been seriously questioned (Brown 1964, 1969a; Maynard Smith 1964; Lack 1966; Wiens 1966; Williams 1966a; Orians 1969, Lewontin 1970). Two general objections have been raised. First, virtually all aspects of animal behavior can be plausibly explained by direct (individual) or indirect (kin) selection without postulating some new and unsubstantiated selective process. Second, interdemic selection can favor altruistic traits only under stringent conditions that do not prevail in most natural populations.

Interdemic selection models

General assumptions. Interdemic selection involves the differential extinction or proliferation of whole populations. While this process could theoretically occur, albeit rather slowly, in most species, it could not favor the spread of an altruistic trait unless populations initially differ with respect to that trait. Some populations must possess the trait while others do not. Herein lies the major theoretical obstacle to interdemic selection. How would altruistic traits initially become established within any of the local populations? Individual selection within each population should eliminate such traits before they could become prevalent. Some process must prevent the elimination of altruistic traits from occurring in at least a few populations, or interdemic selection would never have a chance of spreading them to other demes.

The usual way this obstacle is overcome in theoretical models is to postulate a population structure amenable to random fixation of traits (Wilson 1973; Wade 1978). When a new population is created by colonization, its genetic makeup represents some sample of the parent population's gene pool. Through chance events gene frequencies may differ considerably from those in the original population. The probability of large deviations occurring is higher in small daughter populations than in larger ones, just as small samples are more likely to deviate from a true population mean than are large samples in any sampling situation. The process whereby random deviations in gene frequencies arise because a few unrepresentative colonizers start a new population is referred to as the **founder effect.**

Chance events can also cause gene frequencies to change between generations within populations. In small populations the genes inherited by progeny may deviate considerably from the population average because accidental mortality or random events affected reproductive success of the parents.

Allele frequencies may therefore change from one generation to the next by chance alone. Several successive increases or decreases in frequency could cause an allele to become fixed or lost, provided the population is small enough (Figure 2–12). Fixation and loss are much less likely to occur in large populations because the chances of a sufficiently disproportionate number of increases or decreases occurring are miniscule.

Figure 2–12

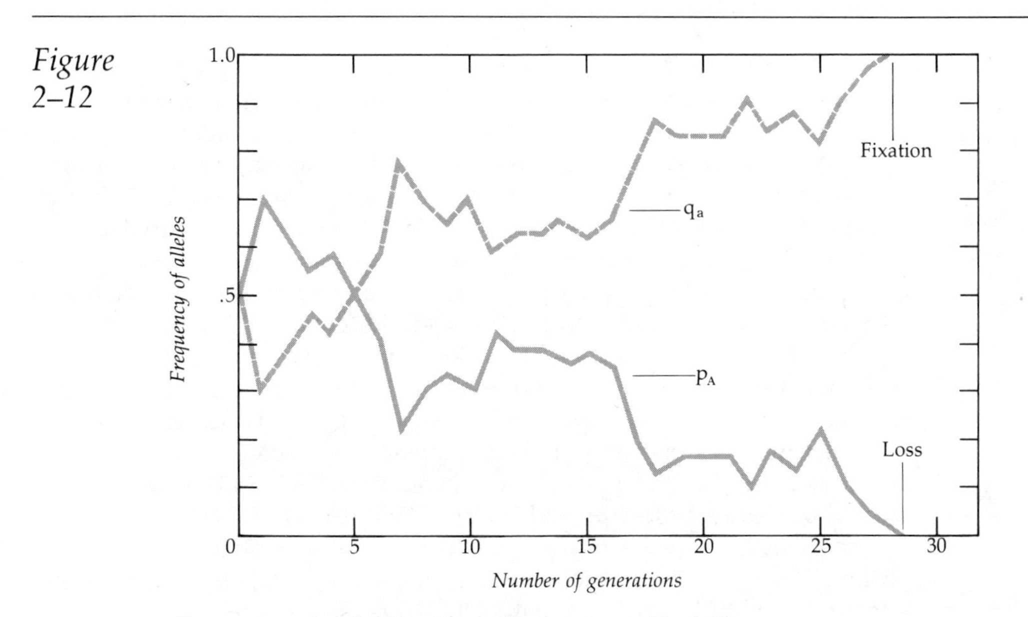

Computer simulation of continuous genetic drift.

Random changes in the frequency of allele a (designated by q_a) and the frequency of allele A (designated by p_A) led to fixation of allele a and loss of allele A in a population of 12 individuals after 28 generations.

Source: From E. O. Wilson and W. H. Bossert, *A Primer of Population Biology* (Sunderland, Mass.: Sinauer Associates, 1971), p. 86. Reprinted by permission.

The process whereby gene frequencies change within a population by chance alone is called **genetic drift.** Genetic drift is not likely to lead to fixation or loss of an allele unless directional selection affecting the allele is weak or absent and gene flow between different local populations is negligible. It has nevertheless been postulated as an important cause of evolutionary changes in protein structure, where at least some amino acid substitutions appear to be selectively neutral (Kimura 1979).

To explain how local populations might vary with respect to an altruistic trait, interdemic selection models assume that species are dispersed as arrays of small demes. Such an array of demes is called a **metapopulation** (after Levins 1970). In models of interdemic selection all members of each deme are assumed to either possess the altruistic trait or lack it. Intermediate frequencies of an altruistic trait are, of course, possible, but then selection between demes becomes mathematically equivalent to trait-group selection (see pp. 69–71 and Wilson 1977).

Models of interdemic selection treat each deme in a metapopulation as a separate entity, with its own proliferation and extinction rates. Interdemic selection occurs when some demes proliferate or become extinct faster than others. Demes containing the altruistic trait are postulated to proliferate faster or become extinct more slowly because the altruistic trait is prevalent. If this process occurs faster than individual selection can reduce the frequency of altruism within demes, it should lead to spread of the altruistic trait.

The Levins model. Two main models of interdemic selection have been devised, the Levins model and the Boorman-Levitt model. The Levins (1970) model is based on a metapopulation consisting of many very small populations, each occupying its own habitat island or patch (Figure 2–13A). At any given time some patches are occupied and some are empty. Empty patches are occasionally colonized by emigrants from extant populations, while extant populations occasionally become extinct, thereby creating new empty patches. A few populations are assumed to possess altruistic traits that were fixed by genetic drift or the founder effect. These traits reduce the probability of the population quickly becoming extinct in newly colonized patches. Under these conditions the altruist genes can increase in frequency within the metapopulation because populations lacking them become extinct faster than populations possessing them.

Several technical difficulties were found in Levins's original mathematical procedure (Boorman and Levitt 1973; Wilson 1973), but most of the problems have since been surmounted by computer simulation models without changing the end result (Levins and Kilmer 1974). Also, a recent experimental study confirms that such a process can occur. Wade (1977) found that differential proliferation of small laboratory populations of flour beetles *(Tribolium confusum)* can increase the frequency of an altruistic trait within a metapopulation possessing the island structure postulated by Levin. Wade's experiment involved differential proliferation of populations rather than differential extinction, but it seems likely that both processes could generate comparable end results.

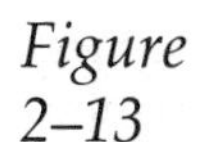

Figure 2–13

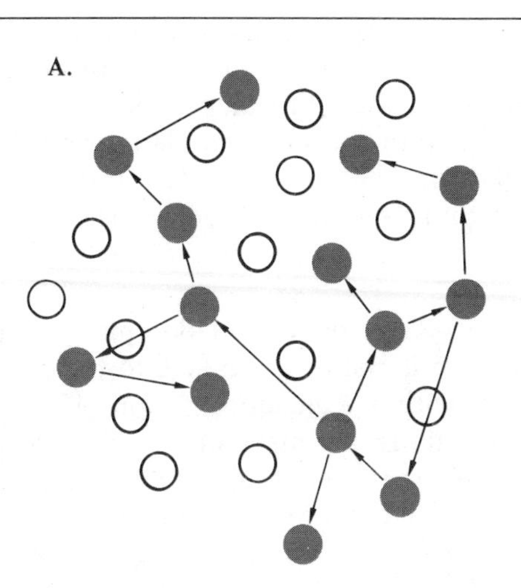

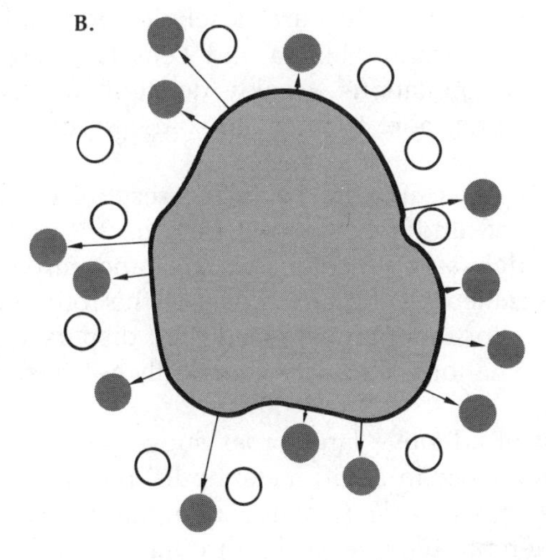

The metapopulations used to simulate interdemic group selection processes.

A. The metapopulation assumed in the Levins model consists of numerous small populations. Habitable sites contain either a small population (solid circles) or are unoccupied but suitable for colonization (open circles). Arrows indicate colonization from parent populations. B. The metapopulation structure assumed in the Boorman-Levitt model consists of a large central population and numerous small peripheral populations. Solid circles indicate habitable sites actually occupied by peripheral populations, and open circles indicate unoccupied habitats that are suitable for colonization. Arrows indicate colonization from the parent population.

Source: Modified from E. O. Wilson, "Group Selection and Its Significance for Ecology," p. 635. Reprinted, with permission, from the 1973 *BioScience*, © American Institute of Biological Sciences.

The major theoretical difficulty limiting applicability of the Levins model is that the model depends on unrealistic assumptions about population structure. Few natural populations are dispersed as small, discrete demes whose members rarely interbreed. Hence few natural populations are susceptible to the differential proliferation or extinction of demes, as postulated by the model.

The Boorman-Levitt model. The Boorman and Levitt (1972, 1973) model assumes a metapopulation consisting of a large, enduring central population surrounded by many small peripheral populations (Figure 2–13B). The peripheral populations are started by colonizers from the central population, and they vary genetically due to a founder effect. Each population begins at carrying capacity and has a different probability of becoming extinct, depending on how frequently the altruistic trait occurs within it. The model shows that an altruistic trait can increase in frequency to values as high as 20–30% within the metapopulation, but only if extinction proceeds fast enough to prevent individual selection from reducing the frequency of altruism within peripheral populations, and only after most peripheral populations have become extinct.

Alternative approaches. The Levins and Boorman-Levitt models suggest that interdemic selection is not likely to lead to the prevalent occurrence of an altruistic trait within a metapopulation. However, both models make restrictive assumptions that make interdemic selection more difficult to demonstrate. Models that relax these assumptions may show that interdemic selection is more likely to occur than previously believed. Several approaches might be taken in relaxing the restrictive assumptions of earlier models.

One alternative approach to postulating random sources of variation among populations is to assume that variability among local populations results from **hitchhiking,** a process by which neutral or weakly deleterious alleles increase in frequency because they are closely linked to an advantageous allele (see Chapter ten and Medina and Petit 1979). If linkage patterns vary among local populations, weakly deleterious alleles promoting altruism might become more frequent in some populations than in others.

A second alternative is to assume that colonizers responsible for starting new populations are not randomly selected from parent populations (Wade 1978). Current models assume colonizers are randomly selected, and hence they are constrained by the limitations of the founder effect. However, if colonizers are not randomly selected (i.e., dispersal patterns are nonrandom), local populations may vary genetically as a result of the dispersal process.

Finally, the concept of adaptive landscapes suggests an alternative scenario to the Levins and Boorman-Levitt metapopulations. According to Wright's (1970, 1977) *shifting balance theory* of evolution, different local populations might reach different adaptive peaks through a combination of chance events, hitchhiking, and individual selection. Interdemic selection could then replace populations on low adaptive peaks with populations on higher adaptive peaks. While this process may not favor the spread of an altruistic trait counter to the pressures of individual selection, it could substantially affect the gene pool. Considerable evidence suggests that populations may in general be structured in a form susceptible to such a process (S. Wright 1978).

Modeling efforts have to date confirmed the logical possibility of interdemic selection as an evolutionary process, but the process has only been substantiated by data from a few natural populations (see Lewontin 1970). Perhaps future studies will demonstrate that interdemic selection is prevalent in nature, but at the present time other selection models are generally relied upon to explain seemingly altruistic behaviors.

Intrademic group selection models

Trait-group selection. Models of group selection that might occur within demes do not rely on improbable assumptions about population structure. One promising model is based on the concept of trait groups. D. S. Wilson (1975, 1977) defines a **trait group** as a subdivision of a population within which each individual breeds, eats, competes, or in other ways interacts with other population members. No individual interacts with every other individual in a population. The network within which each individual does interact is that individual's trait group. Trait groups may be discrete entities such as mosquitos inhabiting a vessel or mouse families inhabiting separate haystacks, or they may be parts of continuous populations in which each individual forms the center of its own trait group, as in territorial or sessile animals.

Wilson's model of trait-group selection can be understood most easily by analyzing the effects of selection on two alleles, one for an altruistic trait and the other for a selfish trait (Wilson 1975). An example might be the use of alarm calls to warn others of a predator's presence. A-type alleles predispose individuals to give the alarm call, while S-type alleles do not. Predation occurs while individuals are living in their trait groups. Individuals with S-type alleles survive within trait groups better than individuals with A-type alleles because they do not attract the predator's attention by giving the call (Figure 2–14A). That is, individual selection is acting in favor of the S-type allele and against the A-type allele. However, at the same time this is happening, trait groups containing more A-type individuals suffer less predation than trait groups containing fewer A-type individuals because individuals in the former trait groups are more frequently warned by alarm callers. If average survival is sufficiently higher within trait groups that contain alarm callers, fewer A-type individuals will die overall, and more will survive to reproduce. The result is a net increase in the frequency of A-type alleles, due to trait-group selection (Figure 2–14B).

Despite the increased frequency of A-type alleles during winter, individual selection would gradually convert trait groups containing some A-type individuals into trait groups containing only S-type individuals unless intermixing occurs. If every individual leaves its trait group (winter flocks in the example above) to breed, produces an average of, say, one mature progeny (the allele affecting alarm calling during winter is assumed to have no effect on reproductive success during summer) (Figure 2–14C), and then settles back into a new trait group the following winter (Figure 2–14D), more trait groups will contain A-type individuals every winter and the frequency of A-type individuals in those trait groups will be increased. The intermixing process prevents individual selection from converting trait groups containing A-type individuals into trait groups lacking such individuals. By repeating this process through several generations, the S-type individual will eventually be eliminated.

Wilson (1975, 1977) and Bell (1978a) have formalized the model mathematically and shown that it can work. The mathematics will not be presented here, but it has some interesting properties. When the frequency

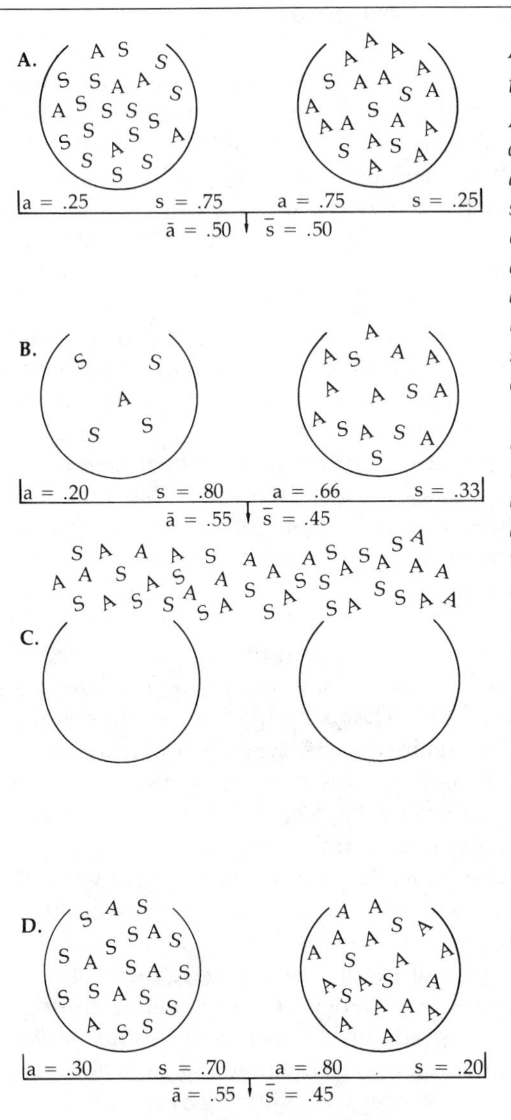

Figure 2–14 *A model of trait-group selection.*

A. Two trait groups in a population contain different frequencies of the alleles coding for altruistic (A) and selfish (S) behaviors. The frequencies of each type are indicated below each trait group by a and s. The average frequency of each type among both trait groups is indicated by ā and s̄. B. Predation leads to lower frequencies of altruist alleles within both trait groups, but overall mortality is higher in the trait group with the lower frequency of altruist alleles. As a result, the average frequency of altruist alleles among both trait groups increases *even though the frequency of those alleles decreases within each trait group. C. Trait groups break apart and intermix randomly or nonrandomly (e.g., during the breeding season). D. The population again sorts into trait groups, with the frequency of altruist alleles within each trait group being higher than in the previous generation.*

Source: From D. S. Wilson, "A Theory of Group Selection." *Proceedings of the National Academy of Sciences* 72 (1975):145. Reprinted by permission.

of an allele is identical in every trait group (i.e., each trait group is genetically equivalent to the entire metapopulation with respect to the specified locus), the model predicts that only direct (individual) selection will occur. When the allele is either present or absent in each trait group (i.e., trait groups are genetically isolated demes), the model predicts that interdemic selection will occur. Finally, when the allele's frequency is variable among trait groups, the model predicts that a type of selection intermediate between individual and interdemic selection occurs. Thus Wilson interprets individual selection and interdemic selection as two extremes of a single continuum. For an understanding of the intermediate situations, the source of genetic variation among trait groups must be explained more clearly.

In intermediate situations the frequency of an A-type allele can take any value between 0–100% in each trait group. One way these situations could arise is through genetic drift or founder effects, which affect allele frequencies in relatively isolated demes. A second and more likely way they could arise is through nonrandom mixing at the time trait groups are formed.

A common basis for nonrandom mixing is limited dispersal by related individuals. If individuals tend to remain in the same general vicinity as their relatives, trait groups would consist of kinship lineage groups. Then Wilson's trait-group selection model becomes mathematically identical to the kin selection model (Maynard Smith 1976a; Wilson 1977, 1979, 1980).

A second possible basis for nonrandom mixing is aggregation of individuals who exhibit similar traits (Wilson 1977, 1979). Unrelated individuals who exhibit altruistic behavior may, for example, prefer to associate with other individuals who exhibit altruistic behavior (Matessi and Jayakar 1976). Trait groups would then consist of unrelated individuals who share particular alleles in common. While this source of variability among trait groups is theoretically possible, an analysis by Charlesworth (1979) indicates that it is not likely to occur in real populations.

Dawkins (1979) argues that trait groups based on such preferences are still subsumed under kin selection theory because individuals are helping genetic "kin" who share the alleles promoting altruistic behavior. However, Wilson's trait-group model does not require the assumption that such individuals share similar genes because they are descended from a common ancestor, while kin selection theory does make that assumption. At the genetic level trait-group selection works in the same manner as kin selection, but it makes less restrictive assumptions about how individuals with similar alleles become associated with each other. Thus Wilson's model is more general than kin selection theory.

Other models. Brown and Brown (1980; Brown 1981) suggest a method for conceptually unifying kin selection theory and trait-group selection theory. The significance of kin selection lies in the fitness gains derived indirectly through progeny of relatives. Similarly, the significance of trait-group selection lies in the fitness gains derived indirectly through progeny of other individuals who possess the same alleles. Thus a general way of conceptualizing fitness at the phenotypic level is to divide it into direct and indirect fitness components. Direct fitness is mediated by propagating alleles to direct progeny, as defined earlier (see p. 62). Indirect fitness is mediated by selectively propagating alleles through the progeny of other individuals who possess the same alleles. These individuals may be relatives who share common ancestry, in which case kin selection is involved, or they may be unrelated individuals, in which case trait-group selection is involved. Consistent with this change, the term *indirect fitness* refers to fitness derived through propagation of alleles via the progeny of other individuals, either relatives or nonrelatives, who possess the same alleles.

Several other models of intrademic group selection have been developed (Cohen and Eshel 1976; Matessi and Jayakar 1976). These models are based on the same general principles as Wilson's model and make the same basic predictions, but they are less useful because they are mathematically less tractable.

Conclusion

Environmental conditions and social contexts are ultimately responsible for shaping the genetic makeup of a species, although they do so only indirectly, by affecting the ability of organisms to reproduce. Natural selection as it affects phenotypes must therefore be distinguished from natural selection as it affects genotypes.

Selection on phenotypes allows individuals or groups possessing some types of alleles to leave more descendants, on average, than individuals or groups possessing alternative alleles. Three types of selection processes may occur at the phenotypic level. Individual selection occurs when some individuals leave more direct progeny than others do. Trait-group selection occurs when some groups within demes leave more descendants than others do. A special case of trait-group selection is kin selection, which occurs when trait groups are comprised of relatives. Finally, interdemic selection occurs when some genetically homogeneous demes leave more descendant demes than others do.

The effect that selection at the phenotypic level has on gene pools is not easily deduced. Selection on phenotypes affects gene frequencies and linkage patterns in a complex manner dictated by the way genotypes interact with biochemical environments during development. Hence no exact predictions can be made about how gene pools change in response to selection acting at the phenotypic level. The complexities of development also preclude simple dichotomies of behavior into instinctual and learned.

The most prevalent forms of natural selection at the phenotypic level are direct (individual) selection and indirect (kin) selection. These two processes are generally believed to be largely responsible for the evolution of animal behavior. The following chapters will explore in more detail how this evolution may have occurred.

Altruism: Does it exist?

In the previous chapter various models of group selection were discussed as possible explanations of altruistic behavior. The focus there was on various selective mechanisms and their plausibility in real populations. Here the focus changes to what actually occurs in real populations. The reality of indirect selection and group selection is most reliably established by demonstrating that altruism actually exists. If altruism does exist in animals, it would imply that Darwin's classical theory of individual selection needs modification. The study of possible instances of altruism is interesting for precisely that reason.

A major theoretical issue is at stake here. The issue is, What types of selection have shaped animal behavior? If altruism has evolved among relatives, indirect (kin) selection must be an important process. If altruism has evolved among nonrelatives, trait-group selection or some other selective process involving nonrelatives must be important. Thus the existence and nature of altruistic interactions form an important cornerstone of sociobiological thought.

The only well-established examples of altruism involve the nonreproductive castes of social insects. Most workers of ants, bees, wasps, and termites devote their entire lives to rearing progeny produced by specialized queens, and they rarely or never produce progeny of their own. The evolution of these castes is a complex problem, and the selective processes involved will not be discussed until Chapter 12.

The present chapter focuses on possible cases of altruistic behavior among vertebrates. The two most thoroughly studied and perhaps most interesting examples are alarm signals used to alert conspecifics of danger and helping behavior directed toward rearing the young of others.

Categories of behavioral interactions

Not all forms of aid-giving behavior are altruistic. Parental behavior is not considered altruistic because aiding direct progeny increases an individual's direct fitness. Cooperative behavior is also not considered altruistic because every participating individual receives direct fitness benefits. To make the distinctions clear, a more rigorous definition of each kind of behavioral interaction is helpful.

From an evolutionary perspective behaviors are conveniently classi-

Chapter three

fied according to how they affect the fitness of individuals exhibiting them (actors) and individuals affected by them (recipients) (Table 3–1). When an actor's behavior increases a recipient's direct fitness at the expense of the actor's direct fitness, the behavior is **altruistic.** When an actor's behavior increases a recipient's direct fitness and also increases the actor's direct fitness, the behavior is **cooperative.** When an actor's behavior increases the actor's direct fitness while reducing the recipient's direct fitness, the behavior is **selfish.** Finally, when the actor's behavior reduces both the actor's direct fitness and the recipient's direct fitness, the behavior is **spiteful** (Hamilton 1964).

Reciprocal altruism

Altruism involves giving aid without expecting to receive any direct benefits in return, while cooperation involves mutual aid among several individuals. Intermediate between these two kinds of interactions is **reciprocal altruism,** which involves reciprocation of benefits following a time lag (after Trivers 1971). During interactions involving reciprocal altruism, an actor aids a recipient without immediate reward, and at some later time the recipient reciprocates. The net outcome is that both actor and recipient benefit from the interaction. However, the time lag introduces an element of risk into the interaction. The actor initially makes a sacrifice without receiving any immediate benefit, and later the recipient makes a sacrifice after the benefits have already been received. Technically, reciprocal altruism is a cooperative interaction, but it does require some special theory because of the risk factor introduced by the time lag.

Theory

Reciprocal altruism can evolve whenever an actor can expect to receive future benefits sufficient for compensating its initial sacrifice. The behavior is advantageous as long as the average benefit gained from the interaction exceeds the average cost for both participants, as is true in any cooperative interaction. The special feature requiring explanation here is *reciprocation.* Why should a recipient ever bother to reciprocate? The recipient has already received the benefits, so why pay the costs?

The answer involves the prospect for benefiting repeatedly from future interactions involving the same individuals (Trivers 1971). Suppose that every individual has many opportunities to help others and be helped in turn. If altruists refrain from helping nonreciprocators, or cheaters, the cheaters will benefit only once or a few times. Meanwhile, altruists who do reciprocate will assist one another and benefit many times. In the long run reciprocators would derive more benefit than cheaters, even though their net benefit in the short run is smaller. If altruists do not or cannot withhold benefits from cheaters, the cheaters will derive more net benefit in both the short and long runs, and reciprocal altruism could not evolve.

Table 3–1 *Classification of intraspecific behavioral interactions based on how the direct fitness of each participant is affected. Neutral effects are omitted.*

Effect of interaction on direct fitness of actor	**Effect of interaction on direct fitness of recipient**	
	Increase	**Decrease**
Increase	Cooperative	Selfish
Decrease	Altruistic	Spiteful

Source: From W. D. Hamilton, "The Genetical Evolution of Social Behavior," *Journal of Theoretical Biology* 7 (1964):1–52. Reprinted by permission.

Thus reciprocal altruism requires that altruists be able to identify cheaters and withhold benefits from them.

The conditions for reciprocal altruism are certainly met in humans, although costs and benefits may be measured in political, social, or economic terms rather than fitness terms. Reciprocation lies at the core of trade agreements, political alliances, and even many friendships. Such arrangements may not have arisen through natural selection, but the theory is the same regardless of how costs and benefits are measured. In animals reciprocation is much less prevalent, but several behavioral interactions have been interpreted as forms of reciprocal altruism. Only one example prominently mentioned by Trivers (1971) will be discussed here.

Cleaning symbioses: An example of reciprocal altruism?

Cleaning symbioses consist of one organism cleaning another usually larger organism, often by entering the host's mouth or gill chambers (Photo 3–1) (Feder 1966; Maynard 1968; Brockman and Hailman 1976; Losey 1978). Cleaning behavior is not itself altruistic, as the cleaner obtains food by picking ectoparasites off a host. What does appear altruistic is the host's failure to eat the cleaner after it has been cleaned. Eating the cleaner seems tempting, since the cleaner is often conveniently positioned in or near the host's mouth. The problem, therefore, is to explain why hosts refrain from eating cleaners.

Trivers (1971) suggests that hosts are reciprocal altruists. He argues that hosts do not eat cleaners because at some future time they will need to visit the same cleaners again. Cleaner organisms are relatively scarce and widely dispersed in their marine environments. Each cleaner defends a regular cleaning station where it services up to several hundred hosts every day (Limbaugh 1961; Limbaugh et al. 1961). Should a host begin eating cleaners, it and all other hosts would soon have difficulty finding hungry cleaners to service them. Thus according to the reciprocal altruism hypothesis, hosts refrain from eating cleaners (a short-term sacrifice) be-

Photo 3–1

Cleaner fish are typically much smaller than their hosts and frequently enter the host's mouth and gill chambers to feed on ectoparasites. *An important question is whether cleaners risk their lives by foraging in such a manner.* Photo by Douglas Faulkner.

cause the benefits of being cleaned again (reciprocation by the recipient) exceed the energetic benefits derivable by eating cleaners.

The reciprocal altruism hypothesis can be evaluated by examining the validity of its assumptions. The first assumption is that refraining from eating cleaners really is a short-term cost for hosts. Several lines of evidence suggest that it is not (Gorlick et al. 1978). Most host species are nonpredatory and hence not likely to attack a cleaner under any circumstances. Cleaners may recognize and avoid predatory hosts who are in search of food, and they are less likely to clean such hosts (Hobson 1971). In addition, some cleaners have special adaptations to reduce the risk of being eaten by predatory hosts. The cleaner wrasse *Labroides phthirophagus,* for example, reduces its risk by avoiding the head and mouth areas of predatory hosts (Potts 1973). Since cleaners are generally very small and agile anyway, such tactics may reduce to zero the value of trying to catch a cleaner. Finally, some cleaners may not be eaten because they are distasteful or toxic (Colin 1975).

On the other hand, some evidence suggests that cleaners are vulnerable to predatory hosts. At least two species, Pederson's cleaner shrimp *(Periclemenes pedersoni)* and the cleaner wrasse *Labroides phthirophagus,* have been found in the stomachs of their hosts (Roessler and Post 1972; Lobel 1976). Hosts have not actually been seen pursuing or attacking cleaners while being cleaned, but such attacks could be rare events as a result of selection against such behavior. The fact that cleaners behave differently when cleaning predatory and nonpredatory hosts provides indirect evidence that cleaners are potentially vulnerable to hosts. Otherwise such discrimination should not have evolved in the first place.

A second assumption of the hypothesis is that hosts benefit from being cleaned. The classic experiment purporting to show a benefit for hosts was done by Limbaugh (1961), who removed all the cleaners from one section of a coral reef and studied the effect on host species. Within two weeks following removal of the cleaners, all but the territorial species of host fish had left the area, and many of the territorial hosts developed ulcerated sores, fuzzy blotches, swellings, and frayed fins. While these results suggest that hosts do benefit greatly by being cleaned, they are inconclusive because attempts to repeat the experiment have yielded only negative results (Youngbluth 1968; Losey 1972a). Subsequent studies indicate that cleaners are able to control ectoparasite infestations on hosts effectively (Hobson 1971; Losey 1974), but the presence of cleaners is apparently not vital for that purpose. Since ectoparasite outbreaks did not occur following removal of cleaners in the studies by Youngbluth and Losey, alternative mechanisms (unidentified) apparently can prevent such outbreaks when cleaners are absent (see Losey 1979). If so, hosts would gain little from being cleaned by cleaners.

A second line of evidence suggests that hosts do benefit from being cleaned. Many hosts actively seek out cleaners, and some visit a cleaner on a regular daily basis (Brockman and Hailman 1976). Although some hosts merely lie still or do no more than fan their tails while visiting a cleaner, others use special displays to solicit cleaning. These displays may consist of a vertical solicitation posture or even changes of color (e.g., Roessler and Post 1972). The prevalence of such behaviors seems to imply that hosts benefit from being cleaned, since otherwise such behaviors should not have evolved, but an alternative interpretation is possible.

If solicitation postures evolved because hosts benefit from being cleaned, one might expect that irritations caused by ectoparasites would elicit them. However, experimental studies show that they do not (Losey 1972a, b, 1979; Losey and Margules 1974). Instead, solicitation postures are elicited by tactile stimulation of the cleaner. At least some cleaners actively seek out hosts and nudge them to induce a cleaning episode, apparently exploiting a tactile reinforcement mechanism in the host (Gorlick et al. 1978). This mechanism may have evolved for reasons unrelated to cleaning, or it may have evolved to facilitate the cleaning interaction. Gorlick and his co-workers have taken the former interpretation and argued that cleaners evolved from purely parasitic ancestors once the opportunity for exploiting such host mechanisms opened a new feeding niche for them to invade.

Other miscellaneous evidence does not greatly clarify the situation. Cleaners have been seen picking over fresh wounds, presumably to the benefit of hosts (Roessler and Post 1972; Brockman and Hailman 1976), but they have also been seen eating fin membranes, scales, and associated dermal and epidermal tissues of their hosts (Gorlick et al. 1978; Losey 1979).

Moreover, some hosts rarely carry ectoparasites. Thus the value of being cleaned is not at all clear. Perhaps some host species benefit while others do not.

Present evidence does not support the hypothesis that cleaning symbioses evolved by reciprocal altruism. The phenomenon is not yet well understood, but in most cases hosts do not seem to incur a cost by refraining from eating cleaners, and they may not even benefit from being cleaned. After carefully reviewing the evidence, Gorlick and co-workers (1978) concluded that "cleaning symbiosis is not the simple, intuitively pleasing picture of interspecific cooperation painted by earlier hypotheses." Perhaps it is not just selfish exploitation of hosts by cleaners either, but only future research can supply the answers.

Alarm signals

Many social animals give signals that alert others to potential danger. How alarm signals may have evolved has stirred considerable debate, and behavioral biologists cannot even agree about whether such signals are altruistic or selfish in nature. Warning others of danger may place a signaling individual in greater jeopardy by drawing the attention of predators, but it may or may not provide any immediate benefits.

Characteristics

Alarm signals have clearly evolved to impart information during periods of danger. This is made clear by special properties that make them especially suitable for that function. For example, small passerine birds, ground squirrels, and some primates typically give high-pitched whistles or squeaks upon detecting a hawk or owl flying overhead (Marler 1955, 1957, 1959; Baldwin 1968; Melchior 1971; Sherman 1977; Vencl 1977). In dense foliage these signals have a ventriloquial effect that reputedly makes the caller more difficult to locate. The same individuals respond to perched hawks or owls and to mammalian predators by giving easy-to-locate chirps or clicks, which in some birds initiate mobbing attacks on predators instead of retreat. Thus high-pitched calls are given during extreme danger, while lower-pitched calls that contain more directional information are given during periods of less danger.

More elaborate differentiation of alarm calls has evolved in some species. Arctic ground squirrels (*Spermophilus undulatus*) give six distinctive alarm calls, with the type and intensity of the call varying with the type, proximity, and behavior of the predator (Carl 1971; Melchior 1971). By noting the type and orientation of a call, a human observer can determine the type of predator (avian or mammalian) and for a mammalian predator its location, whether it is lying, standing, or moving, and, if moving, its rate of travel through the colony. Vervet monkeys (*Cercopithecus aethiops*) give different calls in response to snakes, minor avian or mammalian predators, and major avian or mammalian predators (Struhsaker 1967a, b). These calls all evoke alertness and slow retreat to cover among nearby conspecifics. An additional call, given when a major predator is very near, evokes headlong flight into dense shrubbery or trees.

Alarm signals are not always given vocally. Many birds signal their intention to take flight by crouching, rapidly flicking their wings or tail, or rapidly spreading and closing their tail feathers (Daanje 1950; Andrew 1961). Such **intention movements** are generally given in conflict situations, when the bird has conflicting tendencies to remain perched or to fly. Hence they often indicate a cause for alarm and frequently occur in conjunction with alarm calls. Sometimes they may serve to coordinate movements of flock members or mated pairs, but at other times they signal the presence of danger.

Many ungulates and rabbits display prominent tail or rump patches when retreating from predators (Figure 3–1), sometimes by elevating the tail or ruffling the hairs around the rump (Guthrie 1971). Gazelles and other small gregarious antelopes exhibit specialized bounding gaits, called *stotting,* that alert other group members to danger (Photo 3–2) (Leuthold 1977; Walther 1977a). Conspicuous alert postures are prevalent in ungulates, primates, ground squirrels, and other mammals (King 1955; Barash 1973; Gautier and Gautier 1977; Leuthold 1977; Walther 1977a). The latter postures may not have evolved as alarm signals, since they allow individuals to survey suspicious circumstances more carefully, but they do have the effect of alerting other group members to potentially hazardous situations at the same time.

Olfactory signals, or **pheromones,** are another common type of alarm signal. Alarm pheromones are widespread among social insects, where they stimulate and coordinate aggressive defense of the colony or precipitate retreat (Wilson 1971). They may be elicited by conspecifics from neighboring colonies, social insects of other species, vertebrate predators, or any other kind of intruder.

Substances with similar effects are also produced by some vertebrates. House mice *(Mus musculus)* show an aversion to areas containing the odor left by a conspecific in stress (Müller-Velten 1966; Carr et al. 1971; Rottman and Snowdon 1972). The odor is often released during urination or defecation, but it is not just an excretory substance and can be released by itself. Adult black-tailed deer *(Odocoileus hemionus columbianus)* release a scent resembling the smell of garlic from their metatarsal glands when fleeing a predator (Müller-Schwarze 1971). Preliminary experiments show that the scent inhibits feeding behavior of other deer, but further experiments are needed to determine whether it also evokes alert postures or flight.

Certain kinds of fish (Ostariophysi, Cypriniformes, and Siluriformes) and tadpoles of various bufonid toads possess special substances in their skin that signal danger to others when released into the water (Pfeiffer 1962, 1963, 1974, 1977; Bardach and Todd 1970). These substances have no known physiological function, though such a function cannot be ruled out.

Figure 3–1

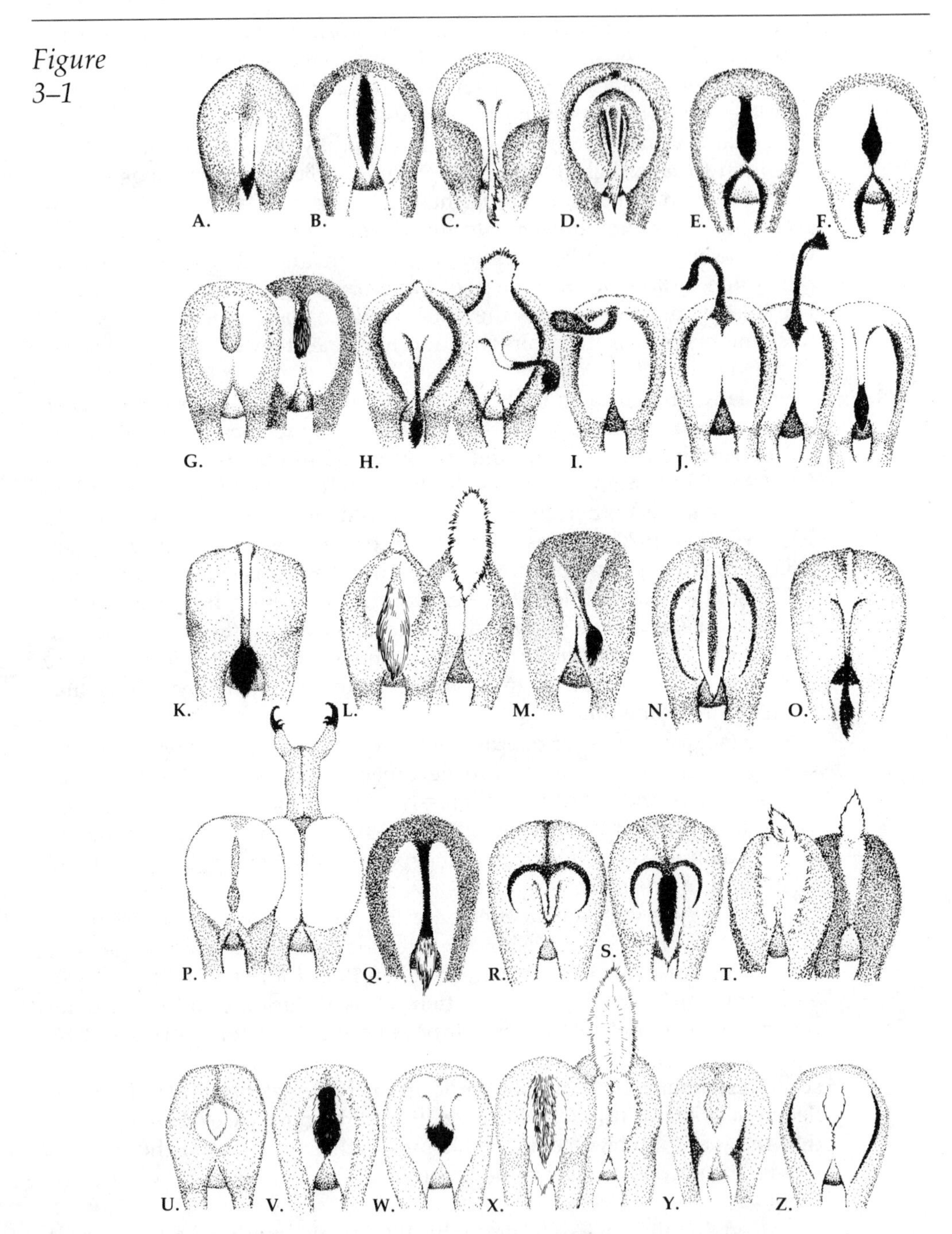

The rump and tail patches of antelopes draw attention to the hindquarters and tail by framing them and accentuating tail movements.

*Guthrie (1971) suggests that this patterning evolved to enhance submissive social displays, but they may have evolved as alarm signals instead. The species illustrated are as follows: A. Uganda kob (*Adenota kob*); B. defassa waterbuck (*Kobus defassa*); C. bontebok (*Damaliscus pygargus*); D. common waterbuck (*Kobus ellipsiprymus*); E.*

Photo 3–2

Thomson's gazelle stotting in response to a predator.

Stotting is a specialized bounding gait given by various ungulates when fleeing predators. Its adaptive significance is not yet well understood, but it has sometimes been interpreted as an alarm display used in conjunction with tail flashing.

Photo from *Hyaena* by Hans Kruuk published by Oxford University Press.

They are contained within specialized epidermal cells lacking direct connections with the surface, and they are released into the water only when the skin is broken (Pfeiffer 1967, 1974, 1977). Conspecific individuals respond to these substances by avoiding areas where they are present or by forming more compact schools. Both responses reduce vulnerability to predation, as will be discussed further below (pp. 121–122).

*stone sheep (*Ovis dalli stoni*); F. bighorn sheep (*Ovis canadensis*); G. blackbuck (*Antilope cervicapra*) (male and female); H. springbok (*Antidorcas marsupialis*) (showing erected tail); I. Thomson's gazelle (*Gazella thomsoni*); J. Grant's gazelle (*Gazella granti*) (male and female); K. banteng (*Bos sondaicus*); L. greater kudu (*Tragelaphus strepsiceros*) (showing erected tail); M. gerenuk (*Litocranius walleri*); N. impala (*Aepyceros malampus*); O. Coke's hartebeest (*Alcelaphus cokei*); P. pronghorn (*Antilocapra americana*) (showing erected rump patch); Q. sable antelope (*Hippotragus niger*); R. sika deer (*Cervus nippon*); S. fallow deer (*Dama dama*); T. caribou (*Rangifer tarandus*) (winter and summer); U. roe deer (*Capreolus capreolus*); V. black-tailed deer (*Odocoileus hemionus columbianus*); W. mule deer (*Odocoileus hemionus*); X. white-tailed deer (*Odocoileus virginianus*) (showing rump patch under tail); Y. red deer (*Cervus elaphus scoticus*); Z. wapiti (*Cervus elaphus canadensis*).*

Source: From R. D. Guthrie, "A New Theory of Mammalian Rump Patch Evolution," *Behaviour* 38 (1971):135. Reprinted by permission.

Aversions to body constituents released into the water do not necessarily indicate that the substances evolved as alarm signals. Sea urchins (*Diadema antillarum*), for example, move rapidly away from areas containing the bodily fluids of crushed conspecifics (Snyder and Snyder 1970). Similarly, aquatic snails escape such areas by dropping from protruding substrates and burrowing into the bottom, while air-breathing snails escape by crawling out of the water (Snyder 1967). Bodily fluids have obviously not evolved to warn others of danger. Escape responses elicited by their presence probably evolved because free body constituents are a sure indication that something is amiss. A signal function is not implicated unless specialized substances or structures with no other identifiable function are involved.

Are alarm signals selfish or cooperative?

In contexts where alarm signals do occur, there are three classes of participants: the signaling individual, the conspecific recipients of the signal, and the predator. The signal may be directed toward conspecific recipients or toward the predator, and it may benefit the signaler, the conspecific recipients, or both. The signal may be intended as a warning, or its role as an alarm signal may be ancillary to other functions. Hypotheses have been proposed on the basis of every possibility.

A first question is whether alarm signals really evolved to warn others. Guthrie (1971), for example, argues that the rump patches of ungulates and various other mammals evolved as submissive signals within social contexts and are given toward predators because prey animals respond to predators as though they were dominant conspecifics. Rump presentation is widely used to appease dominant individuals of either sex during aggressive interactions and to appease males during courtship interactions. Conspicuous markings on the rump and hind legs might therefore have evolved to enhance the effectiveness of submissive displays. Guthrie offers in support of his hypothesis the observation that rump patches have evolved in both gregarious and solitary species. He argues that in solitary species they could not serve as alarm signals because individuals rarely associate with one another. However, this argument is not convincing, because rump patches may allow mothers to alert dependent offspring to danger even in solitary species.

Guthrie's hypothesis might conceivably explain static markings on the hindquarters, but it cannot explain displays revealed by flashing tail markings. If tail-flashing displays increase a prey's vulnerability to predation and are primarily used as submissive displays, selection should favor concealment of the displays if at all possible during encounters with predators. In the case of white-tailed deer (*Odocoileus virginianus*), just the opposite happens. White-tailed deer depress their tails, thereby concealing their white tail patches, when signaling submissiveness, and they raise their tails to flash their tail patches when bounding away from a disturbance (Hirth and McCullough 1977). Thus at least tail flashing evolved in alarm contexts.

One hypothesis for explaining tail patches is that they facilitate social cohesion during flight (McCullough 1969; Hirth and McCullough 1977). Deer may benefit from remaining together in relatively open areas, especially when endangered by predators, because several characteristics of group responses could reduce the vulnerability of every individual (see pp. 111–122). White-tailed deer, for example, are not as solitary as sometimes assumed, since both doe groups and buck groups sometimes occur. Nevertheless, this hypothesis appears untenable because tail flashing is exhibited by solitary individuals as well, not just by individuals in groups. If tail flashing entails a risk, selection should favor its use only in appropriate or beneficial contexts. The fact that solitary individuals flash their tails therefore implies that social cohesion is not the only reason for the display.

Another hypothesis is that tail flashing evolved as a signal directed at the predator. This hypothesis is suggested by the fact that tail patches appear oriented toward the predator, not toward conspecifics. Prey animals could conceivably benefit from signaling to a predator that surprise has been lost because the signal might induce a predator to attack prematurely while the distance is still too great for an attack to be successful (Smythe 1970a, 1977). In effect the signal might force a predator to reveal its intention to attack while the prey animal has all the advantages on its side. Rump patches may even dissuade a predator from attacking altogether. Once a predator has been detected, it has little chance of success. A signal that surprise has been lost might therefore persuade a predator to give up and leave to hunt elsewhere, thereby allowing the prey animal to resume other activities.

Rump patches and tail flashing do seem directed at a predator, and they are often accompanied by specialized bounding gaits, or *stotting*, which make the prey extremely conspicuous (see Photo 3–2). However, closer inspection of the evidence does not support Smythe's pursuit-invitation hypothesis (Coblentz 1980). Stotting by Thomson's and Grant's gazelles *(Gazella thomsoni* and *G. granti)* is continued after a predator has initiated an attack and often appears responsible for the predator capturing the stotting individual. Attacks initiated after stotting displays are given often result in kills (as often as 85% of the time), so stotting is not restricted to contexts when prey individuals have all the advantages. As a general rule, stotting and tail flashing are given in response to every attack, even when predators were not detected early, suggesting that they do not just serve to initiate premature attacks.

The hypothesis that tail flashing and stotting dissuade predators from attacking seems unlikely. Predators usually attack when they approach within some critical distance, and it seems unlikely that this critical distance is affected by such signals. Since tail flashing and stotting are given even after an attack is initiated, they cannot serve solely to dissuade predators from attacking. They also do not seem to discourage predators from continuing an attack, though this possibility has not been carefully examined. A more likely explanation for stotting is that bounding high in the air gives a prey animal a better view of its immediate surroundings. The behavior might therefore enable prey to take a more effective escape route than would otherwise be possible.

Although rump patches and tail flashing probably did not evolve as deterrents to predatory attacks, some other alarm signals may have evolved for that purpose. One possible example is the alarm-duetting behavior of klipspringers *(Oreotragus oreotragus),* a small monogamous antelope of open, rocky habitats in Africa (Photo 3–3) (Tilson and Norton MS). Mated male and female klipspringers always remain near each other, and they are quickly warned of danger by alert postures or sudden movements of the other pair member. When threatened by a predator, they typically give one short alarm call and then flee up a rock escarpment. The calls cannot serve to warn young or the other pair member because every individual has already responded to the predator's presence by fleeing. They also are not directed toward possible relatives on neighboring territories, as pairs give the calls even when located out of earshot of neighboring pairs. Apparently the best explanation is that the alarm duets announce that the predator has been detected, thereby encouraging it to leave the area, so that the klipspringers can resume foraging sooner.

The same hypothesis may also explain the alarm calls given by small passerine birds. In one experimental study barn owls *(Tyto alba)* could

Photo 3–3

A mated pair of klipspringers standing on a rock outcrop above the Kuiseb Canyon in Namibia (South West Africa) after fleeing a predator. Photo by Ronald L. Tilson.

localize the high-pitched alarm calls of songbirds just as easily as other sounds, but they responded to the alarm calls less often than to lower-pitched atonal sounds (Shalter and Schleidt 1977). This result suggests that barn owls do not bother attacking prey animals once they have lost the element of surprise. One problem with interpreting the results, however, is that barn owls are nocturnal predators and do not ordinarily hunt small birds, except possibly at dusk. Their greater responsiveness to atonal sounds may simply reflect their reliance on such sounds to detect small mammals rummaging in leaf litter or vegetative undergrowth.

A second study resolves this problem. The same experimental procedure was repeated using goshawks *(Accipiter gentilis),* a large accipiter hawk that normally hunts songbirds (Shalter 1978). The goshawks were also able to localize the high-pitched alarm calls given by songbirds, and they too showed low responsiveness to the calls. Thus the alarm calls of passerine birds may have evolved to dissuade predators from attacking rather than to warn conspecifics of danger. This conclusion is only tentative, however, because the experiments do not show that goshawks can localize high-pitched calls when the songbirds are in dense foliage. Also, the calls may confer several advantages at once. Some additional possible benefits will be discussed below.

Animals may derive more direct benefits from warning others because the simultaneous responses of many individuals create havoc and may direct the predator's attention away from the caller (Charnov and Krebs 1975; Owens and Goss-Custard 1976). Scattering by many individuals at once may confuse a predator enough so that the caller's risk of being caught is actually reduced. Also, an individual retreating alone to cover may increase its susceptibility to attack by leaving the group. By warning other group members, a caller can remain with the group while retreating. Finally, the high-pitched ventriloquial alarm calls of small birds may cause the predator to search in the wrong place for the caller (Perrins 1968). The effect might increase the predator's search time, thereby encouraging it to give up, or it might divert the predator's attention toward other group members. One possible example is alarm chorusing by common bushtits *(Psaltriparus minimus).* When attacked by accipiter hawks, bushtits fly to cover and then give continuous high-pitched calls while remaining hidden from sight (Grinnell 1903; Miller 1922). The calls do not serve a warning function since all group members have already been alerted, but they could perhaps serve to encourage the hawk to leave. These ideas have not been carefully examined, and the extent to which they apply remains to be determined.

Are alarm signals altruistic?

Several hypotheses are based on the assumption that alarm signals really are altruistic. Three general hypotheses have been suggested. Alarm signals may have evolved by indirect (kin) selection (Maynard Smith 1965), selection for reciprocal altruism (Trivers 1971), or group selection (Wilson 1975; Matessi and Jayakar 1976).

Widely cited support for the indirect (kin) selection hypothesis comes from an eight-year study of Belding's ground squirrel *(Spermophilus beldingi)* (Photo 3–4), which involved observations of over 1800 individually marked animals (Sherman 1977). Alarm calling by ground squirrels does appear risky, because individuals giving calls were stalked and killed by predators more often than were individuals not giving calls (Sherman 1977, 1980).

To evaluate the indirect (kin) selection hypothesis, one must know the genetic relatedness of callers and recipients. In Belding's ground squirrels adult females remain sedentary once they have successfully reared young, and their daughters usually breed nearby if suitable habitat is available. In contrast, sons usually emigrate to seek receptive females in other areas. Thus females are usually related to other females in the vicinity, while males are not.

On the 102 occasions when ground squirrels were observed while a predator was present, females gave alarm calls significantly more often than males (Figure 3–2) (Sherman 1977). In addition, females with relatives nearby gave alarm calls more often than did females without relatives nearby. Similar results have also been obtained for round-tailed ground squirrels *(S. tereticaudus)*(Dunford 1977a; Leger and Owings 1978), Sonoma chipmunks *(Eutamias sonomae)* (S. F. Smith 1978), and black-tailed prairie dogs *(Cynomys ludovicianus)*(Hoogland 1980a). More frequent calling when

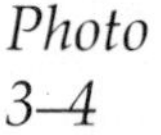

Photo 3–4

A Belding's ground squirrel in an alert posture.

Ground squirrels become alert and often give alarm calls whenever they are disturbed by a predator or any unidentifiable movement or sound.

Photo from P. W. Sherman, "Nepotism and the Evolution of Alarm Calls," *Science* 197 (23 September 1977):1247.

Figure 3–2

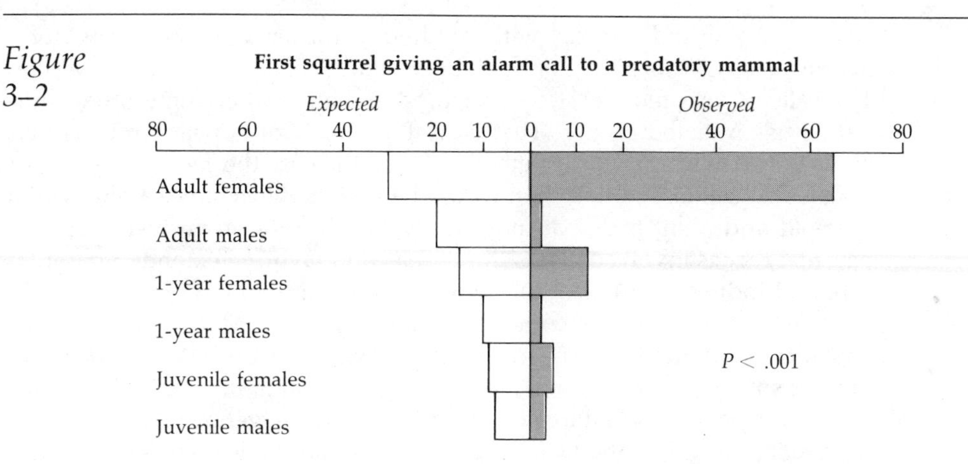

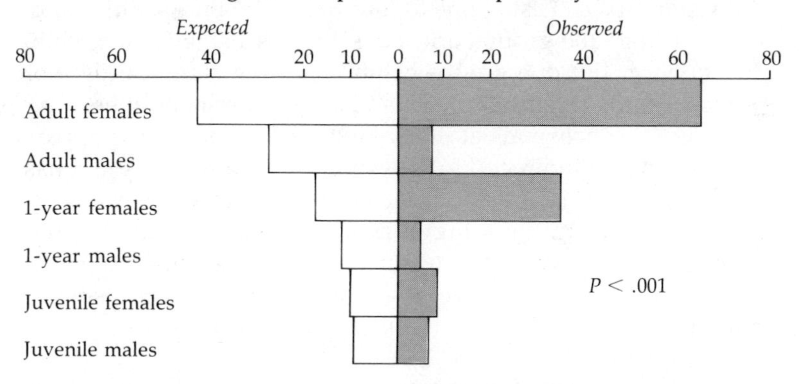

Expected and observed frequencies (given along the horizontal axis) of alarm calling by Belding's ground squirrels.

Data were analyzed separately for each age and sex class. Expected values were calculated by assuming that animals called in direct proportion to the number of times they were present when a predatory mammal appeared. Observed values were obtained from 102 actual interactions between ground squirrels and predators. P = probability that the overall difference between expected and observed values could have arisen by chance alone.

Source: From P. W. Sherman, "Nepotism and the Evolution of Alarm Calls," *Science* 197 (23 September 1977):1246–53. Copyright 1977 by the American Association for the Advancement of Science. Reprinted by permission.

relatives are present suggests that alarm-calling behavior evolved as altruistic aid of relatives, but the data must be interpreted with care.

Closer inspection of Sherman's (1977) data for Belding's ground squirrel shows that many of the female calls are given when daughters are present. One might therefore argue that alarm calling is basically a form of extended parental care, making direct (individual) selection rather than indirect (kin) selection the most likely basis for the behavior (Shields 1980).

However, year-old females with mothers or sisters, but not daughters, nearby also call more frequently than would be expected by chance alone. The calls of year-old females occur predominantly after some other individual has initiated calling. That is, calling behavior appears to be contagious, especially for young females. Nevertheless, the fact that year-old females do call more often than would be expected by chance alone, while year-old and adult males do not, suggests that more than just extended parental care is involved. Thus direct selection favoring extended parental care and indirect (kin) selection may both be important.

Additional evidence comes from a study of white-tailed deer (Hirth and McCullough 1977). The responses of white-tailed deer to predators involve snorting as well as tail flashing. Again, evaluating the indirect (kin) selection hypothesis requires information about the genetic relatedness of signalers and recipients. Doe groups of white-tailed deer consist of either solitary does or young does who are still associating with their mothers. Buck groups consist of one to four unrelated bucks. Hirth and McCullough found that doe groups and buck groups exhibit tail flashing with equal frequency, but doe groups exhibit snorting significantly more often than buck groups (Figure 3–3). This result is extremely interesting, since it implies two conclusions at once. First, alarm snorting is apparently favored by direct and indirect (kin) selection; second, tail flashing has not evolved in response to the same selective factors as alarm snorting. One might argue that alarm snorting is just a form of extended parental care, not involving indirect selection, since doe groups consist of mothers and their female offspring. However, daughters also give alarm snorts, which warn mothers and sisters, and this behavior cannot be attributed to direct selection.

The indirect (kin) selection hypothesis only makes sense when the recipients of alarm signals are nondescendant relatives. Many ground squirrel colonies, most primate groups, and some bird flocks consist of such relatives, but many ungulate herds and bird flocks probably do not (see Trivers 1971; Leuthold 1977).

A second hypothesis that assumes alarm signals are altruistic is based on the theory of reciprocal altruism (Trivers 1971). Callers might benefit by reducing a predator's likelihood of (1) returning to the same area at a future time, (2) developing an increased tendency to hunt that species of prey, or (3) gaining information about the local area or about hunting techniques that would improve future hunting success. These benefits are reciprocal if different individuals call on each occasion. Predators often do hunt near sites where they were previously successful (Curio 1976) and they do develop tendencies to hunt certain kinds of prey (see p. 224), but the effect that warning signals have on a predator's hunting behavior has not been ascertained.

The major problem with this hypothesis is that reciprocal altruism is only possible if benefits can be withheld from cheaters. The benefits proposed by Trivers would be obtained by both signaling and nonsignaling

Figure 3–3

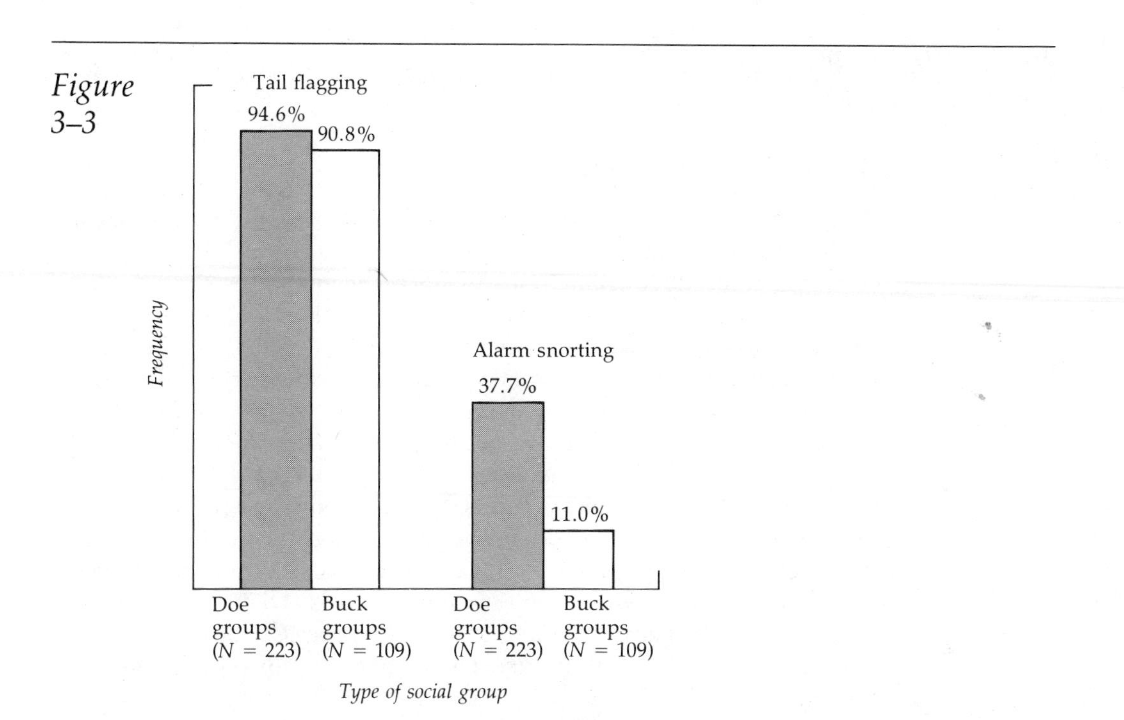

Observed frequencies of tail flagging and alarm snorting within doe groups and buck groups of white-tailed deer.

Doe groups consisted of mothers and daughters, while buck groups consisted of unrelated males. The value given above each bar indicates the percentage frequency of occurrence within groups distributed by a potential predator. N = number of groups observed for each category.

Source: Based on data from Hirth and McCullough 1977.

individuals, and the former have no obvious way to withhold benefits from the latter. Hence the reciprocal altruism hypothesis is not likely to be valid.

The benefits proposed by Trivers could also plausibly favor the evolution of alarm signals through trait-group selection. If individuals do benefit in the ways suggested by Trivers, trait groups that contain individuals who warn others would suffer less predation than trait groups that contain only individuals who do not. Of course, the benefits need not be so indirect for trait-group selection to work. The recipients could derive immediate benefits at the expense of signaling individuals (Wilson 1975; Matessi and Jayakar 1976). Unrelated altruists may possibly tend to associate more with each other than with nonaltruists, thereby creating the variability among trait groups required for trait-group selection to occur (see p. 71), but relevant evidence for evaluating that possibility is not yet available.

Underutilization of resources and reproductive curtailment

Theoretical considerations

The hypothesis proposed by Wynne-Edwards (1962) that most social behaviors evolved to regulate population density was introduced and discussed in the preceding chapter. In essence the hypothesis states that animals voluntarily underutilize resources or curtail reproductive output to prevent the deleterious consequences of overpopulation. Such behavior would be altruistic in that individuals would be lowering direct (individual) fitness to benefit the population as a whole. This hypothesis has been rejected by most biologists largely because known interdemic selection processes cannot lead to fixation of altruistic traits and most social behaviors can be explained by direct (individual) or indirect (kin) selection.

The development of plausible trait-group selection models and the possible prevalence in populations of altruism favored by indirect (kin) selection have reopened the issue. Wynne-Edwards (1977), for example, accepts the evidence against interdemic selection processes but argues that indirect (kin) selection could favor social mechanisms for regulating population density.

Before this argument can be evaluated, a clear distinction must be made between behaviors that evolved to regulate population density and behaviors that evolved for other reasons but secondarily affect population density. Much of the skepticism about Wynne-Edwards's hypothesis stems from Wynne-Edwards's own failure to make this distinction. The definitions of altruism and selfishness given earlier provide suitable criteria for distinguishing the two possibilities. If a behavioral trait decreases direct fitness but increases inclusive fitness by regulating population density, it probably evolved as a population regulatory mechanism. If not, the trait probably evolved for other reasons, and its consequences on population density should be construed as secondary effects.

Most social behaviors should be interpreted as selfish or cooperative behaviors, since the individuals performing them derive direct fitness benefits. Some behaviors may be altruistic, having evolved through indirect (kin) selection or perhaps through trait-group selection, but whether any of them evolved specifically to regulate population density is debatable.

Red grouse territoriality: A case study

Perhaps the best evidence for a trait that may have evolved to regulate population density comes from long-term studies of red grouse *(Lagopus lagopus scoticus)*, a subspecies of willow ptarmigan found on Scottish moors (Photo 3–5). Red grouse defend large multipurpose territories where they forage, mate, nest, and rear young. In summer months when food is abundant and adults are still caring for broods, territory defense is relaxed and occurs only during early morning hours (Watson and Jenkins 1968).

Photo 3–5

A male red grouse on its territory in Scotland.
Males that fail to obtain territories in autumn usually die during winter, but nevertheless they do not fight to the death for territories. Why?
Photo by Nicholas Picozzi.

Family groups break up between August and September, and then older cocks begin to vigorously defend territories and court hens (Watson 1977). Shortly thereafter, between September and November, young cocks establish their first territories and also begin courting hens.

Not all cocks acquire and hold territories. Some old cocks are evicted by stronger newcomers, and many young cocks never succeed in establishing territories. These birds are forced into undefended areas where food and cover are scarce. Similarly, many hens fail to acquire mates, and they too are forced into undefended areas. Most territories attract only one hen, but a few larger territories attract two (Watson and Miller 1971; Miller and Watson 1978).

The territorial behavior of red grouse limits population density (Watson and Moss 1971, 1972). This effect is shown by removing cocks or hens from territories during autumn. Birds removed from their territories are quickly replaced by previously nonterritorial individuals (Watson and Jenkins 1968). Also, territoriality adjusts population density to food availability, as territory size changes according to both quantity and nutritive quality of heather, the main food of red grouse (Watson and Moss 1972; Moss et al. 1975; Lance 1978b; Miller and Watson 1978). Since virtually all birds forced into undefended areas die over the winter (Watson and Moss 1971, 1972), territoriality limits overall population size as well as breeding density.

The main mortality agent is predation, but high susceptibility to predation is probably caused by lack of cover and the poor health or weakened physical condition of nonterritorial birds. The few nonterritorial birds who do survive to spring generally die of malnutrition or disease (Watson 1977).

Wynne-Edwards (1977) contends that territorial behavior evolved in red grouse to regulate population density. The main support for this hypothesis is that territoriality does limit both breeding density and overall population size. However, territorial behavior also reduces competition for food and mates. Thus individuals benefit from the behavior, and the parsimonious view is to interpret it in terms of individual selection (Watson 1977). The effect territoriality has on population density is probably a secondary consequence lacking adaptive significance.

A more interesting problem concerns the behavior of nonterritorial individuals. Why do these individuals accept their fate instead of fighting more persistently for a territory? Since they are doomed to die without a territory, why not fight to the death, if necessary, to win territorial disputes? A more vigorous effort might allow these individuals to squeeze onto small territories, thereby enabling them to survive the winter, even if resident owners cannot be displaced entirely. Here, then, is a possible example of birds altruistically leaving an area to prevent overexploitation of resources (Wynne-Edwards 1977).

The key issue is whether giving up without more of a fight can be explained as a selfish behavior. Watson (1977) argues that it can. He suggests that a nonterritorial male would have a better chance of replacing a territorial male who later becomes weak or dies by not making a serious challenge against healthy territorial males. A male who fights violently for a territory risks being badly injured, and even if he wins against the original territory resident, his weakened condition after such a fight would probably prevent him from holding the territory against subsequent challenges. Moreover, the chances for victory are slim. Nonterritorial males lack territories in the first place because they lost contests with resident territory owners earlier in the season, and escalating those contests would probably not alter the outcome.

One problem with Watson's hypothesis is that nonterritorial males *can* win a territory once they are administered small implants of testosterone (Watson 1970). One therefore wonders why they fail to manufacture more testosterone on their own. Perhaps if nonterritorial males produced more testosterone, they would induce something like an arms race among males, since territorial males could respond in kind (Sievert Rohwer, personal communication). Then territorial disputes might escalate in severity without leading to any change in the eventual victors. Increased testosterone production might give nonterritorial challengers a temporary advantage, but it would probably not improve their chances of holding a territory in the long run. Escalated fighting would not reduce a nonterritorial male's fitness if he were certain to die over the winter, but to the extent that it reduces a male's chances of finding a vacant territory or replacing a weakened territory resident, it would reduce his fitness.

The situation for females is much less clear. More than one female can breed on at least some territories, yet many territories contain only one female. Perhaps there really are no vacancies open to unmated females, but the situation is unclear. Removal experiments suggest that females are excluded and cannot find vacancies, but the reasons why they are excluded have not been identified. Resident females are occasionally aggressive toward intruding females, but female aggression occurs so infrequently that it does not appear responsible for the exclusion of unmated females (Watson and Jenkins 1964). Wynne-Edwards would maintain that females voluntarily refrain from breeding to reduce or prevent overpopulation. However, detailed evidence regarding the plausibility of indirect (kin or trait-group) selection operating in red grouse populations will be necessary before his hypothesis can be taken seriously.

Helping others reproduce

Helpers at the nest

In many birds certain individuals help others rear young instead of or in addition to producing young of their own. This phenomenon, widely known as *helping at the nest* (Skutch 1935), usually involves individuals other than parents who attend nests and feed young. Nest helpers have been reported in over 100 species of birds, and many more instances will probably be uncovered with future research (Skutch 1961; Grimes 1976; Rowley 1976; Woolfenden 1976; Zahavi 1976). Helping behavior occurs in troglodytid and malurid wrens (Selander 1964; Rowley 1965), African bee-eaters (Fry 1972a, b, 1975; Emlen 1978), New World jays (Woolfenden 1973, 1975; Brown 1970, 1972, 1974; Hardy 1976; Raitt and Hardy 1979), Old World babblers (Zahavi 1974, 1976; Brown and Balda 1977; Brown and Brown 1980), neotropical blackbirds (Orians et al. 1977), and many other species.

Three general hypotheses have been advanced to explain helping behavior (Emlen 1978). The behavior may be altruistic (helpers sacrifice direct fitness to benefit others), cooperative (helpers and recipients both benefit), or selfish (helpers benefit at the expense of recipients). To evaluate these hypotheses, one needs two sorts of information. First, who benefits from helping behavior? Second, how are the benefits obtained?

Is helping altruistic?

First hypothesis: Indirect (kin) selection. A widely accepted hypothesis is that helping behavior is an altruistic trait favored by indirect (kin) selection. The role of indirect (kin) selection has been implicated because in most species for which data are available nest helpers are prior offspring or occasionally more distant relatives of breeders (reviewed by Orians et al. 1977; J. L. Brown 1978; Emlen 1978). Helpers may be offspring from earlier broods produced the same year, or they may be from broods produced in preceding years. Previous offspring may continue to help breeders for three to six years, as occurs, for example, in kookaburras *(Dacelos gigas)* (a kingfisher of Australian dry forest and savanna), white-winged choughs *(Corcorax melanorhamphus)* (a crowlike bird of the plains of Australia), grey-crowned babblers *(Pomatostomus temporalis)* (a thrasherlike bird of Australian woodlands) (Parry 1973; Rowley 1978; Brown and Brown 1980), and yellow-billed shrikes *(Corvinella corvina)* (a predatory songbird of African woodland savanna) (Grimes 1980).

The relevance of indirect (kin) selection theory requires a cost-benefit analysis. Does the ratio of benefit to cost *(B/C)* for the helper justify a preference for giving help rather than becoming a breeder? Data for two species help answer this question.

In Florida scrub jays *(Aphelocoma coerulescens)* the benefits of giving help are substantially less than the costs, showing that offspring should rear their own young whenever possible rather than helping their parents

Table 3–2 *Relative genetic contributions to the next generation of a scrub jay that helps its parents rear offspring compared to one that can find a territory vacancy and rear its own offspring.*

Parameter	Empirical value of parameter	Number of nests in sample
N_2 = No. of young produced by an experienced breeder with assistance from a helper	2.20	81
N_1 = No. of young produced by an experienced breeder without assistance from a helper	1.62	45
$\overline{H}$ = average number of helpers present in groups with helpers	1.86	81
N_0 = No. of young produced by pairs in which at least one parent is breeding for the first time	1.36	55
N_0^1 = subsample of N_0 in which no helpers were present	1.03	37
N_0^2 = subsample of N_0 in which both parents were breeding for the first time and no helpers were present	1.50	6
r_p = coefficient of relatedness between breeder and its own offspring	.5	—
r_h = coefficient of relatedness between helper and offspring cared for by helper	.5	—

Equation	Meaning of equation
$\frac{N_2 - N_1}{\overline{H}} \times r_h$	No. of genes propagated by helping parents rear offspring (benefit to helper)
$N_0 \times r_p$	No. of genes propagated by rearing own offspring (average for all new breeders) (potential cost to helper)
$N_0^1 \times r_p$	No. of genes propagated by rearing own offspring (average for new breeders without assistance of helpers) (potential cost to helper)
$N_0^2 \times r_p$	No. of genes propagated by rearing own offspring (average for new breeders who pair with other new breeders and do not receive assistance from any helpers) (potential cost to helper)

Prediction based on indirect (kin) selection hypothesis

General case	$\frac{N_2 - N_1}{\overline{H}} \times r_h > N_0 \times r_p$
Special case	$\frac{N_2 - N_1}{\overline{H}} \times r_h > N_0^1 \times r_p$
Restrictive case	$\frac{N_2 - N_1}{\overline{H}} \times r_h > N_0^2 \times r_p$

Empirical test: Case	Genetic contribution as helper	Genetic contribution as parent	Prediction upheld?
General	.16	.68	No
Special	.16	.51	No
Restrictive	.16	.75	No

Source: From S. T. Emlen, "Cooperative Breeding," in J. R. Krebs and N. B. Davies, eds., *Behavioral Ecology: An Evolutionary Approach* (Sunderland, Mass.: Sinauer Associates, 1978), p. 256. Reprinted by permission of Blackwell Scientific Publications Ltd. Data provided by Glen Woolfenden.

rear young (Table 3–2) (Emlen 1978). That is what they do. Helping behavior is exhibited only by subadults who cannot acquire territories of their own. Only for those individuals is helping a better option than not helping.

The data for superb blue wrens *(Malurus cyaneus)* give a similar result, though with one difference (Table 3–3) (Brown 1975a; Emlen 1978). In this species males have no problem acquiring territories, but they do have a problem finding mates. Because the sex ratio is skewed toward surplus males, many males who establish territories are unable to breed. Although males who acquire mates are most successful by producing progeny of their own, average year-old males cannot do so unless the availability of females is unusually high. Males are clearly cognizant of their mating

Table 3–3

Relative genetic contributions to the next generation of a superb blue wren that helps its parents rear offspring compared to one that can find a territory vacancy and rear its own offspring. Parameters, equations, and predictions are the same as those in Table 3–2.

Parameter	Empirical value of parameter	No. of breeding groups in sample
N_2	2.83	12
N_1	1.50	16
$\overline{H}$	1.08	12
N_0 assumed $= N_1$	[1.50]	—
N_0 assumed $= .7N_1$	[1.05]	—
r_h	.5	—
r_p	.5	—

Empirical test (for male helpers): Case	Genetic contribution as helper	Genetic contribution as parent	Prediction upheld?
General	.62	.75	No
General, adjusted for skewed sex ratio	.62	.52	Yes

Source: From S. T. Emlen, "Cooperative Breeding," in J. R. Krebs and N. B. Davies, eds., *Behavioral Ecology: An Evolutionary Approach* (Sunderland, Mass.: Sinauer Associates, 1978), p. 258. Reprinted by permission of Blackwell Scientific Publications Ltd. Based on data from Rowley 1965.

opportunities, as more males help breeders in years when females are in short supply than in years when females are more available (Emlen 1978). Females never help breeders rear young because they never have any trouble finding a mate and can therefore always do better by producing young of their own.

The data for superb blue wrens raise an interesting possibility. If parents benefit from being helped, they might improve their chances of receiving help by rearing more sons than daughters. The origin of the skewed sex ratio is uncertain and may result from higher mortality among breeding females than among breeding males, but perhaps it results instead from parental manipulation of natal sex ratios (see Trivers and Willard 1973; see also pp. 411–414). It would be interesting to know whether skewed sex ratios are already evident during the parental care period or arise after individuals reach adulthood.

The type of cost-benefit analysis given above assumes that helping behavior entails some cost for helpers, where *cost* refers to the extent that direct fitness of helpers is reduced by giving help to others. Feeding or otherwise caring for nestlings does involve sacrifice in terms of commodities or services, but in virtually no case can this sacrifice be translated into a fitness loss (Brown and Brown 1980). Giving food to a nestling does not represent a fitness cost if the helper has no other way of using the food to enhance its direct fitness.

There is good reason to believe that helpers have no alternative reproductive options open to them. Helping behavior has evolved primarily in sedentary populations with densities at or near carrying capacity (Brown 1969a, 1974, 1978; Harrison 1969; Fry 1972a, 1975, 1977; Orians et al. 1977; Emlen 1978; Gaston 1978; Rowley 1978; Smith and Robertson 1978; Stacey 1979). Such species typically exhibit low fecundity, deferred breeding, high adult survival, and low dispersal (MacArthur and Wilson 1967; Pianka 1970), and most species in which helping occurs show these attributes. When a population is at or near carrying capacity, few suitable breeding habitats are vacant and young males have little opportunity to acquire territories (e.g., Brown 1969a, Parry 1973; Woolfenden 1975). Subadult males mainly acquire territories by replacing previous territory owners who have died, and subadult females mainly find mates by replacing resident females who have died. Thus opportunities for immediate breeding are severely limited. Indirect (kin) selection may be important to the extent that helpers benefit by propagating genes through relatives, but helping behavior may not be altruistic, as it apparently engenders little or no cost in terms of a helper's own reproductive potential.

One potential problem with the indirect (kin) selection hypothesis is that helpers are not always related to breeders. Helpers are occasionally or regularly unrelated to the breeders who receive help in green woodhoopoes *(Phoeniculus purpureus)* (Ligon and Ligon 1978a, b), piñon jays *(Gymnorhinus cyanocephalus)* (Balda and Bateman 1971; J.L. Brown 1978), common murres *(Uria aalge)* (Tschanz 1979), Arabian babblers *(Turdoides squamiceps)* (Zahavi, unpublished seminar), and possibly barn swallows *(Hirundo rustica)* (Myers and Waller 1977). Such evidence does not negate the potential importance of indirect (kin) selection when helpers are related to breeders, as there is no reason to believe that helping behavior has evolved for the same reason in every species, but it does suggest that other hypotheses are relevant for at least some species.

Second hypothesis: Curtailment of reproduction. A second hypothesis based on the assumption that helpers are altruists is that nest helpers are voluntarily curtailing reproductive output to prevent overpopulation (Wynne-Edwards 1962). This viewpoint has been taken by several authors (e.g., Skutch 1961; Parry 1970, 1973; Fry 1972a, 1975). The general problems arising from the need to postulate interdemic selection processes to produce such curtailments have been discussed earlier. The specific problem here is why helpers aid breeders at all if they really are trying to curtail reproduction. Logically, nonbreeders should merely reside on the parental territory without providing help, or they should depart and live in undefended habitat not suitable for breeding. One might argue that the activities of helpers actually reduce the reproductive success of breeders, but current evidence does not support this supposition (see pp. 101–104). The fact that helping behavior has evolved largely when alternative opportunities for breeding are absent further suggests that reproductive curtailment by helpers is involuntary, and Morton and Parry (1974) have shown for kookaburras that reproductive output is not curtailed by young birds helping their parents instead of breeding on their own.

Third hypothesis: Reciprocation. A third hypothesis is based on the idea of reciprocation. Helpers may provide help in order to establish familial ties with nestlings, which later reciprocate by assisting those that earlier gave help to become breeders (Ligon and Ligon 1978a, b). In green woodhoopoes breeding groups usually arise when kin groups emigrate to new territories. Breeding status in these groups is based on high dominance status. Since kin groups are often derived from several successive broods, help given to nestlings might later be reciprocated if the former helpers achieve high status through the aid of those nestlings that had survived.

The only support for this hypothesis comes from observations that helpers compete for opportunities to rear nestlings and frequently preen and vocalize toward older nestlings. Competing to feed nestlings cannot be explained by indirect (kin) selection, because an individual should prefer having someone else rear its relative's progeny whenever possible. Competition to provide help could only evolve if providing help confers direct benefits. The nature of these benefits is unclear, but reciprocation as proposed by Ligon and Ligon is one possibility.

A major problem with the reciprocation hypothesis is that matured nestlings have no reason to reciprocate by aiding former helpers. They have already received all the help they can ever expect, and by reciprocating, they are less likely to become breeders themselves. In addition, an alternative explanation for why helpers compete to feed and preen nestlings can be suggested. This explanation is based on the handicap principle, a controversial topic that will be discussed more fully in Chapter 10 (see pp. 432–435). According to the handicap principle, individuals incur a direct fitness cost (here, helping rear progeny of others) to communicate competitive superiority in other contexts as an aid to gaining higher social status (see Zahavi 1975, 1977a).

Is helping cooperative or selfish?

Helping behavior may be cooperative or selfish rather than altruistic. To assess these possibilities, one must determine the true beneficiaries of the behavior.

Benefits to helpers. Helpers could benefit from aiding breeders rear young in several ways (Brown 1969a, 1974, 1978; Orians et al. 1977; Emlen 1978; Gaston 1978; Stallcup and Woolfenden 1978; Woolfenden and Fitzpatrick 1978). They may gain valuable experience that later improves their own reproductive success. They may be better able to find food or avoid predators by remaining in familiar habitat on their natal territories. Both helpers and breeders might benefit from cooperative foraging or mutual defense against predation. Helpers may be better able to acquire a suitable territory by waiting to inherit the one held by their parents or by helping to increase the parental territory's size enough to partition off a piece of it. If helpers derive such benefits, they may help feed young primarily because it allows faster increases in territory size (see below) or because it encourages parents to tolerate their presence on the territory. In that case helping behavior would be cooperative, not altruistic.

Little evidence is available for evaluating direct benefits derived by helpers. The principal evidence pertains to territory inheritance by male helpers. A minority of male Florida scrub jays (39%) acquire territories by inheriting the parent's territory after death of the father or by budding (partitioning) off a small territory from the original territory (Figure 3–4) (Woolfenden and Fitzpatrick 1978). Budding is possible, of course, only after the parental territory has grown in size. Territory size in scrub jays is positively correlated with family size (breeders plus helpers), and changes in territory size are apparently caused by changes in family size. Larger families result in small part from the help provided by a helper in rearing nestlings, but they mainly result from efforts of the parents. Thus a male helper may benefit because his feeding of nestlings increases group size, which in turn leads to slightly increased territory size and hence a somewhat greater prospect of budding a new territory from the parental territory. However, the magnitude of this benefit seems small on average.

Female helpers could not benefit in the same way because they do not ordinarily inherit mates once their mothers die. They usually find vacancies by making forays in spring to search for unmated males. Helping behavior by females must therefore have evolved for other reasons.

An important benefit for male helpers in several species may be control over greater amounts of resource by the lineage group. That is, male helpers may help defend larger territories, making more resources available for production of nondescendant relatives. Since controlling more resources mainly provides indirect fitness benefits, this effect, if important, would implicate indirect (kin) selection processes. The size of breeding territories is positively correlated with group size in Australian magpies *(Gymnorhina tibicen)* (Carrick 1972), striated jungle babblers *(Turdoides striatus)* (Gaston and Perrins 1975), kookaburras (Parry 1973), and acorn woodpeckers *(Melanerpes formicivorous)* (Stacy 1973; MacRoberts and MacRoberts 1976). However, these studies do not make clear whether territories increase in size because groups are larger or groups become larger because larger territories are more suitable for nesting.

Benefits to breeders. Helping is probably not selfish behavior because present evidence indicates that it benefits breeders. The majority of studies show that breeding pairs rear more young on average when assisted by helpers than when nesting without help (J. L. Brown 1978; Emlen 1978). This evidence alone is not conclusive, however, because a positive corre-

Figure 3–4

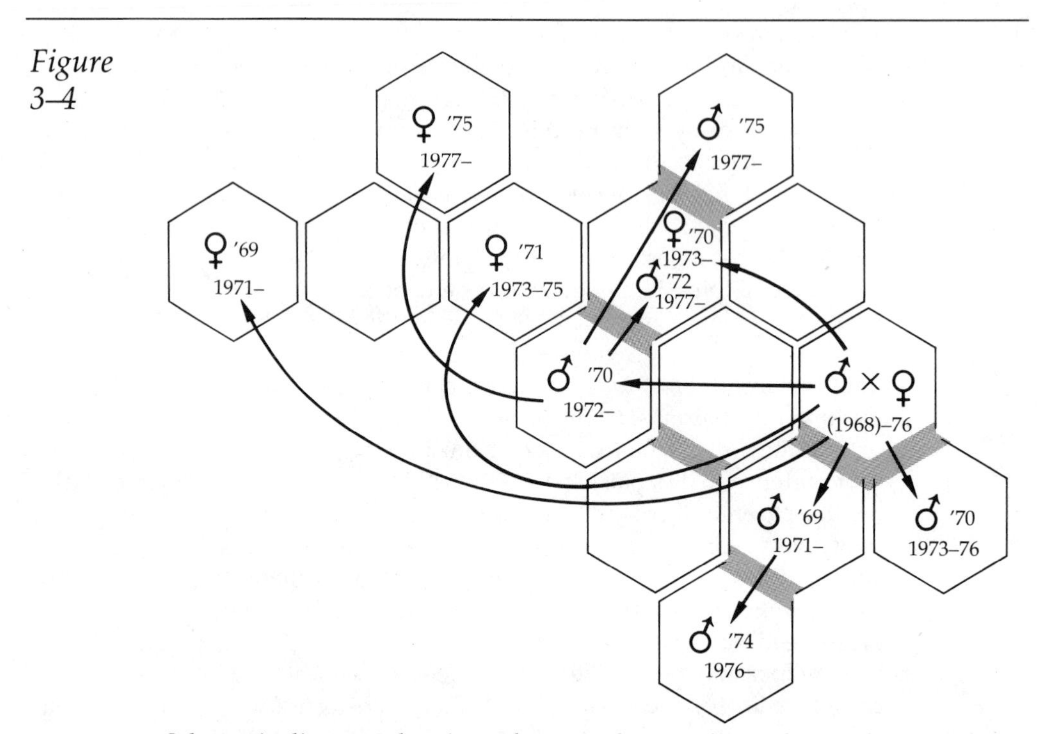

Schematic diagram showing where ten known descendants of one Florida scrub jay pair eventually bred.

Hexagons represent individual breeding territories. Shaded borders demarcate territories that formed by budding from parental territories. Arrows indicate dispersal of helpers from natal territories to breeding territories. Sex and hatching year are shown for each helper on its breeding territory. Dates of first breeding and (for four individuals) last breeding are shown for each helper below its sex and hatching year.

Source: From G. E. Woolfenden and J. W. Fitzpatrick, "The Inheritance of Territory in Group-breeding Birds," p. 107. Reprinted, with permission, from the 1978 *BioScience*, © American Institute of Biological Sciences.

lation between breeding success and number of helpers could arise if some third variable is correlated with both. For instance, older breeders are more experienced and have produced more previous progeny. Their higher success may therefore be due to their greater experience, not to the greater assistance received from helpers (e.g., see Woolfenden 1975). Similarly, higher-quality territories may allow higher breeding success and support larger numbers of helpers, with higher success resulting from higher territory quality and not the greater assistance provided by helpers (e.g., Brown and Balda 1977). To resolve such issues, one must obtain more detailed information about causal relationships.

The study that best disentangles causal relationships involves the grey-crowned babbler (Brown et al. 1978; Brown and Brown 1980). Fledging success of babbler groups was initially found to correlate with group size, age of mother, and two vegetational indices of territory quality. Any of the four independent variables, alone or in combination, could explain the higher fledging success observed in groups containing helpers. One way to isolate the most important (but not necessarily causal) variable is stepwise multiple regression analysis, which first tests the effect of each variable in turn while controlling the effects of all remaining variables and then isolates the variable having the greatest effect. The results indicate that higher fledging success stems from larger group size, not older age of mother or higher quality of territory. To further test this conclusion, Brown and his co-workers matched several breeding groups according to group size, territory quality, and maternal age and then removed helpers from half the groups. Fledging success was significantly lower in the experimental group than in the control group, and the magnitude of difference was similar to that predicted by stepwise multiple regression analysis. The results therefore show that helpers do increase the reproductive success of breeders, at least in grey-crowned babblers.

Some further analysis involving over 30 variables of possible significance reveals how helpers help (Brown et al. 1978). The rate at which food was delivered to nestlings by all group members combined was not correlated with group size. Consequently, nestlings were not fed any more often in larger breeding groups, and hence the number of fledglings produced per egg and per nest was not correlated with group size. However, parents delivered food less often in larger groups because helpers contributed a greater proportion of the food. As a result, parents could produce more broods. Thus larger groups produced more fledglings because they raised more broods, which was made possible by earlier initiation of first broods and shorter intervals between broods.

Possible detrimental effects of helpers. The presence of helpers could conceivably be detrimental to the fitness of breeders. Helpers may deplete resources on the territory, attract predators, or increase conspicuousness of nests (Emlen 1978). To the extent that parental care is affected by previous experience, incompetence of helpers might reduce rather than enhance nestling survival. Male helpers may copulate with breeding females, leading to reduced chances of paternity for dominant males. Such copulations do in fact occur among red-throated bee-eaters *(Merops bulocki)* (Fry 1977) and acorn woodpeckers (Stacy 1973; Koenig and Pitelka 1979). Finally, since additional progeny on a territory might compete with helpers for food or the chance to later inherit the territory, helpers may deliberately disrupt reproductive activities of breeders to reduce future competition with siblings (Zahavi 1974; but see Brown 1975b for rebuttal).

Some evidence indicates that the presence of helpers does not increase annual reproductive output of breeders in red-throated bee-eaters (Fry 1972a), white-throated bee-eaters *(Merops bulockoides)* (Emlen 1978), Arabian babblers (Zahavi 1974), striated jungle babblers (Gaston and Perrins 1975), or green woodhoopoes (Ligon and Ligon 1978b), suggesting that helping behavior is not beneficial to breeders. However, this evidence is far from conclusive. Unpublished data for red-throated bee-eaters indicate that helpers do increase the output of breeders (Jerram L. Brown, personal communication). The data for Arabian babblers do not show a significant increase if a two-tailed statistical test is used, but they do if a one-tailed test is used (Jerram L. Brown, personal communication). The sample sizes for striated jungle babblers were too small to detect a significant increase. Helpers in green woodhoopoes do not enhance the success of breeders in some years, but they do in others. In addition, assistance of helpers should reduce the time and energy parents must spend caring for nestlings, and this effect could allow parents to survive longer than would otherwise be possible (J. L. Brown 1978; Stallcup and Woolfenden 1978; J. L. Craig 1979). When increased survival is an important advantage of receiving help, annual reproductive rate is a poor measure of fitness consequences.

The behavior of helpers should not be detrimental to breeders because deleterious consequences of that sort should cause parent birds to eject helpers from the territory. The existence of such consequences in other species probably explains why helping behavior has not evolved in many species that otherwise fulfill the conditions usually associated with its occurrence (Orians et al. 1977). Colonial seabirds and polygamous marsh-breeding blackbirds, for example, experience shortages of breeding habitat, exhibit high site fidelity that could maintain contact between relatives, and are characterized by deferred breeding of subadults for one or more years. Helping behavior may have failed to evolve in these species because the costs of competing with helpers for food or mates are higher than any advantages that might be gained from receiving help.

Helpers in mammals

Helping behavior has evolved in several social mammals. Female elephants *(Loxodonta africana)* regularly nurse calves other than their own within social groups (Douglas-Hamilton and Douglas-Hamilton 1975). Female lions *(Panthera leo)* nurse cubs other than their own and baby-sit them while the remaining lionesses are out hunting (Schaller 1972; Bertram 1975a, b, 1976). Female brown hyenas *(Hyaena brunnea)* sometimes nurse cubs other than their own, and all females, including those without cubs, bring food to the den and help guard cubs (Owens and Owens 1979a, b). Female banded mongooses *(Mungos mungo)* nurse cubs indiscriminately within social groups, and both males and nonlactating females guard all cubs in the den while lactating females are away foraging (Rood 1974). Subordinate adult female and yearling dwarf mongooses *(Helogale parvula)* guard and bring food to young produced by the dominant female (Rood 1978). In African wild dogs *(Lycaon pictus)* and wolves *(Canis lupus)* all pack members help feed pups even though only one, or at most two, females gave birth to them (Kühme 1965a, b; Mech 1970; Lawick-Goodall and Lawick-Goodall 1970; Fentress and Ryon 1981). Prior offspring sometimes remain with parents and help feed subsequent litters in red foxes *(Vulpes vulpes)* and black-backed jackals *(Canis mesomelas)* (MacDonald 1979; Moehlman 1979). Thus helping behavior is by no means restricted to birds.

Most of the hypotheses and theoretical arguments discussed above for birds are equally applicable to mammals. Helping behavior in mammals may involve altruism, cooperation, or possibly selfishness. Evidence for evaluating these alternatives is inadequate for every species.

Helping behavior in mammals is usually interpreted in terms of indirect (kin) selection because it generally involves relatives. Female elephants and lions within each social group are nearly always genetically related, usually as siblings, mothers, daughters, or cousins (Schaller 1972; Bertram 1975a, b, 1976; Douglas-Hamilton and Douglas-Hamilton 1975). African wild dog packs and wolf packs are family groups consisting of a dominant breeding pair and their prior offspring (Mech 1970; Frame and Frame 1976; Frame et al. 1979). Nonbreeding helpers are also prior offspring of the breeding pair in black-backed jackals and probably in red foxes (MacDonald 1979; Moehlman 1979). Most helpers in dwarf mongooses are prior offspring of the dominant pair, though occasionally unrelated immigrants also provide substantial help (Rood 1978). Genetic relationships are unknown in brown hyena and banded mongoose groups. The mere existence of genetic relatedness among helpers does not demonstrate the importance of indirect (kin) selection, however. A careful analysis of all costs and benefits accrued by both helpers and breeders is necessary before helping behavior can be properly understood (e.g., see Rood 1978).

Alloparental behavior in primates

Female primates other than the mother commonly hold or handle infants, a behavior often called *aunting*. The term derives from the British auntie, a close female friend of the family who helps care for children (Rowell et al. 1964). However, to avoid possible confusion with genetically related aunts, a better term for females performing such behaviors is **allomothers** (*allo* meaning "other") (after Wilson 1975a).

Like other forms of helping behavior, allomothering could be interpreted as altruistic, cooperative, or selfish. Again, assessing each alternative requires an assessment of who benefits and who suffers from the behavior.

Benefits to allomothers. Allomothering behavior is often interpreted as altruistic, on the assumption that time and energy invested in another female's infant reduces the allomother's direct fitness. However, there is no good evidence supporting this assumption, and some evidence supports just the opposite conclusion.

The most likely benefit for allomothers is a gain in mothering experience (Gartlan 1969; Lancaster 1971; Blaffer Hrdy 1976). Maternal care is complex in primates and requires considerable experience before females become proficient at it. By handling infants of other females, inexperienced females might gain valuable mothering experience and thereby make fewer mistakes with their own infants. Experimental studies with rhesus monkeys *(Macaca mulatta)* show that females deprived of mothering experience during their own infancy grow up to become inferior mothers (Harlow et al. 1966), and that females who have previously reared infants are better mothers than females who have not (Seay 1966). Under field conditions first infants survive to the age of six months only 50% of the time, compared to 90% for fourth or subsequent infants of experienced mothers (Drickamer 1974a). In Hanuman langurs *(Presbytis entellus)* and Japanese monkeys *(Macaca fuscata),* older allomothers are less awkward at handling infants than young allomothers, though this difference could be due to maturational changes in coordination or motivational states rather than differences in previous mothering experience (Blaffer Hrdy 1977a; Kurland 1977).

If allomothering evolved as a way for females to gain experience with infants, it should be exhibited mainly by young females. Most evidence supports this prediction (Blaffer Hrdy 1976, 1977a). Lancaster (1971), for example, found that 295 of 347 allomothering episodes in vervet monkeys were initiated by females without previous mothering experience. The other 52 episodes were initiated by older females who had previous mothering experience. A similar preponderance of allomothering behavior by inexperienced females has been shown for rhesus monkeys (Spencer-Booth 1968), Japanese monkeys (Kurland 1977), Hanuman langurs (Sugiyama 1965), Nilgiri langurs *(Presbytis johnii)* (Poirier 1970), and several other primates (Blaffer Hrdy 1977a).

Not all allomothering is performed by inexperienced females. Some cases involve experienced adult females who are neither pregnant nor lactating. Blaffer Hrdy (1977a) suggests that allomothering is a way for adult females to gain information and become familiar with new troop members (the newborn infants). Her interpretation is consistent with the widespread behavior of allomothers holding newborn infants upside down, sniffing and peering at them, and inspecting their genitals.

Benefits to mothers and infants. Allomothering may benefit mothers in several ways. In Nilgiri langurs, patas monkeys *(Erythrocebus patas),* and vervet monkeys, mothers commonly place their infants near another female before wandering off to forage (Poirier 1968; Lancaster 1971; Blaffer Hrdy 1976). Females of these species forage on the ground but depend on vigilance and quick retreat up a tree to escape predators. By placing their infants near another female, mothers can feed farther from the safety of trees than would otherwise be safe. Similar behavior has also been reported for Japanese monkeys, but it has only been observed near artificial provisioning sites (Kurland 1977). It may therefore be an artifact in that species.

A second potential benefit may be increased infant survival. Lactating females who recently lost their own infants occasionally adopt orphans who would otherwise have died (Blaffer Hrdy 1976). Nonlactating females may also adopt infants, and in one laboratory study of rhesus monkeys, such females began producing milk after infants were forced on them (Hansen 1966). Adoption of orphaned infants has been reported under natural conditions for Japanese monkeys, rhesus monkeys, and chimpanzees *(Pan troglodytes)* (Itani 1959; Rowell 1963; Sade 1965; Lawick-Goodall 1968). Females may be more prone to adopt orphaned infants if they have had previous experience allomothering them, but no direct connection has ever been demonstrated between the two events. Similarly, females with allomothering experience may be more likely to rescue infants from hazardous situations, though again no connection has ever been demonstrated.

Costs to mothers and infants. Some evidence suggests that allomothering may not benefit mothers or their infants. Mothers often resist attempts by other females to hold infants. Rhesus monkey mothers, for example, are initially aggressive toward prospective allomothers and only allow them to handle infants after the allomother has groomed them or sidled up unobtrusively while pretending to forage (Rowell et al. 1964). The Hanuman langur is exceptional in that mothers are very tolerant of other females (Jay 1965). They allow allomothers to hold infants within a few hours of birth, and as many as eight females may hand an infant back and forth on the first day.

Allomothers often hold infants in awkward positions or treat them roughly (Photo 3–6) (Blaffer Hrdy 1976, 1977a). Rough handling generally occurs when the allomother is trying to take the infant away from its mother, when she is trying to keep it after the mother wants to take it back, when she is trying to get rid of it, when she is inspecting it, or, rarely, when she is punishing it. Probably for this reason mothers supervise allomothers closely while their infants are being handled, and they are quick

Photo 3–6

A Hanuman langur allomother sitting on another female's infant and pushing its head into the rock.

Such abuse is a risk taken whenever a mother allows another female to take her infant.

Photo by Sarah Blaffer-Hrdy/Anthro-Photo.

to retrieve infants from incompetent allomothers. Sometimes attempts to retrieve infants lead to pulling contests, with the infant caught in the middle. Allomothers may even kidnap infants, and kidnapping by dominant females occasionally leads to their starvation (Bourliere et al. 1970; Mohnot 1974; Blaffer Hrdy 1976, 1977a). Such effects and even minor injuries caused by rough handling are very rare, however, which makes the deleterious effects of allomothering seem small (Mohnot 1974; Blaffer Hrdy 1977a). Perhaps the benefits derived by mothers and their infants exceed these costs, although this conjecture is not yet supported by empirical evidence.

If allomothering is detrimental to mothers and infants, tolerating it would be altruistic behavior on the part of mothers unless allomothers were previous offspring of the mothers. Altruism by mothers could evolve in this context provided that allomothers are usually relatives and the experiences gained by allomothers are enough to offset the costs to infants. The relevance of indirect (kin) selection to allomothering behavior has recently been evaluated for Japanese monkeys. Genealogies are now known for several troops near Kyoto, Japan, which have been studied for more than two decades, and genetic relationships among mothers and allomothers can therefore be analyzed. Kurland (1977) found that only 18 of 140 allomothering episodes involved relatives. Thus indirect (kin) selection does not seem to be implicated. The most reasonable interpretation is that allomothering is a cooperative interaction, with both allomothers and mothers benefiting from it.

Conclusion

Altruism may involve either relatives or nonrelatives. When it involves relatives, indirect (kin) selection is implicated, provided that fitness benefits derived through the progeny of relatives multiplied by the coefficient of relatedness between altruist and recipient exceed the fitness costs incurred by sacrificing personal survival or reproductive output. When it involves nonrelatives, trait-group selection among groups of nonrelatives or selection for reciprocal exchange of benefits are implicated.

Several behavioral interactions that occur in vertebrates are widely viewed as altruistic. The two most prevalent of these are alarm signaling and helping behavior. Additional examples will be encountered in the next chapter. Present evidence suggests that alarm signals have sometimes evolved to aid kin. The best evidence supporting an important role of indirect (kin) selection comes from studies of ground-dwelling rodents and white-tailed deer. For other species the evidence for altruism is weak, and a growing body of literature is suggesting that alarm calling evolved to promote selfish interests. Likewise, the available evidence indicates that helping behavior has usually evolved as a cooperative interaction. Helping behavior may be altruistic in some mammals and birds, but in many species both helpers and recipients derive direct benefits.

Despite these conclusions, the importance of indirect (kin) selection cannot be entirely ruled out. Individuals who help nondescendant relatives in cooperative interactions ordinarily derive larger net benefits than individuals who help nonrelatives, because they benefit indirectly through the relatives' progeny as well as directly through personal fitness gains. Thus aid-giving behavior may frequently have evolved through direct (individual) selection, but indirect (kin) selection may still have determined who receives the help.

Cooperation

Altruism has great theoretical significance for understanding selective processes, but it is not a major organizing influence in most social systems. A much more prevalent form of aid-giving behavior is cooperation. Behaviors involved in thwarting predators or finding food often confer mutual benefits. Such behaviors are of fundamental importance in sociobiology, because they are instrumental in bringing animals together as social groups.

Cooperation should evolve whenever the individuals who benefit least average higher fitness than individuals who do not cooperate at all. It may be advantageous because it enhances adult survival, adult fecundity, or offspring survival. These benefits may be derived through mutual defense against predators, enhanced foraging efficiency, or reduced costs of remaining together in groups. This chapter evaluates each potential benefit in turn.

Cooperative defenses against predation

Mutual vigilance

One of the most prevalent advantages of social behavior is an improved ability to detect predators. Galton (1871) rather colorfully described the advantages of mutual vigilance while writing about the social behavior of African cattle:

> *To live gregariously is to become a fibre in a vast sentient web overspreading many acres; it is to become the possessor of faculties always awake, of eyes that see in all directions, of ears and nostrils that explore a broad belt of air; it is to become the occupier of every bit of vantage ground whence the approach of a lurking enemy might be overlooked.*

The same description could have just as easily been written about many other social animals in addition to cattle.

The importance of mutual vigilance is made clear by observations of how readily social animals respond to the alertness and alarm signals of other group members. Many animals are even responsive to behavioral indications of danger shown by other species, and in some cases they may actively associate with other species who possess complementary sensory capabilities. For example, baboons, zebras, gazelles, and other antelopes often forage together, and each species responds to warning signals given

Chapter four

by the others when predators are detected (Devore and Hall 1965). Baboons probably benefit from the ungulates' superior sense of smell, while ungulates probably benefit from the baboons' superior vision. It is tempting to conclude that baboons and ungulates associate with each other to enhance vigilance, although the associations could conceivably result instead from similar habitat preferences.

Powell (1974) has experimentally demonstrated the value of mutual vigilance by studying the response times of caged starlings *(Sturnus vulgaris)* to a stuffed hawk model flown over them on a monofilament line. He found that groups of ten starlings responded more than 20% faster to the stuffed hawk than did solitary individuals. Single birds could not look up while eating grain from the bottom of the cage, but birds in flocks could rely on the vigilance of other flock members while foraging. Powell's experiment has been repeated, using caged red-billed queleas *(Quelea quelea)* and a trained chanting goshawk *(Melierax metabetes)* that frequently preys upon them, and the results again show that increased flock size leads to increased probability of detecting a hawk flying overhead (Lazarus 1979).

The responses of individuals to the presence of a predator further confirm the importance of mutual vigilance. When a trained Harris hawk *(Buteo unicinctus)* is released over an area where yellow-eyed juncos *(Junco phaeonotus)* are foraging, flocks become larger and the time each individual devotes to surveillance increases (Caraco et al. 1980). This result indicates that flocking behavior provides protection from predators and that mutual vigilance is likely to be an important advantage of flocking.

A gain in alertness may often be decisive in escaping an attacking predator. Many predators rely on the element of surprise to capture prey, and when surprise is lost, their chances of success are greatly diminished (Rudebeck 1950; Schaller 1972; Elliott et al. 1977; Kenward 1978). That flock members do gain survival advantages has recently been documented. Page and Whiteacre (1975) found that the mortality rate of wintering shorebirds averaged over three times higher for solitary individuals than for flock members. The difference resulted primarily from the hunting efforts of merlins *(Falco columbarius),* a pigeon-sized falcon that hunts small birds by diving from above and catching them unawares. Kenward (1978) found that the capture success of a trained goshawk was negatively correlated with the size of wood pigeon *(Columba palumbus)* flocks and attributed the correlation partly to improved vigilance in larger flocks (Figure 4–1). Neither study proves that mutual vigilance is responsible for reducing vulnerability to predators, as flocking behavior may confer other forms of protection as well, but they do show that flocking behavior has survival advantages.

Using a simple mathematical model, Pulliam (1973) has identified four key variables that should affect the value of mutual vigilance. These are (1) group size of the prey, (2) the frequency that prey individuals look about for predators, (3) the time required for a predator to make its final rush from concealment, and (4) the number of times a group is attacked. Because the model assumes that vigilant members of a prey group have no chance of spotting a predator during the stalking phase but are certain

Figure 4–1

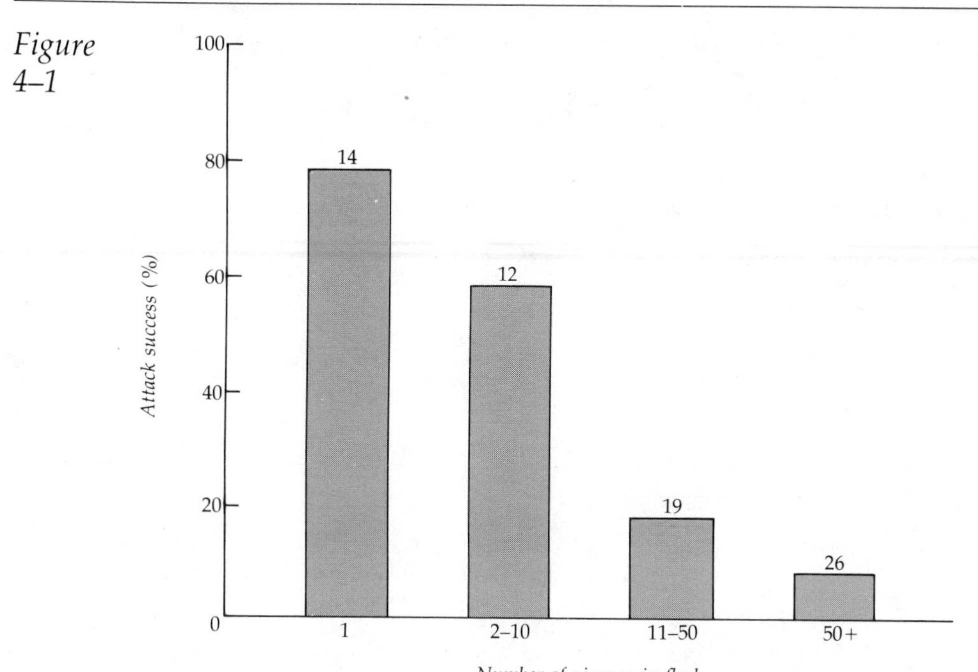

Attack success of an avian predator decreases as flock size of the prey species increases.

The data were obtained by releasing a trained goshawk in the vicinity of wood pigeon flocks. Numbers above the bars of the histogram indicate the number of experimental trials conducted for the indicated range of flock sizes.

Source: From R. E. Kenward, "Hawks and Doves: Attack Success and Selection in Goshawk Flights at Woodpigeons," *Journal of Animal Ecology* 47 (1978):453. Reprinted by permission of Blackwell Scientific Publications Limited.

of spotting it during the final rush, it leaves out several additional variables that should be important. These include the stalking tactics of the predator, the number of predators involved in the attack, topography, vegetation structure, and physical condition of the prey, all of which influence a prey animal's chances of detecting a predator during both phases of a hunt.

Pulliam's model illustrates several general features of mutual vigilance. For any given predatory attack the probability of detecting a predator quickly approaches 1.0 as flock size increases (Figure 4–2). The asymptote is reached faster if individuals look up more often or if predators take longer to make their final rush. The amount of time spent on surveillance affects the probability of detecting predators, but it also reduces the time left available for foraging. Selection should therefore favor some optimal allocation of time between the two activities. When prey animals are harder pressed to find food, they should rely more on vigilance of others so they can spend more time foraging. However, this necessitates larger group sizes, which in turn increases competition for food. Thus a delicate trade-

Figure 4–2

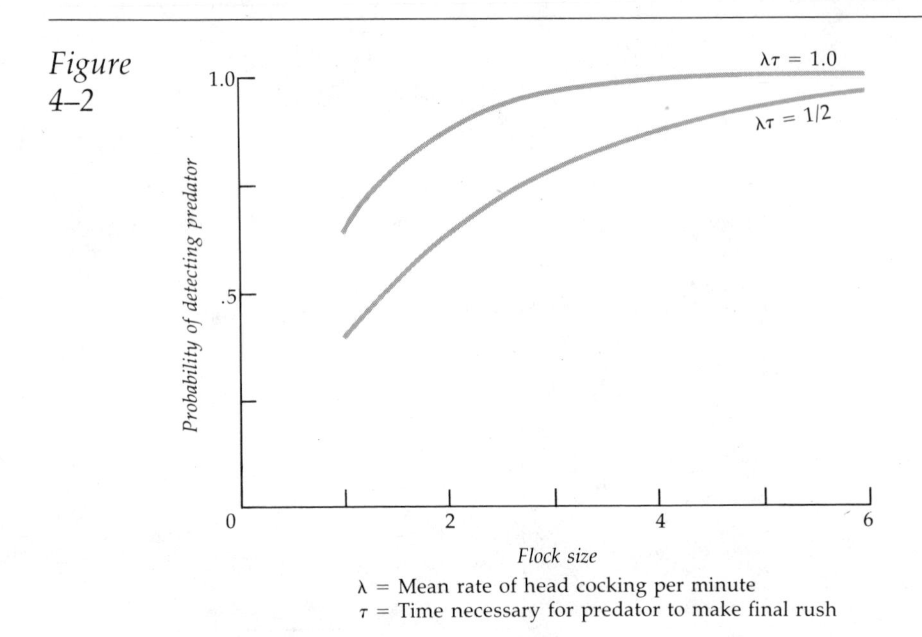

The probability that a prey will detect a stalking predator as a function of prey group size.

As prey group size increases, the probability of detecting the predator approaches (but never quite reaches) 1.0. The steepness of the curve depends on the rate at which prey animals look about for predators (λ) and the time necessary for a predator to make its final rush (τ). The curves are hypothetical and are based on a mathematical model of predator-prey interactions.

Source: Based on data in Pulliam 1973.

off exists between the advantages gained from mutual vigilance and the disadvantages suffered from increased competition.

The duration of a predator's attack phase is shorter in dense cover and for faster predators. Also, prey animals are less able to detect a stalking predator in dense cover; when the predator is small, cryptically colored, or stealthy; or when the predator ambushes prey instead of stalking them. Thus the value of mutual vigilance decreases when vegetation is dense or the predator is difficult to detect before an attack.

The curves shown in Figure 4–2 suggest that prey animals gain little additional advantage from mutual vigilance once group size reaches some moderately small level. However, the probabilities of detecting a predator were graphed for single attacks. The probability of detecting a predator on every one of many independent attacks is much smaller and is calculated by multiplying the probabilities for each attack. For example, when $\lambda\tau = 1/2$ (see Figure 4–2), the probability of detecting a predator during the stalk phase on each of ten independent attacks is $.95^{10} = .60$ for a flock of six birds, compared to $.39^{10} = .00008$ for a single bird. When even more attacks are considered, the ability to consistently detect stalking predators is substantially higher for large flock sizes as compared to moderately sized flocks. Thus the advantages gained from mutual vigilance can favor very large group sizes.

Active defense and mobbing

Flight is the most common defense for prey once a predator is detected, but some animals set up a defensive screen or counterattack. Either individuals or groups may counterattack. When social groups cooperatively attack, harass, or pursue a potential predator, the behavior is called **mobbing**.

The epitome of active defense occurs in social termites, ants, bees, and wasps. Soldiers or workers defend colonies by collectively attacking any intruders that threaten them, often relying on massed attacks to overwhelm or repulse intruders much larger than they are (Wilson 1971). In the myrmicine ant (*Pheidole dentata*) recruitment of soldiers is narrowly specific (Wilson 1975c, 1976b). Workers lay down the pheromone trail responsible for recruiting soldiers only when they encounter fire ants (*Solenopsis geminata*) or other species of the same genus. The fire ant is a large, aggressive species and is the most formidable competitor of *Pheidole*. In other ants and in bees, wasps, and termites, massed attacks are also stimulated by pheromones released when workers are strongly disturbed or attacked, but they are elicited by a broader range of stimuli (e.g., members of rival colonies, insect or vertebrate predators, hive robbers) (Wilson 1971).

Many insects congregate together to increase local concentration of noxious secretions or sprays. Groups of red wood ants (*Formica rufa*) collectively eject acid sprays toward intruders, and groups of nasutitermitine termites coat intruders with odorous resins (Eisner 1970). Bombardier beetles (*Brachinus*), which emit small puffs of an irritant substance, frequently congregate for the same reason (Wautier 1971). Similarly, numerous unpalatable tropical butterflies aggregate at night, and although they do not emit any chemical toxins, they may enhance the effectiveness of their warning colors or scents by roosting together (Cott 1940; Benson and Emmel 1973). (Active defense is not exhibited by these butterflies, but the role of cooperation does parallel that of species that do show active defense.)

A rather unusual mode of defense is shown by the social owlfly (*Ascaloptynx furciger*) (Henry 1972). Eggs are laid in clumps and after they hatch larvae congregate into a bristling mass of jaws whenever they are disturbed (Figure 4–3). This combined effort makes it difficult for an avian predator to grab any of the larvae from behind the jaws.

A classic example of active defense among vertebrates is shown by the muskox (*Ovibus moschatus*), a large bovid found in North American Arctic tundra (Tener 1954, 1965). When attacked by wolves, adult female muskoxen form a defensive ring around their calves, while the adult male remains outside the ring and commonly charges the wolves in an attempt to drive them off (Photo 4–1). Like many animals, muskoxen respond differently to different kinds of predators. When approached by humans,

Figure 4–3

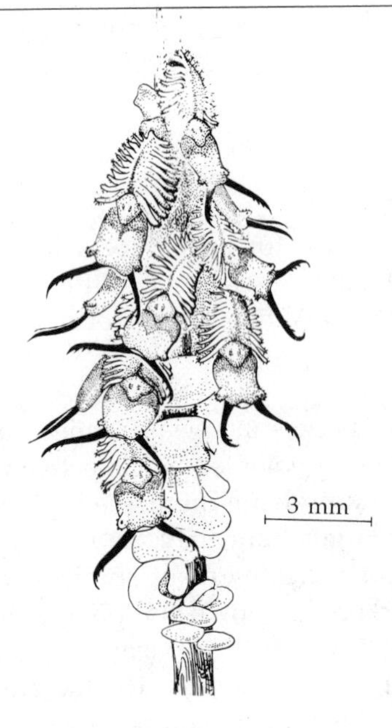

Social owlfly larvae form a bristling mass of jaws when confronted by an avian predator.

Source: From C. S. Henry, "Eggs and Repagula of *Ululodes* and *Ascaloptynx* (Neuroptera: ascalaphidae): A Comparative Study," *Psyche* 79 (1972):18. Reprinted by permission.

they flee instead of forming a defensive circle. Similar defensive formations also occur in elephants (Sikes 1971; Douglas-Hamilton 1972), water buffalos (*Bubalus bubabis*) (Eisenberg and Lockhart 1972), and African Cape buffalo (*Syncerus caffer*) (Kruuk 1975).

Some published reports suggest that primates may counterattack against certain predators. In experimental settings recently captured chimpanzees will charge a stuffed leopard (*Panthera pardus*), brandish or hurl sticks at it, and generally create a ruckus (Photo 4–2) (Kortlandt 1962, 1965, 1967, 1972; Kortlandt and Kooij 1963). Mobbing is most prevalent among chimpanzees that live on open savanna and is less prevalent among chimpanzees that live in forests (Kortlandt 1972). Chimpanzees respond most intensely when presented with a stuffed leopard that has a stuffed chimpanzee infant in its paws. After a few minutes of attack responses, the chimpanzees seem to show "increasing doubts" about the leopard's authenticity (Kortlandt 1967, 1972). Their rage dissipates, and they begin showing some curiosity, suggesting that they behaved earlier as if the leopard were alive. Even so, they continued showing attack behavior sporadically for as long as several hours (Kortlandt 1972).

Despite these results, the situation is artificial in the sense that the stuffed leopard did not attack or retreat as a real one might. Wild chimpanzees have been observed brandishing clubs and throwing stones at other chimpanzees but not at live leopards (Goodall 1963a; Kortlandt 1965;

Photo 4–1 *Defensive circle of muskoxen.*
When threatened by wolves, female muskoxen form a defensive circle, with horns directed outward. The male (on right) remains outside the circle and attacks any wolves that approach too closely.
Photo by Ted Grant/National Film Board Photothèque.

Photo 4–2

Chimpanzee attacking a stuffed leopard with a club.
When a stuffed leopard is first presented, chimpanzees mob it, brandishing clubs and hurling sticks. The reaction is strongest when a stuffed infant chimpanzee is placed in the leopard's paws.
Photo by Adriaan Kortlandt.

Reynolds and Reynolds 1965). An experiment using live leopards would help resolve this problem. Unfortunately, a planned experiment with a live, caged leopard failed because a brush fire engulfed the observation area (Kortlandt, personal communication). Hence the reality of mobbing behavior under natural conditions is still in some doubt.

Baboons have rarely been seen actually attacking large predators. In one instance involving olive baboons (*Papio cynocephalus*), males and even females with clinging infants lunged at a leopard who sprang out of bushes into their midst (Altmann and Altmann 1970). In another instance olive baboons attacked and lethally wounded a serval cat (*Felis serval*) (Teleki 1973a). Chacma baboons (*Papio ursinus*) have been observed throwing and knocking stones off cliffs in the general direction of humans (Hamilton et al. 1975a, 1978a). Such behavior is less risky than direct counterattacks, but it indicates a propensity to actively defend against predators. When presented with a stuffed leopard, adult male olive baboons scream and lunge at it, eventually slashing its hindquarters and dragging it off once the leopard fails to react (Devore 1972; Kortlandt 1980). Dominant male chacma baboons respond to dog packs by immediately interposing themselves between the rest of the troop and the dogs, with a single adult male not uncommonly maiming or killing several large dogs (Stoltz and Saayman 1970). When approached by humans, however, adult males break and run, fleeing in front of their troop instead of behind it. Nevertheless, the rarity of such observations suggests that mutual vigilance and flight are more typical defenses against predators. Of course, the very presence of a human observer may make mobbing a less likely response than usual, which could explain why mobbing is only rarely seen. Popularized versions of baboons attacking humans and other predators often leave the impression that such events occur frequently, but the available evidence suggests that such reports are greatly exaggerated.

Although baboons rarely counterattack large predators, they may threaten them. Occasionally adult olive baboons approach or run toward wild cheetahs (*Acinonyx jubatus*), leopards, or lions, displaying their canines and barking at them (Devore and Washburn 1963; Hall 1963; Altmann and Altmann 1970; Baenninger et al. 1977; Bertram 1978a), and adult males regularly become interposed between the troop and a pursuing predator during retreats (Altmann and Altmann 1970; Devore 1972). This positioning frequently arises because adult males start fleeing later, run more slowly, and stop and turn sooner than other troop members. Rowell (1966), for example, states that adult males in her study troop took a vanguard position primarily when "the cause of alarm was so slight that the more confident males did not respond to something that set the juveniles running." She found that stronger stimuli (e.g., her own movements) provoked precipitous flight, suggesting that males often lag behind because they feel less threatened and hence are less responsive to most predators. According to Rhine and Owens (1972; also Rhine 1975), adult males are more likely to enter hazardous clearings first and to lead or bring up the rear during troop progressions. Such positioning could provide protection for other troop members. Altmann (1979), however, found that the order of troop progression is essentially random, although adult males are sometimes found in the front and back thirds of the troop in potentially dangerous situations. Thus the extent to which male baboons take risks to protect other troop members is not yet clear. The taking of such risks would not represent a form of altruism because adult males do have a vested interest in other troop members. Many infants in the troop are their offspring, and the adult females are their best source of future matings.

Mobbing does occur in several other social mammals. Coatimundis (*Nasua narica*), agoutis (*Dasyprocta punctata*), and various primates mob snakes (Janzen 1970a; Smythe 1970b; Oppenheimer 1977; Owings and Coss 1977), and large ungulates sometimes charge in concert to repulse lions, hyenas, and African wild dogs (Kruuk 1972; Schaller 1972; Leuthold 1977).

Many species of colonial or flocking birds mob potential predators (S. A. Altmann 1956). Mobbing behavior varies greatly among species and ranges from simply approaching a predator while giving distinctive calls to swooping down from behind and pecking at the predator's head or body. Mobbing is most effective in gregarious species because many individuals can be recruited to the scene, but even in nongregarious species parent birds sometimes gang up on predators that threaten their eggs or young (Curio 1975). A perched hawk is most likely to elicit a mobbing response when it is "sharp set," a posture indicating that the bird is hungry and on the alert for vulnerable prey (Photo 4–3) (Hamerstrom 1957). Thus birds are apparently most prone to mob when the predator poses a real threat.

Prey animals sometimes approach predators without actually mobbing them. Such behavior occurs, for example, in gulls and in various African antelopes (Kruuk 1964, 1972; Kruuk and Turner 1967; Walther 1969; Schaller 1972). Kruuk (1976) suggests that prey animals approach predators to collect information about potential sources of danger, which may be useful in avoiding predators at some future time. Approaching a predator singly would be a hazardous venture, but approaching as a group reduces the risk for every individual.

Because mobbing has risks associated with it, prey animals should be less prone to mob dangerous predators. A study of mobbing behavior in California ground squirrels (*Spermophilus beecheyi*) supports this prediction (Owings and Coss 1977). Ground squirrels mob snakes by approaching them, kicking sand on them, or pouncing on them. They also wave their tails sideways ("tail flagging") while confronting a snake. In experimental tests ground squirrels from areas where rattlesnakes are absent give these responses commonly toward nonpoisonous snakes (but not toward rattlesnakes). Ground squirrels from areas where rattlesnakes are present, however, give them in low frequencies to all snakes. Thus ground squirrels mob snakes (except those making threatening sounds) where poisonous snakes are absent but do not mob any snakes where poisonous snakes are present. Similarly, pied flycatchers (*Ficedula hypoleuca*) vigorously attack woodpeckers and squirrels by swooping down and pecking at them, but they remain at a safe distance from accipiter hawks capable of catching small birds on the wing (Curio 1975).

A. B.

Photo 4–3 *Birds can determine when a perched hawk is ready to hunt by its posture.*

*A. A red-tailed hawk (*Buteo jamaicensis*) perched in a relaxed posture. B. A red-tailed hawk perched in a sharp-set posture, ready to attack a vulnerable prey.*

Photo from F. Hamerstrom, "The Influence of a Hawk's Appetite on Mobbing," *Condor* 59 (1957):193.

Some potential benefits of mobbing behavior are driving away or deterring a predator by harassing it or interfering with its hunting behavior, discouraging the predator from returning to the same general area on future occasions, diverting the predator's attention away from relatives or offspring, and announcing to the predator that remaining in the area is futile because surprise has been lost (Curio et al. 1978; Harvey and Greenwood 1978). Note that these proposed benefits closely parallel those proposed to explain alarm signals, but evidence for evaluating them is largely lacking. Mobbing by colonial birds is known to effectively deter predators (references in Hoogland and Sherman 1976), but its effectiveness for other species is not well documented.

Erratic flight and explosive scattering

Fleeing along an unpredictable path is a common defensive tactic once a predator attacks a prey (Humphries and Driver 1967, 1970; Driver and Humphries 1970). Both solitary and social animals respond to danger with erratic flight, but simultaneous erratic scattering of many individuals is likely to be more confusing to a predator than erratic flight by a single individual. Predators typically single out a particular prey individual, often one that appears odd, weak, or unwary (e.g., Mueller 1971, 1975). Scattering presents the predator with multiple moving targets, which may make the task of singling out a particularly vulnerable prey more difficult. Scattering by many individuals at once also makes the predator's task of maintaining sensory contact with a prospective victim more difficult. In addition, the sudden explosion of many individuals moving in different directions at once may startle a predator and cause it to hesitate or jump back. Such effects can break sensory contact between the predator and its intended victim or provide enough additional time for the prey animal to reach safety. The margin between success and failure in predator-prey interactions is often small, making even slight delays potentially important.

The effectiveness of creating commotion has been clearly demonstrated experimentally. Neill and Cullen (1974) showed that predatory squid (*Loligo vulgaris*), cuttlefish (*Sepia officinalis*), pike (*Esox lucius*), and perch (*Perca fluviatilis*) have lower capture success when attacking shoals of fish than when attacking solitary individuals. The behavioral data show that this results, in part, because scattering by multiple targets disrupts the predator's attack behavior. Major (1978) subsequently confirmed these results in an experimental study of jack (*Caranx ignobilis*), a predatory fish that preys upon Hawaiian anchovies (*Stolephorus purpureus*) and other small fish.

Geometrical effects

Predators are most likely to attack individuals on the periphery of a group, because vulnerable individuals are more easily spotted and more easily approached when on the periphery. Prey animals should therefore gain protection by remaining in or fleeing toward the center of a group (Williams 1964; Hamilton 1971; Vine 1971). The idea is that an animal is safer when it can keep another individual between itself and potential predators, especially when living in open environments where other forms of cover are lacking. Seeking the center of a group for protection is selfish behavior, but individuals still derive mutual benefits because they all gain opportunities to shield themselves behind other group members. This tactic may be responsible for the tight bunching together seen in fish schools, bird flocks, and mammal herds during predatory attacks.

Gaining safety by seeking the center of social groups has been documented by Milinski (1977), who studied predatory attacks of three-spined sticklebacks (*Gasterosteus aculeatus*) on water fleas (*Daphnia magna*). His data show that water fleas on the periphery of swarms are subjected to more attacks than are water fleas in the center. They also show that water fleas on the periphery of swarms are attacked less often than isolated water fleas. The latter condition is not necessary, however, for swarming to be advantageous. If every water flea spends an equal proportion of time in the center, swarming will be advantageous as long as the average survival rate of all water fleas in a swarm exceeds that of isolated water fleas.

Another example of bunching occurs in starling flocks. Starlings fly in loose formations when they are above a falcon but bunch together when they are below it (Lorenz 1943; Tinbergen 1951; Mohr 1960). Falcons normally attack birds in flight by swooping down from above and striking them in midair. Thus starlings bunch together when they are vulnerable to the falcon but not when they are invulnerable. According to one interpretation, bunching together creates a physical hazard for the falcon should it decide to attack. A falcon stooping (diving) into a tightly bunched flock risks a midair collision that could injure its wings or other soft body parts, and this risk might dissuade it from attacking even the most peripheral members of the flock as long as the flock remains tightly bunched. A second interpretation is that bunching results from every individual seeking the flock's center. In loose flocks some peripheral individuals might be picked off by the falcon, and centripetal movements by those individuals would automatically produce tightly bunched flocks.

Although the shielding effect usually reduces the average individual's vulnerability to predators, bunching together can sometimes make a predator's task easier. For example, lions gain an advantage under certain circumstances when ungulate herds are large and tightly bunched. Animals in front or in the center of herds at water holes are sometimes unable to flee until others have done so, giving lions valuable seconds that could mean the difference between success and failure (Schaller 1972). While prey species are normally less vulnerable to lions when bunched together, there are costs associated with the behavior, just as there are with every behavior.

Cooperative enhancement of foraging efficiency

Mutual vigilance

One way cooperation improves foraging success has already been mentioned. Mutual vigilance for predators allows each individual in a group to spend more time foraging without increasing vulnerability to predators. In Powell's (1974) experimental study of starlings, for example, each flock member spent less time looking up and more time foraging than did solitary individuals (Table 4–1). The same effect has been demonstrated under seminatural conditions for wood pigeons foraging in clover fields (Murton et al. 1971), great-tailed grackles *(Quiscalus mexicanus)* foraging on lawns (J. N. M. Smith 1977), yellow-bellied marmots (*Marmota flaviventris*) foraging on talus slopes (Svendsen 1974), and Olympic marmots (*Marmota olympus*) foraging with other members of extended family groups (Barash 1973). Caraco and co-workers (1980) found that the amount of gain in foraging time is greater for members of junco flocks when a predator is present than when no predator is present. This result suggests that the foraging advantage gained from mutual vigilance increases with the frequency that predators are encountered.

Table 4–1 *Surveillance versus foraging time as a function of flock size in starlings and tricolored blackbirds* (Agelaius tricolor).

Flocks	A (1 starling)	B (5 starlings)	C (10 starlings)	D (1 starling, 9 blackbirds)
Number of flocks[a]	6	6	6	7
Number of observations per flock	61	58	117	54
Percentage of foraging time spent feeding[b]	53 ± 13	70 ± 12	88 ± 9	85 ± 9
Percentage of foraging time spent on surveillance	47	30	12	15
Frequency of surveillant observations per second[c]	.39 ± .13	.30 ± .89	.19 ± .13	.27 ± .16

[a]Each flock composed of naive birds.

[b]Significantly different for groups A, B, and C (*P*<.01 analysis of variance); groups C and D not significantly different (*P*>.01 analysis of variance).

[c]Significantly different for groups A, B, and C (*P*<.01 analysis of variance); groups C and D are significantly different (*P*<.01 analysis of variance).

Source: From G. V. N. Powell, "Experimental Analysis of the Social Value of Flocking by Starlings *(Sturnus vulgaris)* in Relation to Predation and Foraging," *Animal Behavior* 22 (1974):503. Reprinted by permission of Baillière Tindall.

The beater effect

Another way social animals enhance foraging efficiency is by hunting prey flushed by other group members (Rand 1954). The **beater effect** capitalizes on the decreased wariness of prey animals who are in headlong flight. When prey are flushed, they may successfully elude one predator only to blunder into the waiting jaws of another. The beater effect was first described for small groups of tropical hornbills (*Bucorvus cafer*), which stalk along the ground attacking insects flushed in their direction by other group members (Swynnerton 1915). Herons and storks occasionally form long, irregular lines and walk across fields flushing insect prey as they go (Rand 1954). Cormorants, pelicans, and mergansers sometimes encircle schools of fish or drive them ahead of the flock, constantly diving in to attack them as they dart about in pandemonium (Bartholomew 1942; Cottam et al. 1942; Emlen and Ambrose 1970; personal observation). In early spring when rivers rise, subadult Nile crocodiles (*Crocodylus niloticus*) form a semicircle where a temporary channel enters a pan, face the inrushing water, and snap up fish emerging from the river (Pooley and Gans 1976). Similarly, Eleonora's falcons (*Falco eleonorae*) form a standing wall into the wind to capture migrating songbirds along coastal cliffs of Mediterranean islands, apparently enhancing their capture efficiency by attacking prey who are preoccupied trying to escape other falcons in the area (Walter 1979). They capitalize on this method of prey capture by breeding in colonies during early autumn when migration along the cliffs is in full swing.

The beater effect does not require cooperation. Individuals can exploit flushed prey no matter how the prey are flushed. For example, neotropical antbirds specialize on flying insects flushed up by moving swarms of army ants, several species of birds follow behind cattle to capture flushed insects, tropical hawks, drongos, hornbills, and various other birds hunt insects flushed by monkeys, gulls often follow behind tractors plowing fields and fishing boats, and small birds often follow behind trains (Rand 1953; Heatwole 1965; Friedmann 1967; Willis 1967; Wahl and Heinemann 1979; Fontaine 1980). Cooperation is clearly not involved here because the success of foraging individuals is not enhanced by the activities of conspecifics.

The beater effect has been proposed as one possible explanation of mixed-species tropical bird flocks. These flocks generally consist of a *nucleus* species, which forms a stable and permanent core to the flock, and various *attendant* species, which associate with the nucleus species on an irregular basis (Moynihan 1962a; Buskirk et al. 1972). The evolution of mixed-species flocks is still not fully understood, but mutual flushing of prey is probably not important because many members of such flocks do not hunt flushed insects (Buskirk 1972, 1976). The nucleus species and at least some attendant species probably flock together as a defense against predation (Willis 1972a; Buskirk 1976; Munn and Terborgh 1979), though some attendant species join flocks as they cross territory boundaries to prevent neighboring conspecifics from entering their territories (Powell 1979).

Overcoming prey defenses

The defensive tactics of prey are often sufficient to greatly reduce the hunting success of individual predators. One method for overcoming these defenses is cooperation. Where one predator acting alone may fail, several acting in concert often succeed.

The most conspicuous examples of cooperative hunting involve social mammals. The social carnivores, particularly lions, spotted hyenas (*Crocuta crocuta*), and social canids, rely on cooperation to kill large prey and increase their chances of killing small prey (Photo 4–4) (Kleiman and Eisenberg 1973; Bekoff 1975; Eaton 1976). The effectiveness of cooperative hunting is well documented and will be discussed more fully in Chapter 13. Once prey are captured, social carnivores may further rely on cooperation to repulse would-be scavengers, especially on African savannas where competition for carcasses is intense. Although cooperative hunting is best known among terrestrial carnivores, it is also an important hunting tactic of killer whales (*Orcinus orca*) and perhaps some dolphins (Fink 1959; Martinez and Klinghammer 1970; Hoese 1971).

Photo 4–4

Social carnivores rely on group effort to capture large prey.
Here a pack of African wild dogs pull down a warthog (Phaccochoerus) *during a cooperative kill.*
Photo by Norman Myers/Bruce Coleman, Inc. Courtesy of World Wildlife Fund.

Ravens (*Corvus corax*) capture eggs from colonies of kittiwake gulls (*Rissa tridactyla*) more successfully when hunting in groups, because kittiwakes are less effective at mobbing several ravens at once (Montevecchi 1979). Kittiwakes breed on steep cliffs of oceanic islands and rely mainly on inaccessibility of nests and mobbing to prevent predation. When ravens hunt alone or in pairs, they are usually chased away by kittiwakes before they can land at a nest. Hence their success at capturing eggs is relatively low. However, groups of three or more ravens are much more successful because some ravens in the group can enter a nest while the gulls are busy mobbing the others.

Group hunting by predatory fish is sometimes an effective way to overcome the social defenses of schooling prey. Major (1978) found that temporary schools of jack are more successful than single individuals in capturing Hawaiian anchovies, which form dense schools to evade predators, because they are better able to isolate individual anchovies from the school.

The Nile crocodile cooperates in tearing apart carcasses that are too large to swallow whole, at least occasionally (Pooley and Gans 1976). Crocodiles ordinarily reduce large prey to a size small enough to swallow by seizing part of the prey in their jaws and rolling over repeatedly until a piece is torn off. However, this maneuver fails for small carcasses because the carcass rotates with the crocodile. With such prey crocodiles have been observed taking the carcass to a second crocodile, who holds the carcass while the first tears a piece off. Alternatively, both crocodiles sometimes grab the carcass and roll on it in opposite directions. Each individual eats what it tears off without hostility toward the other. Some evidence suggests that crocodiles may, on rare occasions, hunt or scavenge prey cooperatively as well. In one instance, for example, two crocodiles were seen walking side by side carrying the carcass of an antelope between them.

Certain spiders cooperate in building large, dense sheet webs to gain protection or to capture prey more effectively (Kullmann 1972; Buskirk 1975, 1979; Burgess 1976; Rypstra 1979). Species that cooperate to build larger webs for capturing prey sometimes also cooperate to overwhelm prey that are too large for a single individual to kill. This kind of cooperation involves mothers and their spiderlings, and it has apparently evolved from prolonged maternal brood care (Kullmann 1972). In other species cooperative web building involves unrelated individuals that have aggregated. Individuals of these species spin their own webs within the aggregation, contributing at most only small amounts of silk to a communal web area, and they do not cooperate in overwhelming prey (Burgess 1976). Such colonies probably evolved to exploit insect-rich habitats not readily exploited by solitary spiders, such as understory openings above small streams (Brach 1977; Buskirk 1979).

In comparable fashion, cooperation can evolve to overcome plant defenses against herbivory. For example, recently hatched larvae of the jack-pine sawfly (*Neodiprion pratti*) are so weak that they have difficulty chewing through the tough outer layer of pine needles (Ghent 1960). When one does succeed, other larvae in the group are immediately attracted to the opening by the release of volatile salivary secretions and plant compounds. The gap is soon widened enough for all larvae to feed along its open edges. After each new breakthrough the successful forager is being parasitized by the others. Nevertheless, every individual benefits over the long run because different individuals are the first to chew through each new pine needle. The foraging groups consist of siblings or perhaps half siblings, as the larvae all developed from eggs laid by a single female, but indirect (kin) selection need not be invoked here because the benefits are mutual.

Enhancing the nutritive quality of food

An unusual way for animals to benefit mutually from feeding in groups is by augmenting the quality of food before it is eaten. In some aphids adults are heavier and more fecund if they have been reared in aggregations, suggesting that they gained mutual benefits during development (Dixon and Wratten 1971; Shearer 1976; but see Hargreaves and Llewellyn 1978). When an aggregation of *Brevicoryne brassica* is reared on one side of a cabbage leaf, an isolated individual reared on the other side still benefits from the aggregation (Way and Cammell 1970). Aphids feed by sucking juices out of the leaf, and while feeding they simultaneously inject saliva (Lamb et al. 1967; Forrest and Noordink 1971). Their saliva contains many substances affecting plant metabolism (Schäller 1968; Miles

1969), and some of these may increase nutritive quality or digestibility of the leaf (see Dixon and Wratten 1971; Forrest 1971). The combined injections of many aphids are evidently more effective in releasing plant nutrients than the injections of a single aphid, which would explain why aphids reared in aggregations grow to a larger size.

Gaining information

Individuals can sometimes increase foraging efficiency by watching other group members forage. An elegant series of experiments show that caged great tits (*Parus major*) learn where to find food by observing where other flock members recently found food items (Krebs et al. 1972; Krebs 1973). The experimental setup consisted of several artificial trees, constructed from wood doweling, each containing one to four different kinds of foraging sites (depending on the experiment) (Photo 4–5). In the first experiment the effect of food dispersion was analyzed. The results show that birds foraging in flocks find clumped foods faster than birds foraging alone or in pairs. This advantage was not evident when food was dispersed. The searching behavior of birds in flocks was observed before and after a food item was found by a flock member. These observations show that flock members are attracted to the particular tree and perch where food has recently been found. They also show that flock members begin to search in locations similar to those where food has been found. Thus birds can derive information about where and how to forage by watching other flock members, and they act on this information if food items are clumped.

Birds can also learn what kinds of food to search for by watching other flock members. For example, titmice became a major nuisance for a while in Great Britain after they learned to open and skim the cream off the top of milk bottles left on porches by milkmen (Fisher and Hinde 1949; Hinde and Fisher 1951). Milk bottles were often attacked within minutes after they were left on a porch, and some flocks even followed milk carts around and opened bottles on them while the milkman was making deliveries! An experimental study involving chaffinches (*Fringilla coelebs*) and house sparrows has subsequently confirmed that flock members observe what others are eating and then begin eating those foods themselves (Turner 1965).

Under natural conditions members of bird flocks often encounter different kinds of food in different localities, with each type of food found in its own particular place. Learning what kinds of food items are present and where they are most likely to be located may often be important in minimizing the time spent searching for food. This is especially true when food is spatially clumped, because then many of the places that might be searched would contain little or no food.

Individuals in social groups might learn where others have depleted resources by watching and remembering where other group members have fed. Foraging efficiency should be increased by an ability to avoid areas already searched by others. This may explain, for example, why Roosevelt

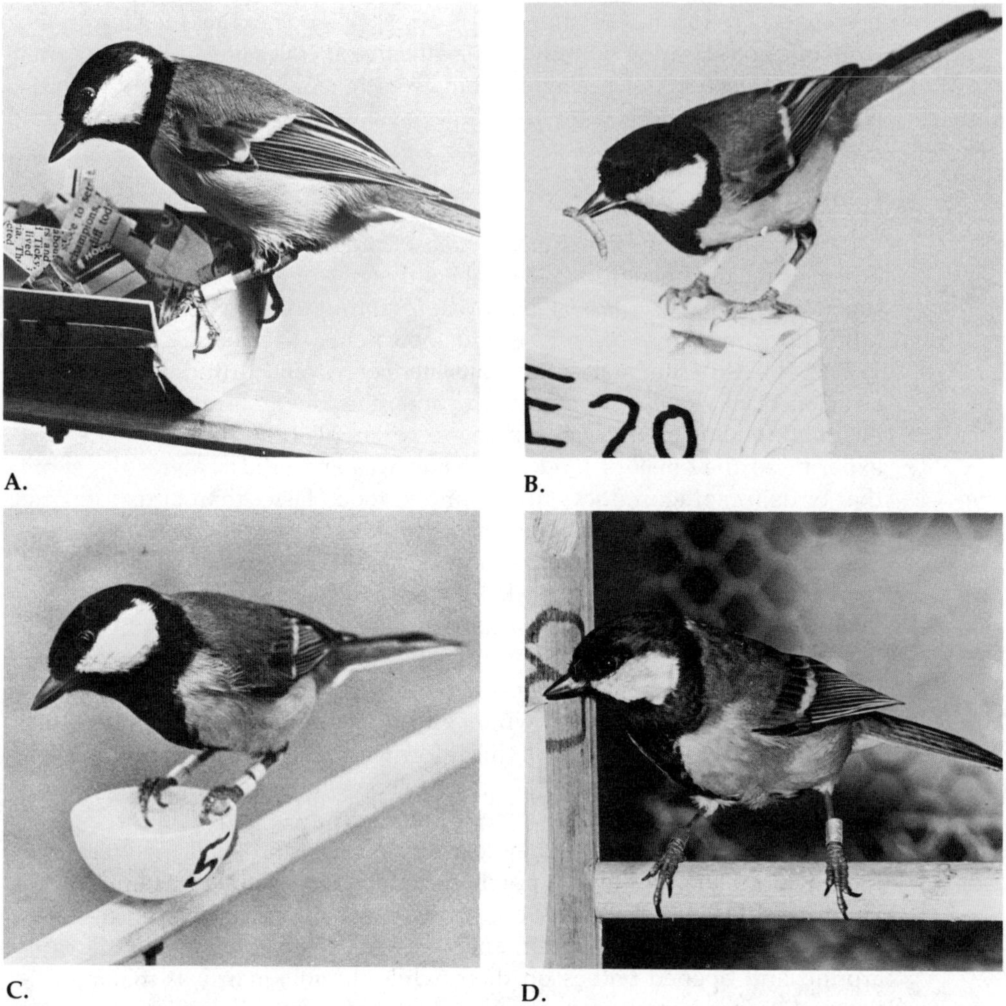

A.

B.

C.

D.

Photo 4–5 *The four foraging sites used to study experimentally the advantages of flocking behavior in great tits.*

A. Food is cached in a plastic hopper filled with cut newspaper. B. Food is cached in wood blocks with a hole drilled in the top and covered with masking tape. C. Food is cached in half a ping-pong ball filled with sawdust. D. Food is cached underneath masking tape stuck to the trunks of the wood-doweling trees.

Photo from J. R. Krebs, et al., "Flocking and Feeding in the Great Tit *Parsus major:* An Experimental Study," *Ibis* 114 (1965):437.

elk (*Cervus canadensis*), herons, storks, and blackbirds sometimes forage along parallel paths (see Rand 1954; M. Altmann 1956; J. N. M. Smith 1977).

Cody (1971a, 1974) has suggested that winter flocks of seed-eating finches avoid previously searched areas after fully depleting them of food. By removing all food in an area before moving elsewhere, flock members may be able to recognize and avoid previously exploited areas. Theoretically, this behavior would minimize the time spent searching any given area for food and hence would be the most efficient way to exploit a nonrenewing food source. However, the actual foraging movements of the flocks observed by Cody do not fit the predictions of his hypothesis (Pyke 1978). Competition evidently induces flock members to forage in the most productive areas available at any given time. This should be the best tactic when flocks do not defend food resources. Flock members trying to gain a long-term advantage by searching efficiently would be less successful than other individuals who always search where food is most abundant. Cody's hypothesis might work if each flock defended its own foraging area, but the finch flocks he studied do not defend foraging areas.

Reducing the costs of grouping behavior

Grouping together is a necessary prerequisite for mutual defense or cooperative foraging, but it entails costs that may in turn be reduced by additional forms of cooperation. One probable cost of grouping behavior is increased exposure to disease organisms and ectoparasites (Alexander 1974). Cooperative defenses against disease transmission seem minimal or absent, except in humans, but **allogrooming** (i.e., the grooming of others) occurs in many animals and may have evolved to reduce ectoparasite infestations or otherwise service the pelage or plumage.

Allogrooming behavior

Allogrooming behavior generally seems to have functional significance for the recipient. Allopreening in birds is usually directed at the top of the head, a part of the body not easily reached by the recipient (Photo 4–6) (Harrison 1965). Allogrooming in primates is concentrated on inaccessible body parts of the recipient, particularly the back, upper arms, armpits, and neck (Sparks 1967). All primates remove and often lick or eat foreign objects and ectoparasites from others, and sometimes they even clean or lick the wounds of others (Carpenter 1940; Simonds 1965). Similarly, lions and other cats lick the head and neck of others reciprocally following a meal, probably for mutual sanitation (Schaller 1972).

The benefits derived by individuals who perform allogrooming are less clear. One obvious benefit is reciprocation, but this cannot be a complete explanation because allogrooming is often a one-sided interaction that is rarely or never reciprocated. The evolution of allogrooming is still poorly understood, but closer inspection of the contexts within which it occurs suggests several potential benefits.

Allopreening in birds occurs most commonly during the breeding season, particularly during courtship (Harrison 1965). Males perform most allopreening and they may allopreen as an inducement for females to accept them as mates. However, there is more to it than that because allopreening continues throughout the nesting cycle in many species. Moreover, females may also allopreen when they are in a temporarily dominant position, such

Photo 4–6

*Male Galapagos penguin (*Spheniscus mendiculus*) preening the head of a female.*

When one bird preens another, it most commonly preens the other bird's head or neck.

Photo from P. D. Boersma, "An Ecological and Behavioral Study of the Galapagos Penguin," *Living Bird* 15 (1976):43–93.

as when they are relieving the male of brooding duties at the nest. Furthermore, allopreening does not occur in every species, and its incidence cannot be explained by phylogenetic heritage alone. A complete explanation must explain why it has evolved in some species but not in others.

One possible explanation is that allopreening has evolved to maintain prolonged pair bonds. Allopreening is most prevalent in species that form lifelong pair bonds, particularly in seabirds that commute long distances for food. Harrison (1965) therefore suggests that one benefit is to discourage individuals from deserting while their mates are off foraging. How allopreening might accomplish this is unclear, however, and allopreening does occur in species that do not forage long distances from their nests.

Another possible explanation is that allopreening serves to maintain dominance relationships. Allopreening is closely associated with dominance behaviors (Harrison 1965). Generally speaking, only the dominant bird or the bird in a temporarily dominant position allopreens, and transitions from allopreening to aggression regularly occur. Allopreening may therefore reinforce dominance status, with dominant individuals exchanging plumage servicing for the benefits derived from high status in other contexts. How this might work is illustrated by cooperative breeding systems, in which breeders may allopreen helpers to help maintain their breeding status (Gaston 1977).

In most primates dominant individuals are the principal recipients of allogrooming, except when it involves maternal grooming of infants (Sparks 1967; Sade 1972; Seyfarth 1976; Stammbach 1978). As a general rule, adult females and juveniles groom adult males (Photo 4–7) and young females groom adult females. Grooming a male is an appeasement gesture that allows a subordinate individual to remain near a dominant one without being attacked or threatened. Subordinates can thereby gain protection from harassment by other troop members and, on occasion, better access to desirable food items. Also, young females may allogroom mothers as a ploy for getting close enough to touch or hold a young infant, thereby enabling them to gain valuable mothering experience (Blaffer Hrdy 1976). Dominant males do perform some allogrooming, primarily when they are interested in a sexually receptive female. Hence allogrooming by dominant males may enhance their copulatory success.

An interesting observation is that allogrooming episodes among female primates predominantly involve relatives (Hunkeler et al. 1972; Kurland 1977; Dunbar 1979a; Wade 1979). The significance of this observation

Photo 4–7

*A female hamadryas baboon (*Papio hamadryas*) allogrooming a male.*
Allogrooming is most commonly directed at less accessible body parts.
Photo by Toni Angermayer, Holzkirchen.

is not yet clear, but it may mean that allogrooming among females is altruistic or that cooperating with relatives is more advantageous than cooperating with nonrelatives.

Male coalitions

Another cost of grouping behavior, at least for males, is increased competition for mates. Female sociality gives dominant males an opportunity to control access to many females at once, making it more difficult for subordinate males to mate. In several social mammals, subdominant males therefore form coalitions to increase their social status or otherwise gain better access to females.

Male olive baboons form temporary coalitions to displace a third male from a consort relationship with a receptive female (Packer 1977; see also Hall and Devore 1965; Altmann and Altmann 1970; Stoltz and Saayman 1970). One male of the coalition acquires the female after a successful displacement, while the other continues fighting with the displaced male. Generally the male who enlists aid is the one who acquires the female. The other male obtains no immediate benefit but does incur the risk of being injured. Males participating in coalitions are often unrelated, but the same individuals do enlist each other's assistance on successive occasions. Hence the aid is reciprocated, and the coalitions apparently represent a form of reciprocal altruism.

Male rhesus monkeys sometimes form stable coalitions to dominate a single stronger male, thereby deriving the benefits of elevated social status (Southwick et al. 1965; Varley and Symmes 1966; Kaufmann 1967). Though each male in the coalition may be individually weaker than another male in the troop, they cannot be beaten when acting in concert. By spending nearly all their time together, they effectively become dominant males. The subdominant males in such coalitions are rarely aggressive toward each other, and they alternate in obtaining benefits such as preferred food items (Southwick et al. 1965). In captive groups coalitions involve relatives more often than nonrelatives (Massey 1977), perhaps because cooperating with relatives increases fitness more than does cooperating with nonrelatives.

In some parts of East Africa male lions form permanent coalitions to take control of female groups (Schaller 1972; Bertram 1975a). All males benefit from the coalitions because all gain nearly equal access to females without resort to dominance disputes or other forms of aggression. Since single males cannot compete against coalitions, cooperation greatly enhances the reproductive success of every participating male. In general, coalitions of three or more males have a higher probability of gaining tenure in female prides, are able to retain tenure for longer periods of time, and produce more surviving progeny per male than single males or coalitions of two males (Bygott et al. 1979). Interestingly, males do not form coalitions in areas of low population density, possibly because smaller female group sizes reduce male competition for mates (Rudnai 1973). Lion coalitions usually consist of relatives (19 out of 21 coalitions whose origins were known), although they occasionally consist of nonrelatives (1 out of 21 coalitions), or relatives plus nonrelatives (1 out of 21 coalitions) (Bertram 1976; Bygott et al. 1979).

Although cheetahs are basically solitary, sibling males form coalitions in some parts of Africa (Eaton and Craig 1973; Eaton 1974). Males copulate with more females when accompanied by another male, but the benefits are not shared equally because only the dominant male copulates with females. Since coalitions consist of brothers, the subordinate males benefit indirectly by helping their brothers obtain mates.

Conclusion

Cooperation is a major organizing influence on animal social behavior. It forms the basis of pair bonding and many kinds of grouping behavior. The most important benefits derived from cooperation are mutual defense, enhanced foraging efficiency, and increased competitive ability within social contexts, but these benefits can take many forms depending on the species involved.

Because cooperation benefits every participant, it is easy to forget that cooperating individuals do not have identical interests. Cooperating individuals interact to promote mutual interests, but they also compete when their interests conflict. The interactions found within social groups provide a good example. Animals form social groups to gain mutual advantages, but within those groups they compete for food, mates, sleeping sites, and other requisites. Competition for these requisites often manifests itself as aggressive behavior, a second major organizing influence on animal social behavior. The next chapter therefore turns to the opposite side of the social coin, namely, aggression.

Aggression

Aggression is a key facet of animal behavior. Animals use aggression to control space, compete for resources, and acquire mates. They use it in parental interactions and in sibling rivalries. They may even use it to thwart predators or members of competing species. Indeed, aggression is a likely response to any situation where the selfish interests of individuals conflict, regardless of whether the individuals belong to the same or different species.

The proximate causation and adaptive significance of animal aggression have great theoretical interest, not only because aggression plays a central role in organizing animal societies, but also because they may shed light on the nature of human aggression. Few subjects have received more attention from behavioral biologists and psychologists, and few have evoked more heated controversy.

Popularized treatments of animal aggression, such as those of Robert Ardrey (1962, 1966) and Konrad Lorenz (1966), have created in the public mind a body of common knowledge that many professional scientists find unacceptable. Many scientists question the validity of drawing comparisons between human and animal aggression, and even among professionals deep-seated differences exist in the ways animal aggression is interpreted. This chapter therefore focuses first on these interpretational differences and what the current evidence allows one to conclude about them. It then goes on to discuss in general terms why animals are aggressive and how they use aggression to promote their selfish interests.

What is aggression?

Definitions

Aggression is a difficult concept to define because even the experts often disagree about whether particular kinds of behavior are or are not aggressive. The concept is loaded with connotations, mostly negative ones, and it is often applied almost indiscriminately to a great variety of actions and emotions (see Kaufmann 1970; Johnson 1972).

Aggressive behavior has been defined in terms of motivational states such as anger or hate, without regard to its consequences, and in terms of specific responses such as injuring or killing others, without regard to

Chapter five

motivational states. Neither approach is entirely satisfactory. Many humans, for example, get extremely angry or hateful without ever attacking or harming anyone, while others are capable of great cruelty and violence without feeling any emotional involvement whatsoever. Attacking other individuals may be used to gain a competitive advantage, in self-defense to gain protection from predators or aggressors, or as an act of predation to obtain food. Thus similar motivational states can lead to very different consequences, and similar consequences can result from very different motivational states. Any definition of aggression must therefore specify something about both underlying motives and behavioral consequences (Johnson 1972; Berkowitz 1974; Moyer 1976; Baron 1977). Here is where the experts disagree. Which motives and which consequences should be included, and which should be excluded? In the final analysis there is no single answer to this question, and aggression must be defined in terms appropriate to specific research objectives.

Aggression is defined in this book as overt behavior directed at harming or threatening to harm another individual with the intent of gaining some advantage. The two central ingredients of the definition are (1) overt attempts to inflict harm and (2) intent to gain some advantage. Both require clarification.

From an evolutionary perspective *harm* refers to a reduction in the recipient individual's fitness. Ideally deleterious consequences of aggression on a recipient's fitness should be verified empirically, but as a practical matter they are often assumed. In studies of proximate causation an evolutionary criterion of harm is impractical. Fitness consequences are often difficult to measure, and in physiological studies they are irrelevant. Moreover, harm in humans may refer to psychological, economic, political, or physical damage that may or may not affect a person's biological fitness. Thus for studies of proximate causation, *harm* refers to any noxious stimulation or consequence to which a recipient shows an aversion (after Moyer 1976).

Intent can only be inferred from overt behavior patterns and the contexts within which they occur. Intentions are not directly observable in either humans or animals, and they are often difficult to infer. Many legal cases, for example, revolve around the issue of aggressive intent precisely because intent is difficult to prove. Despite this difficulty, intent must be an essential part of the definition; otherwise accidently inflicting harm on others or inflicting harm while trying to help others (e.g., medical malpractice, dental work, parental discipline of offspring) would not be excluded from the concept of aggression.

The criterion for establishing intent is goal-directedness. Any attempt to gain physical, psychological, economic, political, or competitive superiority by inflicting harm has aggressive intent. Inferring aggressive intent is most ambiguous for human behavior, because aggressiveness in humans is easily confused with ambition and assertiveness. Working hard to excel or to outcompete another for recognition is not aggressive unless behaviors specifically directed at harming another are employed. Becoming more knowledgeable about a job or being friendly with a superior to gain favorable recognition is not aggressive, but trying to gain the same advantage by denigrating or humiliating another is. Certainly many people do use aggressive behavior in competitive situations, but not all behaviors used in such situations are aggressive. Similarly, asserting one's personal rights, feelings, or preferences is not necessarily aggressive, though it may be interpreted that way. Assertiveness is a positive expression of personal rights or points of view, while aggressiveness is a means used to achieve personal goals by putting down, demeaning, humiliating, or otherwise hurting others (Moyer 1976; Alberti and Emmons 1978). When interpreting animal behavior, this distinction is less relevant because there are usually few social or selective pressures to spare another individual's feelings.

The terms *aggressiveness* and *aggression* should be carefully distinguished (Moyer 1976). *Aggressiveness* is a predisposition or tendency to behave aggressively, while *aggression* is an overt behavioral response. Aggressiveness is an internal state that may or may not lead to overt aggression.

Aggressive behavior usually entails risks, as anyone who has gotten into a fight or has experienced an effective riposte knows. For this reason attack behavior is often closely associated with self-protective behavior, even to the extent that threat postures usually consist of both attack and withdrawal components (see Marler and Hamilton 1966; Hinde 1970). Hence behavioral biologists generally include behaviors entailing threat, attack, submission, and withdrawal under the single concept of **agonistic behavior** (after Scott and Fredericson 1951). The concept of agonistic behavior is particularly useful when discussing interactions between two or more individuals, as each individual may alternate between aggressive and submissive behaviors during a single interaction.

Kinds of aggression

Aggression occurs in numerous environmental and social contexts. It generally results from different ultimate causes in different contexts, and it may also result from different proximate causes. Classification systems of aggression vary considerably, depending on whether they are based on ultimate causes, proximate causes, or a combination of both (see Moyer 1976). The classification system adopted here is modified from Wilson (1975a) and is based largely on ultimate causes to facilitate evolutionary analyses. Table 5–1 defines the principal forms of aggression and describes the contexts within which they normally occur. The list is not exhaustive, but it includes most forms of aggression occurring among nonhuman animals.

Is predation a type of aggression?

Predation is in many ways fundamentally different from other forms of aggression. Predatory behavior is elicited by different kinds of external

Table 5–1 *Types of animal aggression, based on an evolutionary classification system.*

Type of aggression	Definition	Social contexts in which it occurs
Intrasexual aggression	Aggressive behavior directed at conspecific individuals of the same sex	Territorial behavior; social dominance
Intersexual aggression	Aggressive behavior directed at conspecific individuals of the opposite sex	Courtship; male herding of females; territorial behavior; social dominance
Parental aggression	Aggressive behavior directed at offspring by parents	Weaning; forced dispersal of offspring; infanticide (rare)
Infanticide	The killing of conspecific infants, who may or may not be direct progeny	Male takeovers of female groups; parental aggression during food scarcity (rare)
Peer aggression	Aggressive behavior of young animals directed at peers or siblings	Fratricide (killing of siblings); aggressive play; social dominance among littermates
Defensive aggression	Aggressive behavior used for self-defense or defense of offspring	Prey animals trapped by predators; subordinate or inferior animals cornered by dominant or superior animals
Redirected aggression	Aggressive behavior directed toward an inanimate object or an inferior individual when the individual evoking the behavior is too strong, too dangerous, or too unapproachable to be attacked directly	Any aggressive interaction involving conspecific individuals
Predatory aggression	Attacking a prey animal to obtain food (not always considered a form of aggression)	Predator-prey interactions

stimuli, is based on different internal states, involves different movement patterns, and is mediated by different neural mechanisms (see Carthy and Ebling 1964; Delgado 1966; Hutchinson and Renfrew 1966; Lorenz 1966; Hinde 1974; Moyer 1976). Predation is a means for capturing food, while aggression is a means for gaining competitive advantages or self-protection. Predatory animals are not necessarily more aggressive toward conspecifics or humans than nonpredatory animals, and often they are less aggressive. To argue, as Ardrey (1962) does in *African Genesis,* that man is innately aggressive because he evolved a killer instinct to hunt game animals is patently absurd. Hunting and murder are entirely different behaviors based on entirely different motives, and a predisposition for the one does not imply a predisposition for the other. Labeling predation as a form of aggressive behavior encourages this kind of illogic, which is one good reason for considering it as something fundamentally different.

Several authorities do consider predation a type of aggression (e.g., Wilson 1975a; Moyer 1976), and certainly there are similarities. Predatory attacks often entail violence and bloodshed, as do many aggressive actions, and they are hostile behaviors directed toward prey. Since the proximate and ultimate causes of other kinds of aggression also differ from one another, the use of differing causation as a pretext for excluding predation from the concept of aggression is inconsistent logic. Finally, antipredatory behavior clearly involves aggressive behavior, and hence predatory behavior should also be considered a form of aggression.

Despite these arguments, predation will not be included as a form of aggression in this book for the following reasons. Although the most dramatic and vivid instances of predation involve carnivorous mammals or birds attacking warm-blooded prey, with victims often disemboweled and devoured piecemeal with much blood in evidence, gruesome acts of predation are by no means typical. Few would argue that the foraging behavior of insectivorous birds or praying mantises is aggressive, and yet such behavior represents an important form of predation. Is predation to be considered aggressive only when it offends our moral sensibilities? Or is aggression to be generalized sufficiently for it to include predation of starfish on oysters and jumping spiders on flies? Theoretical and empirical studies of predation and aggression have little in common, and combining the two into a single concept facilitates misleading and fallacious assertions. Reinforcing the conceptual differences between predation and aggression seems more important than reinforcing their similarities, which are perhaps already emphasized more than they should be.

A controversy: Opposing motivational models

Hydraulic model

Konrad Lorenz (1966) and many other ethologists argue that animal aggression springs from an innate fighting instinct. Lorenz views aggressiveness as an innate primary drive, on a par with hunger, thirst, and sex, which spontaneously accumulates motive force or "aggressive energy" through time. This wellspring of "aggressive energy" is presumed to accumulate until eventually it demands release through overt aggression, making aggression an inevitable aspect of animal behavior. The "aggressive energy" postulated by Lorenz has often been likened to a hydraulic system, with the pressure (motivational energy) accumulating until it forces open a valve (inhibitory block), to be discharged as overt aggression. As the time elapses following an aggressive episode, the animal is assumed to become more and more predisposed to behave aggressively, until finally it begins seeking out appropriate stimulus situations for releasing its "aggressive energy."

The motivational mechanism involved in accumulating "aggressive energy" is postulated to be a single generalized drive and is assumed to be structured largely along genetically determined guidelines. It is assumed to accumulate energy spontaneously without external stimulation, and this energy is assumed to act as a motive force responsible for causing aggressive behavior to occur. Thus Lorenz and others view aggressiveness as an inevitable consequence of an animal's genetic heritage, regardless of upbringing (see Lorenz 1966; Eibl-Eibesfeldt 1974), a view espoused in several popularized books on human behavior (e.g., Ardrey 1962, 1966; Storr 1968, 1972; Tiger and Fox 1971).

Frustration-aggression model

Many social psychologists reject the idea that aggression is an innate primary drive (e.g., Berkowitz 1969, 1974; Feshbach 1970; Bandura 1973). They argue that aggressiveness itself is elicited by environmental conditions, with genetic influences being extremely indirect. They believe that learning plays a crucial and overriding role in its development and expression, making social factors largely responsible for its occurrence.

The most influential theory in this vein has been that aggressiveness stems from frustration (i.e., the thwarting of goal-directed behavior) (Dollard et al. 1939). The frustration-aggression hypothesis initially stated that frustration always leads to aggression and aggression always stems from frustration. However, the facts fail to fit this simple relationship, and the hypothesis has consequently evolved into something quite different. Because frustration does not always evoke overt aggression, the hypothesis was modified to state that frustration leads to increased aggressiveness, which may or may not be expressed through overt aggression (Miller 1941). In addition, because aggression does not always stem from frustration (it may be caused by physical attack, pain, and various other provocations as well), it was further modified to state that aggressiveness is produced by "noxious stimulation" of any sort (Buss 1961). Finally, because aggressiveness does not always lead to overt aggression, "eliciting stimuli" specific to the behavioral response must act in conjunction with noxious stimulation to "release" aggressive behavior (Berkowitz 1965, 1969). Thus the hypothesis now states that noxious stimulation leads to heightened aggressiveness, which in conjunction with appropriate external stimuli leads to overt aggressive behavior.

Many social psychologists who support the modified frustration-aggression hypothesis still believe that aggressiveness represents an accumulation of drive-specific energy (see Baron 1977). This view again makes aggression seem inevitable, as frustration and other noxious stimuli are unavoidable. Proponents of this view are therefore logically forced to the same conclusion reached by proponents of Lorenz's hydraulic theory, with the only principal difference being that aggression is ascribed to environmental rather than genetic causation.

Switchboard model

Many behavioral biologists and psychologists reject the view that motivation energizes behavior. Instead they consider motivation to be primarily a routing mechanism, something akin to a telephone switchboard (e.g., Hinde 1970, 1974; Bandura 1973; Zillman 1978). It routes incoming calls or stimuli to appropriate response mechanisms but does not possess energy of its own for driving the responses evoked. The routing is accomplished by changes in neural sensitivity of the various motivational networks. At any given moment each network is sensitized to a different degree, and the most sensitized network will evoke an overt response. The routing mechanism or switchboard is structured according to previous experience and genetic heritage. Hence both learning and natural selection play important roles in determining which responses are given in any given situation.

The switchboard model has very different implications concerning the inevitability of aggression. It implies that a behavior can be inevitable only if its underlying motivational network becomes progressively and irreversibly more sensitized during periods between overt responses. Feeding behavior is a good example of this process, as hunger becomes progressively stronger and cannot be diminished without food intake until the body is pathologically weakened by lack of food. A similar process does not seem to occur for aggressive behavior. Attack tendencies can increase or decrease as stimulus situations change, without ever being expressed through overt aggression. Thus the switchboard model predicts that aggression is not inevitable and can theoretically be avoided altogether by preventing oversensitization of the underlying motivational mechanisms. In human societies this could be accomplished by reducing economic and social causes of heightened aggressiveness, by teaching people alternative ways of coping with frustration, competition, and stress, and by improving therapeutic treatments of pathologically aggressive individuals.

Two sources of controversy

The controversy over proximate causes of aggression centers around two key questions: First, is aggressiveness a genetically determined and hence inevitable drive? And second, is it a general drive comparable to hunger or thirst that must be satisfied through acts of violence?

The answers to these questions have more than academic significance. They prescribe how we look at human beings and, more importantly, how we go about being human. If aggression is an inevitable consequence of our heredity, society cannot hold individuals responsible for violent actions and must devise ways to channel aggression into harmless outlets. Pathologically or criminally aggressive individuals cannot be treated and must be removed from society without hope of rehabilitation. If, however, aggression is not an inevitable consequence of our heredity, education and rehabilitation programs should be intensified in an effort to teach people nonaggressive ways of solving problems and reaching goals. Pathologically and criminally aggressive people could be treated, and sanitariums and penitentiaries should be structured with that objective in mind. At a more personal level, each person's view of aggression affects how he or she responds to the aggressive behavior of others and how willing he or she is to learn nonaggressive ways of solving problems and reaching goals.

The theoretical issues involved are clearly important for coping with human aggression, as each theory suggests a different solution. They are also important in interpreting animal behavior and in designing research for elucidating the underlying physiological mechanisms involved. Because the social implications are important and because aggression is an illustrative case study for proximate analyses of behavior, the relevant evidence concerning proximate causation will be discussed in detail before the adaptive significance of aggression is examined.

Developmental origins of aggressiveness

Evidence for a genetic influence

One indication that genetics influences aggressiveness comes from inbred strains of laboratory and domesticated animals. Inbred strains of laboratory mice vary considerably with respect to aggressive behavior. One strain in particular (C57BL10) consistently outfights other strains when members of the different strains are matched against each other (Ginsburg and Allee 1942; Scott 1942; see also Scott 1966). To eliminate the possibility that differing maternal behavior is responsible for this difference, Lagerspetz and Wuorinen (cited in Lagerspetz 1969) placed the offspring of aggressive strains with mothers of nonaggressive strains and vice versa. They found that mice of aggressive strains remained significantly more aggressive than those of nonaggressive strains even though they were cross-fostered by parents of the opposite strain. Hence the strain differences stemmed from differing genetic heritage and not differing early experiences. These experiments do not conclusively establish that genetics influences aggressiveness per se, however, because the difference between strains may have resulted from superior fighting ability rather than heightened aggressiveness. Similar differences can be shown within natural populations. For example, the white-striped morph of the white-crowned sparrow *(Zonotrichia leucophrys)* is more aggressive in intraspecific interactions than the tan-striped morph, in all sizes of winter flocks (Figure 5–1) (Ficken et al. 1978).

Selective breeding of strains with differing temperaments provides another line of evidence supporting a genetic influence on aggressiveness. Aggressive strains of white mice and laboratory rats have been produced under controlled conditions by breeding the most aggressive individuals each generation (Hall 1938; Hall and Klein 1942; Lagerspetz 1969). Aggressive temperaments have been bred into dog breeds such as German

Figure 5–1

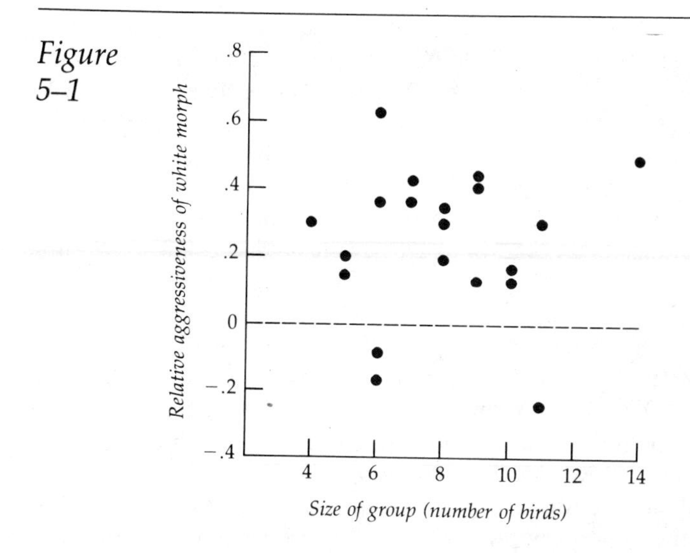

The aggressiveness of white-striped morphs of the white-crowned sparrow compared to that of tan-striped morphs in nonbreeding flocks.

The morphs are easily distinguished by the color of the stripes on the crown of their heads. Deviations from the zero line indicate differences between expected and observed frequencies of aggression. Note that flock size does not affect the relative aggressiveness of the two morphs.

Source: From R. W. Ficken, M. S. Ficken and J. P. Hailman, "Differential Aggression in Genetically Different Morphs of the White-throated Sparrow (*Zonotrichia albicollis*)," *Zeitschrift für Tierpsychologie* 46 (1978):48. Reprinted by permission.

shepherds, Doberman pinschers, pit bulls, terriers, and others to make them better guard dogs, fighters, or hunters (Scott and Fuller 1965). In one cross-fostering experiment litters of fox terrier and beagle pups were divided shortly after birth and reared together in mixed litters (James 1951). Despite similar rearing experiences the terriers consistently monopolized both food and females once they reached adulthood, even though terriers are smaller than beagles. Thus the difference in temperaments can be attributed to genetic differences between strains rather than different early experiences.

Siamese fighting fish *(Betta splendens)* have long been pitted against each other by the natives of Thailand, much as fighting cocks and pit bulls have been in other countries (Smith 1945; see also Thompson 1969). For many centuries the winners of such contests have consistently been used as breeding stock, with the result that domestic strains of fighting fish are much more aggressive than their wild ancestors and must be kept apart to prevent them from fiercely attacking each other.

Aggressiveness can also be bred out of animals, as has been done in the domestication process. The dog is a domesticated version of the wolf and has been greatly pacified through many centuries of careful breeding (Scott 1968). The laboratory rat is a domesticated descendant of the brown Norway rat *(Rattus norvegicus)* and is much less aggressive than its wild ancestor (Karli 1956). This difference persists when first-generation descendants of both types are reared in social isolation, showing that it results from genetic rather than environmental factors (Price 1978). The paradise fish *(Macropodus operculus),* a distant relative of the Siamese fighting fish, has been bred for docility instead of aggressiveness, and domestic strains sold commercially to fish hobbyists are much less aggressive than wild strains (Ward 1967).

A genetic influence on human aggressiveness has been inferred from studies of men with abnormal numbers of sex chromosomes. A normal male possesses one X and one Y chromosome, but some males possess an additional Y chromosome. Studies of prison populations show that the XYY, or "supermale," syndrome is much more prevalent among criminally insane, antisocial, and criminal males incarcerated in institutions than among males in the general populace (e.g., Court-Brown 1967; Price and Whatmore 1967; Daly 1969; Jarvik et al. 1973; Witkin et al. 1976). Such studies are often cited as proof that genetic factors influence human aggressive behavior, even though a direct causal link between the XYY abnormality and heightened aggressiveness has not been established (Shah 1970, 1976). However, a recent study of Danish men shows that the relationship is not causal. This study compared behavior of XYY and normal males in the general populace for the first time and showed that XYY males do not commit violent crimes or any other type of criminal activity more often than normal males (Witkin et al. 1976). XYY males did score more poorly on standardized tests measuring intellectual proficiency, however, which suggests that the higher percentage of XYY males incarcerated for aggressive behavior may arise because XYY males are more easily caught and punished. Thus, earlier evidence notwithstanding, the XYY syndrome evidently does not cause heightened aggressiveness in humans.

A more indirect link between genetic factors and aggressiveness is mediated by the sex hormones. The presence of male sex hormones in young animals before or soon after birth is necessary for a genetically male individual to actually develop into a phenotypic male. External genitalia, neurological development, and adult sexual behavior are all dependent on the types of hormones present during a critical period early in infancy (Harris and Levine 1965; Saunders 1968; Beach 1971). The same is true for male aggressive behavior. Neonatally castrated male mice and rats show female-like patterns of fighting behavior and reduced aggressiveness when compared to normal males (Bronson and Desjardins 1968, 1969; Edwards 1968, 1969; Conner and Levine 1969; Peters and Bronson 1971; Peters et al. 1972; Miley 1973). Conversely, female mice and rhesus monkeys can be made more aggressive by injecting them with male sex hormones during infancy (Edwards 1968, 1969; Goy 1968). Thus to the extent that hormone production is governed genetically, development of aggressiveness is indirectly affected by the genome.

Evidence for learned influences

One important variable affecting the development of aggressiveness is the type of mothering received by a young animal. For example, if mice from a passive, inbred strain are reared by a foster mother from an aggressive strain, they become more aggressive as adults (Southwick 1968). Similarly, mice reared by foster rat mothers fight more frequently as adults than mice reared by mouse mothers (Hudgens et al. 1967), and rhesus monkeys reared by hostile or hyperaggressive mothers are themselves hostile and hyperaggressive when compared to those reared by normal mothers (Photo 5–1) (Sackett 1967; Boelkins and Heiser 1970). Aggressive human parents do not necessarily produce aggressive offspring (McCord et al. 1961), but excessive permissiveness and the use of severe physical punishment are correlated with heightened aggressiveness of children (Sears et al. 1957; Feshbach 1970). Human children also become more aggressive when they are deprived of maternal care, affection, or both (Ain-

Photo 5–1

A rhesus monkey mother showing hostile neglect of its infant.

Such behavior and more extreme forms of overt aggression are typical of mothers who were reared in social isolation. Such behavior is especially prevalent when the infant is the mother's first one.

Photo by Harry F. Harlow, University of Wisconsin Primate Laboratory.

sworth 1962; Rutter 1972). The extent young infants are handled affects many traits, including "emotionality" and aggressiveness, in mice, rats, rhesus monkeys, and humans (Johnson 1972; Hinde 1974).

Early experience with peers also affects the development of aggressiveness. Infant rhesus monkeys reared by their own mothers but denied access to peers become hyperaggressive, with the magnitude of the effect depending on how long the isolation period lasts (Harlow and Harlow 1969). In contrast, infants reared on cloth surrogate mothers (Photo 5–2) but allowed access to peers show essentially normal levels of aggressive behavior.

Severe social deprivation has pronounced effects on aggressive behavior. When individuals are reared in complete social isolation, aggressiveness usually increases in red jungle fowl *(Gallus gallus)* (Kruijt 1964), mice (Valzelli 1969, 1974), rats (Bevan et al. 1951; Seitz 1954), rabbits (Wolf and Haxthausen 1960), dogs (Kuo 1967), and rhesus monkeys (Harlow and Harlow 1965; Arling et al. 1969), though it remains nearly unchanged or decreases in some strains of mice (King and Gurney 1954; King 1957; Hutchinson et al. 1965; Valzelli and Garattini 1968), rats (Valzelli and Garattini 1968), and dogs (Fisher 1955; Scott and Fuller 1965; Fuller and Clark 1966). However, complete isolation drastically affects all facets of behavior and physiology. Hence one possible interpretation of these observations is that complete isolation generates abnormal fear responses, which in turn cause animals to overreact to moderate stimulation. Such overreactions could cause animals reared in isolation to fight when normal animals would not.

Photo 5–2

An infant rhesus monkey clinging to its cloth surrogate mother.

Infants reared on such substitute mothers are essentially normal as adults provided that they are allowed to interact with peers during development.

Photo by Harry F. Harlow, University of Wisconsin Primate Laboratory.

Another interpretation might be that animals reared in isolation simply fail to recognize and respond appropriately to appeasement gestures and threats because they never learned to do so during early development.

Animals can be conditioned to become more or less aggressive. They are often trained to be more aggressive by rewarding them with food (e.g., Reynolds et al. 1963; Ulrich et al. 1963) or by exposing them to a continuous electric current that is shut off each time they begin to fight (Miller 1948). Rhesus monkeys become more aggressive when given a mild shock while becoming less aggressive when punished with a strong shock (Ulrich and Symannek 1969). Therapeutic methods to control human aggressiveness are based on rewarding nonaggressive behavior and punishing aggressive behavior (Moyer 1976). These methods have often been used successfully to treat individuals with abnormal aggressive tendencies. Many studies show that human aggression can increase or decrease as a result of imitating parents, friends, heroes, or television episodes (e.g., see Bandura 1973; Liebert 1974; Goldstein 1975). Thus there is clearly a learned component to the development and expression of aggressive behavior.

Conclusion

Aggressiveness is affected by both heredity and experience. No phenotypic trait, including aggressiveness, is entirely preprogrammed in the genome; nor is any trait entirely learned apart from genetic heritage. Every trait develops from the fertilized egg and inevitably results from complex interactions between genetic and environmental factors during every stage of development. To argue that aggressiveness is either entirely innate or entirely learned is simplistic and misleading. Aggressiveness is not a fixed and immutable consequence of heredity; nor is it just a pathological response resulting from undesirable learning experiences. The capacity to behave aggressively is genetically based and has adaptive value. What may sometimes be maladaptive or pathological is the expression of aggressive behavior in inappropriate contexts.

Proximate causation

The second issue raised on page 141 is whether aggression stems from a single generalized motivational state or from several differing motivational states. Since motivational states are postulated to represent underlying physiological mechanisms, the best approach to use in resolving the issue is to examine those mechanisms. An enormous amount of research has been devoted to identifying the proximate causes of aggression in vertebrates, and the available information can only be summarized here. The main objective in the following sections is to demonstrate the variety of different mechanisms involved. Fuller discussions concerning the ways these mechanisms work are given in Marler and Hamilton (1966), Clemente and Lindsley (1967), Garattini and Sigg (1969), Hinde (1970), Eleftheriou and Scott (1971), Johnson (1972), Fields and Sweet (1975), Moyer (1976), Smith and Kling (1976), Brain (1979a, b), and Rodgers (1979).

Stimuli that elicit aggression

Aggression is normally elicited by the presence of another individual. In fact, the various types of aggression are defined according to the type of individuals (e.g., male, female, offspring, peer, predator, etc.) eliciting an aggressive response. As a general rule, the intensity of the stimulus depends on the distance between the individuals involved. Close proximity and crowding ordinarily increase the frequency or intensity of aggressive encounters (Marler 1976).

Aggressive responses are often evoked by very specific features of the target individual. For example, territorial male European robins *(Erithacus rubecula)* recognize intruding males by their red breast feathers rather than by an overall perception of the opponent (Lack 1939a). They will threaten a tuft of red breast feathers placed on a post almost as much as they will an entire stuffed male robin (Photo 5–3). Similarly, male three-spined sticklebacks recognize rival males by their red bellies (Tinbergen 1951). When models of various shapes, sizes, and colors are introduced into a territory, a male stickleback will attack objects of nearly any shape if they are red underneath but will not attack realistically shaped models that lack red bellies. Territorial birds attack tape recordings of male songs

Photo 5–3

A European robin threatening a tuft of red breast feathers on a post to the left and above the photograph.

A tuft of red feathers is nearly as effective as an entire stuffed robin in eliciting an aggressive response.

Photo by F. V. Blackburn, courtesy *British Birds.*

played on their territories, but only if the songs contain specific elements characteristic of the species' song (Falls 1963, 1969). Members of ant colonies all share a specific colony odor, and they will attack any insect entering the colony that lacks that odor (Wilson 1971). Thus animals are often responsive to only certain stimuli, a phenomenon referred to as **stimulus filtering,** and this selective responsiveness greatly influences which behavior pattern an animal exhibits at any given time (see Marler and Hamilton 1966; Hinde 1970). When a response is elicited only by very specific stimuli, the stimulus responsible for eliciting it is usually referred to as a **sign stimulus** or **releaser.**

Stimulus filtering results from sensory, neural, and hormonal mechanisms that developed under the control of both genetic and environmental factors. It focuses the animal's attention on the most relevant stimuli in the environment at the time, and it is responsible for eliciting appropriate responses in each behavioral context. The selective factors determining how much stimulus filtering occurs are unclear, but the degree of filtering certainly varies among species. Aggressive behavior in birds, lower vertebrates, and invertebrates is primarily a response to very specific sign stimuli, while in mammals it is more frequently a response to generalized stimulus patterns.

Endocrine controls

Hormones are part of an integrated physiological system underlying aggressive behavior. The hormones most directly involved are the sex hormones, especially those of the male (Wilson 1975a; Leshner 1978; Brain 1979a). Also involved are epinephrine and norepinephrine, which help an animal respond to sudden danger, and adrenocortical hormones, which prepare an animal to withstand prolonged periods of stress (Wilson 1975a; Leshner 1978; Brain 1979b; Rodgers 1979). The sources of these hormones and a summary of their effects are shown in Figure 5–2.

Male sex hormones. Aggressiveness during social interactions is closely associated with male sex hormones, or *androgens*. The most important of these hormones are dihydrotestosterone and other steroid metabolites of testosterone (Lloyd and Weisz 1975).

The earliest evidence that male androgens affect aggressive behavior came from castrating domestic animals. Castration has long been known to transform wild stallions into relatively tame geldings and savage bulls into plodding oxen. In 1849 Arnold Berthold showed experimentally that roosters stop crowing and fighting after they have been castrated and regain these behaviors when testes from normal roosters are implanted into their abdominal cavities (Wilson 1975a). More recent experiments have shown that injecting adequate amounts of testosterone is sufficient to restore crowing and fighting behavior without implanting entire gonads. Indeed, ovariectomized hens take on the external appearance and pugnacious behavior of a male when injected with the appropriate androgens (Benoit 1950).

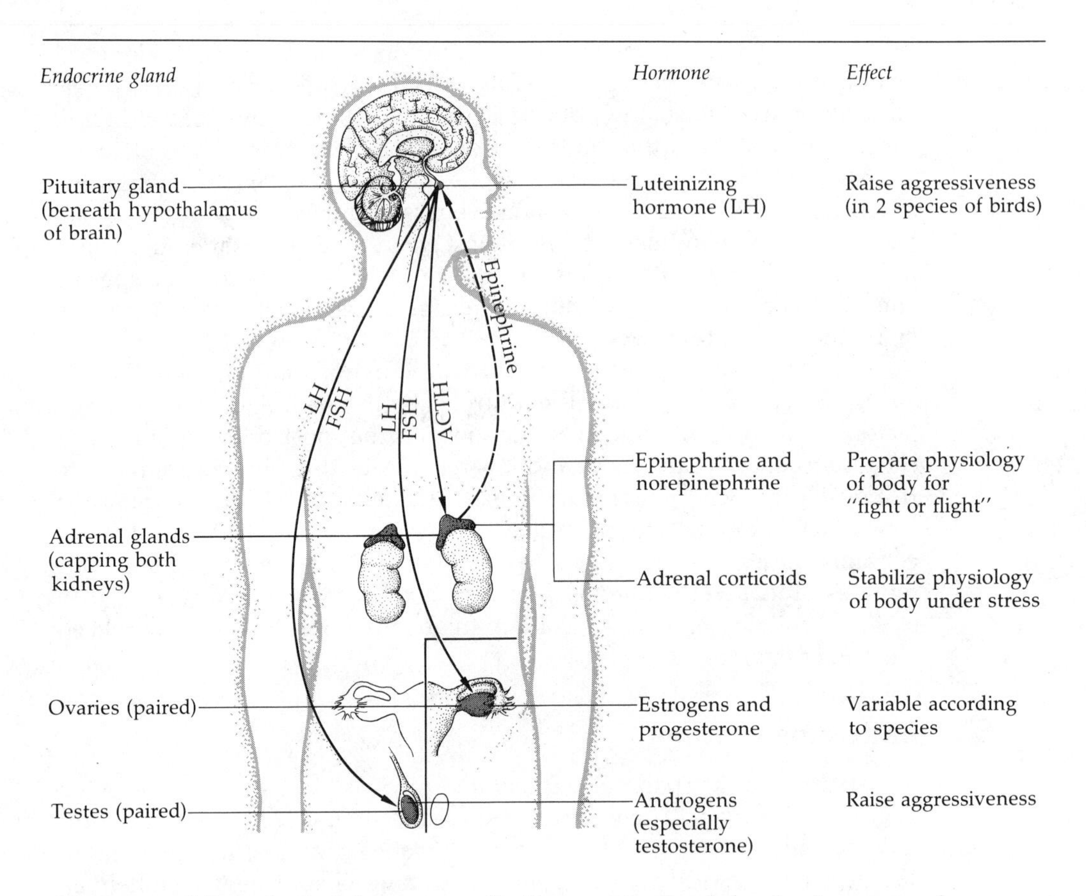

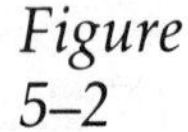

Figure 5–2 *The principal hormones affecting aggressive behavior in mammals.*

The anterior pituitary gland, which is controlled by nerve impulses from the brain and to a lesser degree by circulating epinephrine, releases adrenocorticotrophin (ACTH) into the bloodstream. ACTH in turn causes the cortex of the adrenal glands to enlarge and stimulates increased production of adrenal cortical hormones. The anterior pituitary gland also releases luteinizing hormone (LH) into the bloodstream. In males LH causes increased production of male sex hormones and, in some species, enlargement of the testes. In females LH acts in conjunction with follicle-stimulating hormone (FSH) to stimulate release of estrogen, a female sex hormone. Epinephrine is produced by the medulla of the adrenal glands, and its release is controlled by the brain. All these hormones affect aggressive behavior, though in different ways. Their specific effects are discussed in the text. (Note that the human system is used here for illustrative purposes, but a similar system operates in all mammals.)

Source: From E. O. Wilson, *Sociobiology: The New Synthesis* (Cambridge, Mass.: Belknap Press of Harvard University Press, 1975), p. 252. Reprinted by permission.

Castration has since been shown to reduce aggressiveness in fish, lizards, turtles, and a wide variety of birds and mammals, including humans, and this effect can usually be reversed by subsequent injections of testosterone (Moyer 1976; Leshner 1978; Brain 1979a).

A second line of evidence showing the effect of testosterone on aggressiveness comes from correlations between temporal changes in plasma testosterone concentrations and concomitant changes in aggressive behavior. The spring resurgence of territorial behavior and male aggressiveness in birds, for example, is closely tied to the seasonal enlargement of male gonads, a change that is stimulated by increasing day lengths each spring (Lofts and Murton 1973; Farner 1975; Farner and Follett 1979). The heightened aggressiveness of males is probably a direct result of increased androgen levels in the blood. This conclusion is supported by evidence that testosterone implants induce aggressive behavior in California quail (*Lophortyx californicus*) prior to gonadal recrudescence in spring (Emlen and Lorenz 1942). Moreover, plasma testosterone levels in male white-crowned sparrows increase just before males change from winter flocking to territorial behavior (Wingfield and Farner 1978a, b). Similarly, an abrupt increase in dominance behavior of squirrel monkeys (*Saimiri sciureus*) at the onset of breeding is correlated with increased production of testosterone (Du Mond and Hutchinson 1967; Nadler and Rosenblum 1972). In laboratory mice male aggressiveness is first exhibited at puberty when testosterone levels are rising (Brain and Nowell 1969; McKinney and Desjardins 1973), and injecting prepubertal mice with androgens induces fighting behavior (Levy and King 1953; Svare and Gandelman 1975).

A third kind of evidence comes from injecting previously subordinate individuals with testosterone. Hens injected with testosterone become more aggressive and fight their way up the pecking order, sometimes going all the way from the most subordinate position to the most dominant position (Allee et al. 1939). They retain their status even after injections are stopped, indicating that more than hormonal factors are involved in maintaining status. One such factor may be "social inertia," as status is rarely challenged once a hierarchy is established (Guhl 1958, 1964, 1968). Injecting testosterone into subordinate male lizards (*Sphenomorphus kosciuskoi*) increases their aggressiveness and allows them to dominate other subordinate individuals, though not subdominate or dominant individuals (Done and Heatwole 1977). Injecting subdominate individuals does enable them to usurp high status from previously dominant individuals. In rhesus monkeys male aggressiveness is correlated with baseline plasma testosterone levels (Rose et al. 1971). However, the highest-ranking males do not always have the highest testosterone levels (Rose et al. 1971), and injecting testosterone does not change status relationships (Mirsky 1955). No correlation between aggressiveness and plasma testosterone levels could be found in male Japanese monkeys (Eaton and Resko 1974). In castrated male chimpanzees testosterone treatments enhance dominance status while estrogen treatments reduce it (Clark and Birch 1945).

Male red grouse implanted with testosterone granules become more aggressive and increase the size of their territories (Watson and Moss 1971). Two previously nonterritorial males in poor condition even regained their health and successfully established territories. Similar results have also been obtained in studies of black-crowned night herons (*Nycticorax nycti-*

corax) (Noble and Wurm 1946), herring gulls (*Larus argentatus*) (Boss 1943), and rock doves (*Columba livia*) (Lumia 1972).

These results raise the difficult question of why subordinate individuals do not produce higher testosterone levels on their own, if by doing so they could compete more successfully with other males. Part of the answer may involve the interacting effects of hormones on their target organs. The rate at which testosterone is secreted and its effectiveness on target organs are both affected by the presence of hormones associated with stress (Lloyd and Weisz 1975). Prior experience with defeat or injury may generate stress in later aggressive interactions, which in turn could maintain testosterone at relatively low levels. Such effects could explain, for example, why chronic subordination of male mice by trained fighters leads to reduced testosterone production among the losers (see Bronson and Desjardins 1971).

Another part of the answer may be that the animal's nutritional state is involved. Blood testosterone is in dynamic equilibrium between a portion bound to protein and an unbound portion in plasma (Lloyd and Weisz 1975). Physiologically active testosterone is the portion not bound to protein, and the concentration of unbound testosterone may depend on how much is being transported by protein carriers. Thus animals in poor physiological condition may produce less binding protein and hence be incapable of increasing testosterone levels.

An evolutionary argument may also be relevant here. As suggested in Chapter 3, a subordinate individual may not gain any long-term advantage from increasing its testosterone level, because dominant individuals could respond by elevating *their* testosterone levels (see p. 94). An escalation of fighting intensity might then ensue, with no ultimate change in who is victorious. Thus the subordinate animal may only increase its risk of injury without improving its chances for winning higher status.

Another hormone linked with male aggressiveness is *luteinizing hormone (LH)*, which is secreted by the pituitary gland and normally stimulates production of testosterone. LH has little effect on plasma testosterone levels in birds during autumn months when gonadal recrudescence and elevated testosterone production cannot be induced by artificially long day lengths. Injecting male starlings and red-billed queleas with LH during autumn leads to increased fighting ability and elevated dominance status (Mathewson 1961; Davis 1963; Crook and Butterfield 1970; Lazarus and Crook 1973). Since the effect could not be attributed to increased testosterone production, LH itself was apparently responsible for the birds' heightened aggressiveness. Perhaps an increase in LH production is responsible for the temporary resurgence in territorial behavior shown by some birds in autumn (e.g., Snow 1958; Orians 1960). However, the evidence for an LH effect on aggression is weak, and a direct role of LH in generating aggressive behavior is controversial (Brain 1979a).

Female sex hormones. Female sex hormones, especially estrogens and progesterone, affect aggressiveness too, but the relationships are less clear. Estrogens implanted in castrated male mice normally do not change previously established levels of aggressiveness (Brain and Poole 1976; Brain and Bowden 1978; Brain 1979a), although they do increase aggressiveness in castrated male hamsters and the scincid lizard *Sphenomorphus kosciuskoi* (Vandenbergh 1971; Done and Heatwole 1977). Their effects on female behavior are variable. Unreceptive female hamsters are normally highly aggressive and dominant over males, but they lose their aggressiveness and become subordinate to males after being spayed (Payne and Swanson 1971). Aggressiveness can be increased in spayed females by injecting them with estrogens at dosages ten times higher than normal physiological levels (Kislack and Beach 1955) but not at dosages near normal physiological levels (Payne and Swanson 1971). In contrast, administering normal dosages of progesterone does increase female aggressiveness and restore female dominance over males.

The results obtained with hamsters may not be typical of other female mammals. Neither ovariectomy nor estrogen injections affect the spontaneous aggressive behavior of female mice (Levy 1954; Tollman and King 1956; Gustafson and Winokur 1960; Edwards and Burge 1971), and ovariectomy does not affect shock-induced aggression in female rats (Conner and Levine 1969). High estrogen levels increase the aggressiveness of female chimpanzees (Birch and Clark 1946), but they have negligible or extremely subtle effects on women (Gottschalk et al. 1961). Different effects of this sort may reflect evolutionary responses associated with differing social organization, but that possibility cannot be evaluated until more species have been studied.

Estrogen and progesterone both affect aggressiveness of female mammals during estrus, but again the effects vary among species. In hamsters estrous females become less aggressive and more tolerant of males who approach them (Vandenbergh 1971; Wise 1974), but in rhesus monkeys they become more aggressive toward males (Carpenter 1942; Michael 1969; Michael and Zumpe 1970). Levels of both estrogen and progesterone are high during estrus, suggesting at least for hamsters that progesterone may counteract the effects of estrogen on aggressiveness (Vandenbergh 1971; but see Wise 1974).

In women aggressiveness and anxiety decrease slightly during ovulation, when estrogen and progesterone levels are high (Ivey and Bardwick 1968), and they increase shortly before menstruation, when progesterone levels are decreasing (Greene and Dalton 1953; Dalton 1964; Hamburg 1966; Hamburg et al. 1968; Weiss 1970). Premenstrual anxiety and aggressiveness are due to a change in the ratio of progesterone to estrogen in blood plasma.

Since the effects of female hormones vary among species, an important factor is the neural substrate upon which the hormones act. It is evidently adaptive for females to become more aggressive during estrus in some species and to become less aggressive in others. Such differences presumably result from the way brain centers respond to hormonal changes occurring during the estrous cycle in each species.

Epinephrine and norepinephrine. *Epinephrine* (also called adrenalin) and *norepinephrine* are primarily responsible for preparing the body for "flight or fight" during dangerous or stressful situations. These two hormones act in conjunction with the sympathetic nervous system to alter metabolic processes and circulatory patterns, thereby mobilizing energy reserves and concentrating them in skeletal muscles and other tissues directly involved in escape reactions and aggressive behavior. The hormonal effects are only temporary, since about half of the epinephrine and norepinephrine released into the bloodstream is inactivated every two or three complete circulations through the body (Mulrow 1973).

Epinephrine is released into the bloodstream almost exclusively by specialized cells, called *chromaffin cells,* in the adrenal medulla (Mulrow 1973). Release of epinephrine is controlled by the sympathetic nervous system, which has direct connections with neural centers in the brain. Epinephrine stimulates stronger and more frequent contractions of the heart, increases blood pressure, constricts blood vessels to the gastrointestinal tract and lungs, dilates blood vessels to the skeletal muscles, dilates bronchi and bronchioli in the lungs, increases metabolic and respiratory rates, and stimulates the release of simple sugars and free fatty acids into the bloodstream.

Norepinephrine is the principal neurotransmitter substance responsible for transmitting nerve impulses across synapses in the sympathetic nervous system. Small amounts are released into the bloodstream by the adrenal medullae and by leakage from sympathetic nerve endings (Mulrow 1973). Norepinephrine acts in generally the same direction as epinephrine, but it affects target tissues in somewhat different ways (Table 5–2). It has little effect on cardiac output and its effect on metabolic processes is less pronounced, but it has about the same effect as epinephrine on blood pressure and coronary blood flow.

Norepinephrine is apparently released more during the actual response to stress or danger, while epinephrine is released more during the anticipatory period just prior to stress or danger (Schildkraut and Kety 1967; Frankenhaeuser 1971). The principal effect of both hormones is to prepare the body physiologically so that any response will be more effective. Neither hormone actually induces aggressive behavior. Both may sensitize brain tissues to stimuli eliciting aggression, but this possibility has not yet been adequately investigated (see Moyer 1976; Brain 1979b).

Adrenocortical hormones and general adaptation syndrome. Physiological preparedness for prolonged periods of stress is under the control of adrenocortical hormones. These hormones, like epinephrine and norepinephrine, may not have any direct effects on aggressiveness. Nevertheless, they are important because they are involved in physiological responses to crowding, which often leads to heightened aggressiveness.

In his landmark studies of stress, Hans Selye (1956) identified three stages of what he called the *general adaptation syndrome (GAS)* to stress. Each stage involves increased production of adrenocortical hormones and progressively more extreme physiological responses.

During the *alarm stage* the pituitary gland releases *adrenocorticotrophin (ACTH),* which in turn stimulates production of cortisone, cortisol, corti-

Table 5–2 *The effects of intravenous infusions of epinephrine in humans compared to the effects of norepinephrine.*

Body system	Epinephrine	Norepinephrine
Cardiac		
Heart rate	+	−
Stroke volume	+ +	+ +
Cardiac output	+ + +	0, −
Arrhythmias	+ + + +	+ + + +
Coronary blood flow	+ +	+ + +
Blood pressure		
Systolic arterial	+ + +	+ + +
Mean arterial	+	+ +
Diastolic arterial	+, 0, −	+ +
Mean pulmonary	+ +	+ +
Peripheral circulation		
Total peripheral resistance	−	+ +
Cerebral blood flow	+	0, −
Muscle blood flow	+ +	0, −
Cutaneous blood flow	− −	+, 0, −
Renal blood flow	−	−
Splanchnic blood flow	+ +	0, +
Metabolic effects		
Oxygen consumption	+ +	0, +
Blood sugar	+ + +	0, +
Blood lactic acid	+ + +	0, +
Eosinopenic response	+	0
Central nervous system		
Respiration	+	+
Subjective sensations	+	0, +

Note: + = increase; 0 = no change; − = decrease.

Source: From R. J. Mulrow, "The Adrenals," in T. C. Ruch and H. D. Patton, eds., *Physiology and Biochemistry*, Vol. III, 20th ed. (Philadelphia: W. B. Saunders Co., 1973), p. 243. (Based on Goodman and Gilman 1970.) Reprinted by permission.

costerone, and other steroid hormones by the adrenal cortex. These hormones lead to relatively long-lasting changes in body physiology. Their effects vary among species, but in general they promote conversion of protein to carbohydrates and the storage of carbohydrates as glycogen in the liver. Thus they cause a general mobilization of energy reserves, which become important when an animal is faced with chronic stress. A second hormone produced by the adrenal cortex, aldosterone, regulates electrolyte metabolism in the body. Aldosterone is not controlled by pituitary gland secretions and is not involved in the GAS.

During the *resistance stage* prolonged stress leads to increased weight of the adrenal glands and greater production of adrenocortical hormones. The hormones, in turn, begin reducing responsiveness to foreign substances by delaying or reducing inflammatory reactions, killing lymphocytes in the lymph nodes, lowering eosinophil (a type of white blood cell) counts in the blood, and suppressing antibody reactions to foreign proteins entering the bloodstream. This stage is often caused by excessive crowding or increased frequencies of aggressive encounters (Davis 1964; Bronson 1967; Christian 1968; Welch and Welch 1969), and the physiological responses border on the pathological. The resistance stage is essentially a last line of defense against chronic stress.

If stress is sufficiently prolonged and severe, the body enters the pathological *exhaustion stage.* It can no longer handle high adrenocorticoid levels without physiological dysfunction. Muscle tissues begin to break down from prolonged conversion of protein to carbohydrates, and immunological responses to disease organisms and foreign substances become severely impaired. The animal becomes weak and sickly, and if stress continues, it is likely to die. Death by stress is the ultimate cost for animals who cannot escape prolonged and excessive crowding.

Subordinate individuals exhibit consistently higher levels of adrenocortical activity than dominant animals in laboratory mice (Southwick and Bland 1959; Louch and Higginbotham 1967), rats (Popova and Naumenko 1972), wolf pups (Fox and Andrews 1973), and rhesus monkeys (Sassenrath 1970). This characteristic is usually interpreted to be a reflection of chronic stress suffered by subordinate individuals. In addition, aggressive rhesus monkeys have higher levels of circulating 17-hydroxycorticosteroids than less aggressive monkeys (Levine et al. 1970) and aggressive strains of laboratory mice have heavier adrenal glands than less aggressive strains (Lagerspetz et al. 1968), possibly for the same reason.

Some evidence implicates a more direct role of corticosteroids in aggressive behavior, apart from stress effects. Removal of the adrenal glands leads to decreased aggressiveness in laboratory mice (Sigg 1969; Brain et al. 1971; Harding and Leshner 1972), and replacement therapy with corticosterone restores their aggressiveness (Candland and Leshner 1974). Injecting intact mice with moderate dosages of corticosteroids increases aggressiveness (Candland and Leshner 1974). Some effects of corticosteroids may be mediated by ACTH. Circulating corticosteroid levels affect pituitary secretion of ACTH via negative feedback loops, and ACTH is known to affect aggressiveness independently of adrenal secretions (Brain et al. 1971; Leshner et al. 1973). The effects of ACTH are primarily long-term, however, and short-term responses are apparently caused by direct actions of adrenocortical secretions rather than ACTH (Leshner and Walker, cited in Leshner 1978).

Neurology of aggression

Neurological control centers that affect aggressive behavior are ordinarily identifed by electrical stimulation, chemical stimulation, or lesioning of selected brain centers. Much research has been devoted to the neurological bases of aggression, and only part of it can be summarized here. The objective is to show the variety of brain centers involved in aggressive behavior and not to explain how the neural mechanisms actually work. It should be realized that the term *brain center* does not imply a well-delineated neurological entity. It simply refers to a specific area of neural integration at the specified location. The various brain centers are all interconnected, either directly or indirectly, but functional pathways are difficult to identify and are not yet well understood. Therefore, the following sections will simply describe the various brain centers involved, without speculating on how they interact. The locations of brain centers referred to below are shown in Figure 5–3. The exact names of brain structures need not be memorized. Listing them is done below simply to show how many are involved. More detailed discussions of current knowledge can be found in Fields and Sweet (1975), Moyer (1976), and Rodgers (1979).

Intrasexual aggression. Brain centers affecting intrasexual aggression are found in the hypothalamus, cerebral cortex, and midbrain. Electrical stimulation of 53 points in *ventrolateral hypothalamus* causes normally nonaggressive rats to attack other males placed in their cages (Woodworth 1971). Lesions in lateral hypothalamus or combined lesions in lateral and ventromedial hypothalamus do not (Adams 1971). Lesions in lateral or medial hypothalamus have no effect on shock-induced aggression or predatory mouse killing by male rats (Adams 1971; Miczek et al. 1974), but lesions in *ventromedial hypothalamus* increase shock-induced aggression (Adams 1971; Eichelman 1971; Grossman 1972).

Electrical stimulation of *lateral hypothalamus* causes submissive male rhesus monkeys to viciously attack a dominant male cage mate (Robinson et al. 1969). A follow-up study shows that stimulation of this nucleus causes males to attack other males more often than it causes them to attack females (Alexander and Perachio 1973). It also causes them to attack subordinate individuals of either sex more often than it causes them to attack dominant males. Similarly, stimulation of lateral hypothalamus causes male opossums *(Didelphius marsupialis)* to attack conspecific males (Roberts et al. 1967). In both studies aggression was elicited only when an appropriate external stimulus (i.e., another individual) was present, indicating that electrical stimulation of hypothalamic nuclei increases the predisposition or motivation to behave aggressively, not overt aggression per se.

Lesions in the *septal area* of neocortex increase aggressiveness of dominant male hamsters, but they increase submissiveness of subordinate males (Sodetz and Bunnell 1967). Septal nuclei have direct neural connections with the hypothalamus and may act as inhibitors or facilitators of activity in hypothalamic nuclei that control aggression. Removal of *prefrontal neocortex* (i.e., posterior region of the frontal lobes) diminishes the incidence of threat displays and increases the frequency of intrasexual aggression in rhesus monkeys (Kling 1975). It also increases hyperactivity and pacing behavior.

Midbrain lesions in the *inferior colliculi* enhance intermale fighting

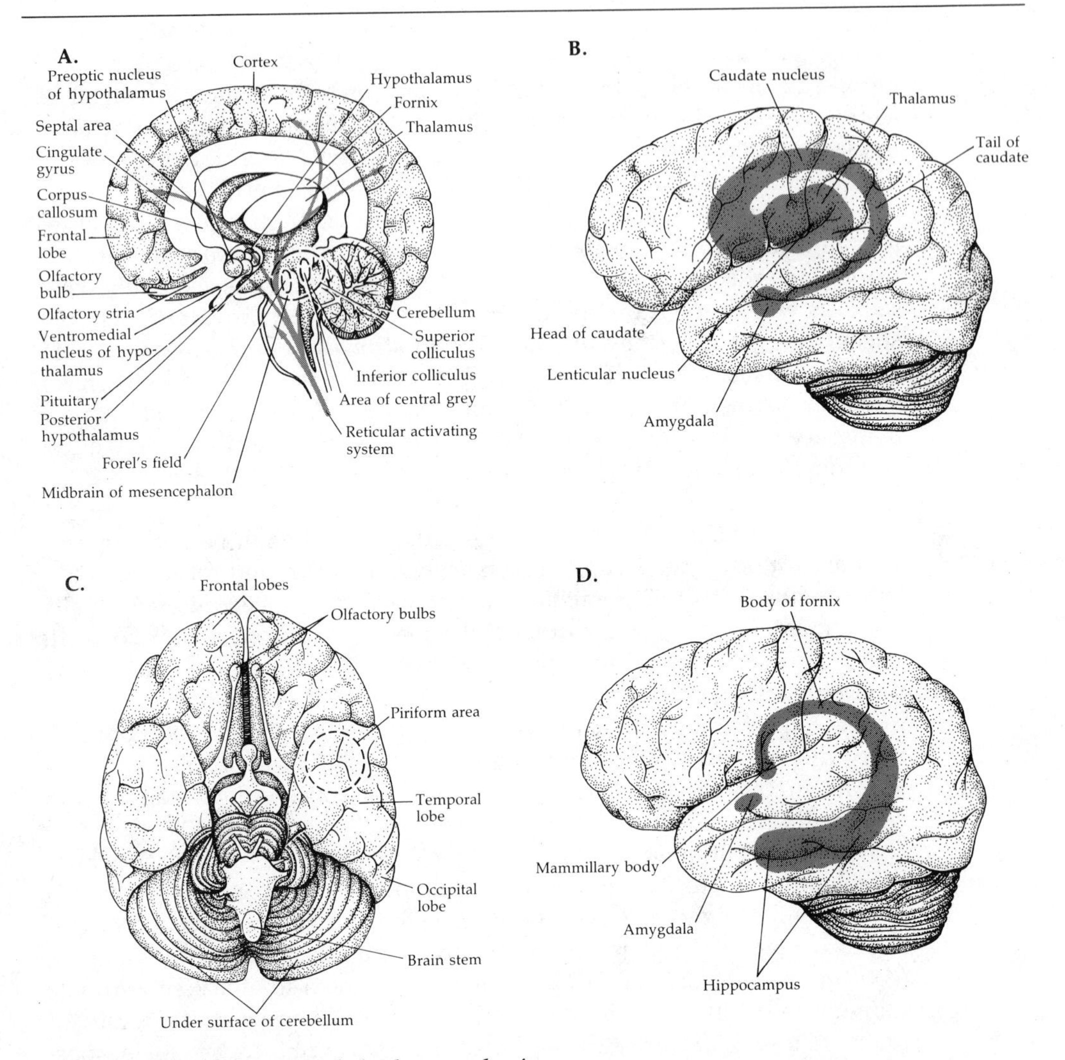

Figure 5–3 *Major parts of the human brain.*

A. Sagital section showing internal structure. B. Location of caudate nucleus and amygdala. C. Ventral view showing lobes of the cerebral cortex. Shaded areas indicate structures below the neocortex (the outer layer of cerebral cortex). D. Location of the hippocampus. Shaded areas indicate structures below the neocortex.

Source: Figures B, C, D and E (pp. 286–287) in *The Psychobiology of Aggression* by K. E. Moyer. Copyright © by K. E. Moyer. Reprinted by permission of Harper & Row, Publishers, Inc.

behavior in rats, but lesions in the superior colliculi do not (Kesner and Keiser 1973). The same lesions have no effect on shock-induced aggression or predatory mouse-killing behavior. The inferior colliculi are part of the midbrain *reticular activating system (RAS),* which generates general arousal states affecting activity levels in all hypothalamic control centers (Moyer 1976). Lesioning nuclei in the inferior colliculi may remove an inhibitory block on hypothalamic nuclei, thereby increasing aggressiveness, but this interpretation has not been definitely proven.

Fear-induced aggression. Numerous brain centers affect aggressive reactions in fearful or stressful situations. Once again, the hypothalamus, neocortex, and midbrain are involved, though the precise neural areas differ from those associated with intrasexual aggression.

Stimulating *anterior hypothalamus* causes a cornered cat to behave aggressively, while stimulating *ventromedial hypothalamus* elicits a generalized rage response (Yasokochi 1960). Lesioning ventromedial hypothalamus of cats may result in extreme savageness, especially when the cat is unable to escape the situation (Wheatley 1944; Kling and Hutt 1958). Stimulating the ventral part of *medial hypothalamus* elicits aggression, while stimulating the dorsal part elicits escape reactions (Romaniuk 1965). This result suggests a dorsoventral organization of hypothalamic areas controlling self-protective behaviors. The strength of stimulation also affects a cat's response. Growling (an attack tendency), hissing (a defensive tendency), and escape can each be elicited by threshold stimulation of nuclei in the *hypothalamus, thalamus,* and *forebrain,* while stronger stimulation of at least the hypothalamus leads to an alternation between these responses (Hunsperger and Bucker 1967; Brown et al. 1969).

Aggressive reactions to fear or stress are drastically reduced by removing the *amygdala* in wild Norway rats, domestic cats, lynxes *(Lynx canadensis),* agoutis, a variety of nonhuman primates, and humans (reviewed by Kling 1972, 1975; Moyer 1976). In rhesus monkeys the effect only occurs if the *medial nuclei* of *amygdala* are included in the lesioned area, indicating that these centers are especially important (Kling 1975). Aggressive responses evoked by stimulating the amygdala can be blocked by lesioning the hypothalamus, but responses evoked by stimulating the hypothalamus cannot be blocked by lesioning the amygdala (see Moyer 1976). The main outflow of the amygdala is to the hypothalamus, particularly to the ventromedial nucleus (Gloor 1975). The effects of hypothalamic stimulation can be blocked by lesions in the thalamus, indicating that hypothalamic output is mediated through the midbrain. Present evidence suggests that the amygdala integrates visual and other sensory input from the neocortex and gives them motivational significance by routing the input to hypothalamic control centers.

Lesioning the *septal area* lowers the threshold for both escape and aggressive responses of rats and mice in stressful situations (reviewed by Moyer 1976). Prolonged stimulation of the *rostral* (anterior) *hippocampus* produces escape attempts and fear-induced aggression in rhesus monkeys (MacLean and Delgado 1953). Removing the hippocampus sometimes reduces aggressive reactions of cats, rhesus monkeys, and baboons (Gol et al. 1963). Stimulating the *caudate nucleus* inhibits aggressiveness of normally hostile rhesus monkeys toward human handlers (Delgado 1960). Bilateral ablation of either the anterior third of the *temporal lobes* of neocortex or the *prefrontal cortex* of rhesus monkeys severely disrupts all forms of social behavior, including fear-induced aggression (Myers 1972; Franzen and Myers 1973). Lesioning the *temporal lobe* reduces fear-induced aggression in the rhesus monkey (Turner 1954). The prefrontal cortex and anterior temporal lobe apparently have direct connections with the amygdala and may be part of a control system that also involves the hypothalamus, subthalamus, and midbrain reticular activating system (Nauta 1962).

Irritable aggression. Irritable aggression includes rage, violent behavior during epileptic seizures, and other forms of aggression that lack escape tendencies (Moyer 1976). It can be elicited by stimulating *anterior* and *medial hypothalamus,* lesioning *ventromedial hypothalamus,* lesioning or chemically stimulating the *hippocampus,* and sometimes by lesioning the *septal area* (reviewed by Moyer 1976). Stimulating or lesioning the *amygdala* can activate or inhibit irritable aggression, depending on what nucleus is treated. Stimulating the *temporal lobe* makes cats irritable and rhesus monkeys vicious and violent (Anand and Dua 1955, 1956). Stimulating the *posterior orbital cortex* (just posterior to the temporal lobe) makes cats vicious (Anand and Dua 1955, 1956). Tumors in the temporal lobe, *frontal lobe, limbic system* (most areas of cerebrum underlying the outer neocortex), and *hypothalamus* lead to extreme aggression and violence in humans (Moyer 1976). Uncontrolled aggressiveness is often associated with epileptic centers in the temporal lobe. Lesions of the temporal lobe, amygdala, *cingulate gyrus, posterior hypothalamus,* several *thalamic nuclei,* and frontal lobe of the neocortex have all reduced aggressive behavior in humans, though none have been effective in every instance.

A general physiological model

Since several types of aggression can be identified, each having different physiological bases, no one model will fit them all in every respect. Moreover, the detailed organization of physiological control systems varies among species, making generalizations even more difficult. The contexts in which aggression occurs vary because species are affected by different selective regimes, and the underlying morphology and physiology of aggression varies because species have different phylogenetic heritages. Nevertheless, a model conforming to the general organizational properties of physiological control systems would greatly clarify motivational concepts. In a series of papers Moyer (1968, 1971, 1975, 1976) has developed a general model to describe how control systems underlying aggressive behavior are organized (Figure 5–4).

The model first assumes that innately organized neural systems within the brain generate aggressive tendencies when activated by appropriate sensory input. These systems consist of brain nuclei or control areas located at several levels in the brain. They have many connections with other neural systems, but they are closely integrated and function as a unit.

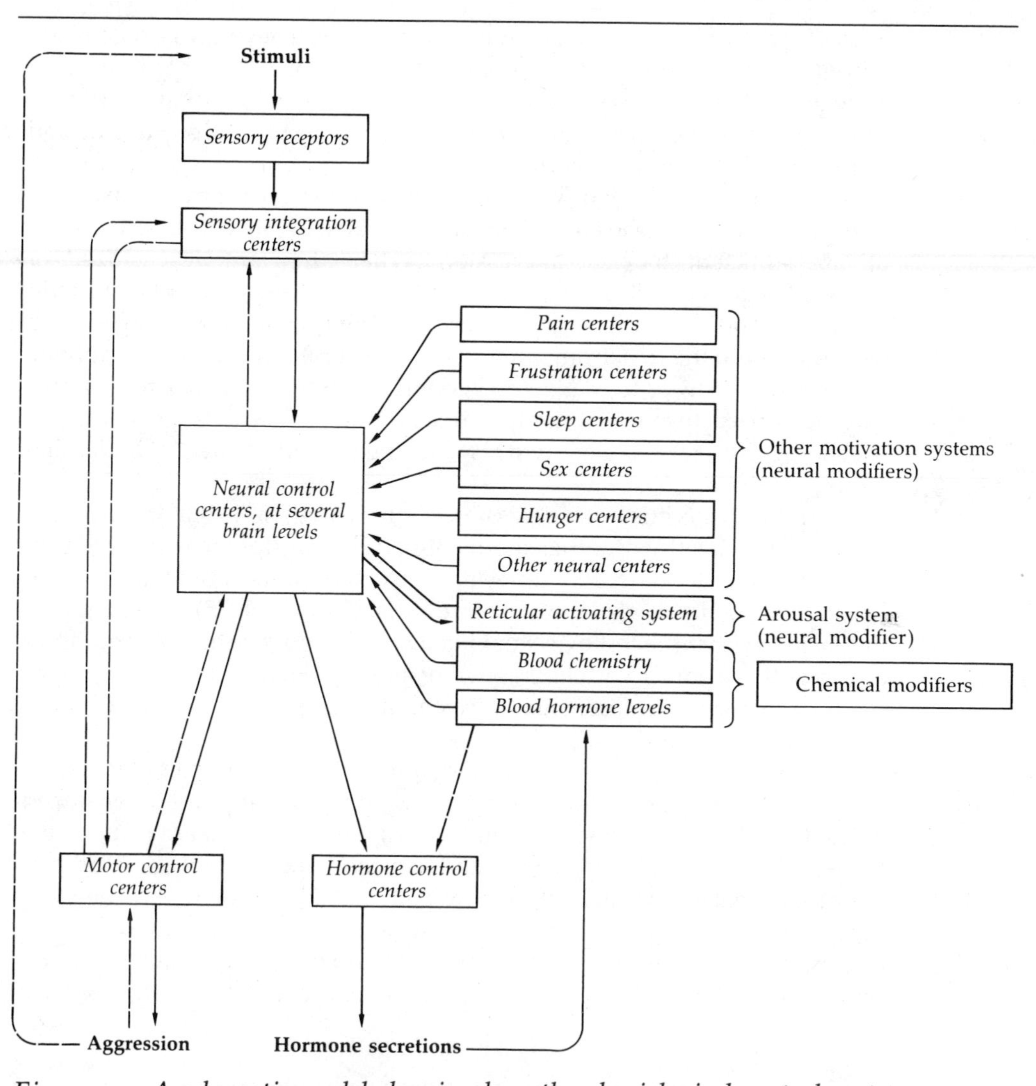

Figure 5–4 *A schematic model showing how the physiological control system underlying aggressive behavior is organized.*

Solid arrows indicate directions of influence or control. Dashed arrows indicate feedback loops. The terms center *and* control center *refer to nondiscrete areas in the brain where neural integration takes place.*

Source: Based on Moyer 1975, 1976.

The functional identity of each system may be based, for example, on lower synaptic resistance to nerve impulses transmitted within the system than to nerve impulses entering or leaving the system.

The neural control areas are activated by sensory input from brain areas responsible for integrating information received by sensory receptors. They generate output to motor control areas, which are responsible for muscular actions and hence behavioral responses. They may also generate output to brain areas controlling hormone production or other physiological responses. As the stimulus situation changes during a behavioral interaction, various feedback loops modify sensory input, motor output, and activity levels in central control areas.

Several systems modify the sensitivity of neural control areas to external stimuli. When the control areas are sensitized, weaker external stimuli are sufficient to elicit aggression. When they are desensitized or inhibited, stronger stimuli are required before aggression can be elicited.

One modifier of these control areas is the reticular activating system (RAS). Input to the RAS from other parts of the brain or from the peripheral nervous system leads to changes in general arousal, which in turn sensitizes or desensitizes the neural control areas associated with aggressiveness. Activating the aggression system also has the reciprocal effect of increasing arousal in the RAS.

Another modifier of aggression control areas is input from neural systems associated with other motivations. The control areas may be sensitized by pain, food deprivation, sleep deprivation, morphine deprivation (in addicted individuals), and frustration (see Moyer 1975). They may be desensitized or inhibited by sexual behavior, foraging behavior, and other activities that conflict with aggressive responses. Finally, the control areas can be sensitized or desensitized by changes in plasma hormone levels, blood chemistry, and drugs.

Thus the control systems underlying aggression are complex neural and chemical networks. They are activated by sensory input, are sensitized or inhibited by physiological systems external to them, and are responsible for evoking aggressive behavior and concomitant physiological adjustments. A separate neural control system, affected by its own set of sensory input and external modifying systems, is responsible for each type of aggression. The various aggression systems probably do not act independently of one another, but the precise interactions between them are unknown.

Implications for motivational models

The physiological evidence does not support the hypothesis that aggressiveness is a single generalized drive sparked by "aggressive energy" specifically associated with it. Aggression consists of diverse behaviors, each mediated by a distinct control system, and these control systems do not energize the behaviors they mediate.

Many motivational states are correlated with aggressive behavior in humans. For example, intrasexual and intersexual aggression can result from stress, irritation, frustration, anger, hostility, or anxiety. Parental aggression can result from irritation, frustration, or anger. Defensive aggression can result from fear or anger. Redirected aggression results from a combination of anger, frustration, and perhaps fear. Similar emotions are not necessarily felt by animals, but animals do sometimes show outward manifestations reminiscent of emotions exhibited by humans. Most biologists and psychologists now accept the conclusion that aggression has multiple causes and cannot be explained by unitary drive concepts (e.g., Tinbergen 1951; Cofer and Appley 1964; Hinde 1970, 1974; Moyer 1976).

Neural control areas are not spontaneously activated by the accumulation of drive-specific motivational energy. Although they do have endogenous levels of neural activity and although their responsiveness to sensory input is modified by other motivational systems, hormones, and blood chemistry, they do not exhibit cumulative increases in neural activity capable of explaining spontaneous behavior. Spontaneous aggression can occur in the absence of external stimuli, but it probably results from desensitization of neural control areas rather than accumulation of "aggressive energy." The energy propelling behavioral responses arises from general arousal of the reticular activating system, not from drive-specific arousal of neural control areas. Exactly how energy is allocated to specific behavioral or physiological responses is determined by the relevant motivational systems. These systems route neural input to the appropriate response mechanisms, but they do not energize motor responses in any sense of the word. Thus the switchboard model of motivational systems is most consistent with the physiological evidence, and conclusions or interpretations based on energy models of motivation appear invalid.

An evolutionary model of aggression

Aggression can be viewed as one of several possible strategies for increasing fitness under competitive conditions. Popp and Devore (1979) recently proposed a simple cost-benefit model of aggression that pinpoints the key variables affecting its evolution. Their approach will be presented in some detail because it provides a general theoretical framework for understanding when aggression should evolve. In the following analysis the focal animal or potential aggressor will be referred to as the *actor,* while its opponent will be referred to as the *rival.*

An actor has some chance of obtaining benefits in a competitive situation without behaving aggressively, but aggression may allow it to obtain additional benefits. For any given situation this net gain in benefits must exceed the costs before aggression can be adaptive.

Four variables are important in computing the net benefit of aggression. These are (1 and 2) the probabilities of successfully gaining access to a disputed resource or mate with and without aggression, (3) the amount that fitness is increased should the actor successfully gain access to the disputed resource or mate (i.e., the benefit), and (4) the effect that aggression has on the rival's fitness if the actor and rival are related.

The net benefit gained from aggression can be computed with a little simple algebra. The procedure is as follows: First, compute the average benefit the actor can expect in a given situation should it compete aggressively. To do this, multiply the benefit derived if the actor's aggression succeeds (B_{a2}) times the probability of success (P_{a1}). Second, compute the average benefit the actor can expect should it compete nonaggressively. To do this, multiply the benefit derived if the actor's nonaggressive competition succeeds (B_{a2}) times the probability of success (P_{a2}). The net benefit gained from aggression is then given by $P_{a1}B_{a1} - P_{a2}B_{a2}$, and aggression will pay when the costs of aggression (C_a) are less than the net benefit:

$$C_a < P_{a1}B_{a1} - P_{a2}B_{a2} \quad (5\text{–}1)$$

The maximum acceptable cost to an aggressor, its *maximum adaptive expenditure* ($C_{a,\max}$), represents the point where aggression becomes maladaptive. Thus

$$C_{a,\max} = P_{a1}B_{a1} - P_{a2}B_{a2} \quad (5\text{–}2)$$

The situation becomes slightly more complicated if the rival is a relative, because any costs suffered by the rival will adversely affect the actor's inclusive fitness. The maximum adaptive expenditure for an actor competing with a relative must therefore include this effect.

In any aggressive interaction the rival may sometimes win. The rival's net benefit from the interaction is computed in the same way as the actor's. Thus the rival's net benefit is $P_{b1}B_{b1} - P_{b2}B_{b2}$, with the subscript *b* denoting that the variables refer to the rival rather than the actor. The rival also suffers costs (C_b), which will exceed the benefits should the actor win. The net reduction in the rival's fitness will therefore be $C_b - (P_{b1}B_{b1} - P_{b2}B_{b2})$. This net cost, multiplied by the degree of relatedness *(r)* between actor and rival, is the amount that the actor's inclusive fitness is diminished because the rival is a relative. Thus aggression toward a relative is adaptive for the actor when

$$C_a < P_{a1}B_{a1} - P_{a2}B_{a2} + r(P_{b1}B_{b1} - P_{b2}B_{b2} - C_b) \quad (5\text{–}3)$$

Equation 5–3 is rather cumbersome and certainly not worth remembering. However, the logic used to derive the equation is worthy of study.

Several useful predictions can be made from Equation (5–3). These predictions are generated by asking what happens to the inequality when one or more of the variables are changed. Thus one can predict that aggression by the actor is more likely to evolve when (1) the cost of aggression for the actor (C_a) is low, (2) the net benefit ($P_{a1}B_{a1} - P_{a2}B_{a2}$) is large, (3) the probability of gaining access to the resource without aggression (P_{a2}) is low, (4) genetic relatedness between rival and actor *(r)* is low, and (5) the net cost to the rival ($C - P_1B_1 + P_2B_2$) is low. In short, aggression becomes more likely whenever conditions increase an actor's maximum adaptive expenditure.

Actual costs and benefits can rarely be measured for actors and rivals. Present methods allow researchers to identify the costs and benefits associated with aggression, but they rarely allow quantification. The model is useful mainly for showing how the key variables are interrelated. The model can also provide insights into the tactics used to win aggressive encounters. But before tactics are discussed, some of the benefits and costs associated with aggression require elaboration.

Benefits of aggression

Intrasexual aggression

Intrasexual aggression is the most widespread and diverse form of aggression. It is generally most prevalent and most intense among males, but it also occurs frequently among females. Intrasexual aggression is associated with defending territories, social status, mates, or resources. No generalizations can be made here about the benefits gained from intrasexual aggression because virtually every conceivable benefit is obtained in some context or in some species. The relevant theory and current evidence will be discussed more fully in later chapters. Territoriality will be discussed in Chapter 7, status in Chapter 13, and defense of mates in Chapters 10 and 11. Suffice it to say here that current evidence is consistent with the predictions drawn from the Popp and Devore model.

Intersexual aggression

Intersexual aggression occurs in several contexts. It may occur during territorial disputes when males and females both defend individually occupied territories, it may occur within social groups during status disputes or disputes over resource items, or it may occur during courtship. The most prevalent context is courtship, and hence that will be the focus here.

Aggression is very prevalent during courtship. It has been thoroughly documented in all types of vertebrates and in some invertebrates by analyzing the postures and movements shown in purely agonistic disputes between males and then showing that very similar postures or movements also occur during courtship (see Andrew 1961; Bastock 1967; Hinde 1970).

A common pattern in vertebrates is for a male to initially threaten or attack an unfamiliar female just as he would a rival male. Gradually the male's aggressiveness subsides, until finally he accepts the female's presence. The female, in turn, is initially wary of the male and only gradually allows him to approach closely enough to copulate. This process is well illustrated by the courtship behavior of gulls (Moynihan 1955, 1962b; Tinbergen 1959, 1960a, b).

In most gulls pairing occurs on a temporary pairing territory, which the male establishes outside but near the nesting colony. Males display on their pairing territories by adopting an oblique body posture and by giving a raucous *long call*. This *oblique-cum-long-call* is an aggressive display used to repel intruders of both sexes. It also serves to attract unmated females.

When a female first lands on a pairing territory, she is threatened or attacked by the male. The male's behavior is very similar to that shown toward intruding males—which is not surprising since in gulls the sexes are very similar in appearance. Unlike an intruding male, however, the female does not return the threat or flee. She adopts an upright appeasement posture, usually facing away from the male (Photo 5–4). The male, in turn, directs his threat postures laterally rather than directly at the female, thereby reducing the intensity of threat. At first the female appears fearful and remains only briefly with the male. She often leaves to alight on another male's territory nearby, but eventually she selects a particular male. She makes the choice by repeatedly returning to the male's territory and remaining there for longer and longer periods of time.

Over a period of several days the male's aggressiveness subsides and is replaced by appeasement postures. The principal posture is *head flagging*, in which the male and female stand with bodies parallel and heads facing away from each other. At the same time, the female becomes bolder and allows closer approach by the male. After spending considerable time together, the female begins begging for food, which the male eventually provides by regurgitation. Soon thereafter the pair begins to copulate, and the pairing process is completed.

Many hypotheses have been advanced to explain male aggression

Photo 5–4

A male black-headed gull (Larus ridibunus) *(left) threatens a female during early courtship.*

The female shows submissiveness by facing away from the male. Male aggression is typical during early stages of courtship in most vertebrates.

Photo by A. Christiansen/Frank W. Lane.

during courtship. Two are proximate explanations, while the others are ultimate explanations. Each will be discussed in turn. The problem has not yet been studied in depth, and hence a definitive explanation is not yet within reach.

Proximate explanations. Male aggressiveness during courtship is commonly interpreted strictly in motivational terms (e.g., Tinbergen 1952; Morris 1956; Etkin 1964; Bastock 1967; Hinde 1970). Unmated males are thought to be initially aggressive toward individuals of both sexes because they are predisposed to defend territories, enforce dominance relationships, or attack prey. Their aggressiveness toward prospective mates is interpreted as an overflow of motivational energy into courtship situations. Male courtship is conceived as being governed by the conflicting drives of attack, withdrawal, and sex, while the female's behavior is interpreted as a means of gradually appeasing the male until his aggressiveness wanes and his sexual drive becomes paramount.

This hypothesis is based on an energy accumulation model of motivation, which was shown earlier to be untenable given present physiological evidence. The different neural control areas involved in each type of aggression and the many feedback circuits affecting an animal's propensity to behave aggressively make a simple overflow of neural activity from one control area to another very unlikely. The overflow hypothesis has little credibility if the switchboard model of motivation is valid.

An alternative hypothesis, based on the switchboard model, is that males initially misidentify unfamiliar females as rival males. Initial sensory input may be misinterpreted by neural control systems, with aggression occurring instead of sexual behavior before a distinction between the contexts of territory defense and courtship can be made. However, continued aggression after several hours or days of courtship certainly cannot be attributed to misidentification of the intruder. By that time males clearly know that the intruder is a prospective mate. Moreover, males are aggressive toward prospective mates even in species where females differ markedly from males in appearance, and it seems very unlikely that males cannot distinguish them from rival males.

Both of the hypotheses above are proximate explanations of intersexual aggression during courtship, and both imply that the behavior has no adaptive significance. Nevertheless, male aggression during courtship may entail some very real fitness costs, since a male's reproductive success often depends on how soon he can become mated or on how many females he can inseminate. Aggression is likely to slow down the courtship process, and it may cause some females not to accept the male as a mate. It is therefore not likely to evolve in courtship contexts unless it provides fitness benefits to offset these costs.

Ultimate explanations. One possible way aggression may be beneficial during courtship stems from the general observation that females are usually selective in choosing their mates (see Chapter 10). A female should increase her fitness by mating with a genetically superior male because her sons would inherit some of the male's superior genes and therefore gain a competitive advantage in the next generation. Hence if a male's reproductive success hinges on his aggressive capabilities, males might be aggressive during courtship to advertise superior competitive abilities in other contexts, and females might prefer mating with aggressive males. Thus aggressiveness during courtship might benefit males by attracting rather than repelling prospective mates.

One prediction of the female choice hypothesis is that males should be less aggressive during courtship when females select mates primarily on the basis of territory quality. Unfortunately, this prediction is difficult to test because the criteria used by females to select mates are not easily determined. Even if females respond to territory quality, they may still respond to male behavior as well.

A second prediction is that females avoid mating with males who are either too aggressive or not aggressive enough during courtship. Evidence for testing this prediction is still rather limited. In one study courtship intensity by polygamous male red-winged blackbirds *(Agelaius phoeniceus)* was positively correlated with density of females nesting on each male's territory (Weatherhead and Robertson 1977). Because the index of courtship intensity was based to a considerable extent on aggressive behaviors, this correlation suggests a female preference for more aggressive males. However, the results must be interpreted carefully. Courtship intensity was negatively correlated with territory size and uncorrelated with *number* of females nesting on the territory (Searcy and Yasukawa 1981; Wittenberger 1981). Hence it may reflect higher aggressiveness among males who possess smaller territories, where competition is most severe, and not a female preference for aggressive males. In a separate study of the same species Yasukawa (1979) found no correlation between intensity of male aggression during courtship and male pairing success.

A third prediction is that males should be less aggressive during courtship if they have already proven superiority over other males. The idea that male aggression during courtship can provide females with information about a male's ability to win fights with other males is questionable. Males who are poor competitors against other males may be just as capable of behaving aggressively toward females as are males who are good competitors. A better criterion for females would be to select mates who have previously won contests with other males. The most superior males should attain higher dominance status or acquire better territories, and females do prefer mating with such males. Why then should males further parade their competitive abilities by attacking or threatening prospective mates, at some risk of causing them to leave? No careful studies have evaluated this prediction, but one gains the impression from the available literature that male aggression during courtship is very prevalent in species where males compete for dominance status or territories of variable quality.

Another hypothesis is also based on female mate selection behavior, but it considers the selection process from a male perspective. An important question for a male when first approached by a female is whether she is physiologically receptive and predisposed to consider him as a mate (Zahavi 1977a). A male can gain a definite competitive advantage over other males if he can avoid wasting time courting females who will not mate with him. By behaving aggressively, a male would increase a female's cost of remaining near him, and this cost should induce her to leave immediately if she is uninterested in him.

Since this hypothesis views aggression as a means used by males to test female intentions, it can readily explain why males are most aggressive during early stages of courtship. A female's persistence in remaining on a territory indicates her intention to mate, and as courtship progresses her intentions become increasingly clear, until finally the male no longer benefits by testing them further.

Two hypothetical examples illustrate how males might benefit from testing female intentions. In territorial birds a female may encroach on a male's territory to exploit resources there, even though she has no intention of remaining to mate. She may, for example, already be mated to another male nearby. If the female is threatened or attacked, the benefits she could derive from such encroachments would not be worth the cost, but if she is not attacked, she would gain free access to resources that would otherwise have been inaccessible to her. By inducing females to leave, males prevent depletion of resources on their territories and avoid wasting time courting unreceptive females. Similarly, in social animals a female may gain access to a desirable food item by making sexual advances toward a dominant male. She might successfully deceive the male and obtain the food item if the male fails to challenge her, but her deception would entail too high a cost if males are initially aggressive and not immediately approachable by sexual advances.

One might predict that males should not be aggressive during courtship when females have no reason to encroach on a male's territory or personal space other than to mate with him. The pairing territories of male gulls, for example, do not contain any resources exploitable by encroaching females. Why, then, should male gulls behave aggressively early in courtship? One possible explanation is that females vary in their readiness to mate. Some females are not yet receptive and enter several territories in succession to gain information about prospective mates. Other females have already completed most of that process and are more nearly prepared to select a mate. By behaving aggressively, territorial males may be able to mate earlier, which in turn may enhance their reproductive success.

Still another hypothesis is based on male mate selection. Trivers (1972) argues that male aggression "may act as a sieve, admitting only those females whose high motivation correlates with early egg laying and high reproductive potential." While males may sometimes be selective in accepting mates, they should be selective only under certain circumstances (Wittenberger 1979a; see Chapter 10). Since males are aggressive during courtship in many species where they should not be selective of mates, notably in polygamous birds and mammals, Trivers's hypothesis fails as a general explanation for intersexual aggression during courtship.

Finally, males may be aggressive during courtship to ensure paternity of the female's offspring (Erickson and Zenone 1978; Zenone et al. 1979). By behaving aggressively, males may delay ovulation in the female long enough for sperm from previous copulations to die. In experimental studies of ring doves (*Streptopelia risoria*), males are far more aggressive during courtship when placed with a female who has recently associated with another male (Erickson and Zenone 1976; Zenone et al. 1979). Females paired with highly aggressive males require about eight days to lay their first eggs (ovulating about 40 hours earlier), while females paired to less aggressive males require about five days to begin laying (Hutchison and Lovari 1976). This delay appears crucial in determining paternity, because sperm remain potent in a female's reproductive tract for about six days. Thus male aggression delays ovulation in the female just long enough to prevent sperm of other males from fertilizing the female's eggs.

The evidence for ring doves strongly supports the paternity assurance hypothesis, but comparable experiments have not been conducted for any other species. Moreover, enhancing paternity cannot be an important selective factor unless females do regularly mate with more than one male in succession. Whether female ring doves do this has not been ascertained. The paternity assurance hypothesis is therefore an exciting possibility, but it cannot be accepted solely on the basis of current evidence.

Parental aggression

A common form of parental aggression in mammals occurs at weaning. When young mammals approach weaning, their mothers become increasingly intolerant of them. Weaning is a relatively prolonged process during which mothers gradually force offspring to forage for themselves by evading or resisting suckling attempts. Mothers may behave aggressively by nipping at their offspring, a behavior that has evolved because the selfish interests of mothers and infants no longer coincide (Trivers 1974). Offspring persist in their efforts to suckle as long as the mother continues lactating, since they improve their survival chances by acquiring the additional food, but at some point these small increments in offspring survival are no longer large enough to make continued nursing advantageous for the mother. That point occurs when mothers can benefit more by diverting the energy used to make milk into body maintenance or new reproductive efforts.

Parents may also use aggression to force offspring out of the social group or territory. Weaned offspring may disperse voluntarily, but they are sometimes forced out by parents or other adults (Photo 5–5). Aggression occurs in this context when the best interests of offspring are served by delaying dispersal, while the best interests of parents are served by forcing dispersal.

Dispersal is generally a period of high mortality. From a young ani-

Photo 5–5

An adult male lion attacking a subadult male while a lioness looks on.
Parental aggression sometimes occurs because subadult males can potentially compete with their fathers for mates.
Photo by Finn Allan/Frank W. Lane.

mal's point of view it should be delayed until a time when chances of surviving dispersal are best, and this time is likely to be when resources are particularly abundant, weather conditions are particularly favorable, or vulnerability to predation is particularly low. However, offspring compete with their parents for resources until they disperse, and older male offspring may even begin competing with adult males for mates. Competition is greatest during periods of resource scarcity and during the mating season. If such periods precede more favorable conditions for dispersal, parents and offspring will disagree over when offspring should disperse. This disagreement may be resolved by parental aggression directed at offspring. Parents should force offspring to disperse when the benefits of avoiding further competition exceed the costs of exposing offspring to higher risks of mortality during dispersal.

A much less common form of parental aggression is deliberate killing of offspring. White storks (*Ciconia ciconia*), for example, occasionally kill young nestlings by throwing them out of their nests or eating them (Schüz 1943, 1957). The nestlings killed by parents are generally small and weak compared to their older siblings. White storks begin incubating eggs as soon as they are laid. Eggs hatch asynchronously, and nestlings differ in age by as much as several days. In the scramble for food the youngest nestlings are often pushed aside by their older and stronger nest mates, with the result that they gradually grow weaker. Finally when they are no longer strong enough to beg for food, they are thrown out of the nest.

One possible benefit of this behavior is that weakened nestlings are removed to sanitize the nest. Weakened nestlings are likely to become diseased or parasitized, which could adversely affect other nestlings in the nest. No data are currently available for evaluating this hypothesis.

A second possible benefit is that parents kill weak nestlings to prevent them from competing with their stronger siblings. It is generally better to produce a few strong, healthy offspring than many weak, sickly ones. By killing nestlings that are likely to die anyway, parents can avoid wasting food on them and can therefore provide more food for nestlings that have a better chance of surviving. If this hypothesis is true, parents should kill nestlings more frequently when they have greater difficulty finding food. Brinckmann (1954) found that nestling storks are most frequently killed in years of food scarcity. Also, most killings are perpetrated by young parents, primarily males, who are less experienced and hence less able to gather food for their young (Schüz 1943). Whether the behavior increases fledging success has not been established, however, so further study is needed.

Haverschmidt (1949) suggests that adults throw nestlings out of the nest because they no longer recognize them as young storks. This hypothesis may be true, as storks habitually throw foreign objects out of their nests, but at best it represents only a proximate cause of the behavior.

Infanticide

Infant killing by adults other than parents occurs regularly in collared lemmings (*Dicrostonyx groenlandicus*) (Mallory and Brooks 1978), Belding's ground squirrels (Sherman 1980), Hanuman langurs (Sugiyama 1967; Yoshiba 1968; Mohnot 1971; Blaffer Hrdy 1974, 1977a, b; Makwana 1979), purple-faced langurs (*Presbytis senex*) (Rudran 1973), and lions (Rudnai 1973; Bertram 1975a, b, 1976). It also occurs, though less frequently, in a wide variety of primates other than langurs (Angst and Thommen 1977; Blaffer Hrdy 1977b; Goodall 1977; Struhsaker 1977).

When infanticide was first discovered among Hanuman langurs, it was thought to be a pathological response to abnormally high population densities (Sugiyama 1967). Infanticide occurred regularly in Sugiyama's study area, where population density was high, but not in an area studied by Jay (1965), where population densities were much lower. Some primatologists still interpret infanticide as pathological behavior (e.g., Dolhinow 1977), but there is now reason to believe that the behavior is adaptive.

In most cases infants are killed exclusively by males who have recently taken possession of female groups. By killing infants fathered by previous males, a new male can induce females in the group to come into estrus earlier than they would have had their infants been spared (Blaffer Hrdy 1974, 1977a, b; Bertram 1975a, b). Langur mothers, for example, give birth only once every two to three years if their infants survive, but they give birth again within a year if their infants are killed (Angst and Thommen

1977). Lionesses give birth every 24 months if their cubs survive, but they give birth within 9 months after their last cub dies (Bertram 1975b). Thus killing infants can greatly reduce the time a male must wait before inseminating the females. This time saving can increase a male's reproductive output substantially, since male tenure in female groups ordinarily lasts only a few years.

According to a recent theoretical model developed from langur studies, the advantages of infant killing depend on the average length of male tenure in female groups (Chapman and Hausfater 1979). The model also predicts that population density should not affect the adaptive value of infanticide. The reason infanticide has been observed less often at low population densities is probably that male takeovers occur much less frequently, making infanticide less frequent as well (see Rudran 1973; Blaffer Hrdy 1977c). Data for testing the model are not yet available, so the model remains conjectural.

A second potential advantage of infant killing, at least for males, is that females in the group all enter estrus synchronously. In lions synchronous births facilitate communal nursing and baby-sitting, and young cubs would not have to compete with older cubs for food (Bertram 1975a, b, 1976). Cubs born synchronously do survive better than cubs born asynchronously, but the precise reasons are unknown.

In Belding's ground squirrels young, prewintering juveniles are killed by nonresident females and one-year-old males (Sherman 1980). Resident males and females do not kill juveniles, which are generally closely related to them. Infanticide by females is not accompanied by cannibalism and may have evolved because it enables females to preempt burrow systems dug by juveniles. Infanticide by males is accompanied by cannibalism and may be a means of increasing fat reserves prior to the onset of winter.

Infanticide reduces the fitness of parents, of course, and parents should attempt to protect their offspring. Female collared lemmings are highly aggressive toward unfamiliar males and often succeed in preventing infanticidal attacks (Mallory and Brooks 1978). Female langurs sometimes try to thwart male attacks on infants by carrying them away, shielding them, or trying to retrieve them from the male's grasp, but their attempts usually fail (Mohnot 1971; Blaffer Hrdy 1977a). Following the infant's death the mother shows many outward signs of distress and is antagonistic toward the male. Nevertheless, her antagonism wanes as she comes into estrus, and she accepts the new male's sexual advances because he represents the best reproductive option remaining available to her.

Infanticide is not restricted to mammals. Both males and females of many colonial gulls and some colonial terns and pelicans attack and kill chicks (Pettingill 1939; Sprunt 1948; Harris 1964; Parsons 1971; Hunt and Hunt 1976). They primarily kill chicks of neighboring pairs who have strayed onto their territories, but gulls do occasionally kill their own chicks.

Chick killing in gulls is generally associated with cannibalism (Photo 5–6) (Harris 1964; Parsons 1971). Unprotected chicks are easy prey and can be captured without flying considerable distances from the colony. Chicks

Photo 5–6

A herring gull cannibalizing a chick that strayed onto its territory. *Cannibalism is a frequent cause of chick mortality in gull colonies. Disturbances caused by intruders (predators or humans) increase the incidence of cannibalism because chicks scatter when frightened.*

Photo from J. Parsons, "Cannibalism in Herring Gulls," *British Birds* 64 (1971):533.

are cannibalized more often in colonies with high nest densities, probably because they are more likely to stray onto neighboring territories when nests are close together (Brown 1967a; Haycock and Threlfall 1975; Hunt and Hunt 1976). The higher incidence of chick killing in dense colonies apparently does not result directly from high competition for food around the colony (Brown 1967a). Food resources away from the colony are energetically more costly to exploit than unprotected chicks, but they are less costly to exploit than chicks protected by their parents. Brown (1967a) argues that chicks are simply easier to protect in less dense colonies, making them less available as prey for other colony members. Of course, more intense competition for food around dense colonies would reduce the time parents could spend protecting chicks.

Peer aggression

The most destructive form of peer aggression is *fratricide*, the killing of siblings. Fratricide regularly occurs among nestling boobies, gannets, skuas, hawks, owls, pelicans, and cranes (Ingram 1959; J. B. Nelson 1964, 1978; Miller 1973; Procter 1975; Johnson and Sloan 1978; O'Connor 1978; Stinson 1979, 1980). Nestlings of these species hatch asynchronously, because incubation begins soon after the first egg is laid, and hence they may differ in age by as much as several days. Asynchronous hatching is thought to be a hedge against uncertain food supplies (Lack 1954). During periods of scarcity the youngest nestlings succumb before they consume much food, and consequently most food goes to nourish nestlings that have the best chance of surviving.

Older nestlings should benefit most from fratricide during periods of food scarcity, but little evidence is currently available for evaluating this hypothesis. In one study Procter (1975) found that fighting among chicks of the South Polar skua (*Catharacta maccormicki*) is elicited primarily by hunger. In another study Ingram (1959, 1962) found that adult short-eared owls (*Asio flammeus*) stockpile extra food alongside the nest where older nestlings can reach it, perhaps to reduce the chances of older nestlings attacking younger ones during temporary periods of hunger. On the other hand, older chicks of hawks and eagles often harass or kill their younger siblings regardless of prey availability (Ingram 1959; Meyburg 1974). One benefit of killing younger siblings is reduced competition for food. By killing young siblings immediately, the older nestlings can obtain all the food instead of just part of it. In hawks, owls, and eagles a second benefit is direct acquisition of food, since fratricide is accompanied by cannibalism. The principal cost is reduced inclusive fitness caused by the death of relatives. A mathematical model based on kin selection theory confirms that increased starvation mortality should favor fratricide (O'Connor 1978). The model also predicts that the alternative option, food sharing coupled with slower development rates, is disadvantageous whenever predation risk is relatively high.

Another form of fratricide involves cannibalism of larvae by workers of social insects. Cannibalism prevents wastage of nutrients and reallocates nutrients to more appropriate use. Ants normally eat all injured eggs and larvae, and they also attack healthy larvae when they are starving (Wilson 1971). Similarly, termites become cannibalistic when experimentally deprived of normal protein sources (Cook and Scott 1933). During food scarcity colonies are hard-pressed to maintain their size, and cannibalism allows workers to stay alive while also preventing emergence of new workers that could not be supported by the available food supply anyway.

The second role of cannibalism in social insects is reallocating nutrients from production of workers to production of reproductives, or vice versa. During the sexual phase army ant workers (*Eciton*) consume most female larvae and convert the protein into hundreds or thousands of male reproductives (Schneirla 1971). Relatively few female larvae are spared, and those develop into very large virgin queens. Conversely, when winged termite reproductives are prevented from dispersing, they are soon pulled apart and eaten by the workers (Lüscher 1952; Ratcliffe et al. 1952). This enables conversion of energy and nutrients initially allocated to reproduction into colony growth and maintenance.

A second type of peer aggression is aggressive play, which is especially prevalent in young mammals (Aldis 1975). Play is widely believed to allow practice of behavior patterns later affecting survival and reproductive success (Dolhinow and Bishop 1970; Bekoff 1974; Fagan 1974; Poirier and Smith 1974; Poirier et al. 1978; E. O. Smith 1978), although some authors question this interpretation (e.g., Welker 1971; Symons 1974, 1978a, b). Aggressive play is much more prevalent among young males than among young females in primates but not in social canids (Poirier 1970, 1972; Lindburg 1971; Baldwin and Baldwin 1974; Bekoff 1974; Poirier and Smith 1974; Moyer 1976). This reflects a parallel difference among adults. Male primates rely heavily on intrasexual aggression to acquire mates and resources, while female primates do not. In contrast, male and female social canids both rely on intrasexual aggression to protect their reproductive interests. Social canids are monogamous, and pair bonds are maintained within social groups by aggressive behavior of dominants directed toward like-sexed subordinates (Wittenberger and Tilson 1980).

A third form of sibling rivalry has clear adaptive significance and involves dominance interactions among offspring still dependent on their mothers. Young piglets compete intensely for teat positions during the first hour after birth by struggling among themselves and scratching each other with their sharp neonatal tusks and incisors (McBride 1963; Hafez and Signoret 1969). Once a teat order is established, it remains intact until weaning. Piglets cannot be conditioned to a new teat order by forcing them to suckle new teats on tranquilized sows, and they defend the teats they claimed when other piglets try to usurp them. The most dominant piglets claim the anterior teats, which produce the most milk. Piglets on the three most anterior teats receive about 84% more milk than piglets on the three most posterior teats, a difference with obvious survival value (Gill and Thompson 1956). A similar teat order exists among kittens, where the preference is for the most posterior teats (Ewer 1959). Teat orders could not be found among infant dogs (Rheingold 1963), grey meerkats (*Suricatta suricatta)* (Ewer 1963a), African giant rats (*Cricetomys gambianus*) (Ewer 1967), or tree shrews (*Tupaia belangeri*) (Martin 1968).

Defensive aggression

Aggressive defense has obvious survival value for a cornered animal lacking alternative modes of defense or escape. A cornered animal must either defend itself or die, and the potential benefit of escaping a predator or aggressor is always worth the risk of being injured. In such situations the cornered animal is normally weaker than its opponent, and its best chance for escape is to adopt a defensive posture rather than attack offensively. Defensive aggression may also have survival value for large prey animals capable of inflicting harm on a potential predator. Large ungulates, for example, are likely to fight back against wolves or other carnivores and they often injure or kill their assailants (e.g., Rausch 1967). Aggressive defense is just one of several alternatives open to a prey animal, and it should not be used unless all less risky alternatives have failed.

Adult animals are sometimes aggressive to protect offspring (Photo 5–7). Colonial birds and some social mammals, for example, mob predators that may attack their young (see pp. 115–120). Even parents of noncolonial birds often swoop down on predators hunting near their nests. The aggressiveness of female mammals, such as bears, who have young nearby is well known.

Photo 5–7 *A wildebeest cow defending its calf, unsuccessfully, from a spotted hyena.*

Large antelopes often aggressively defend offspring from predators when not greatly endangered themselves.

Aggressively defending offspring is not necessarily adaptive. Some predators are more difficult to fend off than others, and some represent a greater threat to the parent's safety than others. Aggressive defense should not evolve unless the chances of thwarting the predator are high enough to offset the risks, in terms of inclusive fitness. Thus animals may never attack predators who threaten their offspring, or they may attack some predators but not others.

Redirected aggression

Aggression redirected toward weaker individuals or inanimate objects often occurs when an animal is highly motivated to attack a foe but is inhibited from doing so by the foe's strength or superiority. Redirected aggression has usually been interpreted by ethologists as a discharge of "aggressive energy" onto a target that cannot retaliate. Its only adaptive significance, according to this interpretation, is that it reduces the animal's likelihood of making a foolhardy attack against a superior opponent. An alternative interpretation, based on the Popp and Devore (1979) model presented earlier, is that redirected aggression evolved as a threat signal (see p. 184).

Costs of aggression

One obvious cost of aggression is that all participants in a physical fight risk being injured. The risks of injury are by no means negligible. A major cause of adult mortality in lions is fighting over recently killed carcasses (Schaller 1972). In many mammals the damaging effects of previous fights are evident in the presence of old wounds. Healed bone fractures, torn ears, and old scars are commonly observed in species as diverse as gibbons, bears, seals, and whales (e.g., Carpenter 1940; Matthews 1964; Cloudsley-Thompson 1965; Bartholomew 1967; Norris 1967; Geist 1977). In mule deer an estimated 5% to 10% of adult males show signs of injury each year (Geist 1974a), while in red deer (*Cervus elephas*) about 23% are injured annually, with 6% showing long-lasting effects (Clutton-Brock et al. 1979). In muskoxen about 10% of adult males die each year as a direct result of intermale aggression during the rut (Wilkinson and Shank 1977). The clashes of bull elk during fights over females occasionally result in a locking of antlers, with both bulls dying because they cannot free themselves.

The risk of being injured is an important consideration in any fight. An individual faced with a clearly superior opponent does better by accepting subordinate status or going elsewhere to seek success rather than forcing a showdown (Tinbergen 1968). Even superior individuals are at risk, because an inferior opponent can inflict injuries before being defeated. For this reason superior animals should rely on threats as much as possible to intimidate weaker rivals.

A second cost of aggression is the time and energy expended during disputes. Aggression reduces the time available for vigilance, thereby increasing exposure of combatants to surprise attacks by predators. It also reduces the time and energy available for finding food, seeking mates, rearing young, and conducting other activities. **Aggressive neglect**, the neglect of offspring while fighting with competitors (after Hutchinson and MacArthur 1959), has been documented or inferred in gannets (*Sula bassana*) (J. B. Nelson 1964, 1965), sunbirds (*Nectarina*) and honeycreepers (*Myzomela*) (Ripley 1959, 1961), and red-winged starlings (*Onychognathus morio*) (Rowan 1966). Neglect of other activities may be equally or more important, but its effect on reproductive success is not easily documented.

Aggression is widespread among animals, but it is not universal or unrestrained. It entails very real costs that discourage wholesale violence. Because all combatants risk injury and expend time and energy during a fight, animals do not fight when they have no hope of winning. Aggression does not occur in every competitive situation, and it should not evolve unless the expected benefits exceed the costs.

Who wins?

Just as military tactics are important in determining who wins a battle, behavioral tactics are important in determining who wins a fight. Other factors being equal, the animal with the best tactics should win. Of course, other factors are not always equal, and an adequate theory should include such factors in addition to behavioral tactics.

One approach to the problem is based on the Popp and Devore (1979) model presented earlier. Recall that aggression is adaptive only when costs are less than expected benefits. Animals should never exceed their maximum adaptive expenditure (C_{max}), and when they reach that expenditure, they should concede defeat. This conclusion suggests that individuals could improve their chances of winning a fight by speeding up the rate at which their opponents incur costs. Thus the ability of each individual to inflict costs on an opponent should be an important determinant of success. This ability will be referred to as the individual's *intrinsic competitive ability* (K).

Now consider a simple aggressive interaction involving two individuals, male a and male b. The time it takes male a to reach his maximum adaptive expenditure (T_a) is determined by his maximum adaptive expenditure ($C_{a,\mathrm{max}}$) and by the rate at which his opponent inflicts costs of aggression upon him (K_b). Thus male a will reach his maximum adaptive expenditure at

$$T_a = \frac{C_{a,\mathrm{max}}}{K_b} \qquad (5\text{–}4)$$

Similarly, male b will reach his maximum adaptive expenditure at

$$T_b = \frac{C_{b,\mathrm{max}}}{K_a} \qquad (5\text{–}5)$$

Male a will win the encounter if male b reaches his maximum adaptive expenditure first. That is, male a wins if

$$T_a > T_b \tag{5–6}$$

Substituting for T_a and T_b, male a wins if

$$\frac{C_{a,\max}}{K_b} > \frac{C_{b,\max}}{K_a}$$

$$C_{a,\max}K_a > C_{b,\max}K_b \tag{5–7}$$

Equation 5–7 predicts that the male with the highest value of $C_{\max}/K$ should win an aggressive interaction. A higher value of $C_{\max}/K$ can come about in two ways. If the maximum adaptive expenditures are approximately equal for both males, the male with the greatest competitive ability should win. If the maximum adaptive expenditure of one male exceeds that of the other, an individual with inferior competitive ability could win (see Maynard Smith and Parker 1976).

There is one problem with the model above. Males should theoretically continue a fight as long as their expected *future* benefits continue to exceed their expected *future* costs. The costs already incurred should not be a deciding factor, except to the extent that they predict what a male's future costs and chances of winning are likely to be. As a general rule, decisions are made according to future prospects, not past levels of investment (Maynard Smith 1977). The model above predicts that one male should give up when the costs *already* incurred exceed expected future benefits. It would be more appropriate if it were reformulated to compare expected future costs to expected future benefits, but this has not been done. Note that similar confusion has cropped up in mate desertion theory. The rationale for prospective rather than retrospective decision making is identical in both contexts (see pp. 417–418 for further discussion).

Tactics

The objective in an aggressive interaction is to obtain the largest possible net gain in fitness. Since benefits to be gained by winning remain unchanged during a dispute, fitness gain is maximized when costs of winning are minimized (Figure 5–5). That animals do try to minimize costs is supported by a study of fighting tactics among female iguanas (*Iguana iguana*) (Rand and Rand 1976). The females fight for ownership of nesting burrows, and disputes are mainly over partially excavated burrows, which require less energy to complete. The principal costs and benefits of aggression are energetic, as female energy reserves are severely limited, and disputes are settled with minimum energy expenditures. The evidence therefore implies that fitness costs are minimized.

Figure 5–5

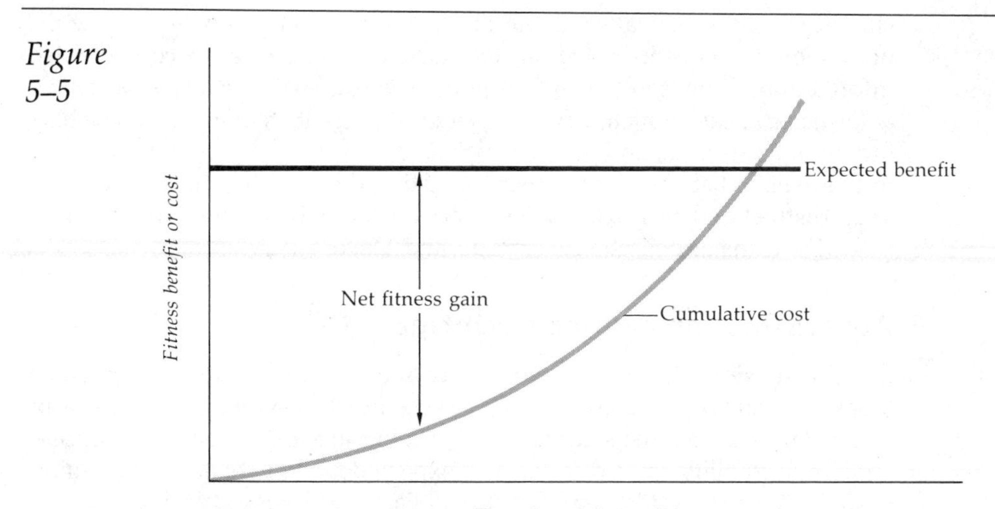

A model of fitness gain during aggressive interactions.
Net fitness gain (expected benefit minus cumulative cost) decreases as duration of aggressive interaction increases. Since expected benefit is fixed, net fitness gain is maximized when cumulative cost (determined by steepness of the cost curve and time elapsed) is minimized.

The loser's strategy should be to minimize costs, but not at the expense of giving up when there is still a reasonable chance of winning. Once a combatant ascertains that the chances of winning are too low to justify further expense, it should give up immediately. A decisive advantage may become evident at any point during a dispute. An opponent may be clearly superior at the outset, in which case it will not be challenged, or a combatant may establish superiority at some stage during the dispute, at which point the other combatant should concede defeat. Thus each participant in a dispute should advertise whatever tactical advantages it might possess over the other, while concealing any weaknesses it might have that could encourage an opponent to initiate or persist in a dispute.

Persuasion versus coordination signals

The classical view of communication behavior assumes that communication has evolved as a cooperative exchange of information (e.g., Morris 1956; Marler 1959, 1961; Klopfer and Hatch 1968; W. J. Smith 1968, 1977). Since communication has been assumed to be mutually advantageous for both signalers and recipients, selection has been assumed to favor accuracy and unambiguity in the messages conveyed (see Wilson 1975a).

Although cooperative interactions do confer mutual benefits, competitive interactions do not. In competitive situations signaling individuals should use communication to gain an advantage. The signaler may withhold some types of information, emphasize others, or even convey false information. Hamilton (1973) first pointed out the distinction between *coordination displays*, which serve to coordinate activities of individuals sharing mutual interests, and *persuasion displays*, which serve to persuade other individuals to behave in a manner benefiting the signaling individual. The same distinction has also been made more recently by several other authors (e.g., Dawkins and Krebs 1978; Popp and Devore 1979).

Advertising competitive advantages

Persuasion signals are used during agonistic interactions to advertise a basic asymmetry in a contest. The two kinds of asymmetries capable of reducing an individual's chances of winning are a difference in intrinsic competitive ability and a difference in expected net benefit at the beginning of or during a fight (Maynard Smith and Parker 1976; Rand and Rand 1976; Popp and Devore 1979). Threat displays are used to advertise competitive ability, while other signals are used to advertise differences in net expected benefits.

A show of superior competitive ability is often sufficient to convince an inferior opponent that it cannot win. An individual possessing larger body mass, horn size, teeth, or other physical attributes indicative of superior fighting ability can often win without a costly fight simply by displaying its strength or weaponry.

The threat displays of animals do just that. A few examples suffice to illustrate the point. Male fiddler crabs (*Uca*) threaten opponents by conspicuously waving their greatly enlarged chelae, or pincers (Hazlett and Bossert 1965; Crane 1967, 1975). Fish threaten opponents by lateral displays that present the maximum body surface toward rivals. Lizards threaten by puffing or engorging their throats, erecting crests, or extending their forelegs to raise the body, all of which increase the surface area visible to an opponent (Photo 5–8) (Carpenter 1977). Nearly all birds display weaponry by extending the head and beak forward toward an opponent, and they increase apparent body size by spreading their wings and fanning their tails (Andrew 1961). Mammals threaten by baring their teeth or displaying their horns or antlers. Such threats are often accompanied by specialized movements and bright colors that make them more conspicuous, and they are given repeatedly to make them more effective (see Dawkins and Krebs 1978). Thus as a general rule, threats enhance apparent fighting ability.

Other signals may be used to advertise an asymmetry in net expected benefits. For example, in breeding groups of Arabian babblers, dominant birds give food to other group members, a behavior that would normally be regarded as altruistic (Zahavi 1977a; Gaston 1978). However, the subordinates reject such food offerings. Zahavi argues that the dominant bird is advertising that he is strong enough to afford giving food away. This view is highly controversial, as there are theoretical problems associated with the idea of accepting handicaps to advertise strength or superiority

Photo 5–8

*The threat posture of a perenty (*Varanus giganteus*), a large monitor lizard of central and western Australia.*

Threats are directed at both conspecifics and potential predators. Note the extended legs, arched back, and raised tail, all of which increase apparent size and make defensive capabilities more visible. The perenty shown here was unusually large, measuring nearly 8 ft (2.4 m) in length.

Photo by Axel Poignant.

(see pp. 432–435), but the rejection behavior of subordinates is difficult to explain any other way.

Established territory owners nearly always win disputes, even against equally matched or perhaps superior opponents. One hypothesis for their consistent success is that residents stand to gain more by winning than do intruders (Dawkins and Krebs 1978). Because territory residents possess more information about safe hiding places and good foraging sites, they may be better able to utilize the territory. Hence their expected benefits are higher, at least in the short run. Of course, an intruder could eventually obtain the same familiarity, but not without first taking some risks. The advantage of prior occupancy is often gained almost immediately upon taking possession of a territory, however, implying that having better information per se is not the reason why territory owners usually win disputes. In those cases prior occupancy is apparently used as a proximate cue for advertising asymmetrical benefits. This is the case in fish and some birds (see Phillips 1971; De Boer and Heuts 1973; Wiens 1973; Krebs 1977).

Exaggeration and deception

Advertisements are more convincing when the selling points are exaggerated. One might therefore expect animals to exaggerate their apparent intrinsic competitive ability and the costs they are willing to incur to win a dispute.

Many threat displays exaggerate apparent body size. The manes of lions and hamadryas baboons *(Papio hamadryas)* (see Photo 4–7) and the facial hair patterns of monkeys have probably evolved to accentuate the size of the head during frontal displays (Hingston 1933; Guthrie 1970; Crook 1972; Schaller 1972). Many mammals also increase apparent body size by erecting hairs along the neck and spine. Many birds threaten by spreading the wings, fanning the tail, and ruffling the body plumage (Photo 5–9), all of which increase apparent body size (Andrew 1961). The same effect is achieved in fish by extending and spreading the fins.

Certain behaviors may have evolved to exaggerate the costs an animal appears willing to suffer in a dispute. Redirected aggression, for example, may exaggerate an animal's willingness to physically assault a rival. Many fish and birds display real or sham nest-building, nest-tending, or incubation behavior during agonistic interactions (e.g., Armstrong 1947; Tinbergen and Iersel 1947; Heiligenberg 1965). Classical ethological theory refers to such seemingly irrelevant displays as *displacement activities* and interprets them as consequences of an overflow of excessive motivational energy into nonadaptive outlets. However, they may have evolved to exaggerate the special benefits a territory owner can derive by successfully expelling intruders (Popp and Devore 1979). When territory residents have

Photo 5–9

A male yellow-headed blackbird Xanthocephalus xanthocephalus) *performing an asymmetrical song spread.* *This threat posture is directed at males perched on neighboring territories or flying overhead. Note the fanned tail and slightly spread wings, both of which increase the male's apparent body size.*

Photo by author.

eggs or nestlings on their territories, they should be willing to incur greater costs than an intruder to win a dispute because they are defending reproductive investments as well as resources. Although displacement activities are often given when nests or eggs are not yet present, they serve as a reminder of how important the territory is to its owner. They may also exaggerate the territory's true value by implying that the territory owner is already defending eggs or offspring. Similarly, sham feeding behavior, another common displacement activity, may emphasize a resident male's willingness to defend his feeding area.

Exaggerations and deceptions are initially favored by selection because they inflate an animal's competitive ability, but they are convincing only as long as the sham goes undetected. Selection should favor ever greater exaggerations and deceptions until the associated costs begin to exceed the benefits. Exaggerations often require morphological changes such as longer hair, broader wings, longer tails, larger fins, or brighter colors. Such changes are likely to hamper other activities, thereby reducing fitness in other contexts. For example, male great-tailed grackles possess longer tails than females, for use in threat displays, but they also suffer greater mortality over winter (Selander and Giller 1961). Though not actually proven, their longer tails perhaps hamper flying ability or in some other way make them less able to withstand the rigors of winter. Thus counterbalancing selective pressures limit the extent to which structures or behaviors can be exaggerated, so that an equilibrium condition is reached in a population.

Exaggerated threats and other deceptions are only likely to fool an opponent when used sparingly. As more and more males employ them, more and more males gain the ability to detect deceptions and devalue exaggerations (Wallace 1973; Parker 1974a). Eventually exaggerations and deceptions lose their effectiveness and no longer provide any advantage. Nevertheless, they persist in populations. One possible reason for this persistence may be males unconsciously devalue the threats of every rival by the same amount, regardless of whether a display is exaggerated or not. If unexaggerated threats are devalued as much as exaggerated ones, any male who fails to give exaggerated threats will be perceived as an inferior opponent. The same argument could also explain the persistent use of displacement activities and other deceptions. Thus exaggerated threats may cost more than they are worth, if rivals devalue them, but must be given anyway to inflate the rival's devalued perception of the threat.

Why not overkill?

Animal conflicts are often settled without using lethal weapons or inflicting serious injury. Consequently, self-restraint is widely held to be an altruistic trait that evolved for the good of the species (e.g., Eibl-Eibesfeldt 1961; Lorenz 1966). However, the selective processes capable of favoring altruism do not seem applicable to the evolution of self-restraint. Self-restraint occurs in disputes regardless of whether contestants are relatives or nonrelatives. Opportunities for reciprocation are minimal because the ability to inflict injury is generally one-sided. Interdemic selection cannot explain the general prevalence of self-restraint because at best it can work only under severe constraints. Trait-group selection could conceivably be involved, but its role cannot be evaluated without further evidence. In addition, altruism should not be assumed until the alternatives have been ruled out. Thus a first step in analyzing self-restraint is to ask who benefits from it.

The objective in an aggressive interaction is to win with maximum fitness gain. As noted earlier, this objective is achieved by minimizing the costs to the victor, not by maximizing the costs to the loser. A victor does not gain any additional benefits by continuing to attack an opponent who has conceded defeat, but he does pay additional costs. Any attempt to inflict bodily harm, either during or after a dispute, is likely to be met with resistance. When bodily harm is imminent, the benefits of defensive aggression are greatly elevated and an inferior opponent is likely to strike back in desperation. While this is not likely to change the outcome of a fight, it is likely to increase the costs for the ultimate victor. Self-restraint is therefore advantageous for the victor because it minimizes his costs of aggression.

Self-restraint has also evolved in aggressive interactions for a second reason. Inflicting superficial injuries is more effective for winning a fight than inflicting deep wounds of a more serious nature would be (Geist 1966, 1977). Superficial injuries are more painful than deep wounds, at least in humans, and hence they should be more persuasive in convincing an opponent to give up the fight. Moreover, inflicting deep wounds may be more risky to the aggressor than inflicting superficial wounds. Long horns thrust deep into an opponent are more difficult to withdraw, and violent struggles by a victim may break the aggressor's neck or skull. Geist therefore suggests that weaponry evolved to maximize trauma, not injury. Trauma may be inflicted by causing painful bruises or lacerations or by making an opponent lose control of its body. For example, the blunt horns of giraffes are used as clubs, while the claws of cats are used to rake an opponent's flesh. Many ungulates lock horns and engage in wrestling contests to throw the opponent off balance. The many points, lumps, twists, prongs, and surface textures of ungulate horns and antlers serve primarily as defensive devices to prevent opponents from inflicting serious injury, and they have relatively little value as offensive weaponry (Geist 1966). Indeed, the evolutionary progression in ungulates has been from taxonomically primitive species that inflict lacerations or puncture wounds to taxonomically advanced species that rely on wrestling and pushing contests to resolve disputes. Thus weaponry is often designed for self-defense and wrestling or pushing contests, not for inflicting death or injury on opponents.

Submission

Submissive displays evolved as a way of avoiding further costs of aggression once an individual has conceded defeat. They have obvious benefits for a loser, and they take advantage of a victor's willingness to desist once victory is attained. The loser often avoids further attacks by fleeing the situation, but when flight is unfeasible, submissive postures are used instead.

Submissive displays are not used solely to concede defeat. They may also signal awareness of inferior status, thereby allowing a subordinate to approach a dominant more closely without evoking an aggressive response. Females use them during courtship to reduce the initial aggression of courting males, and juveniles use them to beg for food. Finally, combatants may use them to de-escalate aggression during a fight to gain some temporary advantage.

Submission is often expressed by adopting a posture that is just the opposite of a threat posture. It is the behavioral antithesis of aggression. Darwin's (1871) description of threat and submission in dogs well exemplifies this **principle of antithesis** (See Figure 5–6.):

> *When a dog approaches a strange dog or man in a savage or hostile frame of mind he walks upright and very stiffly; his head is slightly raised, or not much lowered; the tail is held erect and quite rigid; the pricked ears are directed forwards, and the eyes have a fixed stare. These actions, as will hereafter be explained, follow from the dog's intention to attack his enemy, and are thus to a large extent intelligible. As he prepares to spring with a savage growl on his enemy, the canine teeth are uncovered, and the ears are pressed close backwards on the head; but with these latter actions we are not here concerned. Let us now suppose that the dog suddenly discovers that the man he is approaching is not a stranger, but his master; and let it be observed how completely and instantaneously his whole bearing is reversed. Instead of walking upright, the body sinks downwards or even crouches, and is thrown into flexuous movements; his tail, instead of being stiff and upright, is lowered and wagged from side to side; his hair instantly becomes smooth; his ears are depressed and drawn backwards, but not closely to the head; and his lips hang loosely. From the drawing back of the ears, the eyelids become elongated, and the eyes no longer appear round and staring.*

Not all submissive postures are based on the principle of antithesis. Submissive animals may appease an aggressive one by eliciting behaviors incompatible with aggression. For example, begging behavior may be used in aggressive contexts to evoke courtship feeding or parental responses, and sexual enticements may be used to evoke sexual responses. By reorienting the attention of an assailant, a submissive animal can sometimes prevent further attacks.

Attack strategies

Animals have many behavioral options open to them during an aggressive encounter. They can threaten an opponent at varying intensities, they can perform various displacement activities, they can attack, or they can retreat. These options should be employed in a way that best enhances an individual's chances of winning the dispute. Exactly how animals do

Figure 5–6

The threat and submissive postures of a dog.
Submission in dogs is the postural antithesis of threat. The upper figure in this 1872 drawing by Darwin shows a threat posture, while the lower figure shows a submissive posture.

this is difficult to analyze. One approach to understanding the problem is to devise computer simulation games that mimic simple aggressive interactions.

Consider, for example, a simple simulation involving two basic offensive tactics (Maynard Smith and Price 1973). "Conventional tactics" involve pushing or wrestling contests and are not likely to cause serious injury. "Dangerous tactics" involve direct assaults or vicious attacks and are likely to cause serious injury. These tactics can be employed, along with retreat, in a variety of ways. The simulation game is constructed by arbitrarily defining the set of rules determining when each tactic is employed by a given combatant. The set of rules for each contestant is its **strategy.**

Maynard Smith and Price (1973) chose to analyze five strategies in their simulation game:

1. "Mouse" never uses dangerous tactics. It retreats when dangerous tactics are used against it but otherwise continues fighting with conventional tactics until the contest has lasted a preassigned number of moves.

2. "Hawk" always uses dangerous tactics. It fights until it suffers serious injury or its opponent retreats. (Note that "hawk" and "mouse" refer here to strategies employed by conspecifics, not to predators or prey.)

3. "Bully" uses dangerous tactics if it makes the first move or if its opponent uses conventional tactics, but it uses conventional tactics if its opponent uses dangerous tactics. If faced with dangerous tactics twice in a row, bully retreats.

4. "Retaliator" uses conventional tactics if it makes the first move or if its opponent uses conventional tactics, but it uses dangerous tactics when its opponent uses dangerous tactics. It retreats if the contest lasts a preassigned number of moves.

5. "Prober-retaliator" usually uses conventional tactics if it makes the first move or if its opponent uses conventional tactics, but occasionally it uses dangerous tactics instead. If the opponent then responds with dangerous tactics, it reverts to conventional tactics. If the opponent responds with conventional tactics, it continues using dangerous tactics. "Prober-retaliator" usually responds to probes (i.e., responds to dangerous tactics of an opponent) with dangerous tactics, but sometimes it responds with conventional tactics.

At the end of each simulated contest, every combatant receives a payoff reflecting the net benefit gained by winning, the costs suffered by being injured, and the costs entailed by expending time and energy on aggression. The costs were predicated on the type of strategy employed. The chances of winning were determined by simulating the contest on a computer. Individuals adopting any given strategy were assumed to have equal fighting ability.

The question is, Which strategy should evolve in a population where individuals can adopt one or another of the strategies above? To answer this question, Maynard Smith and Price (1973) defined an **evolutionarily stable strategy,** or **ESS,** as a strategy that, if adopted by most members of a population, leads to higher fitness payoffs than does any other strategy a rare mutant individual might adopt against it. A payoff matrix for the simulation game shows how the ESS criterion works.

In a population consisting mostly of "hawks," an individual using any given strategy will almost always be fighting a "hawk." The payoff for a "hawk" fighting a "hawk" is −19.5, which is less than the payoff for a "mouse" fighting a "hawk" or a "bully" fighting a "hawk" (Table 5–3). That is, a rare mutant adopting the "mouse" or "bully" strategy would

Table 5–3 *Fitness payoffs for various strategies in a simulation game of aggressive interactions.*

Contestant receiving the payoff	Opponent				
	"Mouse"	"Hawk"	"Bully"	"Retaliator"	"Prober-retaliator"
"Mouse"	29.0	19.5	19.5	29.0	17.2
"Hawk"	80.0	−19.5	74.6	−18.1	−18.9
"Bully"	80.0	4.9	41.5	11.9	11.2
"Retaliator"	29.0	−22.3	57.1	29.0	23.1
"Prober-retaliator"	56.7	−20.1	59.4	26.9	21.9

Note: Fitness payoffs depend on the strategy adopted by the opponent. See text for definitions of terms.

Source: From J. Maynard Smith and G. R. Price, "The Logic of Animal Conflict," *Nature* 246 (1973):16. Reprinted by permission.

have higher relative fitness than a "hawk" strategist. Hence the "mouse" and "bully" strategists would become more prevalent in the population, while the "hawk" strategists would become less prevalent. Similarly, the "mouse" strategy is not an ESS because "hawks," "bullies," and "prober-retaliators" all average higher fitness payoffs when most opponents are "mice," and the "bully" strategy is not an ESS because "hawks," "retaliators," and "prober-retaliators" average higher fitness payoffs when most opponents are "bullies." In contrast, the "retaliator" strategy is an ESS because no strategy offers higher payoffs when most opponents are "retaliators." The "prober-retaliator" strategy is almost an ESS because only "retaliators" average higher payoffs when most opponents are "prober-retaliators."

For this simple situation the most successful strategy is conventional tactics mixed with dangerous tactics. The simulation shows that continued attempts to inflict injury, the "hawk" strategy, is not necessarily the best way to win an encounter. Thus self-restraint and conventional tactics can evolve as selfish traits.

Conflicts among real animals generally involve more complex strategies than those simulated by Maynard Smith and Price (1973). In a recent study of territorial spiders, for example, Riechert (1978) found that a female's strategy varies according to her competitive ability and her energetic investment in the contended resource. Most females use a "retaliator" strategy, but females with a weight advantage use a "hawk" strategy instead. Unusually small females use the "mouse" strategy if they are visiting another female's web but use the "retaliator" strategy when defending their own webs against the "hawk" strategy.

Elaborate models have been devised to simulate contests where the benefits or competitive abilities of combatants are asymmetric (e.g., Maynard Smith 1974a; Parker 1974a; Maynard Smith and Parker 1976). These models will become very useful once they are refined sufficiently to predict outcomes of real contests. Some first steps in that direction have already been achieved by Norman and co-workers (1977) and Caryl (1979).

ESS as a general approach

The concept of evolutionarily stable strategies should prove useful in constructing models of many evolutionary processes because it provides a general criterion for determining which of several alternative strategies should become fixed in a population. The approach has already been used with some success in evaluating the evolutionary origins of sex (Maynard Smith 1976a, 1978), some mating strategies (Charnov 1979), and parental strategies (MacNair and Parker 1979; Parker and MacNair 1979), among others. The ESS approach is a more formal way of describing selective processes on an adaptive landscape, and it is useful because it provides a mathematical criterion for deciding when a population has reached an adaptive peak. ESS models are particularly useful for making predictions that can be tested empirically and for clarifying what kinds of behavioral observations are not yet understood (Maynard Smith 1979). Thus they are useful tools for interpreting the evolutionary significance of observed behavioral interactions.

Despite its attractiveness, the ESS approach does have limitations. Assumptions must be made about the costs and benefits associated with each strategy. These assumptions often cannot be validated empirically, and yet the parameter values used in simulations may be crucial in determining what strategy constitutes an ESS. The models must be built on mathematical relationships used to define how different strategies affect each other during interactions, and these too may be built on unverifiable or oversimplified assumptions. In addition, all models to date have assumed haploid genetics and ignored the complex genetic consequences of diploidy and polygenic inheritance. The ESS approach is an entertaining and insightful way of showing what is possible, but it is not a valid way of showing what is. Nevertheless, it is likely to have great theoretical impact in coming years.

Conclusion

Aggression is a multifaceted phenomenon. It cannot be understood with simplistic notions of instinct or aggressive energy. The underlying physiological mechanisms involve complex networks of sensory, neural, and hormonal systems not amenable to analysis in those terms. Aggression should not be considered as just one kind of behavior. It occurs in many contexts, and a different physiological control system is responsible for its expression in each context. At the proximate level aggression is affected by genetic and experiential processes, it is suppressed or triggered by sensory input, and it is channeled, modified, and constrained by underlying neural and hormonal controls. It is not inevitable, and it is not the only way an individual can compete.

The social significance of aggression cannot be overemphasized. Aggression is a major organizing influence in animal social systems, and it has clear adaptive significance. It is intimately involved in spacing behavior, mating behavior, and even parental behavior. Animals are not aggressive just to be mean or vicious. Quite the contrary—animals are aggressive to pursue selfish interests in competitive or threatening situations, and they usually minimize actual violence within the constraints imposed by winning tactics. "Nature red in tooth and claw" is often greatly exaggerated, especially with respect to intraspecific aggression.

Environmental bases of behavior

Behavior has evolved as a means for solving problems. To survive and reproduce an animal must solve a set of ecological problems and a set of social problems. The ecological problems involve finding a place to live, acquiring resources, and avoiding predation. The social problems involve finding mates, rearing young, and, sometimes, attaining high social status. These two sets of problems are not mutually exclusive. For instance, acquiring resources may be prerequisite to attracting mates, and attaining high social status may be prerequisite to acquiring resources as well as attracting mates.

Before the behavioral solutions to problems can be analyzed, the problems themselves must be understood. This chapter focuses on ecological problems and where they come from; later chapters will discuss social problems and how they are overcome.

An overview of ecological problems faced by animals

A useful way of conceptualizing the ecological problems faced by animals is to diagram their underlying causes and behavioral solutions (Figure 6–1). The three major problems are represented by rectangles in the center of the diagram. Their solutions can be seen by tracing the arrows downward. The problem of finding a place to live is solved by selecting a suitable site and competing successfully for it. The problem of acquiring resources is solved by overcoming plant or prey (i.e., resource) defenses and competing successfully to acquire them for use. The problem of avoiding predation is solved by defending against predators. Each solution is mediated by a set of behavioral strategies appropriate to both the environmental and social contexts in which they occur. Note that the term *strategy* is used here in the same sense as it was defined in Chapter 5.

Returning to the center of the diagram, the causes of the three problems can be seen by tracing the arrows upward. The problem of finding a place to live results from limited habitat availability, which in turn results from competition and the particular habitat requirements of each species. The problem of acquiring resources results from limited abundance and availability of resources, which in turn results from competition and successful plant or prey defenses. The problem of avoiding predation results directly from the threat of predation, which in turn is the means of resource acquisition for predatory species.

Chapter six

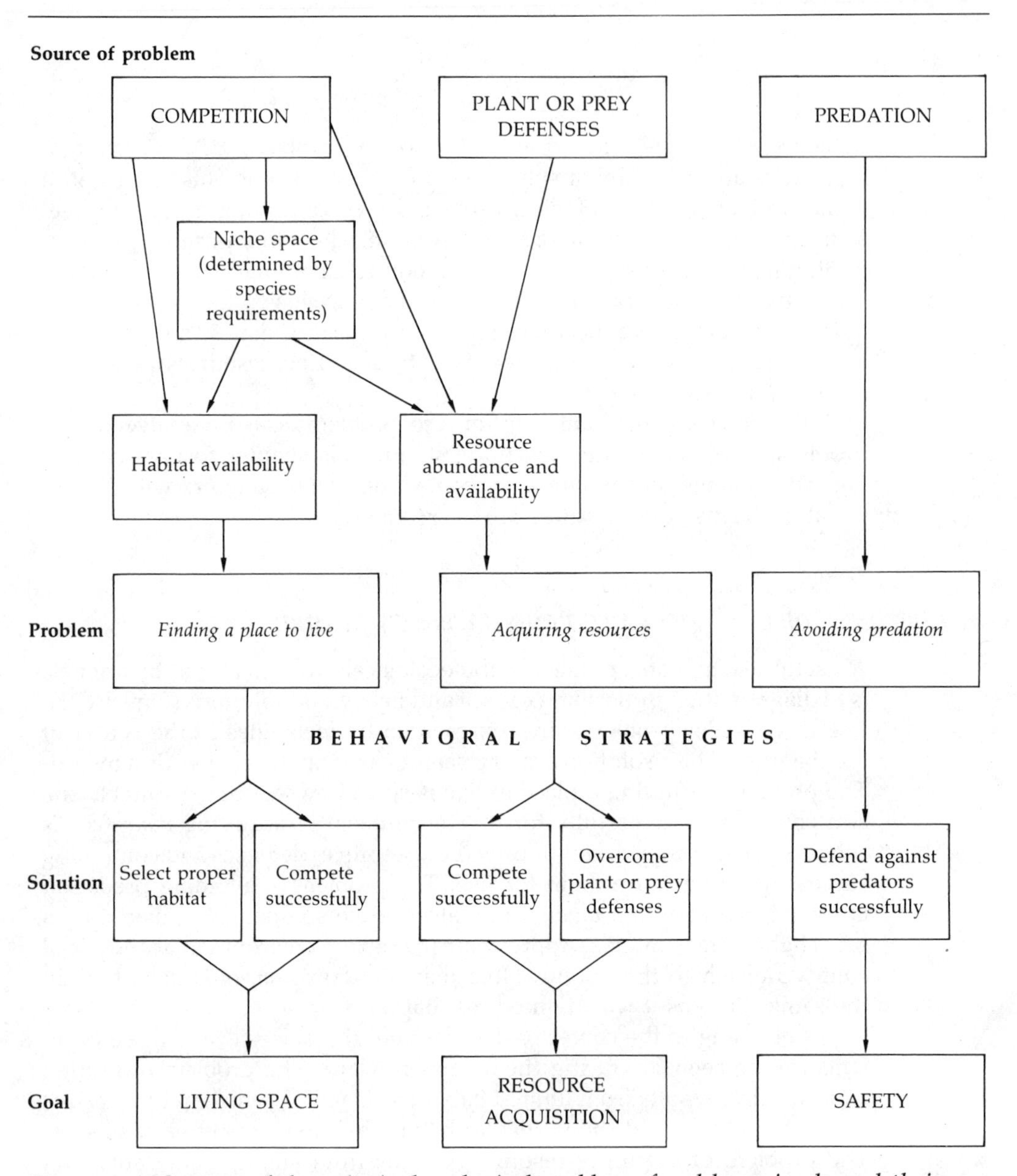

Figure 6–1 *Sources of the principal ecological problems faced by animals and their general behavioral solutions (see text).*

The diagram shown in Figure 6–1 essentially outlines the presentation in this chapter. The chapter first explains in general terms how competition constrains niche dimensions and then goes on to discuss habitat selection as a behavioral solution to the problem of finding suitable living space. Following that it discusses the distributional patterns of resource abundance and the determinants of resource availability. Then it turns to plant defenses against herbivores, with emphasis placed on chemical defenses, and how those defenses are overcome. Finally, it describes the tactics used by predators to capture prey and the countertactics used by prey to evade predators. As we shall see in later chapters, the problems of acquiring resources and evading predators are major selective factors shaping animal behavior. Hence those topics in particular are fundamentally important for evaluating the adaptive significance of animal behavior.

Competition and niche space

Definitions

To consider the effects of competition on a species' habitat and resource requirements, some precise terminology is necessary. **Competition** occurs when use of a resource by one consumer reduces the amount of resource available to another, either of the same species **(intraspecific competition)** or of a different species **(interspecific competition).** It manifests itself through either resource depletion or direct interference with competitors. The former is *scramble competition,* while the latter is *contest* or *interference competition* (after Nicholson 1957). In scramble competition individuals rush to acquire or utilize a resource sooner or more efficiently than their competitors. In contest competition individuals prevent competitors from exploiting a resource by defending or controlling it aggressively.

Specialists versus generalists

Because competition reduces an individual's fitness by constraining lifetime reproductive success, it creates an impetus to reduce competition. The intensity of competition can decrease only if demand for the limiting resource is decreased. Demand may decrease as a result of factors either extrinsic or intrinsic to the population. The extrinsic factors include adverse weather, increased predation pressure, and parasite outbreaks, all of which reduce population density. The principal intrinsic factor is utilization of a broader resource spectrum by some or all population members. Wynne-Edwards (1962) has argued that reduced population density can also be an intrinsic (i.e., evolved) response of populations to increased intraspecific competition, but the earlier discussions in Chapters 2 and 3 explain why this view is unlikely.

One consequence of increased generalization within a species is usually intensified interspecific competition. Because most resources are already exploited by one species or another, any new resources that a species begins to utilize will already be in demand. Thus there is a general tendency for interspecific competition to oppose intrinsic responses of a population to intraspecific competition (Svärdson 1949). The result is resource partitioning, with each species specializing to some degree on its own set of resources.

As a general rule, species tend to specialize on particular sets of resources. Other factors being equal, specialists should always outcompete generalists, because generalists are "jacks of all trades but masters of none" (MacArthur and Levins 1964, 1967). As long as more specialized members of a population leave more surviving descendants than do less specialized individuals, selection will continue to favor increasing specialization. This process is ultimately halted by limitations in resource availability. Highly specialized individuals cannot compete as well for resources lying outside their specialty areas, and as they become increasingly specialized, they have greater difficulty acquiring sufficient resources to reproduce as well as less specialized conspecifics. Hence an equilibrium degree of specialization will eventually be reached.

A major factor affecting how specialized a species becomes is the spatial distribution of the relevant resource (MacArthur and Pianka 1966; Levins 1968; Emlen 1973; Pulliam 1974). When a food resource, for example, is widely dispersed in small packets, animals must forage in a wide variety of different microhabitat types to obtain sufficient food. They may specialize on some microhabitat types, taking their chances in the others, or they may become generalists capable of competing in all microhabitats about equally well. The extent to which they specialize should depend on how much the various microhabitat types differ from one another. If the microhabitats are very different, specialization will not pay because individuals specializing to compete in one type will not be able to compete in the other types. If microhabitat types are similar, specialization will begin to pay because specializing to compete in one type will detract less from competitive ability in the other types. In the limiting case all microhabitat types are identical, and individuals spend all their time in a single microhabitat. A parallel argument can be made with regard to resource types (e.g., food sizes or prey types). Thus extreme specialization to maximize competitive ability is most likely to evolve in species that concentrate on single microhabitat or resource types.

Of course, spatial variability is not the only parameter affecting the range of resources exploited by an animal. Temporal variability must also be considered (MacArthur and Levins 1964, 1967). Animals cannot afford to specialize on resource types that are periodically scarce or absent unless they can survive periods of scarcity without acquiring any resources. The time span of resource fluctuations is therefore critical. Animals may circumvent short-term shortages by exploiting resources during peak periods of availability and then resting or conducting other activities during periods of scarcity. Longer-term shortages, such as seasonal depressions in resource availability, are more difficult to avoid, but many species circumvent them by producing resistant egg or larval stages, becoming dormant, or migrating. Thus the resource requirements and reproductive biology of a species are both important in determining whether it can specialize on a fluctuating resource.

If specialization is so effective for enhancing competitive ability, why are some species extreme generalists? The answer is that some resources cannot support specialists of their own. A substantial body of theory on optimal foraging behavior predicts that species should become more generalized as resource availability declines (Emlen 1966, 1968; MacArthur and Pianka 1966; Schoener 1971; MacArthur 1972; Charnov 1973, 1975, 1977; Hamilton et al. 1978b). When a particular food resource is scarce, a consumer cannot afford to bypass less desirable food items, because average search time for the most desirable items is too high. Hence some food resources are not heavily utilized by species specialized to exploit them, making them available to generalist species (Figure 6–2). Specialists may exploit such resources to some extent, but their competitive abilities are hindered by specialized morphological structures or relatively inflexible behavioral repertoires designed to exploit other resources. Thus generalists have a competitive advantage when exploiting resources lying outside the specialty areas of specialist consumers. This is one reason why every biotic community supports a few food generalists, or *omnivores.*

Niche space

The net effect of specialization is resource partitioning. Some species are specialized to exploit some resources, while other species are specialized

Figure 6–2

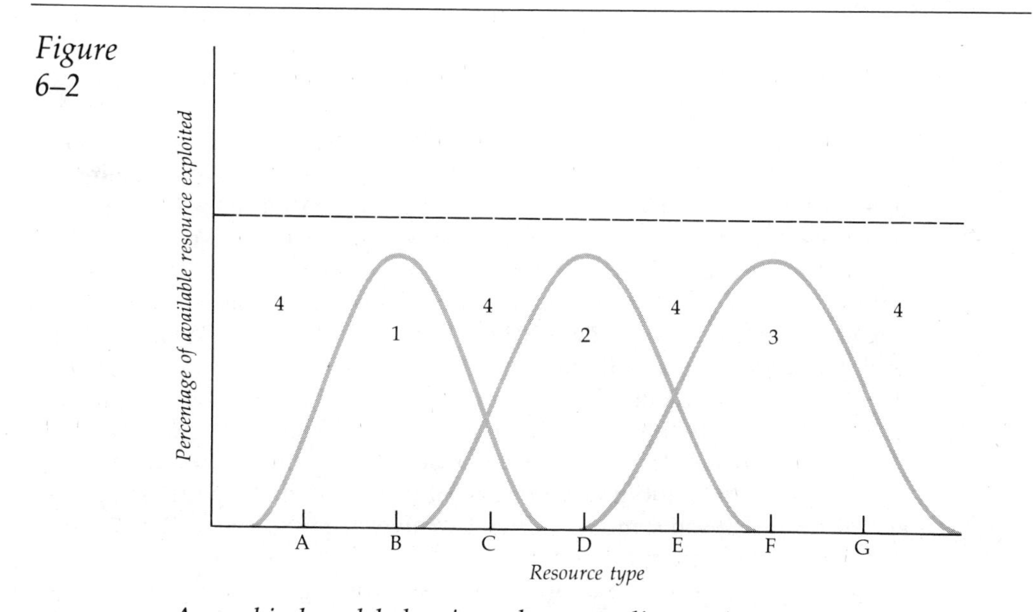

A graphical model showing why generalist species exist in every ecological community.

Specialist species 1, 2, and 3 consume most of resources B, D, and F but cannot compete effectively with a generalist species (4) for resources A, C, E, and G. No species are specialized to exploit resources A, C, E, or G because overall abundance, spatial distribution, or temporal availability of those resources preclude specialization on them. The dashed line indicates the maximum percentage of each resource that can feasibly be utilized.

to exploit different resources. Species vary considerably in their degree of specialization, and the resource spectra they exploit often overlap. This overlap is the basis of interspecific competition. Nevertheless, overlap is never complete. Each species concentrates on a different part of the overall resource spectrum. That is, specialization is the essence of resource partitioning.

Definitions. To study resource partitioning, ecologists have developed the concept of niche space. The classical definition of an **ecological niche** is an animal's "place in its biotic community, its relations to food and enemies" (Elton 1927). According to this definition, niche space includes dietary preferences, foraging methods, and predator defense tactics. It represents the functional role an animal plays within a biotic community. In contradistinction to niche is **habitat**, an animal's spatial or physiognomic position within a biotic community.

Ecologists have never been able to agree on any one definition of niche, though most agree that the concept is useful. Some have included position in a community as part of the niche concept (e.g., Grinnell 1917, 1924; Allee et al. 1949; Odum 1971), and Dice (1952) even went so far as to say that "the term does not include, except indirectly, any consideration of the function the species serves in the community." Others have excluded position from the niche concept and maintained a careful distinction between niche and habitat. Thus niche has had two separate meanings in ecology, the "functional niche" and the "place niche" (Clarke 1954).

The hypervolume model of niche space. Modern concepts of niche space have been strongly influenced by the work of G. Evelyn Hutchinson (1957, 1965, 1967). Hutchinson's model of niche space is based on the fact that each animal species can tolerate only a limited range of conditions along any given environmental gradient and can compete effectively only within a limited part of the overall resource spectrum. Fitness with respect to each environmental gradient or resource spectrum is generally highest under intermediate conditions and decreases toward the extremes. All environmental gradients and resource spectra affecting an animal can be plotted on their own axes, giving rise to a multidimensional portrayal of the species' niche. The *n* dimensions delineating this multidimensional *hypervolume* correspond to the number of variables affecting fitness in that species. Hutchinson's model is most easily visualized by picturing each dimension of a niche space separately. Along any given dimension one can plot average fitness of the population as a function of some environmental gradient or resource spectrum that varies along that dimension. For each dimension average fitness is greater than zero only within the zone where the species can survive and reproduce. This zone is referred to as *niche width* or *niche breadth* along that dimension. Note that the relative fitness of individuals is not appropriate for delineating niche boundaries. Fitness here refers to average fitness for the entire population. That is, niche boundaries are defined by the conditions that would cause a local population of the species to become extinct.

The niche boundaries along each dimension can be combined to obtain an overall picture of niche space. These boundaries can be connected together into one large, multidimensional graph to generate a niche hypervolume. Note that *niche hypervolume* refers to a volume delineated by more than three ecological gradients. An entire hypervolume may be based on hundreds or thousands of dimensions. About the only way to visualize space in that many dimensions is to take cross sections along 1–3 dimensions at a time.

Average fitness is not everywhere identical within a niche space. Fitness varies along each dimension of a hypervolume. A plot of fitness curves along every axis results in a fitness density superimposed on a niche space to produce a multidimensional surface representing average fitness at each point in the niche. A simple two-dimensional example is shown in Figure 6–3. The fitness surface for each species has a single peak, generated by the bell-shaped fitness curves along each environmental gradient. Note that the surface is very similar to an adaptive landscape. It differs primarily with respect to the way in which the horizontal axes are defined. This similarity is more than fortuitous. An organism's response to environmental variation is a phenotypic trait, and each axis of a niche hyperspace reflects a separate trait. In effect, a niche space is delineated by the phenotypic traits associated with each relevant environmental variable, and therefore it *is* an adaptive landscape (see MacArthur 1958).

The *fundamental niche* of a species is the n-dimensional polygon whose

Figure 6–3

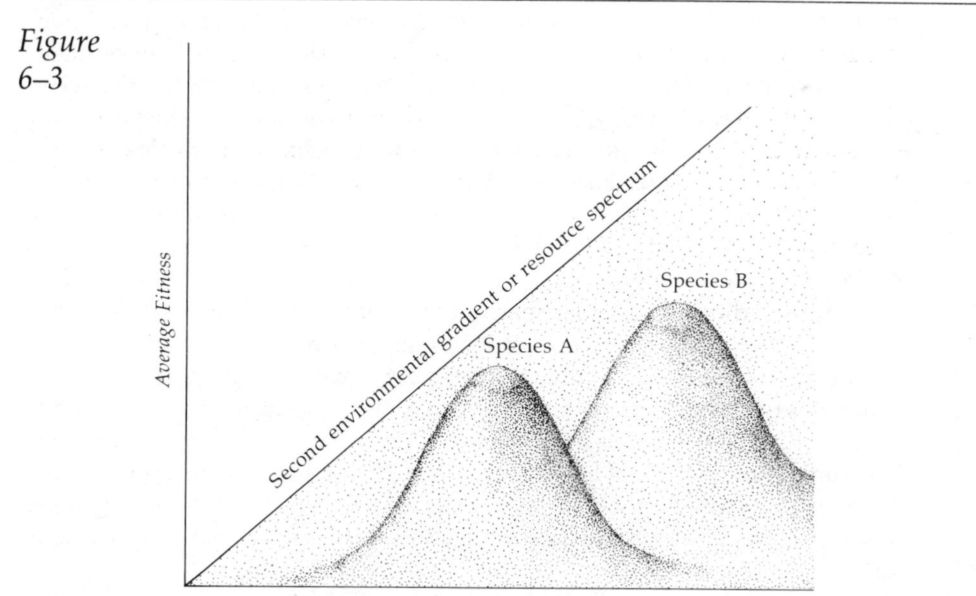

A two-dimensional niche space.

Average fitness at each point in an environmental mosaic (horizontal plain) is plotted along the vertical axis. Note the similarity to adaptive landscapes.

sides extend to the boundaries demarcating niche width along every dimension of the hypervolume. It corresponds to the entire range of environmental conditions within which the species can persist. The *realized niche* is a subset of the fundamental niche. It represents the range of environmental conditions actually occupied by a real population of the species. Realized niches are usually smaller than fundamental niches (they obviously cannot be larger), since species seldom utilize their entire fundamental niches at any given time and place.

Hutchinson's original examples of niche dimensions represented tolerance ranges along microclimatic or physiognomic gradients. His model therefore emphasized a species' place in biotic communities rather than its functional role. More recent theorists, however, have shifted emphasis toward resource utilization spectra, which focuses the niche concept more on a species' functional role in biotic communities (see Pianka 1976).

Response surfaces. As a practical matter, average fitness cannot be measured with respect to environmental gradients or resource spectra. Therefore, a quantitative picture of niche space is obtained by plotting variables that should reflect fitness. One approach is to plot response probabilities along a given niche dimension. The responses shown by a species in a given situation should, on average, be adaptive, and hence they should indirectly reflect fitness. Consider, for example, the response surface of blue-gray gnatcatchers *(Polioptila caerulea),* a small insectivorous songbird, with respect to prey size and height of prey above ground level (Figure 6–4) (Whittaker et al. 1973). The surface was obtained by plotting the proportion of prey items taken within each size class and at each height above ground level. Ideally the proportions should be calculated as a percentage of the relative numbers present, since peaks on the surface might otherwise result from greater abundance of prey within certain size classes or at certain heights, but this goal is usually difficult to achieve in practice. Notice that the degree of specialization along any one dimension can be easily discerned from a response curve. Narrower, more peaked curves indicate greater degrees of specialization; broader, less peaked curves indicate lesser degrees of specialization.

By including all variables relevant to a species' foraging behavior, one could obtain a complete picture of its foraging niche. Such a picture could then be combined with similar pictures for a competing species to give some idea of niche overlap. An extension of the procedure to all competing species in a biotic community would allow a vivid, though admittedly oversimplified, picture of how resources are being partitioned in that community. Unfortunately, obtaining the necessary data for plotting response surfaces is costly and time-consuming, so the procedure has only limited value.

Response surfaces allow quantification of niche overlap between two or more competing species, simply by plotting comparable data for each species along the same dimensions. It is tempting to equate niche overlap with intensity of interspecific competition, but the relationship between

Figure 6–4

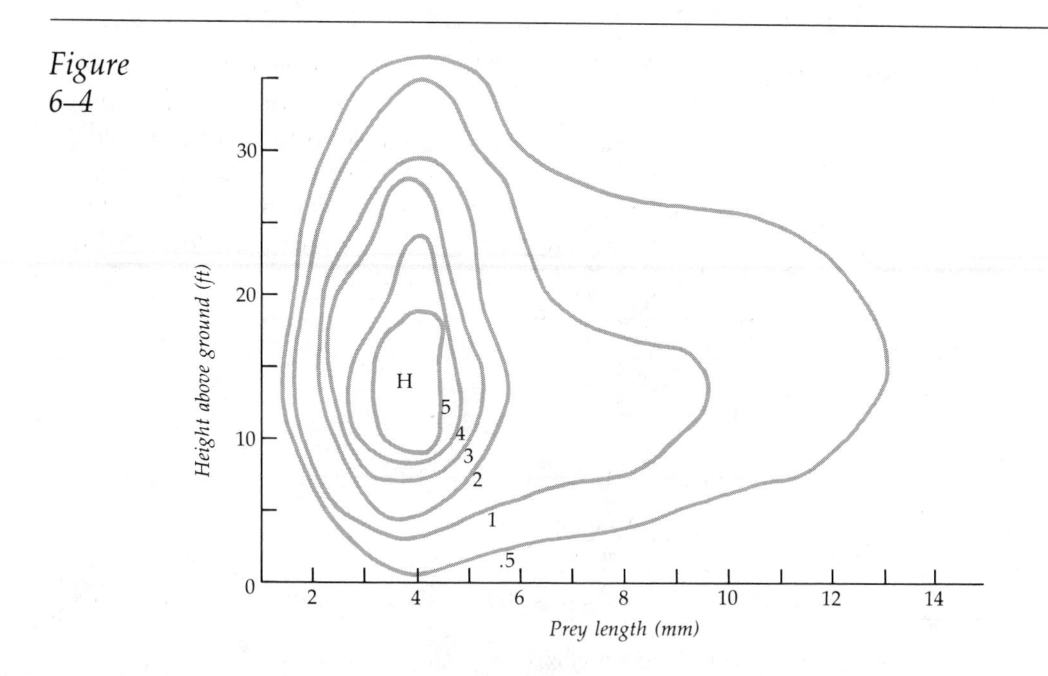

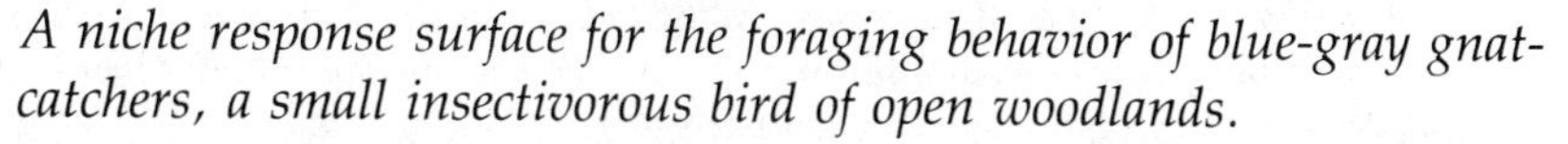

A niche response surface for the foraging behavior of blue-gray gnatcatchers, a small insectivorous bird of open woodlands.

The contour lines map the percentage of dietary items (indicated by the number along the contour) taken of each prey length and at each height above ground level during the incubation period. Each concentric ring encloses all points with percentages equal to or greater than the value indicated along that contour. H refers to values higher than 5%.

Source: Reprinted from R. H. Whittaker, S. A. Levin and R. B. Root, "Niche, Habitat and Ecotype," *American Naturalist* 107 (1973), p. 332, by permission of The University of Chicago Press. Copyright © 1973 by The University of Chicago Press.

niche overlap and competition intensity is too complex for that (see Colwell and Futuyama 1971). Competition is affected by resource availability as well as degree of niche overlap, and niche models do not take this fact into account. Two species may both obtain much of their food from a single, abundant resource without competing much at all. Two species may also obtain only a small proportion of their food from a scarce resource and yet compete intensely for that resource. In addition, niche overlap may be high along one dimension but be minimal when several dimensions are considered together (Pianka 1974, 1976). A simple two-dimensional diagram shows how this result is possible (Figure 6–5). Two species may, for example, both exploit the same type of prey but concentrate on different prey sizes, or they may both exploit the same prey sizes but concentrate on different prey types or different foraging sites. Hence they would not experience high competition despite high niche overlap along the prey-type or prey-size dimension.

Figure 6–5

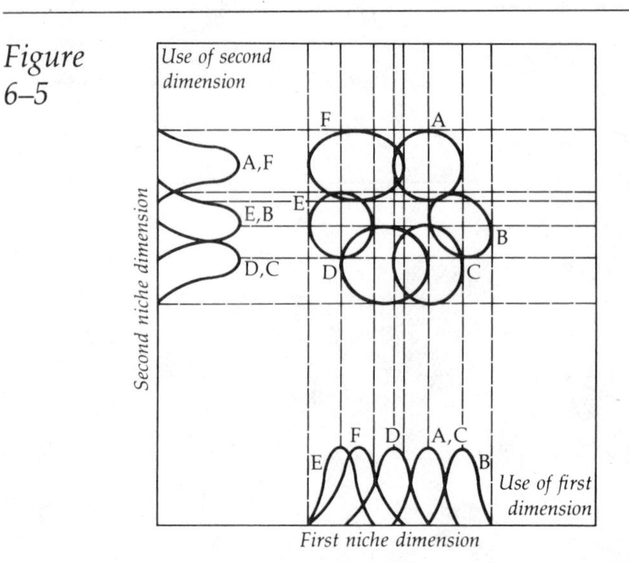

Schematic diagram showing how high niche overlap along one dimension can be much reduced by low overlap along a second dimension when both dimensions are considered together.

The letters A–F refer to different species. The bell-shaped curves along each axis represent niche breadth and frequency of resource use along that niche dimension. Circles and ellipses indicate the niche spaces of each species, as determined by the combined niche breadths of both niche dimensions. Note the high overlap of species F with species E and D along the horizontal dimension, and the low overlap between those same species once both dimensions are considered together.

Source: Figure 6.7 (page 198) from *Evolutionary Ecology* by Eric R. Pianka. Copyright © 1974 by Eric R. Pianka. Reprinted by permission of Harper & Row, Publishers, Inc.

Habitat selection

One way animals satisfy their niche requirements is to settle in appropriate habitats. By being in an appropriate habitat, they can hunt appropriate prey and evade predators in the manner best suited to their particular specialization. Thus animals often choose their habitats, a behavior referred to as **habitat selection.**

Correlations between an animal's abundance and particular habitat characteristics do not prove that habitat selection has taken place (Klopfer 1969). Nonrandom settling patterns may result from physical or biotic dispersal. Camouflaged animals, for example, may be found mainly on appropriate matching backgrounds because individuals selectively choose to settle on those backgrounds or because all individuals settling on other backgrounds are quickly eaten by predators. Of course, if differential mortality is the reason why most individuals are found on appropriate backgrounds, it would create a strong selective impetus for behavioral selection of suitable backgrounds. Thus many such correlations observed in nature are likely to result from behavioral habitat selection, even though conclusive evidence is often difficult to obtain (see Partridge 1978 for further discussion).

Habitats clearly differ in intrinsic quality. That is, some habitats satisfy niche requirements better than others. Nevertheless, being in an intrinsically better habitat is not necessarily advantageous. As population density increases, higher intrinsic quality becomes offset by increased competition intensity. It is therefore possible for individuals in poor habitats to achieve the same fitness as individuals in superior habitats (Fretwell and Lucas 1969). The trade-off between habitat quality and competition intensity is an important consideration in studies of habitat selection behavior, and more will be said about it in Chapter 7.

Having taken competition intensity into account, animals should prefer settling in the best habitats they can find. However, they cannot afford to search indefinitely for the best available habitat because continued search is costly. Searching for better habitats requires time and energy and entails higher risks of predation than are experienced once an individual gains familiarity with a chosen site (Levins 1968). A habitat should be accepted when the costs of continued search are no longer offset by the chances of finding a better habitat than the best one already discovered and available for settling.

Animals must choose habitats according to proximate stimuli, many of which differ from the ultimate factors affecting their fitness (Lack 1954; Hildèn 1965a). Ultimate factors are frequently unsuitable as proximate cues, either because they are too costly to assess or because they are absent at the time habitats must be chosen. Reproductive success in many bird species, for example, depends on food availability during the nestling period, but habitats are often chosen largely according to vegetational or physical features because insects fed to nestlings are absent until relatively late in the breeding season (Lack 1933, 1944; Miller 1942; Svärdson 1949; Hildèn 1965a; Klopfer 1969). Presumably the proximate cues used to select habitats are to some extent predictive of food availability later in the season, but low predictability may often make identification of the best habitats difficult.

The specific cues used to select habitats should depend on the costs and benefits associated with using them. The benefits depend on how well the cues predict future habitat conditions affecting fitness. The costs depend on the length of time required to assess the cues in each prospective habitat, the amount of energy expended while assessing them, and the amount of risk involved in assessing them. These costs depend in turn on the sensory and locomotory capabilities of the animal involved. They also depend on the nature of the cues themselves. Gross vegetational features can often be assessed rather quickly, while assessments of food availability may require extensive foraging in each prospective habitat. In addition, physical and vegetational features are often more predictable than food availability, which would make them more suitable for assessing future habitat conditions than direct assessments of food availability. Several different proximate cues may be used by a species, and the combination of cues used probably represents a compromise between the information value of each cue and the costs entailed in assessing it.

Resource characteristics

Much of animal behavior involves resource acquisition. Hence a clear picture of resource characteristics is essential. Of particular importance are spatial and temporal patterns of resource distribution and the determinants of resource availability.

The spatial and temporal distribution of resources, in particular, is a major factor determining how animals space themselves, seek or choose mates, and otherwise interact with one another. Precisely quantifying resource distribution patterns is therefore crucial in analyzing the adaptive significance of animal behavior. At present, distribution patterns can be described much more precisely in theory than is usually achieved in practice. Quantitative data for evaluating resource distributions are all too frequently absent in sociobiological studies, thus forcing many investigators to rely heavily on superficial assessments or subjective impressions of resource characteristics when interpreting behavioral phenomena. Adequately quantifying resource characteristics is one of the most pressing problems facing sociobiology today. For this reason a more sophisticated scheme for classifying resource distribution patterns will be presented here than is currently necessary for understanding most of the existing sociobiological literature.

Definition

A **resource** is a substance, object, or energy source required for normal body maintenance, growth, and reproduction (Ricklefs 1979). The chief resources required by animals are food, water, shelter, and breeding sites. Food resources are especially important in shaping animal social organization, and they will be given particular attention here.

Classifying resources into discrete types is often difficult, especially for food resources. One approach is based on determining whether resource types are harvested or exploited independently of one another (MacArthur and Levins 1967). More specifically, the food resources exploited by two species are different if depletion by one species has no direct effect on resource availability for the other species. This approach is useful for studies of competition, since resources are considered the same when competition exists. However, it does not always give an intuitively reasonable basis for classifying resources. Two species may exploit the same resource without competition if that resource is sufficiently abundant. For most purposes resource types are therefore classified according to other criteria, such as prey size or prey species.

Distributions in space

The spatial distribution of a resource can be *random*, *regular*, or *clumped* (Figure 6–6). These are idealized patterns rarely encountered in nature. As a practical matter, statistical procedures are used to determine whether a

Figure 6–6

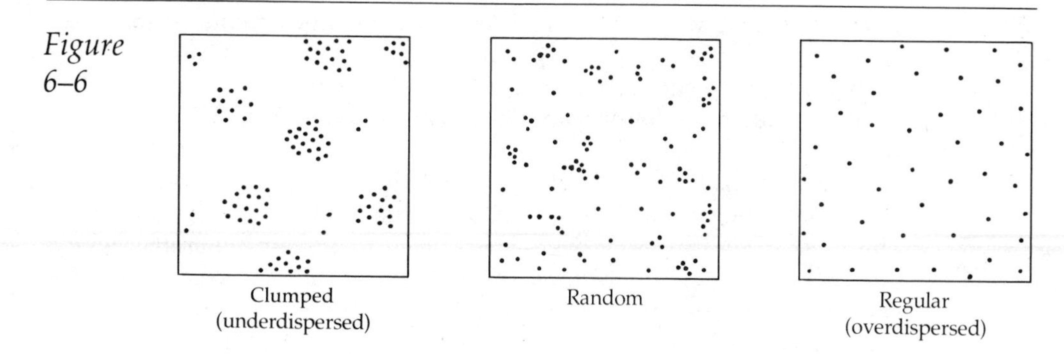

Three patterns of resource distribution are clumped, random, and regular. All possible intermediate dispersion patterns can also occur.

given pattern departs significantly from randomness toward regularity (i.e., *overdispersion)* or toward clumping (i.e., *underdispersion*) (see Greig-Smith 1964; Kershaw 1964; Iwao 1968; Goodall 1970; Pielou 1977). The exact procedures used for this purpose need not concern us here.

Random spacing. A resource is randomly distributed if each location has an equal probability of containing a given amount of resource. Physical features of the environment such as soil types, minerals, water, microrelief, and topographical features are usually not randomly distributed. As a result, most plant populations are also not randomly distributed. For example, Clapham (1936; modified by Blackman 1942) found that only 3 of 44 species of prairie plants were randomly distributed, and those were among the least common species studied. Random spacing is most frequently observed in plants that occur at low densities, possibly because rarity makes departures from randomness more difficult to detect statistically (Archibald 1948). Random spacing is also found among species growing in recently disturbed homogeneous environments. Many weeds found in freshly plowed fields, for example, are randomly distributed (Singh and Chalam 1937; Singh and Das 1938, 1939).

Because physical features and plant populations are not usually distributed randomly, neither are most animal populations. Random dispersion patterns in animals are most likely to result from larvae settling at random in homogeneous environments, individuals moving at random within a population, or resources being distributed at random in the environment. Some examples of the first possibility are various species of bivalve clams collected from small homogeneous mudflats (Connell 1955; Jackson 1968), certain species of woodland spiders that colonize the undersides of boards placed uniformly across a forest floor (Cole 1946), and flour beetle larvae living in laboratory containers of flour (Park 1933). Some examples of the second possibility are certain motile deposit-feeding clams (e.g., *Nucula proxima*) that live in silt-clay sediments of ocean bottoms (Levinton 1972) and bark beetles (e.g., *Ips grandicollis*) that attack weak or physically damaged trees (Mason 1970). Rare or uncommon animals are generally underdispersed rather than randomly distributed, unlike plants, probably because such species exploit rare microhabitats that are not likely to be distributed randomly throughout a biotic community (Hairston 1959).

Regular spacing. Regular or uniform spacing patterns usually result from strong negative interactions among members of a population. The most likely causes of overdispersion in plants are *allelopathy,* which involves the release of chemical toxins into the soil, and competition for water or light. In many cases the actual cause of overdispersion is controversial (see Harper 1977). Many, and perhaps most, plants release chemical toxins or growth inhibitors into the soil (Rice 1974). The release of toxins by certain plants found on terpentine soils, for example, is allegedly responsible for the bare zone found around each plant or stand of plants. If such bare zones are large enough, they could lead to regular spacing, though this has not actually been demonstrated. Competition for water is evidently responsible for the regular-spacing patterns often observed among desert plants such as creosote bush (*Larrea*), rabbitbrush (*Chrysothamnus*), and sagebrush (*Artimesia*) (Woodell et al. 1969). Competition for light may occasionally lead to regular spacing of trees in closed canopy forests, though this is exceedingly rare (e.g., see Greig-Smith 1952; Jones 1955–1956).

Territoriality and mutual avoidance are common causes of regular spacing in animals, but these behaviors do not necessarily lead to overdispersion, especially when food, shelter, or breeding sites are unevenly distributed. Regular spacing is most clearly evident within dense colonies of breeding seabirds, where breeding pairs are packed together on small nesting territories. Mutual avoidance is responsible for regular spacing of trypetid fly larvae (*Dacus tryoni*) among loquat fruits, since females generally avoid ovipositing in fruits that have already been parasitized by other females (Monro 1967).

Clumped spacing. The vast majority of both plant and animal populations exhibit clumped spacing. Clumping can result from underdispersion of physical features, microhabitats, or resources in the environment; vegetative growth of plants by means of rhizomes or comparable structures; nonrandom dispersal patterns of seeds, larvae, or other propagules; or mutual attraction among mobile individuals. Resource clumps—or **patches,** as they are often called—can be treated as entities which themselves can be spaced randomly, regularly, or in clumps. Likewise, individuals within clumps can be spaced randomly, regularly, or in clumps. For this reason the dispersion pattern detected by statistical methods depends strongly on the scale of the sampling procedure (Skellam 1952; Greig-Smith 1964), an important consideration when designing a sampling program. Similarly, the dispersion pattern perceived by an animal depends on the animal's size and scale of movements. Organisms of differing size or mobility may perceive a resource quite differently, depending on whether their movements are always confined within a single resource patch or pass through many patches in succession.

Intensity and grain. *Intensity* and *grain* are useful attributes for characterizing the way resources are perceived (Pielou 1977). Imagine a resource as a mosaic of areas, each of which differs in resource abundance or density. Areas of high density are resource patches, and areas of low density are interstices between patches. The size of each patch is arbitrarily determined by statistical criteria on a scale appropriate for the particular organism under study. For example, patch size in a mosaic representing a particular insect population may be defined as an area containing a specified density of insects if the foraging behavior of an avian predator is being studied, but it may be defined instead as the size of a single insect if a parasite of the insect is being studied.

The *intensity* or **patchiness** of a resource measures the difference in resource density between the interior of resource patches and the interstices lying between them. *Patchy* resources are clumped resources with relatively high intensity. *Homogeneous, uniform,* or *dispersed* resources are resources with relatively low intensity or a regular or random distribution pattern.

Grain refers to the size and spacing of resource patches. Four general patterns of grain are commonly encountered. *Fine-grained* resources occur as small patches situated relatively close to one another; *coarse-grained* resources occur as large patches situated relatively far apart; *fine-grained, spaced* resources occur as small patches situated relatively far apart; and *coarse-grained, packed* resources occur as large patches situated relatively close to one another (Figure 6–7). Once again patch size and closeness are defined relative to the size and mobility of a particular organism. Fine-grained and uniformly distributed resources are often loosely referred to as dispersed resources, while coarse-grained and patchy resources are often referred to as clumped resources.

Distributions in time

The temporal distribution of a resource refers to its abundance through time in a given locality. To the extent that resources consist of plant or animal populations, temporal distribution patterns result from details of the species' life cycle, population dynamics, density-dependent regulating mechanisms, and density-independent perturbations caused by weather or other environmental conditions. A large part of population ecology is concerned with these topics, and several excellent textbooks provide full treatment of them (e.g., Andrewartha and Birch 1954; Odum 1971; Emlen 1973; Pianka 1974; Krebs 1978; Ricklefs 1979).

In a given locality resource abundance can be either relatively constant or variable. A *continuously available resource* is one that is always present in about the same abundance or in amounts adequate to support a particular population at a constant level, while a *fluctuating resource* is one whose abundance varies sufficiently that an animal exploiting it must change its

Figure 6–7

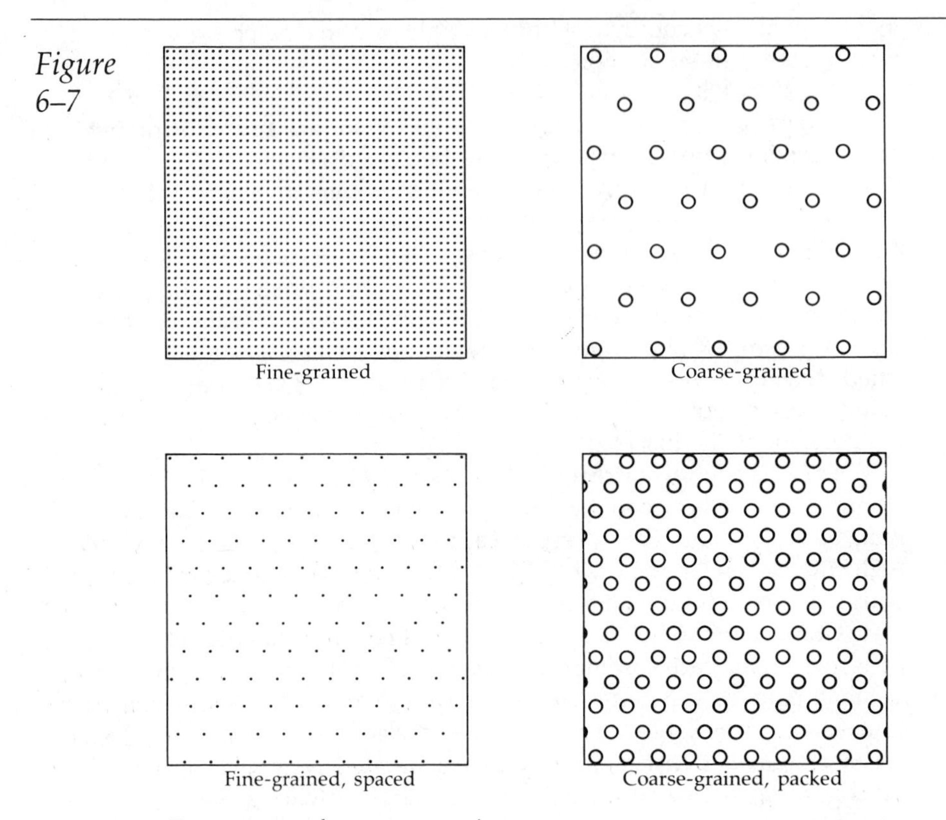

Four types of resource grain.

Grain refers to the size and spacing of resource patches. Points represent small resource packets or patches, and circles represent large or dense resource patches.

usage pattern on a temporal basis. Fluctuating resources can vary on a regular basis, with nearly constant time intervals between peaks or troughs, or it can vary irregularly, with the length of this interval varying randomly about some mean value. The former pattern is referred to as a *cyclic* or *periodic fluctuation,* while the latter is referred to as a *random, aperiodic,* or *unpredictable fluctuation.* A *predictable* resource either is continuously available or fluctuates in a periodic manner. Resource abundance in separate localities can peak at about the same time during each cycle *(in phase),* at consistently different times *(out of phase),* or at times unrelated to each other *(aphasic).*

An important variable is the time scale of resource fluctuations. Animal responses, in the form of movements to new patches or switches to other resource types, depend on whether resource abundance varies hourly, daily, seasonally, or annually.

Spatiotemporally clumped resources are spatially clumped, temporally fluctuating, and aphasic or out of phase among the various patches exploited by a given local population. This pattern of resource distribution is regularly associated with certain kinds of animal social organization, as will be shown in later chapters.

Resource availability

Resource availability refers to the amount of resource an animal can actually capture, utilize, or otherwise exploit. Two determinants of availability are resource abundance relative to population size and the resource requirements of each individual within a consumer population. Other determinants of availability are renewability of the resource and the ease of discovering, gaining access to, or capturing the resource.

Renewability. **Renewability** refers to the time required for a food resource to be replenished. A renewing resource is replenished more or less continuously, while a nonrenewing resource is replenished only on a seasonal, annual, or longer-term basis. Of course, rate of replenishment may vary considerably during the course of a year, and this variability may have an important effect on how animals explit a resource. Some examples of renewing resources are summer insect populations, new foliage during vegetative growing seasons, and detritis washing ashore along coastlines. Some nonrenewing resources are seed crops produced at the end of a growing season and overwintering populations of insect larvae.

Depleting a renewing resource does not cause cumulative increases in competition intensity through time, except when the resource is overexploited, but depleting a nonrenewing resource always does. Renewability is actually one determinant of resource abundance and hence affects availability by affecting abundance.

Accessibility. Ease of access or discovery is a major determinant of food availability. Foliage may be too high, too far out on small limbs, or too well protected by dense thickets or thorns for an herbivore to reach. Prey animals may be physically inaccessible to a predator, or they may escape detection by remaining camouflaged and behaving secretively. Since plant parts and prey individuals are not equally inaccessible or equally capable of escaping detection, not all individuals in a given population are equally available as food for a given consumer species. Plant or prey individuals may be available to some consumer species but not to others because each consumer species has its own specializations for overcoming plant or prey defenses. For example, foliage high in a tree is inaccessible for an antelope but may be readily accessible to an arboreal monkey. A prairie dog hidden in a burrow is inaccessible to a coyote but is easy prey for a ferret.

Ease of capture. Prey individuals vary greatly in their ease of capture. Juveniles and inexperienced subadults are usually easier to capture than adults. Prey that depend on early detection of predators and flight to safety are difficult to capture when healthy and alert but become vulnerable when old, poorly nourished, injured, or asleep. Prey are often more vulnerable to some hunting tactics than to others, making them more vulnerable to some predators than to others.

Availability versus abundance. Understanding the determinants of resource availability is important when evaluating the way a resource affects an animal's social organization. The problem of measuring availability is often circumvented by using abundance as an index of availability, but this approach must be used cautiously when evaluating the effects of competition on behavior. Indications of food abundance in excess of consumer needs can be misleading because many apparently suitable food items may be unavailable to the animals exploiting them. How well sampling procedures reflect food availability, as opposed to food abundance, is therefore an important consideration when evaluating relationships between food resources and behavior.

Resource quality

Resources generally vary in quality, with *quality* measured by the effect a given unit of resource has on an animal's fitness. The quality of water resources depends primarily on how contaminated they are with pollutants or toxins. The quality of shelter depends on how much protection it affords from adverse weather or predators. The quality of breeding sites often depends on how well the sites provide protection for adults or offspring and on how near they are to food or other resources. The quality of food resources depends on nutrient content, energy yield, digestibility, and the concentration level of toxins. Because the quality of food resources is especially important in studies of animal behavior, it requires more detailed analysis.

Quality of food resources. The nutritive quality of food resources is exceedingly difficult to measure or evaluate. It depends on both nutrient content of the food and nutritional requirements of the consumer. The latter are poorly known except for humans, domestic animals, and laboratory animals, though much information is also available for animals kept in zoos. Unfortunately, nutritional requirements in a zoo are often different from those in the wild because energy demands and stress levels are normally substantially different. In addition, a complicating factor requiring consideration is the fact that an animal's nutritional requirements vary with reproductive condition. Pregnant and lactating female squirrels, for example, require greatly increased calcium intake, making them prone to chew on bones (Bakken 1952). Many granivorous (grain-eating) and frugivorous (fruit-eating) birds consume insects during the breeding season to obtain the protein necessary for producing eggs but do not consume them at other times of the year.

The caloric content of a food item depends in part on its composition. Foods consisting largely of carbohydrates or fats have higher caloric content than foods containing large amounts of protein. Foods high in structural components such as bone, cellulose, or chitinous exoskeletons average lower caloric content by weight and are also more difficult to digest. This fact helps explain, for example, why most herbivorous animals prefer young, growing foliage to mature leaves and why most insectivorous birds prefer larvae to mature insects.

The nutritive quality of food also depends on an animal's ability to digest or assimilate it. Specialized diets are generally accompanied by masticatory or digestive specializations. The spotted hyena, for example, has an unusually powerful chewing apparatus for crushing the largest and hardest bones of prey animals, unlike other carnivores, which lack strong enough teeth and muscles for the task (Photo 6–1) (Kruuk 1972). Horseflies *(Chrysops* and *Tabanus)* feed on plant juices as well as blood, and they produce both carbolytic and proteolytic enzymes (Wigglesworth 1929, 1931). In contrast, tsetse flies *(Glossina)* feed only on blood and lack carbolytic enzymes. Herbivorous mammals are much more capable of digesting foliage than omnivores or carnivores because they possess intestinal flora necessary for breaking down cellulose (Jennings 1965). However, they are less capable of digesting meat because they possess fewer and weaker proteolytic enzymes. Digestive specializations can even affect assimilation efficiencies among animals with similar diets. Horses digest cellulose by caecal fermentation and are about 70% as efficient as ruminants, which digest cellulose in fore-stomach fermentation chambers (Phillipson and McAnally 1942; Heinlein et al. 1966; Vandernoot et al. 1967). On the other hand, digestion is slower in the ruminant system because the rate at which food can be processed in the stomach is slower (see Balch and Campling 1965; Heinlein et al. 1966; Vandernoot et al. 1967). A parallel difference also exists in primates, where some species digest cellulose by fore-stomach fermentation, while the majority digest cellulose by caecal fermentation (Clutton-Brock and Harvey 1977a). Caecal fermentation is apparently a specialization for more fibrous diets, as species with fibrous diets must process larger volumes of food per unit time to ingest adequate amounts of nutrients (Janis 1976).

Finally, digestibility is affected by digestibility-reducing substances and toxins present in food. Plant secondary compounds and toxins are important as antiherbivory agents and will be discussed in the next section. Similarly, some potential prey animals produce chemical toxins to deter predators, a topic to be discussed in the following section on predator-prey interactions. Consumers must either avoid such foods, accept the ill effects, or somehow neutralize such substances.

Plant-herbivore interactions

Overview of plant defenses

Many plants, especially annuals, depend on patchy distributions and short growing seasons to escape herbivores. An ephemeral and unpredictable existence reduces each plant's chances of being located and eaten by an herbivore before it can reproduce, and it prevents any herbivore

Photo 6–1

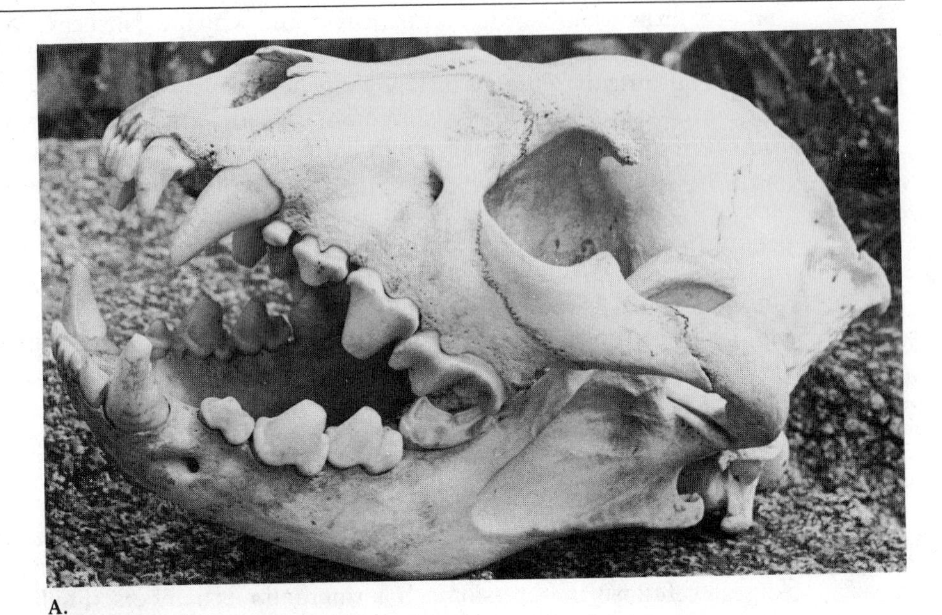

A.

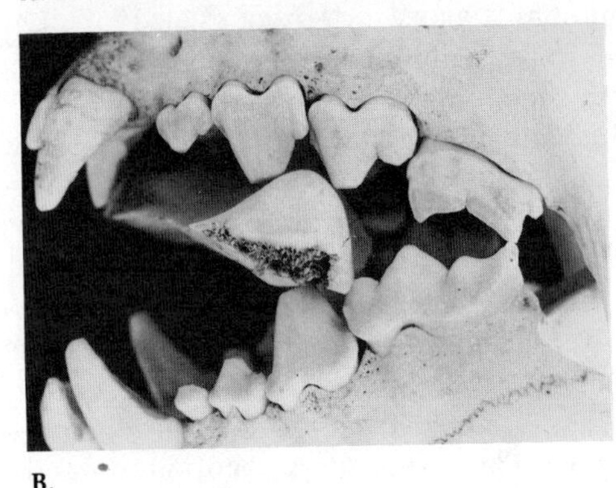

B.

The unusually powerful chewing apparatus of spotted hyenas, which is capable of breaking even the hardest bones.

This capability allows hyenas to exploit parts of carcasses that cannot be eaten by any other predators or scavengers, an important advantage when food is scarce. A. Hyena skull, showing heavy jaw bones and massive teeth. B. Close-up view of the large molar teeth, which are specialized for crushing even the hardest bones of mammal carcasses.

Photos from *Hyaena* by Hans Kruuk published by Oxford University Press.

from becoming specialized on that plant species (Harper 1977). The same tactic is also common for reducing predation on seeds (Janzen 1970b, 1971). Seasonal or annual variations in seed production cause periods of glut and famine for seed predators. Animal populations are limited by periods of food scarcity and hence cannot fully exploit the available food during periods of glut. This circumstance allows many seeds to escape until they can germinate. Seed-caching mammals must store enough seed to survive the worst winters, and in average years many cached seeds are superfluous to the animal's food requirements. These also escape consumption and are left to germinate. Another form of temporal defense is the production of foliage after female insects have finished ovipositing or during seasons unfavorable to herbivore activities (Janzen 1970b; Opler 1974). Some plants provide living sites or food for predaceous or omnivorous insects, especially ants, which in turn provide the plant with protection from herbivores (Janzen 1966, 1967a, 1972, 1974). Other plants depend on hairs, spines, thorns, or impenetrable foliage to deter large herbivores. In many ways such defenses parallel prey defenses against predators, but fuller discussion of this parallel cannot be included here. The interested reader should refer to Harper (1977).

In addition to the defenses listed above, many plants produce chemicals to reduce digestibility or to poison herbivores not specialized to detoxify them. The prevalence of such antiherbivory agents is only now being appreciated, and the existence of these substances has important implications for evaluating food availability and analyzing the food preferences of herbivores. Food resources exploited by herbivores often seem very abundant, but in reality the rate at which these resources can be exploited is severely limited by the ability of herbivores to detoxify them. Thus we cannot hope to measure food availability of herbivores without first learning about plant toxins and how herbivores respond to them.

Plant secondary compounds as antiherbivory agents

Plant secondary compounds are substances not involved in the basic processes of cell metabolism. Over 10,000 low-molecular-weight secondary metabolites are known in fungi and higher plants, and these represent only a small fraction of the probable total number present in all plants (Swain 1977). Quite possibly the total may reach or even exceed the total number of plant species (about 400,000) currently known to science (Raven et al. 1976).

Some plant secondary compounds are implicated as storage compounds or regulators of metabolism or growth, but the great majority have ecological significance beyond those functions. Some serve primarily to attract animal pollinators or seed dispersers, while others deter animals that do not effectively disperse pollen or seeds from consuming them (Rhoades 1979). Nevertheless, many plant secondary compounds have apparently evolved as antiherbivory agents to deter animals from eating particular plant species or plant parts.

Several lines of evidence indicate that plant secondary compounds act as antiherbivory agents. First, a wide variety of these substances have been shown experimentally to deter herbivores as diverse as slugs, insects, birds, and mammals from eating plants containing them (Bowers et al. 1976; Janzen et al. 1976; Levin 1976a; Rhoades and Cates 1976; Rhoades 1979). In some cases the mode of deterrence has been identified (see pp. 217–220). Second, many plants respond to parasite infestations or herbivore attacks by increasing production of secondary compounds (Levin 1976a), implying that such compounds serve a defensive function. For example,

Pease and his co-workers (1979) have shown that many deciduous plants grazed by snowshoe hares *(Lepus americanus)* respond to heavy browsing by producing resins and other compounds toxic to hares, with a 2–3-year time lag. The result is a 10–11-year cycle in plant chemistry that parallels the 10–11-year cycle in snowshoe hare abundance. Third, alternative functions can often be ruled out. Plant secondary compounds cannot be metabolic waste products because waste products are structurally simpler than starting materials, while plant secondary compounds are not (Solomon and Crane 1970; Swain 1977). They also cannot be storage compounds because storage compounds should be readily synthesized and degraded, while plant secondary compounds are not. Finally, the structural complexity and variety of plant secondary compounds have changed through evolution, and such changes cannot be explained by changes in excretory functions, storage requirements, or basic metabolic processes (but see Seigler and Price 1976; Seigler 1977; Jones 1979).

Types of antiherbivory compounds

Plants produce two general types of protective chemical agents: digestibility-reducing substances and toxins. Some compounds, such as tannins, are both (Goldstein and Swain 1965; Feeny 1969; Sumere et al. 1975).

Digestibility-reducing substances are either protein and starch complexers or inhibitors of specific proteolytic enzymes (Rhoades and Cates 1976). Protein and starch complexers react with protein or starch molecules to form larger, less digestible compounds. Proteolytic enzyme inhibitors deactivate specific enzymes within an herbivore's digestive tract, thereby blocking digestion of proteins.

Toxins disrupt metabolic or developmental processes of herbivores. Compounds such as alkaloids, pyrethrins, rotenoids, glycosides, saponins, mustard oils, and cyanide interfere with liver, kidney, and heart functions, act as nerve poisons, disrupt muscle actions, or interfere with basic metabolic processes (Jacobson and Crosby 1971; Freeland and Janzen 1974; Levin 1976a; Rhoades and Cates 1976; Rhoades 1979). Plant analogues of insect hormones or antihormones interfere with insect metamorphosis and development (Sláma 1969, 1979; Heftman 1970; Williams 1970; Rees 1971; Nakanishi et al. 1972; Bowers et al. 1976), while analogues of vertebrate estrogens interfere with implantation and cause abortions (Allen and Kitts 1961; Loper 1968; Braden and McDonald 1970). Toxic amino acid substitutes become incorporated into animal proteins in place of normal dietary amino acids and lead to defective protein function (Fowden et al. 1967; Bell 1972; Rehr et al. 1973). A variety of other side effects and physiological dysfunctions have also been described, and undoubtedly more will be uncovered with future research.

Plant chemical defense strategies

The production of chemical defenses is energetically costly to a plant, making indiscriminate allocation of them to the various plant parts unfeasible. The reality of this cost is well documented by studies showing that growth rate and crop yield are negatively correlated with concentrations of secondary compounds in plant tissues (Rhoades 1979). Since production costs are high, plants should concentrate chemical defenses in their most valuable tissues. They should also use the most cost-effective form of defense. Digestibility-reducing compounds and toxins are not equally expensive to produce, nor are they equally effective. The type of defense a plant exhibits should therefore reflect the relative advantages and costs of each.

Cost-benefit analysis. The effectiveness of digestibility-reducing substances depends on dosage. The higher the concentration in a tissue, the better the tissue is protected (Feeny 1976; Rhoades 1979). At high concentrations these substances are highly effective against both generalist and specialist herbivores because their high molecular weights make them difficult to degrade or neutralize, even with specialized digestive physiology. However, high-dosage requirements and high molecular weights make digestibility-reducing substances energetically expensive to produce.

In comparison, toxins are energetically cheap to produce because they are low-molecular-weight compounds that are capable of crossing cell membranes readily and they are effective even in low concentrations. However, they provide little protection against herbivores capable of detoxifying them. Toxins are a poor defense against specialist herbivores because specialists can afford the costs of detoxifying the specific toxins produced by the few plants they eat. In fact, specialist herbivores often rely on plant toxins to identify suitable food plants, with the result that toxins often act as attractants for specialist herbivores (e.g., Smith 1966; Fraenkel 1969; Rees 1969; Emden 1972; Kogan 1977; Staedler 1977). In contrast, toxins are a good defense against generalist herbivores because generalists cannot afford the many chemical mechanisms that would be necessary to detoxify the large variety of toxins they encounter in the plants they eat (Freeland and Janzen 1974).

Chemical defenses against generalist herbivores. As pointed out earlier, herbivores cannot specialize on ephemeral plants or plant parts because they must shift to alternative foods during periods when the ephemerals are absent. Hence ephemerals are primarily vulnerable to generalist herbivores. Although both digestibility-reducing substances and toxins deter generalist herbivores, toxins are cheaper to produce. One might therefore predict that ephemerals should usually rely on toxins rather than digestibility-reducing substances to defend against herbivores (Rhoades and Cates 1976). Present evidence supports this prediction. Toxins are most prevalent in ephemeral herbaceous plants and in the actively growing parts of woody plants (McKey 1974; Levin 1976b; Rhoades and Cates 1976).

The importance of seasonal availability is made evident by the chemical defense tactics used by plants that produce both toxins and digestibility-reducing substances. In such species a seasonal shift from one type of chemical defense to the other often occurs. For example, the young, actively growing tissues of bracken fern *(Pteridium aquilinum)* contain high concentrations of cyanogenic glycosides and low concentrations of tannins in early spring, but this changes progressively toward low concentrations of glycosides and high concentrations of tannins as the fronds mature (Figure 6–8) (Lawton 1976). Thus the ferns shift from reliance on toxins during the short-term growing period to reliance on digestibility-reducing substances during the long-term period of maturity. The latter substances are a defense against specialist herbivores, as will be discussed more fully in the following section.

Toxins are not always deployed against generalist herbivores. Producing toxins does require some energy, and in some cases that energy may be better used for other purposes. Nicotine toxins in tobacco plants (*Nicotiana tabacum*), for example, are least concentrated in seedlings, become more concentrated in immature plants, and are most concentrated in fully mature plants (James 1950). Young, undifferentiated tissues may be metabolically more costly to defend than mature leaves (McKey 1974; Orians and Janzen 1974), or a greater proportion of the available energy may be best allocated to growth during early stages of development when growth rates are highest. In any case, plants are not equally well protected during every stage of growth. This characteristic, along with the higher nutrient

Figure 6–8

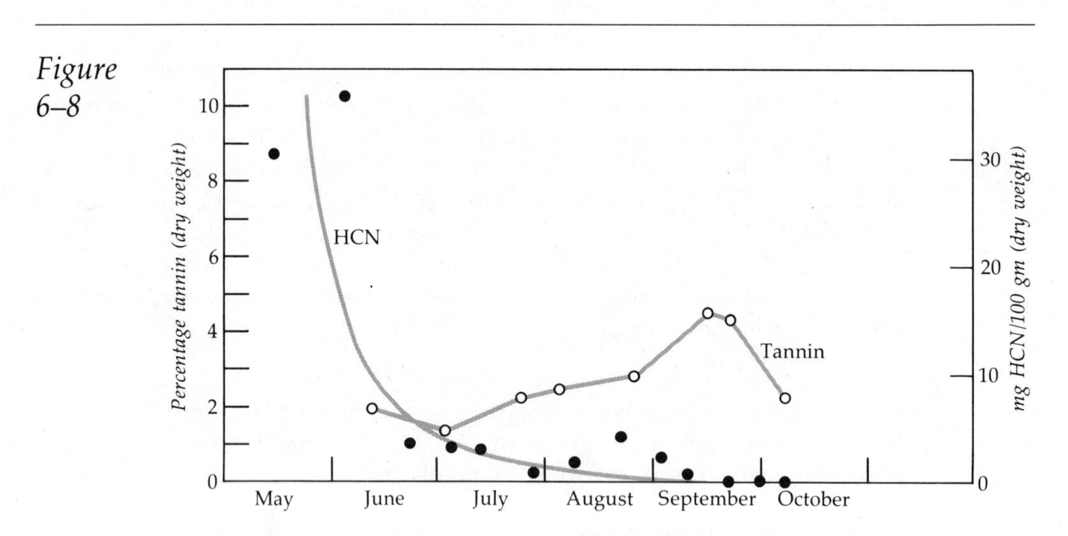

Seasonal changes in cyanide (HCN) and tannin content of bracken fern fronds.

Note that cyanide content is highest in young developing fronds, while tannin content is highest in older mature fronds.

Source: From D. F. Rhoades and R. G. Cates, "Toward a General Theory of Plant Antiherbivore Chemistry," in J. W. Wallace and R. L. Mansell, eds., *Biochemical Interaction between Plants and Insects: Advances in Phytochemistry,* Vol. 10 (New York: Plenum Press, 1976), p. 183. Modified from J. H. Lawton, "The Structure of the Arthropod Community on Bracken (Pteridium aquilinium)," *Botanical Journal of the Linnaean Society* 73 (1976):187–216. Reprinted by permission.

content of young tissues, may explain why herbivores usually prefer young, growing plant parts over mature ones.

Many ephemeral plant species produce few or no toxins because most of their energy is devoted to rapid growth and reproduction (Cates and Orians 1975). Such species are typical of early successional stages, and they depend on colonizing newly opened habitats for success. They therefore rely on escape in time and space instead of toxins to escape herbivores.

Chemical defenses against specialist herbivores. Mature leaves are vulnerable to both specialist and generalist herbivores. Specialization is possible because mature leaves persist for prolonged periods each growing season. One might therefore predict that mature leaves should be protected by digestibility-reducing substances, and indeed, concentrations of those substances are highest in such tissues (Rhoades and Cates 1976).

Striking evidence that duration of availability rather than tissue age is responsible for the high concentration of digestibility-reducing substances in perennial plants comes from a comparative study of oak trees (*Quercus*) and creosote bushes. Digestibility-reducing tannins are most concentrated in the mature leaves of oaks, as they are in most other plants that produce tannins, but in creosote bushes digestibility-reducing phenolic resins are most concentrated in the youngest leaves (Rhoades and Cates 1976). This difference results from a difference in the growth patterns shown by the two species.

Oaks leaf out rapidly in spring, and young leaves are only available to herbivores for a few weeks (Feeny 1970). Hence young leaves are essentially invulnerable to specialist herbivores. Mature leaves, on the other hand, are present during nearly the entire growing season and represent a predictable resource for specialist herbivores. Just the opposite pattern is true for creosote bushes (Rhoades and Cates 1976; Rhoades 1977). During wet periods creosote leaves grow in sequence from the growing tips of sprays outward, with younger leaves continually replacing older ones on the outer tips of the branches. When a dry spell sets in, the inner mature leaves are shed first, leaving only young leaves on the plant. Mature leaves are therefore present only during ephemeral periods of rainfall, while young leaves persist through the long dry season. Thus it appears that length of growing season, not age of leaves, determines which type of chemical defense is most advantageous.

Further evidence showing the importance of seasonal availability comes from studying the way digestibility-reducing substances are distributed across species. Tannins are most prevalent among evergreen trees (87% of the species studied) and deciduous trees (79% of the species studied); they are least prevalent among annuals (17% of the species studied) and herbaceous perennials (14%) (Bate-Smith and Metcalfe 1957). Trees are more vulnerable to specialist herbivores than either annuals or perennials because availability of foliage is more prolonged and predictable for trees. Again, vulnerability to specialist herbivores is associated with use of digestibility-reducing substances. However, this pattern is not true for all types of digestibility-reducing substances, since proteolytic enzyme inhibitors have been found only in annuals and perennials (Rhoades and Cates 1976). Why the latter substances are restricted to plants that seem relatively invulnerable to attack by specialist herbivores is still unknown.

Chemical defenses against seed dispersers. Foliage is not the only part of plants to contain plant secondary compounds. Unripe fruits contain high concentrations of tannins, which give them an astringent taste and makes them less digestible (Goldstein and Swain 1965). As fruits ripen, the tannins polymerize and lose their effect on palatability and digestibility. This mechanism prevents seed-dispersing animals from eating fruits before seeds have matured. Similarly, large amounts of tannins are found in the seed coats of many plants, even in species otherwise lacking tannins (Bate-Smith and Riberéau-Gayon 1959; Janzen 1971). Tannins are probably present in seed coats to deter seed predators who are incapable of removing seed coats before digestion, which in some species is necessary in order for seeds to germinate.

Detoxification systems of herbivores

Plant-herbivore interactions are coevolved systems. Herbivores continually evolve new strategies to overcome plant defenses, while at the same time plants continually evolve better defenses to counteract those new strategies. Since every herbivore species has its own means of overcoming plant defenses, the defenses exhibited by each plant species should be a compromise between the high cost of gaining complete protection from every herbivore species and the high cost of foregoing protection altogether.

Common methods used by herbivores to circumvent plant defenses are avoidance of unusually toxic plant tissues, specialization on relatively few plant species to limit the number of different toxins ingested, and development of specialized digestive systems capable of neutralizing or ameliorating the effects of digestibility-reducing substances (Freeland and Janzen 1974). The metabolic costs of detoxification are particularly important determinants of food availability because they limit the rate at which herbivores can ingest toxic plant materials.

One mode of detoxification is to break down foreign chemicals by means of microsomal or drug enzymes (Schuster 1964; Parke 1968; Mandel 1972; Adamson and Davies 1973). These enzymes are concentrated in the liver but are also found in the intestinal mucosa, kidneys, lungs, skin, testes, and thyroid glands. They function to metabolize excess steroid hormones as well as to break down foreign chemicals. Detoxification by microsomal enzymes is an energetically expensive process. It entails absorbing the toxin into the bloodstream, producing a different high-molecular-weight enzyme for breaking down each toxin encountered, and excreting the metabolic waste products resulting from degradation of the toxin (Freeland and Janzen 1974). These costs limit the rate at which toxins can be processed and hence the rate at which foods can be ingested.

A second mode of detoxification involves a change in gut chemistry. Digestion in most mammals occurs in an acidic stomach, followed by further digestion in an alkaline small intestine, caecum, and colon. Plant toxins remain inert until they are activated by enzymes present in the small intestine (see Freeland and Janzen 1974). However, they cannot be degraded before they reach the small intestine because stomach acids interfere with the degradation process. Hence many toxins are absorbed into the bloodstream before they can be degraded in the digestive tract. In ruminating mammals such as deer and cattle, however, this is not true. Because these species have alkaline stomachs, toxins can be degraded before they are absorbed into the bloodstream (Moir 1968).

Ruminant stomachs contain many strains of bacteria and protozoa, each capable of degrading specific plant toxins. The rate at which a toxin is degraded depends on the composition of the gut flora, which in turn depends on how much of the toxin has been previously ingested. The herbivore's diet acts as a strong selective agent in determining which strains of bacteria and protozoa flourish. Strains capable of degrading toxins that are only rarely ingested persist in the gut flora, but only in low concentrations. Ruminants can gradually develop a capability for degrading new toxins, but because the process is slow, they maintain steady diets consisting of foliage with prolonged availability rather than changing diets continually to exploit ephemeral plant materials.

Both microsomal and gut detoxification systems are unable to handle a new toxin in large quantities. Herbivores cannot afford the energetic costs of maintaining detoxification systems not currently being used. Production of the necessary enzymes or gut flora must first be induced by ingesting small amounts of the toxin. For this reason herbivores are generally cautious in sampling new types of food, are generally capable of learning to eat or not eat foods by consuming them in very small quantities, and generally have a good memory for tastes associated with particular plants or plant tissues (Freeland and Janzen 1974). Since only a certain amount of toxin can be processed per unit time, generalist herbivores should be limited by the rate of detoxification rather than the availability of each food type. Thus food intake of herbivores that exploit seemingly abundant foods such as mature deciduous leaves or conifer needles may be limited by detoxification capabilities rather than food abundance. This is an important point. It has been suggested as an important factor affecting grouse mating systems (Wittenberger 1978a) and may well have important implications for other herbivore social systems as well.

Predator-prey interactions

Predator-prey interactions parallel plant-herbivore interactions in being coevolved systems. Natural selection continually modifies the defensive mechanisms of prey animals and the corresponding capture techniques and assimilation mechanisms of predators. Several general factors are important in shaping prey defenses. They include (1) hunting tactics and capture capabilities of the principal predators threatening each prey species, (2) structural characteristics of the prey animal's environment, and (3) limitations imposed by the prey animal's size, morphology, physiology, sensory capabilities, and behavioral repertoire.

Predator hunting tactics and their effect on prey defenses

Filter feeding. One method of capturing prey is to *filter* them out of the environment. This may be accomplished with the help of a specialized filtering apparatus or by systematically sifting through substrate materials. Some examples are bivalve molluscs filtering microorganisms from water currents, earthworms filtering invertebrates out of the soil, orb-building spiders filtering insects from the air, sandpipers filtering invertebrates out of intertidal mudflats, and baleen whales filtering plankton from large volumes of ocean water. The prey of filter feeders have no effective defense other than avoiding areas where predators are likely to be found. Since predators tend to concentrate where food is most abundant, selection should favor wide spacing of prey that are mainly vulnerable to filter feeders. Nevertheless, such prey often become concentrated in local areas by air or water currents, nutrient concentrations, or specialized habitat requirements.

Ambushing and aggressive mimicry. A second method of capturing prey is to *ambush* them. It may entail sitting and waiting for prey to come within striking distance, or it may entail stealthy approach toward unwary prey. Sessile marine organisms like coral polyps, sea anemones, and tubeworms wait for prey to drift within reach of their tentacles. Ant lion larvae (*Myrmeleon*) ambush ants by digging a small pit in sandy soil (Wheeler 1930). When a victim falls into the pit, it cannot scramble out because the sandy sides of the pit keep sliding downward. Many spiders ambush prey by lying in wait inside a burrow or in some other hiding place, where they can jump out on an unsuspecting prey as it walks by (Bristowe 1958; Buchli 1969). Praying mantises sit motionless in vegetation, relying on their resemblance to twigs or bark to escape notice, until an unwary victim comes within reach (Photo 6–2). Frogs sit quietly in wait for flying insects. Many snakes lie coiled in wait for small rodents and birds. Flycatching birds perch quietly on exposed branches or other vantage points until they spot an unwary flying insect and then dart out to capture it. North American cuckoos (*Coccyzus*) wait quietly in thick foliage of a tree for large insects, amphibians, or reptiles to come within reach (Hamilton and Hamilton 1965). Cats often wait quietly along game trails and pounce upon unsuspecting birds, rabbits, deer, or other prey. They also sneak up stealthily on prey and then pounce suddenly to make their capture.

Although ambush predators usually depend on concealment or stealth to escape notice of prospective prey, a few gain close approach by resembling harmless animals. This tactic is commonly referred to as **ag-**

A.

B.

Photo 6–2 *Praying mantises rely heavily on camouflage, both to escape avian predators and to capture insect prey.*

A. Phyllocrania paradoxa *of Ghana. B.* Theopompella westwoodi *of Ghana.*

Photos from M. Edmunds, *Defence in Animals: A Survey of Anti-Predator Defences* (Essex: Longman Group Ltd., 1974), p. 110.

gressive mimicry. One example of an aggressive mimic is the sabre-toothed blenny (*Aspidontus taeniatus*), a fish that mimics the cleaner fish *Labroides dimidiatus*. When a host fish allows the mimic to approach closely, expecting to be cleaned, the mimic takes a bite from one of the host's fins (Eibl-Eibesfeldt 1959; Randall and Randall 1960). Another example is the zone-tailed hawk (*Buteo albonotatus*) of southwestern North America, which resembles black vultures (*Coragyps atratus*) and soars among them (Wickler 1968). Because of the close resemblance, prey animals pay no attention to the hawk until it swoops down for a strike.

A few species go a step further and actually lure prey toward them. The ant-mimicking thomisid spider *Amyciaea* of Asia and Australia behaves like an ant in distress and pounces on any ant approaching to investigate (Hingston 1927; Mathew 1954; Clyne 1969a). Bola spiders (*Dicrostichus, Cladomylea, Mastophora*) twirl a sticky ball of silk impregnated with chemicals to attract certain species of moths (Wickler 1968; Eibl-Eibesfeldt 1970). The chemicals are the same as those released by female moths to attract mates. The angler fish *Phrynelox scaber* projects a worm-like appendage from its mouth and lures prospective victims nearer by vibrating it (Pietsch and Grobecker 1978). When the victim is near enough, the angler fish swallows violently, engulfing the prey and a large volume of water into its stomach. Another species, the batfish (*Ogcocephalus*), angles downward to lure small crustaceans within reach (Wickler 1968). Deep-sea angler fish (Ceratiidae) have bioluminescent lures of various colors, either in their mouths or suspended from long appendages, to attract prey (Wickler 1968; Eibl-Eibesfeldt 1970). The alligator snapping turtle *(Macroclemmys temmincki)* of the southern United States wiggles a worm-like appendage on its tongue to attract small fish. Such deceptions cannot work if the prey are not fooled. Hence predators relying on deception to lure prey depend on camouflage, unusual body shapes, or concealment to make the deceptions work.

Prey animals use a variety of defenses against ambush predators. Many rely on chemical defenses that make them toxic or unpalatable to predators. They possess wary habits and try to avoid potentially hazardous ambush sites. Mammalian prey, for example, are extremely wary when visiting water holes or salt licks, and they are often leery of entering dense cover or open clearings. Such areas cannot be avoided entirely, but mutual vigilance and other cooperative defenses may be used to reduce the risk. The deceptive tactics used by some predators favor improved discrimination abilities among prey. Improved discrimination often results from learning, but prey probably inherit a propensity to learn such discriminations.

Search and pursuit. A third method of capturing prey is *search and pursuit*. Many predators search likely areas for prey and pursue vulnerable individuals that they detect. The duration of pursuit may be very short or quite prolonged, depending on the relative agility and speed of predator and prey.

Search-and-pursue predators often form search images for specific types of prey (Tinbergen 1960). A *search image* is simply a preconceived notion of what is being sought. Predators are thought to form search images based on key stimuli unique to particular kinds of prey. The basic idea of search images is that predators look for particular prey types, based on their previous experience. Search images are formed after a prey type has been captured several times in succession, and they aid in detecting additional prey of the same type. Forming a search image is most useful when hunting clumped prey; it has little value when hunting widely dispersed prey.

Prey animals evade search-and-pursue predators by escaping detection or, once detected, escaping capture. Many tactics are used to those ends, and they will be discussed more fully in later sections.

Cooperative hunting. A fourth method of capturing prey is *cooperative hunting* (see pp. 123–126). Cooperative hunting is especially advantageous when prey are larger or more agile than the predator. It does not evolve unless it increases a predator's capture success enough to offset the concomitant increase in competition at kills. Counterattack, mutual vigilance, and various other social defenses are commonly employed to escape cooperative hunters. Since cooperative hunting basically entails search-and-pursue tactics, sometimes along with a stalking phase (e.g., lions), nearly all prey defenses used against search-and-pursue predators may also be used against cooperative hunters.

Other predator characteristics affecting prey defenses. Two other aspects of predator hunting tactics can affect the defenses employed by a prey species: the sensory modalities used to detect prey and the range of prey sizes hunted by a predator. Camouflage and mimicry based on visual cues, for example, are effective against visual hunters but not against hunters relying on other sensory modalities to detect prey. All predators prefer prey of certain sizes and usually will not attack prey that are too small or too large (see Charnov 1977). Prey animals might therefore evolve larger sizes to escape some of their more important predators (see Wiley 1974). The importance of body size is most evident in animals with indeterminant growth. Small fish fry, for example, are highly vulnerable to predation, but as they grow larger, their vulnerability greatly decreases. This makes rapid growth highly advantageous and places a premium on any behaviors that enhance growth rate. Whether predation affects equilibrium body size in species with determinate growth is less clear because many other factors also affect optimal body size (see Searcy 1979, 1980; pp. 428, 430).

Spatiotemporal escape mechanisms

Ephemeral habitats. Prey animals living in ephemeral habitats often depend on escape in space and time, just as ephemeral plants do. These so-called "fugitive species" (after Hutchinson 1951) or "*r*-selected species" (after MacArthur and Wilson 1967) always stay one step ahead of their predators by continually colonizing newly created ephemeral habitat patches. They reproduce rapidly in the new habitat before predators have much chance of finding them, and their progeny soon disperse as propagules to colonize new habitats. This life cycle is possible only because competition is virtually absent in newly colonized habitats. The sorts of habitats colonized by fugitive species include temporary pools of water, fresh dung or carrion, and habitat patches in early stages of vegetational succession.

Temporal synchrony. Some prey animals periodically swamp the exploitative capacities of predators because their abundance undergoes high-amplitude fluctuations. Since predator densities are limited by food availability during periods of scarcity, predators cannot fully exploit sudden surges in prey abundance. Prolonged surges in abundance allow predator densities to increase, but shorter-term surges do not.

Periodical cicadas (*Magicicada*) are a good example of how this effect works. Adult cicadas emerge synchronously from underground nymphs once every 13 or 17 years, depending on the species (Figure 6–9) (Lloyd and Dybas 1966). When the adults emerge, they emerge in very large numbers. During cicada peaks the forests reverberate with their calls. Although avian predators concentrate heavily on cicadas during outbreaks, they cannot capture more than a small proportion of them before the cicadas

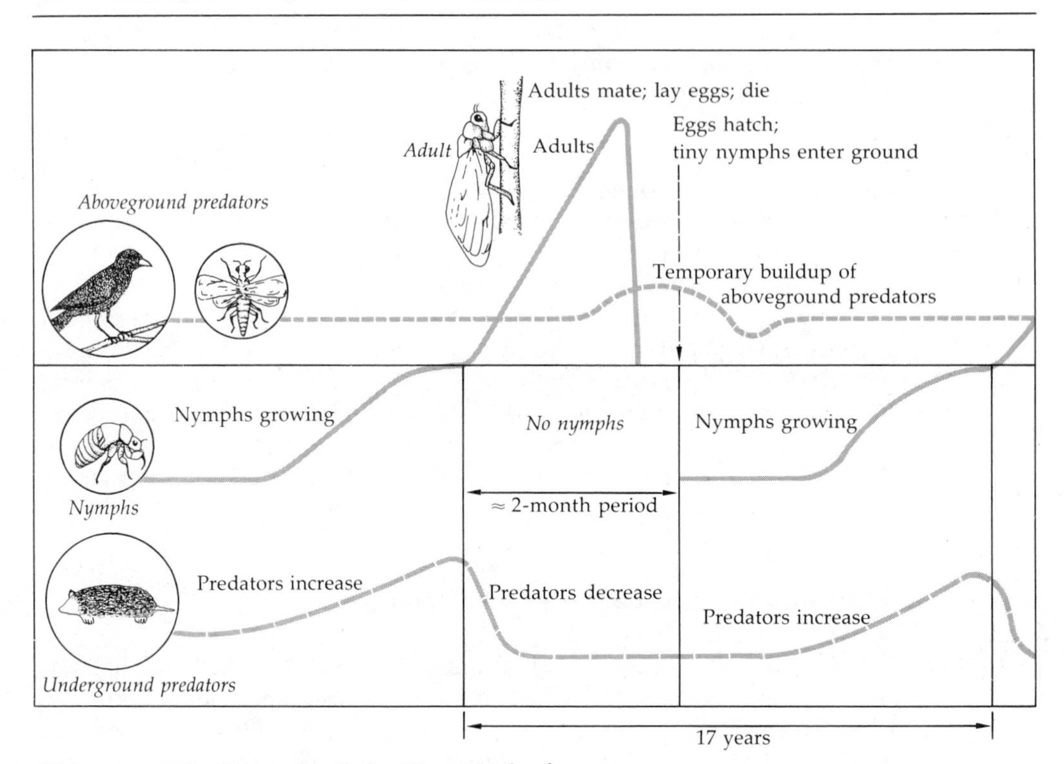

Figure 6–9 *The life cycle of the 17-year cicada.*

The cicada life cycle probably evolved as a defense against predation. Adult cicadas emerge from underground only once every 17 years to lay their eggs. At that time they become extremely abundant for a brief period before they all die. The populations of insectivorous birds and parasitoid wasps, the chief enemies of cicadas, cannot increase fast enough to exploit the sudden superabundance of prey. As a result, most adult cicadas escape predation. Similarly, larvae escape predation by moles to some extent because mole populations decline every 17 years when all the cicada larvae emerge as adults. Mole populations remain relatively low for several years following each cicada emergence, until developing larvae of the next cycle become large enough to be worth eating.

Source: From M. Lloyd and H. S. Dybas, "The Periodical Cicada Problem, I. Population Ecology, II. Evolution," *Evolution* 20 (1966):139. Reprinted by permission.

finish reproducing and die. During most of their life cycle cicadas remain underground as nymphs, where they are inaccessible to aboveground predators. Note that the length of each cycle is ordinarily a prime number, that is, a number not divisible by any smaller number other than 1. Peaks of emergence therefore do not consistently match the peaks of any shorter-term population cycles of predatory species.

A variety of other insects have similar life cycles, with periodicities ranging from two to five years (Bulmer 1977). A few species, such as wood crickets (*Nemobius sylvestris*), have two-year life cycles but are not periodic because a separate, reproductively isolated year class emerges each year in the same geographical region (Ragge 1965). The evolution of periodic emergence patterns is not well studied. Mathematical models suggest that they should evolve only when competition between larval year classes is greater than competition within year classes (Bulmer 1977). Predation pressure should favor periodical emergence by reinforcing the effects of high competition between year classes, and it probably explains why emergences are synchronous.

Synchronous emergence on a seasonal basis can also deter predators effectively. Damselflies and dragonflies emerge from aquatic larvae in several synchronized pulses, probably to swamp the exploitative ability of blackbirds and other avian predators (Corbet 1962; Orians 1980). Some species of frogs metamorphose synchronously to reduce predation by fish and birds (Arnold and Wasserug 1978). Synchronous nesting reduces the number of eggs destroyed by predators in many colonial birds (Hoogland and Sherman 1976). The common vole (*Microtus arvalis*) shows a diurnal activity pattern, with most individuals foraging synchronously for short periods throughout the day (Daan and Slopsema 1978). Voles that are active in phase with the population are less likely to be killed by kestrels (*Falco tinnunculus*) than voles that are active out of phase with the population, indicating that synchronous foraging swamps the ability of kestrels to capture prey. In East Africa nearly all wildebeest (*Connochaetes taurinus*) give birth during a brief two-week calving season early in the rainy season (Estes 1976). Calves born early or late are more frequently captured by predators than calves born at the peak, apparently because predators cannot fully utilize all the calves present during peak production. This effect is further enhanced by aggregating into larger herds, as large nursery herds contain a higher proportion of calves than do smaller nursery herds, suggesting that calf mortality is lower in large herds (Estes 1966; Estes and Estes 1979).

Spatial clumping. Spatial clumping can have the same effect as temporal clumping, provided that predators remain dispersed (Taylor 1976, 1977). As Brown (1975a) puts it, prey animals are safer in an area where predators have full stomachs than in areas where they have empty stomachs. The importance of this effect, which Brown terms the "gluttony principle," may be considerable, especially for gregarious animals. Predators often remain dispersed despite clumping of prey (e.g., Calaby 1951). The spatial distribution of predators is attuned to the distribution of all important prey species, as well as to the distribution of other requisites, and it cannot change in response to clumping by a single prey species, especially if that prey is only a minor part of the predator's diet.

Swamping the exploitative ability of predators may explain why virtually the entire eastern North American population of monarch butterflies (*Danaus plexippus*) congregates in a few small woodlots in Mexico each winter (Urquhart 1976; Calvert et al. 1979). The butterflies are packed so densely that they are frequently trodden by cattle, and their combined weight is sufficient to break off tree branches 3 inches thick. The butterflies are attacked by birds, mice, shrews, and ants, but most escape to migrate north the following spring.

Clumped prey may also be safer because dispersed predators would detect clumps less frequently than would dispersed individuals. Mathematical models show, for example, that fish are safer in schools because schools are detected less frequently by predators (Brock and Riffenburgh 1960; Cushing and Harden-Jones 1968; Pitcher 1973). Certain ducks may aggregate in crèches (i.e., groups of several or many broods) for the same reason. Munro and Bédard (1977) found that common eider ducklings (*Somateria mollisima*) are safer in crèches than in dispersed broods when attacked by single gulls, apparently because crèches are detected less often than scattered broods. Ducklings are less safe in crèches when attacked by groups of gulls, but such attacks are relatively infrequent, and crèching is, on average, advantageous to duckling survival.

Spatial and temporal clumping are effective defenses against predation only after prey densities more than saturate the exploitative abilities of all predators combined. Clumping is likely to attract predators and hence is likely to increase vulnerability of prey until the saturation point is reached. Since degree of clumping is limited by competitive interactions among prey, selection may favor dispersed spacing of prey and other forms of defensive tactics when food is dispersed. Escape through oversatiation of local predator populations is feasible primarily for prey that are unconstrained by competition until they reach high densities.

Morphological defenses of prey

Some animals, like some plants, use armor or spines to thwart predators. Circumventing such defenses is often not worth the effort for predators that have alternative prey available to them. Gastropod snails, bivalve molluscs, wood lice (*Armadillium*), millipedes, terrapins (i.e., freshwater turtles), tortoises, armadillos (*Tolypeutes*), and pangolins (*Manis*) (a peculiar group of arboreal and terrestrial mammals of Africa and India that eats exclusively ants and termites) retreat into hard shells or roll up into tight balls shielded by body armor when threatened by a predator (Edmunds 1974). The hard exoskeletons of many insects may provide similar protection, at least against some invertebrate predators. Hermit crabs do not grow shells of their own, but they live inside empty gastropod shells and move from smaller to larger shells as they grow larger.

Spines and stiff body hairs are common defenses against predation. Grasshoppers have strong spines on their forelegs that may pose a threat to small predators such as lizards (Edmunds 1974). Caterpillars often have stiff bristles or body hairs that, in conjunction with distastefulness or toxicity, make them unappealing to avian predators (Cott 1940). Many fish have sharp spines on their dorsal fins that make swallowing by large fish difficult or even hazardous (e.g., Hoogland et al. 1956–1957; Hartmann 1979; Keenleyside 1979). Surgeonfish of tropical coral reefs are so named because they possess a single scalpel-sharp spine just anterior to the caudal fin on each side of the body. Any attempt to grab or swallow a surgeonfish can cause severe injury to a would-be predator. A variety of lizards have sharp spines on their bodies as well. For example, the thorny devil (*Moloch horridus*), a tiny lizard of Australia, has spines all over its body to discourage predators from grabbing it (Poignant 1965). Various mammals ward off predators with loosely attached body spines, including porcupines (*Erethizion, Atherurus*), hedgehogs (*Erinaceus*), echidnas (*Tachyglossus*), and streaked tenrecs (*Hemicentetes semispinosus*) (Photo 6–3). A predator approaching any one of these animals is likely to get a noseful of painful spines.

Morphological defenses are not without cost, as they usually hamper locomotory capabilities. They are generally found on slow-moving animals as one of several defense mechanisms, and they are most effective against generalist predators that have not evolved specialized hunting tactics for circumventing them. Nearly all morphological defenses can be overcome by predators specialized to cope with them. Snails, for example, are swallowed whole and then dissolved internally by the opisthobranch sea slug (*Aglaja inermis*) (Paine 1963). Mollusc shells are cracked open with rows of flat teeth by plaice (*Pleuronectes*) and rays (*Raia*), with notched chelae by crabs (*Calappa*), and with stones by song thrushes (*Turdus philomelos*) and sea otters (*Enhydra lutris*) (Edmunds 1974). Starfish pry open bivalve clams with powerful hydraulic tube feet and insert their stomachs to digest the clams' internal body parts. Oystercatchers (*Hematopus*), a type of shorebird, pry open clams and snip the musculature that holds the shell shut, using their specially shaped bills. Banded mongooses break open the hard exoskeletons of glomerid millipedes by throwing them repeatedly against a stone (Eisner and Davis 1967; Eisner 1968). Egyptian vultures (*Neophron*

A. B.

Photo 6–3 *Body spines are one means of defending against predators.*

Two species that rely on spines for escaping predation are shown here: A. the streaked tenrec and B. the echidna or spiny anteater.

A. Photo reproduced by permission of the National Zoological Park, Smithsonian Institution, Washington, D.C. B. Photo by Axel Poignant.

percnopterus) crack open the nearly impregnable shells of ostrich eggs by hurling rocks at them (Lawick-Goodall and Lawick 1966). Thus most morphological defenses are not invincible, but they do discourage all but the most persistent or most specialized predators.

Camouflage

Camouflage is one of the more common defensive tactics employed to deter predators (Photo 6–4). Animals that resemble part of their environment and are normally overlooked by predators as a result are said to be *cryptic*.

An obvious and effective form of crypsis is to simply blend into the background coloration of the substrate (Kettlewell 1973; Kaufman 1974; Wicklund 1975; Curio 1976; Endler 1978). To do this, animals must either select appropriate backgrounds or change colors to match the backgrounds they happen to be on (see Frisch 1973 and Edmunds 1974 for examples). A few species modify their environments to improve crypsis or even carry camouflage items about with them. Sandwich terns (*Sterna sandvicensis*),

Photo 6–4

*This grasshopper (*Zabilius aridus*) from Ghana gains protection from predators by closely resembling a leaf.*

Such camouflage tactics are commonly used as a defense against predation.

Photo from M. Edmunds, *Defense in Animals: A Survey of Anti-Predator Defenses* (Essex: Longman Group Ltd., 1974), p. 110.

for example, defecate and scatter eggshells around in their colonies, possibly to reduce conspicuousness of eggs and chicks to crows (Croze 1970), and various anthropods carry debris around with them to enhance their resemblance to the background (Edmunds 1974).

Blending with background coloration is only one way to achieve crypsis. Some animals take on the appearance of inanimate objects or plant structures to break up body contours and make them less recognizable against patterned backgrounds. *Countershading,* the occurrence of dark colors above and light colors below, is common among aquatic animals to eliminate shadows created by overhead illumination. Many marine plankton are transparent, making them virtually invisible in the open ocean.

Crypsis based on visual stimuli is, of course, only effective against visual hunters. Many predators detect prey by means of auditory, olfactory, or tactile cues (e.g., see Marler and Hamilton 1966), and some, such as sharks, rays, and the freshwater catfish *(Ictalurus nebulosus),* can even detect prey by using electric potentials generated by their bodies (Kalmijn 1971; Peters and Bretschneider 1972). Sounds and smells could conceivably be masked by background noise or odors, but the extent to which prey animals conceal nonvisual signs of their presence has never been properly investigated.

Crypsis is an effective antipredator defense, but only in the right circumstances. In exposed areas cryptically colored animals must remain essentially motionless to stay protected. Predators often rely on motion to detect prey, and even slight movements may reveal a prey animal's presence. Since remaining motionless is often incompatible with essential activities such as foraging and reproduction, cryptic animals are frequently active primarily at night when movements are less readily detected. Species active during the day usually behave furtively and stay in dense cover to avoid detection. Animals occasionally rely on crypsis while performing some activities but not while performing others. Many birds, for example, depend on crypsis while incubating eggs but not while foraging or defending territories.

Camouflage loses much of its effectiveness when prey animals become too common. Predators may then develop a search image for the prey and find many or most prey individuals despite their camouflage. Tinbergen and his co-workers (1967) experimentally demonstrated the effect of spacing by laying out painted chicken eggs in a 3 x 3 grid, with 8 of the 9 eggs in each trial partially hidden under soil or vegetation. They found that many more eggs were eaten by carrion crows (*Corvus corone*) when eggs were only 50 cm apart than when they were 800 cm apart. Croze (1970) repeated the experiment, using artificial insect larvae made from colored flour and lard, and found that the proportion of prey items eaten by crows increased with prey density. Thus crypsis is most effective when prey are widely spaced and least effective when prey are clumped.

Finally, counterselection for alternative color patterns may preclude crypsis (Hamilton 1973). Intense competition for females or territories may favor bright colors or conspicuous markings on males to enhance courtship or threat displays. In certain thermal environments black or white coloration is important for maintaining optimal body temperature (Hamilton 1973; Henwood 1975; Dmi'el et al. 1980). Such colors make crypsis less effective, and when the net advantages of possessing them exceed the advantages of being cryptic, alternative modes of defense are employed to escape predation.

Fleeing predators

An important alternative to crypsis is flight. Many cryptic species resort to flight when their camouflage fails to fool a predator. Flight is also a primary means of escape for species lacking cryptic coloration and behavior, and in those species vigilance is an important component of their defense strategy.

Flight may entail erratic evasion tactics, attempts at outrunning the predator, or rapid escape to a place where the predator cannot follow. Safe retreats may be to a new milieu, as when a bird takes flight to escape a mammalian predator or a frog jumps into a pond to escape an avian predator, or they may be inaccessible places like cliff faces, crevices, small outer limbs of trees, or burrows previously dug into the substrate. The effectiveness of evasive tactics depends, of course, on the prey's speed, agility, and stamina compared to those of the predator. It also depends on how far prey animals wander from safe retreats while conducting daily activities. Thus morphological, ecological, and social factors often dictate whether flight will be a successful line of defense.

Inaccessibility of prey

Some animals avoid predators by living permanently in safe or inaccessible places. Many marine invertebrates live in tubes or burrows dug into the substrate (Edmunds 1974). They never leave their burrows, although they must often emerge partially to filter microorganisms out of water currents. Hence they are vulnerable to predators capable of approaching closely without being detected. A few animals gain protection by residing close to well-protected species. Anemonefish (*Amphiprion, Gobius,* etc.) live near the mouth openings of sea anemones and are protected by the anemones' stinging nematocysts, which are specialized cells for inflicting injury on potential predators (Abel 1960; Davenport 1962, 1966; Mariscal 1970a, b, 1972; Schlicter 1970; Allen 1975). An anemone tolerates the presence of an anemonefish after an initial period of habituation, possibly because the fish brings some food to it (Mariscal 1972; but see Allen 1975). Similarly, harvestfish (*Peprilus alepidotus*), butterfish (*Poronotus triacanthus*), whiting (*Gadus merlangus*), and man-of-war fish (*Nomeus gronowi*) hide beneath the umbrellas of various jellyfish to derive protection from the jellyfish's nematocysts (Mansueti 1963; Horn 1970; Rees 1966; Gotto

1969). Moles live underground in tunnels and feed on soil invertebrates, where they are unreachable by aboveground predators. Many marine invertebrates, earthworms, burrowing lizards, some birds, and mammals besides moles also escape predators by living part or most of their lives underground (see Edmunds 1974).

Living permanently in safe places is feasible only if the prey animal can satisfy all requirements for survival and reproduction there. This constraint forces most animals to leave safe places during at least part of their lives. Thus safe retreats are often employed primarily while animals are resting, grooming themselves, sleeping, or lying dormant.

Chemical defenses

Like plants, some animals rely on chemical defenses to avoid being eaten, though animals deploy chemical defenses in more diverse ways.

Sprays and secretions. Some animals, such as bombardier beetles (Photo 6–5), certain ants, spitting cobras (*Naja, Hemachatus*), and skunks (*Spilogale, Mephitis*) spray noxious chemicals when threatened by potential predators (Buckley and Porges 1956; Eisner 1970). Many gastropods secrete noxious substances that diffuse through the water and repel predatory fish (Thompson 1960a, b; Fretter and Graham 1962; Edmunds 1968). Noxious sprays are prevalent among arthropods, and some species can spray them with considerable force (Eisner 1970). For example, a predaceous assassin bug (*Platymeris rhadamantus*) of Zanzibar and East Africa can spray its cobra-venom-like saliva up to 30 cm (Edwards 1960, 1961), and a whipscorpion (*Mastigoproctus giganteus*) of the southwestern United States can spray an acidic secretion up to 80 cm (Eisner et al. 1961). These distances are not the effective distance of defense, as most arthropods do not discharge sprays until they are physically contacted, but they give some idea of how forcefully the sprays are ejected (Eisner 1970).

Some species also use nonnoxious sprays as chemical defenses. Cuttlefish release black ink clouds as a smoke screen for making good their escape (Holmes 1940; Boycott 1958), and deep-sea squid (*Heteroteuthis*), certain shrimp (*Acanthephyra*), and searciid fish discharge luminous clouds that probably serve the same purpose (Nicol 1971).

Another method of administering toxic secretions is by rubbing them on would-be assailants. Some aphids secrete a waxy substance when touched by a parasitic wasp (Eisner 1970). The substance quickly hardens and may entrap the wasp. The daddy-long-legs (*Vonomes sayi*) secretes a chemical irritant and rubs it against attacking ants (Eisner et al. 1971). Females of the tropical wasp (*Mischocyttarus drewseni*) rub ant-repellents from their abdomens onto pedicels used to suspend their nests from overhangs (Photo 6–6) (Jeanne 1970, 1972). Although such defenses are prevalent in arthropods, they are relatively rare in vertebrates. The more usual form of chemical defense in vertebrates is injected venoms.

Photo 6–5

A.

B.

C.

*A bombardier beetle (*Brachinus crepitans*) emitting a noxious spray at an ant.*

An approaching ant (A) bites the beetle's hind leg (B) and is sprayed by a noxious solution of quinones (C).

Photos from T. Eisner, "Chemical Defense against Predation in Arthropods," pp. 156–217 in E. Sondheimer and J. B. Simeone, eds., *Chemical Ecology* (New York: Academic Press, 1970), p. 188. Photos by Thomas Eisner and D. Aneshansley (Cornell University).

Photo 6–6

A social wasp of the species Mischocyttarus drewseni *rubbing ant-repellent from an abdominal gland onto the pedicel of her nest.*

This behavior helps prevent ants from attacking the wasp's brood.

Photo from R. L. Jeanne, "Chemical Defense of Brood by a Social Wasp," *Science* 168 (19 June 1970):1465. Copyright 1970 by the American Association for the Advancement of Science.

Venoms. A great diversity of animals inject poisons by means of stings, spines, nematocysts, fangs, or other structures. All coelenterates inflict wounds by discharging poison-laden nematocysts, though the majority are not toxic to humans. The scyphozoan sea wasp *Chironex fleckeri* is the most poisonous marine organism known, and its sting can kill a human within 3 to 8 minutes (Halstead 1971a). One family of starfish (Acanthasteridae) and two families of sea urchins (Echinothuridae and Diadematidae) inflict extremely painful wounds with brittle spines that cover their bodies (Halstead 1971b). Among marine polychaetes bristleworms inject toxins with chitinous bristles called setae and bloodworms inject toxins with sharp fangs (Halstead 1971b). Octopuses inflict poisonous bites with their sharp beaks when handled by humans (McMichael 1971), and tiny planktonic sea slugs (*Glaucus atlanticus* and *Glaucilla marginata*) employ poisonous nematocysts obtained by foraging on the chondrophores (defensive structures) of Portuguese men-of-war (*Physalia utriaulus*) (Thompson and Bennett 1969).

Wasps and bees are well known among arthropods for their poisonous stings. Honeybees (*Apis mellifera*) have specialized stingers designed to detach when pushed into soft vertebrate skin, at the cost of the bee's life (Frisch 1955; Maschwitz and Kloft 1971). The poison sac detaches with the stinger and continues pumping venom into the prospective hive robber after the bee has died.

Stinging structures are greatly reduced on most higher ants, possibly because the main predators of ants are other arthropods, but ants can still ward off assailants with stinging bites and some ant groups do possess functional stingers (Cavill and Robertson 1965; Eisner 1970; Maschwitz and Kloft 1971). The stings of some ants are extremely painful. Janzen (1972) writes, for example, that "while hundreds of stings of *Pseudomyrmex* can be tolerated if there is a compelling reason to invade the tree, one to five *Pachysima* stings were enough to drive me away from a tree, leaving me very reluctant to return." The chief enemies of *Pseudomyrmex*, an acacia tree ant, are other invertebrates, but the chief enemies of *Pachysima* are probably large herbivorous mammals that eat foliage occupied by the ants. Another ant with a potent sting is the imported fire ant (*Solenopsis saevissima*), a species introduced inadvertently into the southern United States from Latin America. The imported fire ant injects a painful paralytic nerve toxin by means of an abdominal stinger that, on rare occasions, can be lethal to humans (Blum et al. 1958). By comparison, the stings of a closely related species, *Solenopsis xyloni*, which is native to the southern United States, have much milder effects (Blum et al. 1961). The most toxic ant venom known is that of the harvester ant (*Pogonomyrmex badius*), whose venom is five times more toxic than that of the Oriental hornet (*Vespa orientalis*) and eight to ten times more toxic than that of the honeybee, the two most toxic insect venoms outside the genus *Pogonomyrmex* (Schmidt and Blum 1978). Its great toxicity probably evolved as a defense against vertebrate predators. Cole (1968) graphically describes the effects of *Pogonomyrmex* venom:

> *The effects of a sting can be very painful. Localized swelling and inflammation ensue rapidly. Soon thereafter a throbbing pain, which may last for several hours, extends to the lymph nodes of the inguinal, axillary, or cervical area, depending on the location of the sting. . . . Multiple stings will produce excruciating pain and may induce systemic disturbances of considerable severity.*

Caterpillars from several families of butterflies and moths are equipped with poisonous hairs or spines, and moths of the neotropical genus *Hylesia* have stingers (tiny venomous spines) along the abdomen (Pesce and Delgado 1971). Other venomous arthropods include some centipedes, most spiders (the majority of which are not toxic to humans), and virtually all scorpions (Balozet 1971; Bücherl 1971a, b, c; Maschwitz and Kloft 1971).

Snakes are probably the best known of venomous vertebrates, even though a majority of snakes are not venomous (see Bücherl et al. 1968; Minton and Minton 1969; Bücherl and Buckley 1971a). Snake venoms are used both to immobilize prey and to thwart predators. The only lizard known to be venomous is the gila monster (*Heloderma suspectum*), which injects venom from poison sacs at the base of its five or six most anterior teeth (Tinkham 1971a, b). Gila monsters use their venom primarily to immobilize prey, but they may gain protection from it as well. A protective function is suggested by their conspicuous black and yellow coloration, which probably warns potential predators to stay away (see p. 241).

Less familiar are various other venomous vertebrates. Horn sharks (Heterodontidae) and spiny dogfish (Squalidae), the only known venomous sharks, inject venom into wounds made by two sharp spines adjacent to their two dorsal fins (Halstead 1971c). Similarly, chimaeras inject venom with a dorsal spine located at the anterior end of their single dorsal fin. Most rays are venomous and inflict their venoms with sharp spines on their tails. The position of venomous spines in Chondrichthyes suggests that venoms are mainly used for defensive purposes.

Venomous bony fish include several catfish, weaverfish, stargazers, scorpionfish, and toadfish (Halstead 1971c). These are all marine, bottom-dwelling species with poisonous dorsal, opercular, anal, or pelvic spines. Most are found in tropical oceans, though a few live in temperate waters. Their reliance on venoms for defense appears to be correlated with sedentary, slow-moving habits and in some cases with lack of safe hiding places. An unusual species is the poison-fang blenny (*Meiacanthus atrodorsalis*), which injects venom with a single long canine tooth at the rear of each lower jaw (Losey 1972c; Springer and Smith-Vaniz 1972). When one of these blennies is taken into a predator's mouth, it inflicts a painful bite that usually causes the predator to reject it unharmed. The reaction is immediate and may be caused by the pain of the initial bite and not by the venom, but the venom is not used to capture food and is solely defensive in function (Keenleyside 1979). Venomous fish are often brightly colored to warn predators away, as in zebrafish (*Pterois volitans*) (Photo 6–7), or well camouflaged to avoid being disturbed, as in stonefish (*Synanceja horrida*) (Photo 6–8).

The only known venomous mammals are platypuses (*Ornithorhynchus paradoxus*) and various species of shrews (*Solenodon, Neomys, Blarina*) (Calaby 1968; Pournelle 1968). In platypuses only the male is venomous, and he injects his venom by means of a movable horny spur on the inside of each hind leg. The adaptive significance of this spur is unclear. Males may use it to attack other males, or perhaps to deter predators, although platypuses have few predators except perhaps large snakes and monitor lizards (Poignant 1965). Shrews are small mouse-like mammals with such high metabolic rates that they must eat almost continuously to stay alive. They primarily eat insects, although water shrews (*Neotomys fodiens*) also eat crayfish, snails, frogs, and small fish, while short-tailed shrews (*Blarina brevicauda*) also eat rodents and moles (Pournelle 1968). The venom of shrews is contained in their saliva and is used mainly to immobilize large-sized prey. It apparently has little or no value in deterring predators, although this possibility has not been entirely ruled out.

Distastefulness. Distastefulness is generally caused by the presence of toxins or emetics. It may result from substances carried in the blood, substances regurgitated or defecated upon handling or swallowing, glandular secretions, or poisons embedded in the flesh or other body tissues.

Distastefulness is especially widespread among arthropods (Eisner 1970). Water beetles secrete a milky substance onto their bodies that is toxic to vertebrate predators. For example, the secretions of *Dytiscus marginalis*

Photo 6–7

The zebrafish or lionfish.

This extremely venomous species uses bright coloration and conspicuous fin structure to warn predators not to attack.

Photo by Richard Rosolek.

narcotize fish and cause fish, amphibians, and small mammals to vomit (Schildtknecht 1971). The surface-swimming gyrinid water beetle *Dineutes discolor* secretes a viscous fluid that makes it unpalatable to fish and newts (Benfield 1972). Various moths exude toxic secretions from specialized glands when poked or touched (e.g., Bisset et al. 1960; Frazer and Rothschild 1960; Ford 1964). Their secretions will poison an avian predator if the moth is ingested. Many species of butterflies have poisons or emetics embedded in their flesh (e.g., Jones et al. 1962; Owen 1971; Brower and Brower 1964; Brower et al. 1967; Reichstein et al. 1968). Eating one of these butterflies is a memorable experience for an inexperienced avian predator (Photo 6–9).

Among vertebrates pufferfish (*Lagocephalus*) are avoided by local fishermen because of their unpleasant taste (Edmunds 1974), and many birds have relatively low palatability for avian predators and cats (Cott 1947). Certain ducks, notably common eiders and shovelers (*Anas clypeata*), defecate on their eggs when flushed from the nest, making them less attractive to predators because of the strong odor (Swennen 1968). Experimental evidence indicates that the feces make otherwise tasty eggs distasteful to mammalian egg predators.

Most toads, frogs, and salamanders secrete toxic or irritant substances

Photo 6–8

The stonefish, another extremely venomous species.
In contrast to zebrafish, stonefish rely on their close resemblance to rocks to avoid being disturbed by predators.
Photo by Richard Rosolek.

from glands in their skin (Luther 1971; Lutz 1971). The secretions of toads are mainly effective against small predators, especially the young of poisonous snakes who have not yet learned to avoid them (Brazil and Vellard, cited in Lutz 1971). The secretions of European bell toads (*Bombino*) are sufficiently toxic to deter turtles and crocodiles placed in a vivarium with them (Gadow 1901).

Constraints on the evolution of chemical defenses. Toxic or noxious substances may be produced intrinsically or taken up from plant tissues (Eisner 1970; Rettenmeyer 1970). In either event the prey animal must avoid poisoning itself. To do this, it must often develop energetically expensive biochemical or morphological mechanisms (Brower 1969). The biochemical costs of synthesizing and storing toxins or the specialized diets required to take up plant secondary compounds for use as toxins may make alternative modes of defense more advantageous than chemical defenses for many species. The danger of autotoxicity or the energetic costs of synthesizing and storing toxins probably explain why developing embryos are rarely unpalatable or toxic (Orians and Janzen 1974).

Chemical defenses are not wholly effective against predators. They work because they reduce the energetic return for a predator and hence

A. B.

Photo 6–9 *The emetics present in some butterflies make them unpalatable to avian predators.*

Here an inexperienced blue jay (Cyanocitta cristata) *eats a monarch butterfly (A), only to discover some unpleasant consequences (B).*

Photos by Lincoln P. Brower.

the range of conditions under which the prey species can profitably be taken (Orians and Janzen 1974). Nearly any defense can be overcome by specialized morphology, physiology, or hunting tactics. A few examples suffice to illustrate this point. The honey buzzard (*Pernis apivora*) feeds almost entirely on wasp and bee larvae that it obtains from hives during summer months (Willis 1972). It gains protection from stings of adult wasps and bees by means of small, close-fitting feathers on its head and by using its small beak to decapitate especially persistent defenders of the hive. Bee-eaters prey upon adult bees by catching them in flight, taking the prey to a perch, and banging and rubbing the prey's abdomen several times to squeeze out the venom (Fry 1969, 1972b). Sometimes they are even able to de-venom bees on the wing. Civets (*Viverra civetta*) and mongooses can kill poisonous snakes because of the quickness of their movements (Hinton and Dunn 1967; Ewer 1973). Secretary birds (*Sagittarius serpentarius*) and short-tailed eagles (*Circaetus gallicus*) are specialized to kill snakes and gain protection from the snake's fangs by means of thick scales on their legs (Edmunds 1974). Many predators, however, cannot afford to specialize on overcoming the chemical defenses of particular prey. When selection favors more generalized diets, such specializations are not worth the cost.

Deceiving predators

A variety of deceptive tactics are used to fool predators into not attacking or into attacking the wrong part of the body. These include mimicry, which depends on the aposematic coloration of many well-defended species, intimidation displays, distraction displays, and death-feigning displays.

Deception always entails some risk because the predator may not be fooled. Hence it is often used as a last resort after all other lines of defense have failed. The effectiveness of deceptive tactics depends on both morphological structures and behavioral repertoires available for perpetrating a deception. It also depends on the predator's inability to see through the deception, and hence it depends on the predator's sensory capabilities.

Aposematism and mimicry. Distastefulness and other chemical defenses are commonly employed in concert with bright coloration or conspicuous morphological structures, which warn predators not to attack. Carpenter (1921) found, for example, that monkeys would eat only 20% of 220 brightly colored insect species offered to them, compared to 73% of 155 dull-colored species. Similarly, Cott (1947) found that brightly colored birds are less palatable to cats and humans than are dull-colored birds. The use of bright colors or striking morphologies to warn predators about noxious or dangerous defenses is often referred to as **aposematism** (for further discussion see Edmunds 1974; Harvey and Greenwood 1978; Baker and Parker 1979).

The association of conspicuous coloration, morphology, and behavior with chemical repellents makes deception by edible or otherwise unprotected species possible as a deterrent against predators. Probably every animal possessing strong repellents or toxins is resembled by one or more species lacking repellents (Wickler 1968; Rettenmeyer 1970). This phenomenon is termed **Batesian mimicry** in honor of the English naturalist Henry Bates who first described it in 1864.

Even slight resemblances to noxious species confer some protection, though of course close resemblances confer more protection than poor ones (see Schmidt 1958, 1960; Brower 1963; Morrell and Turner 1970; Brower et al. 1971; Ford 1971; Pilecki and O'Donald 1971). The best mimics are not immune to predation, and a few predators even specialize on mimics. The West African wasp *Pison xanthopus* and the Malayan wasp *Trypoxylon placidum*, for example, sometimes stock their nests almost entirely with ant-mimicking salticid spiders (Richards 1947; Edmunds 1974). They apparently concentrate on these spiders by forming specific search images for them. Without specialized detection methods, however, predators are usually unable to distinguish mimics from models, thus conferring protection on the otherwise poorly protected mimics.

The effectiveness of mimicry depends on how noxious the model species is and on how abundant the mimic is relative to the model. As mimics become more common, more of the warningly colored individuals encountered by a predator will lack chemical defenses and can be safely eaten. Consequently, the predator takes longer to learn the association between the warning coloration and chemical defense of the model and hence eats more mimics and models before it learns to avoid them. Also, when a mimic is relatively common, it becomes more worthwhile for predators to develop an ability to discriminate between mimics and models. Thus mimicry is most effective when mimics are relatively uncommon and models are not already mimicked by other species. Mimics are sometimes common, but in those cases they are often polymorphic, with different local populations mimicking different models (e.g., see Wickler 1968; Owen and Chanter 1969, 1971).

Intimidation displays. Mimicry is not the only way prey animals deceive predators. Some species bluff predators by increasing their apparent size. Stick insects (Phasmidae), for example, often respond to predatory attacks by displaying their wings (Robinson 1968a, b, c). Pufferfish (Tetraodontidae) and porcupinefish (Diodonidae) inflate themselves with water when attacked by large predatory fish. Toads often inflate their lungs when threatened (Lutz 1971), and one Brazilian species (*Physalaemus nattereri*) combines this effect with large abdominal eyespots that make it look like a large animal's head (Photo 6–10). The bearded dragon (*Amphibolurus barbatus*) threatens predators by inflating a large pouch under its lower jaw, depending largely on bluff to scare off would-be predators (Poignant 1965).

Eyespots are used by many prey species to intimidate would-be predators. They are most often found on structures that can be flashed suddenly in a predator's face. Structures containing eyespots include the forewings and hindwings of moths and butterflies, the abdomens of various insect larvae, the flanks of certain fish, the display plumes of peacocks (*Pavo cristatus*), and the shells of Burmese soft-shelled turtles (*Trionyx hurum*) (Cott 1940; Wickler 1968; Edmunds 1974). Eyespots are often effective in startling birds and other predators, giving the prey a chance to escape (Blest 1957; Coppinger 1969, 1970).

A striking example of an intimidation display is that of snake caterpillars, the larvae of several species of neotropical moths. Upon being disturbed, snake caterpillars inflate their abdomens and display posterior eyespots in a manner strikingly similar to a snake's head (Photo 6–11) (Wickler 1968; Brower 1971). They even "strike" at objects touching them, thereby making the illusion even more convincing.

Distraction displays. Distraction displays are another ruse used against predators. Many snakes and lizards twitch their tails to divert attacks away from their heads (Wickler 1968; Greene 1973). The warning rattles of rattlesnakes probably evolved to further enhance this behavior. Tail twitching is also used to divert attention of prospective prey, allowing

Photo 6–10

The Brazilian toad Physalaemus nattereri.

This toad looks deceptively like a large animal's head peering out of the darkness whenever it is threatened by a would-be predator. The deception is effected by inflating the lungs and displaying two large abdominal eyespots.

Photo by Ivan Sazima.

snakes and lizards to catch prey off guard. The tails of some lizards are fragilely connected to their bodies and break off easily when grabbed by a predator. The tail may even writhe about on the ground following detachment, while the lizard makes its escape (Cott 1940). Similar detachable autonomous body parts have also evolved in various molluscs, marine polychaete worms, and arthropods (Edmunds 1974).

Photo 6–11

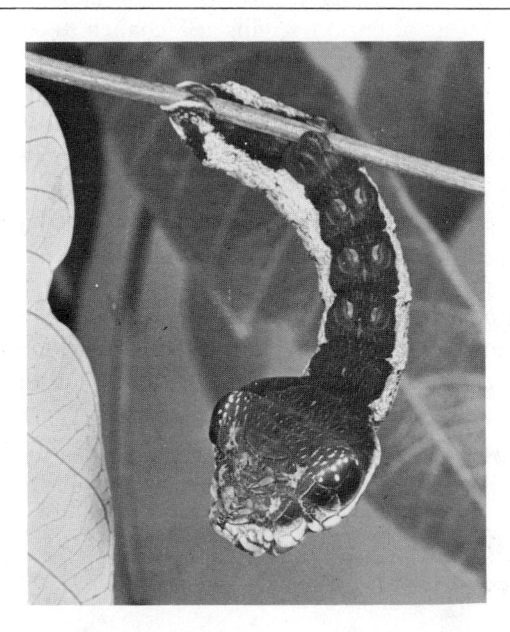

Snake caterpillars gain protection by looking and behaving like a snake's head.

The deception is achieved by drooping and inflating the abdomen, displaying large eyespots, and swaying the end of the abdomen back and forth.

Photo by Lincoln P. Brower.

Other devices are also used to deflect predatory attacks away from vulnerable body parts. Some butterflies have false antennae or small eyespots on their wings to misdirect attacks, while various fish have small eyespots on their fins for the same purpose (Wickler 1968). The plant hopper (*Ancrya annamensis*) of Thailand has a complete false head on its abdomen to deceive predators into attacking the least vulnerable end of its body (Photo 6–12) (Wickler 1968; Alcock 1975a).

Ground-nesting birds commonly use distraction displays to lure predators away from concealed nests (Simmons 1952; Harvey and Greenwood 1978). The parent birds feign injury by flopping through the grass, making them appear to be easy victims for the predator. Once the predator is a safe distance from the nest, however, the bird flies off leaving the predator empty-mouthed. Similar behavior is employed by freshwater bowfin fish (*Amia*), which thrash about as if they were injured when large predatory fish are near their young (Lagler et al. 1962).

Feigning death. A few animals escape predators by feigning death. The North American opossum is well known for employing this tactic (Hamilton 1963). Some tropical toads and bromeliad frogs turn over on their backs and lie stiff and motionless for up to 30 minutes after defensive postures fail to intimidate a predator (Lutz 1971). Hognose snakes (*Heterodon*), when annoyed, raise the head and neck, flatten the neck, hiss, and otherwise display in a fearsome manner (Leviton 1971). If this bluff fails to dissuade a potential predator, they roll over on their backs, limply protrude their tongues, and expose their bellies as if they were dead. Some spiders, beetles, bugs, grasshoppers, stick insects, and mantids become inert when attacked, with legs either outstretched or pulled close to their bodies (Edmunds 1974). By feigning death, a prey animal may cause a predator to leave or relax its attention, giving it a chance to escape.

Overview of prey defense

Prey animals are often faced with many different predators, and no one defense is effective against them all. They may therefore rely on several lines of defense and respond differently to each type of predator. Species faced with specialized predators frequently have specialized defenses to thwart them. For example, gastropods evade most predators by withdrawing into calcareous shells, but this defense is ineffective against starfish. When touched by a starfish, gastropods instead flex and extend their foot, enabling them to leap away (Feder 1972). Even with specialized defenses, however, virtually no combination of defense tactics is completely successful in deterring predators. As fast as a new and more effective defense can evolve, some countertactic will arise in the predator population. Thus predation is nearly always a strong selective pressure affecting an animal's morphology and behavior.

Photo 6–12

The plant hopper Ancrya annamensis *of Thailand.*

The plant hopper deflects predatory attacks away from its vital organs by displaying a false head, complete with large false antennae, on its abdomen. Note the small head and camouflaged eyes at the bug's real anterior end (toward the right side of the photograph).

Photo by Edward S. Ross.

Conclusion

The reproductive and survival advantages to be gained by acquiring resources, avoiding being eaten, and coping with the physical vagaries of ever-changing environments are major selective pressures affecting behavior. The best ways to exploit resources depend on the spatiotemporal distribution and availability of those resources, on plant defenses against herbivory or prey defenses against predation, and on the intensities of intraspecific and interspecific competition. Defenses against being eaten are shaped by morphological and physiological constraints imposed by other niche requirements of the organism, by the foraging tactics of the consumer species, and by habitat structure. Coping with ever-changing environments requires behavioral plasticity or special life history cycles. The behavioral repertoires of animals have evolved to meet these challenges, and the objective in coming chapters will be to show why specific behaviors have evolved in response to them.

Territoriality

The spacing pattern exhibited by a species results from cooperative interactions that bring individuals together and competitive interactions that spread them apart. Spatiotemporal distribution of resources and vulnerability to predation are major factors affecting the net value of forming social groups to cooperate in capturing food or defending against predators. The same two factors also affect the net value of spreading apart by mutually avoiding conspecifics or aggressively defending space. Spacing behavior is an enormously varied phenomenon not easily covered in a single chapter. The present chapter therefore focuses on aggressive defense of space, while Chapter 8 focuses on the origins of clumped spacing and Chapter 13 discusses spacing behavior among and within social groups.

The central issue in the present chapter is why animals defend space. Aggression entails several costs, as was discussed in Chapter 5, and it cannot evolve unless it confers benefits. Thus the first problem is to identify the benefits derived from defending space. The next problem is to identify the factors making a resource worth defending. A related problem involves the degree to which resources should be defended. Finally, there is the problem of why space is defended only against conspecifics in some species, while in others it is also defended against competing species.

The concepts of home range and territory

Home range

Nearly every animal spends its life in a circumscribed area, called its **home range,** except while dispersing or migrating (Burt 1943; Jewell 1966). In field studies home ranges are mapped by following an animal about and plotting its movements on a map or by plotting the various locations where an animal has been sighted (Figure 7–1). The outermost points on such a map are then connected to form the smallest possible convex polygon, which is operationally considered the animal's home range.

Home range size, as measured by the convex polygon method, is greatly affected by how long the animal is observed. With a large number of observations home range size will approach an upper limit or asymptote, at least during any given season (Odum and Kuenzler 1955). Sample size bias can be avoided by obtaining large numbers of observations, directly

Chapter seven

Figure 7–1

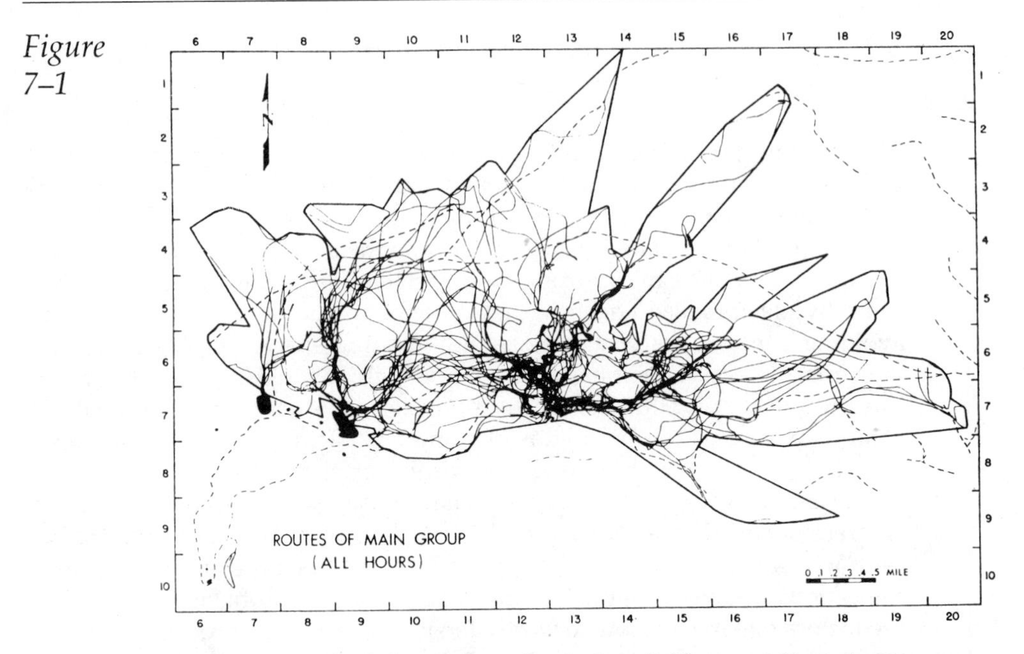

Home range of a baboon troop in Amboseli National Park in Kenya.

The map was obtained by recording troop movements for a period of 139 days. Note that a few areas in the home range are used intensively, while most areas are not.

Source: From S. A. Altmann and J. Altmann, *Baboon Ecology: African Field Research* (Chicago, Ill.: University of Chicago Press, 1970), p. 115. Reprinted by permission of the authors.

observing locations of defended boundaries, or using statistical procedures based on elliptical estimates of home range area (see Jennrich and Turner 1969; Koeppl et al. 1975; Madden and Marcus 1978; Baker and Mewaldt 1979).

Producing a single home range map is generally insufficient for characterizing an animal's use of space. Home range has a temporal component as well as a spatial component. That is, home ranges change through time. The largest changes occur on a seasonal basis. Home ranges obviously change for migratory animals, which inhabit different geographical regions at different times of year, but they may also change substantially for sedentary animals if the resource base changes seasonally. In addition, smaller-scale changes occur on a daily and weekly basis as resources fluctuate in availability. Thus home range boundaries are rarely fixed, and a home range map is just a cross-section of the way an animal uses space.

Home range is more than just an area within which an animal lives and reproduces. It is an area where the animal can become intimately familiar with its surroundings. Indeed, familiarity is a major reason why animals restrict their activities to circumscribed areas. In familiar surroundings an animal knows where to look for food and where to find water. It knows where to hide or seek shelter when threatened by predators or adverse weather. It knows how to avoid accidents, and it knows where to mate and rear offspring. Animals who leave familiar surroundings must relearn this information about the new area. That is one reason why animals do not leave familiar areas unless habitat changes or competitors force them to disperse and why mortality is generally higher during dispersal.

Home ranges have internal structure. Some parts of a home range fulfill habitat requirements better than others, and animal movements reflect this heterogeneity. Notice, for example, that not all parts of the baboon home range shown in Figure 7–1 were used with equal intensity. Areas of intensive use, or **core areas,** are centered around sleeping sites, high-quality food patches, and water holes, which are the principal resources required by a baboon troop. The remaining parts of the home range are used much less intensively and often primarily while traveling between core areas.

The home ranges of many animals, particularly mammals, consist of relatively few core areas interconnected by a network of narrow pathways (Hediger 1949; Adams and Davis 1967; Ewer 1968; Bailey 1974). When polygonal maps of such home ranges are drawn from field observations, the ranges of neighboring individuals or groups often appear to overlap. However, closer inspection reveals that most of the overlap is illusory. Although travel routes frequently intersect, core areas generally do not overlap. In addition, the overlap of travel routes shown on a set of home range maps may be even less than it appears, because neighboring individuals may use them at different times of day or avoid routes recently used by another (Leyhausen 1965). Thus home range maps based on polygonal or elliptical methods do not convey a true picture of how space is being used.

One way to circumvent such problems is to conceptualize home range as an *activity field* consisting of a rectangular grid system, with the value for each grid location defined as the proportion of time the animal spends at that location (Waser and Wiley 1979). The extent to which each location is used exclusively by a given individual can then be measured by superimposing an *isolation field* over the activity field. The value at each location on an isolation field is the ratio of time spent at that location by a given individual divided by the total time spent there by all individuals who utilize it. Thus an activity field quantifies the way an individual utilizes space, while an isolation field quantifies the extent to which an individual monopolizes that space.

Territory

Territorial behavior is extremely diverse and difficult to define or categorize. It can be recognized by three attributes, not all of which apply to every kind of territory (see Hinde 1956; Brown and Orians 1970). Territorial animals (1) restrict some or all of their activities to a defended area, (2) advertise their presence within that area, and (3) maintain essentially exclusive possession of all parts of that area. When defining territoriality, different authors emphasize these attributes to different degrees, partly because their emphasis depends on what research questions are being asked.

Territory is most commonly defined as any defended area (Noble 1939). Noble's definition is based on the first attribute listed above and implies that the other two attributes need not be true. Because it is very broad, some authors have tried to restrict the definition by including self-advertisement as a second necessary characteristic of territoriality (e.g., Nice 1933; Mayr 1935; Lack 1939a; Brown and Orians 1970). However, this change has not gained wide acceptance because it narrows the definition too much.

Several problems are inherent in Noble's definition of territoriality. One problem is that not all defended areas have discrete boundaries. Animals sometimes exclude intruders to differing degrees, depending on how closely the intruder approaches the resident's core area. Some biologists therefore redefine territory as any area where an animal exhibits site-specific dominance (Hediger 1949; Emlen 1957a; Willis 1967; Murray 1969; Leuthold 1977). They view a territory as that part of a home range where the resident animal wins aggressive interactions with intruders more often than it loses them. Since the behaviors used to dominate intruders are similar to those used by related species to defend exclusive territories, these biologists feel that the territoriality concept should not be restricted to exclusively defended areas.

Many other biologists reject this view. The concept of defense has traditionally implied *successful* defense and hence exclusive possession of an area (Brown and Orians 1970; Brown 1975a). If territoriality is redefined in terms of dominance interactions, the concept becomes too general. For example, it would then include overlapping home ranges whenever neighboring individuals or groups show hostility upon making contact. Behaviors associated with such spacing patterns are often distinctively different from those associated with overt territory defense, and they should not be subsumed under the same definition. As an alternative, Brown (1975a) refers to nonexclusively defended areas as **dominions.**

Another problem with Noble's definition is the ambiguous meaning of the word *defense.* Threats and overt attacks directed at intruders clearly constitute aggressive defense of an area, but mutual avoidance mediated by scent-marking, long-distance vocalizing, or visual sightings is more difficult to interpret. Although such behaviors are traditionally not viewed as aggressive, Krebs (1971) argues that the mere existence of mutual avoidance implies some sort of underlying threat; otherwise, animals should not go out of their way to avoid one another. One reply to Krebs's argument is that animals avoid each other to avoid competition. Areas recently visited by others contain fewer resources because they have recently been exploited by another. Interpretations are especially difficult for mammals because overt defense of core areas may be important even though it is infrequent and rarely observed. Since home ranges of most mammals are relatively permanent, previous encounters between neighbors can be remembered and respected for long periods of time. Hence overt aggression would rarely be necessary to maintain spacing. Moreover, aggressive encounters are often unlikely to be detected by an observer because movement patterns and spacing of many mammals are studied by trap-recapture methods, radio tracking, or intermittent sightings rather than continuous observation.

The most controversial alternative to Noble's definition is the idea that a territory is any exclusively occupied area (Pitelka 1959; Schoener 1968). This definition is the ecological equivalent of Krebs's (1971) definition, but it denies the relevance of aggressive behavior altogether. Pitelka (1959) argues that "the fundamental importance of territory lies not in the mechanism (overt defense or any other action) by which the territory becomes identified with its occupant, but in the degree to which it is in fact used exclusively by its occupants." Brown (1975a) objects to Pitelka's definition because territoriality is a behavioral concept that should be defined at least partially in behavioral terms. Pitelka (1959) and Schoener (1968) reject Noble's definition because territoriality is an ecological concept that should not be defined in behavioral terms. There is no obvious resolution to this controversy, and the appropriate definition must depend on the particular questions being asked. If the question concerns why animals defend space, Noble's definition of territoriality is most appropriate. If the question concerns why animals occupy exclusive spaces, Pitelka's definition becomes more appropriate.

History of the territory concept

The fact that many animals defend space has been known at least since the time of Pliny and Aristotle, but the modern study of territorial behavior began as recently as 1920 when H. Eliot Howard published his classic book *Territory in Bird Life*. Howard's book had a major impact on behavior research in England but failed to spark much interest elsewhere. The significance of his work was not fully appreciated in North America until an amateur ornithologist, Margaret Morse Nice, reported the results of her 8-year study of song sparrows *(Melospiza melodia)* (Nice 1937). Nice's work provided the standard format for life history studies during the ensuing 20 years. These studies, which dealt with all aspects of an animal's behavior and ecology, became exceedingly popular and laid the groundwork for more sophisticated research on the adaptive significance of animal behavior in general.

Howard's (1920) ideas foreshadowed most of the subsequent research on avian territoriality (Stokes 1974). Howard recognized the importance of territoriality as a means of defending resources, pointed out that birds behave in similar ways while defending territories and courting prospective mates, and suggested that territoriality might serve to limit local population densities. These ideas have since expanded into very broad and active areas of behavioral research.

Although Eliot Howard is generally credited with rediscovering territoriality in birds, his work was preceded by two earlier studies. Bernhard Altum (1868) was the first biologist to recognize the significance and prevalence of territorial behavior, but his results failed to gain an international audience until Mayr (1935) translated them into English. C. B. Moffat (1903) was the first English-speaking naturalist to realize the importance of territorial behavior, but he published his ideas in *The Irish Naturalist*, where they occasioned little notice.

Much of the early research on territoriality sought to determine how individuals benefit by defending territories. Altum (1868) and Howard (1920) based their studies mainly on birds that forage on territories and concluded that territory defense evolved to ensure an adequate food supply. However, subsequent authors soon pointed out that not all birds forage on their territories. Colonial birds defend nest sites but not feeding grounds, and most promiscuous birds defend only a display arena used to attract mates (Nice 1941; Hinde 1956; Lack 1968). Since the type of resource being defended did not suggest any general benefit derived from defending a territory, several other hypotheses were advanced. Courtship and mating usually occur on territories, suggesting that territories may have evolved to attract mates, maintain pair bonds, or prevent conspecifics from interfering with copulations (Hinde 1956; O'Donald 1963; McLaren 1972; J. M. Emlen 1978). Others suggested that territoriality functions to reduce aggression (Hinde 1956; Geist 1974b; Owen-Smith 1977), prevent epidemics, protect nests or offspring (Hinde 1956), space out individuals to make concealed nests more difficult for predators to locate (Lack 1954, 1966; Crook 1965), or regulate population densities to prevent overexploitation of food resources (Kalela 1954; Wynne-Edwards 1962). All of these hypotheses appeared plausible for at least some species, but none provided a general explanation of territoriality or any criterion for distinguishing among selective advantages, secondary consequences, and coincidental correlates of territoriality. Jerram L. Brown (1964) finally surmounted these problems by proposing a general theory based on a cost-benefit analysis.

Why exclusive territories evolve

Brown's (1964) theory for explaining territoriality is derived from the costs and benefits of aggressively competing for space. Brown argued that territory defense is costly in terms of reduced survival and reproductive potential and cannot evolve unless those costs are offset by compensatory benefits. Patrolling territory boundaries and chasing out intruders requires time and energy that could be spent foraging, attracting mates, caring for offspring, or performing other activities. Aggressive defense entails risks of injury or death and reduces the time available for antipredator surveillance. Since boundary patrols and aggression have no intrinsic benefits, they should not evolve unless they enable individuals to acquire requisites that could not be obtained more cheaply in another way.

If every individual could obtain as much of every resource as it could possibly use, there would be no point in defending any resource. Competition is therefore the single most important factor favoring territoriality (Figure 7–2). In the real world all animals compete for resources, although the resources under contention vary from one species to the next. Individuals or groups may compete for mates, food, cover, shelter, breeding sites, space, particular habitat configurations (such as proximity of breeding sites to food resources), or any other requisite that influences survival or reproductive success. Males nearly always compete for mates because their reproductive success is generally limited by the number of females they can inseminate, while both sexes may compete for other resources that limit reproductive output or influence survival. Because competition is virtually universal, every species could potentially show territorial behavior. Nevertheless, not all do. Thus some second factor must also be important. That factor is net benefit, which varies with the intensity of competition.

Competition intensity depends on the number of individuals encountering a resource per unit time and the relative availability of that resource compared to the requirements of individuals competing for it (Figure 7–2). The intensity of competition affects both the costs and benefits

Figure 7–2

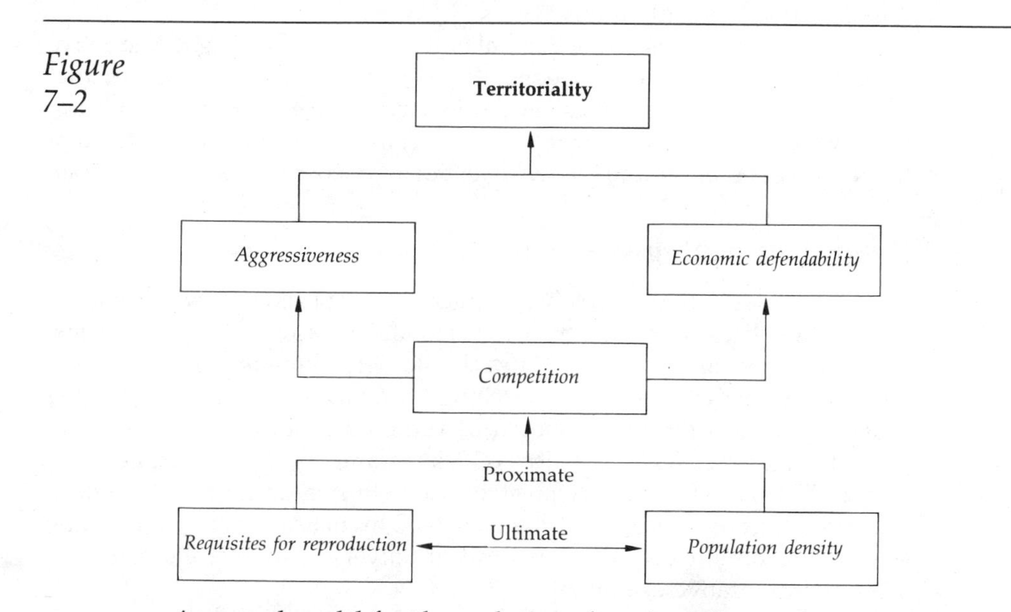

A general model for the evolution of territoriality.

Resource requirements and population density determine the intensity of competition. An intermediate degree of competition, in turn, makes resources worth defending and hence makes aggressive defense of resources advantageous. The result is territoriality.

Source: From J. L. Brown, "The Evolution of Diversity in Avian Territorial Systems," *Wilson Bulletin* 76 (1964):161. Reprinted by permission.

of resource defense. When too many individuals compete for a resource, the cost of excluding them all becomes excessively high. When few individuals compete for a resource, the benefits gained by excluding them become vanishingly small. Thus territoriality should not be advantageous unless competition intensity falls within some intermediate range, and resources falling within that range are referred to as **economically defensible resources** (after Brown 1964). Resources are economically defensible if individuals defending them leave more progeny than do individuals not defending them, and territoriality should evolve only when animals are competing for economically defensible resources.

What is defended?

The costs of territory defense stem from essentially the same sources in every species, but the potential benefits are diverse. The main theoretical issues therefore revolve around the benefits of territoriality. Each potential benefit can be considered as one hypothesis for explaining territoriality. No one hypothesis will explain territoriality in every species; nor will all hypotheses apply to any given species. The initial problem, therefore, is to identify the hypothesis or hypotheses applicable to each species and from that information deduce the environmental contexts within which each hypothesis should apply.

To solve this problem, a general criterion for identifying the adaptive benefits of territoriality is needed. Brown's (1964) model provides that criterion: namely, competition. A particular resource cannot favor territoriality unless individuals compete for it. Ideally, one should also show that the resource is economically defensible, but in practice that is rarely feasible.

Defense of mating stations

Territories defended solely to attract or control access to mates should evolve when the sites most attractive to females are economically defensible while other resources are not. Males should defend mating territories only if alternative activities such as searching for females do not lead to higher fitness, and such territories should be centered in identifiable areas that predictably attract females (Alcock 1975b; Campanella 1975; Emlen and Oring 1977). The economic defensibility of mating stations has yet not been evaluated for any species, but the reasons why males compete for mating stations are often clear. These reasons are diverse and best illustrated by some representative examples.

Insects. Males of the damselfly *Argia apicalis* defend pond edges where females emerge from aquatic larval stages (Bick and Bick 1965). When a female emerges, a territorial male soon latches onto her to form a tandem pair. The male remains with her for a prolonged period while copulating to prevent other males from inseminating her before she oviposits (Parker 1970a). Males compete for pond edges because pond edges are the best places to encounter newly emerged females and not enough shoreline exists to accommodate every male (Corbet 1962). The reason newly emerged females are most abundant along shorelines has to do with the basic biology of odonates. When aquatic larvae are ready to metamorphose, they swim ashore and emerge on vegetation or other shoreline substrates (Corbet 1962; Orians 1980). There they metamorphose into adults, dry their wings, and then fly.

Not all territorial odonates defend emergence sites. Male damselflies of the genus *Calopteryx* defend emergent vegetation along stream margins where females lay their eggs. When a female arrives on a territory, the male forms a tandem pair and remains with her until she oviposits. Why some males defend oviposition sites while others defend emergence sites is unclear. One important factor may be the length of time elapsing between female emergence and oviposition. Another may be a difference in how much it costs to defend the two types of sites.

Males of most territorial dragonflies also defend oviposition sites (Jacobs 1955; Ito 1960; Kormondy 1961; Corbet 1962; Johnson 1964; Moore 1964; Campanella 1975). Campanella (1975) argues that males should be territorial when females oviposit in a limited number of predictable locations but not when females oviposit in a wide variety of unpredictable locations. In the latter event males should establish some type of dominance hierarchy or patrol widely in search of females. Evidence for evaluating the extent to which males can predict where females oviposit is available for very few species as yet, but the limited evidence currently available does support this hypothesis.

Males of some wasps and bees defend territories where females can be predictably found. Males of the cicada killer wasp (*Specius speciosus*) and bumblebee wolf (*Philanthus bicinctus*) defend territories in sandy areas where female emergence holes are most common (Lin 1963; Gwynne 1978). In the bumblebee wolf males scent-mark territories by dragging their abdomens on vegetation. The scent may repel intruding males or perhaps attract females, but its precise function is unknown. Since not all males can defend territories, some must patrol in search of females. This allows a comparison of male success for each strategy. In both species territorial males gain better access to newly emerging females than do nonterritorial males.

In many other sphecid wasps males defend entrance holes of female nest burrows (Alcock 1975b). A territorial male waits inside the nest burrow until the female returns with nest provisions prior to ovipositing and then attempts to copulate with her. Nest burrows are not the only places where females copulate, but they are the most likely places to find receptive females.

The males of several bee species defend mating territories centered around food plants favored by females (Haas 1960; Cazier and Linsley 1963; Jaycox 1967; Pechuman 1967; Raw 1975; Eickwort 1977). Competition for especially attractive plants is difficult to demonstrate, but in *Anthidium manicatum* only the largest males are successful in defending territories and obtaining mates (Haas 1960). In some species of carpenter bees *(Xylocopa)* males defend small territories around bushes where females feed (Janzen 1964; Linsley 1965; Velthius and Camargo 1975a, b; Anzenberger 1977), while in other species they defend entrance holes of nests built into the side of buildings, where they try to intercept arriving and departing females (Cruden 1966; Gerling and Hermann 1978).

Some males of the primitively eusocial halictine bee (*Lasioglossum rohweri*) defend entrances to nest burrows dug into the ground, while other males patrol around food flowers and nesting areas in search of females (Barrows 1976). Territorial males may leave their territories to patrol for females, but some males never defend territories. Why some males defend nest entrances while others patrol is unknown.

In the desert clicker (*Ligurotettix coquilletti*), a grasshopper of the Sonoran desert, males defend territories centered around small creosote bushes (Otte and Joern 1975). Only certain-sized bushes are likely to harbor females, and the ones chosen by territorial males lie within that size range. Each male defends a separate bush, which may attract numerous females. Each bush provides ample food and oviposition sites for accommodating many females, so females do not compete for breeding sites. However, males do compete for bushes because there are fewer bushes than there are males. Not every small bush is defended by a territorial male, but unoccupied bushes may be unsuitable due to unfavorable plant chemistry.

Males of the cactus fly (*Odontoloxozus longicornis*) defend territories around saguaro cacti (*Carnega gigantea*), where females oviposit (Mangan 1979). Females who land on a territory copulate with the male and then usually oviposit immediately. They may copulate and oviposit several times before departing. Some males do not acquire territories and instead patrol in search of receptive females. They copulate with females in undefended areas, and these females then fly elsewhere to oviposit. During adverse weather, when it is hot or windy, males leave their territories and search for females who have taken refuge in cracks or holes in the cacti. Upon finding a female, the male courts and guards her, copulating frequently.

Territory defense benefits male cactus flies in at least two ways (Mangan 1979). Females who copulate with territorial males oviposit immediately, which prevents other males from fertilizing that batch of eggs, and females who visit territories average higher fertility than females found by patrolling males. These benefits can be obtained, however, only when females are actively visiting male territories. When adverse weather immobilizes females, males achieve greatest success by abandoning their territories and seeking out females who have taken refuge.

Oviposition sites are also defended by some male butterflies (Baker 1972). Males of the small tortoiseshell (*Aglaius urticae*) defend vertical edges of walls or hedges, along with nearby nettles where females oviposit. Because females tend to follow edges while searching for oviposition sites, males defending territories that contain walls or hedges mate more often than males defending isolated nettles in open fields. Similarly, male peacock butterflies (*Inachis io*) defend rows of trees near oviposition sites. Their territories do not contain nettles, where females oviposit, but females prefer nettles near forest edges.

Not all territorial butterflies defend oviposition sites. Males of the speckled wood butterfly (*Pararge aegeria*) defend patches of sunlight on the forest floor (Davies 1978a). Sun-specks enhance the males' color and may aid in attracting females. However, the relative mating success of males in sun-specks compared to males in shadows has not been measured.

Males of many butterflies congregate on hilltops, where they defend mating territories against other males (Shields 1967). These territories are not associated with favorable food sources, oviposition sites, or any other known resource. Why males congregate and defend territories on prominent terrain features is poorly understood. Shields suggests that hilltopping behavior facilitates the sexes finding each other when individuals would otherwise be widely dispersed, as hilltopping is most prevalent in rare species (see also Maynard Smith 1976b). However, since hilltopping does occur in a few relatively common species, other factors may also be involved.

Defense of mating stations is a prevalent form of territoriality among insects. The most commonly defended locations are female emergence sites, entrances to female nests, oviposition sites, and food plants visited by females. These are all predictable locations where females can be found. Such sites are generally limited in availability, and defense of them probably increases a male's mating success.

Vertebrates. The selective factors favoring defense of mating stations in vertebrates are poorly understood. Although many vertebrates mate on territories that are also used for other purposes, relatively few defend territories used exclusively to attract or control access to mates.

Territories used exclusively for mating are unusual in fish, but Popper and Fishelson (1973) describe one instance in the coral reef fish *Anthias squamipinnis.* This small reef fish lives in swarms permanently associated with particular rocks or coral heads (Figure 7–3). Prior to spawning some males leave the swarm to establish territories on the horizontal surface of the rock or coral head. Meanwhile, many other males remain in deeper water with the swarm. At spawning time the swarm moves into shallower water where the territorial males are located, and after considerable display both sexes broadcast their spawn into the water. Nonterritorial males remain with the swarm during spawning, even though territorial males continually chase them away. The advantages of territory defense are unclear. The behavior is associated with spawning, but the relative reproductive success of territorial and nonterritorial males is unknown.

Many promiscuous birds defend dispersed mating stations at traditional sites, each of which is occupied by a single male every year (Lack 1968; Gilliard 1969; Hjorth 1970). In forest-dwelling grouse male territories are used to obtain food as well as mates, but defense is focused primarily around the perches where males advertise for females (e.g., Blackford 1958,

Figure 7–3

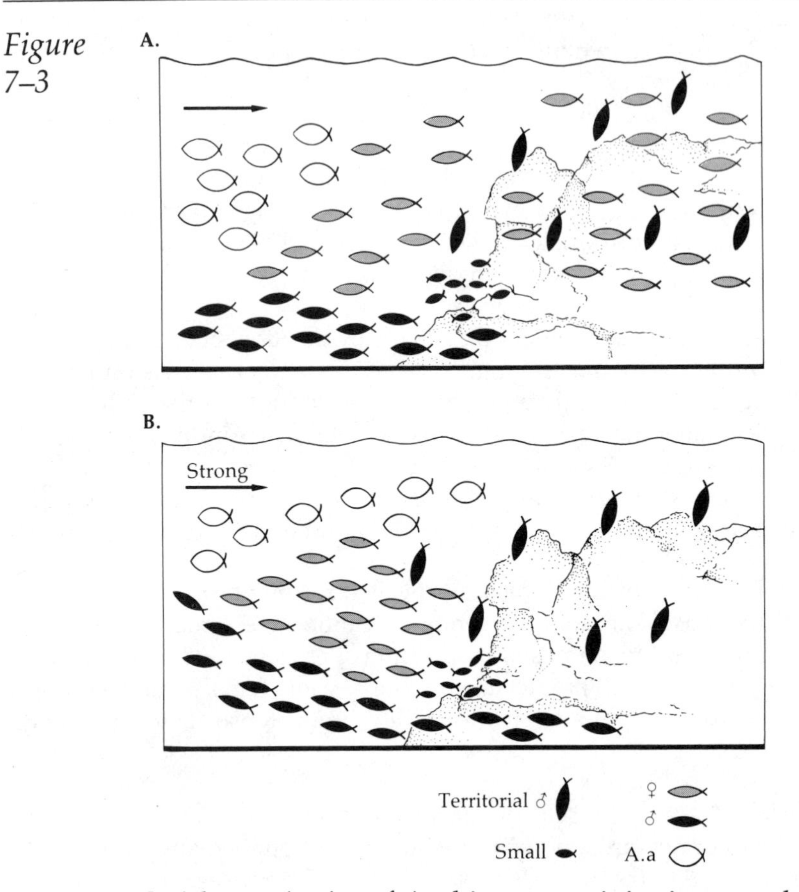

Social organization of Anthias squamipinnis, *a coral reef fish from the Red Sea.*

Prior to spawning, some males begin defending territories and exclude the remaining males from shallow reef areas. Nonterritorial males remain at the bottom of swarms in deep water, moving into the shallows only when females spawn. Juveniles remain consistently in deep water at the bottom edge of the reef. The direction of current is indicated by an arrow. Females forage over and near the reef when the current is weak (A) but move out to deeper water when the current is strong (B). A.a. = Abudefduf azysron, *a reef fish that usually associates closely with* Anthias squamipinnis *swarms.*

Source: From D. Popper and L. Fishelson, "Ecology and Behavior of *Anthias squamipinnis* (Peters, 1855) (Anthiidae, Teleostei) in the Coral Habitat of Eilat (Red Sea)," *Journal of Experimental Zoology* 184 (1973):413. Reprinted by permission.

1963; Gullion and Marshall 1968; Lance 1970; Ellison 1971, 1973). Males clearly compete for display sites, but the reasons why some sites are more attractive to females than others are unknown.

One hypothesis is that females are safer at certain types of sites, either because they are less easily detected by predators there or because they can detect stalking predators more easily there. The display sites of male grouse and birds-of-paradise are usually situated in relatively open areas in their forest habitats (Gilliard 1969; Hjorth 1970), possibly to enhance the ability of both sexes to detect predators.

Another hypothesis is that females are safer if they mate at familiar sites (Wittenberger 1978a). Familiarity with traditional sites might allow females to travel to and from male display sites more safely, just as familiarity with home ranges affords greater protection from predators. However, no evidence is available for evaluating this hypothesis.

A variety of promiscuous birds display together on leks instead of at dispersed display sites, and males ordinarily defend territories on those leks (Photo 7–1) (Lack 1968; Wittenberger 1979a). Similar display territories have also evolved in two antelopes, the Uganda kob (*Kobus kob*) and the blesbok (*Damaliscus dorcas*) (Buechner and Schloeth 1965; Leuthold 1966; Lynch 1974). Of interest here is why males defend territories on leks. The evolution of lek displays themselves is a separate issue, which will be discussed in Chapter 11.

The advantages of defending territories on leks are not immediately obvious. In some birds and mammals males holding central territories on a lek typically obtain more mates than males holding peripheral territories (Hogan-Warburg 1966; Hjorth 1970; Wiley 1973; Floody and Arnold 1975; Shepard 1975). In other species this is not the case. Females of the latter species may be attracted to particular territories rather than to particular

Photo 7–1

*Male sage grouse (*Centrocercus urophasianus*) displaying toward a group of females on a lek.*

Each male holds a small territory on the lek and attempts to attract as many females as possible by means of elaborate display behavior. The males shown here hold central territories, which are preferred by females.

Photo from R. Haven Wiley, "Territoriality and Non-Random Mating in Sage Grouse, *Centrocercus urophasianus*," *Animal Behavior Monographs* 6 (1973):85–169.

mates, but the reasons why some territories might be more attractive than others is unknown (see Chapter 11). The important point here is that any female preference for particular territories or positions on a lek would strongly favor male defense of locations preferred by females.

Males of several tropical bats defend mating territories centered around preferred roosting or mating sites (Bradbury and Emmons 1974; Bradbury 1977a, b). In *Saccopteryx bilineata* males defend the vertical trunks of large buttressed trees, where females prefer to roost during daytime. Good roosting sites are evidently scarce, as unmated males live solitarily on territories adjacent to those of mated males or in separate trees. In *Phyllostomus bastatus* males defend crevices in caves where 10 to 100 females and their young cluster tightly together. Territory ownership shows a high turnover rate because competition for territories is severe. In a related species, *P. discolor,* males defend small groups of females within large roosting congregations found inside hollow trees, while in *Tylonycteris pachypus* and *T. robustula* males defend female groups within hollow cavities of bamboo stalks. Finally, in *Hypsignathus monstrosus* males defend territories on leks, which are located in treetops along streams.

Defense of breeding sites

Many animals defend breeding sites where they spawn, nest, or rear young, while foraging largely outside territory boundaries. Since the mating stations of many insects are centered around oviposition sites, they might be considered as defended breeding sites. A distinction between mating stations and defended breeding sites can be made, however, according to whether the site is used solely for attracting females or is also used for rearing offspring.

Fish. Many species of cichlid fish defend spawning sites (Fryer and Iles 1972). Eggs are deposited in nest scrapes that vary in elaborateness from mere depressions in the bottom to built-up mounds of sand. Either males, females, or both guard the nest site until eggs hatch. In *Cichlasoma maculicauda* spawning substrates appear unlimited in availability, since many previously used sites are unoccupied at any given moment and new arrivals never have difficulty obtaining a place to spawn (Perrone 1975). This may be typical, as many species spawn on sand or gravel bottoms that appear unlimited in availability.

Similarly, spawning sites appear unlimited for many marine fish. The gobiid fish *Signigobius biocellatus* spawns in burrows dug into sandy bottoms, with each pair defending from one to six sealed burrows until the eggs hatch (Hudson 1977). Male painted greenlings defend all-purpose territories in rocky reefs throughout the year, but spawning sites and other resources do not appear to be in short supply (De Martini 1976). Many pomacentrid reef fish defend spawning burrows in seemingly unlimited expanses of sandy bottom (Abel 1961; Turner and Ebert 1962; Myrberg et al. 1967; Albrecht 1969; Rasa 1969; Fishelson 1970).

In species where spawning substrates appear unlimited, defense of spawning sites might be explained in at least two ways. One possibility is that microhabitat variations make some spawning sites better than others. Then competition for the best sites might be intense, and territoriality would allow control of the best sites. A second possibility is that individuals are defending eggs rather than territories. If aggressive behavior is directed mainly at predators, it should be interpreted as parental behavior, not territorial behavior. In the pomacentrid fish cited above, territory defense begins shortly before spawning and continues only until eggs hatch. It is also directed primarily at egg predators, not at competitors for spawning sites. Thus the defensive behavior of these species apparently represents a form of parental care.

Spawning sites do appear limited in some fish. Most freshwater sunfish and bass defend territories around their nest sites (Breder 1936; Miller 1963). Males of the pumpkinseed sunfish (*Lepomis gibbosus*) defend nests only in certain depths of water, preferring to nest near submerged logs or other objects in shallow water. At least the most preferred sites appear limited in availability, as many less preferred sites are also utilized. Male mottled sculpins (*Cottus bairdi*) defend holes dug into the bottom of freshwater streams or crevices under rocks, where they guard egg masses spawned by one or more females (Bailey 1952; Ludwig and Norden 1969). The number of suitable sites appears limited, since artificial breeding sites introduced into a stream are soon defended by previously nonterritorial males (Downhower and Brown 1979).

Male pupfish (*Cyprinodon*) defend heterogeneous bottom substrates containing limestone embankments, rocks, or mats of rooted vegetation, where gravid females spawn (Kodric-Brown 1977, 1978). Several lines of evidence demonstrate that males compete for spawning sites. First, clear differences exist in the attractiveness of available spawning sites, with some sites attracting many more females than do others. This circumstance implies that males compete for the most attractive sites. Second, all suitable sites, including those of marginal quality, are defended, and a surplus of nonbreeding adult males exists as schools in undefended areas unsuitable for spawning. Finally, vacant territories created by removing territorial males are quickly occupied by new arrivals, as are newly suitable territories created by experimentally adding materials to undefended substrates.

Among marine species male striped blennies (*Chasmodes bosquianus*) defend empty oyster shells, where breeding occurs (Phillips 1977a). Males compete intensively for the relatively scarce shells with small openings, which are the only shells that attract gravid females. They also defend open shells that fail to attract females, probably for use as shelter sites. Male beaugregories (*Eupomacentrus leucostictus*) defend nest sites consisting of empty conch shells, tin cans, drain pipes, and other openings found along rocky shorelines and shallow reefs (Brockmann 1973). Nest sites are likely to vary in quality, with at least the best sites being limited in availability. Little is known about the spawning behavior of many marine fish, and additional examples of defended spawning sites will undoubtedly surface during future research.

Amphibians. Many amphibians defend sites where females lay eggs (Wells 1977a). In salamanders male defense of oviposition sites has been reported in several species having external fertilization (Kerbert 1904; Bishop 1941; Thorn 1962) but not in any species having internal fertilization. Males of some frogs defend tree holes or burrows that females use as oviposition sites and males use as daytime shelters (Pengilley 1971; Wiewandt 1971). Males of other species defend calling sites or floating vegetation, where males attract mates and females oviposit (Wells 1977a, b). Females do not oviposit on male territories in every species, and in such cases male territories may serve primarily as mating stations (see Wells 1977a).

Competition for oviposition sites has not been demonstrated for most amphibians, but it probably exists because sites are likely to vary in quality. For example, preferred oviposition sites in bullfrogs (*Rana catesbeiana*) are areas where water temperatures are high enough to permit rapid development but not so high that they cause developmental abnormalities (Howard 1978a). The quality of sites may also depend on vegetation density, which may affect vulnerability of egg masses to leeches (*Macrobdella decora*). The largest males are most successful in competing for oviposition sites preferred by females, and as a result they control the sites where embryo mortality is lowest and hence are most likely to attract a female (Figure 7–4) (Howard 1978b).

Birds. Hole-nesting birds generally defend nest sites against individuals of both their own and other hole-nesting species. Several lines of evidence show that holes are in short supply. Aggression around holes is prevalent (e.g., Bourne 1974); holes often vary in suitability, with many poor holes occupied by nesting pairs (e.g., Curio 1959); and introducing artificial nest boxes usually leads to marked increases in breeding density (e.g., Lack 1958; Berndt and Sternberg 1968). Birds that excavate their own holes may not compete for nest sites, unless suitable substrates for digging holes are scarce or variable in quality, but they must still defend them to prevent usurpation of their holes by competitors. In effect, excavated holes must be defended to protect the time and energy invested in making them.

Territoriality in many colonial birds also stems from competition for suitable nest sites. Auks and murres are specialized to nest in various-sized holes in cliff faces, on cliff ledges, on top of rock stacks, on steep slopes above cliffs, or on flat terrain as a response to interspecific competition for nest sites (Lack 1934). In red-billed tropicbirds (*Phaëthon aethereus*) and two African cormorants, only part of the population can breed at any given time because many pairs cannot find suitable nest sites (Marshall and Roberts 1959; Stonehouse 1962; Snow 1965a). Males of polygamous colonial birds compete intensively for habitats that attract the most females, and they attract multiple mates primarily because many males are excluded from suitable habitats (Wittenberger 1976a, 1979a; see Chapter 11).

Nesting substrate is not in short supply for all colonial birds. Nest sites are readily available in seabird colonies situated on flat terrain and

Figure 7–4

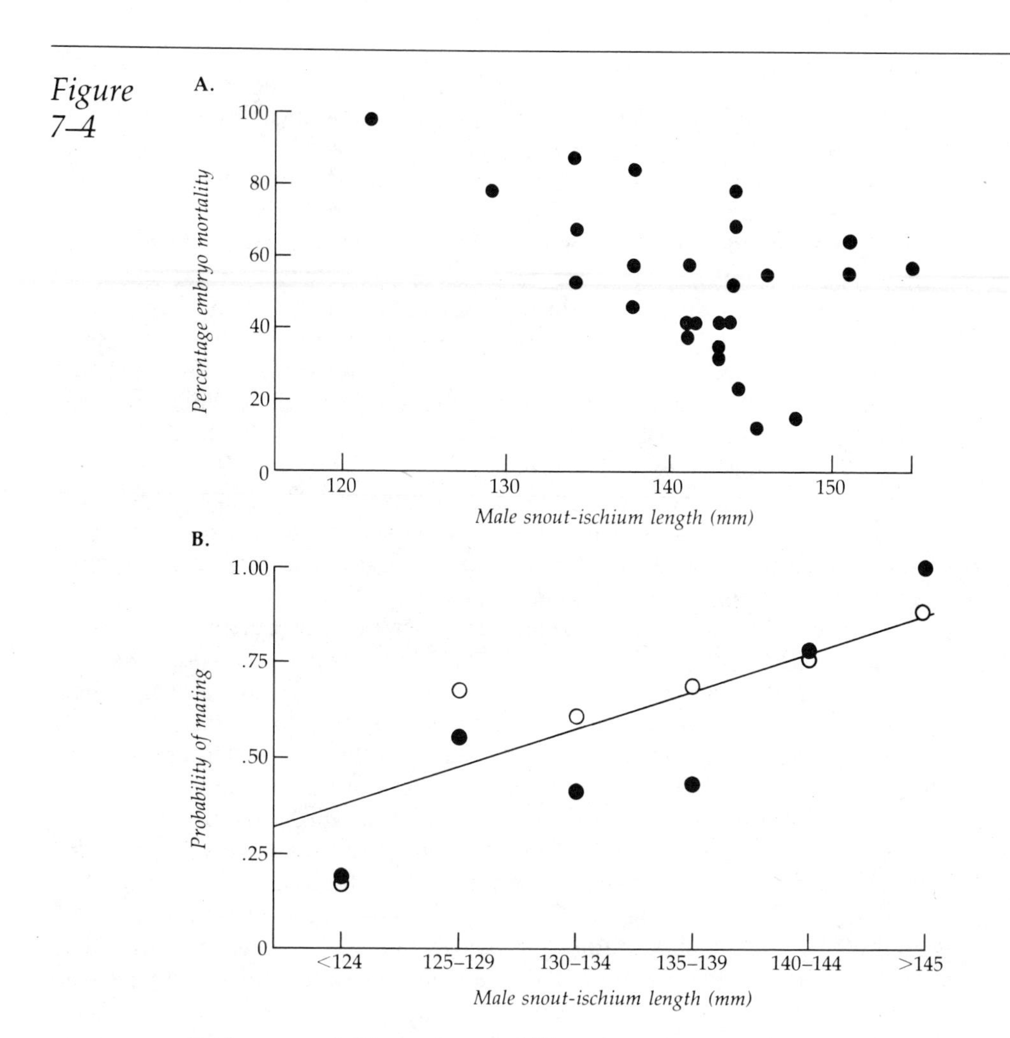

Embryo mortality and probability of mating as a function of male size in bullfrogs.

The largest males, as measured by snout-ischium length, control the sites where embryo mortality is lowest. As a result, they are more likely to attract a female than are smaller males. Open and closed circles in B refer to different years.

Source: From R. D. Howard, "The Evolution of Mating Strategies in Bullfrogs, *Rana catesbeiana*," *Evolution* 32 (1978):863. Reprinted by permission.

for most monogamous colonial birds (see Wittenberger and Tilson 1980). Territory defense in these species may reserve central or otherwise favorable locations within a colony, reduce interference with copulations, protect nesting material, or reduce injury to chicks (see Tinbergen 1956, 1960; Hoogland and Sherman 1976; Wittenberger and Tilson 1980).

Protecting time and energy investments

Some animals defend territories to protect resources that have been constructed or accumulated. Such resources are frequently defended because competitors can often enhance their own fitness by stealing or usurping time and energy investments of others. Defending an investment is advantageous whenever the costs of replacing the investment are less than the costs of defense.

Two species of mantis shrimps (*Cloudopsis scorpio* and *Oratosquilla inorta*) probably defend burrows because digging new burrows is costly (Dingle and Caldwell 1975). These species depend on burrows dug into intertidal mudflats for protection from predators, especially while molting exoskeletons, and it takes several hours to construct a burrow. Evicted individuals would be exposed to high predation risk, as well as having to spend time and energy constructing new burrows. The most aggressive species is the one that digs the longest and most elaborate burrow systems, suggesting that intensity of defense may be correlated with cost of constructing a new burrow.

Male field crickets (Gryllidae) defend crevices or burrows that they dig under food plants (Alexander 1961). Their crevices and burrows are used either to gain protection while foraging or to store or deposit eggs. Evicted individuals would be temporarily left without protection or would be forced to lay eggs in less safe places. The intensity of defense shown by a species is again correlated with the extent of burrowing behavior.

Several species of benthic fish defend spawning burrows in sandy bottoms, which they use as protection from predators (Clark 1972; Colin 1973). Garden eels (*Gorgasia*), for example, live in colonies and forage on plankton, a nondepletable food resource. Suitable bottom substrate appears unlimited in availability. Although burrows may be defended during the spawning season to protect eggs, they are also defended during nonspawning seasons. One reason for year-round defense may be that the fish rely on burrows for protection and loss of a burrow would leave the occupant defenseless until it could dig a new burrow.

The garibaldi (*Hypsopops rubicunda*), a pomacentrid reef fish, defends spawning sites that males must spend time and energy creating (Clarke 1970). Females spawn only in dense mats of red algae, which males maintain by removing competing plant growth and herbivores. Two years of maintenance are required before algae beds are sufficiently dense for spawning. Males use the same beds every year, and the few new beds started each year are rarely maintained long enough to attract females. When a resident male dies, his nest site is quickly claimed by a previously nonterritorial male. Territory defense is mainly to prevent takeover of spawning beds, but males do defend their food supply of sponges as well, apparently to preserve an adequate year-round food supply near the algae beds.

Males of two South American tree frogs (*Hyla boans* and *H. rosenbergi*) construct elaborate mud nests along shallow edges of rivers and ponds, and they cannot attract any mates until they possess a nest (Lutz 1960,

1973). Females lay their eggs in one of the artificial ponds formed by a male's nest, where their tadpoles will gain protection from currents and most predators until they mature. Nest construction represents a significant investment of time and energy, and males defend their nests to protect that investment.

Certain species of herons and storks nest on top of massive nesting platforms constructed out of sticks (Haverschmidt 1949; Lowe 1954; Cottrille and Cottrille 1958; Palmer 1962). Platforms are reused, with minor alterations and additions, year after year. Males usually cannot obtain mates until they acquire a nest platform, and males with nests must continually repulse intruding males who lack nests of their own. Following pair formation both sexes vigorously defend their nests against unmated males seeking to acquire nest sites and neighboring pairs attempting to steal nest materials.

Acorn woodpeckers of California live in extended family groups and store acorns by pounding them into holes dug into the sides of trees, telephone poles, and fence posts (Photo 7–2) (MacRoberts and MacRoberts 1976). They defend their entire home ranges, but defense is strongest around storage granaries. Territory defense serves at least partly to protect stored acorns, since it is directed against squirrels and jays as well as other acorn woodpeckers, but it probably also serves to maintain control of the granary itself. The holes used to store acorns are themselves a valuable resource. A granary may contain as many as 11,000 holes, and each hole requires 30 to 60 minutes to dig. Most holes are drilled during the latter part of the storing season or in winter, when drilling interferes least with breeding and storing activities. That the holes themselves are defended is shown by the fact that acorn woodpeckers continue defending empty granaries after acorn stores are exhausted. In contrast, acorn woodpeckers abandon territories as soon as acorn stores are exhausted in areas where acorns are stored in natural crevices or under loose bark instead of in excavated holes (Stacey and Bock 1978).

The beaver (*Castor canadensis* and *C. fiber*) is a classic example of a mammal that defends an energy investment (Bradt 1938; Tevis 1950; Wilsson 1968, 1971; Hodgdon and Larson 1973). Beavers expend considerable time and energy constructing a lodge and dam and caching food for winter, and they vigorously defend their holdings against intruding beavers. Similarly, muskrats (*Ondatra zibethicus*) defend the elaborate lodges they construct within marshes (Banfield 1974; Lowery 1974). Loss of these investments to a usurper would represent a major cost, as replacement costs are high and chances of avoiding predation and surviving winter would be greatly diminished without a lodge and, in the case of beavers, food stores.

Defense of shelter sites

Defending shelter sites is similar to defending burrows. However, there is one conceptual difference. Shelter sites are defended because displaced individuals are vulnerable to predation while *searching* for a new

A.

B.

Acorn woodpeckers store large numbers of acorns in holes excavated in dead trees or telephone poles.

C.

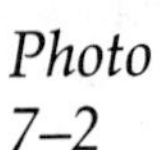

Photo 7–2

The woodpeckers rely on these acorn stores for food during winter, when insects and other food resources are seasonally scarce. A. An acorn woodpecker putting an acorn into a previously drilled hole. B. Close-up of a storage granary, showing how tightly acorns are wedged into holes. Such a tight fit makes it difficult for squirrels and jays to steal acorns from the granary. C. A large storage granary, showing the immense number of acorns stored by some extended-family groups.

Photos from M. H. MacRoberts and B. R. MacRoberts, "Social Organization and Behavior of the Acorn Woodpecker in Central Coastal California," *Ornithological Monographs* 21 (1976):1–115.

shelter, while burrows are defended because displaced individuals are vulnerable to predation while *digging* a new burrow.

Many mantis shrimps (Stomatopoda) defend natural cavities in detritus-covered rocks or rubble piles (Figure 7–5) (Dingle and Caldwell 1969; Dingle et al. 1973; Caldwell and Dingle 1976; Caldwell 1979). They leave their cavities only briefly while darting out for food, probably because they are prey to many carnivorous fish. All suitable cavities are occupied, and territory defense evidently secures them a safe retreat.

The decapod crayfish *Oronectes virilis* also defends safe retreats, though many other crabs and crayfish are nonterritorial (Bovbjerg 1953). Spiny lobsters (*Jasus lalandei*) are normally nonterritorial, but they will defend shelter sites under laboratory conditions if suitable sites are limited in availability (Fielder 1965).

Pomacentrid anemonefish reside as permanent pairs inside sea anemones, where they defend a territory against conspecific intruders (Fricke 1974; Allen 1975; Ross 1978). Anemonefish gain protection by remaining near the tentacles of sea anemones, and anemones are in short supply for at least some species (Ross 1978). They may also gain reduced competition for food by excluding competitors from the space around the anemone, and they may gain access to better foraging locations if some anemones are situated in better places for obtaining food than others.

Figure 7–5

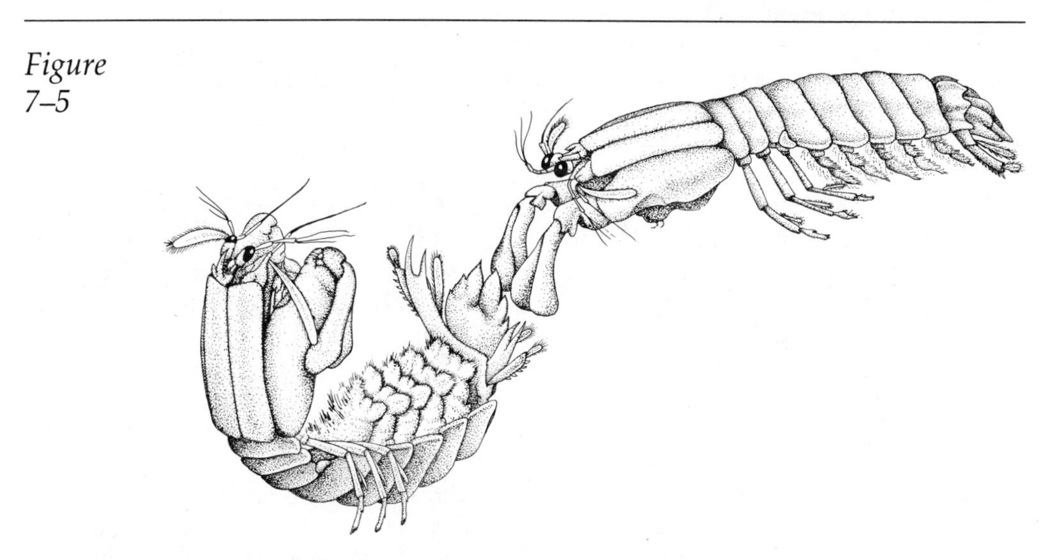

The agonistic behavior of stomatopod crustaceans.

The stomatopod on the left is in a defensive posture, presenting its armored telson (terminal abdominal segment) toward its opponent. The opponent on the right is lunging to strike. The two opponents alternate coiling, lunging forward to strike, and taking a defensive posture in a highly stereotyped sequence until one opponent concedes defeat.

Source: From Roy L. Caldwell and Hugh Dingle, "Stomatopods," *Scientific American* 234 (January 1976):84.

Similarly, many species of gobies reside as monogamous pairs inside burrows dug by alpheid shrimp and keep other gobies away from the burrow entrance (Karplus et al. 1972, 1974; Preston 1978; Karplus and Tuvia 1979). The interaction between gobies and shrimp is symbiotic. Gobies gain safe shelters while shrimp derive protection from the early-warning signals given by gobies. In a few cases gobies are also cleaned by the shrimp (Karplus et al. 1972). Gobies are often observed without shrimp burrows while shrimp burrows are rarely observed without gobies (Preston 1978). This suggests that gobies compete for a limited number of shrimp burrows.

The sedentary mussel blenny (*Hypsoblennius jenkinsi*) of California defends territories in boring-clam *(Serpulorbis)* burrows, snail tubes, and mussel beds (Stephens et al. 1970). There is strong competition for safe holes, especially ones large enough to house larger individuals. A related nonsedentary blenny, *H. gilberti,* wanders over rock intertidal areas and does not defend territories. Another blenny, the Hawaiian rockskipper *(Istiblennius zebra),* defends natural cavities in the splash zone of rocky shorelines (Phillips 1977b). Suitable crevices are evidently in short supply, since not all individuals possess one. Similarly, the redlip blenny *(Ophioblennius atlanticus)* of Caribbean reefs defends small individual territories on rocky substrates, which they use for shelter sites (Nursall 1973; Itzkowitz 1974).

In two species of freshwater trout the largest individuals defend feeding positions nearest to shelter from currents and predators (Jenkins 1969). Large individuals are dominant over small ones and can displace them from their territories. Hence large individuals can move relatively freely within a stream, while small individuals cannot. Similar spacing behavior has been described under laboratory conditions for the green sunfish (*Lepomis cyanellus*) (Greenberg 1947). The spacing behavior of these species seems intermediate between territoriality and dominance, suggesting that the two behaviors are closely related. This affinity is not surprising, since both kinds of behavior are based on aggression. Consequently some authors view territoriality and dominance as opposite ends of a behavioral continuum (e.g., see Itzkowitz 1977; Leuthold 1977).

Defense of food resources

Invertebrates. Defense of food resources is often difficult to prove because animals usually use their territories for purposes additional to foraging. The evidence is clearest when animals feed on territories during nonbreeding seasons. Even then territory defense may have evolved to reserve breeding sites for the following reproductive period. This is probably the case for male garibaldis (see p. 264) and perhaps also for some nonmigratory birds (Lack 1954, 1968). Since territories defended in nonbreeding seasons often provide shelter or safe foraging sites as well as food, defense of these commodities is often intertwined. The purpose of such territories can be interpreted as defense of food resources when food is limited and shelter sites are not and as a defense of shelter sites when shelter is limited and food is not.

In the marine amphipod *Erichthonius braziliensis*, individuals attach to settlement sites and build tubes that they defend against wandering conspecifics (Connell 1963). Tubes are uniformly spaced, and each is surrounded by a bare area that has been grazed clean. Neighbors do not fight, since their grazing areas do not overlap, but residents repulse wanderers aggressively to prevent them from settling. Only occasionally are wanderers successful in driving out residents and taking over their tubes. Whether territories are defended to protect an energy investment (the tube), a favorable grazing area, or both is unclear.

Owl limpets (*Lottia gigantea*) defend grazing areas in intertidal zones by dislodging barnacles and sea anemones (Stimson 1970, 1973). Territories contain thick films of algae, which owl limpets use for food, and experimental removal of the limpets soon results in disappearance of these films. Owl limpets are larger than most other limpets and cannot crop algal films as closely as their smaller competitors. Territory defense allows sufficient growth of algae for owl limpets to forage efficiently, and it keeps the grazing area free of other encrusting organisms that compete with both algae and owl limpets for attachment space.

Sedentary limpets of the genus *Patella* defend foraging territories, while congeneric migratory species move shoreward or seaward with the tide instead (Branch 1975). The sedentary species defend algal beds by ousting limpets and other invertebrates that graze on algae. They also remove undesirable species of algae, much like weeding a garden. Following experimental removal of territory residents, the algal beds are soon grazed down by other limpets or become overgrown by other algae.

Many social ants and termites defend foraging areas around their nests (Brian 1965; Brian et al. 1965, 1966; Wilson 1971; Carroll and Janzen 1973; Baroni Urbani 1979). Such territories are not usually comparable to those found in vertebrates because they are usually not exclusively occupied. Exploited areas within defended boundaries often consist of trails and points rather than well-defined surfaces (Baroni Urbani 1979). Nevertheless, some species do defend boundaries around their foraging arenas. The form of territory defense depends on the spatiotemporal distribution of food resources, and three different patterns have been identified (Hölldobler 1979a).

Territory defense among social insects was first reported for red wood ants (*Formica rufa*), which defend aphid-bearing trees (Elton 1932; Pickles 1935, 1936). Wood ants from neighboring colonies are especially hostile in overpopulated areas where food is scarce (Marikovskiy 1962). Under extreme conditions of deprivation, workers even hunt for members of other colonies and cannibalize them. Territory defense occurs in many other species of aphid-tending ants as well (Brian 1955; Tsuneki and Adachi 1957; Dobrzańska 1958; Waloff and Blackith 1962; Yasuno 1965). The African weaver ant (*Oecophylla longinoda*), which tends a variety of homopteran insects and eats the honeydew secreted by them (Way 1954), is particularly aggressive in defending territories (Hölldobler 1979b). These ants defend relatively fixed territory boundaries, which are marked by colony-specific odors, and when intruders are detected, a mass attack is mobilized by releasing recruitment pheromones.

Defense of relatively fixed territory boundaries results from evenly distributed food resources. The ants studied by Dobrzańska (1958) in Poland partition foraging space in two different ways, depending on how food is distributed. Aphid-tending species are territorial and defend permanent food sources against all other ant species, while species that exploit temporary food patches on an opportunistic basis are nonterritorial. Similarly, the food resources defended by African weaver ants are also evenly distributed (Hölldobler 1979a).

The second pattern of resource defense occurs in harvester ants (*Pogonomyrmex*), which exploit patchy seed resources (Hölldobler 1974, 1976a, b, 1979a). These ants exploit seeds by establishing trunk trails that irradiate from the colony nest to relatively permanent food clumps (Figure 7–6). Aggression between workers from neighboring colonies is normally infrequent because trunk trails ordinarily do not intersect, but intense territorial fighting occurs when artificial food sources bring them together. The intensity of aggressive confrontations depends on how close intruders are to a colony nest. When colonies are established too near each other, aggression by workers of the stronger colony eventually leads to elimination of the weaker colony. The foraging pattern of harvester ants varies with food density, but it is not yet clear how this variability affects territorial behavior (Bernstein 1975).

The third pattern of defense occurs in honeypot ants (*Myrmecocystus mimicus*), so named because some workers (the honeypot caste) are specialized to store food in greatly distended abdomens (Hölldobler 1976c, 1979a). A main food source is termites, which represent a spatiotemporally fluctuating food resource because exploited colonies are eventually wiped out. Territory defense arises when foraging workers detect workers from a foreign colony near a food source. When an encounter occurs, some workers return to the colony and recruit an "army" of 200 or more workers, which swarm over the alien colony and engage its workers in an elaborate display tournament. Actual physical fighting and hence fatalities are rare, but the swarming behavior of workers from the first colony prevents workers from the second colony from gaining access to the food source. Once the food source is exhausted, the invading "army" retreats and the tournament ends. The tournaments sometimes last for several days and are interrupted only at night, when the workers are inactive. Neighboring colonies also conduct tournaments between the respective colonies frequently. These contests prevent workers from entering the foraging zones of their respective neighbors. If one colony is substantially weaker than the other, tournaments are quickly ended and workers from the stronger colony raid the nest of the weaker colony. Such raids result in death of the queen and enslavement of the surviving workers from the weaker colony. The

---

Source: From B. Hölldobler, "Home Range Orientation and Territoriality in Harvesting Ants," *Proceedings of the National Academy of Sciences* 71 (1974):3275. Reprinted by permission of the author.

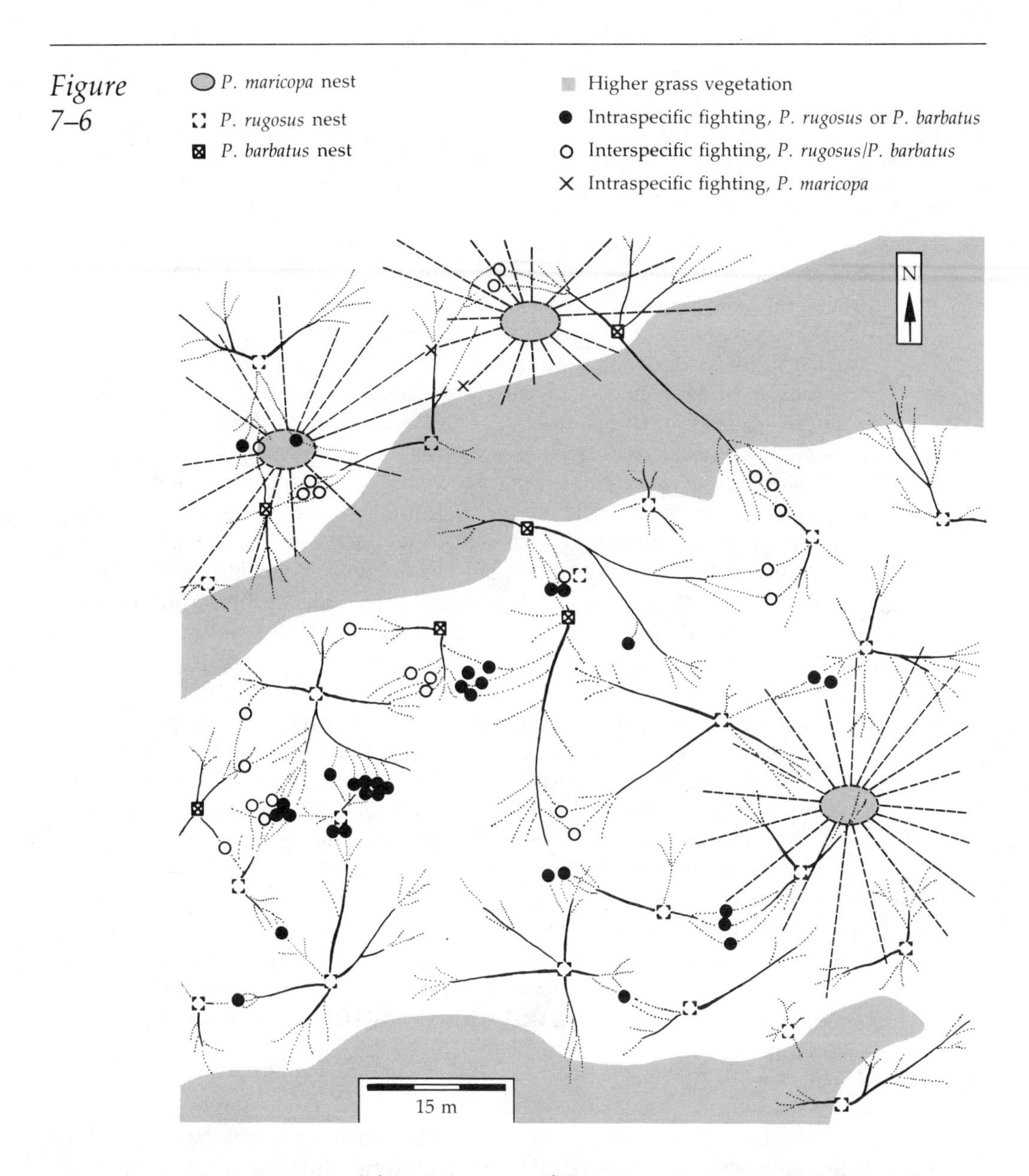

Figure 7–6

The colonies and foraging areas of Pogonomyrmex *ants in an Arizona desert.*

Colony locations are indicated by the symbols shown in the legend. Solid lines emanating from colony sites indicate trunk lines used by large numbers of workers. Dotted lines branching from the trunk lines indicate foraging paths taken by individual workers in search of new food sources. Dashed lines indicate the general foraging area of P. maricopa*, which lacks trunk trails. Note that each colony controls its own foraging area, with workers attacking intruders that they detect within that area.*

locations of tournaments between colonies are not fixed but change as the locations of foraging sites change.

An interesting characteristic of territorial ants is that they are all *monogynous* (Hölldobler 1979a). That is, each colony contains only one queen. *Polygynous* ants, which have more than one queen per colony, are never territorial. Hölldobler suggests that this relationship can be explained by indirect (kin) selection theory. In polygynous ants mating often involves a high degree of inbreeding, with virgin queens mating with males produced in the same colony. Those queens then establish colonies near their natal colonies, with the result that neighboring colonies often consist of relatives. In contrast, monogynous ants are strongly outbred and new colonies are established at considerable distances from the queen's natal colony. Aggression directed against relatives reduces the net benefit obtainable from resource defense (see p. 164), which may explain the absence of territoriality in polygynous ants. However, it remains possible that monogynous ants are territorial because their foraging ecology differs from that of polygynous ants in some consistent, though as yet unidentified, manner.

The mound-building Australian termite *Coptotermes brunneus* defends territories around the few eucalyptus trees growing in their arid environment (Greaves 1962). Mature colonies are spaced about 100 m apart, and although no boundary defense has been observed between them, new colonies cannot persist within the foraging radius of mature ones because workers quickly invade and destroy them. That *Coptotermes brunneus* is territorial because it competes for trees is revealed by comparing them with the related species *C. acinaciformes*, which is nonterritorial and lives in subtropical rain forests, where trees suitable for infestation abound.

Territoriality has been reported in several other termites as well. Competition for food is evident in all these species. The mound-building termites *Hodotermes mossambicus* and *Trinervitermes trinervoides* of South Africa live in pastures, where they severely deplete new growth of grass and create problems for local ranchers (Nel 1968). The mound-building termite *Odontotermes redemanni* of India also defends foraging areas, and in that species territory size is correlated with colony size, suggesting that colony size is limited by the amount of food available on the territory (Banarjee 1975).

Vertebrates. Many vertebrates defend territories where they forage, but such territories are often used for other purposes as well. In most cases territories are defended for several reasons and are best treated as multipurpose territories. However, a few vertebrates defend territories largely or entirely to control food resources.

Many coral reef fish defend territories even though they broadcast their gametes into the pelagic zone (Feddern 1968; Barlow 1974; Ehrlich 1975; Reese 1975; Ehrlich et al. 1977). Their territorial behavior is therefore not associated with defense of spawning sites or care of eggs or young. Among pomacentrids territory defense is directed at food competitors but not at egg predators (Low 1971), suggesting that it evolved mainly as a defense of food resources. In comparison, pomacentrids that spawn on their territories exclude egg predators as well as food competitors (Myrberg 1972; Brockmann 1973; Myrberg and Thresher 1974; Ebersole 1977). Territoriality in the latter species evidently evolved for the dual purpose of protecting eggs and reducing competition for food. The eight species studied by Sale (1977) defend rubble piles, which are used as foraging sites and as protective shelter sites. They may defend territories to preserve food resources, control shelter sites, or both. In *Pomacentrus lividus* territory defense allows green algae on coral heads to grow into thick mats (Vine 1974). Undefended areas are kept closely cropped by other algal grazers on the reef, showing that territoriality is important in maintaining the food supply.

As in social insects, spatiotemporal distribution of food is correlated with territoriality. Territorial species of butterflyfish and surgeonfish graze on evenly distributed and continuously available coral polyps, which are relatively easy to defend (Barlow 1974; Reese 1975; Keenleyside 1979). In contrast, nonterritorial species of surgeonfish exploit temporary clumps of detritus or filter food from unlimited expanses of sandy bottom.

Many birds defend territories during the nonbreeding season. Some may defend winter territories to reserve a breeding territory for spring (Lack 1954, 1968), but that cannot be the case for migratory species. Food is generally scarce for birds during winter, but not all species are territorial. An important factor favoring territoriality is the spatial distribution of food. For example, Zahavi (1971a) induced dominant members of white wagtail (*Motacilla alba*) flocks to defend small temporary territories by providing them with small piles of food. When he dispersed the same amount of food over a wider area, dominant individuals did not show territorial behavior and the wagtails fed as flocks. Local variations in food distribution also induce some individuals to defend territories under natural conditions (Davies 1976a). The same situation occurs among wintering shorebird flocks (Myers et al. 1979a; Myers and Myers 1980).

Many birds forage on their territories while breeding, but it is usually hard to identify exactly what is being defended. In exceptional cases feeding areas are defended independently of nest sites, and one can then infer that food resources are being defended. Avocets (*Recurvirostra americana*), for example, defend two spatially separate territories, a small one around the nest site and a larger one where they forage (Gibson 1971). Hawks and owls usually nest in inaccessible sites and range widely in search of food (Brown and Amadon 1968). Some species, such as the red-tailed hawk (*Buteo jamaicensis*), defend both nest sites and hunting grounds (Fitch et al. 1946; Craighead and Craighead 1956). Red-tails defend nests most vigorously while they contain young and hunting grounds most vigorously before eggs are laid and after young fledge.

Most territorial mammals conduct all their activities on permanent territories, making it difficult to identify what is being defended. Nevertheless, at least some mammals clearly defend food resources. Tree squirrels (*Tamiasciurus*) live in coniferous forests and defend individual territories all year (Smith 1968). In winter territories are centered around food caches, and territory size is adjusted to food abundance. These territories are not defended just to reserve a breeding site for spring, as many of them are abandoned before summer because they are unsuitable for breeding (Kemp and Keith 1970).

The territories of some antelopes and primates are also associated with food resources. Male vicuñas (*Vicuna vicuna*) control access to female groups by defending the best grazing areas (Koford 1957; Franklin 1974). Food is scarce in the semiarid grasslands where vicuñas live, and the fact that territory size is correlated with quantity and quality of available forage shows that territory defense is related to food resources. Male pronghorn antelopes (*Antilocapra americana*) defend grazing areas and water holes used by female herds on a temporary basis (Kitchen 1974). Most nongregarious African antelopes are territorial and forage on their territories (Owen-Smith 1977). Defense of food resources per se has not been demonstrated, but indirect evidence indicates that food is limited for these species (Jarman 1974). Many arboreal and semiterrestrial primates defend territories (Bates 1970), but again defense of food resources per se has usually not been demonstrated. Struhsaker (1967b) found that vervet monkeys defend territories when faced with strong competition for food but not when competition is relaxed. Baboons ordinarily do not defend territories, but chacma baboons defend water holes in Namibia (South West Africa) and probably food resources in Botswana (Hamilton et al. 1975b).

Defense of multipurpose territories

The complexities of territoriality are nowhere more evident than for animals that defend multipurpose territories. Because such territories provide all requisites for survival and reproduction, the environmental factors favoring their evolution are not easy to identify. A wide variety of lizards, birds, and mammals defend multipurpose territories (Eisenberg 1966; Lack 1968; Verner and Willson 1969; Brattstrom 1974; Stamps 1977). The adaptive significance of these territories is controversial. The controversy centers around two competing hypotheses, which have been evaluated primarily with data on avian territories.

Lack (1954, 1966) contends that multipurpose territories evolved solely to space out individuals, thereby reducing vulnerability of nests and fledged young to predation. He argues against their role in preserving food resources for two reasons. First, a number of birds do not defend food resources while feeding young, even though competition for food should be most severe at that time. If territory defense evolved to alleviate competition for food, it should be strongest during the nestling period rather than earlier in the season. Second, bird populations are limited by winter rather than summer food availability, and hence they should not experience much competition during summer.

The alternative hypothesis is that multipurpose territories evolved as a defense of limited food resources. The fact that territory defense wanes during the nestling period does not negate its role in alleviating competition for food. If local breeding densities are limited by territorial behavior, territory defense in spring would reduce competition for food throughout the summer because it would force many competitors to breed elsewhere. Once competitors have settled elsewhere, continued defense becomes less advantageous because local competition for food is reduced. Furthermore, the cost of continued defense increases during summer because the time and energy spent defending territories could be spent feeding nestlings. That is, alternative reproductive activities conflict more with territory defense during the nestling period than they do earlier in spring. Waning of territorial behavior in summer indicates a change in the trade-off between territoriality and alternative activities but does not imply that spring territoriality has no effect on competition for food during summer. With regard to Lack's second point, competition cannot be assumed to be absent in summer even if populations are limited by winter food supplies. Although population density may not be limited by summer food availability, reproductive success is. The number of offspring parents can successfully rear depends on how fast offspring can be fed, and any depletion of food resources is likely to reduce food delivery rates. Hence minimizing the depletion rate by forcing some competitors to breed elsewhere should increase an individual's reproductive success.

One indication that multipurpose territories evolved to protect food resources is that territory size is often correlated with food abundance. For example, arthropod abundance on ovenbird (*Seiurus aurocapillus*) territories is negatively correlated with territory size, so that territories of every size contain about the same amount of food (Stenger 1958). Similarly, insect abundance on winter wren (*Troglodytes troglodytes*) territories, which was sampled by capturing insects with sticky pads, is negatively correlated with territory size (Cody and Cody 1972). In golden-winged sunbirds *(Nectarina reichenowi)* and rufous hummingbirds *(Selaphorus rufus)* winter territories each contain about the same amount of nectar regardless of territory size, and the amount on each territory approximately equals the territory resident's metabolic energy requirement (Gill and Wolf 1975; Gass et al. 1976; Pyke 1979). The size of red grouse territories is negatively correlated with abundance of heather, the main food plant of red grouse, in areas where heather is sparse but not in areas where heather is abundant (Lance 1978b; Miller and Watson 1978). In the latter areas territory size is negatively correlated with the nitrogen content of heather, suggesting that it may be adjusted to quality rather than quantity of available food (Lance 1978b). In Townsend's solitaire (*Myadestes townsendi*) seasonal changes in mean territory size are negatively correlated with abundance of juniper berries, their main food resource during winter (Salomonson and Balda 1977). Finally, the density of isopods, the main prey of wintering sanderlings (*Calidris alba*) along coastal sandy beaches, is negatively correlated with the size of sanderling territories (Myers et al. 1979b). This correlation resulted from higher intrusion pressure by competing sanderlings in areas of high prey abundance, not from direct adjustments of territory size to food availability.

Similar results have been obtained for a few nonavian species. Territory size in the iquanid lizard *Sceloporus jarrovi* is inversely related to food abundance and changes quickly and reversibly when food abundance is manipulated experimentally (Simon 1975). Territories of resident tree squirrels contain one to three times the squirrels' annual food requirements, with smaller territories sometimes containing somewhat more food than do larger ones (Smith 1968).

Using a more indirect approach, Yeaton and Cody (1974) argued that food availability should decrease as number of species competing for food increases. They found a positive correlation between size of song sparrow territories and number of competing species, implying that song sparrows adjust territory size to intensity of competition for food (Figure 7–7). However, areas with few competing species were always islands, so the results might be due to some factor other than competition intensity that correlates with island community structure.

In a more general analysis Schoener (1968) found that average territory size for birds is correlated with body weight and diet (Figure 7–8). Body size is also correlated with territory size in lizards (Turner et al. 1969) and with home range size in mammals (McNab 1963). Larger animals probably do not need to space themselves farther apart to gain protection from predators, but they may well require larger territories or home ranges to satisfy their energy requirements. Nevertheless, population densities are usually lower for large animals than for small ones, which would automatically result in larger average territory sizes among large animals.

The fact that territory size is correlated with food abundance does not prove that food is being defended. Higher food abundance would attract and support more individuals per unit area, which would automatically reduce territory size even if territories serve primarily to space individuals apart. The best way to distinguish between the spacing and food defense hypotheses is to determine whether territoriality limits local population densities in the best habitats (Fretwell and Lucas 1969). If it does, it probably evolved to ameliorate competition for food. If not, it probably evolved to space individuals or nests as a defense against predation.

Effect of territoriality on population density

Density levels model

Territoriality can affect populations by spacing individuals out, by limiting density in the best habitats, or by controlling total size of the breeding population (Brown 1969a, b). According to Lack's hypothesis, multipurpose territoriality only serves to space individuals out in a habitat and does not limit local population density in any habitat. According to the food defense hypothesis, multipurpose territories limit local population density in at least the best habitats.

A simple way of looking at how territoriality affects population density

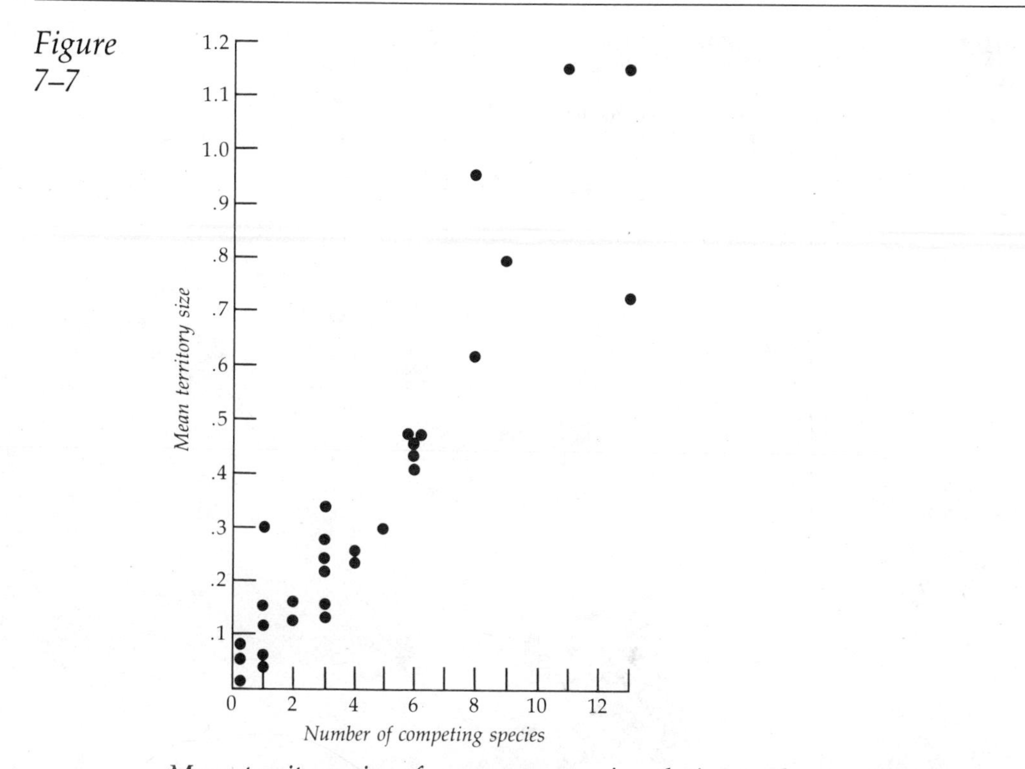

Figure 7–7

Mean territory size of song sparrows in relation to the number of competing species present.

Each point represents a different locality in Wyoming or the Pacific Northwest. The localities with few competing species are all islands. The correlation suggests that mean territory size decreases with increasing interspecific competition for food.

Source: Modified from R. I. Yeaton and M. L. Cody, "Competitive Release in Island Song Sparrow Populations," *Theoretical Population Biology* 5 (1974):50. Reprinted by permission.

is to define three critical density levels (Figure 7–9) (Brown 1969a, b). At *level 1* a population is sufficiently sparse that every individual can breed in favorable habitat, and territorial behavior serves primarily to space individuals out. At *level 2* some individuals are excluded from favorable habitat by the territorial behavior of prior residents, but excluded individuals are still able to breed in poorer habitats. Territorial behavior limits local population density in the best habitats and probably evolved as a defense of food resources. At *level 3* some individuals are excluded from all breeding habitats, thereby creating a surplus of nonbreeding *floaters* in the population. Territorial behavior then limits overall size of the breeding population and again probably evolved as a defense of food resources.

Some relatively uncommon territorial birds may breed at level 1 densities. The Kirtland's warbler (*Dendroica kirtlandii*) is an endemic wood warbler that breeds only in jack-pine forests of central Michigan (Mayfield

Figure 7–8

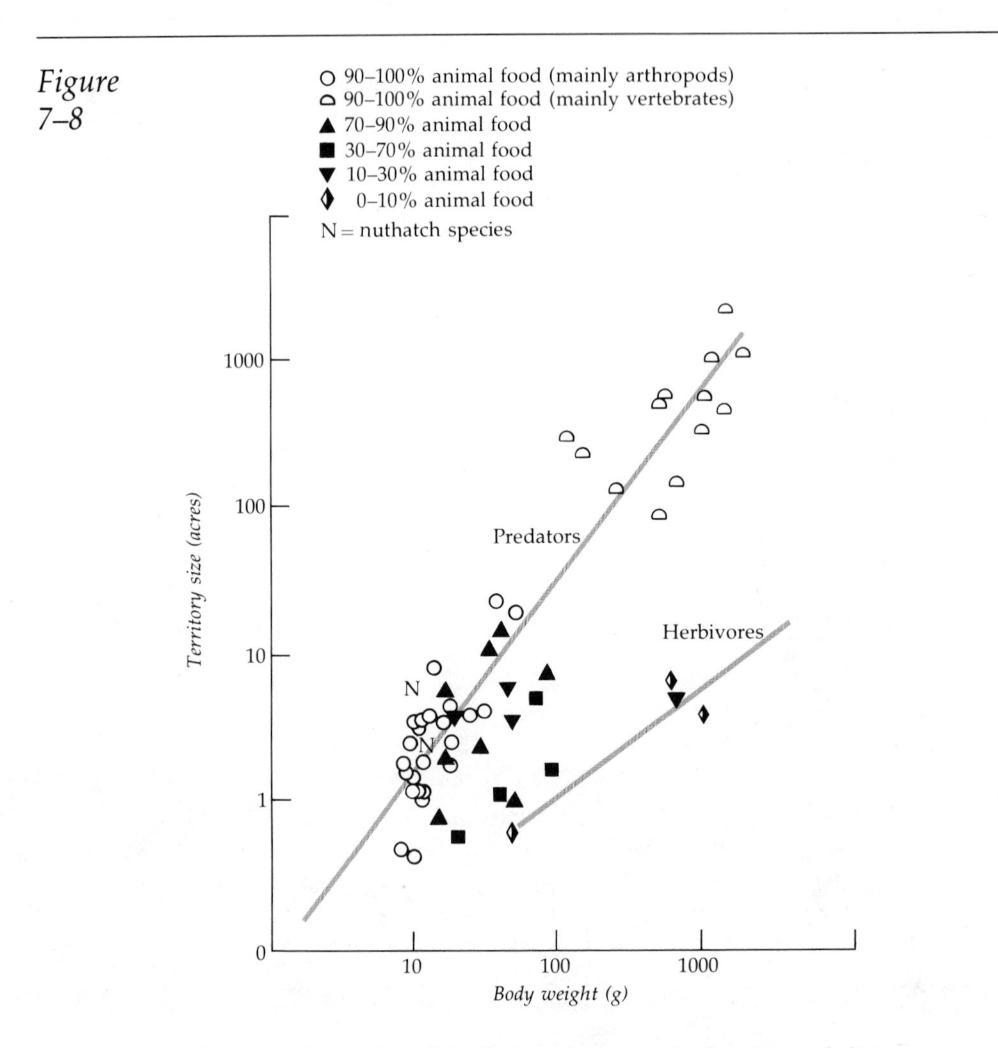

Mean territory size of birds in relation to body size and diet.

Each point represents a different species. The correlation suggests that mean territory size increases with body weight, with the rate of increase and absolute territory size depending on diet.

Source: From T. W. Schoener, "Sizes of Feeding Territories Among Birds," *Ecology* 49 (1968):127. Reprinted by permission of Duke University Press. Copyright 1968 by the Ecological Society of America.

1960). Since not all apparently suitable habitat is occupied, every pair can apparently find a suitable place to breed. The tree sparrow (*Spizella arborea*), which nests in subarctic Alaskan and Canadian tundra, is another possible example. Weeden (1965) concluded that local densities of Alaskan tree sparrows were not limited by territorial behavior because "seemingly suitable habitat" remained unoccupied during each of three breeding seasons

Figure 7–9

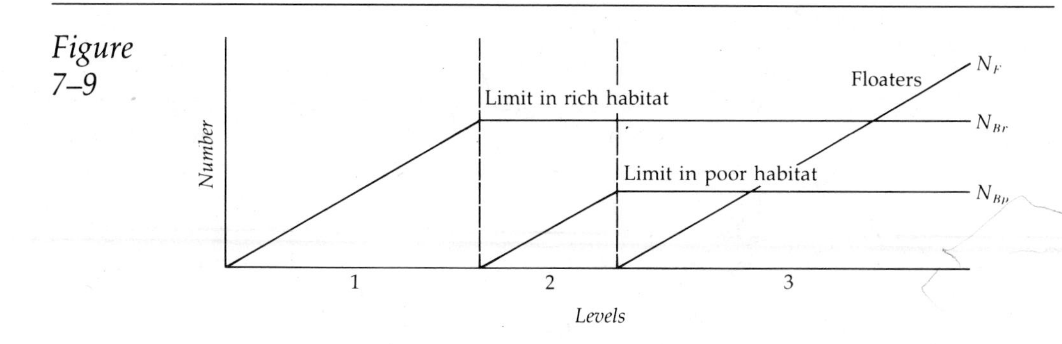

A simple model of how territoriality might affect population density. *At level 1 all individuals can settle in rich habitats, and territoriality does not affect population density. At level 2 a limited number of individuals (N_{Br}) can settle in rich habitat, with the remainder forced into poor habitat. At level 3 a limited number of individuals (N_{Br}) can settle in rich habitat, and a limited number of additional individuals (N_{Bp}) can settle in poor habitat. Remaining individuals are nonbreeding floaters (N_F).*

Source: Reprinted from J. L. Brown, "The Buffer Effect and Productivity in Tit Populations," *American Naturalist* 103 (1969), p. 348, by permission of The University of Chicago Press. Copyright © 1969 by The University of Chicago Press.

and territories never reached a minimum size that might limit local breeding density, even in years when overall population densities were high.

Although these species may exist at level 1 densities, interpreting the evidence is not straightforward. The essential question concerns whether individuals compete for resources. If they do, territoriality could have evolved to defend resources from competitors. According to the simple density levels model, competition for habitats and resources is absent at level 1 densities. However, this implication of the model is not necessarily valid. When habitats vary in quality, even slight variations in quality can favor defense of the best areas regardless of population density. Although the species discussed above occupy habitats characterized by relatively uniform stands of vegetation, the possibility that sites within those stands vary in quality cannot be dismissed because insect abundance can vary to a surprising extent even within uniform stands of vegetation. Thus population density alone is insufficient evidence for determining whether all individuals have access to the best habitats. Distinguishing level 1 densities from level 2 densities requires an assessment of population densities *relative to habitat quality*. The studies cited above provide no evidence regarding the variability or homogeneity of habitat quality, and hence conclusions cannot be drawn regarding the intensity of competition that exists at the population densities observed. Indeed, the density levels concept is inadequate for evaluating the effects of variable habitat quality, and some additional theory is required.

Habitat distribution model

High-quality habitats contain more or better resources and hence attract more individuals than do poorer habitats. This higher intrusion pressure leads to increased competition for food and space and increases the cost of territory defense. Although individuals gain more benefits by living in high-quality habitat, they also incur more costs. If the gain in benefits is completely offset by the increase in costs, individuals will not gain any net benefit by settling in a high-quality habitat. Despite the intrinsic variability in habitat quality, no habitat would be better than any other for surviving and reproducing. Under such conditions local population densities are not limited by territoriality unless some individuals are prevented from breeding altogether. Instead, local population densities are limited by the costs of competition in each habitat.

Fretwell and Lucas (1969) devised a simple way to distinguish between populations limited by competition and those limited by territoriality. Consider first the case where every individual is free to enter any habitat it pleases. Territory defense by resident birds does not prevent individuals from entering a habitat; it merely spaces out those that do. Individuals should first enter the highest-quality habitats available. As population density increases in those habitats, competition for food increases and average reproductive success decreases. Eventually, settling in the best habitat is no longer advantageous because competition reduces reproductive success below that possible in poorer unoccupied habitats. At that point individuals should begin settling in the poorer habitats. In short, individuals should always settle where expected fitness is highest. The end result should be that individuals in every habitat reproduce about equally well, even though local population densities may vary considerably between habitats (*net benefit 1* in Figure 7–10). Fretwell and Lucas termed this scenario an **ideal free distribution** of breeding densities.

In the alternative scenario territorial behavior does limit the number of individuals entering good habitats. Some individuals are forced into poor habitats even though they would reproduce more successfully if they could settle in better habitat. Because the costs of competition do not increase as fast as the benefits of breeding in better habitats, reproductive success should be positively correlated with population density (*net benefit 2* in Figure 7–10). Fretwell and Lucas termed this scenario an **ideal dominance distribution** of breeding densities.

The habitat distribution model provides a way of distinguishing between the spacing hypothesis and the food defense hypothesis. If a population fits the ideal free distribution, multipurpose territories probably evolved to space out nests. If it fits the ideal dominance distribution, they probably evolved to reduce competition for food.

Effect of territoriality on local population densities

Relatively few studies have attempted to test the habitat distribution model. Fretwell and Calver (1969) performed one test in a study of dick-

Figure 7–10

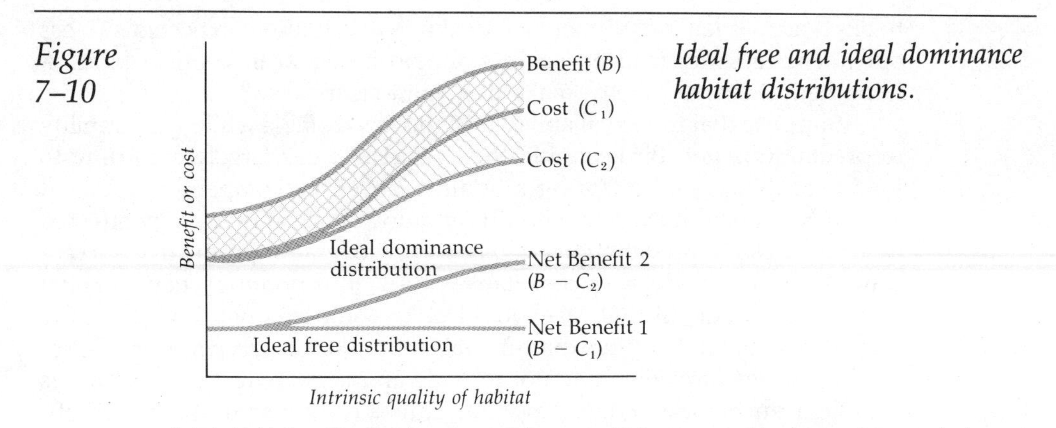

Ideal free and ideal dominance habitat distributions.

In an ideal free distribution competitive costs (C_1), generated by increasing population density, increase at the same rate as benefits (B) as habitat quality increases. The result is equal net benefit ($B - C_1$) for individuals in every habitat. In an ideal dominance distribution competitive costs (C_2) increase more slowly than benefits (B) as habitat quality increases. The result is greater net benefit ($B - C_2$) for individuals in the best habitats.

cissels (*Spiza americana*), a polygamous grassland bird in which males pair with up to several females concurrently. They found that male breeding success is positively correlated with population density, as predicted by the ideal dominance distribution. This result should be generally true for males of polygamous animals when resource-based territories are defended, because the ability of males to monopolize more than their share of resources or nesting habitat is responsible for polygamy in the first place (see Chapter 11).

Fretwell (1969a) also tested the model using a monogamous species. He compared the reproductive success of field sparrows (*Spizella pusilla*) breeding at different densities in two structurally different habitats and found that reproductive success was lowest where population density was highest. Since population density was not positively correlated with reproductive success, the distribution of field sparrows across habitats does not fit the ideal dominance distribution. The data also do not fit the ideal free distribution, since reproductive success was not equal in all habitats, but Fretwell nevertheless concluded that they fit the ideal free distribution better than the ideal dominance distribution. To explain why the results deviated from the ideal free distribution, he suggested that the birds were unable to make perfect habitat decisions or else reproductive success was measured inaccurately.

Interpreting Fretwell's (1969a) results is difficult for a rather subtle reason. Since the best that birds can do is try to predict what their actual success will be, they should select habitats according to the reproductive success they can *expect* to achieve in any given habitat. They cannot select habitats according to the success they actually achieve, which is what biologists normally measure when testing theoretical models. Thus when habitats are settled according to an ideal free distribution, expected success but not necessarily realized success should be equal in every habitat. A hypothetical example helps make this point clear.

Suppose that food availability is highly predictable while vulnerability to predation is not. Birds should then select habitats largely according to food availability, and offspring mortality caused by competition for food should be similar in every habitat. If offspring survival is strongly affected by unpredictable predation events, however, overall reproductive success could easily vary widely between habitats. An appropriate test of the habitat distribution model in that case would be to compare offspring starvation rates across habitats rather than overall mortality or survival rates. Also, nestling starvation alone may not provide an adequate test if starvation is prevalent during the fledgling period. Thus a careful analysis of both offspring mortality and the criteria used to select habitats is necessary before any reasonable conclusions can be drawn about the habitat distribution a particular species exhibits.

Research on hole-nesting European titmice provides some of the best evidence that territorial behavior can limit local population densities in superior habitats. The evidence is nevertheless equivocal.

In Holland great tits (*Parus major*) and coal tits (*P. ater*) breed at relatively high and constant densities in mixed woodlands every year, while breeding at lower and more variable densities in nearby pine woodlands (Kluijver 1951; Kluijver and Tinbergen 1953). Annual variations in total population size have little effect on breeding density in mixed woodlands, but they have a marked effect in pine woodlands. On the basis of this evidence Kluijver and Tinbergen concluded that territorial birds in mixed woodlands maintain a constant density by excluding nonresidents, with a variable number of surplus individuals being forced into less suitable pine woodlands each year.

Unfortunately, the results of Kluijver and Tinbergen are open to an alternative interpretation. Titmice are hole-nesting birds that defend nonexclusive dominions around their nest holes (Hinde 1952; Gibb 1956). Since numerous experiments with artificial nest boxes show that populations of hole-nesting birds are often limited by nest site availability, the availability of suitable nest holes must be included in the analysis. If holes are limited in availability, population densities may remain constant in mixed woodlands because a relatively fixed number of holes can be occupied each year. Territorial behavior would then play no role in limiting population density other than preventing more than one pair from occupying each hole, except to the extent that some suitable holes remain vacant because they are too near occupied holes. Hence population densities in mixed woodlands may have remained constant because they were limited by nest hole availability, not because they were limited by territoriality.

A 1971 study, conducted near Oxford in England, helps clarify the situation. Krebs studied a population of great tits that nested largely in artificial nest boxes placed in mixed woodlands. Since not all nest boxes were occupied, population density was not limited by nest site availability.

Krebs experimentally removed several established breeding pairs from the stable population in mixed woodlands and found that the pairs were rapidly replaced by individuals that had previously held territories in nearby hedgerows. Hence in Krebs's study area, territoriality limited breeding density in the mixed woodlands, not nest site availability.

The above evidence allows two possible conclusions. Either Kluijver and Tinbergen interpreted their results correctly, or different factors limit population density in the two areas. The habitat distribution model helps resolve the two possibilities. If nest hole availability is responsible for limiting population density, the population should exhibit an ideal free density distribution. If, on the other hand, territoriality is responsible for limiting population density, the population should exhibit an ideal dominance density distribution. The relevant data have been analyzed by Brown (1969b), with the following results: In Holland reproductive success was just as high in pine woodlands (7.1 young per pair) as in mixed woodlands (6.8 young per pair) (Brown 1969b, based on Kluijver 1951), suggesting a settling pattern that best fits an ideal free distribution. However, in England reproductive success was higher in mixed woodlands (8.3 young per pair) than in hedgerows (6.8 young per pair) (Krebs 1971), suggesting a settling pattern that best fits an ideal dominance distribution. Thus the most reasonable conclusion is that different factors limit population density in the two study areas.

Effect of territoriality on overall population size

Territoriality would limit the overall size of breeding populations (level 3 of Brown 1969a, b) if some potential breeders are prevented from settling in any suitable breeding habitat. The effect would also limit total population size if territoriality forces some individuals into habitats where they cannot survive. Such effects should be considered secondary consequences of territoriality, not adaptive attributes (Brown 1969a). That is, they probably do not act as selective factors promoting the evolution of territoriality.

One indication that territoriality limits the size of breeding populations is the existence of nonbreeding floaters. The presence of nonbreeders is usually documented by removing territory residents. If resident birds are quickly replaced by new arrivals midway through a breeding season, a nonbreeding contingent of floaters might be inferred. However, new arrivals in a habitat may be individuals who previously held territories in nearby marginal territories, as was shown for great tits in England (Krebs 1971). Replacements following removal of territory residents therefore do not prove the existence of a nonbreeding contingent. In addition, some males may not hold territories for long because they cannot acquire mates. Hence the size of a breeding population cannot be considered limited by territorial behavior unless females as well as males are prevented from breeding.

Numerous removal experiments have been conducted, and many have led to replacement of resident individuals (Brown 1969a). In the most frequently cited studies, attempts were made to shoot as many passerine birds breeding in an eastern deciduous forest as possible (Hensley and Cope 1951; Stewart and Aldrich 1951). The number of males shot was about twice as many as were originally present on territories, but the number of females shot was similar to the original number known to have been breeding on the study area. The results therefore suggest that nonbreeding males but not nonbreeding females were present prior to the experiment. However, surplus females would be more difficult to detect because females are less conspicuous and hence less likely to be shot. In addition, females are more prone to desert the area following death of their mates. Unless a strongly skewed sex ratio is postulated in the overall population, it seems likely that nonbreeding females were also present. In several other studies resident females were replaced following removal, showing that nonbreeding females are probably present in at least some populations (Kendeigh and Baldwin 1937; Snow 1958; Ribaut 1964; Watson and Jenkins 1968; Knapton and Krebs 1974).

The existence of nonbreeding males and females has been documented in a few populations by direct observations. Australian bell-magpies (*Gymnorhina dorsalis* and *G. tibicen*) defend group territories around small groves of trees in agricultural districts (Wilson 1946; Robinson 1956; Carrick 1963). Population densities are very high due to abundant food provided by agricultural crops, and many individuals of both sexes fail to find suitable breeding habitats. Nonbreeding birds of both sexes form large flocks, which forage in open cultivated areas and roost in extensive woodlands. Breeding appears possible only in the relatively few isolated groves of trees, as the few birds attempting to breed in extensive woodlands never succeed because other nonbreeders interfere. Mixed-sex flocks of nonbreeding floaters have also been described for red grouse and dunlins (*Calidris alpina*) (Jenkins et al. 1963; Holmes 1970).

In rufous collared sparrows (*Zonotrichia capensis*) some males and females live in an "underworld" of nonbreeding floaters (S. M. Smith 1978). Each underworld female skulks around on the territory of a particular breeding pair. When the resident female disappears, the underworld female surfaces and pairs with the resident male. Each underworld male skulks around on the territories of several resident pairs, and male home ranges often overlap those of other underworld males. In the areas of overlap underworld males establish dominance relationships. When a resident male disappears, the most dominant underworld male in that area usually replaces him. Both males and females in the underworld are extremely furtive and difficult to observe. They avoid territory residents and spend at least part of their time in undefended areas. Resident birds probably tolerate the presence of underworld individuals primarily because the costs of excluding them are too high relative to the benefits. Another factor may be that underworld individuals are at least sometimes prior offspring of resident pairs. Parents who force prior offspring out of the underworld might reduce their success in contributing offspring to the next generation. Some anecdotal evidence suggests that an underworld of nonbreeding floaters exists in other species as well (Brown 1969a), but the behavior of such individuals has not been studied.

Defensibility theory

Economic defensibility depends to a large extent on the spatiotemporal distribution of resources. Some evidence cited above suggests that evenly distributed resources are more defensible than spatiotemporally clumped resources, but additional theory is needed to show why.

A convenient starting point is defensibility of food resources. Food is a major determinant of spacing behavior, and theory based on food resources is readily extended to other types of resources.

Defensibility of food resources is often measured in terms of energy yield, because energy yield is a convenient currency for measuring both costs and benefits of defense. It is important to realize, however, that not all costs and benefits can be measured in that currency. An important benefit may be higher nutrient yield, and important costs may be increased risk of predation or decreased ability to feed offspring. Benefits of territory defense must offset both energetic and nonenergetic costs before territoriality can evolve. Thus one should bear in mind that the proper currency of defense economics is lifetime reproductive output, not energy.

The amount of resource a territorial animal can control depends on territory size and the extent to which competitors are excluded. Defensibility therefore has two components: the size of area required to obtain adequate resources and the ability of defenders to exclude competitors from areas of that size.

A model of optimal territory size

For evenly distributed resources the cost of territory defense should depend on the number of intruding competitors that must be expelled per unit time, the time and energy required to detect each intrusion, and the difficulty of expelling each intruder once it has been detected. These factors should all increase with territory size. Number of intruders should increase with territory size because a larger territory contains a greater fraction of the total resource, which creates greater demand for the defended resources. Detecting intrusions requires systematic patrolling of defended areas, and patrolling larger areas requires more time and energy. Once an intruder is detected, it must be expelled, and expulsion should become more difficult as demand for resources on the territory increases. The difficulty of expelling intruders should depend on how persistently the intruder remains, and intruders should be more persistent when a greater fraction of the total resource is contained on the territory.

The benefits of territory defense should also increase with territory size, but not as a direct proportion. Animals derive little benefit from defending food resources in an area too small to satisfy their minimum requirements unless they can make up the shortage in undefended areas without losing their ability to exclude intruders from their territory. The benefits of territory defense should therefore be low until a threshold size is reached, and then they should increase dramatically. For this reason territories often appear to have an irreducible minimum size (Huxley 1934). At the other end of the scale, the additional benefits derived by controlling ever larger territories gradually diminish. More distant resources are more costly to harvest and defend, and little is gained by defending surplus resources that cannot be utilized. Thus territories tend to contract when competition increases until they reach an irreducible minimum size, and they tend to expand when competition is relaxed until they reach a maximum size.

A simple model for predicting optimal territory size can be devised from the considerations above (Figure 7–11). The shape of the benefit curve reflects the low value of defending too small an area and the diminishing returns gained from defending ever larger areas. The shape of the cost curve is assumed to be linear because defense costs are assumed to be directly proportional to territory area. The actual shapes of the benefit and cost curves have not been determined for most species, and they are likely to vary depending on vegetation structure and the type of resource being defended.

Optimal territory size lies at the point where benefits (*B*) minus costs (*C*) is greatest. This is the point where an individual's fitness is maximal given local habitat conditions on the territory. Each territory has its own unique set of cost and benefit curves, which reflect local habitat conditions, and hence each has its own optimal size. Although a territory is economically defensible as long as benefits exceed costs, an animal can defend a smaller- or larger-than-optimal territory only at some cost to its fitness. Thus animals should try to predict optimal territory size and defend territories near that size. The extent to which they actually do that is not easy to evaluate.

Figure 7–11

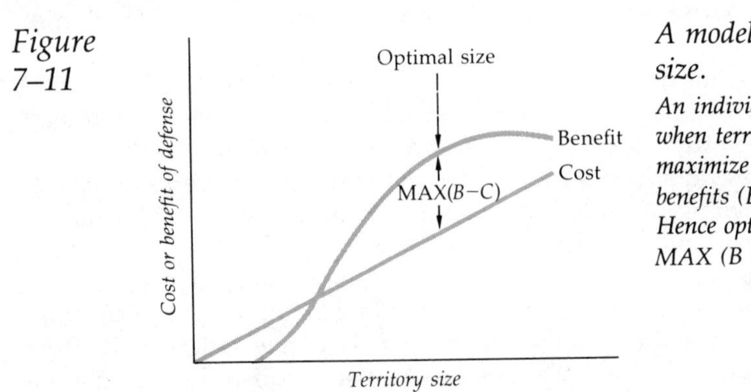

A model of optimal territory size.

An individual's fitness is maximal when territory size is adjusted to maximize the difference between benefits (B) and costs (C) of defense. Hence optimal territory size occurs at MAX (B − C).

Factors affecting optimal territory size

Spatiotemporal clumping of resources. The above discussion assumed that food resources were evenly distributed and continuously available. Under those conditions an animal can find adequate food in a relatively small area throughout the year. However, when food is spatially clumped and the locations of clumps change on a short-term basis, an area of the same size would contain more than enough food at some times and less than enough at other times. An animal exploiting such food resources would therefore have to range over a larger area during the course of a year, and hence it would have to defend a larger territory.

The costs of defending habitat patches that contain large amounts of food are high because clumping increases the number of individuals competing for each clump. Intruders are also more likely to be persistent in remaining in a good habitat patch because areas outside patches contain much less food. At the same time, the benefits of defense are less if the food in each clump is relatively ephemeral because some of the food eaten by competitors would be surplus that could not have been utilized anyway. As degree of spatiotemporal clumping increases, the net benefit of resource defense decreases, until finally costs exceed benefits and the resource becomes indefensible (see Horn 1968).

The same effect determines how many nest sites can be defended. When suitable nest sites are evenly distributed, as in ground-nesting birds, a large number of suitable sites can be defended, and optimal territory size depends on the costs and benefits associated with defending other resources. When suitable nest sites are spatially clumped, as they are for colonial birds, only sufficient space for one or a few nests plus a surrounding buffer zone can be economically defended.

Type of resource defended. Another factor potentially affecting optimal territory size is that males and females may have different resource requirements or may defend different resources altogether. For example, in sheet-web spiders (*Linyphia triangularis*) each sex defends the webs built by females, but for different reasons (Rovner 1968). Females defend their orbs from other females to protect the time and energy invested in constructing the orb. The size of a female's orb should maximize the difference between the benefits of having a large prey-capturing surface and the costs of defending a larger area. Males wander about seeking sexually receptive females, and when they find one, they defend both the female and her orb from other males. Males acquire no food from the female's orb and are therefore not directly concerned with size of the prey-capturing surface. They cut away the outer strands of female orbs to make them smaller, probably to reduce the cost of defending females. One might predict that males leave an area of web just sufficient to prevent females from deserting, a hypothesis that has not been tested. Thus optimal territory size is evidently smaller for males than for females, apparently because males defend access to mates while females defend prey-capturing surfaces.

Stage of breeding cycle. The stage of an animal's breeding cycle influences both costs and benefits of territory defense. When reproductive success is limited by the amount of time and energy available for parental care, territorial behavior becomes more costly during the rearing period. In seasonally territorial animals the benefits also decline because competition for breeding sites decreases after individuals settle elsewhere to breed. The actual point where territories are no longer worth defending should depend on how many broods are reared each season, the extent to which males and females shift to new areas following failed breeding attempts, the number of nonbreeding floaters remaining in the general vicinity, and the role territory defense plays in protecting offspring from predators or conspecific intruders.

Other factors. Several additional factors influence the size of territories actually defended. Territories tend to be smaller when established simultaneously than when established in succession (Assem 1967; Knapton and Krebs 1974). Individuals vary in their ability to defend a given area because they differ genetically and have had different previous experiences. Even more importantly, animals must adjust territory size according to proximate cues that are only partially predictive of future habitat conditions. The ability to assess current habitat conditions and predict future conditions undoubtedly varies among environments and among individuals. One would expect animals to defend large enough territories to provide a cushion against uncertainties, especially when habitat predictability is low (Brown and Orians 1970), and the little evidence available on this point suggests that they do (C. C. Smith 1968; Gill and Wolf 1975). However, correlations between territory size and resource availability do not necessarily imply that such adjustments result from animals assessing food abundance and then changing the area defended accordingly (Myers et al. 1979b). Animals may simply defend the largest area possible given the intensity of competition that they encounter. Then territories would be smaller where food is most abundant solely because competition intensity is higher there. If that is true, difficulties in assessing food availability would not be a factor. While little evidence is available yet on this point, variations in intrusion pressure are sufficient to explain most of the variability in territory size among wintering sanderlings (Myers et al. 1979b) and western gulls (*Larus occidentalis*) (Ewald et al. 1980).

Testing the territory size model

The costs and benefits of territory defense can rarely be measured, but the feeding territories of certain nectivorous birds are ideally suited for the purpose. Data are now available for species from three different taxonomic families, and in general they support the model.

The first quantitative test of defensibility theory was achieved by Gill and Wolf (1975), who analyzed the energetics of territory defense for golden-winged sunbirds, an African equivalent of New World hummingbirds. Sunbirds defend individual feeding territories against competitors of both their own and different species. They obtain nearly all of their energy from nectar, which they extract from flowers on their territories. By defending flowers, sunbirds can maintain average nectar levels higher than those in undefended areas (Figure 7–12). The hypothesis tested by Gill and Wolf is that this difference enables sunbirds to obtain a net energy gain despite the costs of territory defense.

Time and energy budgets were computed to estimate the caloric benefits and costs of territory defense. The energetic benefit of defense equals the amount of energy conserved during foraging as a result of maintaining high nectar levels on the territory. When nectar levels are high, foraging time decreases because fewer flowers are visited each day. This decrease in foraging time translates into reduced energy expenditures because less time is spent flying and more time is spent resting. On the other side of the equation, chasing intruders out of the territory represents an energy expense. This cost was estimated from the time spent chasing intruders by using published estimates for the caloric cost of flight. Thus both benefits and costs could be computed in caloric terms.

The amount of benefit gained from territory defense depends on the difference in nectar levels between defended and undefended areas. Given the known costs of defense, a threshold difference necessary for defense to be beneficial could be calculated. The results from five days of sampling nectar levels show that the observed difference does exceed this threshold value. Moreover, on the one day when it was below threshold, the sunbirds did not defend territories until mid-afternoon. The data therefore show that sunbirds do obtain a net energy gain by defending territories.

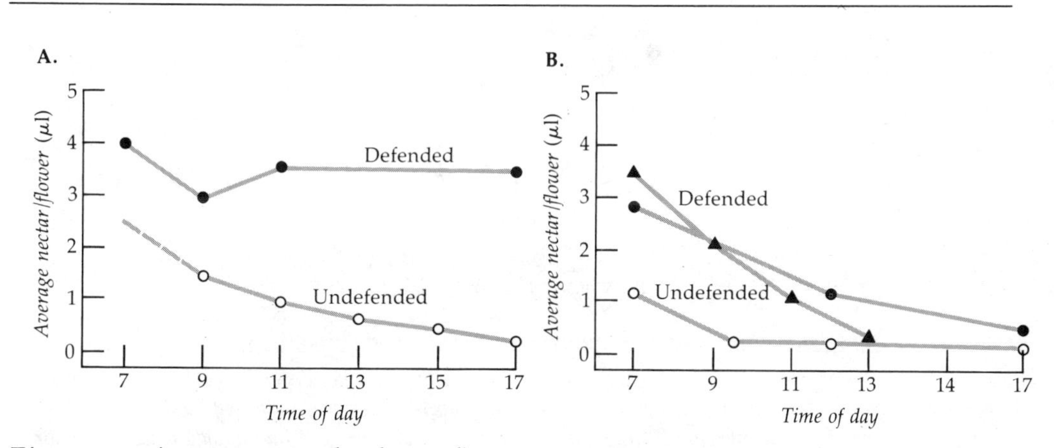

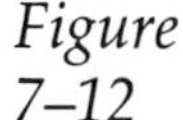

Figure 7–12 *Average nectar levels per flower on territories of golden-winged sunbirds and in nearby undefended areas.*

For the most part, defended areas contain higher nectar levels than undefended areas, presumably because competitors for the nectar have been excluded from defended areas.

Source: From F. B. Gill and L. L. Wolf, "Economics of Feeding Territoriality in the Golden-winged Sunbird," *Ecology* 56 (1975):340. Reprinted by permission of Duke University Press. Copyright 1975 by the Ecological Society of America.

In a second study Carpenter and MacMillen (1976) analyzed the costs and benefits of territory defense for the iiwi (pronounced e-e-vee) *(Vestiaria coccinea)*, a nectivorous Hawaiian honeycreeper. Using a simple algebraic model, they predicted that iiwis should obtain a net energy gain when defending intermediate numbers of flowers but not when defending higher or lower numbers of flowers. By estimating the parameter values of the model, they could predict the range of flower numbers that should be worth defending. The results show that iiwis were territorial in areas where the number of flowers fell within the predicted range but not in areas where the number of flowers fell outside that range (Figure 7–13).

Although the data for iiwis support defensibility theory, they must be interpreted with caution. The range of flower numbers predicted to be economically defensible was calculated by estimating several different parameter values, and some of the estimates may be inaccurate because sample sizes were small. Even small errors in some of the estimates would give markedly different predictions, which could reverse the conclusions drawn from the study. Admittedly, the relevant parameters are difficult and time-consuming to measure, but larger sample sizes will be necessary before the results can be confidently accepted. Also, the model is structured in such a way that it does not compare the benefits of territory defense to the benefits of alternative foraging strategies (Davies 1978b). Finally, the model

Figure 7–13

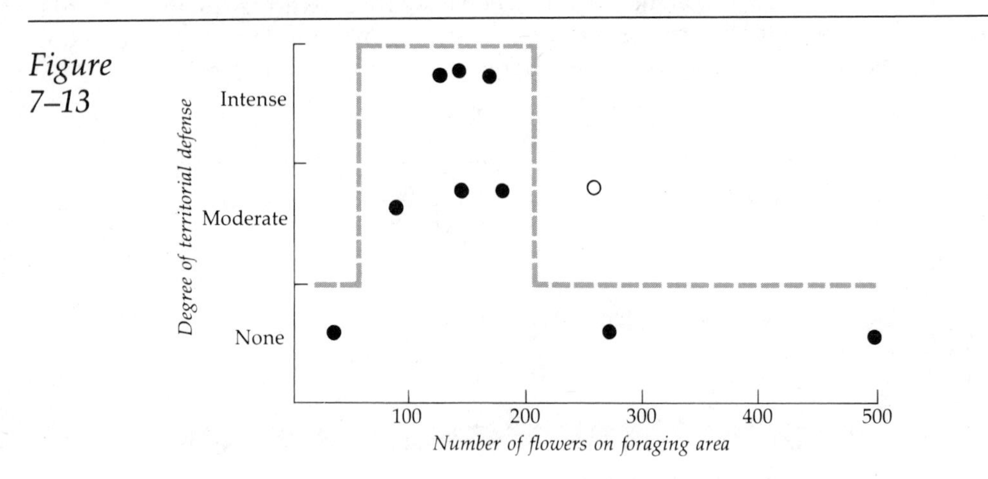

Predicted and observed degrees of territory defense in the iiwi.
The predicted level of defense, indicated by the dashed line, depends on the number of flowers present on the territory. The actual levels of defense, indicated by open and closed circles, are shown for 10 individuals. Closed circles represent individuals whose behavior fits the model, while the open circle represents an individual whose behavior did not fit the model.

Source: From F. L. Carpenter and R. E. MacMillen, "Threshold Model of Feeding Territoriality and Test with a Hawaiian Honeycreeper," *Science* 194 (5 November 1976).

is structured differently than Brown's (1964) original cost-benefit model. It includes basic cost of living and yield to nonterritorial individuals in the equation along with cost and benefit of territory defense. Hence the data would not have negated Brown's model even if they had not supported the predictions derived by Carpenter and MacMillen (see Pyke 1979).

To test defensibility theory experimentally, Ewald (1980) measured the energetic costs and benefits of territory defense for wintering Anna's hummingbirds *(Calypte anna)* in an area where hummingbirds obtain most of their food from artificial feeders. By using feeders, Ewald could experimentally manipulate food availability on territories, a significant advantage over previous studies.

Two models of defensibility were tested. The first corresponded to Brown's (1964) hypothesis that a resource is defensible when the benefits of defense are greater than the costs. The second corresponded to Wolf's (1978) hypothesis that a resource is defensible when net energy intake is higher for territorial individuals than for nonterritorial individuals. The two models could be distinguished because they predict different thresholds for the occurrence of territorial behavior.

To test the models, Ewald manipulated sugar concentrations at the artificial feeders and recorded presence or absence of territorial behavior following the manipulation. The results show that territorial behavior began to occur near the lower threshold predicted by the first model (Table 7–1). Hence they support Brown's hypothesis that a resource is defensible when the benefits of defense exceed the costs.

The super-territory hypothesis

The theory described above assumes that optimal territory size is determined by the point where fitness gain of the territory owner is maximal. However, an alternative theory is possible based on the concepts of relative fitness and spite.

A trait should evolve if it increases an individual's genetic contribution to future generations relative to that of individuals with alternative traits. A trait could theoretically evolve because it enables individuals exhibiting it to produce more surviving descendants than individuals not exhibiting it, or because it reduces the number of surviving descendants produced by individuals exhibiting it less than it reduces the number of surviving descendants produced by individuals not exhibiting it. That is, a trait could have high relative fitness if it has a stronger negative effect on competitors than it has on individuals exhibiting it.

Using this argument, Verner (1977) proposed that animals should defend more space than they require solely to prevent competitors from reproducing. According to Verner's super-territory hypothesis, animals should defend the largest area possible even though defense of surplus resources entails a net cost for territory owners. By defending surplus resources, territory residents should gain higher relative fitness because the success of competitors is lowered.

Table 7–1 *Presence or absence of territoriality in Anna's hummingbirds at various levels of energy availability.*

P	*T*	*P*	*T*	*P*	*T*
.00	−	.11	−	.17	+
.01	−	.13	−	.18	+
.02	−	.14	+	.54	+
.02	−	.15	+	∞	+
.07	−	.16	+	∞	+

Note: P = proportion of an individual's 24-hour energy expenditure available from artificial feeder per day. T = presence or absence of territoriality with, + = territoriality present and − = territoriality absent (each symbol represents one hummingbird).

Source: From P. Ewald, "Energetics of Resource Defense: An Experimental Approach," *Acta XVII Congressus Internationalis Ornithologicus,* 1980. Reprinted by permission.

The hypothesis that animals defend surplus resources is not easy to test because an apparent surplus may be defended as insurance against sudden and unpredictable periods of resource depression. Some data are available regarding the amount of food present on territories, and they suggest that territories may not contain surpluses. The evidence supporting this conclusion was reviewed on pages 275–276.

A more important issue, however, is whether the logic underlying Verner's (1977) hypothesis is sound. For a trait to evolve, the average fitness of individuals possessing the trait must exceed the average fitness of individuals not possessing it. Consider now a trait that reduces the fitness of competitors at some expense to carriers of the trait, and assume that the reduction is greater for competitors than for carriers. Not all competitors will be adversely affected because not all will interact with carriers. The trait cannot increase in frequency unless the average effect on all competitors is greater than the average effect on all carriers. The adverse effect on the relatively few competitors who interact with carriers must therefore be very high before the trait can evolve (Rothstein 1979).

A simple algebraic computation makes this point clearer (Pleasants and Pleasants 1979). Let 1 = average fitness of all individuals before the trait arose, C_c = reduction in fitness of competitors adversely affected by carriers, C_t = reduction in fitness of carriers, and p = proportion of competitors adversely affected by carriers. Then the average fitness for competitors is

$$\overline{W}_c = (1 - p) + p(1 - C_c)$$

and the average fitness for carriers is

$$\overline{W}_t = 1 - C_t$$

For the trait to confer higher relative fitness on carriers, $\overline{W}_t$ must be greater than $\overline{W}_c$, or

$$1 - C_t > (1 - p) + p(1 - C_c)$$

Simplifying,

$$C_c > \frac{C_t}{p}$$

That is, the trait will be advantageous if the cost to competitors is $1/p$ times larger than the cost to carriers. Suppose that the trait arises by mutation at a frequency that affects 1 out of every 10,000 competitors. Then the cost to competitors must be 10,000 times larger than the cost to carriers before the trait can be advantageous. Clearly, the trait is not likely to evolve unless the initial frequency is high. That is most likely to occur if a species consists of small disjunct populations or if gene flow is very low (Rothstein 1979). The latter situation might lead to selection for the trait in a local area, creating an epicenter that could then spread slowly outward, but as a general rule the trait is not likely to evolve.

The above analysis is oversimplified because it does not include the cost that carriers of the trait inflict on each other. If this cost is included, the threshold value of C_c becomes even higher (Rothstein 1979).

A remaining question concerns whether inhibitory effects become important once a trait reaches high frequencies in a population. Many traits have beneficial effects on carriers and concomitant deleterious effects on competitors. Can these deleterious effects increase the strength of selection for such traits above that generated by the beneficial effect on carriers? If the deleterious effects are inflicted on equal proportions of carriers and noncarriers, the answer is no. If for some reason noncarriers are affected more frequently than carriers, the answer is yes. Remember that relative fitness is determined by the average effect on all carriers and all noncarriers. The outcome therefore depends on what proportion of each is affected.

Exclusiveness of territory defense

When territory defense becomes more costly, an animal could defend a smaller area, abandon its territory altogether, or maintain less exclusive control of the same-sized area. The best alternative depends in part on how the costs and benefits of territory defense vary as a function of the degree to which intruders are excluded.

Intruders should be completely excluded when the benefits of territory defense increase faster than the costs as degree of exclusion increases (Figure 7–14A). Optimal territory size decreases as the slope of the cost curve increases, but optimal degree of exclusion should not change unless the shape of the cost or benefit curve also changes. If conditions cause the cost curve to become steeper than the benefit curve, the territory should

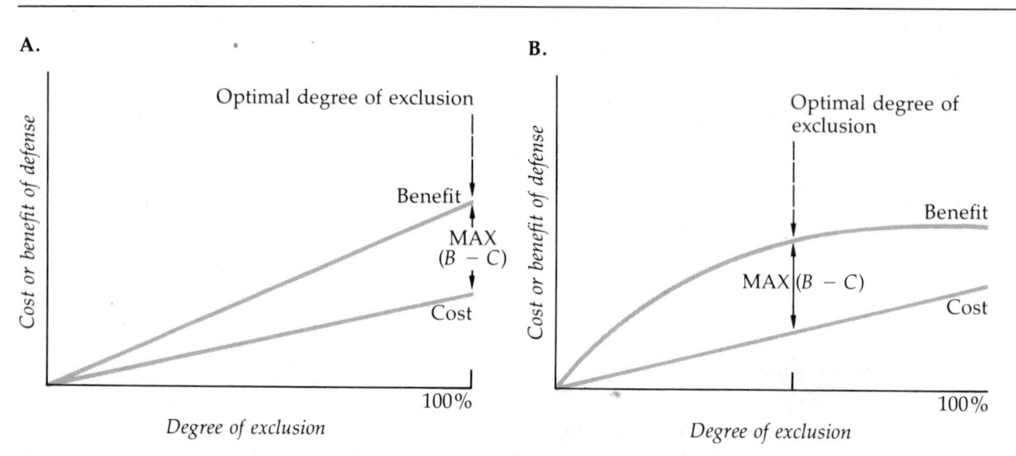

Figure 7–14 ***The optimal degree to which competitors should be excluded, depending on the shapes of the cost and benefit curves for territory defense.*** *Optimal degree of exclusion occurs at the point where benefits minus costs is maximal, which is indicated by MAX (B − C). A. When the cost and benefit curves are parallel, 100% exclusion is optimal. B. When the cost and benefit curves converge beyond an intermediate exclusion level, an intermediate degree of exclusion is optimal.*

be abandoned entirely, because then the costs exceed the benefits for all degrees of exclusion.

Complete exclusion should be advantageous when intruders threaten to take possession of breeding sites, harass resident females, steal copulations, or injure offspring. It should also be advantageous when resource depletion by each intruder costs more than expelling the intruder.

Intermediate degrees of exclusion should be optimal when the benefit curve is steeper than the cost curve at low levels of exclusion but not at high levels of exclusion (Figure 7–14B). Then the net benefit of territory defense is maximal when some of the intruders are excluded some of the time.

Intermediate degrees of exclusion should occur when moderate resource depletion by intruders costs a territory resident less than does excluding intruders more completely. The most important variable is the effect resource depletion has on a resident's foraging efficiency, and not food abundance per se. Depletion of surplus food may have a strong effect on foraging efficiency by making food more difficult to find (Charnov et al. 1977). Hence the existence of a food surplus does not necessarily favor incomplete exclusion of intruders.

Dominions

The optimal degree of exclusion should vary with distance to resources because more distant resources are more costly to harvest. More distant resources are also more difficult to defend, since the defender must travel further to meet intruders and intruders are more likely to persist in trying to exploit resources nearer their own core areas. Thus optimal degree of exclusion should decrease with increasing distance from a defender's core area, giving rise to the concentric zones of dominance characteristic of dominions (Figure 7–15).

Dominions should evolve when much of the food taken by intruders is surplus and depletion by intruders does not make the remaining resources more difficult to harvest. These conditions might arise if breeding densities are limited by factors other than food availability or if food occurs in rich, temporary patches.

Most dominions defended by birds are associated with conspicuous, spatiotemporally clumped food resources. Blue jays *(Cyanocitta cristata)* and Steller's jays *(C. stelleri)* exploit fruiting trees, acorn crops, and the remains of human picnics (Bent 1946; Hardy 1961; Brown 1963, 1974). American robins *(Turdus migratorius)* exploit fruits, along with more evenly distributed earthworms and other invertebrates (Young 1951, 1956). Mockingbirds *(Mimus polyglottos)* normally eat dispersed invertebrate prey and defend exclusive territories, but exclusion becomes incomplete when artificial bird feeders are placed on their territories (Michener 1951). The principal exceptions to this pattern are European titmice, which defend dominions and are insectivorous (Hinde 1952; Gibb 1956). Because breeding densities of titmice are limited by nest hole availability, only a small fraction of the available insects are exploited (Betts 1955). Hence depletion of food resources by intruders probably has minimal effect on the foraging efficiency of residents.

Who should be excluded?

Defending resources against conspecific intruders is prevalent because conspecifics have identical resource requirements and hence compete strongly with territory residents. Defending resources against members of other species is less frequent because interspecific competition is less severe. Nevertheless, some animals do defend territories against other species. The problem here is to explain why some species are interspecifically territorial while others are not.

Two competing hypotheses have been proposed to explain interspecific territoriality. The first hypothesis is that territory residents exclude members of other species because they mistake them for conspecifics. The other hypothesis is that they exclude interspecific competitors because the benefits of excluding them exceed the costs. The misidentification hypothesis implies that interspecific territoriality is either maladaptive or nonadaptive, while the resource defense hypothesis implies that interspecific territoriality is adaptive.

Misidentification hypothesis

Interspecific territoriality has often been attributed to cases of mistaken identity (Hinde 1952; Murray 1971; Becker 1977; Gochfeld 1979a). The added costs of defending a territory against members of another species

Figure 7–15

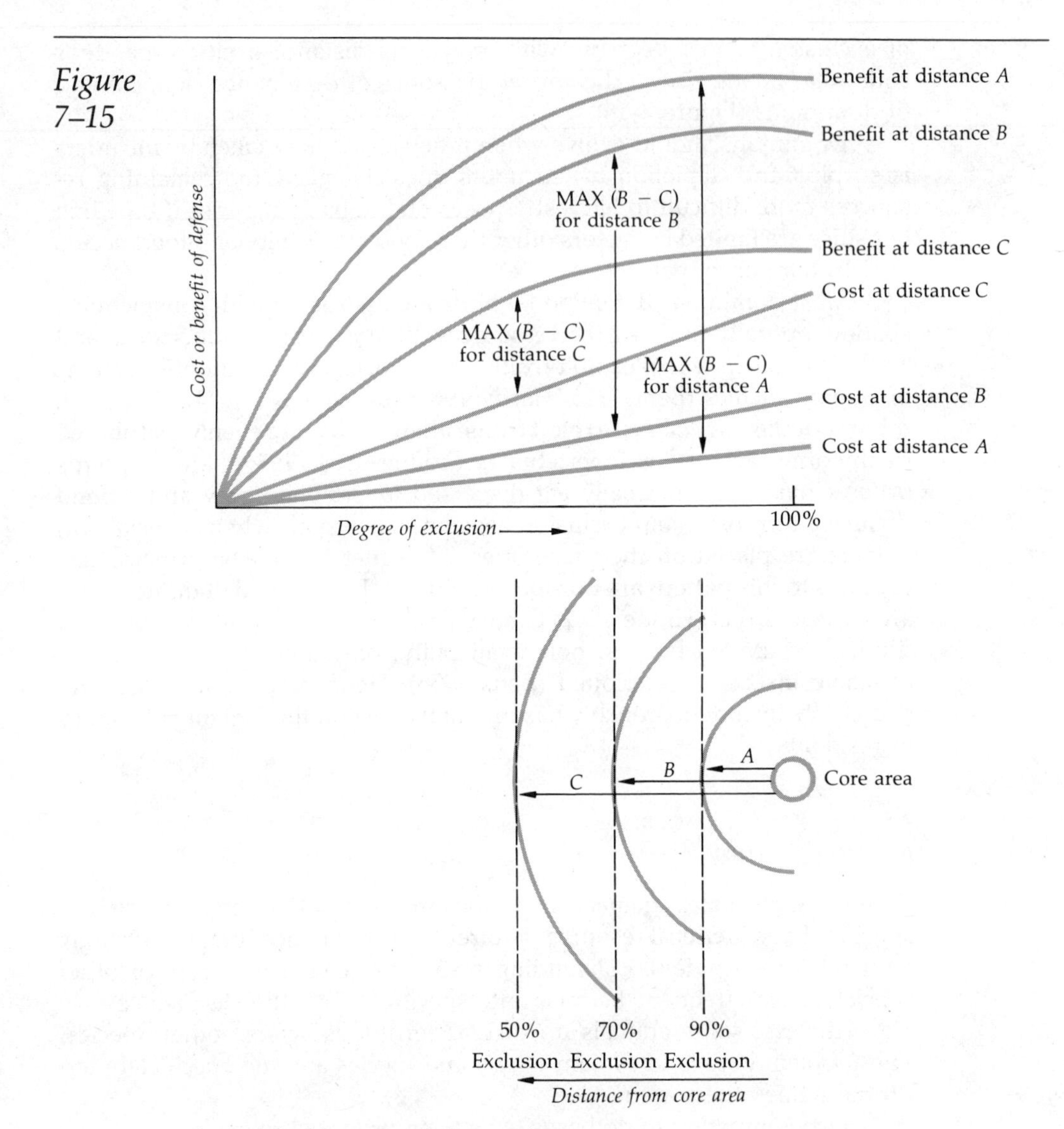

A model of how dominions arise.

When intermediate degrees of exclusion are optimal, the degree of exclusion should decrease with increasing distance from the core area if benefits decrease and costs increase with distance from the core area.

should quickly select for an ability to discriminate between conspecifics and members of other species. Nevertheless, Murray (1971) contends that interspecific territoriality based solely on mistaken identity can persist indefinitely if relatively few members of the population express it. He argues that gene flow from other parts of the population would swamp selection against interspecific territoriality whenever geographic ranges or habitat requirements of the two species involved are only narrowly overlapping.

In line with the misidentification hypothesis, interspecific territoriality does often occur among species with similar plumage or songs (Murray 1971). However, it is not clear whether similar species are interspecifically territorial because they misidentify each other or are similar to facilitate interspecific territoriality (Cody 1969, 1971b, 1973a). In support of the latter interpretation, Cody (1969) points to several examples of distantly related species that are interspecifically territorial and possess strikingly similar plumage. Since the similarity among such species cannot be attributed to common ancestry, it apparently evolved to enhance interspecific threats. Unfortunately, the evidence is not clear-cut. Many of the species cited by Cody are more closely related than he believed, and for most of his examples interspecific territoriality was inferred without adequate documentation (Murray 1976). Evidence that interspecifically territorial species are more similar in zones of geographical overlap than outside those zones is lacking, and an attempt to demonstrate character convergence among two interspecifically territorial (but also closely related) wren species failed (Brown 1977). Cody's point is well taken, but the existence of character convergece among interspecifically territorial species remains to be demonstrated.

One problem with the misidentification hypothesis is that gene flow from outside the zone of overlap is not likely to prevent selection from favoring an ability to discriminate between conspecifics and members of similar species. Gene flow from such areas may affect the rate of evolution, but it should not affect the direction. Moreover, some interspecifically territorial species have broadly overlapping geographical ranges and hence could not be greatly affected by gene flow from areas outside the zone of overlap. Red-eyed vireos *(Vireo olivaceous)* and Philadelphia vireos *(V. philadelphicus),* for example, defend territories against each other even though they share broadly overlapping home ranges and breed in similar habitats (Rice 1978). Also, Philadelphia vireos were shown to be capable of discriminating between conspecifics and red-eyed vireos, though a comparable discriminatory ability could not be demonstrated in red-eyed vireos.

Another problem is that some interspecifically territorial species are not at all similar (e.g., see Orians and Willson 1964; Ebersole 1977; Catchpole 1978; Moore 1978). Misidentification is difficult to believe when the species involved are strikingly different in either coloration or song characteristics.

Finally, the misidentification hypothesis cannot explain why interspecifically territorial species are usually competitors (e.g., see Simmons 1951; Orians and Willson 1964; Low 1971; Myrberg 1972; Brockmann 1973; Myrberg and Thresher 1974; Thresher 1976; Ebersole 1977; Catchpole 1978; Cody 1978; Moore 1978; Walters 1979). The main exceptions to this pattern are fish that exclude egg predators from their territories (e.g., Albrecht 1969; Clarke 1970; Keenleyside 1972a; Myrberg 1972; Brockmann 1973; Myrberg and Thresher 1974; Thresher 1976; Ebersole 1977). According to the misidentification hypothesis, interspecific territoriality should evolve among similar species regardless of whether they compete with one another. Instead, it has evolved among competing species regardless of whether they are similar to one another. A few isolated instances of interspecific territoriality may arise from mistaken identity (e.g., see Lill 1976), but in most instances it probably evolved to reduce interspecific competition for food or to protect eggs from predators.

Food competition hypothesis

Interspecific territoriality is one of several possible ways an individual can compete with members of another species. Alternative options for reducing interspecific competition include foraging or nesting in different microhabitats, specializing on food types or food sizes not easily captured by competitors, or specializing in the way food is captured. Interspecific territoriality is usually costly, and ecological specialization or divergence should be better alternatives unless they are precluded by habitat structure or interspecific competition (Orians and Willson 1964).

Opportunities for foraging in specialized microhabitats, on special types of prey, or in specialized ways are limited within structurally simple habitats like marshes, grasslands, shorelines, and deserts. Competing species of birds in those habitats often exhibit interspecific territoriality (Orians and Willson 1964). In some cases individuals of one species usurp the best habitats aggressively, while members of the other rely on more generalized nesting and foraging capabilities so that they can breed in the poorer habitats left open to them. Yellow-headed blackbirds *(Xanthocephalus xanthocephalus),* for example, defend exclusive territories against red-winged blackbirds in the most productive cattail and bulrush marshes, often by harassing nesting female red-wings until they are forced to abandon their nests. As a result, red-wings nest primarily in marshes unsuitable for yellow-heads, except in pockets where their densities become high enough to prevent yellow-heads from forcing them out. Male yellow-heads can always win aggressive interactions with red-wings because they are larger in size, but their larger size restricts them to the outer edges of highly productive reedbeds where food is most abundant (Orians 1980). Thus yellow-heads gained the ability to monopolize the most productive habitats but lost the ability to exploit less productive habitats.

Opportunities for ecological divergence are limited when several species are specialized to forage in the same strata or zones of structurally complex habitats (Orians and Willson 1964). Several interspecifically territorial species fit this situation, including some hawks that hunt ground-dwelling vertebrates, some woodpeckers that forage on tree trunks or along large branches, and some hummingbirds that exploit nectar-producing flowers. Similarly, opportunities for specializing on particular kinds or sizes of nest holes are limited, and most hole-nesting birds defend their holes from other species (Haartman 1971; Bourne 1974).

Two recent studies show that the degree of interspecific territoriality varies with degree of dietary overlap. Ebersole (1977) used an elaborate statistical procedure to compute competitive overlap between the beaugregory, a pomacentrid reef fish of the West Indies, and 40 species that compete with it for food. He found a significant correlation between the frequency that beaugregories attacked trespassers of other species and his index of competitive overlap. Moore (1978) studied a population of mockingbirds that primarily eat fruit during winter and defend winter territories centered around fruit-laden trees and shrubs. The mockingbirds defended their territories against other frugivorous species but not against insectivorous species, and the degree of defense, as measured by the percentage of intrusions provoking an attack, was directly correlated with the percentage of fruit in the intruding species' diet (Figure 7–16).

Conclusion

Spacing behavior is a central component of social organization, and for that reason territoriality often forms the social fabric within which animals live and reproduce. It constrains opportunities for exploiting resources, restricts entry into the best habitats, and often affects the mating options open to each sex. As we shall see in later chapters, territoriality has far-reaching consequences on other aspects of social organization, and hence its evolution has great theoretical importance.

Figure 7–16

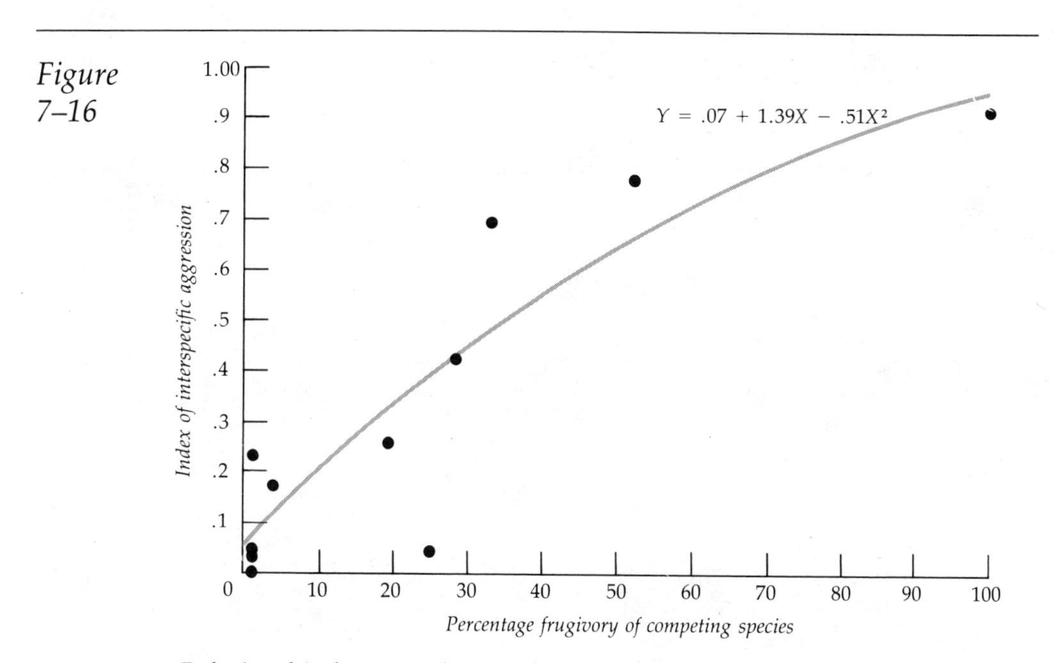

Relationship between degree of interspecific aggression by territorial mockingbirds and percentage of dietary overlap with birds competing for fruit on mockingbird territories.

Source: Based on data in Moore 1978.

Territoriality is only one of several possible ways that animals compete for resources, and it is often not the best way. Defending resources entails time and energy expenditure, exposes animals to possible injury during aggressive encounters, and increases the risk of predation. These costs must be offset by compensatory benefits before territoriality can evolve. Animals gain a wide variety of benefits through territoriality. They defend favorable places for finding mates, favorable breeding sites, time and energy investments, suitable shelter sites, and high-quality food resources.

Not all resources are worth defending. Some are sufficiently abundant that animals can acquire enough resources without having to defend them. Others are too costly to defend because local competition for them is very high. A resource that is worth defending, in terms of fitness, is said to be economically defensible. When the benefits of defense exceed the costs, defending the resource increases fitness and should evolve. Several quantitative studies have now shown that defended food resources are economically defensible. These studies confirm our basic theories about territoriality and provide some of the strongest evidence currently available that animal behavior is adaptive.

Coloniality

Spacing behavior results from the opposing forces of mutual repulsion and mutual attraction. Mutual repulsion arises from aggressive defense of territories or mutual avoidance, and it leads to wider dispersal or more regular spacing of individuals. Mutual attraction arises from the attraction of many individuals to clumped resources or to one another, and it leads to spatial clumping of individuals. The previous chapter discussed the selective factors favoring aggressive defense of resources. The discussion now turns to the reasons why individuals are mutually attracted to fixed locations or colonies. As we shall see in Chapter 13, similar issues will arise again when trying to explain how mobile social groups evolve.

Types of clumped spacing patterns

Three general types of clumped spatial organization can be usefully distinguished: aggregations, social groups, and colonies. They can be distinguished by differences in duration, internal structure, and the factors responsible for their formation. Many gradations between these categories exist, which often makes specific grouping patterns difficult to categorize. Nevertheless, the classification helps focus on the evolutionary origins of different kinds of spacing behavior.

Aggregations

Aggregations are temporary assemblages of individuals brought together by physical forces in the environment or the attraction of many individuals to external stimuli or resources. They never result from mutual attraction based on cooperative behavior, though individuals may use the presence of others as cues for locating resource patches. Coordinated activities and integrated social relationships are typically absent in aggregations, and individuals usually arrive and depart separately unless their movements are controlled by external physical agents. Some examples of aggregations are aerial insects brought together by wind currents, marine plankton brought together by ocean currents or upwellings, seabirds attracted to fishing boats by offal discarded over the side, and migratory hawks channeled by local terrain features (Photo 8–1).

Chapter eight

Photo 8–1

An aggregation of migratory broad-winged hawks (Buteo platypterus) *over Panama City, Panama.*

The hawks migrate in large concentrations because their movements are channeled by the narrow isthmus of Panama and the chain of mountains running lengthwise through it.

Photo by Neal G. Smith.

Social groups

Social groups arise through mutual attraction of individuals to gain cooperative benefits. They usually persist on at least a seasonal basis, though membership within groups may change more frequently. Behavioral activities are coordinated by a communication network of varying complexity. Social groups often, though not always, contain internally organized subgroups, dominance hierarchies, or mating consortships. Fixed spatial relationships are not maintained among group members, in contrast to the internal spatial organization of colonies. Some examples are fish schools, bird flocks, whale pods, ungulate herds, and primate troops.

Colonies

The main distinction between colonies and social groups is spatial organization. The activities of colony members are centered around a fixed location or *central place,* and/or fixed spatial relationships are maintained among colony members. The factors favoring coloniality are sometimes similar to those favoring social groups. Colonies may form because individuals or mated pairs come together to obtain cooperative benefits, but they may also form because individuals or mated pairs are independently attracted to the same location.

Examples of colonial invertebrates are sponges, some hydrozoans, bryozoans, entoprocts, and eusocial insects (Wilson 1971, Boardman et al. 1973). With the exception of hydrozoans and eusocial insects, colony members are sessile as adults. Hence spatial positions are fixed within each colony. Colonial hydrozoans, such as the Portuguese man-of-war, are free-floating superorganisms whose members interact like organ systems. Different colony members function as feeding appendages, digestive organs, reproductive organs, protective tentacles, and buoyancy organs. Since colony members are permanently attached to one another, positions within each colony are fixed. Eusocial insects do not maintain fixed spatial positions with respect to each other, but they center their activities around a fixed nest site.

Examples of colonial birds include seabirds, pelicans, gulls, terns, flamingos, herons, swallows, some blackbirds, and some weaver finches (Lack 1968). Examples of colonial mammals include prairie dogs, ground squirrels, sea lions, fur seals, and elephant seals (Eisenberg 1966). Vertebrate colonies are always centered around a fixed location, and individuals within colonies often, though not always, maintain relatively fixed spatial relationships by means of territory defense. The remainder of this chapter will focus entirely on vertebrate coloniality.

Costs of coloniality

Crowding together into breeding or roosting colonies always entails costs. The high density of individuals in a colony intensifies competition for nearby food resources, attracts predators, facilitates spread of disease organisms and ectoparasites, and exacerbates the deleterious effects of neighboring individuals interfering with reproductive activities (Alexander 1974). Some of these costs are directly observable and well documented.

Evidence for increased competition

Several lines of evidence show that competition for food around colony sites is often severe. Occasionally depletion of food resources can be directly measured, but more frequently evidence for competition is indirect. A few examples suffice to illustrate the kinds of evidence available for demonstrating intense competition around colonies.

The red-billed quelea is an African weaver finch that breeds and roosts in colonies of up to several million individuals (Ward 1965a, b). Its normal diet during the dry season consists of small grass seeds, but as the season progresses its diet shifts to larger seeds of other plants because small seeds become increasingly scarce. If the shift occurs early enough, it coincides with ripening of cereal grains and leads to severe crop damage. Roosting colonies are typically small and numerous early in the dry season, but many colonies are soon abandoned because nearby seed crops are exhausted. The birds from abandoned colonies join nearby colonies where food is more abundant, ultimately leading to a few very large colonies.

Food availability around colonies has not been measured by direct sampling, but indirect evidence shows that food becomes increasingly scarce. Both sexes lose weight as the dry season progresses, indicating that food is increasingly difficult to obtain. Also, adult sex ratios become increasingly skewed toward males because males dominate females in dry-season flocks and consequently gain better access to food (Ward 1965c). Females obtain less food each day than males, as shown by the smaller amounts of food present in their crops, and consequently many more females than males die during the dry season (Ward 1965a).

Small clutch size is often an indicator of intense competition around colonies. The majority of seabirds lay only one egg per clutch (Lack 1968), and species that lay two eggs often cannot rear more than one chick to independence (Nelson 1970, 1978). Experimental additions of a second egg to single-egg clutches show that parents of several species ordinarily cannot provide enough food to support more than a single chick (Rice and Kenyon 1962; Harris 1966; Lack 1968; Nelson 1969). One-egg clutches are rare among birds, and most occur in colonial species. Since birds are believed to produce as many offspring as they can rear, the more frequent occurrence of one-egg clutches among colonial species implies that food is more difficult to obtain than in most noncolonial species. Evidence that one-egg clutches result from intense competition for food comes from comparing species that lay one-egg clutches to related species that lay larger clutches. Most boobies lay one or occasionally two eggs per clutch and can rear only one chick to independence, but blue-footed boobies (*Sula nebouxii*) and Peruvian boobies (*S. variegata*) lay clutches of three or four eggs (Nelson 1970). The latter two species breed near unusually productive ocean upwellings where food is particularly abundant, while most boobies do not.

Two attributes of chick development are frequently indicative of food shortages. As adaptations to food scarcity, chicks may develop very slowly and possess a capability for withstanding prolonged periods of fasting. Both attributes are typical of most seabird chicks (Ashmole 1963; Stonehouse 1963; Nelson 1967a, 1970, 1978; Boersma and Wheelwright 1979), and both are unusual among noncolonial birds.

Another indicator of food shortages around colonies is prevalent starvation of chicks or mass abandonment of colonies. In tricolored blackbirds young fledged successfully in only 7 of 14 colonies because 7 colonies were abandoned en masse, apparently in response to food shortages around the colonies (Payne 1969). Food shortages also cause mass abandonment of breeding colonies in Peruvian boobies (Vogt, cited in Nelson 1964), white pelicans (*Pelecanus erythrorhynchus* and *P. onocrotalus*) (Brown and Urban 1969; Johnson and Sloan 1978), and red-billed queleas (Jones and Ward 1979). The typical pattern among winter roosting colonies of passerine birds is abandonment of small roosts midway through winter, with the birds moving to larger roosts nearby (Neff and Meanley 1957; Wynne-Edwards 1962; Hamilton and Gilbert 1969; Ward and Zahavi 1973). Small roosts are apparently associated with local concentrations of food, and when those are depleted, the roosts are abandoned (Hamilton and Watt 1970).

Long foraging flights by colony members usually imply that food is scarce near a colony. Distant food resources should not be exploited unless nearer ones are unavailable or heavily depleted, because traveling long distances for food requires considerable expenditures of time and energy (Hamilton and Watt 1970). Long foraging trips are the rule for large breeding and roosting colonies. The now extinct passenger pigeon (*Ectopistes migratorius*) often flew 50 miles or more from breeding colonies and much farther from roosting colonies (Schorger 1955). Many colonial seabirds forage far offshore and have time for only one foraging trip per day (Lack 1968). Starlings and some blackbirds regularly fly 50 miles or more from winter roosts (Hamilton and Gilbert 1969). Female Alaskan fur seals (*Callorhinus ursinus*) leave their pups for up to a week at a time while foraging at sea (Bartholomew and Hoel 1953). Such long foraging expeditions not only limit the rate offspring can be fed but also entail long absences that leave offspring more exposed to predation and other hazards.

Interspecific competition for food is evidenced by partitioning of food resources or foraging zones. For example, the eight species of seabirds breeding on Christmas Island all exploit the same prey but capture them in different proportions (Nelson 1970). In addition, the five species of terns breeding there either specialize on different prey sizes or forage at different distances from the colony (Ashmole 1968). Among auks, puffins, and murres each species specializes on different prey types (plankton, fish, marine invertebrates, or some combination thereof) and different prey sizes (Bédard 1969a). The planktonic feeders even capture different proportions of each prey species because they possess different bill morphologies (Bédard 1969b). Some evidence suggests that these species may further partition food resources by foraging at different distances from mixed colonies (Cody 1973b), but the evidence is based on small sample sizes and is not corroborated by other studies (Bédard 1976). Similar partitioning of food resources has also been documented for many other seabirds and colonial landbirds (e.g., Fisher and Lockley 1954; Nelson 1970; Lack 1971).

Evidence for high predation in colonies

Large colonies are very conspicuous and contain large numbers of potential prey. Consequently they are likely to attract numerous predators. One indication that predation pressure is important comes from the kinds of places where colonies are situated. Most breeding colonies of birds are found in relatively inaccessible places such as marshes, isolated trees, cliffs, or islands, where many predators have difficulty gaining access to them (Lack 1968). Similarly, breeding colonies of sea lions, fur seals, and elephant seals are usually situated on islands, which cannot be reached by terrestrial predators (Bartholomew 1970).

The parents of many colonial birds attend young nestlings continuously to protect them from predators. This is particularly true of larger species such as murres, pelicans, flamingos, herons, storks, and gulls, since adults of these species are large enough to chase off many potential nest predators (see Wittenberger and Tilson 1980). That predation pressure is high in colonies is well documented by numerous studies, many of which will be cited later in the chapter (see Ricklefs 1969).

Evidence for increased disease and ectoparasitism

Evidence that group living increases disease or ectoparasitism is usually difficult to obtain without sophisticated analyses of animal health or tedious counts of ectoparasites. Certainly research on human epidemiology suggests that clumped spacing in animals should increase the incidence of disease and ectoparasitism, but not much direct evidence is available on this question. The best evidence to date comes from a study of black-tailed and white-tailed prairie dogs (*Cynomys ludovicianus* and *C. leucurus*) (Hoogland 1979a).

The most serious disease of prairie dogs, as well as other ground squirrels, is sylvatic or bubonic plague, which has been known to wipe out entire colonies of both black-tailed and white-tailed prairie dogs (Barnes et al. 1972; Clark 1977). The importance of plague during the past evolution of prairie dog coloniality is unclear, since it may or may not be native to North America (see Pollitzer 1951; Olsen 1970), but coloniality does make prairie dogs particularly vulnerable to plague (see below). Since plague is transmitted almost exclusively by fleas, the effects of coloniality on its spread can be evaluated indirectly by studying the incidence of flea infestation.

Increased transmission of communicable diseases and ectoparasites presumably results from more frequent contact between individuals. Hoogland's (1979a) data suggest that such contacts are more common in large prairie dog wards (colony subsections) than in smaller wards. They also suggest that infestations are more common in the dense colonies of black-tailed prairie dogs than in the less dense colonies of white-tailed prairie dogs. For both species the number of fleas per burrow entrance was greater in large wards than in smaller ones. In addition, the number of fleas per burrow entrance was significantly higher in black-tail colonies than in white-tail colonies. Finally, the number of fleas and lice counted on both adults and juveniles was greater for black-tails than for white-tails (Table 8–1). Number of fleas or other ectoparasites may or may not be correlated with susceptibility to diseases or other fitness costs, but the data do suggest that increased disease and ectoparasite transmission represents one potential cost of coloniality.

Evidence for interference with reproductive activities

Probably the severest form of interference within colonies is killing offspring of other colony members. Killing and cannibalizing young chicks regularly occurs in colonies of herons, pelicans, and large gulls (references in Wittenberger and Tilson 1980). Pup mortality in grey seals (*Halichoerus grypus*) and northern elephant seals (*Mirounga angustirostris*) frequently

Table 8–1 *Comparison of ectoparasite infestations on adult and juvenile prairie dogs of two species.*

Age and time of sample	Fleas	Lice	Mites and ticks	All ectoparasites
White-tail adults, during breeding	8.47 ± 16.6 .778 (*N* = 45)	.250 ± .577 .188 (*N* = 16)	.022 ± .149 .022 (*N* = 45)	8.58 ± 16.5 .822 (*N* = 45)
Black-tail adults, during breeding	9.34 ± 15.2 .875 (*N* = 128)	.171 ± .811 .085 (*N* = 129)	.016 ± .177 .008 (*N* = 128)	9.51 ± 15.2 .883 (*N* = 128)
White-tail adults, at first juvenile emergences	.951 ± 1.60 .393 (*N* = 61)	.097 ± .301 .097 (*N* = 31)	.105 ± .344 .093 (*N* = 86)	1.81 ± 1.97 .677 (*N* = 31)
Black-tail adults, at first juvenile emergences	1.36 ± 2.01 .516 (*N* = 64)	5.52 ± 7.67 .762 (*N* = 63)	.016 ± .126 .016 (*N* = 63)	6.65 ± 7.52 .903 (*N* = 62)
White-tail young, at their first emergences	.469 ± 1.40 .232 (*N* = 177)	3.35 ± 5.83 .463 (*N* = 82)	.136 ± .487 .093 (*N* = 236)	4.17 ± 5.90 .683 (*N* = 82)
Black-tail young, at their first emergences	.930 ± 2.18 .349 (*N* = 86)	80.7 ± 33.6 .943 (*N* = 88)	.057 ± .278 .045 (*N* = 88)	81.8 ± 33.5 .965 (*N* = 86)

Note: Values are mean numbers (with standard deviations) of ectoparasites found per individual, proportion of individuals parasitized, and the sample size *(N)*.

Source: From J. L. Hoogland, "Aggression, Ectoparasitism, and other Possible Costs of Prairie Dog (Sciuridae, *Cynomys* spp.) Coloniality," *Behaviour* 69 (1979):18–19. Reprinted by permission.

results from males crushing pups during fights over females (Coulson and Hickling 1964; Le Boeuf et al. 1972; Le Boeuf and Briggs 1977). Seal pups also die frequently because they become separated from their mothers. Orphaned pups usually die of starvation, but many additional pups are severely bitten by adult females while their mothers are at sea foraging.

A second kind of interference is theft of nest materials. Many seabirds, gulls, and herons regularly steal nest materials from their neighbors. In mixed colonies of herons and anhingas (*Anhinga anhinga*), anhinga nests are frequently torn completely apart by herons (Meanley 1954). Gathering nest materials consumes much time and energy, especially when nest materials are scarce or nests are large and bulky, and loss of nest materials can be a significant cost for a nesting pair.

Breeding in colonies may reduce a male's probability of fathering his mate's offspring because coloniality may increase the prevalence of forced copulations or the likelihood that females will solicit copulations from neighboring males. In red-winged blackbirds, for example, a high proportion of females lay fertile clutches after their mates have been vasectomized prior to mating (Bray et al. 1975; Roberts and Kennelly 1980). The proportion of fertile clutches decreases with increasing distance to the nearest fertile territorial male, indicating that females often copulate with neighboring territory owners. Copulations between females and males other than their mates are prevalent in colonial birds (Beecher and Beecher 1979; Gladstone 1979; Mock 1980), but whether they are more prevalent than in noncolonial birds is unknown. Many of these copulations are forced, but at least some are not. The extent to which they lead to fertilization of eggs is unknown, except for red-winged blackbirds.

Some extremes in colony sizes

The costs of breeding in colonies are likely to increase with colony size and density. Competition for resources is directly proportional to number of individuals exploiting them; attractiveness of a colony for predators is directly proportional to number of potential prey found in them; and likelihood of interference from neighbors is directly proportional to density within the colony. A few extreme examples of colony size and density therefore give some idea of just how high the costs of coloniality can be. Several examples are listed in Table 8–2, and a few are discussed more fully in text.

The guanay cormorant (*Phalacrocorax bouganvillei*) of Peru breeds in colonies containing about 3 nests per square meter of space (Photo 8–2) (Coker 1919). A 15-acre colony contains 180,000 breeding pairs and deposits guano at the rate of 11 cm per year. The accumulated guano deposits have been the basis of a rich fertilizer industry for over a century. Some of the larger colonies of this species contain upward of 4–5 million birds (Murphy 1936).

Landbird colonies are usually smaller, but some rival seabird colonies in size. Breeding colonies of queleas contain up to several million pairs and cover as much as 30 to 75 acres of thorn trees (Ward 1965b). Large trees harbor hundreds of nests, and even small bushes hold 10 to 50 nests. One colony of the now extinct passenger pigeon covered an area of 2200 km^2 (850 mi^2) and numbered 136 million birds, while a few colonies reportedly contained up to 2 *billion* birds (Schorger 1937)!

Roosting colonies can be equally impressive. The largest winter roosts of the passenger pigeon covered several thousand acres and contained upward of 2 billion birds (Schorger 1955). Pigeons were so dense in such roosts that they perched as much as a meter deep on top of one another and their combined weight often broke large branches off trees, killing hundreds of birds underneath. An unusually large roost of bramblings (*Fringilla montifringilla*) occupied two 15-acre woodlots in Switzerland and contained an estimated 72 million individuals during the winter of 1950–1951 (Mühlethaler 1952; Schifferli 1953). According to one estimate, these roosts contained nearly two-thirds of the entire European brambling population that year (Newton 1972).

Table 8–2 *Some extremes in colony size.*

Species	Estimated size	Source
Breeding colonies		
Wilson's storm petrel *(Oceanodroma oceanicus)*	1 million +	Fisher and Lockley 1954
Great shearwater *(Puffinus gravia)*	2 million	Rowan 1965
Guanay cormorant *(Phalacrocorax bouganvillei)*	4–5 million	Murphy 1936
Common murre *(Uria aalge)*	Several million	Fisher and Lockley 1954
Thick-billed murre *(Uria lomvia)*	1.6 million	Krasovskii, cited in Nelson 1970
Common puffin *(Fratercula arctica)*	1 million	Fisher and Lockley 1954
Little auk *(Plautus alle)*	5 million	Fisher and Lockley 1954
Wide-awake tern *(Sterna fuscata)*	3/4–1 million	Chapin 1954; Ashmole 1963
Passenger pigeon *(Ectopistes migratorius)*	136 million +	Schorger 1937
Red-billed quelea *(Quelea quelea)*	Several million	Ward 1965b
Roosting colonies		
Passenger pigeon *(Ectopistes migratorius)*	2 billion +	Schorger 1955
Blackbirds and starlings *(Agelaius, Euphagus, Sturnus vulgaris)*	4–15 million	Hamilton and Gilbert 1969; Welty 1975
Brambling *(Fringilla montifringilla)*	72 million	Mühlethaler 1952; Schifferli 1953
Mexican free-tailed bat *(Tadarida brasiliensis)*	4–30 million	Moore 1948; Davis et al. 1962

Note: These are maximum sizes and not necessarily typical of the species listed.

Clearly the costs of joining such colonies must be high. For coloniality to be adaptive, the attendant benefits must be equally high. Few attempts have been made to measure the benefits of coloniality, but a variety of possible benefits have been suggested.

Benefits of coloniality

Several hypotheses have been advanced to explain why vertebrates breed or roost in colonies. Coloniality may enable more efficient exploitation of food resources, provide better access to scarce habitat patches, or confer greater protection from predators. Each hypothesis is supported by some evidence, and each may apply to some species. However, no hypothesis

A.

Photo 8–2

B.

Many seabirds typically breed in very dense colonies.

A. A breeding colony of the guanay cormorant on an island off the Peruvian coast. B. Close-up view of a guanay cormorant incubating eggs. Note how close nests are to their neighbors.

Photos courtesy of the American Museum of Natural History.

provides a general explanation of coloniality. Coloniality is a diverse phenomenon with multiple causes, and the selective factors favoring its evolution vary considerably among species.

Food distribution hypothesis

A geometrical model. Distribution of food resources influences the cost of harvesting food. Travel costs to and from food patches can be minimized by dispersed breeding when food is uniformly distributed and by colonial breeding when food is spatiotemporally clumped (Horn 1968). A simple geometrical model shows why.

Figure 8–1 *A geometrical model showing how colonial nesting can increase the energetic efficiency of exploiting spatiotemporally clumped food resources.*

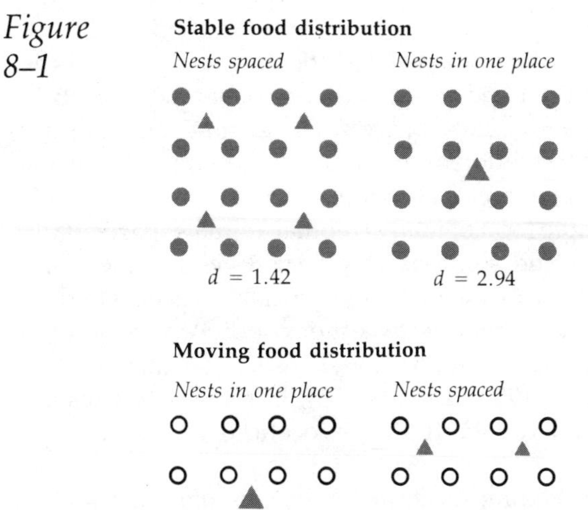

Solid circles represent continuously available food sources, each of which contains one-fourth of the food required by a single nesting pair. Open circles represent spatiotemporally clumped food sources that contain food one at a time and at that time contain enough food to satisfy the food requirements of four nesting pairs. Small triangles represent locations of single nests. Large triangles represent locations of colonies containing four nests. d = the average distance traveled to obtain food, based on the assumption that each adult visits the nearest available food source on each trip. Note that d is minimized by dispersed nesting when food is evenly distributed and continuously available but is minimized by colonial nesting when food is spatiotemporally clumped.

Source: From H. S. Horn, "The Adaptive Significance of Colonial Nesting in the Brewer's Blackbird *(Euphagus cyanocephalus)*," *Ecology* 49 (1968):690. Reprinted by permission of Duke University Press. Copyright 1968 by the Ecological Society of America.

Suppose that an area contains just enough food to support four nesting pairs of birds. When this food is distributed evenly in 16 food patches and is continuously available in every patch, each nesting pair need visit only 4 patches to obtain adequate food. Some simple calculations show that mean travel distance is lowest if nests are evenly dispersed throughout the foraging arena (Figure 8–1). In contrast, when all the food is first in one patch and then in another, with patch locations varying between the same 16 sites on a daily basis, mean travel distance is lowest if nests are placed in the center of the foraging arena. Thus dispersed nesting should be the most efficient way to exploit evenly distributed food resources, while colonial nesting should be the most efficient way to exploit spatiotemporally clumped resources.

Food distribution patterns encountered by colonial animals. The distribution of food resources exploited by colonial animals is often difficult to measure. Many types of food are inherently difficult to sample, and the various food types captured by a species may differ considerably both in distribution and importance. Indirect evidence suggests that some but not all colonial animals do exploit spatiotemporally clumped food resources, as predicted by Horn's (1968) geometrical model.

Aerial insects are the staple food of many colonial animals, particularly swallows, swifts, many bats, and some hawks (Lack 1968; Dalquest and Walton 1970; Bradbury 1977a), and they are often locally concentrated by wind movements and updrafts. Changing weather conditions cause temporal fluctuations in overall abundance, as fewer insects fly during cool, wet, and windy weather (Lack 1956; Bryant 1975), but such changes are not likely to produce temporal variability in the locations of the best foraging areas.

Temporal variability in feeding locations is an essential component of Horn's (1968) model. For aerial insect populations such variability could arise if changes in wind direction concentrate insects at different locations each day (Emlen and Demong 1975). At present, few data are available to evaluate temporal variability of aerial insect populations. The principal evidence comes from a study of tropical emballonurid bats. The abundance of insects exploited by these bats varies seasonally but not to any great extent on a shorter-term basis (Bradbury and Vehrencamp 1976). Seasonal changes in roost locations can be attributed to this variability, but coloniality within seasons cannot.

Fruit is another food resource commonly exploited by colonial birds and bats (Lack 1968; Bradbury 1977a). The spatial distribution and temporal availability of fruit varies considerably among plant species, especially in the tropics. Fruits often ripen synchronously on any given plant, but different individual plants frequently produce fruit at markedly different times, even within the same species. Some plant species fruit synchronously over a short fruiting season to reduce losses to seed predators (Janzen 1967b, 1971; Smythe 1970c), while others fruit over a period of weeks or months to reduce intraspecific competition for faunal seed dispersers (Frankie et al. 1974). The various plant species also fruit at different times of year to reduce interspecific competition for seed dispersers (Snow 1965b, 1971a; Smythe 1970c; Frankie et al. 1974; Janzen 1974; Howe and Estabrook 1977; Thompson and Willson 1979). As a result, some fruits are fairly evenly distributed while others are spatiotemporally clumped (see Klein and Klein 1975).

A relationship between spatiotemporal clumping of fruit and coloniality has yet to be demonstrated. The relationship does not hold for at least one colonial bird, the phainopepla (*Phainopepla nitens*), of desert canyons in the southwestern United States, which exploits insects and mistletoe berries that seem to be uniformly distributed (Walsberg 1977). An analysis comparing food distributions for colonial and noncolonial frugivores has not yet been conducted, but such an analysis would greatly clarify the relationship, if any, between food distribution and coloniality.

Colonial marsh-breeding blackbirds generally exploit damselflies, dragonflies, and other insects that have recently emerged from aquatic larval stages (Orians and Horn 1969). Most of these insects emerge along shorelines or on the outer edges of reedbeds, and they emerge primarily in the morning (damselflies) or at night (dragonflies) (Orians 1980). Many fewer damselflies emerge in the afternoon, and by then most tenerals (recently metamorphosed adults) have dispersed into the surrounding uplands. Within each marsh the density of emerging insects varies greatly from one place to another, probably due to local differences in currents, water temperatures, bottom characteristics, and fish foraging intensity. Emergence also varies greatly from one day to the next because it is cyclic during most of the blackbird breeding season and because it is suppressed by cloud cover, rain, and cold weather (Orians 1980). Nevertheless, the relative abundance of emerging insects at each location is less variable, with some locations being consistently better than others (Wittenberger, unpublished data).

In eastern Washington the foraging behavior of colonial blackbirds reflects the spatiotemporally variable distribution of aquatic insects. Brewer's blackbirds (*Euphagus cyanocephalus*) nest in clumps of greasewood or sagebrush near ponds or marshes. During morning hours they forage for tenerals along stream or pond margins, but in the afternoon they shift to upland prey (Horn 1968). This shift reflects the daily cycle of aquatic insect emergence and results in a diurnal cycle of both the prey types delivered to nestlings and the directions taken by adults when leaving colonies to forage (Figure 8–2). Red-winged and yellow-headed blackbirds, which nest in marshes and exploit recently emerged aquatic insects, often show a similar pattern of foraging behavior (Willson 1966; Orians and Horn 1969).

Colonial nesting by great blue herons (*Ardea herodias*) is also associated with spatiotemporal clumping of food resources (Krebs 1974). Individuals that breed in colonies primarily hunt invertebrates and fish along the margins of shallow ponds. Distribution of these prey can be inferred from the distribution of foraging herons because herons soon leave a feeding site if no food is captured there. Herons from breeding colonies feed in different locations each day, and the number feeding at any given location varies from hour to hour. This pattern suggests that prey is spatiotemporally clumped. In contrast, nearby herons that breed in widely spaced trees and defend large multipurpose territories forage along irrigation ditches in pastures, where they capture mice and other evenly distributed prey.

Not all colonial birds exploit spatiotemporally clumped food resources. Piñon jays breed in colonies near abundant sources of piñon pine cones (Balda and Bateman 1971). The cones are spatially clumped, but the best foraging locations do not vary much on a short-term basis. Similarly, crossbills (*Loxia*) wander widely and breed opportunistically in colonies whenever a flock locates a sufficiently abundant supply of spruce cones (Newton 1970). The extinct passenger pigeon bred near local concentrations

Figure 8–2

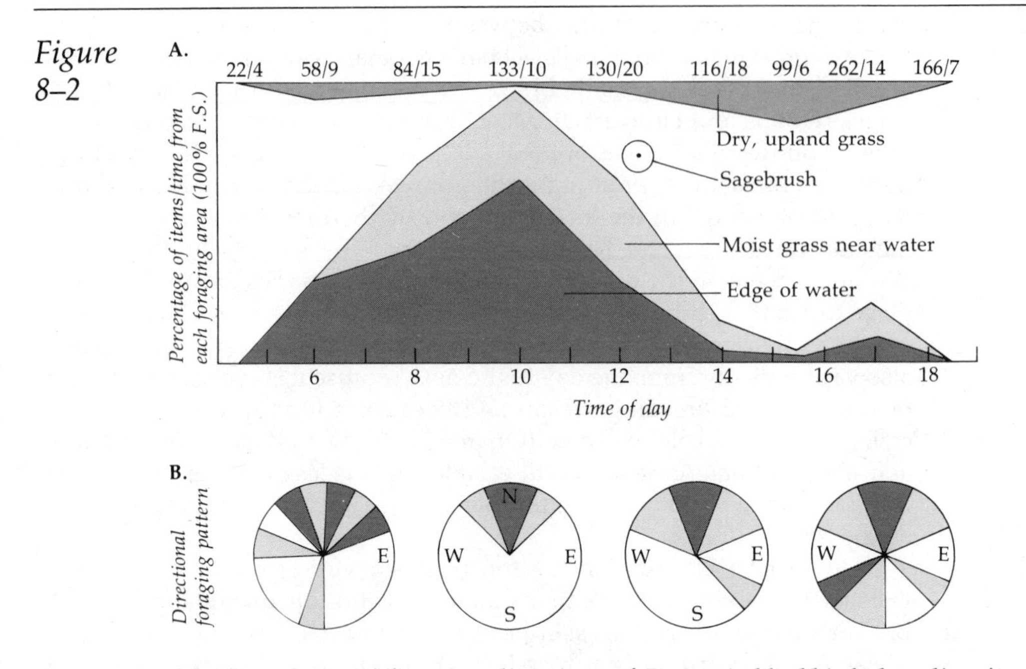

Nestling diet and foraging directions of Brewer's blackbirds breeding in greasewood uplands near cattail marshes.

A. The upper diagram shows the percentage of food items obtained in each major habitat type as a function of time of day. For a given time of day each plotted point represents the percentage of food items of a given type that were delivered to nestlings. Numbers above the graph are total numbers of food items in the sample and numbers of samples taken for each time of day. B. The lower diagram shows the foraging directions taken by birds departing the colony at the times of day indicated. Data were collected by observing 12 individuals for a half hour each and plotting the directions taken during that interval on a circular diagram. The data were then pooled for all individuals observed at each time of day, and the sectors were shaded according to how many trips were taken in each direction. Dense shading represents the most commonly taken directions; lighter shading represents less frequently taken directions; lack of shading represents directions not taken by any individuals.

Source: From H. S. Horn, "The Adaptive Significance of Colonial Nesting in the Brewer's Blackbird *(Euphagus cyanocephalus)*," *Ecology* 49 (1968):691. Reprinted by permission of Duke University Press. Copyright 1968 by the Ecological Society of America.

of beech mast, which was their main food during the breeding season (Schorger 1955). Cone and nut crops fluctuate greatly between years in any given area, causing species that exploit them to breed in different localities each year, but during any given breeding season the best foraging locations do not change much on a short-term basis.

Problems with the hypothesis. One problem with Horn's (1968) central-place model is that it does not easily explain the formation of very large colonies. Resources near large colonies are soon depleted, forcing

Figure 8–3

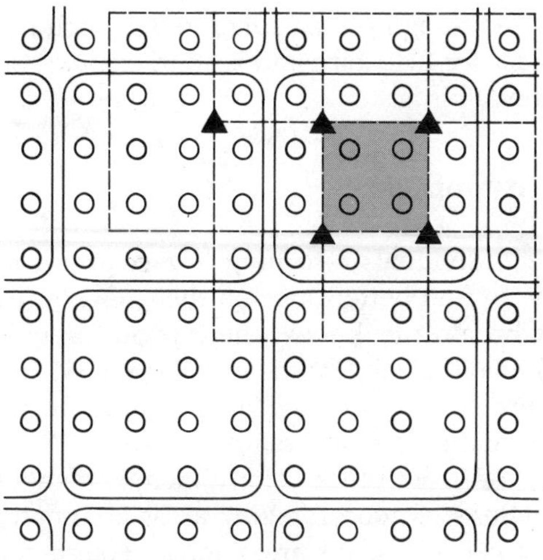

An expanded foraging arena eliminates the energetic advantage of coloniality in cases where food is spatiotemporally clumped.

$d = 2.94$

Open circles, triangles, and d are defined as in Figure 8–1. Each of the boxes outlined with solid lines represents a foraging space equivalent to that shown in the lower half of Figure 8–1. Boxes outlined by dashed lines indicate the foraging arenas around the nests centered within them. Note that nests can be dispersed and still be situated at the center of their own sets of food resources. The boxes surrounding adjacent nests overlap. The shaded area shows a segment of the entire region where food sources are exploited by four nesting pairs. If more nests were added to the diagram, most of the region would be exploited to the same degree, as it is in Horn's original model.

Source: Matthew P. Rowe, personal communication.

individuals to commute long distances for food. From an energetic standpoint individuals should be able to forage more efficiently by segregating into several smaller colonies, each centered in its own foraging arena.

A second problem is that coloniality does not minimize travel distances to spatiotemporally clumped food resources unless the colony is centrally located. Colonies are usually situated at sites that impede access by predators, and such sites are often not conveniently available in a central location. Many colonies are therefore not located in the center of their foraging arenas. It is not yet clear how acentric a colony site can be before the advantages of shorter average commuting distances are lost, but many colony sites probably do not fall within that limit.

Finally, Horn's geometrical model depends on the bounded nature of the food distribution (Matthew P. Rowe, personal communication). In Figure 8–1 dispersed individuals must commute farther than clumped individuals when food is spatiotemporally clumped because their nests lie near the edge of an isolated foraging area. However, if the foraging area is not isolated, each pair can nest in the center of its own food distribution and still remain dispersed (Figure 8–3). Dispersed individuals then have the same average commuting distance as clumped individuals. The model therefore works only if the foraging area surrounding each colony is isolated. This may be the case for blackbird and seabird colonies but probably not for many other colonial birds.

Information center hypothesis

The hypothesis. Individuals returning to a colony from poor foraging areas may be able to find better ones on their next outward trip by following individuals who already know where good foraging areas are located (Ward 1965a; Zahavi 1971b; Ward and Zahavi 1973; Emlen and Demong 1975). Thus one potential benefit of breeding or roosting in a colony is to acquire information about foraging sites.

Acquiring information from other colony members should be most beneficial when the locations of good foraging areas are relatively unpredictable. Food should persist in good areas long enough for returning individuals to still find food there but not for so long that individuals rarely need to search elsewhere for new food sources. Food should also be sufficiently concentrated that several individuals can exploit good foraging sites profitably. Colonies would therefore be most useful as information centers when food resources are spatiotemporally variable on a time scale of hours or days. The evidence discussed above indicates that at least some colonial birds do exploit such resources.

Coloniality might also facilitate relocating food sources when landmarks are absent, as in offshore-foraging seabirds, because individuals could follow the steady stream of other colony members flying to and from them (Ward and Zahavi 1973). However, species that commute long distances for food often show remarkably accurate homing ability based on solar cues alone (Matthews 1968), so an absence of terrestrial landmarks does not necessarily imply that individuals can benefit by following others.

The evidence. A preliminary question in evaluating the information center hypothesis is whether individuals have an opportunity to learn where food is located by following other colony members. Such opportunities should be especially prevalent when colony members feed together, and in fact most colonial birds do (Lack 1968; Newton 1972; Ward and Zahavi 1973). Some exceptions are herons, hawks, and vultures, which breed or roost in colonies and forage alone, but Ward and Zahavi (1973) argue that those species might still benefit by learning better places to search for food.

Identifying a likely individual or flock to follow from a colony is not as difficult as it might seem. Several potential cues can be suggested, though none have been studied to determine whether they are actually used. Individuals returning from good feeding areas should have full crops or large amounts of food in their bills, both of which might be monitored by neighboring individuals. Consistently successful foragers should have better-fed chicks, and in winter they should maintain heavier body weights. Even in the absence of such cues, a bird that left a poorer-than-average feeding area on its previous trip might increase its chances of finding a better area simply by following a randomly chosen individual or flock on its next trip. However, individuals adopting such a tactic may not find better foraging areas than they would by striking out in a random direction of their own.

Proof that individuals actually follow other colony members to food is not easily obtained. Birds depart winter roosts in consecutive waves that show up on radar as *ring angels* (Photo 8–3) (Eastwood et al. 1962; Eastwood 1967), and in some species individuals often lag behind each wave, vacillating between leaving the roost and returning to it. Ward and Zahavi (1973) suggest that these stragglers are birds that left poor foraging areas the previous evening and are trying to decide which flock to follow. However, no relationship between hesitancy in leaving the roost and poor foraging success the previous day has been shown, and such hesitancy may result from high vulnerability to predation during departure (see pp. 335–337).

Birds in breeding colonies arrive and depart amidst a teeming hubbub of activity, and movements from the colony are seldom organized into flocks. Nevertheless, direct observations show that individuals do sometimes follow others from colonies of tricolored blackbirds, Brewer's blackbirds, bank swallows (*Riparia riparia*), and phainopeplas (Orians 1961b; Horn 1968; Emlen and Demong 1975; Walsberg 1977). In addition, offshore-foraging seabirds regularly commute to and from distant feeding grounds

---

Photo 8–3

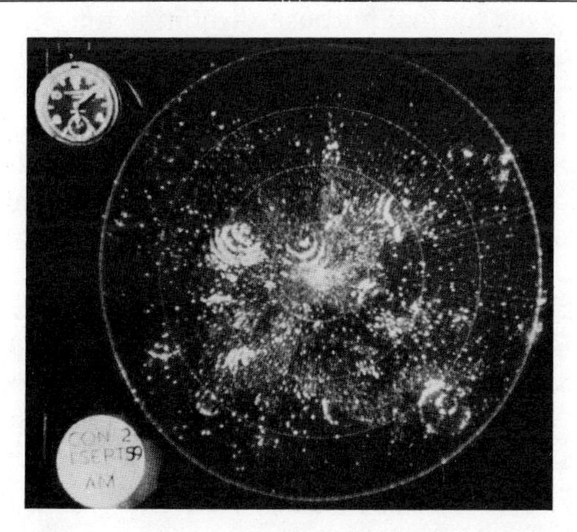

Ring angels on radar screens, once mysterious apparitions for radar operators, result from massed departures of birds leaving colonial roosts.

The concentric rings arise because several successive waves depart from the roost. Rings disappear from radar screens as the birds in each wave disperse some distance away from the roost. Note that the roost is often not centrally located around the principal foraging grounds, as indicated by the absence of rings in some directions around most roosts.

Photo from E. Eastwood, *Radar Ornithology* (London: Methuen and Co., Ltd, 1967).

in streams, while oilbirds and herons typically depart colonies in small flocks (e.g., Snow 1961–1962; Krebs 1974; Custer and Osborn 1978). These observations suggest that some individuals may be following others to their foraging grounds, but no studies to date have shown that less successful foragers follow more successful foragers. Nor have any studies shown that individuals actually find better foraging sites by following others. Until data are available to show that individuals really do find better feeding grounds by following others, the information center hypothesis remains unproven.

Problems with the hypothesis. One problem with the information center hypothesis is that it fails to explain why individuals or flocks return to a central roost after leaving superior foraging areas. They would have no use for obtaining information that day, and they would very likely recruit competitors the following morning. A better strategy would be to roost near the food source until it is exhausted and then visit the colony to seek information about new feeding grounds. Of course, if every individual adopted that policy, colonies would provide little information for anyone. One possible reason successful foragers might return to the colony and share information is that they expect others to reciprocate in the future. However, reciprocal altruism requires the ability to discriminate against cheaters, and nothing prevents cheaters from returning to a colony on days when they fail to find good foraging areas and remaining away on days when they do find good foraging areas. For breeding colonies the information center hypothesis is more tenable because individuals must return regularly to feed offspring.

Even if individuals do acquire information by following others, coloniality may not have evolved for that purpose. Acquiring information may have evolved as a secondary adaptation once coloniality evolved for other reasons. Separating secondary adaptations from primary selective factors is not easy, but it can be done with systematic comparative studies. Acquiring information is probably a secondary adaptation if it is exhibited by only a subset of all colonial species with similar breeding biology. As an example, suppose that some cliff-nesting seabirds acquire information from other colony members, while others do not. Every species nests on cliffs and forages offshore. One might then conclude that coloniality evolved because suitable nesting cliffs are scarce, with some species secondarily evolving tactics for acquiring information as a response to the way their particular food resources are distributed. To date no comparative studies of this sort have been conducted, so it is not yet possible to distinguish between primary selective factors and secondary adaptations.

Clumping of suitable breeding habitat

Colonial social organization may be imposed on a species by clumping of suitable breeding habitat. This can occur in two ways. First, suitable breeding sites may be patchily distributed, either because environmental conditions favor breeding at sites that are relatively inaccessible to predators or because breeding is restricted to relatively little land mass (islands) within large bodies of water. Second, the suitability of breeding habitat may depend on habitat modifications maintained by previous occupants, with colonization of potentially suitable but unoccupied habitat being difficult because time and energy must be expended to modify newly settled habitat. Each possibility will be discussed in turn.

Hypothesis based on clumped breeding sites. The two primary defenses against nest predation in birds are use of relatively inaccessible nest sites and concealment. Many birds build nests on the outer tips of tree limbs, in dense thickets, or in thorny bushes, where predators cannot easily reach them. Such sites are numerous and dispersed throughout woodlands and some scrub habitats, and most birds that utilize them are noncolonial. In more open grasslands and other two-dimensional habitats, inaccessible nest sites are less available and most noncolonial birds rely on concealment to deter nest predators.

Several constraints limit the effectiveness of concealment. One is large body size. Large birds are more easily detected by predators, even when sitting quietly on a nest, and hence they require more cover to remain concealed. Their young often require large bulky nests, which are not easily concealed, as in herons, crows, and hawks. Bulky nests are unnecessary in ground-nesting species, but large adults have greater difficulty approaching or leaving nests unnoticed. Thus many large birds nest in treetops, on cliffs, or in other relatively inaccessible places to escape predators, often relying on continuous nest attendance and defensive aggression to ward off predators that do reach their nests.

Another constraint is population density. Concealment is most effective when nests are more widely dispersed and relatively uncommon, as several studies have experimentally demonstrated (see p. 231) (Murton 1958; Tinbergen et al. 1967; Croze 1970). High local population densities should make concealment less effective because large numbers of individuals entering and leaving an area attract predators. High density may therefore make reliance on inaccessible nest sites a better defense against predation than concealment. Coloniality might then evolve because inaccessible nest sites are clumped and relatively scarce compared to population size.

One cause of high local breeding densities is clumping of food resources. As food resources become more clumped, so also do populations exploiting them. At the critical density when use of inaccessible sites is safer than concealment, local populations should coalesce into colonies.

Evidence for clumped breeding sites. Clumping of suitable breeding sites may well explain many instances of coloniality in landbirds. Most colonial species exploit clumped food resources, breed at high local densities, and nest in relatively inaccessible sites (Lack 1968). The critical question, though, is whether potentially suitable nest sites really are scarce enough to cause high local breeding densities.

For many species they apparently are. Colonial weaver finches breed at very high densities near abundant seed sources (Crook 1964; Ward 1965a). They could conceivably nest on the ground, relying on nest concealment for protection, but breeding densities are too high for concealment to be effective unless many individuals commuted long distances for food. Instead, they nest in the relatively few isolated trees near their foraging grounds. Similarly, large numbers of blackbirds nest in marshes or reedbeds and breed at high densities because the availability of marshes is limited compared to population size (see Wittenberger 1976a, 1979a). In most colonial birds not all potential colony sites are actually occupied, but once a critical density is reached, additional factors cause dense packing within established colonies. Such factors include information exchange between colony members (see previous section) and defensive tactics used to deter predators (see pp. 328–335). The important point, though, is that space is insufficient in these species for all individuals to effectively conceal their nests.

Some colonial landbirds do not seem to lack suitable nest sites near their foraging grounds. For example, great blue herons nest in the tops of trees, often in isolated groves or on islands. The number of suitable sites in many areas appears unlimited, suggesting that colonial nesting probably evolved to enhance foraging efficiency (Krebs 1974). A variety of passerine birds breed in small, loosely spaced colonies, with many apparently suitable nest sites remaining unoccupied (Lack 1968). Thus clumping of nest sites may explain how large colonies evolve in very abundant species, but it may not explain why smaller colonies evolve in relatively uncommon species.

The clearest evidence for nest site shortages comes from studies of marine birds. The foraging arena of pelagic seabirds such as albatrosses, shearwaters, petrels, alcids, and some terns is extremely large, covering vast expanses of open ocean, and the land area available for nesting is limited to relatively few islands and continental coastlines. In most places the land area is much too small for nests to be spaced far apart, and in any case many seabirds are too large to effectively conceal nests. There can be little doubt that colonial nesting among seabirds results from clumping of safe nest sites (Nelson 1970; Ashmole 1971), even though additional factors must be invoked to explain why nests are densely packed within colonies (see pp. 328–335).

A similar argument explains why fur seals and sea lions breed in colonies (Photo 8–4) (Bartholomew 1970; Stirling 1975). These species breed on islands or remote coastlines, which are relatively scarce compared to the large breeding populations supported by the vast food resources available in the open ocean and continental shelf areas. The number of suitable breeding sites is limited by several factors (Bartholomew 1970). Adaptations for marine foraging preclude the evolution of good locomotory ability on land, thereby limiting terrestrial breeding to flat coastal areas or pack ice. Breeding is restricted to islands or pack ice, where few predators are present, because young pinnipeds are defenseless against most terrestrial pred-

Photo 8–4

A breeding rookery of elephant seals on Año Nuevo Island off the coast of southern California.

The entire beach shown here is occupied by seals, but this is not true at all breeding rookeries. Even so, the scarcity of accessible beaches on islands is probably responsible for colonial breeding in elephant seals, fur seals, and sea lions.

Photo by B. J. Le Boeuf and K. T. Briggs.

ators (Orr 1965; Peterson and Bartholomew 1967; Bartholomew 1970). Breeding is further restricted to areas where fish and other food resources are sufficiently abundant offshore. At latitudes where pack ice is absent, breeding must occur on relatively few islands or coastal beaches. An absence of ice forces fur seals and sea lions to breed on islands, and that is why they are colonial. In contrast, seals that breed in polar regions, where land-fast ice and pack ice are extensive, breed either in loose colonies or in widely scattered locations (Stirling 1975). Species that breed on land-fast ice are loosely colonial, while species that breed on pack ice, which is more extensive, are noncolonial.

Habitat clumping may explain why some bats and wintering songbirds roost in large colonies. These species are very abundant and live in open grasslands or agricultural districts where roosting cover is sparse. A combination of open habitats and high local population densities would make suitable roosting sites relatively scarce, and dispersed roosting to enhance concealment would lose its effectiveness if a high proportion of the suitable sites are occupied every night. Under conditions where concealment is relatively ineffective, clumping together in large roosting colonies to swamp the exploitative abilities of predators might well be a safer alternative.

Ground squirrels and prairie dogs. Several hypotheses have been invoked to explain coloniality in diurnal rodents. One is based on shortage of suitable breeding sites. A second is based on the costs associated with colonizing habitats away from natal breeding areas. A third is based on the benefits gained from mutual vigilance. The first two hypotheses involve shortages of suitable habitat and will be discussed in this section. The third hypothesis involves cooperative defense against predators and will be discussed in the next section.

The hypothesis that coloniality arises from shortages of breeding sites is difficult to evaluate. Colonial ground squirrels live in tundra, alpine meadows, or recently disturbed areas such as pastures, fallow fields, and lawns (e.g., Carl 1971; Yeaton 1972; Dunford 1977b; Sherman 1977), and such habitats often appear to be very abundant. Nevertheless, many of the unoccupied portions of such habitats may be unsuitable for breeding because they lack subtle habitat characteristics required for successful breeding. Arctic ground squirrels, for example, avoid areas where visibility and mobility are hindered, and consequently they breed only in certain types of vegetation patches on open slopes (Carl 1971). They are further restricted to well-drained areas with relatively deep permafrost layers, where they can dig adequate burrow systems. Similarly, California ground squirrels are restricted to short-grass habitats with relatively short herbaceous vegetation (Evans and Holdenreid 1943). They are found mainly in scattered areas that are heavily grazed by cattle or other ungulates. Though such areas abound today, they may have been scarce during most of the period when ground squirrel social organization was evolving.

The situation for marmots is equally unclear. Yellow-bellied marmots (*Marmota flaviventris*) inhabit subalpine meadows on forested slopes or rock slides in Great Basin deserts, alpine marmots (*M. marmot*) inhabit alpine meadows on forested slopes, and hoary marmots (*M. caligata*) inhabit rocky talus slopes near alpine or Arctic treelines (Downhower and Armitage 1971; Barash 1974, 1976; Holmes 1979). The availability of talus slopes and mountain meadows is limited, but not all such habitats are occupied. One possible reason why some habitats remain vacant is that some apparently suitable habitats are in reality unsuitable because they do not contain adequate food or lack soil characteristics suitable for burrowing (see Svendsen 1974). A second possible reason is that colonizing unoccupied habitats may be too costly. Colonization involves travel over hazardous or inhospitable terrain, and when suitable unoccupied habitats are few and far between, the risks of such travel may well be more costly than remaining in occupied habitats and accepting the increased costs of competition entailed by colonial breeding.

In prairie dogs potentially suitable habitat is almost certainly not limited in availability (Hoogland 1980b). Colonies gradually expand into previously unoccupied peripheral areas, and many similar habitats further from the colony remain unoccupied. However, prairie dogs live in early successional stages of prairie vegetation, and they depend on the burrowing and foraging activities of other colony members to maintain habitat suitability (King 1955, 1959; Koford 1958). Unoccupied areas are covered by taller vegetation, and colonization of these areas requires considerable clipping of tall vegetation, along with the need to dig new burrow systems. Clipping of vegetation is important, both to increase visibility of mammalian predators and to facilitate growth of weedy food plants. Colonizing the fringes of an existing colony is probably easier than colonizing more distant areas because vegetation is already clipped in adjacent areas and burrows can be dug before the home territory is entirely abandoned. These advantages alone would create an impetus for colonial breeding. Thus prairie dogs may be colonial because the costs of colonizing distant habitats are greater than the costs of remaining in an established colony.

Mutual defense hypothesis

Several cooperative defenses against predation could make breeding or roosting in colonies beneficial. Coloniality may allow individuals to detect predators more quickly, mob predators more effectively, or swamp the ability of predators to exploit all available prey. One problem in evaluating the mutual defense hypothesis is that cooperative defenses may have evolved as secondary adaptations after coloniality evolved. A similar problem was encountered earlier while discussing the information center hypothesis.

Some evidence indicates that mutual vigilance for predators may be one advantage favoring coloniality in prairie dogs, ground squirrels, and marmots (Hoogland 1979b, 1980b; Wittenberger 1979a, 1980a). Females of most colonial rodents warn others in the colony by adopting conspicuous alert postures and giving specialized alarm calls (King 1955; Armitage 1962; Waring 1970; Carl 1971; Smith et al. 1973; Nel 1975; Dunford 1977b; Sherman 1977; Owings and Virginia 1978). Prairie dogs spend less time being vigilant in larger and denser wards than in smaller and less dense wards (Hoogland 1979b, 1980b). Vigilance is also more effective in larger wards, as measured by the time elapsing between presentation of a stuffed mammalian predator and the first alarm call (Figure 8–4).

Olympic marmots, which breed as dispersed family groups, spend less time on vigilance when foraging in groups than when foraging alone (Barash 1973). Similarly, female yellow-bellied marmots spend less time on vigilance when breeding in colonies than when breeding in isolated areas (Svendsen 1974). Relying on the vigilance of other colony members may therefore allow more time for foraging, as explained in Chapter 4 (see pp. 122–123).

Armitage and Downhower (1974) observed no predation in colonies of yellow-bellied marmots but found some evidence for predation on isolated individuals, suggesting that individuals are safer in colonies. However, this difference could be an artifact resulting from the more frequent

Figure 8–4

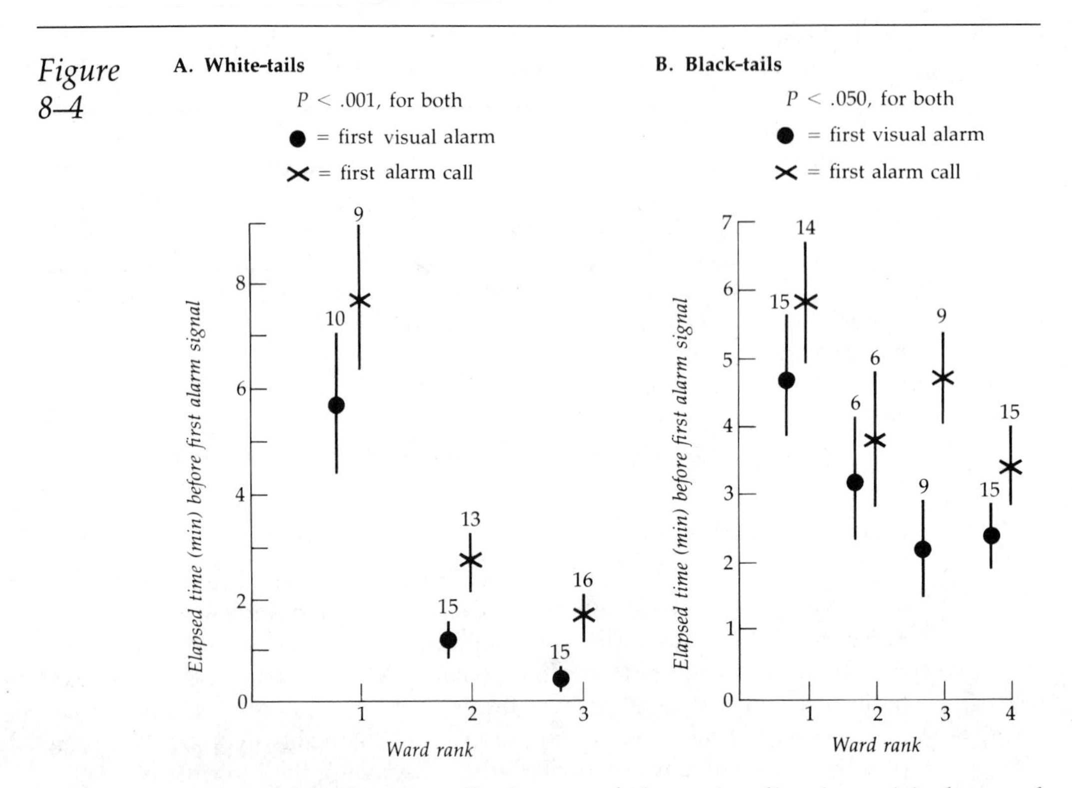

Relationships between effectiveness of alarm signaling in prairie dogs and ward (colony subsection) size.

Ward rank is a relative measure of ward size. Numbers over each plotted point represent sample size. The data show the average amount of time elapsing from the moment a stuffed badger was placed on a ward to the first alert posturing and alarm calling by prairie dogs in the ward. Quicker response times enhance the ability of prairie dogs to escape intruding predators. Since response times decrease as ward size increases, larger wards should confer greater protection from predators than smaller wards. A. Results for white-tailed prairie dogs. B. Results for black-tailed prairie dogs. Vertical lines surrounding each point represent standard errors.

Source: From J. L. Hoogland, "The Evolution of Coloniality in White-tailed and Black-tailed Prairie Dogs (Scuridae: *Cynomys leucurus* and *C. ludovicianus*)," *Ecological Monographs* (1980), in press. Reprinted by permission of Duke University Press. Copyright 1980 by the Ecological Society of America.

presence of human observers near the colonies they studied.

Seasonal changes in spacing behavior suggest that mutual vigilance is one factor favoring coloniality in Arctic ground squirrels (Carl 1971). Arctic ground squirrels are colonial from June to August when their main predator is the red fox *(Vulpes fulva)*, but they become solitary from August to November when their main predator is the grizzly bear *(Ursus chelan)*. Foxes are ambush hunters best deterred by effective vigilance, while grizzly bears are burrow excavators best eluded by avoiding detection.

Clustering to conserve heat

Various birds and mammals roost or hibernate in small groups to conserve heat. Heat loss is proportional to the surface-to-volume ratio of an irradiating heat source, and by clustering together, animals can reduce their surface-to-volume ratio.

Some songbirds in temperate regions sleep in clusters during cold winter nights (Photo 8–5) (Armstrong 1955; Löhrl 1955; Welty 1975). Such clusters usually involve species with small body size and, hence, high surface-to-volume ratios. They usually contain a dozen or so individuals although Tucker (1943) found one cluster containing over 50 long-tailed tits *(Aegithalos caudatus)*. Similar sleeping clusters have also been reported among tropical birds (e.g., Pycraft 1910; Skutch 1944a). Prevention of heat loss can be equally important in the tropics because nighttime temperatures often drop considerably below daytime temperatures.

Prevention of heat and water loss is at least one advantage of clustering in bats, particularly among temperate species that hibernate during winter (Bradbury 1977a). Although most temperate zone bats hibernate in clusters, a few, such as the lesser horseshoe bat *(Rhinolophus hipposideros)*

Photo 8–5

*A sleeping cluster of about 15 European tree creepers(*Certhia familiaris*).*

A variety of small songbirds sleep in clusters on cold winter nights to conserve body heat. Photo by Hans Löhrl.

(Matthews 1937; Daan and Wickers 1968), hibernate solitarily, suggesting that other factors can outweigh the thermoregulatory advantages of clustering. Among tropical bats predictably stable roost temperatures and temporary food shortages may induce clustering and torpor. The body temperatures of the spear-nosed bat *(Phyllostomus discolor)*, for example, drop lower when the bats are exposed to typical ambient temperatures than when exposed to unusually cold temperatures (McNab 1969). Moreover, body temperatures of central individuals inside a cluster drop lower than those of individuals on the periphery, apparently because central individuals experience more stable temperatures than peripheral individuals. In the long-tongued bat *(Glossophaga soricinia)* a single night of food deprivation induces clustering and torpor (Rasweiler 1973). These responses reduce metabolic energy requirements and presumably help the bats survive temporary food shortages. The thermoregulatory advantages of clustering help explain why bats cluster together in small daytime roosts or form clusters within larger colonies, but they do not explain why spatially segregated clusters clump together to form large roosting colonies.

Effects of predation on colony organization

Increased density

A common attribute of breeding colonies is that not all suitable space in and around them is occupied. Many tropical seabirds, particularly those nesting on flat terrain, breed at exceedingly high densities of up to several nests per square meter but occupy only part of the available space (Nelson 1970). Similarly, elephant seals crowd together into dense pods on breeding rookeries rather than disperse more widely across beaches (Bartholomew 1952). Cliff swallows *(Petrochelidon pyrrhonota)* are restricted to nesting on a limited number of cliffs, but many portions of the available cliffs contain no nests. As a general rule, breeding densities within colonies are much greater than would be expected by uniform or random spacing throughout the available space, even though prior residents often aggressively resist dense packing.

One hypothesis for dense packing in colonies is that peripheral nests are more vulnerable to predation than central ones (Tenaza 1971). Predators encounter peripheral nests first, and they are exposed to less severe mobbing on the periphery because fewer individuals are recruited to mob them. Tenaza's hypothesis is similar to Hamilton's (1971) shielding hypothesis for fish schools and other social groups, but it differs in that prey occupy fixed locations in colonies, unlike the constant changing of positions within social groups.

Peripheral nests are more vulnerable to predation than central nests in adelie penguins *(Pygoscelis adeliae)* (Eklund 1961; Taylor 1962; Reid 1964; Penney 1968), cattle egrets *(Ardeola ibis)* (Siegfried 1972), white pelicans (Schaller 1964; but see also Knopf, 1979), black-headed gulls *(Larus ridi-*

bundus) (Patterson 1965), kittiwake gulls (Coulson 1968), sooty terns *(Sterna fuscata)* (Feare 1976), cliff swallows (Emlen 1952a), bank swallows (Emlen 1971; Hoogland and Sherman 1976), piñon jays (Balda and Bateman 1972), and Brewer's blackbirds (Horn 1968). Thus, for at least some species, packing into the center of a colony confers protection for eggs or chicks. This advantage leads to competition for centrally located nest sites, which in turn causes dense packing of nests in the colony center.

Tenaza's hypothesis is not tenable for all colonial birds. In gannet colonies peripheral nests are not more vulnerable to predators than central nests (Nelson 1970). Indeed, gannets have no predators other than humans because they nest on top of extremely inaccessible rock stacks (Nelson 1966a, b). Nevertheless, gannets nest at very high densities, and late arrivals invariably try to crowd between established pairs instead of placing themselves a meter or two away where they could avoid much of the aggression directed at them. Nelson (1970) writes:

> *Successful nesting depends on a site large enough to accommodate two large adults and their chick, to provide an adequate landing space for a heavy bird and to permit displays. Encroachment on this minimum is bitterly resisted and a gannetry takes on its characteristic quality of orderly density allied with intense hostility. Territory is established and maintained by strenuous and damaging fighting, usually required at least once, and often very much more, from every male, and by extremely frequent, vigorous and season-long display.*

Clearly, crowding into the center of a colony instead of settling on the periphery entails a high cost, and yet peripheral nests are just as successful as central ones. The benefits gained by crowding into the center of a colony are therefore far from clear for gannets.

Increased breeding synchrony: The Fraser Darling effect

The hypothesis invoked by Nelson (1970) to explain high breeding densities in gannet colonies is that social stimulation from neighboring pairs induces earlier breeding and increased breeding synchrony, which in turn lead to higher reproductive success. This hypothesis was first proposed by F. Fraser Darling (1938) to explain coloniality in gulls and is commonly referred to as the **Fraser Darling effect.** The hypothesis hinges on three points: (1) Does local breeding synchrony occur in colonies? (2) Does earlier breeding and increased breeding synchrony lead to higher nesting success? (3) Does increased colony size or nest density stimulate earlier breeding and increased breeding synchrony?

Evidence for local breeding synchrony. To support the first point, Darling (1938) pointed out that courtship behavior in colonial gulls is infectious and often spreads rapidly among many pairs in a colony, a fact since confirmed by more recent studies (e.g., Hickling 1959; Brown and Baird 1965). In addition, copulations within a colony often occur synchronously, at least on a local scale (Brown 1967b). Contagious mating of this sort should increase breeding synchrony.

Local breeding synchrony within subsections of colonies has been documented for white pelicans (Behle 1944; Schaller 1964), herring gulls (Paynter 1949; Parsons 1976), brown-hooded gulls *(Larus maculipennis)* (Burger 1974a), swallow-tailed gulls *(L. furcatus)* (Hailman 1964; Snow and Snow 1967; Nelson 1968), sooty terns (Feare 1976), sandwich terns *(Sterna sandvicensis)* (Veen 1977), and black skimmers *(Rhynchops nigra)* (Gochfeld 1979b). Among songbirds local breeding synchrony has been demonstrated in cliff swallows (Emlen 1952a), bank swallows (Hoogland and Sherman 1976), black-headed weaver finches *(Ploceus cucullatus)* (Hall 1970), and tricolored blackbirds (Lack and Emlen 1939; Orians 1961b). In great frigatebirds *(Fregata minor)* pairs display in groups away from their nesting colonies, and these groups later give rise to synchronized nesting groups within colonies (Nelson 1970).

Breeding synchrony and nesting success. Increased breeding synchrony could enhance reproductive success by swamping the ability of predators to exploit the sudden surge in abundance of eggs or chicks (see pp. 226–228). It could also increase reproductive success if peak food abundance is of short duration or if unpredictable weather conditions later in the season affect postfledging survival of late-hatching chicks. In black-headed gulls, sooty terns, and sandwich terns a smaller proportion of nests are taken by predators during peak nesting periods (Figure 8–5) (Patterson 1965; Feare 1976; Veen 1977), while in several other gulls chicks are less frequently cannibalized during peak nesting periods (Weidmann 1956; Brown 1967a; Parsons 1971). In great frigatebirds asynchronous breeding reduces egg and chick survival because nonbreeders interfere with the reproductive activities of breeders (Nelson 1968). Finally, in gannets and common puffins *(Fratercula arctica)* late-hatching chicks survive less well because food abundance declines after the breeding peak (Nelson 1970).

Correlations between earliness or synchrony of breeding and nest success do not necessarily imply a causal relationship. Late breeders are often younger and may be less successful because they are inexperienced, not because they bred later in the season (Lack 1943; Armstrong 1947). Younger age is at least partially responsible for the lower success of late breeders in colonies of kittiwake gulls and shag cormorants *(Phalacrocorax aristotelis)* (Coulson and White 1956, 1960; Snow 1960). Age may also be an important factor in storm petrels *(Oceanodroma castro)* (Harris 1969). Late breeders may be less successful because they are newly established pairs rather than pairs that were formed in a previous year. Experienced gull pairs are more successful than newly established pairs, and they also breed earlier in the season (Coulson 1966; Mills 1973). Perhaps their higher success does result from earlier breeding, but it may result instead from their greater previous experience together as a pair. Late breeders may also be

Figure 8–5

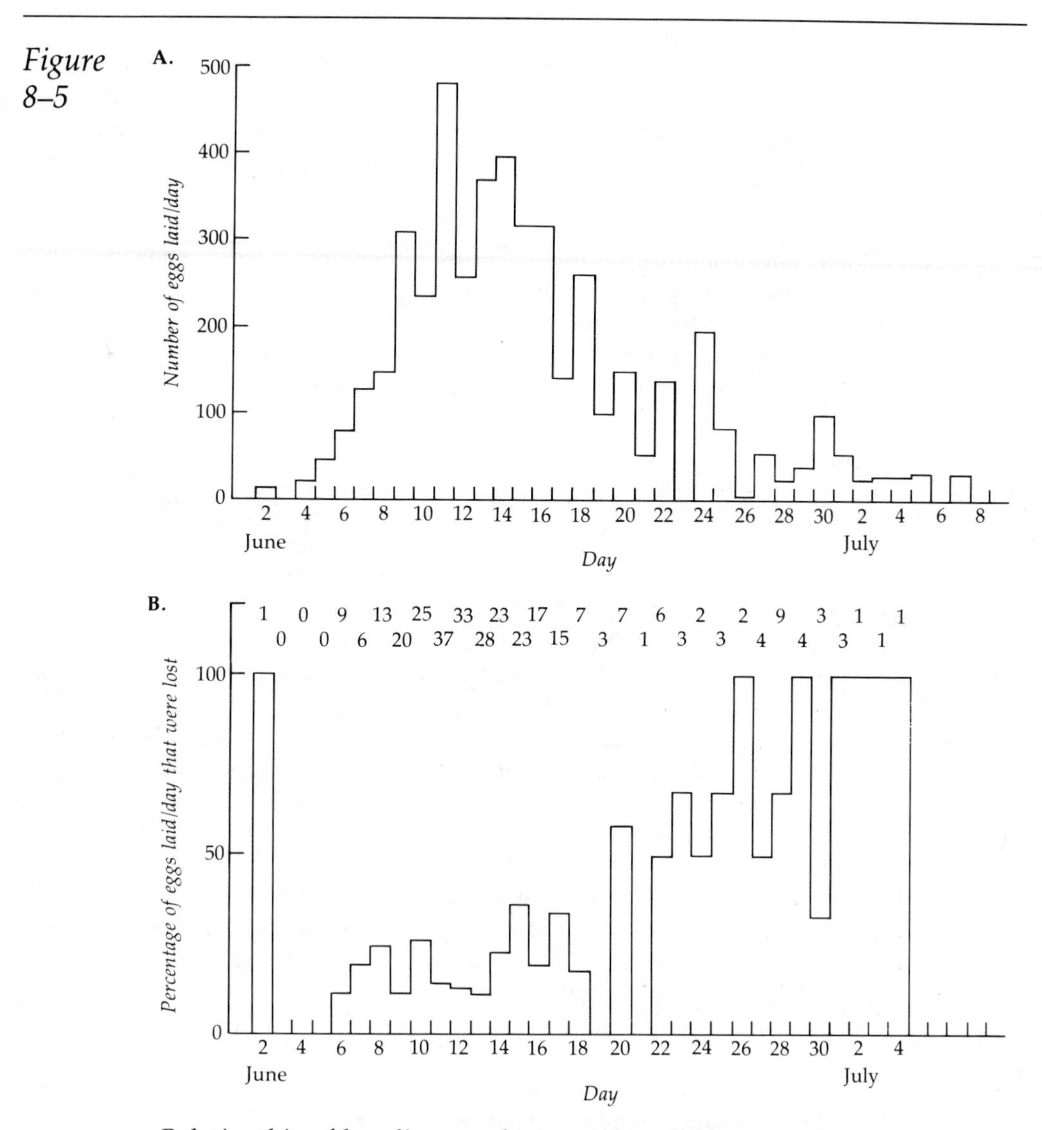

Relationship of breeding synchrony and nesting success in sooty terns.

A. The number of eggs laid per day within the study colony on the Seychelles Islands. B. The percentage of eggs laid each day that were lost to predators. Numbers at top of histogram refer to the number of eggs on which egg mortality data were based. By comparing the top and bottom graphs one can readily observe that more eggs were lost when laid on days when few eggs were laid than were lost on days when many eggs were laid. That is, synchronous laying was associated with a lower rate of egg loss.

Source: From C. J. Feare, "The Breeding of the Sooty Tern *Sterna fuscata* in the Seychelles and the Effects of Experimental Removal of Eggs," *Journal of Zoology, London* 179 (1976):317–360. Reprinted by permission.

less successful because they are nesting for the second time that season and have already suffered an energetic drain from their first nesting attempt (Patterson 1965). When degree of breeding synchrony is correlated with colony size, less synchronous breeders may be less successful because they are in smaller colonies, where a higher proportion of all nests are on the periphery (Figure 8–6, Photo 8–6) (Tenaza 1971). Similarly, higher nest success in dense colonies may result from closer spacing of nests and hence a lower proportion of peripheral nests rather than from greater breeding synchrony. Finally, colonies exhibiting higher breeding synchrony may be more successful because more food is available or because nest site quality is higher, not because synchronous breeding confers an advantage (Orians 1961b; Robertson 1973). Many factors contribute to nest success in colonial birds, and, consequently, the importance of any one factor is difficult to demonstrate.

Synchronous breeding may be advantageous for some species but not for others. The adaptive value of synchronous breeding is difficult to evaluate for any given species because many variables are involved. Individuals may benefit from breeding more synchronously because offspring are less vulnerable to predation or cannibalism or because food is only abundant for a brief period. On the other hand, individuals may not benefit if synchronous breeding intensifies competition for food and nest sites. The trade-off between these benefits and costs should determine when each individual breeds. If benefits exceed costs for most individuals, breeding synchrony should be high. If not, breeding synchrony should be low.

The best time to breed is likely to vary among individuals. Young individuals and recently formed pairs may be more successful by starting relatively late because they are less able to compete for food or nest sites at the breeding peak. Breeding synchrony is an aggregate of many individual nesting cycles and should be analyzed as an array of individual choices. No study has yet determined what the optimal breeding time is for each class of breeding pair within a colony; nor has any study determined whether pairs adjust their breeding schedule in light of that optimum.

Relationships among colony size, nesting density, and breeding synchrony. Current evidence suggests that increased nesting density or colony size leads to increased breeding synchrony and/or earlier laying in some colonial birds, though not in others. Breeding is more synchronous in the denser parts of gannet colonies and for larger breeding groups within a colony (Nelson 1966a). Laying is earlier in larger colonies of herring gulls, and breeding is more synchronous until colony size reaches about 250 pairs (Burger 1979). In larger colonies breeding is not more synchronous overall, but it probably is more synchronous within subsections of the colony. Laying is earlier in larger and denser colonies of kittiwake gulls (Coulson and White 1956, 1958, 1960, 1961; Coulson 1966). Overall breeding syn-

Figure 8–6

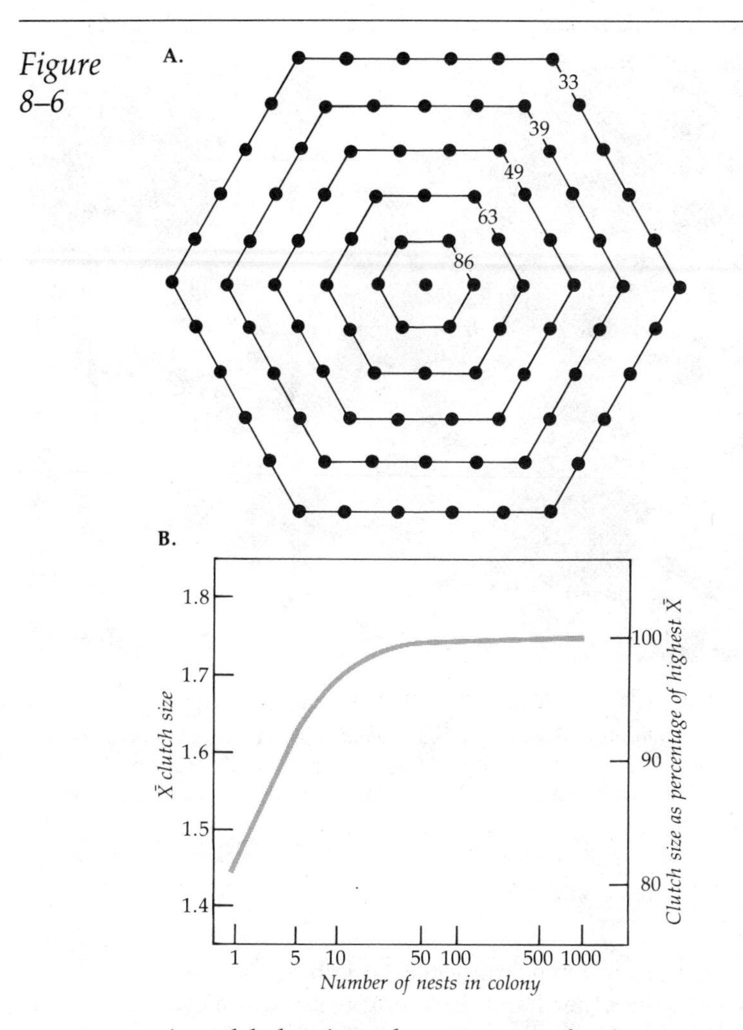

A model showing why mean reproductive success increases with colony size when peripheral nests are less successful than central nests in a breeding colony.

A. Each concentric hexagon represents the periphery of a hypothetical colony with maximum nest density. Numbers indicate the proportion of all colony nests that are located on the periphery for colonies bounded by each size of hexagon. Note that a lower proportion of nests are peripheral when colony size is larger. B. A hypothetical curve showing how mean clutch size ($\bar{X}$) in a colony should increase with colony size. The curve is obtained by multiplying mean clutch sizes of peripheral and central nesters by the proportion of each type of nester in the colony. The value for colonies of size 1 is based on mean clutch size of isolated nesters. For adelie penguins mean clutch size was 1.76 eggs for central nesters, 1.69 eggs for peripheral nesters, and 1.45 eggs for isolated nesters.

Source: From R. Tenaza, "Behavior and Nesting Success Relative to Nest Location in Adelie Penguins *(Pygoscelis adeliae)*," *Condor* 73 (1971):85. Reprinted by permission.

Photo 8–6

Nesting adelie penguins in the center of a breeding colony.
Note that the penguin indicated by the arrow is surrounded by six neighbors in an approximately hexagonal pattern. This configuration is the basis for Tenaza's model explaining why nesting success increases in larger colonies (compare with Figure 8–6).
Photo by Richard R. Tenaza.

chrony is lower, but local synchrony within subsections of colonies was not evaluated. On the other hand, breeding is not more synchronous in the denser parts of lesser black-backed gull *(Larus fuscus)* colonies (Weidmann 1956), glaucous-winged gull *(L. glaucescens)* colonies (Vermeer 1963), or mixed herring gull and lesser black-backed gull colonies (MacRoberts and MacRoberts 1972). Breeding is also not more synchronous in larger colonies of silver grebes *(Podiceps occidentalis)* (Burger 1974b), Rolland's grebes *(Rollandia rolland)* (Burger 1974b), white pelicans (Knopf 1979), sandwich terns (Langham 1974), red-winged blackbirds (Mayr 1941; Smith 1943; Orians 1961b), or tricolored blackbirds (Orians 1961b).

Correlations between increased breeding synchrony and larger colony size or greater nest density must be interpreted cautiously. Larger or denser colonies may contain more old individuals than do smaller or less dense colonies, and older individuals are likely to breed earlier or more synchronously than are inexperienced first-time breeders (Lack 1943b; Armstrong 1955; Coulson and White 1956, 1958, 1960; Hailman 1964). In addition, since breeding can be synchronized by numerous external factors, including day length, weather conditions, and resource availability (Immelmann

1973), comparisons based on geographically separated colonies must control for these variables. The work with gannets, herring gulls, and kittiwakes has taken these factors into account, and they provide the best evidence that increased breeding density leads to earlier or more synchronous breeding.

Social stimulation and breeding synchrony. Two experimental studies show that social stimulation could increase breeding synchrony. Introducing a third individual of either sex into the cage of a breeding pair induces ovulation in budgerigars *(Melopstittacus undulatus)* (Brockway 1964), and sounds from the other breeding pairs in nearby cages accelerate ovarian development in female ring doves (Lott et al. 1967). These studies demonstrate that a physiological basis for social stimulation exists, but they do not show that high breeding density is more stimulating than low breeding density within a colony.

The proximate mechanisms responsible for synchronous breeding are difficult to identify. Although social stimulation is one possible mechanism, present evidence does not prove its importance in actual breeding colonies. External factors may well synchronize breeding events of entire colonies. However, such factors seem less important in generating local synchrony within colonies. Although local factors such as tidal rhythms or vegetational structure may sometimes promote local breeding synchrony by making certain subsections of a colony better than others at any given time, such factors can often be ruled out (see Burger 1979). Thus the existence of local breeding synchrony within colonies suggests that the Fraser Darling effect is important, at least in seabird colonies. Its importance in landbird colonies remains to be established.

Massed flights and choruses

Birds that roost in large colonies do not just fly to the roost in late afternoon or early evening. They typically gather together in open assembly areas or circle continually over the roost site. Usually they produce a nearly continuous cacophony of sounds, which crescendos shortly before the birds enter their final roost. Massed flights above assembly areas are common, though individuals often vacillate considerably before joining the flights. Frequently these massed aerial maneuvers eventually culminate in a final en masse plunge into the roost just before dusk.

The excitement evident in preroosting flocks of starlings is well described by Wynne-Edwards (1962), who wrote:

> *Going to roost is accompanied by excited community singing, alternating with bickering and rivalry over perch sites. Moreover, on fine evenings, especially early in the autumn, either a part or sometimes the whole of the noisy company not infrequently rises with a great roar of wings to engage in the most impressive aerial maneuvers over the site: the massed flock may extend in a tight formation sometimes hundreds of yards in length, changing shape and direction like a giant amoeba silhouetted against the sky.*

Similar behavior, with minor variations, occurs in most other colonially roosting birds and is observable by visiting a blackbird roost shortly before dusk.

Wynne-Edwards (1962) interpreted preroosting flights and choruses as epideictic displays for assessing population density. Since his views depend on interdemic selection, which has been criticized on theoretical grounds (see pp. 63–64), this hypothesis seems untenable. An alternative hypothesis is that massed flights and choruses are tactics for reducing vulnerability to predation at the roost.

Individuals risk encountering a hidden predator whenever they enter dense foliage, and the safest way to enter such areas is to follow behind another individual (Hamilton 1971). The probability of a predator lying in wait at a roost site may seem low, but the costs of waiting for another colony member to enter the roost may well be even lower. There is certainly some risk involved in entering a roost, as evidenced by a Cooper's hawk (*Accipiter cooperii*) I once observed regularly waiting at dusk inside a small Brewer's blackbird roost in Stockton, California.

According to Hamilton's (1971) "you-first" hypothesis, individuals congregate in preroosting assembly areas to wait until other individuals venture into the roost. Evening choruses in preroosting assembly areas may serve to attract additional birds to the assembly area, thereby spreading the risk of predation during the waiting period across more individuals and increasing the number of individuals that might enter the roost first. Massed flights to and fro just before dusk may represent vacillations in which every individual is following the others with the expectation that eventually some will descend into the roost. They may also reduce vulnerability to predation while waiting to enter the roost. Individuals may often hesitate in joining massed flights because there is some small risk involved or because it represents an energy drain.

Descriptions of birds entering roosts fit the you-first hypothesis very well. Rudebeck (1955), for example, describes the preroosting flight of swallows as follows:

> *With incredible agility the birds were moving promiscuously around each other against the rapidly darkening sky. . . . Finally, the birds alighted in the reeds. Some few of them dropped suddenly, and then it was a question of* seconds [*my emphasis*] *until the others followed. Like a torrential shower of rain they hurled themselves down more or less vertically, and disappeared instantaneously. . . . Within a short time—I would estimate it at two or three minutes . . .—all the swallows* [*an estimated one million of them!*] *descended.*

A description of chimney swifts *(Chaetura pelagica)* entering a roost is remarkably similar (Pickens 1935):

> *Sometimes small groups passing above the chimney drop toward its mouth in a gesture of seeming salute but the urge of the ring above seems to draw them back. It is dip, dodge and pass on. At last, however, with increasing darkness, some drop from the ring and settle within the chimney and a living line follows like a thread from a rapidly revolving spool or reel.*

Such observations are typical of colonial landbirds that roost in dense cover. The rapid descent of so many birds after the first few individuals enter the roost certainly suggests that they were all waiting for some individual to enter the roost first.

Massed flights into and away from roosts probably reduce vulnerability to predation by swamping the ability of predators to fully exploit all available prey. The reality of the threat to both birds and bats is made clear by a few examples. Accipiter hawks and peregrine falcons (*Falco peregrinus*) aggregate above the bat roost at Ney Cave and kill bats as fast as they can eat them for the five hours it takes the colony to depart each evening (Moore 1948). Several birds of prey commonly visit the bat caves studied by Davis and co-workers (1962) in Texas, including great horned owls (*Bubo virginianus*), red-tailed hawks, and Cooper's hawks. Only one or two of each were seen at a cave at any one time, but their regular appearance represented a clear threat to the bats. Several species of hawks hunt around dickcissel roosts in Trinidad, and one species, the merlin, apparently specializes on dickcissels going to roost (ffrench 1967). In tropical Africa and Southeast Asia a specialized crepuscular hawk, the bat-eating buzzard (*Machaeramphus alcinus*), subsists almost entirely on bats and swifts that share roosting caves on an alternating schedule (Pryer 1884). Roosting colonies represent a reliable source of abundant prey, and it is not surprising that some predators take advantage of them.

Swamping predators may be one advantage of arriving and departing synchronously, but it is not necessarily the only one. Roosting individuals may exhibit synchronous movements because they respond to the same external cues, namely, sunrise and sunset, to fully utilize daylight or nighttime hours. Nevertheless, the fact that individuals often vary in their arrival or departure times and sometimes vacillate before joining waves departing from roosts suggests that more than just external timing factors are involved.

Effects of competition on colony organization

Competition may influence the distances individuals travel to find food, and it may be an important factor underlying the mass abandonment of colonies. To understand how competition affects the behavior of colonial animals, some new theory is required. The theory is basically one of economics, and it can be formulated in terms of either energy or money.

Central-place systems

Most colonial animals, along with many noncolonial animals, utilize space in relation to a fixed point in space. Such **central-place systems** are characterized by the rhythmical movement of individuals or groups to and from a fixed location (Lösch 1954; Hamilton and Watt 1970; Covich 1976; Orians and Pearson 1978). Horn's (1968) geometrical model for colonial breeding in the center of a foraging arena is a simple example of a central-place system. Other examples of central-place systems are social insect colonies, vertebrate breeding or roosting colonies, nests of noncolonial birds, dens of noncolonial mammals, and human cities.

Central-place systems are recognizable by their geometry (Figure 8–7) (Hamilton and Watt 1970). The *core area* is where individuals in the system focus their activities. It encompasses the most intensively utilized part of the system and may include nest sites, denning sites, sleeping sites, or water holes. The degree of environmental modification or resource depletion in a core area depends on the density of individuals using it. Areas surrounding the nests or dens of noncolonial animals are often hardly modified at all, while areas under large bird and bat roosts contain large accumulations of droppings (e.g., Loefer and Patten 1941; Schorger 1955; Davis et al. 1962) and human cities cover the landscape with pavement and buildings.

The *trampling zone* is an area where resource availability is depressed by pathways leading to and from the core area. Since traffic is most concentrated near the core area, trampling effects are accentuated there. The trampling zone is minimal for animals that fly unless significant amounts of droppings accumulate along flight paths. The trampling zone of mammals includes game trails and denuded areas around water holes and sleeping sites. In human technological societies it is especially large and includes the extensive road networks, roadside developments, railroads, canals, and power lines that interconnect cities and towns.

The *biodeterioration zone* includes the regions surrounding a core area where resource availability has been depressed. It is an area where desirable food items have been heavily depleted. In grazing animals it consists of areas where overgrazing has led to replacement of palatable plant species with unpalatable ones. In humans it includes all overexploited and polluted areas outside residential and industrial districts. The biodeterioration zone blends imperceptibly into the principal resource acquisition zone or *arena*. In human technological societies most exploited areas lie within the biodeterioration zone, and a distinctive arena zone may be largely absent.

Competition and use of space

The biodeterioration zone expands outward as the number of individuals using a core area increases and as slowly renewing resources become depleted. Individuals must therefore travel further to find food as colonies become larger and as resource availability near the colony becomes depressed. However, not all individuals travel the same distance from a colony to forage. Some individuals travel much farther than others, and by doing so they expend considerably more time and energy to reach their foraging areas. These costs seem to place individuals traveling far from the colony at a disadvantage, which raises the question of how some individuals can afford to forage further from the colony than others.

Figure 8–7

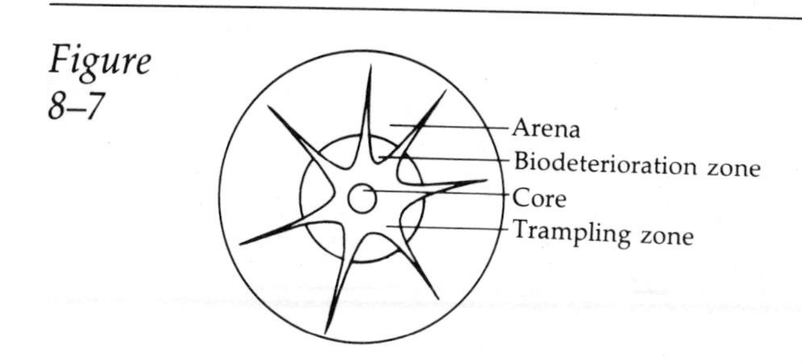

Schematic diagram of a central-place system.

The various components of the system are described in the text. The relative size and shape of each component depend on the biology of the species involved and on the number of individuals utilizing the system.

Source: From W. J. Hamilton, III and K. E. F. Watt, "Refuging." Reproduced, with permission, from the *Annual Review of Ecology and Systematics*, Volume 1, p. 264. © 1970 by Annual Reviews Inc.

One hypothesis for answering this question is that all travel distances are energetically equivalent (Hamilton et al. 1967). If competition for food decreases with increasing distance from the core area, the higher energetic costs of traveling to more distant feeding grounds can be offset by opportunities for acquiring food at a faster rate. Individuals might then obtain the same net energy gain from a foraging trip no matter how far they travel from the core area (up to some outer limit).

Hamilton and co-workers (1967) tested their hypothesis by estimating the relative energetic value of foraging at different distances from a large starling roost in central California. They calculated an index of net energy gain for each distance by estimating energy expenditures during flight, the density of starlings exploiting potential foraging habitats at each distance, and the relative availability and quality of each type of potential foraging habitat. Despite their rather crude estimates, the results show that the net energy gain achieved at different distances from the roost is relatively constant for distances up to 40 miles, the maximum distance flown by starlings from the study roost (Table 8–3). The data suggest that more starlings should have foraged 10 miles from the roost than were actually observed, since net energy gain appeared somewhat higher at that distance, but Hamilton and his colleagues believe they underestimated competition intensity for that distance.

The hypothesis was tested further by measuring the net energy return of honeybees foraging at different distances from a hive (Gilbert 1975). An experimental hive was placed in vegetationless desert, where the bees were obliged to forage entirely at artificial feeders, and special feeding trays were designed in such a way that food resources were depleted in proportion to the number of bees visiting the feeder. Gilbert placed feeding trays at 400, 800, and 1600 m from the hive and weighed the stomach contents of individually marked workers returning from feeders at each distance. The

Table 8–3 *Relative suitability of foraging habitat at various distances from a large starling roost in California.*

Distance from roost (mi)	Relative suitability
10	3.10
20	2.86
30	2.86
40	2.82
50	2.66
60	2.52

Note: Relative suitability is an indirect index of the net energetic gain achieved by flying a given distance from the roost.

Source: From W. J. Hamilton, III, et al., "Starling Roost Dispersal and a Hypothetical Mechanism Regulating Rhythmical Animal Movement to and from Dispersal Centers," *Ecology* 48 (1967):825–833. Reprinted by permission of Duke University Press. Copyright 1967, by the Ecological Society of America.

data show that bees visiting more distant feeders return with the same net caloric value of food as bees visiting feeders closer to the hive. This result is not obtained with the usual feeders used in bee studies because crowding around feeders prevents bees from depleting feeders near the hive enough to simulate competition under natural conditions.

Additional data come from a study by Boch (1956) in Germany, who measured stomach contents of bees foraging up to several kilometers from the hive. Gilbert (1975) reanalyzed Boch's data and showed that the net caloric return for bees feeding 2–6 km from the hive is essentially constant on any given day (Figure 8–8). The net caloric return of bees foraging at distances less than 2 km was much higher, probably because crowding around feeders prevented those feeders from being depleted. Thus under conditions of resource depletion, bees adjust their foraging distance to food availability in such a way that greater flight costs are compensated by increased food intake rates.

Hamilton and co-workers (1967) based their hypothesis solely on energetic considerations. They did not consider other potential costs, such as increased risks of predation while traveling to and from a food source and higher risks of nest predation or mate desertion caused by longer absences from nests. In some cases such costs may be more important than energetic costs, and they too must be compensated by relaxed competition or reduced predation pressure at more distant foraging grounds if the equivalence hypothesis is valid.

Mass abandonment of colonies

According to central-place theory, the foraging arena around a core area represents a **commons** (see Hardin 1968). That is, all individuals uti-

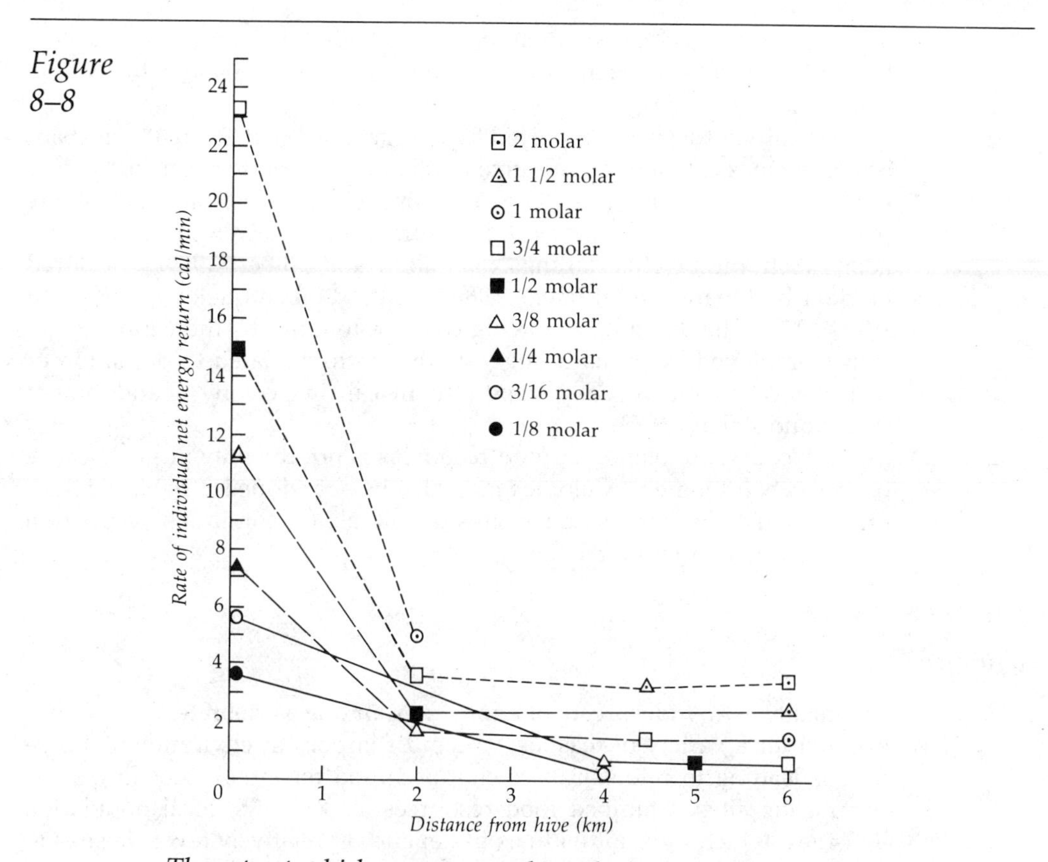

Figure 8–8

The rate at which sucrose was brought back to a hive (cal/min) by honeybees foraging at different distances from the hive.

Each set of connected points on the graph represents a series of feeders placed at varying distances along a single flight line. Sucrose concentrations, measured in terms of molarity, varied among feeders, but calculations of caloric delivery rates took this variability into account. Note that the net rate of energy return was constant in any given direction for distances between 2–6 km, but the return rate varied with direction from the hive.

Source: Redrawn from Gilbert 1975; based on data from R. Boch, "Die Tanze der Bienen bei nahen und fernen Trachquellen," *Zeitschrift für vergleichende Physiologie* 38 (1956):136–7. Reprinted by permission.

lizing the arena suffer equally from its continued exploitation, a conclusion that follows directly from the hypothesis that foraging success is energetically equivalent at all distances from the core area. This characteristic of central-place systems may explain the previously inexplicable mass movement of entire roosting colonies to new sites and the occasional mass abandonment of breeding colonies.

Mass action by colony members is understandable when one realizes that all individuals in the system are equally affected by resource depletion. Once the energetic costs of traveling to a foraging arena exceed the benefits for one individual, they exceed the benefits for all. Of course, differences in experience cause some variation among individuals, but nevertheless the commons effect should create a fairly discrete threshold for colony abandonment.

This hypothesis is supported by observations that colonies are abandoned en masse following resource depletion or periods of unusually low food availability in Peruvian boobies (Vogt, cited in Nelson 1970), white pelicans (Brown and Urban 1969; Johnson and Sloan 1978), passenger pigeons (Schorger 1955), starlings (Hamilton and Gilbert 1969), tricolored blackbirds (Orians 1961a; Payne 1969), and red-billed queleas (Jones and Ward 1979). The difficulty of forcing colonies to move by the various scare tactics employed by humans suggests that factors related to predation or disturbance are not usually important in causing colony abandonment (Ward and Zahavi 1973).

Of course, depletion of food resources is not the only possible cause of mass abandonment. Colonies may also be abandoned if habitat deterioration makes formerly suitable sites unsuitable for continued occupation (e.g., Morris and Hunter 1976).

Conclusion

Coloniality is a widespread phenomenon among vertebrates, and it has evolved for a variety of reasons. The most important environmental conditions leading to coloniality are clumped food resources, breeding sites, or roosting sites. Clumped food resources lead to high local population densities, which force individuals to depend on relatively few safe sites to breed or roost. A few animals must breed in colonies because terrestrial breeding combined with marine foraging greatly restricts the availability of suitable sites. Very abundant species may be unable to disperse sufficiently to rely on concealment, causing them to breed or roost in colonies as a defense against predation. Acquiring information about good foraging locations may be one advantage of coloniality, but this has not yet been shown to be an important selective factor. It remains very possible that acquiring information by following others evolved secondarily after other factors led to coloniality. The costs of coloniality are high, especially for members of large colonies, and the benefits must be correspondingly high before coloniality can evolve.

Once coloniality has evolved, predation and competition strongly affect colony organization and lead to a variety of behavioral tactics designed to ameliorate their effects. High vulnerability to predation leads to breeding in the most inaccessible or safest sites available. The higher vulnerability of peripheral nests favors dense packing of nests within the central parts of many bird colonies. Synchronous breeding is often an effective defense against predation but may also arise when breeding is limited by short seasons of food availability. In predatory and omnivorous birds cannibalism is often a profitable way to obtain food because vulnerable eggs or chicks are present at high densities nearby. The high risks of predation and cannibalism cause many parents to attend offspring continuously during early stages of development. Colony members often arrive or depart en masse to reduce risks of predation, especially at roosting colonies. Finally, individuals adjust travel distances and foraging tactics to avoid competition as much as possible. As a result, competition affects all colony members about equally, and when food resources become too depleted to sustain their needs, they often abandon the colony en masse. Thus coloniality is a complex form of social organization, with individuals adopting a variety of behavioral tactics to minimize the high costs associated with it.

Life history patterns and parental care

Animals have at their disposal only a finite amount of time, energy, and nutrients. These should be channeled into a combination of activities that best enhances the rate at which genes are propagated to the next generation. To that end, animals can use their time and resources in three basic ways: to enhance their own survival, to produce direct progeny or progeny of relatives at a faster rate, or to enhance survival and reproduction of progeny they have already produced. Since time and resources are limited, allocations made to any one type of activity necessarily reduces the time and resources left available for the other types of activities. A three-way trade-off exists, and natural selection should lead to an optimal balance among the three activities. Parental care is one way of enhancing offspring survival. Its evolution must therefore be couched within the broader context of survivorship and fecundity strategies. The way an animal apportions time and resources to survivorship, reproductive effort, and parental care is referred to as its **life history pattern.**

Life history patterns

Life tables

Patterns of time and resource allocation are reflected by survival and reproductive rates at each age. Age-specific survivorship and fecundity are analyzed by constructing a **life table,** which tabulates the proportion of individuals surviving to each age and the average reproductive rate achieved by those individuals. Complete data sets for constructing life tables are difficult to obtain, and in many cases data are available only for survivorship. The survivorship table constructed by Deevey (1947) for Dall sheep *(Ovis dalli)* provides a good example for illustrative purposes (Table 9–1). Data for the table were obtained by assessing age at death for 608 skeletal remains collected in Mount McKinley National Park by Murie (1944). The table shows that immature and very old sheep survive less well than sheep of intermediate ages, a pattern typical of large mammals. A complete life table could not be constructed for Dall sheep because no data were available on age-specific reproductive rates. Such data are much more difficult to obtain because live sheep cannot be aged without capturing them, and sheep of known ages must be studied intensively before reliable data on reproductive success can be obtained.

Chapter nine

Table 9–1 *Partial life table for Dall sheep constructed by assessing age at death of 608 skeletal remains.*

Age interval (years)	Number surviving at beginning of age interval out of 1000 born	Number dying in age interval out of 1000 born	Mortality rate per 1000 alive at beginning of age interval
0–1	1000	199	.199
1–2	801	12	.015
2–3	789	13	.017
3–4	776	12	.015
4–5	764	30	.039
5–6	734	46	.063
6–7	688	48	.070
7–8	640	69	.108
8–9	571	132	.231
9–10	439	187	.426
10–11	252	156	.619
11–12	96	90	.937
12–13	6	3	.500
13–14	3	3	1.000

Source: From E. S. Deevey, Jr., "Life Tables for Natural Populations of Animals," *Quarterly Review of Biology* 22 (1947):289. Reprinted by permission.

The most convenient way to visualize survivorship data is to plot them graphically. The survivorship curve for Dall sheep is shown in Figure 9–1, based on the data in Table 9–1. Note that survival rate is represented by the steepness of the curve. The steep slope for young and old age groups is indicative of the poorer survival occurring at those ages.

Survivorship patterns. Generalized survivorship curves can be classified into three types (Figure 9–2). Each type is typical of certain types of organisms, as will be explained below. These idealized curves resemble the survivorship patterns of many real populations, but intermediate curves can and do occur.

A type I survivorship curve arises when juvenile survival is relatively high and adult mortality is most prevalent among older age groups, after senescence has set in. Adults in prime condition have the highest survival rate because they are least susceptible to malnutrition, disease, and predation. Species with type I survivorship usually have low fecundity because they use most of their time and resources to promote adult and offspring survival. Type I survivorship is typical of large mammals such as Dall sheep.

A type II survivorship curve arises when juvenile survival is relatively

Figure 9–1

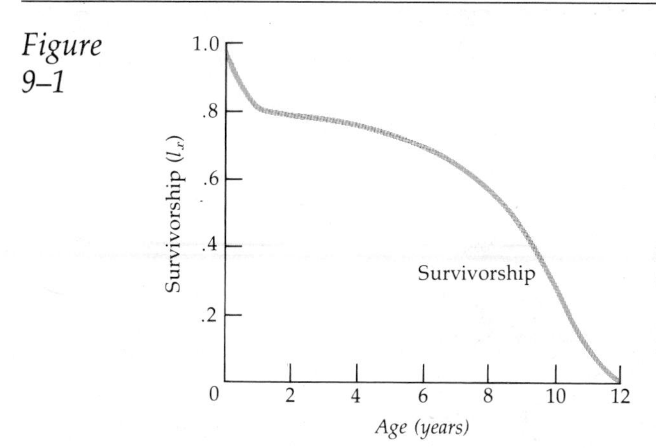

Age-specific survivorship curve for Dall sheep of Mount McKinley National Park, Alaska.

The curve gives the proportion of an initial age cohort surviving to each given age.

Source: Based on R. E. Ricklefs, *Ecology* (Portland, Oregon: Chiron Press, 1972), p. 395. (Data from Deevey 1947.)

high and adult mortality is nearly constant for all age groups. Usually, juvenile survivorship is lower than adult survivorship, creating a steep initial drop in the curve (not shown in Figure 9–2) before the curve levels off to a steady decline. Since adults of all ages are equally susceptible to malnutrition, disease, and predation, no well-defined period of senescence is evident. Species with type II survivorship again devote considerable time and resources to promoting adult and offspring survival, though perhaps they allocate proportionally more time and resources to fecundity than species with type I survivorship. Type II survivorship is typical of most birds but also occurs in a few other animals.

A type III survivorship curve arises when juvenile survival is very low and adult survival is relatively high. Juveniles are often propagules that disperse as seeds, eggs, larvae, or other early developmental stages, and their survivorship often depends on where they settle following dispersal. Predation and adverse consequences of competition for food or space affect juveniles much more than they affect adults. Species with type III survivorship allocate large amounts of time and resources to fecundity and use relatively little to promote offspring survival. Type III survivorship is typical of plants, invertebrates, and most lower vertebrates.

Fecundity patterns. The two basic fecundity patterns are repeated reproduction **(iteroparity)** and reproduction in one suicidal burst **(semelparity).** Semelparity, or as Gadgil and Bossert (1970) describe it, "big bang" reproduction, is well exemplified by migratory salmon, which live in the ocean until maturity and then migrate up their natal streams to spawn and die. The fecundity pattern of iteroparous animals can take several forms. If experience improves reproductive performance, young individuals reproduce at a lower rate than fully mature individuals. If not, reproductive rate is relatively constant for all age groups until senescence sets in. Senescence generally leads to a decreased reproductive rate. Experience is most important in birds and mammals, while senescence is important primarily in animals exhibiting type I survivorship.

Figure 9–2

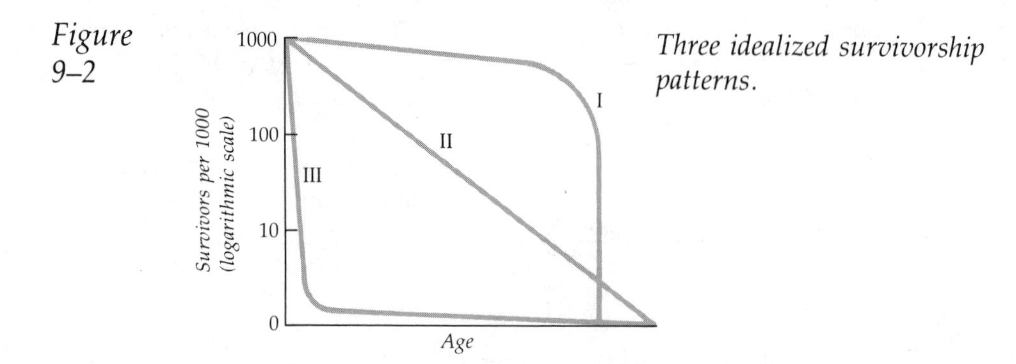

Three idealized survivorship patterns.

Type I survivorship arises when mortality is low until senescence sets in. Type II survivorship arises when mortality occurs at a constant rate regardless of age. Type III survivorship arises when juvenile mortality is high and adult mortality is low. In real populations juvenile mortality is typically higher than adult mortality. The curves for type I and type II survivorship do not reflect this characteristic, and a more realistic representation would be to include a steep dip in each curve for the youngest age groups (see Figure 9–1, for example).

Source: Modified from E. S. Deevey, Jr., "Life Tables for Natural Populations of Animals," *Quarterly Review of Biology* 22 (1947):285. Reprinted by permission.

Iteroparity is generally associated with high longevity, while semelparity is associated with low longevity (insects) or very high reproductive costs (salmon). Several mathematical models have been developed to explain the evolution of iteroparity and semelparity, but these are beyond the scope of the present discussion. Interested readers should refer to Stearns (1976).

Reproductive trade-offs

Trade-off between fecundity and growth. When adults grow continually throughout life, a pattern referred to as **indeterminate growth,** large individuals generally achieve higher fecundity than small individuals. For females larger individuals are typically more fecund because they have more abdominal space for carrying egg masses. For males larger individuals are typically more fecund either because they can mate with larger females or because they are better able to compete for matings. Thus a trade-off exists between immediate reproductive output and growth, which translates into greater reproductive output in the future provided that the animal survives. The amount of resources allocated to each alternative should depend on how good prospects are for continued survival (Williams 1966b). If chances for survival are high, more resources should be allocated to growth. If chances for survival are low, fewer resources should be allocated to growth and more should be allocated to immediate reproduction.

Evidence for a trade-off between fecundity and survival comes from several sources. Among 14 species of lizards, for example, the mean survival rate of adult females is negatively correlated with mean annual reproductive rate (Figure 9–3) (Tinkle 1969; see also Tinkle et al. 1970; Huey et al. 1974). This correlation suggests that either increased fecundity can be achieved only at the expense of reduced survival rate or increased survival rate can be achieved only at the expense of reduced fecundity. Many lizards build up fat reserves or grow larger during periods unfavorable for reproduction but not during periods favorable for reproduction (Licht and Gorman 1970; Sexton et al. 1971; Andrews and Rand 1973). This enables lizards to make the most of their resources at all times and implies that the energetic costs of reproduction preclude growth during breeding periods. Among desmognathine salamanders adult females survive less well than adult males (Organ 1961). Since immatures of both sexes survive about equally well, the lower survival of adult females probably stems from the greater time and energy expended by females on reproduction. Hence a trade-off seems to exist between reproductive effort and survivorship, with the optimal allocation for each activity differing for the two sexes.

Figure 9–3

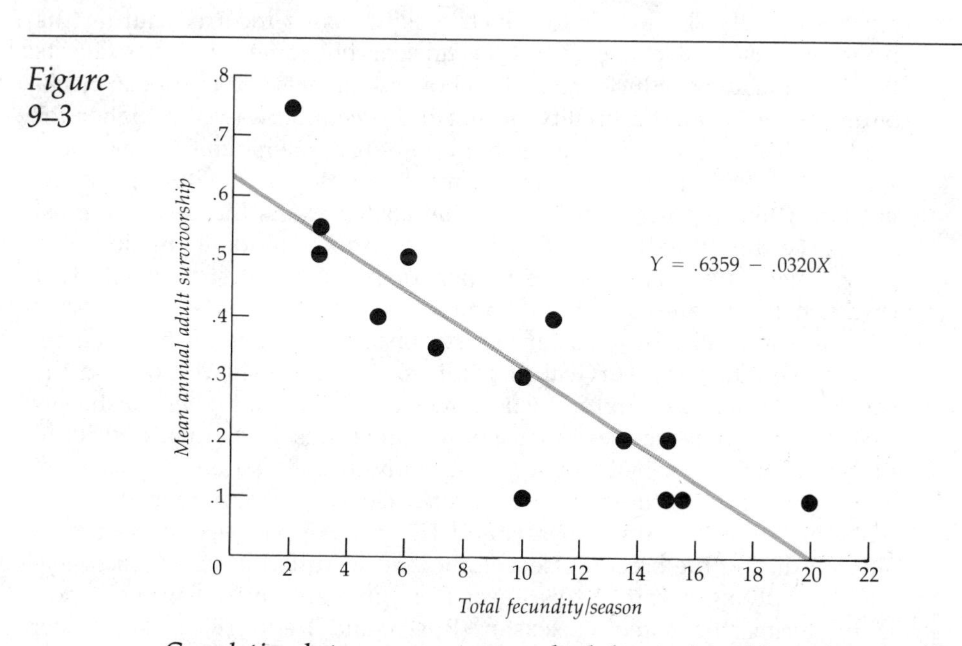

Correlation between mean annual adult survivorship and season-long fecundity among 14 species of lizards.

Each plotted point represents a different species.

Source: Reprinted from D. W. Tinkle, "The Concept of Reproductive Effort and its Relation to the Evolution of Life Histories of Lizards," *American Naturalist* 103 (1969), p. 508, by permission of The University of Chicago Press. Copyright © 1969 by The University of Chicago Press.

Trade-off between reproduction and self-maintenance. When adults grow to a predetermined size and then reproduce (a pattern which is referred to as **determinate growth**), increased reproductive effort reduces the resources left available for self-maintenance and exposes adults to greater risks of predation. Female snow buntings *(Plectrophenax nivalis)*, Lapland longspurs *(Calcarius lapponicus)*, and house martins *(Delichon urbica)*, for example, gradually lose weight during incubation and nestling periods because they expend large amounts of time and energy on parental care (Hussell 1972; Bryant 1979). Furthermore, females that rear large broods lose more weight than females that rear smaller broods. Male song sparrows and bullfinches *(Pyrrhula pyrrhula)* lose 5% to 10% of their body weight during the nestling period (Nice 1937; Newton 1966). Female blue grouse *(Dendragapus obscurus)* and ring-necked pheasants *(Phasianus colchicus)* lose nearly 20% of their body weight during incubation and do not recover it until after young reach independence (Breitenbach and Meyer 1959; Redfield 1973). Weight losses during incubation have also been documented in least auklets *(Aethia pusilla)* and several species of penguins (Richdale 1957; Bédard 1969b; Johnson and West 1973; Boersma 1976; Hunt 1980).

Such weight losses have little evolutionary significance unless they affect survival. The extent to which weight losses incurred during the breeding season affect later survival is unclear, but some evidence suggests that breeding does reduce survival. In house sparrows and European blackbirds *(Turdus merula)* mortality during the breeding season is higher for adults that breed than for adults that do not (Summers-Smith 1956; Snow 1958). In great tits and house martins most mortality occurs during winter, but nevertheless annual mortality is higher for adults that rear the most young (Perrins 1965; Bryant 1979). Similarly, annual mortality in king penguins *(Aptenodytes patagonica)* is higher for parents that successfully rear chicks than for parents that fail (Stonehouse 1960).

Sex ratio data provide further evidence for a trade-off between reproductive effort and survival. If adult sex ratios are skewed toward the sex that invests less in reproduction, mortality must be higher for the sex that invests more since sex ratios among immatures are normally 50:50. In ducks and grouse females rear young without male parental assistance, and adult sex ratios are skewed toward excess males (references in Wittenberger 1979a and Wittenberger and Tilson 1980). The sex ratios become skewed during the breeding season, at least in ruffed grouse *(Bonasa umbellus)*, implying that the skew arises from greater female than male mortality during the breeding season (Rusch and Keith 1971). The lower survival rate of females apparently results from the high cost of producing eggs and caring for young (Lack 1954; King 1973). In quail and some songbirds males contribute substantially less parental care than females, and again sex ratios are skewed toward excess males (references in Wittenberger and Tilson 1980).

A comparable trade-off has been documented for males. Male ruffed grouse who display for females on logs that are used year after year attract more mates than males who display on logs that are used only sporadically, but they also suffer higher mortality from predation (Gullion and Marshall 1968). Adult male elephant seals enjoy relatively high survival until they attain high dominance status in a breeding rookery, but once they attain high status, they rarely survive to breed the following year (Le Boeuf and Peterson 1969; Le Boeuf 1974). Dominant males obtain most of the females, but they sustain a severe energy drain while breeding because they cannot leave the rookery to forage. Living off stored body fat for several months apparently weakens them so much that they usually cannot survive the next year.

Trade-off between fecundity and parental care. The above evidence shows that a trade-off exists between immediate reproductive effort and continued survival. With respect to immediate reproductive effort, a second trade-off exists between fecundity and parental care. When offspring are given extensive parental care, less energy is available for producing eggs and fewer offspring can be cared for at one time. Consequently, extensive parental care is associated with reduced fecundity. For example, among marine bottom-dwelling molluscs, worms, crabs, sea urchins, and starfish, species that bear live young or provide extensive parental care lay 20 to 100 eggs at a time (Thorson 1950). Species that produce large eggs or provide some parental care lay 100 to 1,000 eggs at a time, and species that produce small eggs and provide no parental care lay 1,000 to 5,000,000 eggs at a time. Live-bearing fish produce as few as 1 to 4 young at a time, while egg-laying fish that provide no parental care produce up to 3,000,000 eggs at a time (Thibault and Schultz 1978). Females of some fish, amphibians, and reptiles reproduce only once every other year, and this reproductive schedule is most commonly associated with extended maternal brood care, live-bearing, or long migrations (Bull and Shine 1979). By not breeding in alternate years, females can more than double their fecundity in years when they do breed because the energy normally used for parental care can be allocated to growth or stored fat to be used the next year for egg production. Birds with precocial young contribute less parental care than birds with altricial young because precocial young hatch fully feathered and are soon capable of foraging for themselves, while altricial young are hatched naked and must be fed by their parents for prolonged periods. Clutch sizes of precocial birds are typically much larger than those of altricial birds, apparently because parents are not capable of caring for as many young at a time when their young are altricial (Lack 1954, 1966, 1968; Klomp 1970).

Optimizing the trade-offs

An animal's overall life history pattern apportions time and resources to fecundity, parental care, and survival, with the amounts allocated to each varying as a function of age. An optimal life history pattern is one that apportions time and energy to those ends in a way that best enhances fitness. A whole constellation of traits is affected by time and energy apportionments, including age of first reproductive effort, egg size, clutch size, frequency of reproduction, types and degrees of parental care, and tactics for enhancing survival. Since these traits often covary, current theories of life history tactics are structured on particular constellations of traits. The bases of current theory are two extreme life history patterns.

At one extreme, opportunistic species are specialized to live and reproduce in short-lived habitats with unpredictable locations, such as dung pats, pools of rainwater, or new forest clearings. The most successful individuals are those that discover new habitats quickly, reproduce rapidly before the habitat deteriorates in quality or competitors find it, and send out the most propagules to colonize new habitats (MacArthur and Wilson 1967; Wilson 1975a). Selection favors good dispersal ability, rapid reproduction, and high fecundity. Prolonged survival of adults has little value because suitable habitats are ephemeral, and once the habitat deteriorates, chances of finding another suitable habitat are minimal. The best way to colonize new habitats before competitors find them is to produce as many propagules as possible. Propagules need only survive a relatively short dispersal phase, with continued survival following dispersal depending largely on the chance occurrence of settling in unoccupied suitable habitat. Investments in offspring beyond those needed to ensure survival of the dispersal phase would be wasted. Hence propagule size should be small, and parental care should be absent.

At the other extreme, sedentary species live in relatively long lasting, stable habitats such as coral reefs or climax stages of vegetational successions. Resident individuals can generally exclude or outcompete new arrivals, and becoming settled in a recently vacated, suitable habitat depends more on competitive ability than on early arrival. The most successful individuals are those that survive for long periods of time and produce highly competitive offspring. Extensive parental care should be advantageous because it makes offspring more competitive and helps them survive long enough to become established. Since considerable time and resources are devoted to survivorship and parental care, fecundity should ordinarily be low.

The expected characteristics of opportunistic and sedentary species are summarized in Table 9–2, and many species conform to one of these two patterns. Opportunistic species have low longevity, immediate breeding, high fecundity, good colonizing ability, low investment per offspring, and no parental care. Sedentary species have high longevity, delayed breeding, low fecundity, poor colonizing ability, high investment per offspring, and extensive parental care.

Much of the research on life history patterns revolves around the question of how opportunistic and sedentary strategies evolve. Two theories have been proposed. The theory of *r* and *K* selection is based on the relationship between population density and mortality rate, while the bet-hedging theory is based on the probabilities and uncertainties of juvenile

Table 9–2 *Life history characteristics of opportunistic and sedentary species, as postulated by MacArthur and Wilson (1967).*

	Opportunistic species	**Sedentary species**
Climate	Variable and/or unpredictable: uncertain	Fairly constant and/or predictable: more certain
Mortality	Often catastrophic, nondirected, density-independent	More directed, density-dependent
Survivorship	Often type III (Deevey 1947)	Usually type I and II (Deevey 1947)
Population size	Variable in time, nonequilibrium; usually well below carrying capacity of environment	Fairly constant in time, equilibrium; at or near carrying capacity of the environment
Colonization of new habitats	Frequent	Rare
Intra- and interspecific competition	Variable, often lax	Usually keen
Selection favors	1. Good colonizing ability 2. Rapid development 3. High r_{max} 4. Early reproduction 5. Small body size 6. Semelparity: single reproduction 7. Low investment per offspring 8. No parental care	1. Poor colonizing ability 2. Slower development, greater competitive ability 3. Low r_{max} 4. Delayed reproduction 5. Larger body size 6. Iteroparity: repeated reproductions 7. High investment per offspring 8. Extensive parental care
Length of life	Short, usually less than 1 year	Longer, usually more than 1 year
Leads to	Productivity	Efficiency

Source: Reprinted from E. R. Pianka, "On r- and K-selection," *American Naturalist* 104 (1970), p. 593, by permission of The University of Chicago Press.

and adult mortality. These two theories have generated a basic issue of considerable importance. Do the opportunistic and sedentary strategies represent end points of a linear continuum, with intermediate life history patterns characterized by intermediate values for all traits, or do life history patterns involve a multidimensional constellation of many independently varying traits? If opportunistic and sedentary strategies are end points of a linear continuum, all life history traits should covary as a unit. If not, intermediate patterns could consist of any combination of traits, without any trait necessarily being exhibited to an intermediate degree. Since r and K selection theory assumes a linear continuum of life history patterns, the answer to this question is crucial in evaluating current theory.

r and *K* selection theory

The growth rate of a population depends to a considerable extent on the intensity of intraspecific competition. In an uncrowded environment, where intraspecific competition is absent, population growth will be exponential (Figure 9–4A). Growth rate can be described by the equation

$$\frac{dN}{dt} = r_m N$$

where dN/dt is rate of population growth, r_m is the maximum possible difference between birth rate and death rate, and N is population size. At any given time t, population size N_t is given by

$$N_t = N_0 e^{r_m t}$$

where N_0 is population size at the beginning of growth.

Exponential growth occurs only when intraspecific competition is absent. In most real populations resources are normally limited, and population members soon begin competing with one another. The effect is to gradually diminish growth rate as local population size increases, until eventually a stable equilibrium size is reached (Figure 9–4B). This equilibrium size is usually referred to as **carrying capacity,** denoted by the constant K. Mathematically, the growth rate for such a population can be described by the logistic equation

$$\frac{dN}{dt} = r_m N \left(\frac{K-N}{K} \right) = r_m N \left(1 - \frac{N}{K} \right)$$

As population size increases, the term $(K - N)/K$ decreases until $N = K$. Then the rate of increase (dN/dt) becomes zero, and population size stabilizes at carrying capacity (K). Note that $(K - N)/K = 1$ when $N = 0$, so the rate of increase is exponential when population growth first begins.

During early stages of population growth, competition and predation are negligible and most mortality is caused by dispersal into inhospitable environments, adverse weather conditions, or deterioration of recently colonized habitats. Mortality rate is not affected by population density (i.e., it is **density-independent**), and population growth is therefore exponential. Some individuals are better able to resist adverse conditions than others, but individuals that expend more resources on growth, self-maintenance, predator protection, or parental care do not appreciably increase survival prospects for themselves or their offspring. Hence selection favors high fecundity.

Each genotype within a population produces descendants at its own rate r_i. Under conditions of density-independent mortality, the number of descendants produced by each genotype increases exponentially, in parallel with overall population growth, with the rate of increase determined by

Figure 9–4

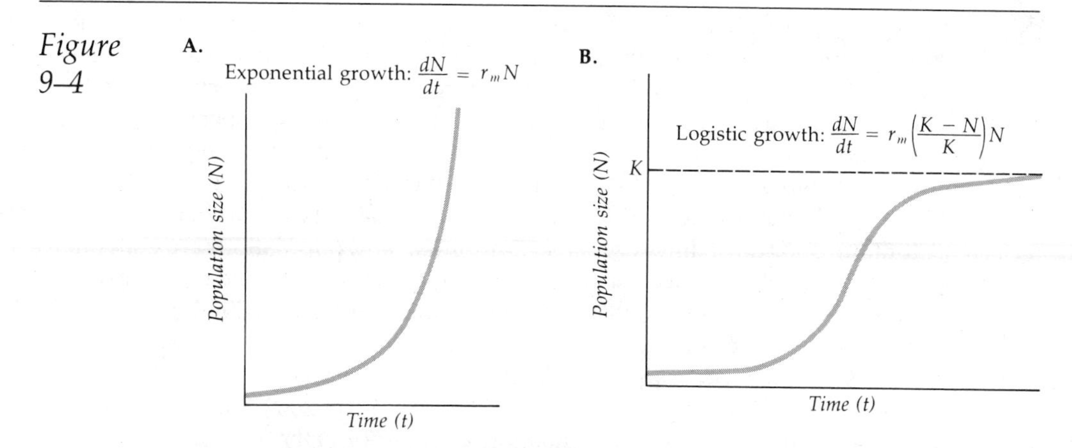

Two models of population growth.

A. Exponential growth leads to rapidly increasing population size. It occurs when competition is minimal or absent. In the real world exponential growth cannot continue indefinitely unless it is interspersed with catastrophic population declines. B. Logistic growth begins with nearly exponential population increase but then levels out as population size nears environmental carrying capacity (K). The leveling out of growth at larger population sizes results from increasingly strong intraspecific competition for resources or density-dependent predation losses.

r_i, just as the rate of overall population increase is determined by r_m. Genotypes with the highest value of r_i will therefore leave the most descendants. For each genotype r_i equals birth rate minus death rate among the genotype's descendants. Since chances of mortality are not influenced by genetic makeup, death rates are similar for all genotypes and r_i is highest for genotypes with the highest birth rates (MacArthur and Wilson 1967; Hairston et al. 1970; Pianka 1972). Because the success of each genotype depends on maximizing r_i, selection under conditions of density-independent mortality is usually referred to as ***r* selection.**

When a population is at or near carrying capacity, competition and predation are much more prevalent causes of mortality, and, consequently, mortality rate varies with density (i.e., it is **density-dependent**). Individuals with poor competitive ability or high vulnerability to predation have little chance of surviving to reproduce. Thus selection favors using time and energy to increase competitive ability or reduce vulnerability to predation. More time and energy should be allocated to growth, body maintenance, and parental care. Producing large numbers of low-quality offspring is not a successful tactic because such offspring have little chance of surviving. Thus density-dependent mortality should lead to increased longevity, reduced fecundity, and higher investment per offspring. Selection under conditions of density-dependent mortality is usually referred to as ***K* selection** because it results from populations existing near carrying capacity (MacArthur and Wilson 1967).

Opportunistic species continually colonize new habitats, and hence they reproduce where intraspecific competition and predation are minimal. Since progeny disperse long before carrying capacity is reached, populations are always in an early stage of growth. Opportunistic species therefore live perpetually under conditions of density-independent mortality, and they should exhibit life history traits predicted for an r selection regime.

Sedentary species live permanently in habitats that remain suitable for long periods of time. Dispersal is minimal, and populations are generally near carrying capacity. Such species therefore live under conditions of density-dependent mortality, and they should exhibit the life history traits predicted for a K selection regime.

One conceptual problem with r and K selection theory is that r is a life history parameter while K is not (Stearns 1976, 1977). That is, selection can affect intrinsic rate of population increase directly but cannot directly affect carrying capacity. One approach for circumventing this problem is to conceptualize the theory in terms of population growth patterns (Green 1980). One might think of r selection as a process generated by rapid population increases and K selection as a process generated by population stability. Then life history parameters can be attributed more directly to selective regimes. However, this approach also has problems. The theory of r and K selection is based on the consequences of density-dependent and density-independent mortality, and mortality is not necessarily density-dependent in stable populations (Wilbur et al. 1974; also below). Probably the best approach is to view the theory directly in terms of mortality patterns and either abandon the terms r selection and K selection altogether or at least continually bear in mind that these terms imply specific relationships between mortality patterns and life history characteristics.

Bet-hedging theory

An alternative to r and K selection theory is that life history traits evolve in response to the survival prospects of adults and juveniles. The theory is based on mathematical models of Murphy (1968) and Schaffer (1974), which are beyond the scope of the present discussion, and on the verbal arguments of Williams (1966a), Wilbur and co-workers (1974), and Hirschfield and Tinkle (1975). A summary of the logic is shown in Figure 9–5.

When adult survival prospects are low and cannot be appreciably increased by allocating more time and energy to survival, selection should always favor immediate and high reproductive effort. If future opportunities to reproduce are minimal, reproducing immediately is better than risking a delay even when prevailing conditions are unfavorable for reproducing. Diverting resources into self-maintenance and growth has little value, and all resources should be used to maximize immediate reproductive output.

When adult survival prospects are potentially high, the energetic costs of immediate reproduction have greater impact on future reproductive

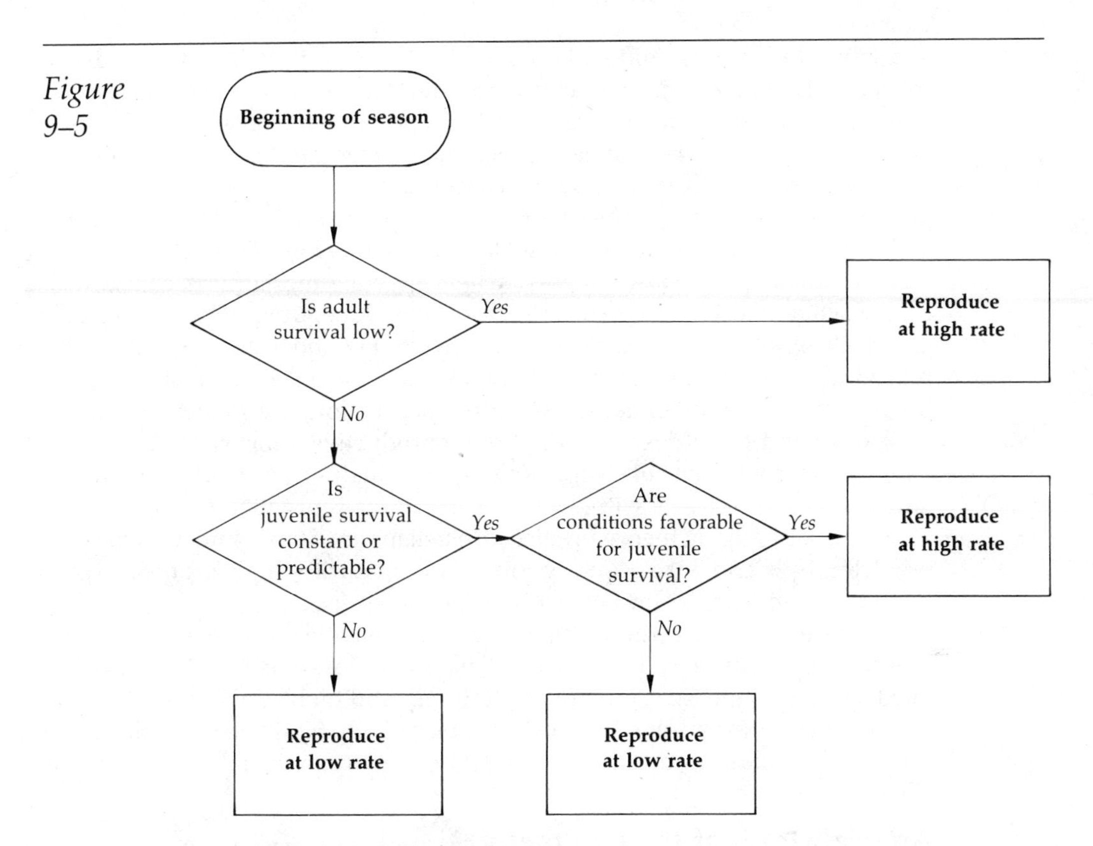

The logic of bet-hedging theory.

The shapes of the symbols have standard flowchart meanings. The oval represents the beginning of the logical argument. Diamonds represent conditional decision points, where the direction of the logic depends on whether a particular condition is true or false. Rectangles represent processes or outcomes. To read the flow diagram, begin at the oval and follow the arrows through the various decision points until each reproductive outcome is reached. The conditions predicated by the decision points leading up to each outcome represent the selective regimes responsible for each outcome.

opportunities. More resources should normally be used to maintain high survivorship, leaving fewer for immediate reproduction. However, the way resources are allocated should vary, depending on the predictability of juvenile survival.

When juvenile survivorship varies annually but is predictable prior to breeding each year, reproductive effort should be high in years favorable for juvenile survival and low in years unfavorable for juvenile survival. High effort may be achieved by ingesting more food to increase the total energy budget or, if that is impossible, by diverting time and resources away from investments in survivorship or growth.

When juvenile survivorship is completely unpredictable prior to breeding, adults should maximize their own survival and reproduce only at a low rate. High reproductive effort in any given year may prove disastrous because it could result in no reproductive output while at the same time greatly diminishing adult survival prospects. Juvenile survivorship is most likely to be unpredictable if development extends across several seasons, if reproductive effort must be committed far in advance of the time juveniles enter the environment, or if juveniles develop in an environment different from the one where adults live (Hirschfield and Tinkle 1975).

In unchanging environments, where juvenile mortality is constant, reproductive effort may be either high or low, depending on the survival prospects of juveniles (Williams 1966a; Hirschfield and Tinkle 1975). High reproductive effort should be advantageous if juvenile survival is high because the sooner juveniles can enter the population, the sooner they can reproduce and leave descendants. Low reproductive effort should be advantageous if juvenile survival is low because high effort would diminish adult survival prospects without providing much reproductive gain.

The bet-hedging theory predicts constellations of traits similar to those predicted by r and K selection theory. Early maturity, high reproductive effort, high fecundity, low investment per offspring, and low adult survivorship are expected when adult mortality is unavoidably high or juvenile mortality is predictably low. Delayed maturity, low reproductive effort, low fecundity, high investment per offspring, and high adult survivorship are expected when adult survivorship is potentially high and juvenile survivorship is either predictably low or completely unpredictable.

An evaluation of the two theories

The theory of r and K selection has gained wide acceptance among biologists, although it has recently been criticized by some authors (e.g., Wilbur et al. 1974; Stearns 1977). Early impetus for the theory came from intercorrelations among life history traits. Comparisons across broad phyletic groups indicate that species with high r_m typically possess traits characteristic of the opportunistic strategy, while species with low r_m typically possess traits characteristic of the sedentary strategy (MacArthur and Wilson 1967; Pianka 1970). Moreover, animals in temperate regions appear to be relatively more r-selected than animals in tropical regions, a difference that may be associated with relatively greater density-independent mortality in temperate regions (Dobzhansky 1950).

In the absence of an alternative theory, such correlations provided tentative support for the theory of r and K selection. However, the bet-hedging theory outlined above now provides an alternative explanation for the same correlations. Therefore, a closer analysis of actual selective regimes is necessary to determine which theory best explains the observed patterns.

The relationship between density-independent mortality and life history patterns has rarely been assessed. In one study increased density-independent mortality of sea stars *(Leptasterias hexactis)*, caused by heavier-than-normal wave action, led to *reduced* female reproductive effort, apparently because it reduced juvenile survival but not adult survival (Menge

1974). This result is contrary to that predicted by *r* and *K* selection theory and consistent with that predicted by bet-hedging theory.

A second series of studies compared the life history traits of three lizard species found in deserts of northern Mexico and southwestern United States (Tinkle 1967, 1973; Turner et al. 1970; Ballinger 1973; Tinkle and Hadley 1975). Although all three species are abundant, it is not clear how closely their densities approach carrying capacity of the environment. Nevertheless, the three species all experience density-dependent mortality. Adult side-blotched lizards *(Uta stansburiana)* have low longevity due to high predation pressure, reach maturity in less than one year, are oviparous (i.e., lay eggs), and lay as many as five clutches of three to five eggs each per year (Table 9–3). Adult Yarrow's spiny lizards *(Sceloporus jarrovi)* have higher longevity, reach maturity in their first year, are viviparous (i.e., bear live young), and produce one litter of about ten young per year. Adult sagebrush lizards *(S. graciosus)* have the highest survival rate of the three species, apparently because they live at higher elevations where predators are less abundant, reach maturity in their second year, are oviparous, and lay no more than two clutches of one to six eggs per season. The proportion of total energy budget devoted to reproduction is highest for sagebrush lizards, intermediate for side-blotched lizards, and lowest for Yarrow's spiny lizards (Tinkle and Hadley 1975).

Two points are of interest in these data. First, all three species experience primarily density-dependent mortality, and yet they exhibit very different life history patterns. Second, the relationship between adult survivorship and reproductive effort is contrary to that predicted by *r* and *K* selection theory. Sagebrush lizards are the most *K*-selected of the three species, and yet they exhibit higher reproductive effort each year than either of the other two species.

Wilbur and co-workers (1974) attribute the life history pattern of sagebrush lizards to their predictably high juvenile mortality rates. However, bet-hedging theory predicts that predictably high juvenile mortality should favor low annual reproductive effort. Hence the life history pattern of sagebrush lizards does not fit that theory either.

Further exceptions to the predicted patterns are common among amphibians. In general, salamanders and at least some frogs with high adult survival have delayed maturity and *high* fecundity. This relationship has been documented for five species of desmognathine salamanders (Organ 1961; Tilley 1968), high- versus low-altitude populations of the mountain salamander *(Desmognathus ochrophaeus)* (Tilley 1973, 1977), six species of ambystomatid salamanders (Husting 1965; Wilbur 1972), the California slender salamander *(Batrachoceps attenuatus)* (Maiorana 1976), and the bullfrog (Wilbur and Collins 1973; Wilbur et al. 1974). The reason is that longer-lived individuals grow to larger sizes, and large size enables higher fecundity. Many long-lived species also lay small eggs, contrary to the predictions of *r* and *K* selection theory, apparently because egg size does not affect vulnerability of newly hatched young to predation or other mortality agents.

Finally, Wilbur and co-workers (1974) point out that prey animals

Table 9–3 *Life history data for three species of lizards.*

Species	Age at maturity (months)	Mode of reproduction	Ratio of clutch or litter weight to adult female body weight[a]	Clutch or litter size	Number of clutches	Proportion of total energy budget devoted to reproduction[a]	Mean annual adult survival rate	Source
Uta stansburiana	9	Oviparous	.20	3–5	4–5	.19	11%	Tinkle 1967
Sceloporus jarrovi	5	Viviparous	.31	7–13 (3–5 for yearlings)	1	.11	33%	Turner et al. 1970 Ballinger 1973
S. graciosus	22	Oviparous	.26	2–6	2	.24	50%	Tinkle 1973

[a]From Tinkle and Hadley 1975.

may exist at a stable equilibrium density well below carrying capacity when population densities are limited by predation rather than food availability. Such prey animals may possess life history traits characteristic of sedentary *K*-selected species even though they live at densities well below carrying capacity, or they may possess traits characteristic of opportunistic *r*-selected species even though they are sedentary and limited by density-dependent mortality. Some populations of the eastern fence lizard *(Sceloporus undulatus)*, for example, have low adult survival and high fecundity even though all populations exist at stable densities and are subjected to density-dependent mortality (Tinkle and Ballinger 1972). Wilbur and co-workers (1974) suggest that *K*-selected traits should evolve if adults possess effective defenses against predation, while *r*-selected traits should evolve if they do not. However, their hypothesis does not easily explain within-species differences such as those reported by Tinkle and Ballinger (1972).

The evidence for lizards and salamanders suggests that traits such as adult longevity, age at maturity, clutch size, frequency of clutches, and extent of parental care do not necessarily covary as a unit. Almost any combination of traits appears possible, though some combinations occur more often than others. The evidence therefore suggests that a linear continuum of traits along a single axis, as implied by both *r* and *K* selection theory and bet-hedging theory, does not adequately characterize the way life history traits interact with one another or with environmental conditions.

A multiple-variable approach

Previous theoretical approaches to life history patterns suffer from two major problems. One is associated with the mathematical models themselves, while the other involves the way field data are used to test the models.

The mathematical models devised to explain life history patterns have generally been based on the trade-off between immediate reproductive effort and adult survivorship. The models do not treat pertinent life history traits separately, and yet predictions are made about those traits. For example, high fecundity, small egg size, and absence of parental care are supposed to covary as a unit, and all are subsumed under the single variable "reproductive effort." Yet species differ in the way they apportion reproductive effort, depending on their reproductive biology and the environmental conditions they face. Reproductive effort may be high because fecundity is high, because parental care is extensive, or because a combination of moderate fecundity and moderate parental care requires high effort. Fecundity itself is a composite variable consisting of clutch size and number of clutches per season. High fecundity may be possible because the resource base enables large clutch sizes or because long seasons enable multiple clutches. These two patterns are clearly very different. In species with indeterminate growth, body size is an important variable affecting both fecundity and survivorship, and it is interrelated with age at reproductive maturity, climatic conditions, competition intensity, predation pressure, and other variables. Current models simply do not allow analyses of how the many relevant variables interact with one another. Each important life history trait should be the subject of a separate modeling effort that evaluates all relevant variables affecting it. Since the various traits do interact, the ultimate goal should be a multiple-variable model showing how all traits interact with one another and with all pertinent environmental variables.

The problem with interpreting field data is that time and energy are poor estimates of fitness costs and benefits. A fixed caloric expenditure on egg production may entail little cost in one species because attempts at improving adult or offspring survival are futile, while in another species the same expenditure may entail a high cost because caloric expenditures on growth or parental care greatly enhance adult or offspring survival. For example, one species of salamander may be very capable of increasing its own survival or the survival of its offspring, so it may invest considerably less resources in fecundity and considerably more in growth and/or parental care. Another species may be incapable of protecting itself or its eggs from predators no matter how much time and energy are expended trying, so it should invest all its resources in fecundity. If the former species suffers high predation despite its best efforts, it may have low survivorship even though it has low fecundity. If the latter species lives where predators are uncommon, it may have high survivorship even though it has high fecundity. Indeed, Tinkle and Hadley (1975) use this argument to interpret the high fecundity of sagebrush lizards. Thus comparisons across species may be difficult to interpret because the impact of time and resource allocations on fitness may be very different for each species.

Parental behavior

Evolution of parental care

Few studies have analyzed the evolutionary origins of parental care outside the context of r and K selection theory, and most authors consider parental care as one component of the life history pattern characterizing K-selected species. However, in view of the problems associated with r and K selection theory, a separate analysis focusing specifically on the evolution of parental care seems desirable.

Like other behaviors, parental behavior offers benefits that increase adult fitness and entails costs that decrease adult fitness. The major benefit is enhancement of offspring survival, while the major cost is reduction of the adult's future reproductive potential. This trade-off is embodied in the concept of **parental investment,** which Trivers (1972) defines as "any investment by the parent in an individual offspring that increases the offspring's chance of surviving (and hence reproductive success) at the cost of the parent's ability to invest in other offspring." However, Trivers's definition is inadequate because parents sometimes invest in several current offspring simultaneously at some detriment to future reproductive potential but not to the detriment of any current offspring. Therefore, parental investment is redefined here to mean any investment in offspring by a parent that enhances the survival prospects of current offspring while reducing the parent's ability to invest in other current offspring or to produce future offspring. Parental investment includes the energy and nutrients used to produce gametes and nurture developing offspring, the time and energy spent protecting offspring, and the risks of predation, accidents, or weather-induced mortality incurred while caring for offspring. **Parental care** is a subset of parental investment. It includes all nongametic investments in offspring following fertilization, while parental investment includes investments in gametes plus investments in parental care.

A graphical model illustrates how the optimal level of parental investment is determined (Figure 9–6). A higher investment in current progeny leads to increased immediate reproductive output at the expense of future reproductive potential. The net benefit of an investment equals the immediate output minus the loss of future output, and the optimal level of investment occurs where net benefit is maximized. Thus any analysis of how parental behavior evolves must consider how much it increases survival of extant offspring compared to how much it decreases a parent's ability to produce additional offspring.

The evolution of parental care is a complex problem. The benefits derived from various types of parental care vary among species and higher taxa because they are affected by morphological constraints, developmental requirements of offspring, and environmental conditions unique to each species. Further complexity arises because the benefits and costs of parental care ordinarily differ for the two sexes. Females can generally be confident that offspring they care for are their own, but males cannot, especially in species with internal fertilization. Low confidence of paternity reduces the benefits males can gain from parental care because it causes males to sometimes unknowingly care for offspring other than their own. The costs of parental care may also differ for the two sexes because alternative reproductive options differ for each sex. The alternatives open to males often include territory defense and seeking additional mates, while the main alternative open to females is acquiring additional energy and nutrients for making more eggs. Thus an analysis of how parental care evolves must consider both the external constraints placed on the species and the differing trade-offs encountered by each sex.

External constraints and the forms of parental care

Parental care can take many forms and is shaped by many external factors, not all of which can be discussed here. Among vertebrates the principal functions of parental care are to protect eggs, nurture developing young, and protect developing young from predators or adverse weather. The ability of parents to provide these kinds of care varies greatly among species.

Figure 9–6

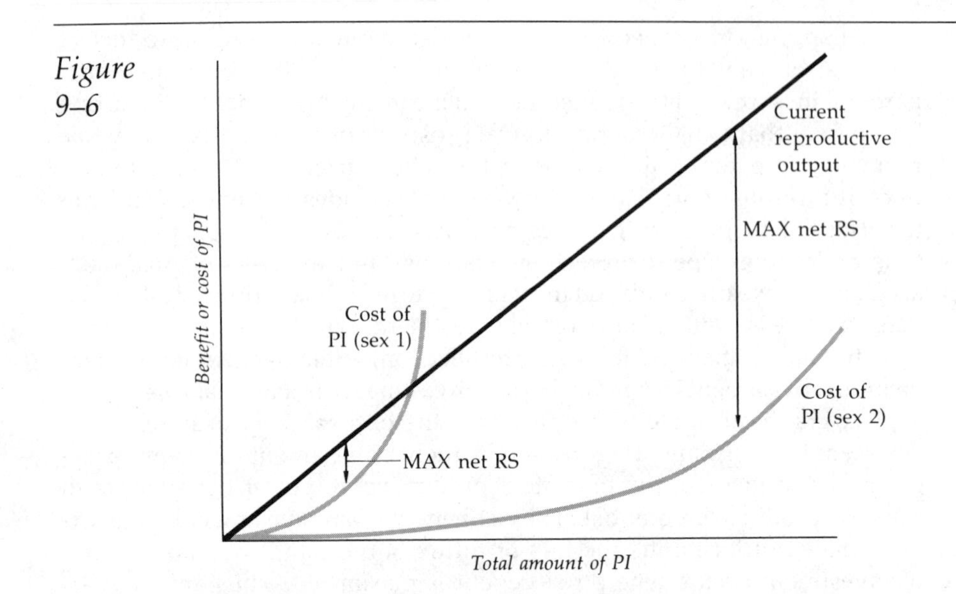

A model of optimal parental investment (PI).

The horizontal axis refers to total amount of parental investment provided by both sexes combined. The vertical axis refers to the benefit or cost of that investment for each sex of parent. A fixed proportion of the total investment is provided by each sex, with sex 1 (normally the female) providing a higher proportion than sex 2. The benefit derived from the total investment, current reproductive output, increases linearly as parental investment increases because more offspring are produced or a higher proportion of the offspring survive to independence. The cost of parental investment increases faster for sex 1 than for sex 2 because sex 1 contributes a higher proportion of the total investment. The optimal level of total investment from the perspective of each sex occurs at the point where the net benefit (benefit minus cost) is maximal for that sex. Note that the optimum differs for each sex, and the optimum for sex 1 limits the total level of investment.

Source: Adapted with permission from R. L. Trivers, "Parental Investment and Sexual Selection," in Bernard Campbell, *Sexual Selection and the Descent of Man* (New York: Aldine Publishing Company, 1972), p. 140. Copyright 1972 by Bernard Campbell.

Enhancing egg survival. In **oviparous** (egg-laying) animals an important factor affecting egg survival is the nature of the surrounding environment. Developing embryos are metabolically active, consuming much oxygen and releasing much carbon dioxide. Eggs therefore cannot survive in environments lacking adequate aeration. Because eggshells and embryonic membranes are permeable to water vapor as well as oxygen and carbon dioxide, eggs are susceptible to desiccation unless they are placed in moist environments or possess special adaptations for preventing water loss. Since embryonic tissues are usually highly palatable (Orians and Janzen

1974), they are highly vulnerable to predation unless placed in safe sites or protected by the parents. Early hatching, made possible by higher ambient temperatures or hatching at earlier developmental stages, reduces the period of vulnerability. Temperature regimes are also important because chilling and overheating can both cause embryonic death. Thus parents have many opportunities for enhancing egg survival. Parents may aerate eggs, incubate eggs, actively defend eggs to prevent predation or cannibalism, safeguard eggs by placing them in safer locations, or safeguard eggs by retaining them internally following fertilization. Whether any of these responses actually evolves depends on the nature of the selective regime and the constraints imposed by morphological limitations on parental capabilities.

Eggs laid in aquatic environments need not be aerated if they are placed in moving water. This is the case, for example, when eggs are simply broadcast into ocean or lake currents or when egg masses are attached to vegetation or under rocks located in currents. However, if eggs are placed in depressions or nests where water currents are minimal, as in many fish, fanning eggs becomes important for aeration. Eggs laid in open terrestrial environments also need not be aerated, but buried eggs must usually be placed in sandy soils or be exposed periodically to the air.

Eggs developing in water must necessarily develop at ambient temperatures. The thermal properties of water make it difficult for parents to maintain higher egg temperatures through incubation, and aquatic oviparous animals are cold-blooded and hence cannot maintain internal body temperatures much above ambient. The only way parents can regulate development rate is to select sites with appropriate local water temperatures (e.g., Howard 1978b).

Eggs laid on land can develop at temperatures above ambient. Reptiles are cold-blooded and cannot keep eggs constantly at high temperatures by incubating them, but they are capable of maintaining body temperatures substantially above ambient for much of the day by absorbing solar radiation. Hence many reptiles are capable of elevating egg temperatures during part of the day, and a few do. Females of some snakes and lizards, for example, coil around eggs to incubate them and possibly also to protect them (Shine and Bull 1979). In addition, reptiles can bury eggs in warm places provided that the soil in those places is moist enough to prevent desiccation.

Incubation has reached its highest pinnacle in birds, where it often involves complex nest-building behavior and carefully adjusted incubation schedules to maintain high and relatively constant temperatures inside the eggs (Drent 1975). In cold environments incubation is generally continuous to prevent chilling of eggs, while in hot environments eggs are shaded during midday to prevent overheating. An unusual form of incubation occurs in three species of megapodes, which are ground-dwelling birds of Australia. Megapodes bury their eggs in a huge mound of decaying vegetation, where their eggs are incubated by heat generated from the decomposition of vegetation in the mound (Photo 9–1) (Frith 1962; MacDonald 1973). Only the male tends eggs, and he maintains proper egg temperatures by removing or adding vegetation to the mound when temperature around the eggs changes.

The ability to protect eggs from predators varies considerably. Adult fish are often as large or larger than potential egg predators, and hence they are frequently capable of chasing predators away. Egg guarding is a common form of parental care in fish, and many species are interspecifically territorial toward potential egg predators. Other species protect eggs by carrying them about until hatching (Balon 1975). In nurseryfish (Kurtidae) egg clusters are carried about on a hook projecting from the male's forehead (Munro 1967). In banjo catfish (Bunocephalidae) they become attached to the ventral body surface of females (N.B. Marshall 1965). In pipefish and seahorses (Syngnathidae) they are carried about inside brood pouches or attached to the ventral surfaces of males (Herald 1959; Breder and Rosen 1966). In cichlids and various other fish they are carried about inside the mouth of one or the other parents (Oppenheimer 1970; Fryer and Iles 1972).

Many fish do not guard eggs, but they often enhance egg survival by spawning where eggs are less accessible to predators. Pelagic species such as herrings and anchovies broadcast gametes into open water, while reef species such as wrasses and surgeonfish broadcast gametes into outward-moving currents that carry fertilized eggs offshore (Breder and Rosen 1966; Barlow 1974). The fertilized eggs of these species become part of the offshore plankton or sink to the bottom where major egg predators are much

Photo 9–1

*A male brush turkey (*Alectura lathami*), a species of megapode in Australia, tending eggs on its mound.*

The male builds and maintains a large mound of vegetation, within which one or more females lay their eggs over a period of several months. Eggs are incubated by fermentation of decomposing vegetation rather than by body heat of the parent, and egg temperatures are controlled by adding or removing vegetation around the eggs. Only the male tends eggs. Once eggs hatch, the chicks tunnel their way to the surface and become fully independent. They receive no food or protection from either parent.

Photo by Axel Poignant.

less common than in continental shelf areas. The spraying characid (*Copenia arnoldi*), a South American species, has the unusual habit of spawning outside of water (Krekorian 1976). Eggs are attached to the underside of vegetation along a shoreline where they are splashed by the male for several days until they hatch. Many other tropical freshwater fish breed early in the rainy season when food is particularly abundant and spawn in recently flooded areas where predators are less common (K. Nelson 1964). Similarly, grunion (*Leuresthes tenuis*) and a closely related species (*Hypomesus pretiosus*) spawn out of the water on sandy beaches of the southern California coast during the highest high tides of each month between February and September (Walker 1952; see also Figure 10–6), presumably as a defense against marine predators. Thus placing eggs out of reach of predators is an alternative tactic when guarding eggs is unfeasible.

Active defense of eggs has also evolved in many semiterrestrial and terrestrial vertebrates. Male salamanders of two species (*Cryptobranchus alleghaniensis* and *Hynobius nebulosus*) defend eggs, especially from the cannibalistic attacks of conspecific females (Smith 1907; Thorn 1962), while females of several other species protect eggs from such attacks (Salthe and Mecham 1974; Houck 1977). Males of the midwife toad (*Alytes obstreticans*) wrap eggs around their hind legs after fertilizing them and then carry them about until they hatch (Salthe and Mecham 1974). Females of the colonial common iguana (*Iguana iguana*) and the marine iguana (*Amblyrhynchus cristatus*) defend nest sites from other females until they are completely filled in to prevent them from being dug up, a behavior that evolved in response to intense competition for nest sites (Carpenter 1965; Rand 1967, 1968). Female Nile crocodiles guard their nests continuously throughout the incubation period (Guggisberg 1972; Pooley and Gans 1976). When their eggs hatch, they dig up their nests and carry the young to shallow water. Digging up the nest is necessary because the ground becomes hardened during incubation. Once transported to water, the hatchling crocodiles remain together and are guarded by both parents for about 12 weeks. Females and possibly males of the king cobra (*Ophiophagus hannah*) also guard nests until their eggs hatch (Oliver 1956; Leakey 1969). Large birds are sometimes capable of attacking nest predators singly, while smaller birds often mob predators.

Nevertheless, protecting eggs from predators is risky in many amphibians, reptiles, and birds because egg predators are frequently larger than the parents. Many species therefore safeguard eggs mainly by placing them in concealed locations or in places where predators have greater difficulty finding them or gaining access to them (Lack 1968; Salthe and Mecham 1974; Ehrenfeld 1979). Amphibians, for example, often place eggs in crevices, tree cavities, or under rocks. Reptiles often bury their eggs underground. Birds often nest on the outer tips of branches, on cliff ledges, in marshes, or on islands. Other birds conceal their nests in tall grass or dense vegetation, while still others have cryptically colored eggs and chicks.

Some fish, sharks, amphibians, and reptiles, and all mammals reduce egg losses by giving birth to live young. **Viviparity** (bearing live young) is an effective means of preventing egg losses, but it drastically limits the number of young that can be produced at one time. Also, retaining eggs internally until they hatch is not a certain means of preventing egg loss, since retained eggs are lost if the parent dies. Such losses represent a potentially important cost of viviparity. The evolution of viviparity is closely tied to an animal's overall life history strategy, and more will be said about its origins in a later section.

Enhancing offspring survival. Parents can enhance offspring survival by protecting them from predators and by nurturing them. They may also lead offspring to safer retreats, better foraging sites, or more suitable thermal environments. The latter behaviors are commonly adjuncts to parental guarding of offspring and normally do not evolve unless parents remain with offspring to protect them anyway.

Fish are often capable of protecting free-swimming fry because they are as large as the potential predators. Nevertheless, guarding fry occurs relatively infrequently (Gross and Shine 1981). Protecting fry is a costly proposition, and alternative reproductive options often make it disadvantageous. The evolution of fry-guarding behavior will be discussed more fully in a later section.

Amphibians and reptiles frequently do not protect young, either because it is too hazardous or because it is energetically too expensive, although male and occasionally female frogs sometimes carry tadpoles on their backs or in specialized pouches (Salthe and Mecham 1974; Wells 1977a; Ridley 1978). Birds rely on a number of defenses to protect young, including nest concealment, placing nests in inaccessible sites, feigning injury to lure predators away from nests, mobbing predators, rapid development of nestlings to reduce the time spent in nests, and cryptic coloration or behavior of newly fledged young. Mammals often protect young by bearing them in relatively safe dens or parturition sites, producing young that are soon capable of fleeing predators, or actively defending young from predators either singly or in groups. Substantial differences occur among species because niche requirements and morphological capabilities are both extremely diverse.

Nurturing young is usually beyond the capabilities of fish, amphibians, and reptiles because the sizes and types of food eaten by adults and young are too different. Adult fish, for example, often hunt relatively large prey, while their fry forage upon plankton and other microorganisms. This difference results at least in part from young being much smaller than adults. Nevertheless, a few fish nurture young by allowing them to feed on mucous produced on their bodies (Photo 9–2) (Barlow 1974).

In contrast, most birds and all mammals must feed their young to keep them alive. Warm-bloodedness, determinate growth, and rapid developmental rates all make feeding of offspring necessary. Only in the most precocial birds are offspring capable of feeding themselves soon after hatching (see Nice 1962), and this is possible only because each egg contains large amounts of yolk to support prolonged development before hatching.

Photo 9–2

*Brood care in the discus fish (*Symphysodon aequifasciata*), a cichlid fish of South America.*

Both parents secrete a proteinaceous slime on their lateral body surfaces that fry use for food during the first few weeks of life. Providing nourishment to fry is an extremely unusual form of parental care among fish.

Photo by A. v. d. Nieuwenhuizen.

The extent to which parents nurture young therefore depends on developmental rates and dietary requirements of young, along with the foraging capabilities and physiological characteristics of the parents.

Shareable versus nonshareable parental investment

Two kinds of parental investment must be distinguished (Figure 9–7). **Shareable parental investment** is any investment in an individual offspring that does not reduce the parent's ability to invest simultaneously in other offspring but does reduce the parent's future reproductive potential (Perrone 1975; Wittenberger 1979a). **Nonshareable parental investment** is any investment that cannot be invested simultaneously in more than one offspring.

Shareable parental investment includes guarding eggs in fish, incubating eggs and brooding young in birds, and being vigilant for predators in birds and mammals. A male fish, for example, can probably guard several nearby egg masses just as effectively as it can guard one, but the time and energy used to guard eggs reduce the time and energy left available for attracting additional mates or growing to a larger size. Birds are capable of incubating several eggs or brooding several nestlings at a time, but they cannot tend more than one nest simultaneously. Incubation and brooding are therefore shareable within broods but not between broods. The alarm calls of adult birds and mammals can alert several offspring to

Figure 9–7

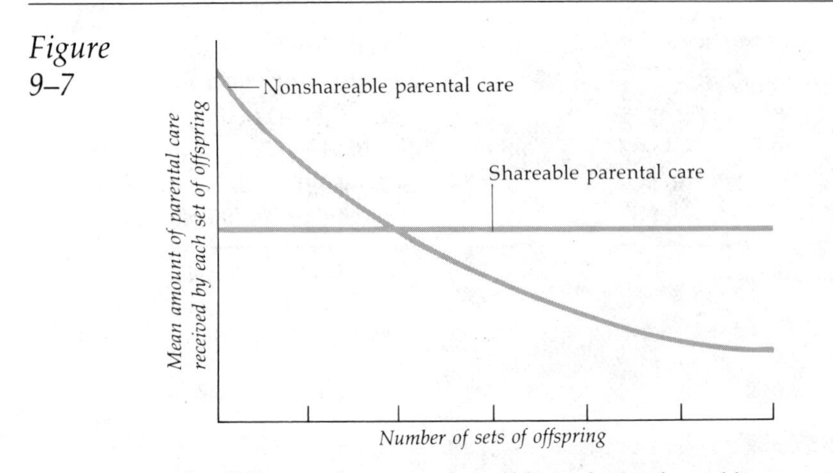

The difference between shareable and nonshareable parental care.

If the presence of additional offspring does not reduce the amount of care received by each, parental care is shareable. If the presence of additional offspring does reduce the amount of care received by each, parental care is nonshareable.

Source: From J. F. Wittenberger, "The Evolution of Mating Systems in Birds and Mammals," in P. Marler and J. G. Vandenbergh, eds., *Handbook of Behavioral Neurobiology, Vol 3: Social Behavior and Communication* (New York: Plenum Press, 1979), p. 312. Reprinted by permission.

danger simultaneously, provided all are near enough to perceive the signal. Adding another offspring to the brood or litter does not reduce the signal's effectiveness unless competition forces some offspring to disperse beyond the range of effective perception.

Nonshareable parental investment includes the energy and nutrients invested in gametes or individual offspring. Food given to one offspring cannot be given to another and cannot be used for self-maintenance or growth. The time spent obtaining food is time that cannot be used to attract mates, defend territories, guard against predators, or conduct other activities. Grooming offspring, an important form of parental investment in primates, is also nonshareable, as is carrying offspring when only one offspring can be carried at a time.

The distinction between shareable and nonshareable parental investment is important because increases in brood size or number of simultaneous broods affect fitness differently for each type of investment. When investments are shareable, cost is fixed regardless of brood size or number of broods, while benefits increase with increasing brood size or number of broods (Figure 9–8A). Hence the net gain from an investment increases with the number of young being cared for simultaneously, and brood size should not be limited by the parent's capacity to care for young. When investments are nonshareable, both costs and benefits increase with increasing brood size or number of broods, though at different rates (Figure 9–8B). The optimal level of investment occurs at some intermediate brood

Figure 9–8

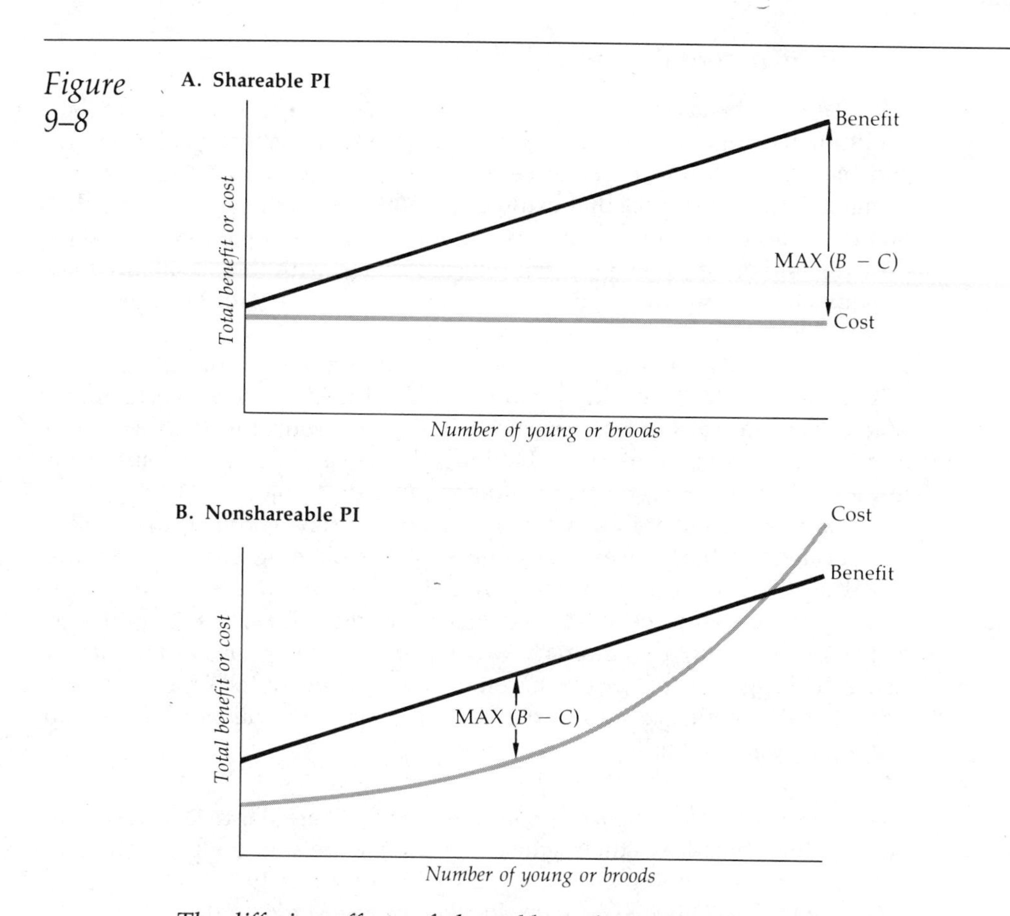

The differing effects of shareable and nonshareable parental investment on optimal brood size or optimal number of broods.

A. If parental investment is shareable, selection should enlarge broods until brood size is limited by other constraints imposed by the species' life history pattern. B. If parental investment is nonshareable, selection should adjust brood size to maximize the net benefit of parental investment unless other life history parameters modify the optimum.

size, where net gain is maximized. Thus brood size should be limited by the cost-benefit relationships of nonshareable parental investments.

The difference between shareable and nonshareable parental investments contributed by males is important when evaluating female mating preferences. An important cost for females who pair with already mated males is reduced male parental assistance (Verner 1964; Orians 1969). However, if male parental investments are shareable, this cost is negligible, and females should be less concerned about a male's mated status. If male parental investments are nonshareable, becoming a second or third mate is costly, and females should be more likely to seek an unmated male. Thus female mating preferences may be strongly influenced by the type of parental investment provided by males.

Origin of parental care in fish

Parental behavior of higher vertebrates is highly evolved, and its evolutionary origins are not readily discernable. The offspring of both birds and mammals require parental care to survive, and comparisons cannot be made between species that exhibit parental care and related species that do not. True, birds and mammals vary in the amount of parental care they provide, and environmental conditions associated with this variability may provide clues about the origins of obligatory parental care. However, the factors affecting the amount of parental care provided may not be the same as those originally favoring the evolution of parental care. Insight into the origins of parental care can be gained from studies of lower vertebrates, where parental care is less prevalent. The best group for analysis is the teleost fishes, where parental behavior is diverse and occurs in a sufficient number of taxonomic groups to allow comparative analyses.

Parental behavior in fish may be absent entirely, confined to one sex, or prevalent in both sexes. Hypotheses for explaining each pattern are based on the trade-offs between parental behavior and alternative reproductive activities (Blumer 1979; Perrone and Zaret 1979). These hypotheses can be evaluated in a preliminary way by comparing species or families of fish exhibiting each pattern of parental care, but eventually more sophisticated cost-benefit analyses of representative species will be necessary to test the hypotheses adequately.

Reasons for the absence of parental care. Parental care is absent in all fish that broadcast their gametes into the pelagic zone of offshore waters or into outgoing currents (Perrone and Zaret 1979). Such species rely heavily on chance events for survival of their eggs, but their eggs do gain some safety by entering the plankton or sinking to the bottom in deep waters, where predators are relatively scarce.

Parental care should be absent when neither parent is capable of enhancing offspring survival enough to offset the attendant costs. This condition can hold if the benefits of guarding eggs are low or if the costs are high. To the extent that eggs are safer in plankton or deep-water zones, the potential benefits of guarding eggs are reduced. Nevertheless, additional factors must be important since many species do not broadcast gametes into the open ocean despite opportunities to do so.

For pelagic (offshore) species parental care is likely to be unusually costly. Eggs normally cannot be deposited on substrates, since the ocean bottom is too deep, and parental care would not be possible unless parents guard floating egg masses or carry eggs about with them in specialized ways. Floating egg masses would attract predators because they would represent a clumped food source, and they would often be hazardous to guard because the parents would be tied to the spot where eggs are located without having any safe shelter sites nearby. In schooling species, for example, it would be extremely difficult, if not impossible, for parents to guard floating egg masses and still remain with a school to derive protection for themselves. Carrying eggs in the mouth or in special structures is a possibility, but such behavior may require unusual preadaptations before it can evolve and in some cases may be incompatible with a species' foraging ecology or social organization. In addition, guarding eggs may be incompatible with food-searching patterns for pelagic species that must roam more widely in search of food. The problem has not yet been given sufficient attention, and more study is needed before the precise disadvantages of parental care will be understood for pelagic species.

For species that breed inshore the reasons why parental care has often failed to evolve are even less clear. One possible reason is that parents are sometimes poorly suited to guard eggs against their most important egg predators, owing to morphological limitations. Another is that guarding eggs is too costly due to the nature of the food supply or predators. The most likely explanation is that species with spatiotemporally variable food resources cannot remain tied to a fixed nest site until eggs hatch. This condition apparently prevails in surgeonfish, for example, which are weakly territorial or defend small, ephemeral feeding territories and broadcast their gametes into outgoing currents (Barlow 1974).

Origin of uniparental care. Parental care should evolve when it enhances egg survival more than enough to offset any reduction in reproductive output entailed by giving care. One parent alone or both parents together could conceivably provide care. When a single parent can enhance offspring survival nearly as effectively as two parents acting together, parental care should be provided by only one parent (Perrone and Zaret 1979). In many circumstances a single parent can probably fan and guard eggs just as effectively as two parents (see Perrone 1975; Williams 1975). Exactly what these circumstances are requires further study, but some factors likely to be important are habitat configuration around the location where eggs are deposited and the nature (size, behavior, and morphology) of egg predators. Under such conditions one parent is likely to desert and leave the eggs for the other to tend alone. The factors determining which sex remains to tend eggs are controversial, and several hypotheses have been proposed.

According to one hypothesis, the sex with the earliest opportunity to desert will do so, leaving the other sex to either care for eggs alone or leave them untended (Dawkins and Carlisle 1976). When the benefits of egg tending are higher than the costs, uniparental care by the deserted sex should evolve.

The only support for this gamete-order hypothesis is that male parental care has evolved mainly in fish with external fertilization (see Ridley 1978; Blumer 1979; Gross and Shine 1981). When fertilization is internal, parental care is usually provided by females. Since females can never desert first when fertilization is internal, such a correlation would be expected were the gamete-order hypothesis true. However, since the same correlation is also predicted by several other hypotheses, as will be shown below, it provides no real support for any one of the hypotheses.

Two predictions of the gamete-order hypothesis have been falsified by comparative evidence (Gross and Shine 1981). First, if gamete order is important, uniparental care by males should not evolve in species with external fertilization when sperm are released before ova. Nevertheless, males build foam nests, release sperm before females spawn, and still provide uniparental care in some species of armored catfish (Callichthyidae) and in the blue gourami (*Trichogaster trichopterus*) (Breder and Rosen 1966; Kramer 1973; Hoedeman 1974). Similarly, females of some mouth-brooding cichlids spawn before males and then remove eggs from male nests to care for them without help from the male (Fryer and Iles 1972). Second, the gamete-order hypothesis predicts that male and female uniparental care should be equally frequent when gamete release is simultaneous. Contrary to that prediction, uniparental care is by males in 36 of 46 species characterized by simultaneous spawning and uniparental care (Gross and Shine 1981).

According to a second hypothesis, uniparental care should be provided by females when confidence of paternity is low and by males when confidence of paternity is high (Ridley 1978; Blumer 1979; Perrone and Zaret 1979). The argument is that males derive fewer benefits from parental care when their relatedness to offspring is reduced by multiple insemination of females. Low confidence of paternity should induce males to desert, causing females to remain and care for eggs.

The strongest support for the confidence-of-paternity hypothesis is the fact that male uniparental care occurs almost entirely in fish with external fertilization. External fertilization allows high confidence of paternity since all eggs are fertilized in the presence of the male. In comparison, confidence of paternity is less certain when fertilization is internal because females can be multiply inseminated prior to spawning. However, as pointed out above, the correlation between male parental care and mode of fertilization is consistent with several hypotheses and therefore does not provide support for any one of them.

The confidence-of-paternity hypothesis seems intuitively reasonable, but in fact it contains a hidden fallacy. Basically the problem is that confidence of paternity cannot affect the evolution of parental care when it is the same for both parental and nonparental males within a population (Gross and Shine 1981). The effect of paternity can be seen with some simple algebra. The number of offspring produced by a nonparental male (N) will equal the number of eggs laid by the females he inseminates (n) multiplied by average survival of those eggs (s) and by confidence of paternity (p). That is, $N = nsp$. The number of offspring produced by a nonparental male (N') will equal the number of eggs laid by the females he inseminates (n') multiplied by average survival of those eggs (s') and by confidence of paternity (p). That is, $N' = n's'p$. Paternal care should evolve when

$$N' > N$$

or

$$n's'p > nsp$$
$$\frac{s'}{s} > \frac{n}{n'}$$

Note that confidence of paternity (p) cancels out and therefore has no effect. Paternal care should evolve when the proportional increase in survival (s'/s) is greater than the proportional decrease in number of eggs fertilized by paternal males (n/n').

Further analysis shows that confidence of paternity can be indirectly related to the evolution of male parental care in the following manner (see Gross and Shine 1981). When confidence of paternity is high, copulations outside the pair bond are necessarily rare and hence territorial males stand to lose few opportunities for achieving such copulations. Consequently, the reproductive cost, in terms of reduced mating success outside the territory, is likely to be low. On the other hand, when confidence of paternity is low, territorial males may frequently mate with females who do not spawn on their territories. Providing parental care may substantially reduce a male's ability to achieve such matings, in which case the cost of parental care would be relatively high. Hence high confidence of paternity may be correlated with the occurrence of male parental care because it eliminates a potential cost of parental care that exists when confidence of paternity is low.

The only way confidence of paternity can directly affect the evolution of male parental care is for it to be different for parental and nonparental males within a population. Such a difference is extremely unlikely to exist in any real populations, except possibly as a secondary consequence that arose after male parental care had already evolved.

A third hypothesis is that parental care is advantageous for one sex but not for the other. One way this could occur arises from the way offspring are physically associated with their parents (Williams 1975; Gross and Shine 1981). Internal fertilization, for example, causes eggs to be associated with females and makes parental care by males impossible or difficult. In some species females retain the eggs internally for part of the developmental period and then lay eggs containing partially developed embryos or bear live young. Males would then have little opportunity for providing parental care. In other species females spawn sometime after being inseminated and then guard eggs from predators. To the extent that males would have to remain with females until spawning, they would pay an increased cost just to gain the opportunity for providing care. These arguments imply that parental care in species with internal fertilization should be provided by females, which is what actually occurs.

When fertilization is external, eggs may or may not be more closely associated with one sex. They are most closely associated with males when spawning occurs on male territories and males aggressively chase females off their territories after spawning is completed. Then females would have little opportunity for providing parental care.

Selection for territoriality should predispose males to guard eggs (Blumer 1979; Borgia 1979; Gross and Shine 1981). If males depend on patrolling to find receptive females, tending eggs is likely to conflict with finding females, and the benefits gained from parental care would more than likely exceed the costs. Then males should not provide parental care, and females could either tend eggs alone or also not provide any parental care, depending on whether they benefit more by providing care or by deserting to produce additional clutches. If, on the other hand, suitable spawning sites are limited in availability or variable in quality, males should defend the best sites and try to attract females to them. Once a male is confined to a territory, caring for egg masses should not interfere much with efforts to attract additional females, and hence the benefits of parental care should exceed the costs for males. Males provide uniparental care of eggs in 45 families of fish, and in 43 of them they defend spawning sites or are at least attached to spawning sites (Ridley 1978; Blumer 1979). The exceptions are male pipefish, seahorses, and nurseryfish, which carry eggs attached to their bellies, in special brood pouches, or attached to bony hooks on their heads, respectively.

A factor that may make parental care disadvantageous for females is length of breeding season (Perrone and Zaret 1979). Prolonged breeding seasons allow females to greatly increase reproductive output by deserting fertilized eggs, because desertion enables them to garner the resources necessary for producing additional clutches (Qasim 1957a; Blumer 1979; Perrone and Zaret 1979). In contrast, deserting eggs does not increase male reproductive output and may decrease it if males require territories to attract mates. Moreover, since egg tending is a shareable form of parental investment, providing parental care may not reduce a male's ability to attract females to his territory. Thus the combination of male territoriality and long breeding seasons may make parental care advantageous for males and disadvantageous for females.

Short breeding seasons and an absence of male territoriality would lead to the opposite outcome. Desertion is no longer as profitable for females because time is too short for them to acquire the resources needed to produce second clutches. A female's entire reproductive effort for the season would be tied up in a single clutch, and protection of that clutch could well be more advantageous than using the extra time and energy for body maintenance or growth. At the same time, when males do not require territories to attract mates, their best recourse may be to roam widely in search of unmated females. Such a strategy would be incompatible with male parental care.

Some evidence supports the hypothesis that length of breeding season affects parental care patterns. In tropical and subtropical blennoid fishes females can produce several clutches of eggs per season because spawning seasons are prolonged, while in temperate and subarctic species females can produce only one clutch per season because spawning seasons are much shorter (Breder 1939, 1941; Qasim 1957b; Reese 1964; Breder and

Rosen 1966; Stephens et al. 1966, 1970). As predicted by the arguments above, egg tending is performed by males in tropical and subtropical species and by females in temperate and subarctic species. Furthermore, uniparental care by males is closely associated with multiple spawning by females as well as with male territoriality. Multiple spawning is the rule in 176 of the 178 species in which males alone tend eggs (Blumer 1979).

The situation is more complex if uniparental care is advantageous for both sexes. In such cases the factors determining which sex provides care are less clear. The magnitude of the net benefit for each sex should not be important because the relative fitness of males to females has no effect on selective outcomes. Selection operates within each sex, not between them. A high net benefit for one sex would not influence the way a lower net benefit affects the other sex. When uniparental care is advantageous for both sexes while biparental care is disadvantageous, various preadaptations and chance events are likely to determine which sex actually provides the care. Once one sex has evolved parental care, selection should not favor parental care by the other sex unless the additional care substantially increases offspring survival above that occurring when only one sex provides care. In the latter event biparental care should evolve.

Origin of biparental care. Biparental care, which involves both sexes caring for offspring, should not evolve unless offspring survival is considerably higher when the second parent participates. Males should not assist females unless the increased survival of offspring resulting from their care is more advantageous than any benefits gained by alternative reproductive activities. The same is true for females. In fish two parents often cannot tend eggs much more effectively than one, but they can guard free-swimming fry more effectively (Perrone and Zaret 1979). Hence biparental care should evolve when conditions favor guarding of fry. Exactly what those conditions are requires further study.

Most species that guard both eggs and fry do exhibit biparental care (Breder and Rosen 1966; Lowe-McConnell 1969; Fryer and Iles 1972; Robertson 1973; Barlow 1974). However, there are exceptions. Mouth-brooding cichlid fish provide protection by allowing fry to swim into their mouths when predators approach, and most mouthbrooders exhibit uniparental care (Fryer and Iles 1972; Perrone and Zaret 1979). The reasons for this pattern are not yet clear. The brood sizes of mouthbrooders that exhibit uniparental care are relatively small, allowing one parent (the female) to effectively protect all their fry without aid, but whether small brood sizes are a consequence or a cause of uniparental care is unclear.

Biparental care has evolved in a few fish that tend eggs but do not guard fry. Anemonefish live as monogamous pairs within the tentacles of sea anemones, where they gain protection from predators, and in some species both sexes help guard eggs (Mariscal 1972; Fricke 1974; Allen 1975). Similarly, the blind goby (*Typhlogobius californensis*) lives as monogamous pairs inside burrows shared with a shrimp, and both sexes help tend eggs (Reese 1964). In both species neither sex can benefit by deserting eggs because both sexes would have to desert the protective confines of their living sites at the same time. Such conditions greatly lower the costs of parental care for both sexes and therefore make biparental care more likely to evolve even when the benefits gained by providing care are relatively low.

Origin of viviparity in reptiles

The evolution of viviparity in reptiles is an interesting problem because bearing live young entails greater female parental investment than does laying eggs. In addition, an understanding of how viviparity evolved in reptiles may clarify how it evolved in mammals, since mammals originally stemmed from reptilian ancestors.

Before viviparity becomes possible, specialized vascular structures in the oviduct must evolve to facilitate gas exchange and decalcification of eggshells (Yaron 1972). Such modifications could evolve when eggs are retained for progressively longer periods of time before being laid (see Dmie'l 1970), but they probably could not arise all at once in a sudden jump from oviparity to viviparity. Hence selection pressures favoring viviparity probably first led to intermediate degrees of egg retention, and they may differ from the selective pressures presently maintaining viviparity (Shine and Bull 1979). If that is true, identifying the selective factors that first led to viviparity may be difficult.

Costs of egg retention and viviparity. Egg retention and viviparity entail several costs for females (Shine and Bull 1979). One or both are most likely to evolve when these costs are low. Identifying environmental conditions that would reduce the costs might therefore help explain how egg retention and viviparity evolved.

One cost is that egg retention reduces the mother's mobility, thereby hampering her ability to forage and escape predation. This cost should be lower in large or venomous species, secretive species, and slow-moving species that do not depend on speed to catch prey or avoid predators (Neill 1964; Fitch 1970). It should also be lower in species where females brood eggs by coiling around them, because then females are immobile during the brooding period anyway (Shine and Bull 1979). The possible relationships of egg retention and viviparity to size, poisonousness, secretive habits, and slow-movement habits have not been systematically evaluated. Since both oviparity and viviparity occur among poisonous and secretive species, these two factors do not adequately account for the origin of viviparity (Tinkle and Gibbons 1977). Whether they have any relevance to the problem at all remains to be determined. Egg retention and viviparity do occur more often than expected by chance alone in squamate reptiles that exhibit maternal brood care (Shine and Bull 1979). Hence maternal brood care may promote the evolution of viviparity (or vice versa).

A second cost is that prolonged egg retention reduces the ability of females to produce additional eggs (Tinkle and Gibbons 1977). Only a small number of eggs can reside in the female's reproductive tract at a time, and the longer eggs are retained, the less frequently new clutches of eggs can be produced. Egg retention and viviparity should therefore be least costly in species with short breeding seasons that could not lay more than one clutch per season anyway. Viviparity is correlated with single-broodedness in snakes and lizards (Tinkle et al. 1970), and it is more prevalent in cool climates where breeding seasons are short (see below). Hence opportunities to reproduce more than once per season may be one constraint that prevents viviparity from evolving in many species, especially in the tropics.

A third cost is that death of a viviparous female during gestation results in automatic loss of offspring (Tinkle and Gibbons 1977). In contrast, death of an oviparous female does not reduce chances for offspring survival except during the period while eggs are still in the female's oviduct. The longer eggs are retained, the higher is the risk of loss resulting from death of the female. One might expect selection to favor egg retention and viviparity more often in species that are less susceptible to predation, but this possibility has not yet been evaluated.

Benefits of egg retention and viviparity. Several benefits have been proposed to explain the evolution of egg retention and viviparity. One possible benefit is that egg retention reduces vulnerability of eggs to predation (Shine and Bull 1979). If predation risk is proportional to the length of incubation, retaining eggs to shorten the exposure period could be advantageous. Of course, this hypothesis presupposes that eggs are generally safer inside a female than buried in the ground, which may or may not be true.

Present evidence does not support the predator defense hypothesis. Although predation is a major source of egg mortality, it is usually highest just after eggs have been laid and possibly just as young emerge (Blair 1960; Auffenberg and Weaver 1969; Moll and Legler 1971; Carr 1973). Intermediate degrees of egg retention would not circumvent the periods of highest vulnerability. If predation is low during the middle of incubation, as present data indicate, longer incubation periods would not appreciably increase predation losses.

A second possible benefit is that egg retention reduces susceptibility to adverse microclimatic effects (Neill 1964). Eggs laid in wet soils may be more susceptible to infection or fungal attacks, while eggs laid in dry soils may be more susceptible to desiccation. High or low moisture content in the nesting substrate may therefore favor viviparity.

A comparison of habitats occupied by viviparous reptiles and squamate reptiles as a whole shows that viviparity is not more prevalent than expected by chance alone in wet or dry climates or in aquatic or arboreal habitats (Table 9–4) (Shine and Bull 1979). Two logical arguments also suggest that microclimatic factors are not likely to explain viviparity (Shine and Bull 1979). First, species occupying wet environments may still be able to find relatively dry nest sites, and vice versa. Adaptations of adults to extreme microclimatic conditions do not necessarily require that eggs be

Table 9–4 *Proportion of viviparous and egg-retentive lizards and snakes in different environments.*

	All squamates[a]	Live bearers in genera with both egg-laying and live-bearing members (n = 38)	Species with egg retention (n = 60)
Wet soil	.12 (.04–.22)	.11	.13
Dry soil	.18 (.03–.35)	.24	.05
Aquatic	.05 (0–.14)	.03	.02
Arboreal	.10 (.02–.28)	.08	.08
Cold climate	.44 (.13–.61)	⩾ .74	.72

[a]Mean for all species, with range of continent means in parentheses.

Source: From R. Shine and J. J. Bull, "The Evolution of Live-Bearing in Lizards and Snakes," *American Naturalist* 113 (1979):915. Reprinted with permission.

laid in similarly extreme conditions. Second, deleterious effects of wet or dry moisture conditions should not favor intermediate degrees of egg retention unless they mainly affect early stages of embryogenesis. Otherwise, shortening the time eggs remain in the nest would not greatly increase egg survival. Thus extreme moisture conditions may help maintain viviparity in a population once it has evolved, but they do not appear capable of favoring intermediate degrees of egg retention.

A third possible benefit is that egg retention facilitates normal development or speeds up the rate of development in cold environments (Greer 1966; Packard 1966; Packard et al. 1977; Tinkle and Gibbons 1977; Shine and Bull 1979). In cold environments nesting substrate may be too cold to allow proper development of embryos, or low ambient temperatures may make development too slow relative to the length of the growing season.

The relationship between climatic regimes and egg retention can be evaluated by comparing environments occupied by egg-retentive and viviparous reptiles to those occupied by squamate reptiles as a whole (Shine and Bull 1979). Since environments where live-bearing species presently live may differ considerably from those where live-bearing species evolved, the best approach is to restrict the analysis to genera that contain both oviparous and viviparous species. Viviparity probably evolved in such genera relatively recently, so this approach minimizes the chances that viviparous species became adapted to new temperature regimes after they became viviparous. The results show that genera containing viviparous species are more often found in cold climates than would be expected by chance alone (Table 9–4). Weekes (1935), Sergeev (1940), and Tinkle and

Gibbons (1977) arrived at the same conclusion by using different methods, so the correlation cannot be attributed to hidden biases in the methodology. However, Tinkle and Gibbons (1977) could find no evidence for faster development in cold climates. Even so, present evidence is inconclusive, and the hypothesis that viviparity evolved to speed up the development rates cannot be rejected.

An alternative interpretation is that egg retention and viviparity evolved to reduce egg losses during periods of uncertainty, which are likely to be more prevalent in cold climates (Tinkle and Gibbons 1977). Eggs may face more unpredictable sources of mortality in cold climates because weather conditions are more variable. Egg retention or viviparity might then be advantageous because development could proceed during periods of adverse environmental conditions and females would be able to select the most propitious time to lay eggs or give birth to young.

Male parental behavior in polygynous birds

The trade-off between parental behavior and alternative reproductive activities is particularly evident for polygynous male birds. A male is considered **polygynous** whenever he forms a prolonged pair bond with two or more females whose nesting cycles overlap in time. Polygynous males typically acquire their mates in succession, with one or more days between each mating, but in some species they occasionally pair with more than one female in a single day.

Males of polygynous birds can either help rear extant offspring of current mates or seek additional mates. Since attracting new mates is a real possibility, males are less likely to care for offspring than males of monogamous species. For this reason males of polygynous birds never help females incubate eggs, in contrast to males of monogamous birds, which frequently do (Verner and Willson 1969).

When breeding seasons are short, females select mates fairly synchronously and males have little chance of attracting additional females once eggs begin hatching in their earliest nests. Males should therefore provide extensive parental care, at least for their first broods. The comparative evidence shows that they do (Table 9–5) (Wittenberger 1979a). When breeding seasons are more prolonged, two patterns of male behavior occur. In some species nesting cycles of successive mates greatly overlap, so that males are courting new females while their earlier mates are feeding nestlings. Males of these species generally do not contribute parental assistance to any of their mates. In a few species nesting cycles of successive mates do not greatly overlap. Mating opportunities are considerably fewer, and males contribute some parental care to at least some of their broods. Thus a clear trade-off exists between mate attraction and parental care. When mating opportunities are relatively good, males allocate most of their time and energy to seeking additional mates. When mating opportunities are relatively poor, males allocate at least some of their time to parental care.

Table 9–5 *Relation between male feeding of young and opportunities for pairing with additional females in polygynous birds.*

Species	Nest(s) where males feed young[a]	Source
Mating periods short and overlap little or not at all with nestling periods		
Pied flycatcher *(Ficedula hypoleuca)*	Primary	Haartman 1951b; Lack 1966
Bobolink *(Dolichonyx oryzivorus)*	Primary (sometimes also secondary)	Martin 1971, 1974; Wittenberger 1980c, unpublished data
Tricolored blackbird *(Agelaius tricolor)*	Primary and secondary(?)	Lack and Emlen 1939; Orians 1961a
Yellow-hooded blackbird *(Agelaius icterocephalus)* (Venezuela)	Primary(?) and sometimes secondary	Wiley and Wiley 1980
Brewer's blackbird *(Euphagus cyanocephalus)*	Primary and secondary	Williams 1952
Lark bunting *(Calamospiza melanocorys)*	Primary	Pleszczynska 1978, personal communication
Mating periods long and overlap substantially with nestling periods		
Australian bell-magpies *(Gymnorhina)*	Usually none	Wilson 1946; Robinson 1956
Great-tailed grackle *(Quiscalus mexicanus)*	None	McIlhenny 1937
Oropendolas and caciques (Icteridae)	None	Chapman 1928; Skutch 1954 Tashian 1957; Drury 1962
Red-winged blackbird *(Agelaius phoeniceus)*	Usually none (primary in some populations)	Orians 1961a; Patterson 1978
Yellow-headed blackbird *(Xanthocephalus xanthocephalus)*	Variable[b]	Willson 1966; Wittenberger, unpublished data
Weaver finches (Ploceinae)	None	Crook 1964
Dickcissel *(Spiza americana)*	None	Zimmerman 1966
Corn bunting *(Emberiza calandra)*	None	Ryves and Ryves 1934
Indigo bunting *(Passerina cyanea)*	None	Carey and Nolan 1975; Carey 1977
Long-billed marsh wrens *(Cistothorus palustris)*	Reduced care of final brood	Verner 1965
Mating periods long but overlap less with nestling periods because nests are staggered		
European wren *(Troglodytes troglodytes)*	Reduced care to all broods	Armstrong 1955
House wren *(Troglodytes aedon)*	Primary and secondary	Kendeigh 1941
Yellow-hooded blackbird *(Agelaius iceterocephalus)* (Trinidad and Surinam)	Primary and sometimes secondary (some males, but not all)	Wiley and Wiley 1980

	Nest(s) where males feed young[a]	Source
Prairie warbler *(Dendroica discolor)*	All broods	Nolan 1978
Ipswich sparrow *(Passerculus sandwichensis princeps)*	All broods	Stobo and McLaren 1975
Savanna sparrow *(Passerculus sandwichensis)*	All broods	Welsh 1975

[a]Primary nest is nest of first-mated female; secondary nest is nest of second-mated female.

[b]Male yellow-headed blackbirds may feed nestlings at primary nest predominantly (Willson 1966), at none of their nests, at primary and lower-ranking nests, or only at low-ranking nests, depending on food availability (Wittenberger, unpublished data).

Brood parasitism

Young birds generally cannot survive without parental care, but in some species adults avoid parental responsibilities by laying their eggs in the nests of other species. The owners of host nests then frequently rear the young as if they were their own (Photo 9–3). The habit of laying in nests of other species is called **brood parasitism.** It has evolved at least seven different times and occurs in about 80 species of birds (Lack 1968). The selective advantages of brood parasitism, the secondarily evolved defenses of hosts, and the tactics used by brood parasites to overcome host defenses raise many interesting evolutionary questions.

Evolutionary origins of brood parasitism

Precursors of brood parasitism. Some precursors of brood parasitism exist in birds that are not normally parasitic. For example, over 100 species of African birds frequently use old nests of other species instead of building their own but remain with the nests and rear their own young (Payne 1977). Some birds are communal breeders and typically lay in nests built by other individuals within breeding groups. It may be more than coincidental that communal breeding occurs in one subfamily of New World cuckoos (Crotophaginae) and in one species of cowbird *(Molothrus badius),* both of which are in families where brood parasitism is prevalent (Brown 1975a). Two species of nonparasitic North American cuckoos occasionally lay their eggs in active nests of other species, with the hosts then rearing the young cuckoos (Nolan and Thompson 1975). This behavior also occurs occasionally in waterfowl, especially in red-headed ducks *(Aythya americana)* (Weller 1959; Joyner 1976). Hamilton and Orians (1965) point out that females of every species occasionally lose their nests during the laying period and have to deposit their eggs somewhere else, with a likely place being

Photo
9–3

White wagtail feeding a fledgling European cuckoo.
Host parents are often smaller than brood parasite fledglings, but that does not deter them from feeding the fledglings.
Photo by F. Sieber.

a nearby nest of another species. Such behaviors provide the variability upon which selection can act.

Benefits of brood parasitism. A major benefit of brood parasitism is release from parental responsibilities. Since adults do not have to provide parental care, they can devote considerably more time and energy to alternative activities such as self-maintenance, egg production, or mate attraction. Moreover, brood parasitism may extend the length of the breeding season because nesting would no longer be restricted to the period when food of the brood parasite is most abundant. The breeding season of brood parasites is limited by availability of host nests, not availability of food resources.

Release from parental responsibilities and extension of the breeding season both allow increased fecundity. High fecundity has been documented in several parasitic birds. Female cuckoos of nine species lay 16 to 26 eggs per season, and female brown-headed cowbirds *(Molothrus ater)* lay about 30 eggs per season (Payne 1973a, b; see also Scott and Ankney 1979). This is undoubtedly more eggs than could be laid by nonparasitic species with comparable foraging habits. Brood parasitism therefore shifts time and resource allocations from parental care to fecundity, and environmental conditions favoring such a shift may promote the evolution of brood parasitism.

A second potential benefit of brood parasitism is that hosts may provide better parental care than could the true parents (Hamilton and Orians 1965; Payne 1977). When a species exploits difficult-to-capture, scarce, highly variable, or ephemeral food resources, nestlings may survive better when cared for by a host that exploits more reliable food resources.

At least some brood parasitic birds do exploit difficult-to-capture, scarce, or uncertain food resources. Parasitic (and nonparasitic) cuckoos capture toxic caterpillars, lizards, and other large prey by perching quietly until a suitable prey ventures within striking distance (Hamilton and Hamilton 1965; Lack 1968). This mode of prey capture reduces food intake rate, and the costs of detoxification may make such food items relatively poor for feeding nestlings. Honeyguides are specialized to eat bee larvae and beeswax, although they also eat other insects, and they often wander widely in search of bee nests (Friedmann 1955, 1960; Lack 1968; Cronin and Sherman 1976). This behavior may hamper the ability of adults to feed offspring adequately. Cowbirds forage around large mammals, picking ectoparasites off them or capturing insects stirred up by their movements. They do not appear to be very adept at obtaining food under other conditions. Since herds of large mammals are often nomadic, cowbird food resources tend to be ephemeral and unpredictable (Hamilton and Orians 1965). If cowbirds had to feed their nestlings, sudden changes in food availability might often result in death of offspring. Parasitic (and nonparasitic) weaver finches exploit spatiotemporally variable seed resources, and most nonparasitic species are adapted to breed very rapidly once conditions become suitable (Friedmann 1960; Crook 1964; Payne and Payne 1977). Breeding seasons are not restricted to short periods of food abundance for parasitic species, and hence they can breed under conditions when nonparasitic species cannot. It should be realized, however, that restrictive diets and unpredictable food resources are not unique to brood parasitic species. Many species that face similar restrictions are not parasitic, indicating that additional factors must be involved.

A third potential benefit is that brood parasitism spreads the risk of egg loss (Payne 1977). When a bird lays all its eggs in one nest, either all the eggs escape predation or all are lost. By laying one egg in each of several nests, the chances of at least one egg escaping predation are increased. Spreading risks is not advantageous of itself, however, because the proportion of eggs lost to predators would be the same in both cases when averaged across all individuals of each phenotype. Risk spreading reduces the variance of egg loss rates but does not affect the mean. Perhaps reduced variance in loss rates is advantageous under some conditions, but no hypothesis has been offered to explain why.

None of the above advantages are uniquely obtainable by brood parasitic species. Many nonparasitic species could benefit from brood parasitism in the same ways, and yet they have remained nonparasitic. Additional constraints or selective factors must therefore be involved. One possibility is that the costs of brood parasitism are lower in parasitic species than in nonparasitic species.

Costs of brood parasitism. An important cost of brood parasitism is reduced survival of eggs and young. Eggs and nestlings of brood parasites are subject to several kinds of mortality unique to the parasitic habit. Eggs may be lost because they are abandoned or ejected from the nest by the host. Eggs deposited after the host has begun incubating may never hatch, or they may hatch too late for the parasite nestling to successfully compete with host nestlings for food. Host parents may not feed parasite nestlings the proper kinds of food necessary for their survival. Finally, when host nests are multiply parasitized, competition among parasite nestlings may prevent any of them from surviving.

Brood parasitism may be less costly for species with particular preadaptations (Hamilton and Orians 1965). For example, short incubation periods would increase survival prospects of eggs laid after the host has begun incubating. Both parasitic and nonparasitic cuckoos have incubation periods of only 10 to 12 days, which are among the shortest of any passerine bird (Hamilton and Hamilton 1965; Lack 1968; Payne 1977). A generalized nestling diet or an ability to specialize on host species with appropriate diets would increase the chances for parasite nestlings to survive on the food provided by a host. Producing large nestlings would improve the ability of parasite nestlings to outcompete host nestlings for food. None of these characteristics are unique to parasitic species, but such characteristics may limit brood parasitism to relatively few species.

Overview. At present there is no adequate explanation for brood parasitism. None of the potential benefits or preadaptations discussed above are unique to parasitic species, and hence none explain why brood parasitism evolved in some species but not in others. Perhaps a theory based on life history strategies would help explain the phenomenon, but to date no such theory has been developed.

Host defenses and countermeasures

Abundant evidence demonstrates that brood parasitism usually reduces reproductive success of hosts (reviewed by Payne 1977). Female cuckoos and cowbirds typically remove a host egg before laying their own, and sometimes they puncture or eat host eggs to induce hosts that have already begun incubating to re-nest. Nestlings of some cuckoos eject host eggs and nestlings from the nest (Photo 9–4), while nestlings of some honeyguides kill host nestlings with specialized hooks on the tips of their bills. Brood parasite nestlings are often larger than host nestlings, allowing them to obtain most of the food and sometimes to crowd host nestlings out of the nest. After fledging, the young parasite may further reduce survival of the host's offspring by monopolizing the host's parental care. Thus the evolution of host defenses against brood parasites is not surprising.

Photo 9–4

A European cuckoo nestling pushing host eggs out of the nest.

Pushing out host eggs allows the cuckoo nestling to monopolize all the food delivered by host parents, but it probably generates some risk that the parents will desert the nest. Such behavior typifies Old World cuckoos but is largely absent in other brood parasitic birds.

Photo by Eric Hosking.

Mobbing brood parasites. One way hosts can avoid brood parasitism is to prevent eggs from being deposited in their nests. To this end, many host species regularly attack or mob brood parasites near their nests (Photo 9–5). Such aggression has been reported primarily for Old World cuckoos and brown-headed cowbirds (Baker 1942; Friedmann 1948, 1963; Rothschild and Clay 1952; Robertson and Norman 1976, 1977).

The deterrence value of host aggression is not well documented. Certainly aggression is not always an effective defense against brood parasitism. Female brood parasites often ignore such attacks if they are larger than the hosts. They also frequently slip onto nests while the hosts are absent. In jacobin cuckoos *(Clamator jacobinus)* the male lures host parents away from the nest, allowing the female to lay while the hosts are preoccupied with chasing the male (Liversidge 1971; Gaston 1976). In Indian koels *(Eudynamis scolopacea)* the male bears a mimetic resemblance to its main host species, the Indian house crow *(Corvus splendens)* (Somanader 1946; Lamba 1963). Because the male resembles an intruding crow, its approach to the crow's nest elicits a territorial response. Again, the female lays while the hosts are chasing off the male. In hawk cuckoos *(Cuculus)* both sexes resemble hawks that prey on small birds (Kuroda 1966; Lack 1968). The males lure hosts away from nests by eliciting a mobbing response that is normally directed toward hawks. In drongo cuckoos *(Surniculus lugubris)* both sexes resemble drongos, which are pugnacious toward small birds (Lack 1968). The resemblance apparently intimidates hosts, thereby scaring them away while the female lays. Hawk cuckoos may also rely on mimicry to scare hosts away from nests, but evidence to show that they really do scare hosts is lacking.

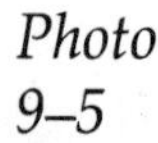

*A nesting pair of nightingales (*Luscinia megarhynchos*) mobbing a stuffed European cuckoo.*

The aggressive behavior of hosts probably deters brood parasites from laying in host nests, but such behavior is often not an effective deterrent against brood parasitism.

Photo by Eric Hosking.

Host aggression should have deterrence value even though countermeasures of brood parasites sometimes circumvent it. Evidence that host aggression has selective value comes from comparing regularly parasitized hosts to hosts that are rarely parasitized. The former species attack stuffed models of brood parasites more frequently and more vigorously than the latter species (Edwards et al. 1949–1950; Smith and Hosking 1955; Robertson and Norman 1976, 1977), indicating that host aggression probably is an evolved response to being parasitized and not just a generalized response to intruders.

Although hosts apparently cannot prevent parasites from laying once their nests are found, they could still benefit by behaving aggressively. At least two benefits are possible. First, aggression by hosts may prevent parasites from finding nests as readily. Robertson and Norman (1977) found, for example, that cowbird parasitism is less prevalent when host densities are high. They attributed this observation to more effective mobbing by hosts. However, their results are not conclusive evidence for showing that aggression deters brood parasites. Intensity of mobbing was not measured as a function of density, and the same correlation would result if high densities of host nests swamp the ability of brood parasites to lay in as high a proportion of all the available nests. Second, aggression by hosts may increase the costs of parasitism, thereby encouraging the parasites to choose nests of less aggressive individuals. No evidence is available for evaluating this hypothesis.

Rejecting parasite eggs. Aggression is not the only means of defending against brood parasitism. A second defense is rejecting parasite eggs after they have been laid, either by ejecting them from the nest or by abandoning the nest altogether. Ejecting eggs is the preferable defense from an energetic standpoint because it does not force the host to lay a new clutch of eggs. However, nest abandonment may be a better defense if parasites replace ejected eggs, which is likely if they have difficulty finding host nests.

Some host species nearly always reject cowbird eggs, while other species nearly always accept them (Rothstein 1975a). Why species differ in this respect is unclear. Rothstein (1975b) suggests that some species are acceptors because the genetic mutations underlying rejection behavior have not yet arisen in the population. An alternative hypothesis is that the costs of rejection are too high. Ejecting eggs is not without cost, since hosts may inadvertently break their own eggs in the process of ejecting parasite eggs. Cedar waxwings *(Bombycilla cedrorum)*, for example, often break their own eggs while ejecting cowbird eggs (Rothstein 1976). Abandoning nests is even more costly, as it requires laying of replacement clutches, which is energetically costly and may delay the nestling period to a less favorable part of the breeding season. Although such costs undoubtedly exist, they cannot explain why some species are rejectors and others are acceptors unless they are generally higher for rejectors. Present evidence is insufficient for comparing the costs among acceptors and rejectors.

Egg mimicry. The most important countermeasure for preventing egg rejection by hosts is egg mimicry (Lack 1968; Payne 1977). Producing eggs that resemble those of the host has evolved mainly in Old World cuckoos. Each female cuckoo probably lays only one type of egg and specializes on the host whose eggs resembles hers. Different females of a given species specialize on different hosts. Hence cuckoo populations consist of subpopulations, or *gentes*, each of which specializes on a particular host.

The effectiveness of egg mimicry is clear. Most hosts of the European cuckoo *(Cuculus canorus)*, for example, reject strange eggs by either abandoning their nests or removing strange eggs from them, but they regularly accept cuckoo eggs resembling their own. The probability of a cuckoo egg being accepted depends on how closely it resembles the host egg (Baker 1942). This relationship creates strong selection for a close resemblance. As a result, cuckoo eggs often resemble host eggs so closely that they can be distinguished only by the thickness of their shells.

Overview. The evolution of host defenses and brood parasite countermeasures is a dynamical process, with selection continually favoring better defenses in host species and better countermeasures in parasite species. In that sense host-brood parasite systems resemble plant-herbivore and predator-prey systems. Hosts employ a diversity of tactics to deter brood parasites. The origins of this diversity are poorly understood. Quite possibly host defenses against brood parasites parallel defenses employed against predators. Mobbing may perhaps be most prevalent in hosts that also mob predators, and it may be least prevalent in hosts that rely on concealed nests to deter predators. Host defenses may also depend on such factors as length of breeding season and number of broods produced per year. Ejecting eggs or abandoning nests may be too costly or unfeasible if breeding seasons are short or many broods can be produced each season, but they may be advantageous if breeding seasons are prolonged and only one brood is produced per year. A variety of such possibilities come to mind, but none have been tested to date.

Parent-offspring conflict

Nature of the conflict

An inherent property of social behavior is that the interests of interacting individuals invariably conflict to some degree. Each individual behaves so as to increase its own inclusive fitness, even when it reduces the inclusive fitness of others. Thus social interactions inevitably entail competition, even during the most cooperative interactions.

The interaction between parents and offspring involves a high level of cooperation, and it has traditionally been viewed as free of conflict. Parents and offspring both have the same objective, namely, enhancement of offspring survival. Nevertheless, parents and offspring are likely to disagree about how important survival of an individual offspring is relative to alternative options open to the parents (Trivers 1974). The fitness gain achieved by an offspring staying alive is higher for the offspring than it is for the parents, and hence offspring should be willing to incur higher costs to stay alive than parents are willing to accept to keep them alive. Thus offspring should in general prefer receiving more parental care than parents are prepared to give.

The region of conflict can be computed from kin selection theory. The cost of parental investment is the same for both parents and offspring because both are related to future offspring of the parents by one-half. However, the benefits are twice as high for offspring because offspring are related to themselves by one, while parents are related to offspring by one-half. Since parents are equally related to present and future offspring, parental investment becomes disadvantageous when the costs, measured in terms of reduced future reproductive output, are less than the benefits, measured in terms of present reproductive output. In contrast, offspring continue to benefit until the parents' cost is twice as much as the parents' benefit. Hence when costs are less than benefits, parents and offspring both benefit from parental investment and there is no conflict (Figure 9–9). When costs are greater than benefits but less than twice the benefits, parents lose from continued parental investment but offspring still gain. Under those conditions a conflict of interest exists between parents and offspring. Finally, when costs exceed twice the benefits, parents and offspring both lose from continued parental investment and there is again no conflict of interests.

Parent-offspring conflict is especially evident while offspring are becoming independent of their parents (Trivers 1974). Weaning in mammals is a prolonged process, during which mothers show increasing intolerance of offspring while offspring continue seeking more parental investment from their mothers (e.g., Harper 1970; Lent 1974). The behavioral evidence shows that a conflict of interests does occur, but it is not sufficient evidence for testing the theory outlined above. For that purpose a careful cost-benefit analysis will be necessary, and to date no such analysis has been conducted.

A similar conflict of interests also occurs in birds. As juvenile birds

Figure 9–9

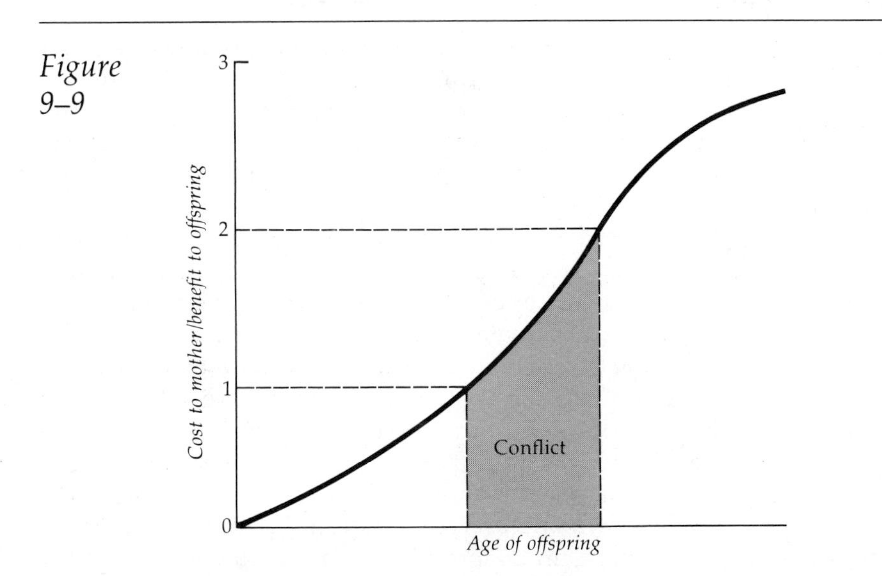

A model of parent-offspring conflict, showing the conditions under which a conflict should occur.

Parent-offspring conflict should exist if the cost of parental investment for the parent divided by the benefit of the investment for offspring is greater than one but less than two. This range of conditions arises because parents are less closely related to offspring than offspring are to themselves (see text).

Source: From E. O. Wilson, *Sociobiology: The New Synthesis* (Cambridge, Mass.: Belknap Press of Harvard University Press, 1975), p. 342 (Based on Trivers 1974.). Reprinted by permission.

grow older, parents become increasingly reluctant to feed them, while juveniles become increasingly more persistent at begging for food (Davies 1976b). Parental resistance makes begging a more costly way for juveniles to obtain food, and eventually self-feeding becomes more profitable. Then juveniles abruptly cease begging and begin foraging entirely for themselves. These results indicate how the conflict is resolved, but again they do not test the theory that a conflict arises only within the range specified by kin selection theory.

Conflicts of interest are not confined to the weaning period. Parents and offspring are likely to disagree over how much should be invested during each stage of development. This should be particularly true during unfavorable periods when food is difficult to obtain. In the most extreme cases, when parents resort to killing their youngest offspring to give older offspring a better chance of surviving (see pp. 171–172), the conflict of interest is clearly evident. However, even under less extreme conditions parents might be most fit by allocating substantial resources to growth or self-maintenance, while offspring would be most fit by obtaining more parental investment than parents are prepared to give.

How are parent-offspring conflicts resolved?

An important question concerns how parent-offspring conflicts are resolved. A variety of theoretical arguments have been made, each generating a different set of predictions, but few empirical data are available. It may turn out that the answer varies with species or context, but the question will not be resolved until better data become available.

Hypotheses. According to one view a compromise is reached, with offspring receiving more parental care than would be optimal for their parents but less than would be optimal for themselves (Trivers 1974). Offspring might obtain this extra parental care by exaggerating their nutritional needs, thereby deceiving parents into providing more care than they otherwise would. Offspring may also obtain extra parental care by making resistance to their demands more costly. Then parents would provide some extra care just to avoid being harassed as much by begging offspring.

A second view is that parents provide no more parental investment than is optimal for their own interests (Alexander 1974). Alexander argues that any behaviors enabling an offspring to extract extra parental investment to the detriment of its parents could not evolve because those same behaviors would eventually reduce the offspring's fitness when it became a parent. If his view is correct, parents should manipulate offspring in conformity with their own best interests, while offspring should accede to parental discipline rather than resist.

Evidence. Two separate modeling efforts indicate that Alexander's (1974) argument does not always hold. A genetic model of parent-offspring conflict suggests that the way conflicts are resolved depends on how high the costs of seeking extra parental investment are for offspring and on how high the costs of resisting such attempts are for parents (Stamps et al. 1979). It predicts that conflicts are resolved in accordance with parental interests if the costs are high for offspring and low for parents but that a compromise will be reached if the costs are high for both offspring and parents. A second model, based on an evolutionarily stable strategy (ESS) approach, predicts that conflicts are resolved in accordance with parental interests if the costs for parents are above some threshold level, while a compromise is reached if the costs are below that threshold level (Parker and MacNair 1978a, b).

One indication of how conflicts are resolved comes from studies of weaning behavior under differing environmental conditions. Under adverse conditions less energy is available for both parental care and self-maintenance of the parent. This result implies that either offspring, parents, or both must suffer when food becomes scarce. If conflicts are resolved in accordance with offspring interests, most of the energy deficit should be suffered by the parents, and offspring should receive nearly as much parental care as they would under better conditions. If they are resolved in accordance with parental interests, most of the energy deficit should be suffered by offspring, not parents.

Although adequate data are not available, most studies show that parental effort of female mammals declines considerably under adverse conditions. Embryos are reabsorbed or aborted, weaning may occur earlier, litter sizes are sometimes reduced, and breeding sometimes becomes less frequent. For example, bighorn sheep *(Ovis canadensis)* lambs are weaned earlier and at lighter weights in deserts than in more productive grasslands, indicating that much of the energy deficit is made up by reduced parental investment (Berger 1979). Crowding generally reduces or eliminates successful reproduction in rodents (e.g., Davis 1951; Christian 1956, 1961, 1963; Christian and LeMunyan 1958; Barnett 1964), and food shortages generally lead to reduced reproductive rates in ungulates (e.g., Preobrazhenskii 1961; Wiltbank et al. 1962; Klein 1970; Lamond 1970; Geist 1971; Sinclair 1977). However, offspring are not the only ones to suffer during adverse conditions. Adult females also lose more weight than usual, indicating that part of the deficit is made up by reducing energy allocations to self-maintenance. Unfortunately, these data alone are not sufficient to determine exactly how parent-offspring conflicts are resolved. Without predictions regarding the optimal allocation patterns from parent and offspring points of view, deviations from parental or offspring optima cannot be detected.

A second indication of how conflicts are resolved comes from sex ratio data in social ants, bees, and wasps. In social hymenoptera workers are more closely related to sisters than to brothers, while queens are equally related to daughters and to sons (Trivers and Hare 1976). As a result, the optimal sex ratio among reproductives produced by a colony should be three virgin queens per male from a worker's point of view, but only one virgin queen per male from the queen's point of view. A comparative analysis of 20 ant species gives an average ratio of nearly 3:1, suggesting that workers control resource allocations to larvae (Trivers and Hare 1976). However, the data are amenable to an alternative interpretation, which will be discussed in Chapter 12, and queens may actually control resource allocations by chemically manipulating worker behavior (Alexander and Sherman 1977). Parent-offspring conflicts that arise in social insects and vertebrates are often treated as comparable phenomena, but they are not. In social insects the offspring (i.e., workers) have physical control of resources and are responsible for performing all care of the parent's (i.e., queen's) subsequent offspring. The queen is placed in a position where she must manipulate offspring behavior to obtain the parental care pattern optimal for her best interests. In vertebrates, on the other hand, parents maintain physical control of resources and perform all parental care themselves, except when they have helpers at the nest or den. Parents do not need to manipulate anyone to determine how parental resources are allocated. Indeed, offspring must manipulate their parents to gain extra parental care from them. Thus parent-offspring conflict in social insects involves parental manipulation of offspring and offspring resistance to being manipulated, while parent-offspring conflict in vertebrates involves offspring manipulation of parents and parental resistance to being manipulated. The answer to how parent-offspring conflicts are resolved therefore depends to a considerable extent on who controls the relevant resources.

Conclusion

Parental care is one aspect of an animal's overall life history strategy, and its evolution depends on how an animal can best allocate limited time and resources to enhance fitness. Two theories have been advanced to explain overall life history patterns, but neither has proven entirely adequate. What is needed is a multiple-variable approach involving a separate cost-benefit analysis for each life history trait.

The evolution of parental care strategies is a neglected topic that has attracted little attention until recently. Some evidence is now available for evaluating how the diverse parental strategies of fish and reptiles may have evolved. However, in higher vertebrates there has been very little analysis of how parents modify the level of parental investment in response to changing environmental or social circumstances.

Perhaps the most significant theoretical advance is the recognition that parents and offspring have conflicting interests. This conflict of interests is indicative of a more general pattern evident in all social interactions. Genetically different individuals never have completely identical interests, and consequently there is always some impetus for manipulating others to pursue selfish interests. The conflict of interests is rather minimal in parent-offspring interactions, but it is extremely prevalent in mating contexts, as will be shown in the next two chapters.

Sex and sexual selection

Sex has such a pervasive influence on behavior that we as humans take its existence largely for granted. After all, what better way is there to reproduce than through sex? A surprising answer for most people is that reproduction without sex appears to be a better way, at least in terms of individual fitness. Several costs are associated with sexual reproduction, and the benefits offsetting them are not immediately obvious. For this reason the evolution of sex poses one of the more challenging problems in evolutionary biology today.

Why sex?

Costs of sexual reproduction

A fundamental disadvantage of sex stems from producing males (Maynard Smith 1971, 1974b, 1978; Treisman and Dawkins 1976). This can be seen by considering the following argument:

Reproductive output is limited by the rate at which resources (i.e., energy and nutrients) can be converted into offspring. An asexually reproducing organism produces progeny without any intervening gametic stage, and all resources devoted to reproduction are converted directly into progeny. In contrast, a sexually reproducing organism converts resources into two gametic types, male and female, which then unite to form progeny. If male and female gametes are produced in equal numbers, all resources devoted to gamete production can be converted into progeny. However, if more male than female gametes are produced, the number of progeny is limited by the number of female gametes produced. Resources used to produce excess male gametes are lost, and hence fewer progeny are produced per unit of resource. In a population consisting of both sexually and asexually reproducing individuals, asexual individuals would produce more progeny per unit of resource and hence would have higher fitness, unless asexual reproduction somehow hampered the ability of these individuals to compete for resources.

For reasons to be discussed later, sexual organisms invest equal amounts of resource into each gametic type (Fisher 1930). When gametes are all the same size, a condition known as **isogamy,** equal numbers of each mating type are produced and gametic wastage beyond natural mor-

Chapter ten

tality is minimal. However, when the male type is smaller than the female type, a condition known as **anisogamy,** more male gametes are produced than female gametes. As the ratio of male to female gametes increases, the proportion of resources wasted on producing excess male gametes gradually approaches one-half the total resources devoted to gamete production. This wastage represents a cost of anisogametic sex and might best be referred to as the **cost of anisogamy.**

The cost of anisogamy is often confused with a supposed *cost of meiosis*. The cost of meiosis results from reduced genetic relatedness between sexually reproducing parents and their offspring (Williams 1975). A sexually reproducing parent is only half related to each offspring, while an asexually reproducing parent is genetically identical with every offspring. Sexually reproducing parents therefore seem to propagate their genes only half as fast as asexually reproducing parents, implying a cost of half of their total reproductive expenditures. However, if sexually and asexually reproducing parents produce the same number of offspring per unit of resource, this cost is nonexistent. Sexually reproducing parents, though only half related to each offspring, would produce twice as many offspring as asexually reproducing parents because they invest only half as much resource into each one (assuming each sex of parent contributes an equal amount of resource for each offspring). Hence the cost of meiosis, as originally defined, is zero. Sexual reproduction is not costly because parents are less related to each offspring; it is costly when resources are wasted on excess gametes of one sex that can never become incorporated into offspring.

A second, though perhaps less significant, cost of sexual reproduction stems from genetic recombination (Williams 1975). Sexual reproduction entails the uniting of two haploid gametes to form a zygote, and any given gamete is likely to possess several deleterious recessive alleles. When the zygote consists of two gametes with the same deleterious recessive (i.e., it is homozygous for that allele), the offspring suffers reduced fitness and may die. Asexually produced offspring are free of this risk because, barring new mutations, they possess exact replicates of all the parent's genes and homozygous individuals have already been selected out of the parental generation. Thus sexually reproducing individuals waste some resources on gametes destined to produce less fit offspring, while asexually reproducing individuals do not. The cost of genetic recombination is an important component of **genetic load,** the fitness reduction caused by homozygosity of deleterious recessives in a population, and it can be substantial (see Wallace 1968; Dobzhansky et al. 1977; Futuyama 1979).

Daly (1978) includes as a third cost of sexual reproduction all the time and energy expended in social interactions to secure a mate and coordinate reproductive activities. However, such expenditures are not costs of sexual reproduction per se because sexual reproduction evolved before such social interactions arose. These expenditures represent costs of the specific social behaviors involved, not of sexual reproduction itself.

The costs of sexual reproduction have not been precisely quantified, but they are clearly substantial. The problem, then, is to identify selective advantages of sexual reproduction large enough to offset those costs.

An interdemic selection hypothesis

A widespread view among population geneticists is that sexual reproduction evolved as a means of increasing the rate at which favorable mutations and new genetic linkage patterns can be incorporated into evolving populations (Crow and Kimura 1965; Kimura and Ohta 1971). New genotypes can be created in asexually reproducing populations only through mutation, a slow process, and they can increase in prevalence only when clones possessing them propagate faster than clones not possessing them (Fisher 1930; Muller 1932). In sexually reproducing populations new genotypes arise much more rapidly because genetic recombination creates new genetic variability much faster than mutations can. Sexually reproducing populations therefore have a long-term advantage in changing environments, as more rapid adaptability to new environmental conditions lessens their chances of going extinct. Over the short term, however, parthenogenetic (i.e., asexually reproducing) females have the advantage because they avoid the costs of sexual reproduction. In a population containing both types of females, individual selection should continually increase the prevalence of asexual reproduction until sexual reproduction is lost.

According to the interdemic selection hypothesis, the faster extinction rate of parthenogenetic strains is a counterselective force preventing the elimination of sexually reproducing forms. Two counterbalancing processes are occurring simultaneously. Local populations with sexually reproducing members go extinct less often and hence spread into habitats previously occupied by asexual populations that became extinct. At the same time, individual selection within local populations that contain sexually reproducing individuals gradually convert those populations into entirely asexual strains. If the former process occurs more rapidly than the latter, sexual reproduction would spread.

The interdemic selection hypothesis is logically plausible because the requirements necessary for interdemic selection to occur are met. Genetically isolated groups (sexually reproducing populations and parthenogenetic strains) could exist as competing entities, and different proliferation or extinction rates among those entities could make one more prevalent than the other.

Group selection cannot favor sexual reproduction unless sexually reproducing populations that give rise to parthenogenetic clones become extinct faster than such clones can become established (Maynard Smith 1978). Otherwise, parthenogenetic clones would increase in frequency relative to sexually reproducing populations. Since extinction of sexually reproducing populations occurs infrequently, the interdemic selection hypothesis requires that new parthenogenetic clones arise only rarely within those populations. Maynard Smith (1978) has reviewed the evidence and concluded that this is probably true, at least for animals.

A strong argument against the interdemic selection hypothesis is based on the stable coexistence of sexually and asexually reproducing individuals within a single population (Williams 1975). If sexual and parthenogenetic forms can coexist, sexual reproduction must confer advantages to individuals sufficient to offset the costs of sex since stabilizing selection requires that the costs and benefits of a trait exactly offset one another. If the costs exceed the benefits, as the interdemic selection hypothesis implies, directional selection should cause sexual reproduction to disappear within local populations rather than coexist in a stable equilibrium. Since stable equilibria involving sexual and asexual modes of reproduction do exist, sexual reproduction must provide benefits at the individual level. In that event, interdemic selection arguments would be largely superfluous.

Although sexual and parthenogenetic modes of reproduction do coexist in several plants and invertebrates, in most cases special circumstances associated with their co-occurrence make rejection of the interdemic selection hypothesis premature. Details of the circumstances cannot be covered here, and interested readers should refer to Maynard Smith (1978).

The interdemic selection hypothesis depends on the assumption that sexually and asexually reproducing populations already exist in a selective regime. Interdemic selection cannot occur when all populations are either one type or the other. Hence interdemic selection cannot explain how sexual reproduction originated from wholly asexual forms in the first place. Genetic drift is the only possible origin of sexual reproduction other than individual or genic selection, and genetic drift could hardly have produced the complex cellular mechanisms involved in meiosis. Thus interdemic selection may help maintain or proliferate sexual modes of reproduction once they arise in a population, but it cannot give rise to them in the first place.

An individual selection hypothesis

In plants and animals that reproduce both sexually and asexually, sexual reproduction always precedes dispersal of propagules (Williams 1975). Dispersal is a period of high uncertainty because dispersing offspring enter new and often unpredictable environments. Here is where sexually produced offspring might have an advantage. Genetic recombination creates new genotypes, some of which are likely to be better suited to a new environment than the parental genotype. When sexual and asexual parents both produce large numbers of propagules, offspring of sexual parents should survive better or compete more successfully in new environments than offspring of asexual parents. The opposite is true if the mode of dispersal ensures that most offspring settle in environments similar to those occupied by their parents. Then genotypes identical to the parental genotypes should be most successful, since the parental genotypes have already proven their superiority in those environments.

Several mathematical and computer simulation models have been used to generate testable predictions based on this hypothesis. The main prediction generated by these models is that genetic recombination should become selectively advantageous in temporally varying environments only if relationships between environmental variables change sign each generation (Charlesworth 1976; Maynard Smith 1978). For example, genetic recombination would be advantageous if hot places are usually wet and cold places are usually dry in one generation, and then hot places become dry and cold places become wet for the next generation. In the absence of such extreme changes, genetic recombination should be selectively neutral or disadvantageous. This prediction has not been tested with empirical data, but such radical changes are not likely to occur for most populations.

The empirical evidence does not support the hypothesis that sexual reproduction evolved as an adaptation to temporally fluctuating environments. Comparative studies show that parthenogenetic strains are most commonly associated with unpredictably fluctuating environments, while sexually reproducing forms are most commonly associated with complex ecosystems in temporally stable environments (Levin 1975; Glesener and Tilman 1978; Maynard Smith 1978). These results are contrary to the hypothesis that genetic recombination evolved because it enables offspring to compete more effectively in changing environments.

An alternative hypothesis is based on spatially varying environments (Maynard Smith 1978). Consider a population, taken as haploid for simplicity, that inhabits two different regions (Figure 10–1). In one region the optimal genotype is AB, while in the other region it is ab. Now suppose

Figure 10–1

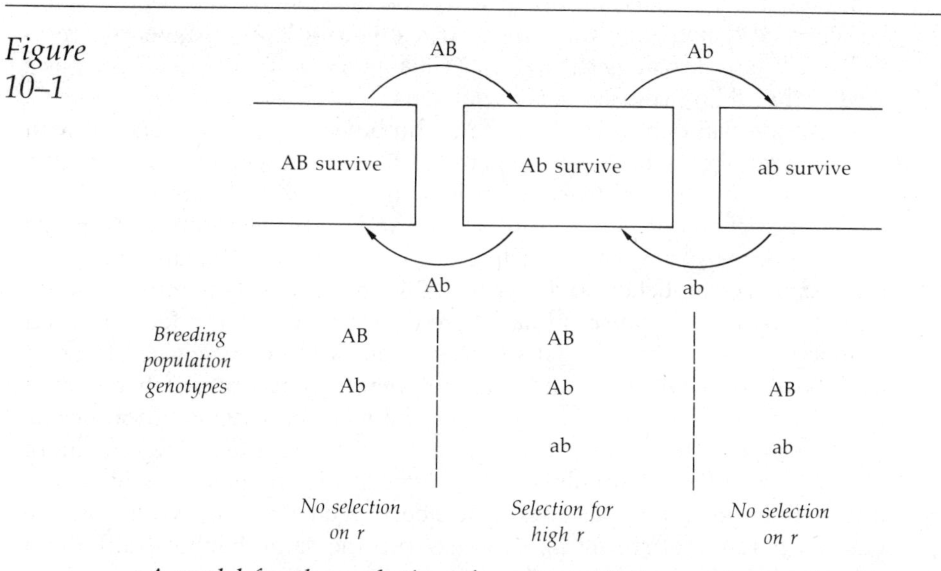

A model for the evolution of sex in spatially varying environments.
If AB and ab genotypes migrate to a region where only Ab genotypes survive, selection will favor high recombination (r) in that region. In regions where only AB or ab genotypes survive, selection will not favor recombination.

Source: From John Maynard Smith, *The Evolution of Sex* (Cambridge: Cambridge University Press, 1978), p. 97. Reprinted by permission.

a new region opens up for colonization, and the optimal genotype in that new region is Ab. Assuming individuals migrate between regions before reproducing, sexual forms have a selective advantage in the central region because genetic recombination allows them to produce Ab offspring at a much faster rate than can asexual AB or ab forms. In the terminal regions there is no selection either for or against genetic recombination.

For the population as a whole, genetic recombination is selectively advantageous if all migrants go to the central region but may not be advantageous if migrants go to both the central region and the other terminal region. That is, genetic recombination is advantageous if AB and ab genotypes migrate only to regions where Ab offspring are most fit but not if AB and ab genotypes also migrate to regions where ab and AB genotypes, respectively, are most fit. The reason is that many of the sedentary sexual individuals in each terminal region will pair with migrants to produce inviable offspring, while sedentary asexual parents will not. The dispersal pattern required by Maynard Smith's model is probably unrealistic for real populations, but models based on more realistic patterns are possible (e.g., see Slatkin 1975). Such models will not say much about the selective advantages or disadvantages of genetic recombination in spatially varying environments until they incorporate dispersal patterns relevant to real populations.

The above models assumed that offspring intermix randomly following dispersal. However, if offspring from a particular parent all enter the same environment and compete with one another, sexual reproduction becomes advantageous under less stringent conditions (Maynard Smith 1976b, 1978). An analogy devised by Williams (1975) helps explain the basic idea of the sibling competition hypothesis.

An asexual parent is like a man who buys 100 raffle tickets, all with the same number, while a sexual parent is like a man who buys fewer raffle tickets (due to the cost of sexual reproduction) all with different numbers. Each raffle ticket represents one offspring, and the ticket number represents the offspring's genotype. If each ticket is for an entirely different raffle (i.e., habitat), the ticket holder with 100 identical tickets is more likely to win at least once because he has more tickets and having identical ticket numbers does not matter. This corresponds to the case where offspring (tickets) disperse randomly and enter different habitats (raffles). In contrast, if all the tickets are for the same raffle, the man with 100 identical tickets clearly has less chance of winning than the man with fewer tickets, all of which have different numbers. Thus genetic recombination, which produces different genotypes (ticket numbers), gains a selective advantage when a parent's offspring all disperse into the same habitat (raffle) and compete for a limited number of resources (prizes).

Siblings are not likely to compete within a single habitat unless parents disperse before reproducing and then produce sedentary offspring. The fact that many sexually reproducing organisms do not show this dispersal pattern therefore poses a serious difficulty for the sibling competition hypothesis. Williams (1975) argues that sexual reproduction may have originated in organisms having the appropriate dispersal pattern, with secondary specializations associated with sex preventing modern organisms that exhibit other dispersal patterns from reverting back to asexual reproduction. However, this argument is untenable for populations exhibiting both sexual and asexual reproduction during different stages of their life cycle.

A genic selection hypothesis

A third approach to the problem is based on the phenomenon of genetic hitchhiking (Maynard Smith and Haigh 1974). Hitchhiking is most clearly exemplified by an experiment with the gut bacterium *Escherichia coli* (Cox and Gibson 1974). Two nonrecombinant strains of *E. coli,* differing only with respect to a mutator gene that greatly increases mutation rates at other loci, were placed together in agar and allowed to compete. The mutator strain started at a low frequency in every trial, but eventually it always replaced the nonmutator strain. This outcome arose despite the fact that mutator genes generate many more deleterious mutations than beneficial mutations.

Since laboratory environments are abnormal for *E. coli,* which normally live in mammalian digestive tracts, some mutations help adapt their phenotypic carriers to the new environment. Favorable mutations are most likely to occur in the mutator strain first, and once they occur, the new genotype will rapidly increase at the expense of all other genotypes. Because no recombination occurs between mutator and nonmutator strains, the new strain that replaces the others will nearly always contain the mutator gene. Hence the mutator gene gets a free ride to fixation. The effect works in spite of the many deleterious mutations caused by the mutator gene because the low fitness of genotypes possessing those mutations has no effect on the high fitness of genotypes possessing favorable mutations until enough time has passed for the latter strains to accumulate deleterious mutations. By that time those strains have already reached fixation.

A similar hitchhiking effect could lead to fixation of genes promoting genetic recombination. Such genes could increase in frequency solely because they generate new recombinant genotypes with higher fitness. Consider, for example, a genotype where a new favorable mutation b is closely linked to an allele A that is deleterious at high frequencies (Strobeck et al. 1976; Maynard Smith 1978). If the gene promoting recombination is closely linked to the Ab linkage group, it could hitchhike a ride to fixation with the b allele as soon as it broke the linkage between A and b.

Two different computer simulation models show that genetic recombination could evolve by hitchhiking (Felsenstein and Yokoyama 1976; Strobeck et al. 1976). The underlying assumptions of the models are not unreasonable, and they do not restrict the processes to only certain kinds of populations (Maynard Smith 1978). Although no empirical evidence currently exists to show that hitchhiking occurs in real populations, it should be universally possible. The hypothesis does have one important weakness. It can explain why a little recombination should be favored over no recombination, but it cannot explain why a lot of recombination should be favored over a little (Maynard Smith 1978). Perhaps hitchhiking is responsible for the origin of sexual reproduction, while other factors are responsible for its spread through a population.

Why two sexes?

The hitchhiking hypothesis may explain how isogametic sex originated, but it cannnot explain how anisogametic sex originated. Isogametic sex provides all the advantages of genetic recombination without entailing the high costs of anisogamy. Quite probably isogamy evolved first, with anisogamy arising in response to additional selective pressures. Several models have been developed to explain how anisogamy could have evolved through disruptive selection (Parker et al. 1972; Bell 1978b; Maynard Smith 1978; Parker 1978b). The following discussion summarizes the main points.

In a primitive isogamous population all gametes were probably motile and just large enough to function as gametes. Such a population is evolutionarily stable unless zygotes produced by the fusion of two small microgametes have very low chances for survival (Maynard Smith 1978). Any mutant genotypes that produce smaller gametes would be less fit because zygotes arising from them would not survive as well, and any mutant genotypes producing larger gametes would be less fit because they could not produce as many gametes per unit of resource. As long as zygotes stemming from larger gametes do not gain any survival advantage, genotypes producing the most gametes will leave the most descendants.

The situation changes when large zygotes survive better than small ones. Then mutant genotypes that produce large gametes could leave more adult progeny even though they produce fewer gametes to start with, since their gametes would give rise to larger-than-average zygotes. Once genotypes producing macrogametes become established in a population, an opportunity for exploiting those genotypes by producing microgametes that selectively fuse only with macrogametes becomes available. Genotypes that produce such microgametes could achieve high fitness because their gametes would give rise to relatively large zygotes without requiring the parents to pay a high cost to produce each gamete.

Gametes of intermediate size would have low fitness in a population containing both microgametes and macrogametes. Genotypes producing gametes of intermediate size would leave fewer descendants than genotypes producing microgametes because their gametes would be less likely to fuse with macrogametes. Microgametes would have two advantages in competing for macrogametes. They would be more abundant because less energy is required to produce each one, and they would be more motile because they have less mass to propel. Genotypes producing gametes of intermediate size would also leave fewer progeny than genotypes producing macrogametes because zygotes formed from fusion of an intermediate-sized gamete and a microgamete would be smaller and hence would not survive as well as zygotes formed by fusion of a macrogamete and a microgamete. Thus a third sex that produces gametes of intermediate size could not become established in a gene pool.

A final question is why macrogametes do not selectively fuse with each other, since the resulting zygotes should have even higher fitness than macrogametes that fuse with microgametes. Parker (1978b) argues that the cost of retaining motility, the risk of failing to find a macrogamete to fuse with, and the risk of inbreeding are too high for this to occur. Fusing with microgametes should evolve as long as the costs of preventing such fusions are higher than the benefits derived by selectively fusing only with other macrogametes.

Sex ratio theory

Why a balanced sex ratio?

Animals generally produce about equal numbers of sons and daughters. This results almost automatically from the genetics underlying sex determination. In most animals the sex of an individual is determined by the sex chromosomes it possesses. If an individual possesses two X chromosomes, it will be a female (except in birds and some insects, where it will be a male). If it possesses one X and one Y chromosome, it will be a male (except in birds and some insects, where it will be a female). As long as gametes carrying X and Y chromosomes are equally likely to form zygotes, equal numbers of each sex of progeny will be conceived. This mechanism for sex determination has evolved by natural selection and should have some adaptive value.

The advantages of producing a 50:50 male:female sex ratio were first elucidated by R. A. Fisher (1930). Suppose that sons and daughters survive equally well to maturity and require equal amounts of parental investment to produce. If females outnumber males at maturity, an average son will produce more progeny than an average daughter because it will contribute gametes to a larger proportion of all the zygotes produced (Figure 10–2). Since sons cost no more to produce than daughters, parents who produce more sons will leave more descendants for each unit of investment than parents who produce fewer sons. Selection will therefore favor genes coding for increased production of sons. The same argument applies if a population contains more males than females, except that then selection would favor increased production of daughters. The advantages gained by producing more of the less common sex gradually disappear as the disparity between number of males and number of females in the population decreases, until at equilibrium sons and daughters are produced in equal numbers.

The sex ratio at conception, or **primary sex ratio,** should be about 50:50 whenever parental investment is limited to gamete production be-

Figure 10–2

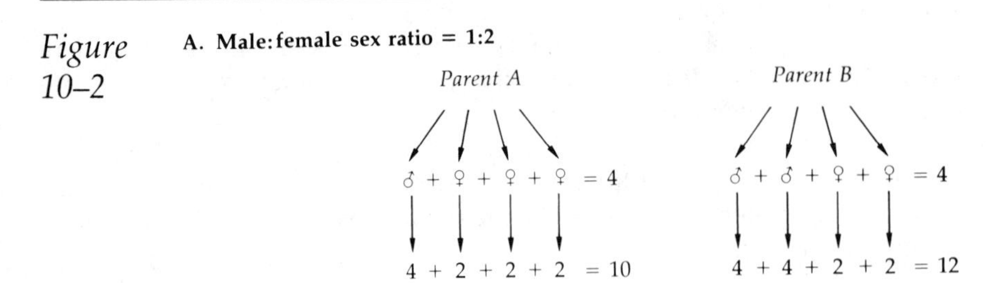

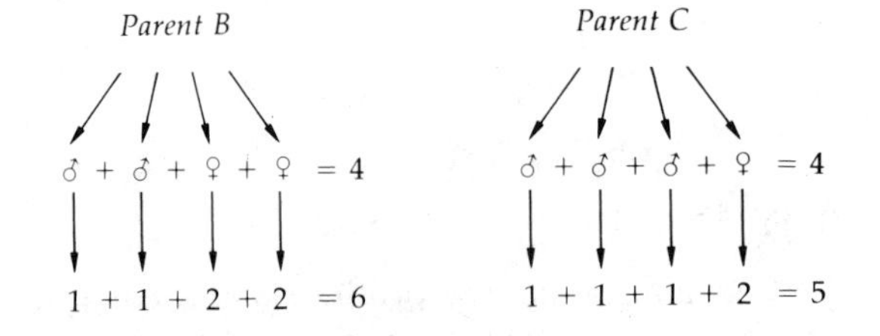

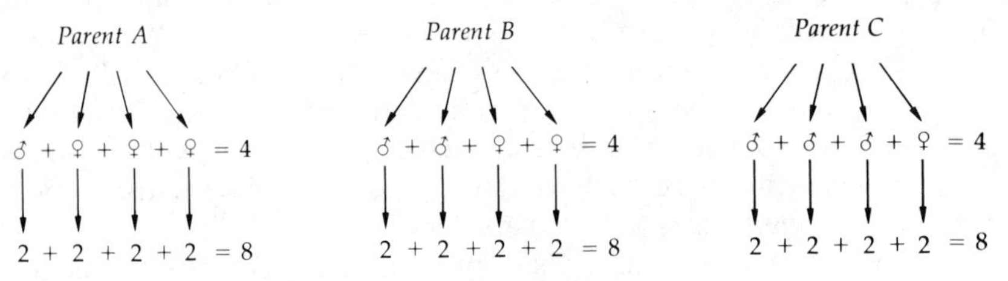

A diagrammatic explanation of why male:female sex ratios are normally balanced.

Consider three types of parents: parent A always produces 3 daughters for every son, parent B always produces an equal number of daughters and sons, and parent C always produces 3 sons for every daughter. The relative number of grandprogeny produced by each type of parent will determine which genotype is most fit. A. If the male:female sex ratio in the population is 1:2, parent A will produce fewer (i.e., 10) grandprogeny than parent B (i.e., 12). The reason is that sons mate with 2 females on average and therefore produce 4 progeny for every 2 produced by daughters. B. If the male:female sex ratio is 2:1, parent B produces more grandprogeny (i.e., 6) than parent C (i.e., 5). The reason is that sons will mate only half the time and hence will produce only 1 progeny for every 2 produced by daughters. C. If the male:female sex ratio is 1:1, no parent has a selective advantage (i.e., all produce 8 grandprogeny). Parents that produce more sons than daughters or more daughters than sons can coexist with parents that produce equal numbers of sons and daughters provided that the adult sex ratio remains balanced.

cause then sons and daughters are equally expensive (or cheap) to produce. However, selection may not favor a 50:50 primary sex ratio (i.e., at conception) if parental care is invested following conception. The general rule is that equal amounts of parental investment should be invested in each sex of progeny (Fisher 1930; Shaw 1958; MacArthur 1965; Leigh 1970; Charnov 1975b; Eshel 1975). Otherwise a unit amount of investment would not yield the same number of descendants when invested in each sex of progeny. Thus when the costs of producing sons and daughters differ, the numerical sex ratio should be adjusted to equalize the total expenditures invested in offspring of each sex.

The cost of producing sons and daughters is about the same in most animals except arthropods, and in those species any deviations from a balanced sex ratio existing within a given age cohort must be a result of higher mortality in the less common sex. If one sex is more likely to die than the other after parental care has ended, the primary sex ratio would not be affected. However, if one sex is more likely to die than the other before parental care ends, the sex ratio will become skewed toward an excess of that sex at conception and an excess of the other sex at independence (Leigh 1970). The reason for this is relatively straightforward. Suppose, for example, that sons are more likely to die before independence than daughters. Then at independence the average son will receive less parental investment than the average daughter because not all sons survive long enough to receive the full investment required to reach independence. For parents to equalize their investment in each sex, they would have to conceive more sons than daughters. At the same time, higher mortality among sons means more parental investment in sons is lost, and this loss must be compensated by higher average reproductive success for the sons who survive. Otherwise the net return per unit of investment would be higher for daughters than for sons, and selection would favor production of more daughters. Surviving sons can achieve higher average success than surviving daughters only if more daughters than sons survive to independence and reproduce. Hence the sex ratio at independence, or secondary sex ratio, must be skewed toward excess daughters.

Deviations from a balanced sex ratio caused by nonrandom mating

Strong deviations from a 50:50 sex ratio can arise from a combination of meiotic drive and nonrandom mating (Hamilton 1967). Consider, first, a population in which males result from zygotes carrying XY sex chromosomes and matings occur at random. Following insemination, X-bearing and Y-bearing sperm compete to fertilize a female's eggs. If a mutation on the Y chromosome enables Y-bearing sperm to always fertilize eggs first, the mutant genotype will always produce sons. The mutation acts solely to influence a sperm's chances of fertilizing eggs; it does not affect a male's chances of inseminating females. This differential probability of some sperm being more likely to fertilize eggs than others is called *meiotic drive*.

Males carrying the above mutation will produce 100% sons, since only Y-type gametes fertilize eggs. Normal males would, of course, produce 50% sons and 50% daughters. In a population where mutant and normal males are equally likely to inseminate females, the mutation will quickly spread because a higher proportion of all the sons produced each generation will carry the mutant Y chromosome (Figure 10–3A). Since more and more females produce entirely sons each generation, fewer and fewer progeny will be daughters. Eventually all progeny will be sons and the population will become extinct. The same thing happens if a similar mutation occurs on the X chromosome, except more generations are required to reach extinction (Figure 10–3B).

Meiotic drive will cause extinction unless a back mutation prevents it from doing so or mating is nonrandom. Nonrandom mating can prevent extinction in the following manner. When each individual can mate with only a restricted and rather permanent set of neighbors, the mutation will cause local extinction where the mutation occurs. This vacant space will tend to be filled by normal immigrants from surrounding areas, thus hampering the spread of the mutation. Local extinctions might then be balanced by immigration to create a stable equilibrium situation, with the sex ratio distorted for the population as a whole. The effect of meiotic drive based on a mutant Y chromosome is probably so strong that the population would probably go extinct despite nonrandom mating, but meiotic drive based on a mutant X chromosome could lead to a stable equilibrium. Cases of X-linked meiotic drive are not uncommon, for example, in wild populations of some fruit flies *(Drosophila)* (Novitski 1947).

Nonrandom mating can produce distorted sex ratios even in the absence of sex-linked meiotic drive (Hamilton 1967). Consider an extreme case in which only brother-sister matings occur. If females lay a fixed number of eggs and a single male can inseminate all of his sisters, a parent who produces only one male in each batch of eggs will leave the most descendants. This situation actually occurs in the mite *Acarophenax triboli*, where females typically produce one son and fourteen daughters in each batch of eggs (Newstead and Duvall, cited by Hamilton 1967). Daughters are born alive, while the single male of each litter hatches, mates with his sisters, and dies before he is born!

In less extreme cases where inbreeding occurs and brothers compete for matings, the expected sex ratio can be shown to equal $(1 - k)/2$, where k = fraction of all females mated by their brothers each generation (Maynard Smith 1978). If $k = 0$ (i.e., mating is random), the expected sex ratio is one-half, as predicted by Fisher's (1930) theory. If $k > 0$ (i.e., inbreeding occurs), the expected sex ratio can be distorted considerably.

Such deviations have been documented in numerous species of in-

►

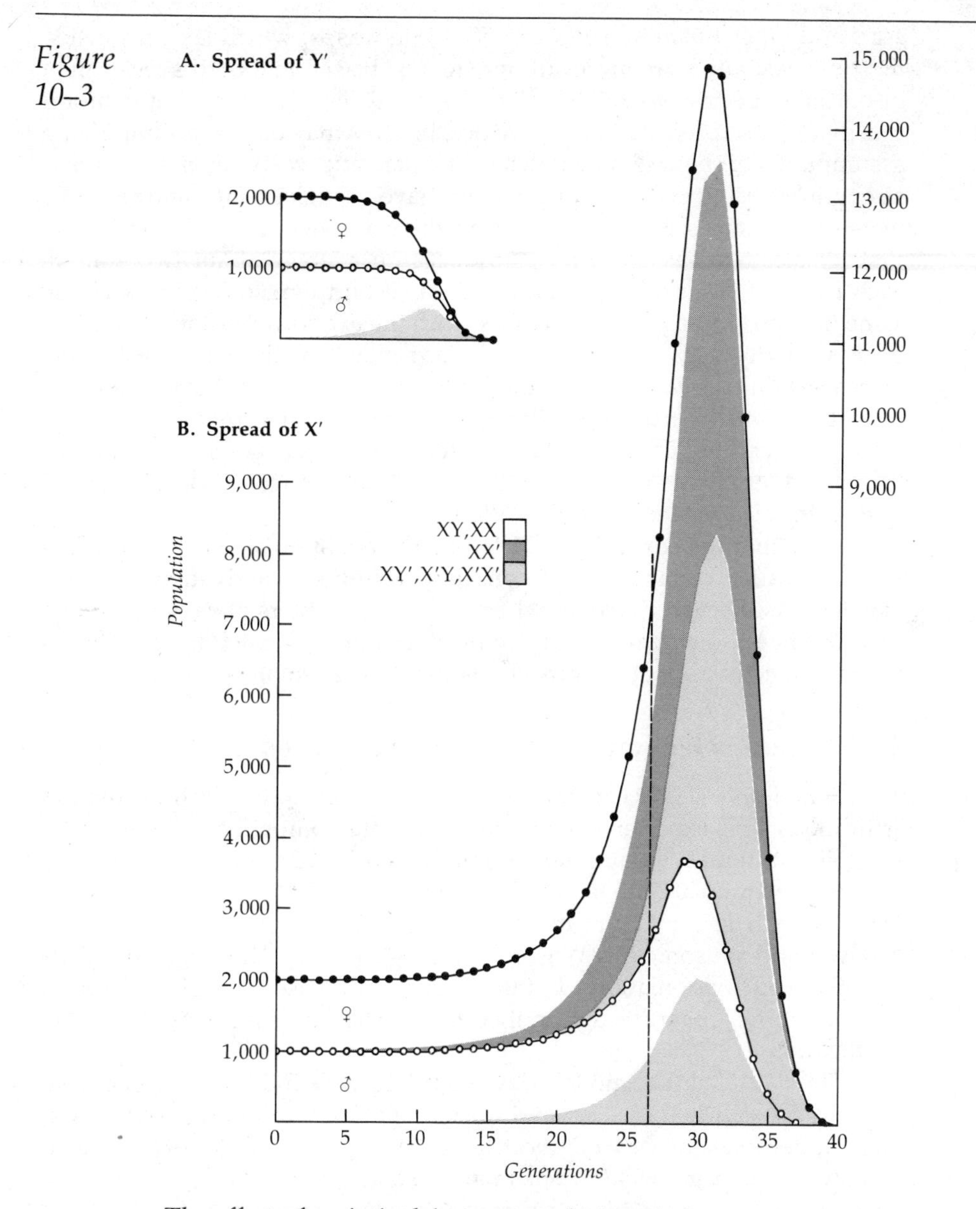

Figure 10–3

The effect of meiotic drive on population size.

The curve for total population size is partitioned into number of females and number of males, as indicated. Shaded areas indicate the proportions of normal and mutant genotypes present within each sex during each generation, where XX and XY are normal genotypes and XX′, XY′, X′Y, and X′X′ are mutant genotypes. A. Y-linked meiotic drive (Y′) leads to increased production of sons, which decreases the number of breeding females and hence causes the population to immediately decline until it becomes extinct. B. X-linked meiotic drive (X′) leads to increased production of daughters, which at first causes population size to increase because the increased number of females allows greater reproductive output. However, it eventually causes the population to decline to extinction once the rarity of males prevents many females from becoming inseminated.

sects and mites (Hamilton 1967, 1979). In fig wasps, which lay eggs inside fig fruits, sex ratios are more strongly female-biased in inbred species than in outbred species (Hamilton 1979). Observed sex ratios do not conform exactly to those predicted by the theory, but this may be due to simplifying assumptions of the original model. In the parasitic wasp *Nasonia vitripennis* strong inbreeding occurs because males have vestigial wings and can only mate with females who developed on the same host (Werren 1980). The first female to parasitize a host produces on average about two sons in each batch of about 29 eggs. However, a second female to parasitize the same host produces many more sons, with the exact proportion depending on how many eggs she lays, to take advantage of the strongly female-biased sex ratio already present on the host. Distortions of the sex ratio are most prevalent in arthropods but may also occur in vertebrates. For example, inbreeding may be responsible for strongly female-biased sex ratios in collared lemmings *(Dicrostonyx groenlandicus)* and wood lemmings *(Myopus schisticolor)* (Stenseth 1978; Williams 1979).

Note that the sex ratio predicted by the equation above is zero when $k = 1$ because certain assumptions underlying the derivation are then violated. As discussed above, when $k = 1$ (i.e., sisters always mate with their brothers), one male should be produced in each batch of eggs unless a single male is incapable of inseminating all the females.

Trait-group selection and distorted sex ratios

Hamilton's (1967) explanation of distorted sex ratios in haplodiploid arthropods that experience extreme local mate competition is eminently sensible if a female's daughters are all mated by her sons. However, attempts to extrapolate the hypothesis to explain distorted sex ratios in populations with less extreme inbreeding (i.e., $0 < k < 1$) are misleading (Colwell and Wilson 1981; Wilson and Colwell 1981). Distorted sex ratios do exist in such populations, but they apparently arise through trait-group selection rather than through individual selection generated by local mate competition.

The mathematical models devised by Colwell and Wilson (1981; Wilson and Colwell 1981) are beyond the scope of the present discussion, but their underlying logic is identical to the trait-group model explained in Chapter 2 (see pages 69–71). The model leads to the following conclusions: First, inbreeding and local mate competition alone do not generate selection for distorted sex ratios. Genes coding for distorted sex ratios are always at a selective disadvantage unless the population is structured into trait groups with respect to sex ratio. Second, the reason female-biased sex ratios are more likely to evolve in small groups is that genetic variance among groups is greater when group sizes are small. Selection for distorted sex ratios is generated by this between-group variance, not by inbreeding per se, which is an incidental by-product of small average group sizes. Third, the sex ratio predicted by Hamilton's (1967) model is correct, but the selective mechanisms implied by the model are incorrect. The reason for this outcome is that both models attribute selection for distorted sex ratios to small group sizes, but for different reasons.

Two lines of evidence support the trait-group model over Hamilton's local mate competition hypothesis (Wilson and Colwell 1981). First, computer simulations of the selective processes involved show that distorted sex ratios no longer arise when genetic variation among trait groups is eliminated and degree of inbreeding (i.e., frequency of matings between relatives) is left unchanged. Second, the trait-group model uniquely predicts that the proportion of females should increase within a population as group size decreases. Field research based on many species of hummingbird mites, which breed in flower corollas and disperse by clinging to the bills of hummingbirds, confirm the validity of this prediction. A reanalysis of data on a predaceous mite (*Amblyseius fallacis*) published by Burnett (1971) yields the same conclusion.

Local mate competition and inbreeding adequately explain distorted sex ratios when offspring disperse every generation and most matings involve siblings (Wilson and Colwell 1981). This situation is a special case of the trait-group model and does not require genetic variation among trait groups for female-biased sex ratios to evolve. Many of the arthropods discussed in the previous section, including *Acarophenax* mites, fig wasps, and various parasitoid wasps, fit that life history pattern. However, when several generations of progeny are produced between each dispersal event, matings are not just between siblings, and distorted sex ratios can arise only through trait-group selection.

Deviations from a balanced sex ratio caused by variations in the reproductive value of sons and daughters

Individuals may sometimes modify sex ratios among their offspring to take advantage of local conditions. For example, if one sex is locally rarer than the other (perhaps due to an environmental perturbation), it would pay to produce offspring of the rarer sex. This might be accomplished by mutations on the sex chromosomes that give either X- or Y-bearing sperm a competitive advantage in fertilizing eggs. The possibility that meiotic drive might evolve in response to locally skewed sex ratios has not been investigated for most animals, but Snyder (1976) has shown that it may occur. He removed approximately half the females from a population of woodchucks (*Marmota marmota*) that initially exhibited a 50:50 adult sex ratio. In the following year the sex ratio among yearlings was 40 males to 89 females, or 31:69.

In polygamous vertebrates relatively few males inseminate most of the females, while many males inseminate few or no females. If the health, vigor, or competitive ability of adult males depends on how much parental investment they received earlier as dependent offspring, parents capable of providing high parental investment should produce more sons, while parents capable of providing less parental investment should produce more daughters (Figure 10–4) (Trivers and Willard 1973). The reason is that the

Figure 10–4

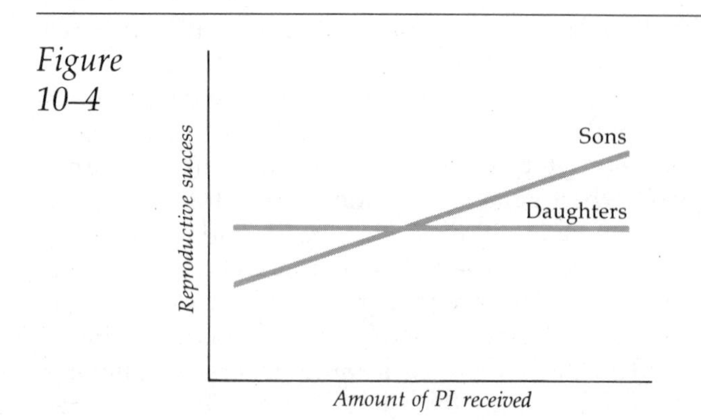

The fitness payoffs for investing parental care in offspring may be different for each sex, depending on how much parental care can be invested.

If the reproductive success of sons (but not daughters) increases with the amount of parental investment (PI) received while young, as shown here, parents who can afford high levels of parental investment should produce sons, while parents who cannot should produce daughters.

former parents are likely to produce successful sons, thereby gaining a high return on their investment in sons, while the latter parents are likely to produce unsuccessful sons, thereby gaining only a low return on their investment in sons. Trivers and Willard cite several studies of deer, seals, domestic animals, and humans to support their claim that mothers in poor nutritional condition produce fewer sons, but a critical review of those studies reveals that no differences in progeny sex ratios were statistically significant and some were in a direction opposite to that predicted by their hypothesis (Myers 1978).

The hypothesis proposed by Trivers and Willard (1973) is most plausible for sexually dimorphic species in which one sex of progeny is more expensive to produce than the other. Relatively little evidence is currently available for testing the hypothesis. One study shows that female common grackles *(Quiscalus quiscula)* produce more daughters early in the season when food is scarce and more sons later in the season when food is more abundant (Howe 1977). Sons probably require higher parental investment than daughters because they fledge at heavier weights. Since parents are less capable of providing parental care when food is scarce, these data provide tentative support for the Trivers-Willard hypothesis. However, in yellow-headed blackbirds just the opposite trend occurs. More sons are produced early in the season when food is scarce, even though sons are considerably larger than daughters when they reach independence (Patterson and Emlen 1980).

The Trivers-Willard effect should be easiest to detect if litter or clutch sizes are small (Williams 1979). Females should then tend to produce either predominantly sons or predominantly daughters, depending on how much energy they can afford to invest in offspring (Figure 10–5). In such species the frequency of all-male and all-female litters should therefore deviate from the binomial distribution predicted by random Mendelian segregation of X and Y chromosomes if the Trivers-Willard effect is important. More

Figure 10–5

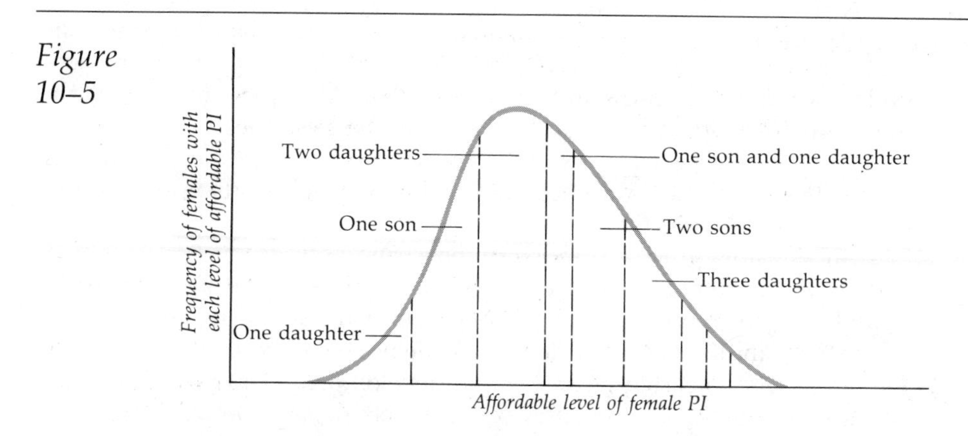

The effect of parental investment (PI) level on composition of litters produced by females in a species that normally produces litters of 1–3 young at a time.

Females that can afford little PI should produce one daughter, but such females should be infrequent in the population. Females that can afford more PI should produce one son and be more frequent in the population. The average female should produce two daughters. Females that can afford above-average PI should produce a son and a daughter, two sons, three daughters, and so on, depending on just how much PI they can afford to invest. Note that the frequency distribution of females does not follow a binomial distribution. If the distribution were binomial, females producing one son and females producing one daughter would be equally common. Similarly, females producing two sons and females producing two daughters would be equally common, while females producing one son and one daughter would be twice as common as females producing either two sons or two daughters. These relationships clearly do not hold in the graph shown here.

Source: G. C. Williams, "The Question of Adaptive Sex Ratio in Outcrossed Vertebrates," *Proceedings of the Royal Society of London, Series B* 205 (1979):570. Reprinted by permission.

precisely, the variance in sex ratio should differ significantly from a binomial distribution.

The best data on sex ratio variance comes from studies of human populations (e.g., Greenberg and White 1967; Edwards 1970). The data closely fit a binomial distribution and hence clearly support the Mendelian theory of random segregation (Williams 1979). Data from other mammals and birds, though limited, also support the Mendelian theory; none provide unambiguous support for the Trivers-Willard hypothesis (Williams 1979; based on data in James 1975, 1976 and Cooke, unpublished).

A variation on the Trivers-Willard hypothesis is that females should produce a disproportionate number of sons or daughters during periods of resource abundance if one sex can translate an early period of rapid growth into higher reproductive success better than can the other sex. The hypothesis is similar to the one discussed above, except here one sex gains a greater advantage than the other by growing during periods of high resource abundance rather than by receiving above-average amounts of parental investment from their mothers.

This hypothesis can be evaluated with sex ratio data for American alligators (*Alligator mississippiensis*), which exhibit skewed primary and secondary sex ratios (Nichols and Chabreck 1980). Alligators are particularly interesting to study with regard to sex ratio because they lack sex chromosomes. Hence primary sex ratios are not a consequence of Mendelian segregation of X and Y chromosomes and may be more susceptible to the effects of natural selection.

In alligators growth is indeterminate, and both age at first breeding and fecundity depend on size rather than chronological age. An early period of rapid growth would allow breeding at an earlier age, and it would enhance fecundity. However, it is not known which sex, if either, benefits the most from an early period of rapid development. The data show that alligator sex ratios are strongly male-biased in wild populations and strongly female-biased in populations on alligator farms (Nichols and Chabreck 1980). Since food is much more abundant on alligator farms, the results indicate that alligators do adjust sex ratios according to food abundance, just as the Trivers-Willard hypothesis predicts. They also indicate that females gain more from an early period of rapid growth than do males. Why this should be so remains to be explained.

A common pattern in parasitic wasps is to lay all-male broods on small or young hosts and to lay broods that contain more females on large hosts (Flanders 1956; Assem 1971). Males are smaller than females, and adult size does not appreciably affect male reproductive output but does affect female reproductive output. Because size of the host affects size of adult progeny, a parent can leave more descendants by producing daughters on larger hosts (Charnov 1979). Once females begin producing more daughters on large hosts, females laying on small hosts should produce more sons than daughters to maintain a balanced sex ratio. The data for parasitoid wasps therefore parallel those for alligators and lend further support to the Trivers-Willard hypothesis in its modified form. As in alligators, natural selection can easily affect sex ratios, though for a different reason. Female Hymenoptera can easily control sex ratios of their progeny because they are *haplodiploid*. Fertilized eggs with two of every chromosome (the diploid condition) give rise to daughters, while unfertilized eggs with one of every chromosome (the haploid condition) give rise to sons. Hence females need only control the release of sperm from storage organs to determine the sex of their offspring.

A few animals have bypassed the usual selection pressure for balanced sex ratios by evolving the capacity to change sex as adults. For example, the tropical Pacific fish *Labroides dimidiatus* forms social groups consisting of one male and a group of females. The tendency for females to change into males is normally blocked by aggressive behavior of the male, but when the male dies, the largest and most dominant female changes into a male and takes over the remaining females in the group (Robertson 1972). Sex reversal occurs in a variety of other fish as well, and in most cases it leads to unbalanced sex ratios with one large dominant male controlling groups of smaller females (Shapiro 1979). This capability allows individuals to reproduce as females during the period when they are too small to compete for female groups, while retaining the opportunity to inseminate many females once they become large enough to maintain dominant status in a group (Warner 1975; Warner et al. 1975).

Monogamous anemonefish (e.g., *Amphiprion bicinctus, A. akallopisos*) are exceptional in that sex reversal tends to maintain balanced adult sex ratios (Fricke and Fricke 1977). Groups of anemonefish generally consist of a mated pair and a number of juveniles whose development is arrested by the presence of adults. Adult females are invariably larger and more dominant than adult males. When two adult males are forcibly paired, the larger one turns into a female. When two adult females are forcibly paired, the larger one wounds or kills the smaller one and then pairs with an outside male or a newly matured male from the group of juveniles. Removal of either the adult female or the adult male results in maturation of a juvenile and, if appropriate, sex reversal by the surviving adult to form a new pair. Hence any local skew in the sex ratio caused by a higher mortality rate in one sex is immediately compensated by the combined effect of sex reversal and maturation of juveniles.

Why male competition and female choice?

In most animals males compete among themselves for mating opportunities, while females take the noncompetitive role of accepting victorious males or choosing from among courting males. Males are usually not very discriminating in their selection of prospective mates. They are prone to court and mount individuals of the wrong sex or species and even inanimate objects that only vaguely resemble females. Females, on the other hand, are generally unwilling to copulate with inappropriate sex partners and are often highly selective in choosing their mates.

Male competition and female choice are direct consequences of the different costs paid by each sex to produce offspring. Consider first the origin of male competition for mates in species lacking parental care. In such species a fundamental bias in the energetic cost of producing gametes is the principal driving force (Bateman 1948; Maynard Smith 1958). Females cannot produce as many gametes as males because ova are larger and hence energetically more expensive to produce than sperm. Female reproductive output is therefore limited by resource availability, not by mate availability. Females ordinarily gain little by mating with more than one male, and enough males are normally available to fertilize all a female's eggs. Consequently, females have no reason to compete for mates. In contrast, male reproductive output is limited by the availability of unfertilized eggs and not by resource availability. Males that acquire less energy may produce fewer sperm, but they still produce many more sperm than needed to fertilize all the eggs produced by females that they can inseminate. Hence the availability of unfertilized eggs (or receptive females) is limiting, and males typically compete to fertilize as many eggs as possible.

Female selectivity in choosing mates stems from the energetic constraints that limit female reproductive output. Females cannot greatly increase their output by mating with many males, and they may even reduce it if they mate indiscriminately. Copulating with a male of the wrong species usually results in infertile eggs or reduced viability of offspring (Mayr 1963), both of which diminish the number of descendants a female leaves in future generations. Accepting a male of inferior competitive ability can have the same effect because the female's offspring will inherit the male's inferior genes and hence leave fewer progeny of their own. Since some males are inferior to others, females risk an irreplaceable part of their lifetime reproductive potential each time they mate with a new male (Orians 1969). They can minimize this risk by selecting mates according to phenotypic traits that reflect male genetic quality (Fisher 1930).

A parallel argument still applies when the costs of gamete production are not the sole consideration. When offspring receive parental care following fertilization, the sex investing less should compete for mates while the sex investing more should be selective in accepting mates (Trivers 1972). In most cases female parental investments are still higher than male parental investments, so male competition and female choice still prevail in species exhibiting parental care. Only rarely have parental investments become higher for males, and then females typically take the active role in competing for mates. What is not clear in such cases is whether a change in the relative parental investments of the two sexes is a cause of such sex role reversal or a secondary consequence of it. More will be said about that question in the next section.

The theory for explaining male competition and female choice has been tested by evaluating two of its key predictions. The first prediction is that male reproductive success should increase substantially when more mates are obtained, while female success should not. The second prediction is that individual reproductive success should vary more among males than among females. This prediction follows from the argument that the best males should mate with many more females than should inferior males, while every female should mate with about the same number of males.

Both predictions were tested by Bateman (1948) in a classical study of mating behavior in fruit flies (*Drosophila melanogaster*). By using genetic markers such as eye color, eye shape, and bristle length to identify progeny, Bateman showed that male reproductive success was more strongly influenced by number of mates obtained than was female reproductive success. He also found that males varied much more than females in their individual success. Since male fruit flies depend on persistent and vigorous courtship to attract mates, the greater success of some males apparently resulted from female choice of the most persistent and vigorous males. Numerous field studies on a variety of other animals have since demonstrated the generality of these results for natural populations (see Trivers 1972; Payne and Payne 1977; Payne 1979; Gibson and Guiness 1980).

Because fruit flies do not provide parental care, Bateman's (1948) study does not test the hypothesis that overall parental investment of the two sexes is responsible for male competition and female choice in species that do provide parental care. Parental investment is notoriously difficult to quantify when parental care is provided, but qualitative assessments suggest that females usually invest more per offspring than males (Trivers 1972). This is certainly true for fish when only females care for young; for egg-retentive and viviparous lizards, in which females retain offspring in the reproductive tract for prolonged periods; and for mammals, where females endure prolonged gestation and provide most or all nutrition for offspring until weaning. It is less true for monogamous birds in which males help incubate eggs, feed young, and brood young, although for most birds male parental contributions appear to be less than female contributions (see Trivers 1972). When males provide most or all parental care, as in some fish, amphibians, and birds, the relative contributions of the two sexes may be reversed. This reversal is associated with female competition for mates and male choice in some species but not in others.

Female competition and male choice

Origin of sex role reversal

Female competition for mating opportunities and male choice of mates is best documented for the few species of fish and birds where sex role reversal occurs. The evolution of sex role reversal is still poorly understood, but it should not evolve unless male reproductive output is limited by time or energy constraints and female output is limited by mate availability.

Female seahorses and pipefish deposit their eggs in specialized brood pouches of males (Fiedler 1955; Takai and Mizokami 1959). The ability of males to brood multiple clutches has not been investigated for most species, but male pygmy seahorses (*Hippocampus zosterae*) can brood eggs from only one female at a time (Strawn 1958). Since males can rear just two broods per month, the availability of males with empty brood pouches is limited. Hence the reproductive output of females is limited by the availability of suitable mates, not by the rate at which they can convert resources into eggs. As a result, females compete for opportunities to mate with non-brooding males. At the same time, reproductive output of males is limited by the time available for brooding eggs. Any time lost because eggs have low viability or because offspring are genetically inferior diminishes their lifetime reproductive output. Males are therefore selective in accepting mates to avoid such losses. Some possible reasons why males brood eggs instead of females were discussed in Chapter 9 (see pp. 372–377), but evidence for evaluating their relevance to species exhibiting sex role reversal is lacking.

Sex role reversal in birds is closely associated with female desertion of mates following egg deposition (Jenni 1974). Trivers (1972) proposed that females should desert when their parental investment prior to desertion is less than that of males. However, this hypothesis cannot explain female desertion in birds because the costs of egg laying for females are always considerably higher than the costs of sperm production and nest construction for males. Several authors were quick to point out a fallacy in Trivers's argument. Deciding between behavioral options (e.g., to desert or not to desert) should in general be predicated on future prospects rather than past investments (Dawkins and Carlisle 1976; Boucher 1977; Maynard Smith 1977). A simple analogy with economic investments makes this point clear. An investor should not continue pouring money into a financially stricken business simply because he stands to suffer a loss if the business fails. Such investments are only justifiable if the prospective gain—namely, saving the business—is worth the risk. Thus mate desertion should not evolve unless the prospective benefits derived from it are higher than the attendant costs.

To test this prediction experimentally, Robertson and Bierman (1979) manipulated clutch sizes of female red-winged blackbirds to obtain two groups of females whose past parental investments were similar but whose expected benefits from continued investment were different. They did this by assuming that females with similar clutch sizes had invested about equally in young of a given age and that females with larger clutch sizes can expect to produce more young on average than females with smaller clutch sizes. Experimental and control groups could therefore be obtained by removing eggs from some nests but not from others, using nests with the same initial clutch size for both groups.

If investment decisions are based on future prospects, females with large clutch sizes after the manipulation should invest more in their broods than females with small clutch sizes. If investment decisions are based on past investments, both groups of females should invest about equally. For redwings Bierman and Robertson found that females with larger clutch sizes following manipulation showed stronger mobbing responses toward potential predators and made more feeding trips to their nests each hour than did females with smaller clutch sizes. However, the increased feeding rate can be attributed entirely to the higher food requirements of larger broods, as the food delivered per nestling did not change. The results therefore provide tentative, though inconclusive, support for the hypothesis that females modify their level of investment according to expected benefits. Since the study did not compare female groups with differing past investments and similar expected benefits, it provides no evidence for evaluating the hypothesis that females modify investment levels in light of past investments.

Females desert their mates either to mate again with new males, a mating pattern referred to as **successive polyandry**, or to enhance their survival prospects by replenishing energy reserves drained by egg production. In most cases of sex role reversal females desert to mate with new males. The conditions favoring such behavior will be discussed in the next chapter when polyandrous mating systems are analyzed. In phalaropes, a group of arctic and temperate zone shorebirds exhibiting complete sex

Photo 10–1

*Sex role reversal in the northern phalarope (*Lobipes lobatus*).*

The male (on top) is dull-colored while the female is bright-colored, a reversal of the usual pattern which arose because females rather than males compete for mates.

Photo by W. James Erckmann.

role reversal (Photo 10–1), females may desert to replenish energy reserves prior to autumn migration (Graul et al. 1977; Wittenberger 1979a). Females are regularly polyandrous in some phalarope populations (Hildèn and Vuolanto 1972; Kistchinski 1975; Schamel and Tracy 1977) but not in others (Höhn 1967; Johns 1969; Ridley 1980). Of course, females may lay a second clutch after dispersing considerable distances, so polyandry may often escape notice unless females are tracked with radiotelemetry.

In most birds female mate desertion cannot evolve because the cost of replacing eggs is too high if the male also deserts. Females should therefore not desert unless conditions strongly predispose males to remain and care for eggs or young following desertion (Wittenberger 1979a). Males should be most prone to care for deserted young if other mating opportunities are rare or absent (Emlen and Oring 1977; Wittenberger 1979a). Confidence of paternity has been suggested as another factor predisposing males to care for deserted clutches (Graul et al. 1977), but subsequent studies show that this factor cannot be important except under unusual conditions (Gross and Shine 1981; see pp. 374–375). Caring for deserted clutches should be the best choice when other mating opportunities are lacking even if confidence of paternity is relatively low, and not caring for deserted clutches is often the best choice when other mating opportunities are prevalent even if confidence of paternity is high. The main factors limiting a deserted male's mating opportunities are short breeding seasons, male-biased sex ratios, and shortages of breeding habitat (Wittenberger 1979a). The importance of these factors will be discussed more fully in the next chapter.

Female competition and male choice without sex role reversal

Female competition for the best males and male choice of mates can theoretically evolve in the absence of sex role reversal, but only under certain conditions (Wittenberger 1979a). Because superior males in a population are likely to attract numerous females, they have the option of choosing between potential mates. They should not usually exercise this option, since they can ordinarily increase their reproductive output by mating with as many females as possible, but they should exercise it when they cannot mate with all the females they attract, unless selecting superior mates delays breeding enough to make the costs too high. Females should compete for mates under two conditions. They should compete if opportunities to mate with superior males are limited or if being the first mate of a superior male is more advantageous than being a later mate.

A monogamous mating system, in which each male pairs with only one female, is the most likely place for male choice and female competition to occur. However, female competition may also occur in polygynous mating systems when order of mating affects a female's fitness. Male choice is less likely to occur in such systems, however, because males can frequently accept as many females as they attract.

Female competition for the best males may be difficult to detect because it may often involve scrambling for prior access to the male rather than overt aggression among females. Male choice of mates may be equally difficult to detect because female choice is occurring at the same time. In a monogamous species several things may be happening at once. Males compete to demonstrate their superiority and females choose the best of those males. Based on those choices, females may compete to mate with the best males, allowing the best males to choose from among several females. A field observer would frequently find it difficult to distinguish between these processes. Males of monogamous animals often court several females in succession before pairing occurs, but it is usually difficult to determine which sex terminated courtship when a female leaves. The female may have left because she was not yet receptive or chose not to accept that male, or she may have left because the male rejected her. Since male aggression toward females is a normal part of courtship in either event, the underlying decision processes are not immediately clear.

Evolution of secondary sexual characteristics

While formulating his theory of natural selection, Darwin (1859) was faced with the difficult problem of explaining a variety of characteristics associated with mating behavior that have no survival value. Darwin conceived of natural selection as a "struggle for survival," and yet these **secondary sexual characteristics** are often detrimental to survival. Many of Darwin's early critics therefore seized on the prevalence of such characteristics to refute the theory of natural selection (see Ghiselin 1974). Darwin, however, recognized that selection can favor traits for enhancing reproductive success as well as traits for enhancing survival. In *The Descent of Man and Selection in Relation to Sex* (Darwin 1871), he argued that secondary sexual characteristics evolve because males possessing them attract either more females or females with superior reproductive capabilities. Darwin referred to this "struggle between members of one sex for the possession of the other sex" as **sexual selection** to distinguish it from natural and artificial selection, which he felt were different processes. Modern evolutionary biologists now realize that sexual selection works in the same way as natural selection, in that both processes favor whatever traits enable individuals to leave more progeny in future generations than do their competitors. Nevertheless, the term *sexual selection* is still useful for referring to selective processes involved in the evolution of secondary sexual characteristics.

Secondary sexual characteristics are recognizable from four basic properties. They are restricted to one sex, usually the male; they usually do not reach full development until sexual maturity; they often, though not always, appear only during reproductive seasons; and they usually do not enhance survival and may reduce it. Some prominent examples of secondary sexual characteristics are increased size, strength, aggressiveness, morphological weaponry, coloration, and ornamentation.

Not all differences between the sexes are secondary sexual characteristics. Genitalia, accessory structures that facilitate copulation (e.g., claspers), and structures involved in parental care (e.g., brood patches, mammary glands) are **primary sexual characteristics**. Also, some morphological or behavioral differences are associated with different feeding specializations or food requirements (Selander 1966, 1972; Ralls 1976; Snyder and Wiley 1976). Such differences stem from competition between males and females for food, not from sexual selection.

Secondary sexual characteristics can aid individuals in competing for mates, or they can result from mating preferences expressed by the opposite sex. **Intrasexual selection** refers to selection stemming from competition for mates, while **intersexual selection** refers to selection stemming from mating choices of the opposite sex.

An important characteristic of sexual selection is that it is necessarily frequency-dependent. That is, the relative fitness of each phenotype depends on the frequency of other phenotypes in the population and not on some external standard imposed by environmental conditions. For example, under intrasexual selection the relative fitness of a male who exhibits a given degree of aggressiveness will depend on the frequency of males who are either more or less aggressive than he is. The male may have high fitness if most males are less aggressive, or he may have low fitness if most males are more aggressive. Similarly, under intersexual selection a male who exhibits a given degree of adornment that is attractive to females may have high fitness if most males are less well adorned, or he may have low fitness if most males are better adorned. An important consequence of this frequency dependence is that the average phenotype exhibited by each sex may not lie at an adaptive peak with respect to environmental conditions (Lande 1980). In other words, frequency-dependent sexual selection can reduce the average fitness of a population below that expected in the absence of frequency-dependent selection.

Intrasexual selection

Male competition for females is especially strong in polygamous animals because some males are able to mate with many females while others are left with none (O'Donald 1973a). Male success is determined by the outcome of both sperm competition and mate competition. **Sperm competition** refers to competition among sperm of several males to fertilize a limited number of eggs, either in an external medium or inside a female's reproductive tract. **Mate competition** refers to competition among males for a limited number of receptive females.

Trivers (1972) has argued that the intensity of intrasexual selection is determined by the ratio between male and female parental investment, with the intensity of selection increasing as this ratio decreases. However, a mathematical analysis of intrasexual selection confirms Bateman's (1948) earlier conclusion that the only important cause of intrasexual selection is higher variability in male reproductive success than in female reproductive success (M. J. Wade 1979). The model also shows that the difference in variance between male and female success is a more appropriate way to measure the intensity of selection than is the parental investment ratio. Finally, a further extension of the model shows that the intensity of total selection on male reproductive success (S_T) equals the intensity of natural selection on females (S_f) multiplied by the adult sex ratio (R) plus the intensity of sexual selection on males (S_s), or $S_T = S_f R + S_s$ (Wade and Arnold 1980). That is, the total effect of selection on males can be partitioned into the additive effects of natural selection and sexual selection.

The intensity of intrasexual selection can be very high in polygamous animals. For example, about 6% of the male elephant seals breeding on Año Nuevo Island off the California coast inseminate about 88% of the females (Le Boeuf and Peterson 1969), and 5% to 10% of the males on sage grouse leks inseminate over 75% of the visiting females (Wiley 1973). Such extreme variability in the success of individual males clearly creates strong selection for superior competitive ability among males. In most polygamous animals males cannot monopolize females to such an extreme degree, but some males still achieve substantially higher success than others.

Sperm competition

When gametes are simply broadcast into the surrounding medium, a male's success hinges on the proportion of eggs fertilized by his sperm. One obvious way to increase the number of eggs fertilized is to produce more numerous and more motile sperm. Both number and motility of sperm increase as sperm size decreases, but smaller size also detracts from survivorship of sperm. Males should therefore produce sperm of some optimal size based on the trade-off between increased number and motility of sperm and decreased viability of sperm. The optimal size of sperm will depend on the relative importance of each factor in the environment where sperm competition takes place.

Males can further enhance their chances of fertilizing eggs by releasing sperm in close synchrony with female release of eggs and by releasing sperm as near to reproductive females as possible. Coordinated release of gametes benefits females as well as males because it reduces the probability of eggs remaining unfertilized. One common method of synchronizing gamete release is to time breeding seasons with reliable external events such as changing day lengths or lunar rhythms (Figure 10–6). Another is to rely on the closely coordinated activities of an elaborate courtship sequence that leads up to fertilization. A third and often more efficient method is internal fertilization (Parker 1970a). For males internal fertilization increases the probability of fertilizing a female's eggs, while for females it reduces chances of wasting gametes.

Mating plugs and sperm deactivation. Once internal fertilization evolves, a variety of new tactics become available to males for improving their success. Sperm competition inside female reproductive tracts can be reduced or eliminated by preventing multiple inseminations of females. In the parasitic acanthocephalan worm *Moniliformis dubius*, males cement shut the female vagina and genital region with secretions from specialized cement glands following insemination (Abele and Gilchrist 1977). They also forcibly copulate with other males and cement shut their vas deferens,

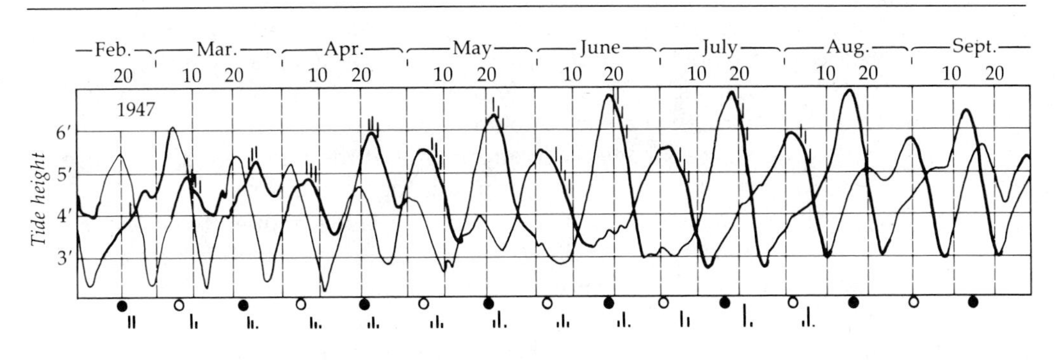

Figure 10–6 *The spawning of grunion, a fish that breeds on the coastal beaches of southern California, is closely tied to the lunar cycle.*

The two curves show the heights of the two high tides each day during the grunion spawning season. Vertical lines just above each tide curve indicate the times of grunion spawning runs. Note that spawning occurs at night (indicated by heavier lines on the tide curves), just after the highest high tides of the month. Peak high tides occur each month during the full moon (indicated by open circles below the graph) and new moon (indicated by solid circles below the graph) of each lunar cycle. Histograms below the graph give the percentage intensity of spawning within each run for the nights when spawning occurs.

Source: From B. W. Walker, "A Guide to the Grunion," *California Fish and Game* 38 (1952):416. Reprinted by permission.

possibly to reduce the number of male competitors in the local population. In many arthropods males insert temporary gelatinous or mucosal mating plugs into the female oviduct following insemination, apparently to obstruct second inseminations by new males and prevent leakage of the male's sperm (Parker 1970a; Ghiselin 1974). Similar mating plugs have also evolved in some mammals, particularly marsupials, rodents, and bats (Rothschild 1955).

A second means of preventing multiple inseminations is the use of chemical agents in seminal fluid to reduce female receptivity (Parker 1970a). This is a common tactic among male insects. To counteract such effects, males of some insects forcibly copulate with females and chemically deactivate previously stored sperm. Deactivation of sperm has been inferred from evidence that the last male to inseminate a female usually fertilizes a disproportionate share of her eggs (Hunter-Jones 1960; Lefevre and Jonsson 1962; Gulger et al. 1965; Parker 1970b), but the chemical or physiological mechanisms involved have not been elucidated. A rather different counterstrategy has evolved in the damselfly *(Calopteryx maculata),* where males actually pump out previously stored sperm by undulating the penis, which is covered with backward-pointing hairs for that purpose (Waage 1979a).

Pregnancy blockage in rodents. Another way to override successful inseminations by a prior male is to chemically block pregnancy in recently mated females. Many studies show that the presence of a new male commonly causes pregnancy blockage in recently mated female mice and voles (Bruce 1966; Bronson 1968, 1971; Stehn and Richmond 1975; Milligan 1976). The block is caused by substances in the male's urine, which stimulates release of gonadotropin and prevents implantation of fertilized eggs (Bruce 1966; Chapman et al. 1970). The same substances induce estrus in the female, allowing the new male to impregnate her soon after he encounters her. The substances have these effects only if the female had been inseminated within four days of encountering the new male.

Although pregnancy blockage in rodents is frequently considered a form of male-male competition (Bronson 1968; Trivers 1972; Wilson 1975a; Rogers and Beauchamp 1976), some evidence suggests that it is actually a mechanism for female mate selection (Schwagmeyer 1979). Presence of the active substances does not automatically block pregnancy, and the pattern of blockage suggests that females use it to abort fertilized eggs when they encounter better mates or better mating contexts. Blockage is most likely to occur if the female is exposed to a new male for two to three days rather than a few hours (Bruce 1961), if the new male is a different genetic strain than the original male (Parkes and Bruce 1961), or if other females are present (Bruce 1963). It does not occur if the new male is castrated (Bronson and Eleftherious 1963; Bruce 1963, 1965; Milligan 1976) or if the original male is still present (Parkes and Bruce 1961). Since pregnancy blockage involves some cost to females, it should not evolve unless females benefit from it or are unable to prevent it (Schwagmeyer 1979). The above evidence shows that females do prevent it under some circumstances. Hence allowing it to occur under other circumstances should be advantageous to females.

Females might benefit from pregnancy blockage in at least two ways. Pregnancy blockage may prevent deleterious effects of inbreeding by allowing an outbred male to override earlier matings (Parkes and Bruce 1961), or it might improve a female's chances of receiving male parental assistance by enabling her to choose a new male when her mate deserts following insemination (Dawkins 1976). The evidence available for evaluating these two hypotheses is currently inconclusive. Pregnancy blockage frequently occurs when the new male is a different genetic strain, regardless of whether he is more or less similar to the female than was the original male. Nevertheless, indiscriminate blockage could still lead to reduced inbreeding on average if females are more likely to encounter related males earlier than unrelated males in natural populations (see Schwagmeyer 1979). Males of many rodents do provide parental care such as nest building, licking pups, and retrieving pups (Horner 1947; Noirot 1969; Gandelman et al. 1970; Rose and Gaines 1976). However, present evidence does not allow a determination of whether pregnancy blockage is restricted to species exhibiting parental care; nor does the evidence allow an evaluation of whether pregnancy blockage actually increases a female's chances of receiving parental assistance from the new male.

Prolonged copulation. Still another way to prevent or delay second inseminations is to copulate for prolonged periods of time. Prolonged copulation reduces the period of time when receptive females can be inseminated by other males and may allow sperm to fertilize eggs prior to a second insemination. Males of some insects rely on prolonged copulation to alleviate sperm competition. House flies *(Musca domestica)* copulate for about an hour even though complete sperm transfer is accomplished in 10 to 15 minutes (Murvosh et al. 1964). Since chemical constituents in male seminal fluid terminates female receptivity (Riemann et al. 1967), prolonged copulations should effectively reduce the likelihood for multiple inseminations. Moths often copulate for 24 hours or more (Richards 1927), and a captive pair of brimstone butterflies *(Rhodocerca rhamni)* copulated continuously for an entire week (Labitte 1919). Some frogs with short breeding seasons remain in amplexus for several days (Wells 1977a), and in certain neotropical species of the genus *Atelopus,* amplexus lasts up to several months (Dole and Durant 1974; Durant and Dole 1974). Prolonged copulations involve a trade-off between preventing multiple inseminations and seeking additional mates. Some compromise should therefore be reached. Parker (1970b) has shown for dung flies *(Scatophaga stercoraria)* that the duration of copulations is adjusted to maximize male success in fertilizing eggs.

Because copulating males are subject to physical displacement by competing males, especially when copulations are prolonged, males often have specialized organs for grasping females. Males of nearly all insects possess prehensile organs fitted with hooks, spines, or claspers for preventing displacements while in copula, and in many species males also have modified antennae, legs, or elytra to serve the same purpose (Richards 1927). Competition to displace males from the backs of females often leads to wrestling contests, with the victor gaining access to the female (Photo 10–2). Males may also use chemical agents to immobilize competitors. For example, males of the armyworm moth *(Pseudaletia unipuncta)* release a pheromone that inhibits approach and copulatory behavior of other males who are attracted by female sex attractants (Hirai et al. 1978). This pheromone effectively prevents competing males from interfering with copulatory behavior of the first male and may have evolved as an adjunct to clasping organs to prevent other males from displacing him during copulation. Since males always benefit by displacing the sperm of other males, such tactics are prevalent, often at considerable cost to the females (see Boorman and Parker 1976; Parker 1979).

Competition for mates

An alternative to prolonged copulation or other means of sperm competition is for males to guard females after insemination (Parker 1974b). Males of some dragonflies remain in tandem with the female until she finishes ovipositing, while in others they guard ovipositing females by chasing away males who approach them (Moore 1960; Corbet 1962; Waage 1979b). Male dung flies do not dismount following copulation but instead remain mounted without genital contact (Parker 1970c). If a paired male is attacked by another male, he rises up on his first and third pairs of legs and raises his middle legs to deflect the intruder away from the female.

Photo 10–2

*Two male tenebrionid beetles (*Onymacris rugatipennis*) of the Namib desert engaged in a wrestling contest over a receptive female.*

The victor will gain access to the female, who is mostly buried in the sand just to the right of the two males.

Photo by William J. Hamilton III.

Should the attacker manage to grasp the female, a fierce struggle follows. Occasionally the intruding male wins, and then he copulates with the female before she continues ovipositing. In many crustaceans males embrace molting females and defend them from other males (Ghiselin 1974). Because copulations are only possible while a female is molting, sequestering behavior gives a male exclusive access to her during the period when she is receptive. The female also benefits from being sequestered because she is especially vulnerable to predators during the molt and gains protection from the male's presence.

Controlling access to females is a common tactic among vertebrates, especially among social mammals. When female groups are small and cohesive, males defend access to entire groups. Since competition for female groups is fierce, this behavior often results in dramatic battles between equally matched rival males. When female groups are large or incohesive, individual males cannot control entire groups, and then they often resort to defending access to one female at a time. Such consort relationships are particularly prevalent in multi-male groups of social ungulates and primates, and they are often closely coordinated with female estrous cycles (e.g., Geist 1971; Hausfater 1975).

Male tiger salamanders *(Ambystoma tigrinum)* have an interesting ploy to counteract attempts to sequester females (Arnold 1976). Males normally sequester females by leading or pushing them away from rival males. When the female nudges a male's tail, he deposits a spermatophore that the female picks up and uses to fertilize her eggs. However, a rival male sometimes intercedes between the female and the first male. He nudges the first male's tail to elicit spermatophore release and then deposits his own spermatophore on top of the first male's spermatophore. Because this makes the first male's spermatophore inaccessible to the female, she picks up the second male's spermatophore instead.

Another tactic in mate competition is to control resources required by females or provide inducements to attract females. Many examples of males controlling resources required by females have already been discussed with regard to territorial behavior. The mating success of males who control resources depends on the distribution and abundance of those resources. This topic will be discussed more fully in the next chapter. Male inducements to attract females cannot work unless females show a preference for males who offer inducements. Hence the inducements used by males to attract mates will be discussed below in the context of intersexual selection.

An important factor potentially affecting male-male competition for mates is the relatedness of competing males. Under conditions of local mate competition among siblings, males may not compete as aggressively because the other males are relatives. For example, in the fig wasp *Blastophaga* competition for females is mainly among genetically identical brothers, and males do not fight over females (Hamilton 1979). In contrast, competition for females is usually among unrelated males in the fig wasp *Idarnes,* and males fight very pugnaciously over females.

Secondary sexual characteristics associated with intrasexual selection

Because competition for mates often entails aggressiveness and fighting prowess, intrasexual selection frequently favors such traits. Many male attributes have evolved to enhance fighting ability, and only a sampling of them can be described here. Numerous additional examples are provided by Darwin (1871), Richards (1927), Hingston (1933), Evans (1968), Wickler (1972), Burton (1976), and Geist (1978).

Body size is determined by the benefits and costs of becoming larger. Energetic constraints and predation pressure are important factors affecting body size, but they ordinarily do not create an impetus for sexual dimorphism in body size because they usually affect both sexes in about the same way (but see Ralls 1976; Snyder and Wiley 1976). Sexual size dimorphism is most likely to evolve when larger size enhances male fighting prowess under conditions of intense intrasexual selection. The importance of intrasexual selection is shown by the more pronounced dimorphism present in polygamous birds and mammals, where the intensity of intrasexual selection is most intense (Amadon 1959; Selander 1965, 1972; Wittenberger 1978a; Alexander et al. 1979). In fact, the degree of sexual size dimorphism is strongly correlated with the extent that successful males can monopolize females in pinnipeds, ungulates, and primates (Leutenegger and Kelly 1977; Alexander et al. 1979). In snakes males are normally smaller than females when males do not fight over them, but they are usually larger than females when they do fight over them (Shine 1978). Similarly, male arthropods are normally smaller than females, but they are usually larger than females when they fight over them (Darwin 1871). Males compete aggressively for mates and are larger than females in many crustaceans, beetles, and dragonflies. Thus aggressive forms of intrasexual selection generally lead to increased size dimorphism.

Another prevalent component of male fighting ability is morphological weaponry. The exaggerated horns and mandibles of some scarabaeid and lucanid beetles (Photo 10–3) and the antlers, horns, tusks, canine teeth, and dermal shields of various mammals are prominent examples (see Geist 1966; Barrette 1977; Eberhard 1979). Such structures are all likely to be favored by intrasexual selection.

Many male secondary sexual characteristics have evolved to intimadate rivals rather than physically subdue them. Structures such as the mane of lions and the beards or other facial hair patterns of ungulates and primates evolved to increase a male's apparent size (Hingston 1933; Guthrie 1970; Crook 1972; Schaller 1972). Bright or conspicuous colors are often associated with anterior body parts or morphological weaponry and may draw attention to attack capabilities, presumably to reduce an opponent's likelihood of initiating or persisting in a fight. However, an extensive multiple regression analysis of coloration patterns in birds indicates that bright coloration has evolved mainly as a defense against predation, even in polygamous species (Baker and Parker 1979). The role bright coloration

Photo 10–3

Some beetles have greatly enlarged mandibles for use in aggressive competition for mates.

*Here two male stag beetles (*Lucanus cervus*) are locked together in an aggressive dispute.*

Photo by Hans Pfletschinger.

plays in enhancing threat displays is debatable, and no unequivocal evidence is presently available.

As a general rule, any attribute that enhances a male's fighting ability can evolve by intrasexual selection, but not all such attributes have. Aggression occurs in many contexts, and competition for mates is only one factor favoring its evolution. The best indication that sexual selection has been important is the existence of sexual dimorphism, but one must bear in mind that sexual dimorphism can result from intersexual competition for resources or differential predation pressure as well as sexual selection.

Intrasexual selection in monogamous animals

Competition for mates is most intense in polygamous animals, where males vary greatly in their mating success. In monogamous animals competition for mates is reduced because each male normally pairs with only a single female. Nevertheless, intrasexual selection can still occur.

Intrasexual selection results from male competition for females, and males of monogamous species compete for females whenever more males than females are ready to breed. This condition could arise, for example, if sex ratios are biased toward excess males (Darwin 1871). In many monogamous birds males appear to be slightly more common than females, probably because the higher parental investment of females reduces their survival below that of males (Lack 1954). In a few birds, particularly ducks, ptarmigan, quail, and cardueline finches, sex ratios are skewed as much as 60:40 in favor of males (references in Wittenberger and Tilson 1980). When sex ratios are skewed that strongly, they generate strong selection for male secondary sexual characteristics.

A shortage of females could also arise if more males than females are ready to breed early in the season. Darwin (1871) argued that females who are ready to breed earliest are in the best physiological condition and are therefore better able to rear young. Pairing with early females may also lead to higher success when predators (such as snakes) become active relatively late in the season or when predators begin breeding too late to exploit early broods. Whenever early breeding is advantageous and not all females begin breeding at the same time, males might compete for females who are ready to breed early (O'Donald 1972a).

The existence of sexual dimorphism in monogamous animals does not necessarily imply that intrasexual selection has occurred. In most cases where sex ratios are about 50:50, males of monogamous animals are territorial (Wittenberger and Tilson 1980). To the extent that territories vary in quality and males compete for the best territories, sexual dimorphism could result from competition for the best territories rather than competition for females.

Consequences of intrasexual selection

Intrasexual selection leads to several important consequences. One is reduced male survival. Larger body size, more conspicuous coloration, and greater pugnaciousness all tend to place males in greater jeopardy of predation or energetic stress. Greater male than female mortality is most clearly evidenced by a skew in the adult sex ratio. A female-biased sex ratio is often indicative of reduced male survival resulting from male secondary sexual characteristics. This contrasts with male-biased sex ratios, which usually result from higher female than male parental investment. Of course, sex ratios are merely indicators of differential mortality, and a proper interpretation can be obtained only by identifying the actual causes of mortality for each sex.

A reduction in male survival has long been postulated as an intuitively likely cost of male secondary sexual characteristics, but a formal theoretical basis for this postulate has only recently been developed. Secondary sexual characteristics generally arise from the effects of numerous genes, and selection operating on polygenic characteristics of this sort has proven mathematically intractable to analyze until Lande (1980) succeeded in modeling the processes involved. Lande's model shows that natural selection acting differentially on the two sexes will result in sexual dimorphism, with the mean equilibrium phenotypes of each sex reaching an optimum with respect to the local environment. When intrasexual selection is operating, the average male phenotype will deviate from the optimum expected under purely natural selection, while the average female phenotype will not. The degree of deviation is determined by the point where the effects of natural selection, as measured by the mortality costs of the secondary sexual characteristics, begin to outweigh the effects of sexual selection, as measured by the reproductive benefits of those characteristics. The effect of the deviation is to reduce average fitness of males in a population, and if the deviation is sufficiently extreme, it could theoretically lead to population extinction.

Empirical evidence demonstrating a survival cost for males is scarce, but some evidence is available. In the polygamous great-tailed grackle males are larger, have longer tails, and are more conspicuously colored than females. Although the sex ratio of nestlings is about 50:50, the adult sex ratio is only 30:70 at the beginning of the breeding season (Selander 1965). The imbalance in sex ratio indicates that male mortality is much higher than female mortality among juveniles, adults, or both. The lower survival rate of males once they leave the nest may be due to their larger size, which might place them at an energetic disadvantage during winter, their longer tails, which might impede flight, or their more conspicuous coloration, which might draw the attention of predators. In grouse sexual size dimorphism is generally most pronounced in polygamous species, and adult sex ratios are negatively correlated with the degree of sexual size dimorphism (see Figure 11–7) (Wittenberger 1978a). This relationship suggests that increased size dimorphism results from intensified competition for females and has the consequence of reducing survival of adult males.

A second consequence of intrasexual selection is delayed breeding by males, especially when male competitive ability is influenced by experience or depends on prolonged maturation of secondary sexual characteristics. If relatively few males monopolize all the females, young males should defer breeding until they can better compete for mates. The importance of intrasexual selection is again shown by a comparison of polygamous and monogamous species. A sexual difference in age of first breeding, or **sexual bimaturism,** is the rule in polygamous birds and mammals but not in related monogamous species (Lack 1968; Emlen 1973; Wiley 1974). Although sexual selection is undoubtedly an important factor favoring sexual bimaturism, it is not the only factor. Sexual bimaturism may also result from sex differences in the amount of risk incurred by breeding, the value of gaining experience prior to breeding, or expected survivorship following breeding (Wittenberger 1979b). Fully understanding sexual bimaturism in any given species requires that a multivariable analysis based on all the relevant factors be performed for that species.

Intersexual selection

Runaway selection

Intersexual selection promotes the evolution of male secondary sexual characteristics as a response to female mating preferences. Darwin (1871) originally conceived of female choice as a preference for certain males based on an aesthetic sense akin to human perceptions of beauty, a view that stirred considerable criticism. While an aesthetic sense may be involved in the proximate stimuli used to select males, it cannot explain the ultimate basis underlying female mating preferences. Mate selection costs time and energy and may expose females to greater risks of predation, and these costs must be compensated by attendant benefits before it can evolve. Male traits preferred by females must somehow increase female fitness before a preference can evolve.

Female mating preferences could promote the evolution of male secondary sexual characteristics if a female's progeny are more fit as a result of inheriting those characteristics (Fisher 1930; O'Donald 1962). Either daughters or sons can benefit, depending on the nature of the trait.

Once a preference becomes established, it could lead to runaway selection for further exaggeration of the preferred traits (Figure 10–7) (Fisher 1930). This runaway process represents a form of intersexual selection. It occurs because females who select males with the preferred traits produce sons who inherit those traits. Their sons, in turn, have an advantage in attracting mates once they mature because they possess the traits preferred by females. The preference therefore strongly biases the type of male that fathers offspring who enter the next generation. Fisher (1930) argued that this runaway process cannot get started unless the male traits confer benefits for females even in the absence of a female preference for males who possess them, since in the early stages of evolution no such preference exists and males possessing the traits do not necessarily attract more mates. However, more recent genetic models, based on both one-gene and polygenic inheritance, show that the male traits need not confer any initial advantage for females (Lande 1981; Kirkpatrick MS). An initial preference for a neutral male trait can theoretically arise by genetic drift or other random events, and once it does, it can trigger a runaway process.

Runaway selection refers to the added advantage males gain by possessing traits preferred by females. It normally results from the ability of preferred males to attract more females, but it can occur in monogamous animals even though each male mates with just one female. When early breeding is advantageous, males with preferred traits will attract females earliest and thereby achieve higher fitness (O'Donald 1972a, b, 1973b). Females who select those males will produce sons with preferred traits, who in turn will achieve higher fitness by attracting mates of their own earlier in the season. The process can still work if breeding is most successful at some intermediate time, with early and late breeders being less successful than individuals who breed at the peak (O'Donald 1972b, 1973b, 1974). If the breeding distribution is symmetrical about the peak breeding period, preferred males have an advantage only if more than 50% of all males possess the preferred traits. If less than 50% possess the traits, preferred males would breed early a disproportionate amount of the time and achieve lower average success than less preferred males.

The handicap principle

The runaway selection hypothesis, as conceived by Fisher (1930), assumes that females select genetically superior males by responding to phenotypic traits reflecting that superiority, but the presence of such signals creates an opportunity for deception by less fit males. By developing traits indicative of genetic superiority to the same degree as the fittest males, less fit males might deceive females into choosing them as their mates. Zahavi

Figure 10–7

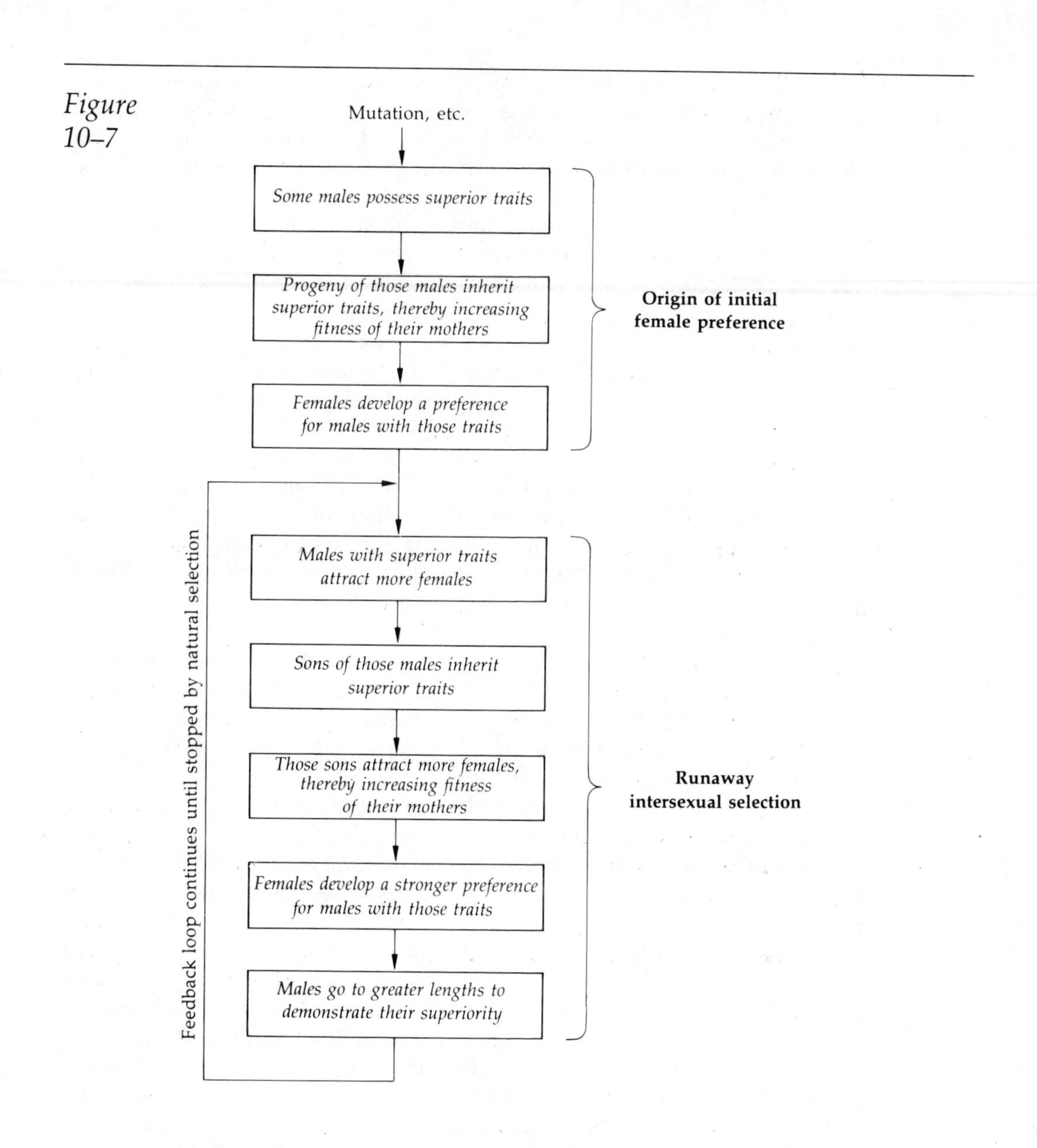

A flow diagram of runaway sexual selection as conceived by R. A. Fisher (1930).

Because males with superior traits produce more fit progeny, selection promotes an initial female preference for males with those traits. Once such a preference arises, males possessing the preferred traits gain an additional mating advantage, which accelerates the rate those traits increase in the population. This acceleration results from runaway sexual selection. Runaway selection leads to continued exaggeration of the preferred male traits until the survival costs associated with the traits offset the increased mating success of males who possess them.

(1975) therefore argues that females should base their preferences on phenotypic traits that cannot easily be possessed by less fit males. The most reliable indicators of genetic superiority would be traits that are too costly for all but the fittest males to afford. The cues used by females to assess male quality should therefore represent a handicap for males. The fittest males would be able to survive and compete for mating opportunities despite such a handicap, but less fit males would not. They would be unable to deceive females without paying the cost of the handicap, and the visible effects of that cost would allow females to unmask the deception. Males not possessing the handicap may be more or less fit than males who do possess the handicap, but females would have no way of discriminating between the more fit and less fit individuals. As long as highly fit males who possess the handicap are more fit than average males who lack the handicap, females should prefer mating with handicapped males. Thus Zahavi argues that male secondary sexual characteristics arising from female choice have evolved precisely *because* they are handicaps for males.

Although Zahavi's (1975) hypothesis appears to be an ingenious explanation for the costly secondary sexual characteristics of males, it faces a major theoretical problem. Females who mate with handicapped males will pass the handicap as well as the male's genes for high fitness on to their sons. The question is, are the progeny of such females more or less fit than progeny of females who do not prefer mating with handicapped males? Genetic models devised to answer this question indicate that progeny of females choosing handicapped males are always less fit (Davis and O'Donald 1976; Maynard Smith 1976c).

Zahavi (1977b) has responded to this criticism by arguing that the assumptions of the genetic models are too simplistic. He suggests that phenotypic expressions of a handicap character vary according to the male's current competitive ability. Young, weak, or nonbreeding males would not develop the character at all or would develop it to a lesser extent than mature males, thereby avoiding the cost of the handicap until they are ready to compete for mates. Any mutation or recombination that causes a male to develop the handicap character to a greater extent than the male could afford would be selected against because the handicap's cost would become too high. Males would develop the handicap character only to the extent that they could afford it. In effect, the handicap character would be an honest indicator of a male's quality.

Maynard Smith's model has been revised to consider this possibility. According to the revised model, when the effect of a handicap is negligible for high-quality individuals but nearly fatal to low-quality individuals, selection will favor a female preference for handicapped males (Eshel 1978). An important constraint is that the actual handicap must be tightly linked to the quality-determining locus for this to occur.

In the last analysis theoretical arguments are insufficient grounds for either rejecting or accepting a hypothesis. The only sound basis for rejecting or accepting any hypothesis is empirical evidence. In a long-term, and as yet largely unpublished, study of Arabian babblers, Zahavi has found evidence supporting his basic argument. Arabian babblers are territorial, group-living birds who exhibit several kinds of apparently altruistic behavior. Individuals give food to other group members, act as sentinels in treetops while other group members forage, and act as nest helpers to rear young other than their own. These behaviors cannot be attributed to indirect (kin) selection because they are just as prevalent in groups consisting of unrelated individuals as they are in groups consisting of related individuals. The most unusual aspect of these behaviors is that babblers *compete* to express them (Zahavi 1977a). Dominant individuals forcibly resist attempts by subordinates to supplant them as sentinels (see also Gaston 1977). They try to give food to subordinates, but subordinates resist accepting such offers. When feeding nestlings, individuals often compete to be helpers at the nest by supplanting others who are trying to deliver food. No theory of altruistic behavior can explain such competition. Zahavi's interpretation is that individuals advertise their high phenotypic quality by behaving altruistically. He argues that individuals pay the costs of altruism to gain the benefits of higher status, which confers the ability to breed within the group.

At present, the handicap principle is not widely accepted as a valid interpretation of how female mating preferences promote the evolution of male secondary sexual characteristics. Runaway selection provides a more parsimonious explanation and is accepted by most biologists. Nevertheless, the possibility that males accept handicaps to advertise high fitness merits further study.

Avoidance of hybridization and inbreeding

Females could benefit in a variety of ways by selectively choosing mates. One possible benefit is reducing the risk of producing infertile eggs or inviable offspring by avoiding hybridization and inbreeding. Many visual, auditory, chemical, and tactile signals have evolved to facilitate species recognition and prevent hybridization (Huxley 1938; Streisinger 1948; Mayr 1963; Smith 1969, 1977), and most animals have species-specific courtship behaviors that serve that purpose (e.g., see Bastock 1967; Marler and Hamilton 1966; Alcock 1975a; Brown 1975a). Since males are usually undiscriminating and prone to court or mount individuals of the wrong species, female choice is largely responsible for the effectiveness of such barriers to hybridization (see Haskins and Haskins 1949 for an exception).

Genetic inbreeding caused by mating with close relatives automatically increases homozygosity of alleles and hence the risk of producing offspring that are homozygous for deleterious or lethal recessive alleles (Lerner 1954). Inbreeding also reduces offspring quality whenever heterozygote superiority (i.e., *heterosis*) is present in a population (Crow and Kimura 1970; Cavalli-Sforza and Bodmer 1971). Male or female dispersal often reduces chances for inbreeding, but female mating preferences are also important.

An experimental study of fruit flies *(Drosophila melanogaster)*, for example, shows a female preference for outbred males (Maynard Smith 1956). Females were given a choice between outbred males, outbred testisless males, inbred males, and inbred testisless males. The mating success of outbred males was significantly higher than that of inbred males, while lack of testes had no effect on male success (Figure 10–8). Females that failed to avoid inbred males produced only about one-quarter as many viable offspring, showing that inbreeding does reduce female fitness for the genetic strains used in the study. Inbred males perform one step of the typical courtship sequence less rapidly than normal males, and female *Drosophila* are less apt to mate with males displaying reduced courtship vigor (Bastock 1956). Thus females may use courtship vigor as a cue for avoiding deleterious effects of inbreeding.

In guppies *(Poecilia reticulata)* males occur as several distinctive color morphs, and females show a preference for rare or novel morphs (Farr 1976, 1977). Exactly what benefits females gain from this preference, if any, are unclear. The preference may have evolved to increase genetic variability

Figure 10–8

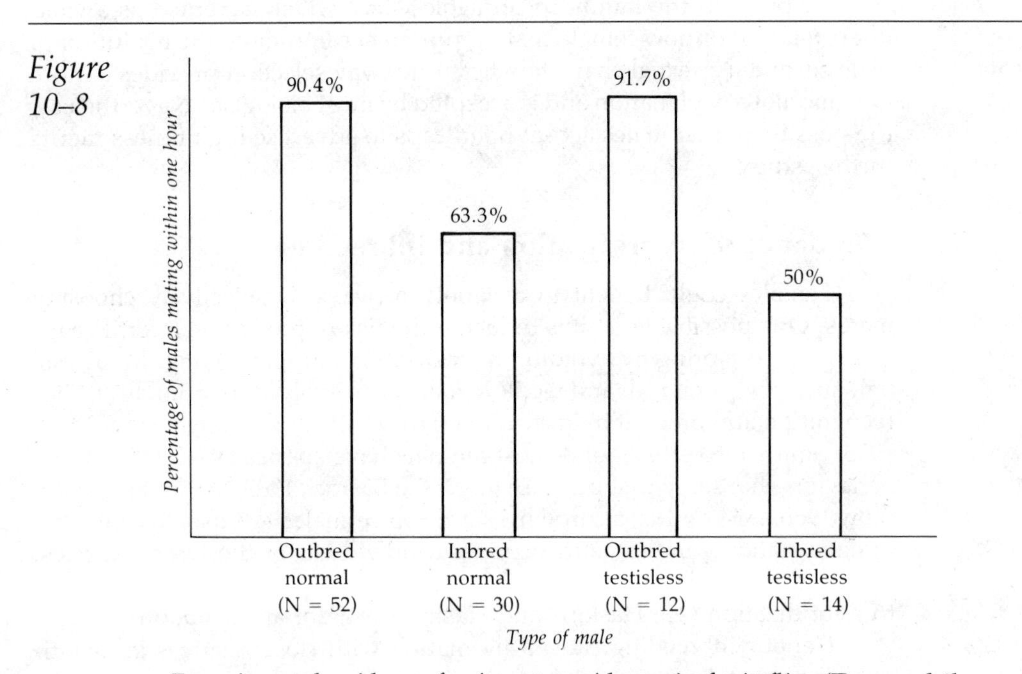

*Experimental evidence for incest avoidance in fruit flies (*Drosophila melanogaster*).*

Each bar of the histogram indicates the percentage of males in each experimental group that had mated one or more times within one hour. N = sample size. Note that outbred males were more successful than inbred males, while males possessing testes had no advantage over males lacking testes.

Source: From J. Maynard Smith, "Fertility, Mating Behaviour and Sexual Selection in *Drosophila subobscura*," *Journal of Genetics* 54:261–79. Reprinted by permission.

among offspring, which might increase offspring survival in newly colonized environments, or it may have evolved to reduce inbreeding effects (see Borgia 1979).

Some female primates may prevent inbreeding by avoiding mates who are close relatives. In free-living rhesus monkeys incestuous matings occur less often than would be expected by chance alone (Sade 1968). One reason is that males tend to leave their natal groups before they breed and hence cannot mate with their relatives. Nevertheless, mother-son matings are inhibited even when males do not leave their natal groups as long as the mother remains socially dominant over her son. Similarly, female olive baboons show a preference for males who transfer into a social group over males who were born in the social group (Packer 1979a). The deleterious effects of inbreeding are well established for humans (Schull and Neel 1965; Adams and Neel 1967; Seemanova 1971) and may help explain the nearly universal existence of incest taboos in human societies (Parker 1976; Demarest 1977; Murray 1980). However, since marriage is often used to cement alliances between unrelated kinship groups, incest taboos may result instead from economic or political considerations unrelated to inbreeding effects.

Ensuring male fertility

In fruit flies *(Drosophila melanogaster)* males continue to court females even after several successive matings reduce their fertility. When virgin and experienced males are placed together with a female, females showed a marked preference for virgin males (Markow et al. 1978). The experienced males used in these experiments were obtained by placing them in vials with females and transferring them into the experimental vials immediately after they had mated several successive times. Virgin and experienced males were of the same genetic strain and did not differ with respect to time before initiating courtship, time spent courting, overall mating speed, or various parameters of courtship behavior. A discrimination on the basis of fertility could be important for females, as insemination causes a reaction with the female's vaginal pouch, which becomes swollen and occluded with an opaque material (Maynard Smith 1956). This material may crystallize into a hardened mass that persists for a week or more. The reaction can occur even if the male lacks testes. (As noted above, however, testisless males do not suffer a mating disadvantage, in contrast to the results for exhausted males.) Thus a female mating with an exhausted male may produce mostly infertile eggs, with substantial loss of fitness. How females make the discrimination between virgin and exhausted males is unknown, but possibly chemical stimuli are involved.

Male genetic quality

Theoretical considerations. Genetic quality of males may be important to females because it affects the genetic quality of offspring. If a female mates with a male who has superior ability to find food or escape predation, her sons and daughters may inherit those abilities. In addition, if the male has superior ability to attract mates, a female mated to him is more likely to produce sons who possess superior ability to attract mates. Hence females should select males who exhibit traits reflective of superior abilities to survive or reproduce (Fisher 1930).

There is one potential theoretical difficulty with this hypothesis (Maynard Smith 1978). The resemblance between parents and offspring depends on *additive genetic variance,* the variance not attributable to allelic dominance or heterosis, and additive genetic variance rapidly decreases to zero under strong directional selection (Falconer 1960). Directional selection for traits that enhance male competitive ability should lead to equilibrium of the trait even in the absence of female mating preferences, and once equilibrium is reached, any remaining differences among males would be nonheritable. Females would therefore not produce more fit offspring by selecting superior males, since additive genetic variance among those males would be virtually absent.

Recurring deleterious mutations and transient polymorphisms arising from new favorable mutations would create some additive genetic variance in populations (Maynard Smith 1978). The amount of additive variance generated from these sources would be small if only one or a few gene loci are involved in the expression of male traits favored by females, but it could be quite large if many loci are involved (see Lande 1976). That is, the problem of low additive genetic variance may have been erroneously predicted from overly simplistic genetic models, in which case it would not represent a genuine obstacle to intersexual selection based on male quality.

Evidence. Detecting female preferences for particular male traits is not easy. In territorial animals genetically superior males are likely to control the best territories, and female preferences for certain males might be based on territory quality rather than male genetic quality. When males contend for dominance status or physical control of females, they monopolize opportunities for mating. Females may then exercise little actual choice in selecting particular males even though they usually copulate with the most competitive males. Despite these problems, some evidence suggests that females may choose mates on the basis of traits reflecting male genetic quality. The strongest evidence comes from an experimental study of fruit flies (*Drosophila melanogaster*) (Partridge 1980). Larvae produced by females who could choose their mates and females who could not were allowed to compete with a standardized strain within culture vials. This procedure makes it possible to measure the relative competitive ability of larvae arising from choice and no-choice matings. The proportion of flies emerging as adults was found to be slightly but significantly higher when females could select mates, thus showing that female choice of mates confers a competitive advantage to offspring. The evidence for vertebrates is weaker but generally supports the same conclusion.

Females of the cyprinodont fish *Nothobranchius guentheri* show a mating preference for the brightest males in choice experiments, even though brighter males are more vulnerable to predation (Haas 1976). Males patrol for females in the turbid waters where this species breeds, and they depend on females detecting them to achieve high reproductive success. Because bright males are visible at greater distances than dull males, they have an advantage in attracting mates. Males vary considerably in brightness, thus creating many opportunities for females to exercise a choice. Females probably prefer bright males because their sons would then be more likely to inherit bright coloration and thereby attract more females once they mature.

Females of several other vertebrates also show preferences for certain types of males, though reasons for their preferences are not always evident. Female Pacific tree frogs *(Hyla regilla)* are preferentially attracted to males who initiate calling bouts, call longer, and call at faster rates (Whitney and Krebs 1975). Whether this preference enhances female fitness is unknown. Female elephant seals sometimes incite a nearby dominant male to attack subordinate males who mount them (Cox and Le Boeuf 1977). Such behavior increases the likelihood of females being inseminated by a dominant male. Whether it improves the likelihood of their male offspring becoming dominant at maturity remains to be investigated. Female primates do not necessarily mate with the most dominant males, and present evidence indicates that they exercise an active choice in selecting their mates (Dixson et al. 1973; Eaton 1973; Hausfater 1975; Lindburg 1975; Seyfarth 1978; Bachman and Kummer 1980). However, the criteria they use and the benefits they gain from such behavior are unknown. As pointed out earlier, such preferences by females need not provide benefits beyond increased mating success of their male offspring. While an initial benefit additional to the mating advantage of sons was previously thought to be necessary for female preferences to evolve, recent genetic models show that an initial benefit is not necessary (Lande 1981; Kirkpatrick MS).

In a variety of birds paired females copulate with males other than their mates (Bray et al. 1975; Gladstone 1979; Mock 1980). Females may benefit from extrapair copulations in at least two ways. Not every female can pair with the best males, and extrapair copulations could allow females to mate with males who are genetically superior to the males they are paired with (Gladstone 1979). Alternatively, females who mate with several males will produce genetically more diverse offspring, which may be advantageous in fluctuating or unpredictable environments (see Janzen 1977). Extrapair copulations have been reported mainly in colonial birds, where females have easier access to males other than their mates, including many monogamous species that lack sexual dimorphism (Gladstone 1979). This suggests that extrapair copulations do not generate much intersexual selection for male secondary sexual characteristics.

Females of many animals show a preference for older males, who have demonstrated better-than-average ability to survive (Trivers 1972). However, such preferences may not result from active mate selection. For example, in polygamous animals older males may attract more females because they are better able to compete for territories or high dominance status, which in turn helps them to monopolize females. Females of many lek-breeding birds prefer males who hold central territories on the lek, and these are usually older males (e.g., see Hjorth 1970). Females may prefer central males because they are older on average than peripheral males, but they may instead prefer central males because central locations on a lek are less vulnerable to surprise attacks by predators (see Wittenberger 1978a, 1979a).

An apparent preference for older males may actually reflect a preference for more experienced males. In the monogamous rock dove, for example, females prefer reproductively more experienced males even when those males are younger than less experienced males (Burley and Moran 1979). They do not show a preference for older males when given a choice between young and old males with similar reproductive experience. Since age and experience are usually correlated, a more superficial study would only reveal a preference for older males, which could easily be interpreted incorrectly.

In animals with indeterminate growth, a preference for older males may actually result from a preference for larger males. Female toads *(Bufo quericus* and *B. americana)* mate preferentially with the largest males, which are also the oldest males (Wilbur et al. 1978). Although Wilbur and coworkers interpret this behavior as a preference for older males, it may be a preference for larger males. Similarly, females of the neotropical frog *Physalaemus pustulosus* mate preferentially with the largest calling males (Ryan 1980). Correlative and experimental evidence suggests that the pitch of male calls may be used by females to assess male size, since larger males give lower-pitched calls. Ryan suggests that females prefer large males because such males have demonstrated their ability to survive longer and acquire resources more successfully. Another possibility is that large males are less easily dislodged during amplexus, which would reduce the risk of female gametes being lost due to interference from other males.

Male parental capabilities

Females may select males because they are better suited as parents. They may often have difficulty evaluating males on the basis of parental capabilities because males should conceal their deficiencies while seeking mates, but criteria for making a choice are sometimes available.

Size can be a good indicator of male parental ability in animals with indeterminate growth. For example, females of the monogamous cichlid fish *Cichlasoma maculicauda* generally mate with the largest available male (Perrone 1975, 1978). Large males are the best parents because they can defend the best foraging areas while accompanying fry. This behavior enables fry to grow faster, which in turn reduces vulnerability of fry to predation and enables both parents to begin breeding again sooner than would otherwise have been possible. Females of the mottled sculpin *(Cottus bairdi)* prefer large males because large males are better at guarding egg masses

(Brown and Downhower 1977; L. Brown 1978; Downhower and Brown 1980). Eggs are more likely to hatch when guarded by large males because smaller males are more prone to eat the female's eggs.

Egg cannibalism by males complicates the mate selection problem for female fish. Each time eggs are cannibalized, the female's reproductive output is reduced. Females should therefore select males in a manner that minimizes the risk of eggs being cannibalized.

One factor affecting female choice under these circumstances should be the number of egg masses already being guarded by the male (Rohwer 1978). If the number of eggs cannibalized is relatively fixed, the probability of egg loss decreases as the number of egg masses guarded by the male increases. Consequently, females should prefer males who are already guarding eggs. Recent data for painted greenlings *(Oxylebius pictus)* confirm this prediction (De Martini 1976). However, males with multiple clutches do not continue attracting females indefinitely. As early clutches begin to mature, males become more likely to cannibalize new eggs laid on their territories (Rohwer 1978). Females should therefore avoid males who are guarding perceptibly advanced eggs, with the result that males with multiple clutches should attract all their mates in a short period of time. This is indeed the case in leaf fish *(Polycentrus schomburgkii)* and painted greenlings (Barlow 1967; De Martini 1976).

A second factor affecting female choice in cannibalistic fish should be male size (Rohwer 1978). Large size enables males to guard eggs from predators more effectively, but larger males have higher energy requirements and hence are likely to cannibalize more eggs. Hence when cannibalism is the most important cause of egg mortality, females should prefer smaller males. Mating preference experiments have not been performed to test this prediction, but some comparative evidence supports it. Males eat far more eggs than do females and are smaller in size than females in three-spined sticklebacks and painted greenlings (Hynes 1950; Semler 1971; De Martini 1976). Similarly, females eat more eggs than do males and are smaller than males in longear sunfish *(Lepomis meglotis)* (Keenleyside 1972b).

Males of most birds provide substantial parental care to nestlings, and the quality of their care could provide a basis for female mating preferences. Cues for evaluating a male's parental capabilities are less available than in fish, but one potential indicator might be the success of previous nesting attempts. Females of monogamous seabirds often pair with the same male in successive years, but they are prone to change mates following nest failure the preceding year (e.g., Richdale 1957; Coulson 1966; Mills 1973). Another possible indicator is male age, as experience is likely to improve parental ability. However, age may be an indicator of better survival ability and not better parental ability. There are presently no data to show that nest success in a previous year or male age really are good indicators of male parental capabilities. Without such evidence female preferences based on these criteria cannot be properly interpreted.

Male inducements for attracting females

Giving food to females during courtship (Photo 10–4) may be used as a signal of willingness to feed the female during incubation or offspring during the parental care period. Courtship feeding may also provide significant energy for making eggs. Royama (1966) has shown that female great tits and blue tits *(Parus caeruleus)* do not have time to gather sufficient energy for making eggs early in the season and that courtship feeding by males is essential for females to breed as early as they do. Krebs (1970) further confirmed that female blue tits benefit from courtship feeding. He showed that females could obtain more food per unit time by soliciting courtship feeding than they could by foraging entirely for themselves. That earliness of breeding is limited by food availability has been shown experimentally for great tits by providing them with extra food. Such artificial provisioning induces great tits to initiate breeding earlier than nearby individuals who were not provisioned (Källender 1974). It behooves females to breed early because survival of nestlings is highest early in the season (Perrins 1965). Females who receive food from courting males can probably breed earlier in spring than females who do not, and hence they are reproductively more successful. This would generate selection for a female preference based on male courtship feeding.

In an earlier study Kluijver (1950) described male great tits feeding females during incubation but discounted its nutritional importance because the behavior occurred very infrequently. However, Kluijver restricted

Photo 10–4

*Courtship feeding by a male red crossbill (*Loxia curvirostra*).*

Giving food to the female is an important part of pair formation in many birds. Its nutritional importance may or may not be significant, depending on the species. Further study will be needed before the evolution of courtship feeding will be understood.

Photo by Eric Hosking.

his observations to behavior at the nest, and Royama found that most courtship feeding occurs away from the nest. Thus careful study is needed to evaluate the nutritional importance of courtship feeding.

With the notable exception of work on pied flycatchers (Haartman 1958; Curio 1959), most early studies regarded courtship feeding as symbolic behavior without nutritional importance (see Lack 1940; Andrew 1961). Since these studies were based largely on observations at nests, this conclusion needs to be reevaluated.

Courtship feeding is prevalent in skuas, gulls, and terns but not among most other marine birds (Hunt 1980). Gulls and terns lay exceptionally heavy clutches relative to body weight, and courtship feeding has major importance because females do not leave the nest for extended periods while laying. Females apparently attend eggs continuously to prevent losses to predation or overheating. Once the clutch is completed, males share in this task. Thus females gain much of their energy from the male during the laying period (Cullen and Ashmole 1963; Brown 1967; Nisbet 1973). In addition, females may gain information about the size or quantity of food a male can provide and the frequency with which he can provide it (Nisbet 1973, 1977). Such information may be indicative of a male's later ability to deliver food to the female during incubation and to nestlings during the parental care period. The extent to which females actually act on such information is unknown.

Many male insects provide females with prey to induce them to mate. Female hangflies *(Bittacus apicalus),* for example, base their choice of mate on the size of prey offered to them by the male (Thornhill 1976a, 1980). By showing a preference for males who provide them with large prey, females obtain enough energy to survive a refractory period and lay a new clutch of eggs without doing much foraging. Females can thereby reduce predation risk, since searching for food increases their vulnerability to predation.

Male balloon flies (Empididae) provide females with prey in many species and apparently rely on provisioning as a means of attracting mates (Kessel 1955). The behavior of males varies among species and can be sorted into an interesting evolutionary progression. In some species males do not deliver any nuptial prey to females and rely instead on other means for obtaining mates. The most primitive use of prey inducements occurs in *Empimorpha comata* and some species of *Rhamphomyia* and *Empis,* where males simply capture an insect and fly about with it in search of females. Once males of these species locate a female, they copulate with her while the female devours the prey. Somewhat more advanced are other species of *Rhamphomyia* and some species of *Hilara*. In these species males capture a prey but do not fly around in search of females. Instead, they join a mating swarm, which females visit to choose their mates. How females make their choices is unknown, but their choices may be based on the size of prey carried by the male. In other species of *Hilara* males wrap their prey inducements in a haphazard network of silken threads, apparently to reduce the struggles of newly captured prey. A more elaborate form of the same behavior occurs in two species of *Empis,* where males completely enclose their prey in silken "balloons." Females of these two species consume the prey during mating, but females of several other species do not. In the latter species males capture only small prey, and they sometimes suck the juices out of prey before seeking mates. Once prey are completely enclosed by an outer covering of silk, males gain the opportunity for attracting females without providing any actual prey. Thus in a few species of *Hilara* males simply weave an empty balloon and use that as an inducement for attracting females.

Females obtain nutriments from males in many other insects as well. For example, females of many orthopterans feed on secretions from male dorsal glands before or during copulation (Alexander and Brown 1963; Chapman 1966), and females of some dipterans feed on secretions from greatly enlarged salivary glands of the male while copulating (Thornhill 1976b). Even spermatophores (i.e., containers for transferring sperm) and mating plugs may have nutritive value for females, although they probably did not evolve as mate attractants. In one study of three butterfly species, for example, radioactive tracers were used to show that nutriments provided by males, probably from their spermatophores, are incorporated into the eggs produced by females (Boggs and Gilbert 1979). In the German cockroach *Blatella germanica* males deposit urates, which are derivatives of uric acid, on spermatophores (Photo 10–5) (Mullins and Keil 1980). About 24 hours after mating, females expel the spermatophores and consume them. Radioactive tracer studies indicate that the urates are metabolized by the embryo early in development. Hence they are apparently used by females to increase reproductive output. Deposition of urates on spermatophores may represent a form of parental investment and have little to do with female mating preferences, but it remains possible that females use urate content on spermatophores as one criterion for selecting mates.

By utilizing nutriments provided by males, females can allocate more energy to egg production than would otherwise have been possible. Hence nuptial feeding is probably always advantageous for females. Whether or not it will evolve depends on how males can best utilize their available energy. In some cases males may best use their energy to attract females through nuptial feeding, while in others they may best use their energy to search for or guard receptive females. It remains for future research to identify the environmental conditions and social constraints promoting the evolution of each alternative.

Conclusion

The origins of isogametic sex are obscure and may have involved several selective processes. Genes promoting genetic recombination may have hitchhiked to fixation by being closely linked to favorable alleles. Perhaps the costs of recombination were relatively low at the outset, allowing positive selection for an ability to produce genetically diverse offspring. Per-

Photo 10–5

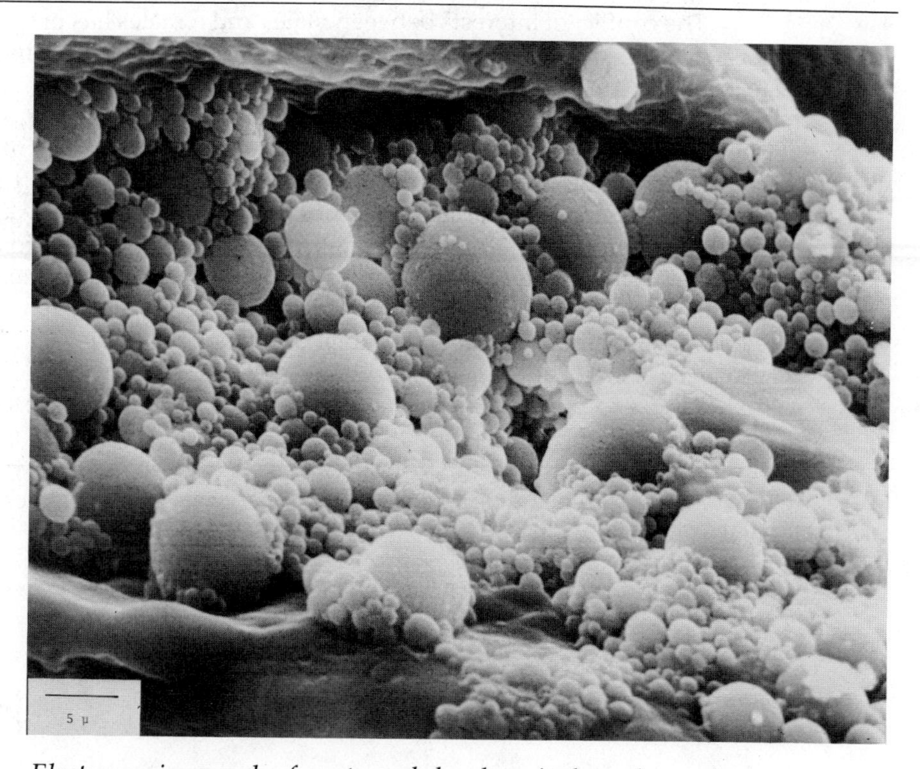

Electron micrograph of urate nodules deposited on the surface of a spermatophore by a male German cockroach.

These nodules may represent a male parental investment, as females use the nutrients contained in them to produce eggs, or they may represent a male inducement for acquiring mates. Scale bar, 5 μm.

Photo from D. E. Mullins and C. B. Keil, "Paternal Investment of Urates in Cockroaches," *Nature* 283 (1980):567.

haps such processes created variability among trait groups or demes, allowing group selection to promote their spread. Whatever the process, the evolution of sex added a whole new dimension to animal social organizations.

Once isogametic sex evolved, disruptive selection led to specialization in gamete size despite the high cost of producing excess male gametes. This difference in gamete size generated a fundamental conflict of interests between the sexes. Males produce many superfluous gametes and are limited by the number of eggs they can fertilize, while females produce no superfluous gametes and are limited by energetic constraints. In higher organisms, where parental care is prevalent, this conflict remains. Females still contribute more to offspring development and continue to be limited by time and energy constraints, while males continue to be limited by the number of eggs they can fertilize. Only in exceptional cases is this pattern reversed.

The conflict of interests between males and females has profound and far-reaching consequences on animal social organization. It causes males to compete among themselves for females, and it causes females to be selective in accepting mates. Males have often evolved large size, strength, aggressiveness, weaponry, exaggerated threats, and ornamentation to compete more effectively for females. They defend resources attractive to females or seek social status to enhance their access to females. Certainly male aggressiveness, size, and strength have not evolved just to compete for mates, but competition for mates predisposes males to be the stronger and more aggressive sex in all social contexts. In contrast, females have evolved coyness, discriminative abilities, and submissiveness to assess prospective mates and to counter the aggressive overtures of zealous males. To be sure, exceptions to these patterns exist, but the patterns are so general that their existence implies a fundamental difference in the way males and females go about enhancing fitness. As we shall see in the following chapters, this difference greatly affects the way males and females interact within animal social systems.

Mating systems

In sexual organisms both males and females face the basic problem of having to form mating relationships in a way that best enhances reproductive success. However, the mating relationship yielding maximal success is not necessarily the same for both sexes. In many species each male tries to fertilize the eggs of as many females as possible, while each female tries to monopolize a single male and the resources he can contribute to her reproductive effort. The result is often a fundamental conflict of interests between the sexes.

The degree of conflict depends on how much males can contribute to female success. If males mainly defend resources and provide vigilance, their contributions may be largely shareable and females have little reason to monopolize mates. However, with the evolution of nonshareable parental care, the situation changes. If females can monopolize males and prevent them from mating with other females, they may obtain substantial nonshareable parental assistance. If females cannot monopolize males or if they accept already mated males as their mates, they may receive much less nonshareable parental assistance. Thus as the potential for males to contribute nonshareable parental assistance increases, the conflict of interests between males and females becomes commensurably greater.

The best way to view mating strategies is to consider the reproductive options open to each sex and analyze the constraints placed on those options by behaviors of the other sex. The basic tenets of current theory stem largely from research on birds. The strongest theoretical impetus has come from studies of polygamous species, even though 90% of all birds are monogamous. Perhaps that is not surprising, as the costs and benefits of reproductive options open to each sex are nowhere more evident than in polygamous mating systems. This chapter will therefore consider first the evolution of polygamous mating systems and then proceed to a discussion of monogamous mating systems.

Some definitions

Mating systems are most commonly classified according to the duration of pair bonding and the number of mates acquired by each sex (Table 11–1). **Monogamy** involves prolonged pair bonding between one male and one female, **polygyny** involves prolonged pair bonding between one male and

Chapter eleven

Table
11–1 *A classification system for vertebrate mating systems.*

General classification	Spatial classification	Temporal classification
I. **Monogamy:** Prolonged association and essentially exclusive mating relationship between one male and one female at a time.	A. **Territorial monogamy:** The monogamous pair shares a common territory. B. **Female defense monogamy:** Each male defends access to a female instead of defending a territory. C. **Dominance-based monogamy:** Females maintain monogamous pair bonds within social groups by dominating more subordinate females.	A. **Serial monogamy:** Individuals of both sexes usually pair with a new mate each year or in each breeding cycle. B. **Permanent monogamy:** Mated pairs usually remain together for life, though sometimes they change mates after failed breeding attempts.
II. **Polygyny:** Prolonged association and essentially exclusive mating relationship between one male and two or more females at a time.	A. **Territorial polygyny:** Some proportion of the males form separate pair bonds with each of two or more females. B. **Harem polygyny:** A single male defends access to each social group of females. C. **Territorial harem polygyny:** Males control access to female social groups by defending group home ranges or habitats where female groups congregate.	A. **Successive polygyny:** Polygynous males acquire each of their mates in temporal succession. B. **Simultaneous polygyny:** Polygynous males acquire all their mates at the same time.
III. **Polyandry:** Prolonged association and essentially exclusive mating relationship between one female and two or more males at a time.	A. **Territorial polyandry:** Several males are paired with at least some territorial females. B. **Nonterritorial polyandry:** Females desert their first males and pair with new males elsewhere.	A. **Successive polyandry:** Polyandrous females acquire each of their mates in temporal succession. B. **Simultaneous polyandry:** Polyandrous females acquire all their mates at the same time.
IV. **Promiscuity:** No prolonged association between the sexes and multiple matings by members of at least one sex.	A. **Broadcast promiscuity:** Gametes are shed into the surrounding environmental medium so that sperm competition is prevalent.	

General classification	Spatial classification	Temporal classification
	B. **Overlap promiscuity:** Promiscuous matings occur between solitary individuals with overlapping home ranges or during brief visits by one sex to the home range or territory of the other.	
	C. **Arena promiscuity:** Males defend a display area or territory that is used exclusively or predominantly for attracting mates.	
	D. **Hierarchical promiscuity:** Males establish dominance hierarchies that affect their ability to inseminate females.	

Source: From J. F. Wittenberger, "The Evolution of Mating Systems in Birds and Mammals," in P. Marler and J. G. Vandenbergh, eds., *Handbook of Behavioral Neurobiology, Vol. 3: Social Behavior and Communication* (New York: Plenum Press, 1979):277. Reprinted by permission.

two or more females, and **polyandry** involves prolonged pair bonding between one female and two or more males. **Promiscuity** involves an absence of prolonged pair bonding and normally implies that successful males mate with more than one female per season. Females may also mate with more than one male, but they often do not. Although the term *promiscuity* often connotes random or indiscriminate mating in everyday usage, promiscuity among animals rarely or never involves random mating. Females are probably always selective in accepting mates, and even when gametes are released into the surrounding medium, sperm competition ordinarily leads to nonrandom fertilization of eggs. Some authors (e.g., Roberts and Kennelly 1980) refer to copulations between individuals other than pair-bonded mates as promiscuous matings within normally monogamous or polygynous mating systems, but a better term for such matings is **extrapair copulations. Polygamy** has been variously defined as prolonged pair bonding with more than one mate at a time (e.g., Selander 1972) or as any nonmonogamous mating relationship (e.g., Wittenberger 1979a). The latter definition will be used here.

The major ambiguity in the definitions above stems from the word *prolonged.* Short-term pairing relationships do occur, and these are often difficult to categorize. Some brood-parasitic birds, for example, form exclusive pair bonds that last for only one day, with females laying each successive egg in consort with a different male (Liversidge 1971). Similarly, males of some arthropods remain in copula or control access to females for prolonged periods to reduce sperm competition (see pp. 425–427). In such cases exclusive mating relationships are maintained until one or more offspring clear each female's reproductive tract or until the females can no longer be inseminated by other males, but the relationship is not prolonged in relation to the length of the breeding season. An unambiguous criterion for prolonged pair bonding is probably impossible, but one reasonable criterion is that the pairing relationship must persist for a substantial fraction (e.g., 20–25%) of the breeding season (Wittenberger and Tilson 1980).

It is often convenient to classify mating systems according to their spatial or temporal organization (Table 11–1). The spatial classification is based on how the sexes come together, and it is more pertinent to current mating system theory than the temporal classification. Nevertheless, the temporal classification is useful for characterizing the exact sequence of events that take place during a breeding season and for evaluating the reproductive options open to individuals of each sex as the breeding season progresses.

Territorial polygyny

Costs of polygyny

Polygynous mating systems involve the conflicting reproductive interests of territorial males, first-mated females, and subsequently mated females. Polygyny entails costs for all three classes of individuals, and it is important to understand what those costs are.

The costs incurred by males result largely from increased competition for females or for habitats that attract females. They are the costs of intrasexual competition and territoriality already discussed in earlier chapters. These costs probably never make polygyny disadvantageous for every male in a population, but they may limit the number of females any given male should try to acquire.

More important are the costs incurred by females. Polygyny is likely to increase competition for food on or near a male's territory, attract more predators to the area, and reduce nonshareable forms of male parental assistance. Competition should be increased because more females are exploiting local food resources while feeding nestlings. More predators may be attracted because polygyny increases the density of nests on a male's territory and hence offers more potential prey for predators. Male parental contributions are reduced because males have greater incentive to devote time and energy to seeking and courting additional mates and cannot increase their overall parental effort in proportion to the number of mates they obtain. Note that only nonshareable male parental assistance is in jeopardy, as males can provide shareable parental assistance to several females just as easily as they can to one female (Wittenberger 1979a).

The costs incurred by females affect *primary* (first-mated) females, *secondary* (second-mated) females, and lower-ranking females, but not necessarily to the same degree. Primary females often receive most or all of a polygynous male's nonshareable parental assistance, while secondary and lower-ranking females receive little or none (see pp. 381–383). Primary females may also be better competitors and better able to occupy the highest-quality nest sites on a territory. Hence the costs of polygyny are ordinarily higher for secondary and lower-ranking females than for primary females. Even so, primary females usually suffer some costs when secondary females breed on the territory, and as a result, they sometimes try to prevent polygynous matings by behaving aggressively toward intruding females.

Primary females may or may not be aggressive toward intruding unmated females, depending on the costs and benefits associated with their aggression. The costs depend on how much aggression increases vulnerability to predation or other mortality agents and interferes with other reproductive activities. They are probably higher in birds than in mammals because female birds must devote substantial time and energy to incubation and brood care (Orians 1980). This difference may explain why primary females are aggressive in birds less frequently than in territorial mammals (see Wittenberger 1979a; Wittenberger and Tilson 1980). The net benefit of aggression by primary females depends on how much the presence of secondary females reduces fitness of primary females and on how difficult it is for primary females to delay or prevent entry of additional females onto male territories. Neither factor has been carefully analyzed for any polygynous animal.

When unmated females pair with already mated males, they incur the costs of secondary mated status plus the costs of overcoming any resistance posed by primary females. Polygyny should not evolve unless females gain benefits from secondary status that exceed both of these costs. Several possible benefits have been proposed.

Skewed sex ratio hypothesis

A prevalent belief among early theorists was that polygyny evolved because a shortage of males forced some females to mate with already mated males. When sex ratios are skewed toward excess females, many females would have to accept secondary mated status or not breed at all. Polygynous mating would provide the excess females with an opportunity to produce young when no other alternative exists.

The skewed sex ratio hypothesis was quickly rejected once polygynous mating systems were studied more carefully. Polygyny has often evolved in birds with 50:50 adult sex ratios, and in many species females pair with already mated males even though unmated males hold territories nearby (Verner 1964; Wittenberger 1976a, 1979a). Female-biased sex ratios do exist among adults of some polygynous birds, but they probably arise as consequences of the severe intrasexual competition among males generated by polygamous mating systems. Skewed sex ratios may cause isolated instances of polygyny in normally monogamous species, but they are not responsible for the regular occurrence of polygyny in any species.

Male emancipation hypothesis

According to a second hypothesis, polygyny evolves in birds when males are not needed to feed offspring, a situation that may arise when food resources are extremely abundant (Armstrong 1955; Crook 1964, 1965). Under such conditions lost male parental assistance and increased competition for food resources might be of little cost to females, and females might improve the genetic quality of their offspring by mating polygamously with genetically superior males.

At best the male emancipation hypothesis can apply to relatively few species. It cannot explain why polygyny evolved in the many species whose food resources are not extremely abundant. Territorial birds and mammals with dispersed breeding distributions commonly defend food resources, and this fact alone implies that competition for food is important (see Brown 1964). The hypothesis also fails to explain why polygyny evolved in the many species where males still contribute parental care, since male emancipation has not occurred in such species.

The male emancipation hypothesis is most plausible for colonial polygynous birds because these species often exploit locally abundant food resources, but even here the hypothesis runs into problems. Loss of male assistance in feeding nestlings should be costly for females no matter how abundant food is. The rate at which females can feed nestlings is still limited by travel time to and from feeding grounds even when the time required to find and capture food is minimal. Male assistance in delivering food to nestlings would therefore increase the rate at which nestlings could be fed even when food is very abundant. Moreover, if food resources are extremely abundant around a breeding colony, colony size is likely to increase until competition for food becomes important, unless a shortage of nest sites limits colony size at a lower level. Finally, the energetic strain of feeding nestlings at a faster rate probably reduces female survival (see Chapter 9), and male parental assistance would alleviate that cost. Thus females probably always incur substantial costs by pairing with already mated males, and the advantages gained by mating with genetically superior males are not likely to be sufficient compensation for those costs.

Polygyny threshold hypothesis

Unmated females may prefer pairing with already mated males because those males defend better territories than those of any remaining unmated males (Verner 1964; Orians 1969). If the quality of territories held by unmated males is low, a female might be able to reproduce more successfully as a secondary mate on a high-quality territory than as a monogamous mate on a low-quality territory. The difference in quality needed to make secondary status the best reproductive option for an unmated female is referred to as the **polygyny threshold** (Verner and Willson 1966).

The polygyny threshold hypothesis is best illustrated with a simple graphical model (Figure 11–1) (Orians 1969). The upper curve in Figure 11–1 represents the reproductive success a female can expect on each territory if no additional females breed there. The lower curve represents the success a female can expect if she becomes a secondary mate on a given territory. Each female should select the best unchosen territory until the polygyny threshold is reached. Then unmated females should begin pairing with already mated males who hold the best territories because they can expect higher success there than on any territories held by the remaining unmated males.

Figure 11–1

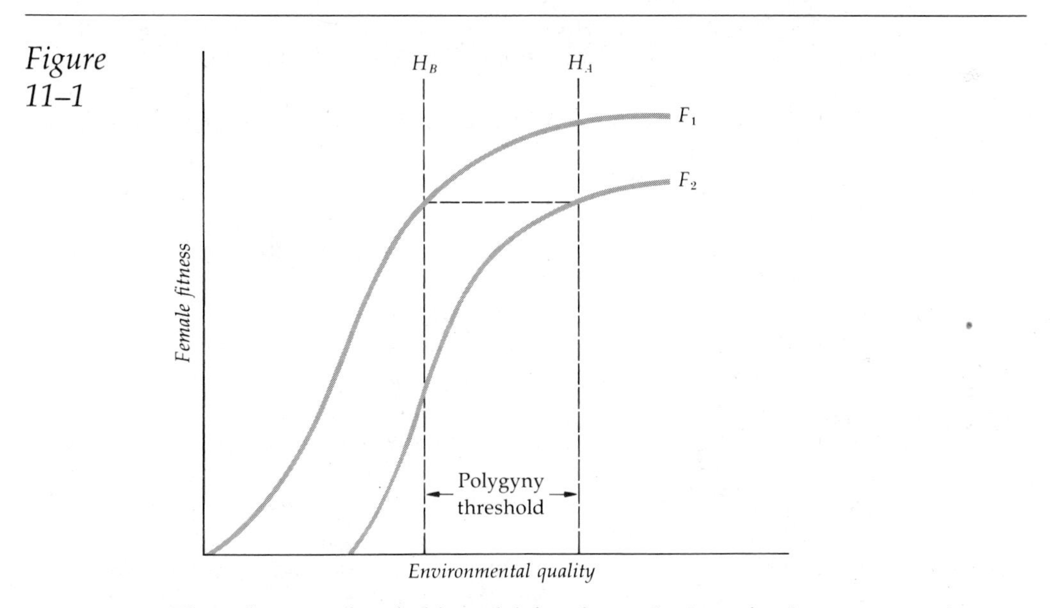

The polygyny threshold model for the evolution of polygynous mating systems.

F_1 = the fitness curve for monogamous females as a function of environmental quality; F_2 = the fitness curve for secondary females as a function of environmental quality; H_A = the best available breeding habitat; H_B = marginal breeding habitat. The polygyny threshold represents the difference in quality between H_A and H_B that would favor polygynous matings. An unmated female should choose to mate with an already mated male in H_A when all males holding territories in habitats superior to H_B are mated.

Source: From J. F. Wittenberger, "The Evolution of Mating Systems in Birds and Mammals," in P. Marler and J. G. Vandenbergh, eds., *Handbook of Behavioral Neurobiology, Vol. 3: Social Behavior and Communication* (New York: Plenum Press, 1979), p. 279. (Modified from Orians 1969.) Reprinted by permission.

Several points about the model require elaboration. First, females should base their mating choices on male quality as well as territory quality (Verner 1964; Orians 1969). A convenient term for the combined quality of a male and his territory is **quality of breeding situation** (Emlen 1957; Wittenberger 1976a). Each male and his territory represent one breeding situation, and unmated females should select the breeding situations where their success will be highest.

Male quality may affect predictions of the model in different ways, depending on whether or not it is correlated with territory quality. If the best males hold the best territories, the rank ordering of breeding situations does not change from the ordering based on environmental quality, but secondary females would not have to produce as many nestlings to be more successful than they could have been as monogamous mates because part of their success would result from higher quality of offspring. To the extent that the polygyny threshold is measured by quantity of offspring, the benefits gained from mating with higher-quality males would lower the threshold (Weatherhead and Robertson 1979). If the best males often hold relatively poor territories, the rank ordering of breeding situations may change from the ordering based on environmental quality (Weatherhead and Robertson 1979; Wittenberger 1979a). Then polygyny might occur on poorer territories than those held by some monogamous males. The effect of male quality will be discussed more fully in a later section (see pp. 459–461).

A second point requiring clarification pertains to the way female fitness is measured. Most studies of polygynous birds estimate female fitness from annual reproductive rate, but in reality fitness depends on lifetime reproductive output. Polygyny can evolve solely because it enhances the annual reproductive rate of secondary females, but this need not be the case. Polygyny may be advantageous even if it reduces annual reproductive rate, provided that it simultaneously increases female survival (Elliot 1975; Altmann et al. 1977). Polygyny may also be advantageous because it increases both annual reproductive rate and female survival. Similarly, polygyny may be disadvantageous even though it increases annual reproductive rate if it decreases female survival at the same time. A general form of the polygyny threshold model must therefore include survival consequences as well as reproductive consequences of secondary mated status (Pleszczynska and Hansell 1980; Wittenberger 1980a).

Finally, since females must decide where to breed in advance of breeding, their mating choices must be based on the success they can *expect* to achieve given the information available to them. Some components of reproductive success may be more predictable than others, and female choices should be based on those components that are most predictable. Tests based on overall success may yield very misleading results, especially if most mortality stems from unpredictable causes. It is entirely possible for females to select territories on the basis of food availability even though most mortality of eggs or nestlings results from predation. If females cannot select nest sites in a way that reduces vulnerability to predation, the best they can do is minimize losses caused by starvation. Thus tests of the polygyny threshold model should be based on predictable components of female success rather than overall success. All studies to date have relied on overall success, and conclusions drawn from those studies should be treated accordingly.

Competitive and cooperative versions of the polygyny threshold model

The females mated with each polygynous male may all suffer reduced fitness because they compete with one another, or they may all gain in fitness because they cooperate with one another (Altmann et al. 1977; Lenington 1977, 1980). Both outcomes are possible, depending on the shapes and relative positions of the fitness curves shown in Figure 11–1.

The graph shown in Figure 11–1 is one form of the competitive model. It assumes that breeding situations are distributed continuously, with no sharp discontinuities separating those of mated and unmated males. The continuous model is appropriate for animals with dispersed breeding distributions but not for animals with clumped breeding distributions. A more appropriate model for colonial polygynous animals assumes a discontinuous habitat distribution, with a sharp discontinuity separating the breeding situations of mated and unmated males (Figure 11–2) (Wittenberger 1979a).

The distinction between the continuous and discontinuous models is important. If the continuous model applies, polygyny evolves when some females can reproduce more successfully as secondary mates on good territories than as monogamous mates on poor territories. Virtually all males can acquire potentially suitable territories, but not all can attract mates. If the discontinuous model applies, polygyny evolves when a fraction of the total male population can control all suitable breeding habitat. Many males cannot acquire potentially suitable territories at all.

In testing the polygyny threshold model one would like to compare the alternative reproductive options open to unmated females who select secondary mated status. One would therefore like to compare the success of secondary females to the success they might have achieved had they bred on territories of unmated males. If the continuous habitat distribution model applies, success on territories of unmated males can be estimated from success of monogamous females who breed on the poorest territories held by mated males or select mates at the same time polygynous matings occur. However, this is not true if the discontinuous habitat distribution applies. Since the polygyny threshold model is often tested by comparing the success of secondary females to that of monogamous females, the distinction between continuous and discontinuous models is crucial when interpreting the empirical evidence.

The cooperative version of the polygyny threshold model assumes that the benefits gained through cooperation exceed the costs of competition on at least some territories (Figure 11–3). Females may benefit from

Figure 11–2

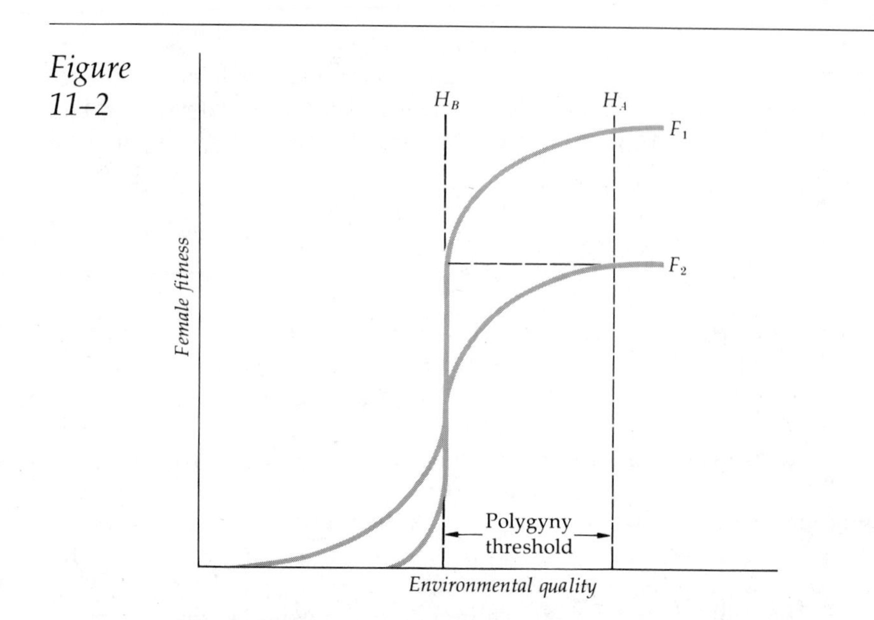

The polygyny threshold model for animals with clumped breeding distributions.

Symbols are as defined in Figure 11–1. All males within suitable habitat (to the right of H_B) attract mates, and many are polygynous because fewer males hold territories than there are females seeking to breed.

Source: From J. F. Wittenberger, "The Evolution of Mating Systems in Birds and Mammals," in P. Marler and J. G. Vandenbergh, eds., *Handbook of Behavioral Neurobiology, Vol. 3: Social Behavior and Communication* (New York: Plenum Press, 1979), p. 283. Reprinted by permission.

mutual vigilance, more effective mobbing of predators, or perhaps other forms of cooperation. Polygyny may be advantageous because cooperation increases the survival of adult females, offspring, or both. Quite possibly cooperation could even increase adult survival at the expense of reduced offspring survival, with polygyny still being advantageous if the gain in future reproductive potential is greater than the loss in immediate reproductive output.

Evidence for testing the polygyny threshold model

The polygyny threshold model has been tested in a variety of ways, mainly by testing various predictions drawn from it (see Wittenberger 1979a; Garson et al. 1981). The most common approach has been to correlate the incidence of polygyny with environmental variables believed to reflect territory quality. Numerous such tests have been conducted, and all show significant correlations (see Wittenberger 1976a, 1979a; Emlen and Oring 1977; Orians 1980). These studies confirm the predicted relationship be-

Figure 11–3

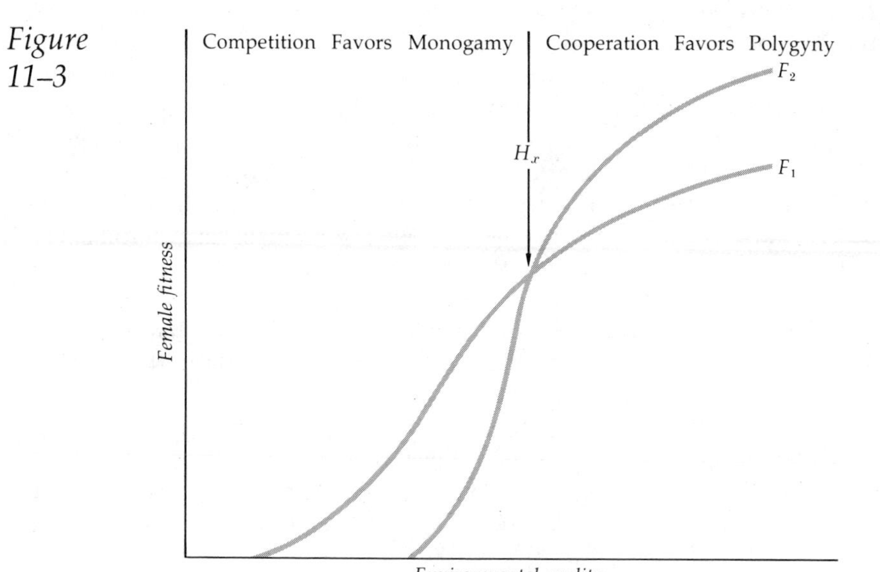

Cooperative version of the polygyny threshold model.

F_1 and F_2 are as defined in Figure 11–1. H_x refers to the level of environmental quality where competition is reduced just enough for the benefits of cooperation to outweigh the costs of competition. Polygyny based on cooperation among females is favored in habitats to the right of H_x, while monogamy is favored to the left of H_x.

Source: From J. F. Wittenberger, "The Evolution of Mating Systems in Birds and Mammals," in P. Marler and J. G. Vandenbergh, eds., *Handbook of Behavioral Neurobiology, Vol. 3: Social Behavior and Communication* (New York: Plenum Press, 1979), p. 282. Reprinted by permission.

tween polygyny and territory quality but do not demonstrate that unmated females do better by accepting secondary status with an already mated male instead of pairing with an unmated male.

The crucial point of the model is the idea that polygyny represents the best choice for unmated females who pair with already mated males. This point can be tested for species with continuous habitat distributions by comparing the success of monogamous and secondary females. Ideally, only predictable components of female success should be used, but present data only allow comparisons based on overall success.

Inspection of Figure 11–1 shows that the polygyny threshold occurs at the point where success of secondary females in the best breeding situations equals that of monogamous females in the poorest breeding situations occupied by monogamous pairs. This is true as long as seasonal factors have little influence on female success. When seasonal factors are important, the success of secondary females should equal that of monogamous females who select mates at the same time.

If male quality does not affect the polygyny threshold, the ratio of success between secondary and monogamous females should equal 1.0 at the polygyny threshold. If male quality is important, the threshold ratio would be less than 1.0 because the smaller number of offspring produced by secondary females would be offset by the higher genetic quality of those offspring. Using a simple model, Weatherhead and Robertson (1979) predicted that the threshold ratio should be about .77 when male quality is important and territory quality is unimportant in determining the quality of breeding situations. A revision of their model to correct for several improper assumptions shows that the threshold can be as low as .50, depending on the breeding sex ratio, the frequency of polygyny in the population, and the average number of females mated to each male (Heisler 1981). If both male quality and territory quality are important, the threshold should lie somewhere between .50 and 1.0.

Data for computing the ratio of success for secondary and monogamous females are available for several noncolonial polygynous birds. The computed ratio is .93 for Ipswich sparrows (*Passerculus sandwichensis princeps*) (Stobo and McLaren 1975; see Wittenberger 1981), .98 for lark buntings (*Calamospiza melanocorys*) (Pleszczynska 1978), and .99 for great reed warblers (*Acrocephalus arundinaceous*) (Dyrcz 1977; see Wittenberger 1981). In addition, data based on the combined success of primary and secondary mates of polygynous males are available for two other species. The success ratio of polygynous females compared to monogamous females is 1.00 for indigo buntings (*Passerina cyanea*) (Carey and Nolan 1979) and 1.18 for prairie warblers (*Dendroica discolor*) (Nolan 1978).

In indigo buntings males do not feed nestlings of either primary or secondary females, and hence the combined success of both give a reasonable estimate for secondary females. If only the data for secondary females are used, the success ratio is .55 (Carey and Nolan 1979). However, secondary females begin breeding later in the season than most monogamous females, and their lower success probably results from seasonal factors. Some males who were unmated at the time secondary females selected mates did acquire mates later. The success ratio of secondary females to monogamous females who mated with those males was 1.58, but sample sizes were small.

In prairie warblers the nestling periods of primary and secondary females do not overlap, and males help feed nestlings in both nests to about the same extent (Nolan 1978). Hence the success of primary and secondary females at any given time should be similar, and combining the data for both helps eliminate the effect of seasonal factors.

The above ratios are minimum estimates for two reasons. First, the success of all monogamous females was used in the computations instead of just those on the most marginal territories. Basing the comparison on all monogamous females is misleading because many monogamous females occupy territories that are higher in quality than any territories available to females who choose secondary mated status (Wittenberger 1981). The data for bobolinks (*Dolichonyx oryzivorous*), for example, give ratios of .61 in Wisconsin (Martin 1974) and .69 in Oregon (Wittenberger 1981). These ratios are poor estimates because in both studies the territories occupied by monogamous pairs were productive wet meadows, while those occupied by unmated males were unproductive dry meadows (Martin 1971; Wittenberger 1976b, 1980b). The existence of a sharp discontinuity in habitat quality between territories of monogamous pairs and unmated males makes the success ratio a poor test of the polygyny threshold. In addition, more secondary than monogamous females are likely to be yearlings. If so, and if success is lower for yearlings because of inexperience, the observed ratio would be reduced by age effects. Then a ratio less than 1.0 could be attributed to differences in age rather than differences in mated status. In Wisconsin bobolinks, for example, most secondary females were yearlings, which may help account for the low ratio observed by Martin (1974).

Present evidence provides strong support for the polygyny threshold model, as it gives values very near the predicted success ratio of 1.0. Since observed ratios are not less than 1.0, they imply that females choose breeding situations largely on the basis of territory quality. Male quality appears relatively unimportant, although this conclusion is only tentative. If secondary females survive less well than monogamous females because of their lower status, they should not accept secondary status unless the success ratio is greater than 1.0. The observed ratio of 1.0 would then not be sufficient to favor polygyny unless the difference is made up by higher genetic quality of offspring.

The strongest criticism of the polygyny threshold model has come from studies of the yellow-bellied marmot. Downhower and Armitage (1971) found that polygynous female marmots reproduce less successfully than monogamous females (Figure 11–4) and hence concluded that polygyny is not advantageous for females. Because these results are often cited as evidence against the model, the yellow-bellied marmot has become something of a test case.

The case of yellow-bellied marmots

Yellow-bellied marmots are diurnal, burrowing rodents that live primarily in colonies situated in open alpine meadows or on talus slopes, although some individuals breed alone in isolated localities (Armitage 1962, 1974; Armitage and Downhower 1974; Svendsen 1974). They are active about five months each summer, depending on climatic conditions, and hibernate for the rest of the year. Males defend territories within each colony, and in Colorado where most studies have been conducted, they pair with one to three females (Downhower and Armitage 1971). Isolated females mate with transient males and breed in small meadows similar in most respects to the larger meadows where colonies are situated (Svendsen 1974).

Hypothesis that polygyny evolved to maximize average fitness of harem members. Three hypotheses have been proposed to explain polygyny in marmots. According to the first hypothesis, polygyny evolved to max-

Figure 11–4

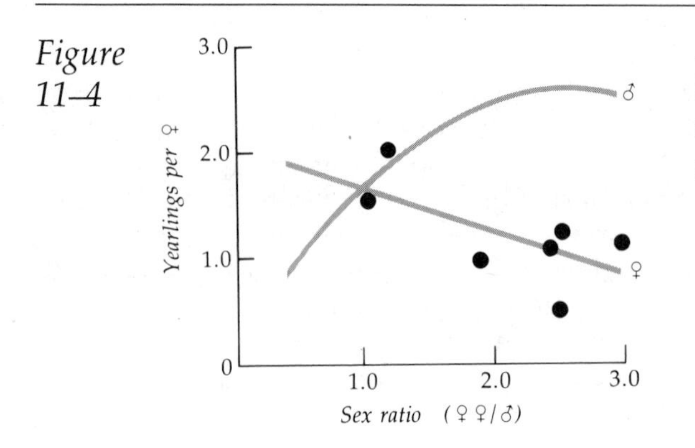

Reproductive success of yellow-bellied marmots in relation to the number of females on each territory.

Sex ratio refers here to number of females mated to a given colony male. The curve for females is a regression line fitted to the data. The curve for males is computed from the curve for females and the number of females mated to each male.

Source: Modified from J. F. Downhower and K. B. Armitage, "The Yellow-bellied Marmot and the Evolution of Polygamy," *American Naturalist* 105 (1971), p. 361, by permission of The University of Chicago Press. Copyright © 1971 by The University of Chicago Press.

imize the *average* reproductive success of each individual on a territory (Downhower and Armitage 1971). Polygyny is presumed to be favored by selection when a male plus two females produce more surviving progeny per individual than a male plus one female. The hypothesis does not assume genetic relatedness among individuals on a territory, and, indeed, individuals are frequently (though not always) unrelated.

This hypothesis is logically unsound because individuals should predicate their behavior solely on the expected reproductive success of themselves and their relatives. They should choose the reproductive option that best enhances their own inclusive fitness. They should not accept an option that reduces their own fitness just because average fitness of all individuals on a territory is still higher than the average fitness of all individuals on other territories.

Downhower and Armitage (1971) developed their hypothesis as an alternative to the polygyny threshold model because their data seemed to contradict one prediction of the model. Orians (1969) predicted that polygynous females should not be less successful on average than monogamous females if the model is correct, and yet the average success of polygynous yellow-bellied marmots is less than the average success of monogamous females. However, for colonial animals the discontinuous version of the polygyny threshold model is applicable, and hence the success ratio of secondary to monogamous females is an invalid test. Monogamous females all occupy territories that are better in quality than any territory (or habitat) remaining available to females who accept polygynous status, and the success achieved by monogamous females is not indicative of the success a low-ranking female could have achieved had she mated with an available unmated male (Wittenberger 1979a). Thus the grounds used by Downhower and Armitage (1971) to reject the polygyny threshold model are invalid.

Hypothesis based on the cooperative version of the polygyny threshold model. The second hypothesis for explaining polygyny in marmots is based on the cooperative version of the polygyny threshold model. According to this hypothesis, polygynous females sacrifice some of their immediate reproductive output to enhance their survival (Elliott 1975). Females might accept the competitive costs of polygyny in return for the vigilance provided by other females on the territory. For a population with zero rate of intrinsic increase (i.e., $r_{max} = 0$), polygyny should be advantageous to females when the percentage increase in their annual survival rate exceeds the percentage decrease in their annual reproductive rate (Wittenberger 1980a). The optimal number of females on each territory occurs at the point where these two percentages are equal, and females should try to become members of breeding groups with that size. Since the intensity of competition may vary among territories, the optimal size of breeding groups may also vary among territories.

Female yellow-bellied marmots may well benefit from mutual vigilance, since they rely on the alarm calls of other females to avoid surprise attacks by predators (Armitage 1962). However, females could theoretically rely on the vigilance of females from neighboring territories rather than the vigilance of other females on the same territory. Mutual vigilance among polygynous females may therefore not confer much of a survival advantage.

Females may also gain mutual energetic benefits because residents on each territory normally hibernate together during winter (Andersen et al. 1976). Clustering together during hibernation reduces heat loss, which could improve overwintering survival and conserve fat reserves. Females rely on fat reserves to breed early in spring, and early breeding is important because it gives juveniles more time to develop and store fat for the following winter (see Armitage and Downhower 1974). However, females do not gain a net reproductive advantage by hibernating in groups. Polygynous females do not produce as many surviving offspring as monogamous females despite the larger size of their hibernation groups. Hibernating in larger groups also does not seem to enhance overwintering survival of juveniles, since juvenile survival is not correlated with group size (Downhower and Armitage 1971). It still remains possible that females survive longer when they hibernate in larger groups, but no data are available on this point. The crucial question of whether females survive better when breeding in larger harems has not yet been adequately studied.

Hypothesis based on the competitive version of the polygyny threshold model. The third hypothesis for explaining polygyny in marmots is based on the competitive version of the polygyny threshold model (Wittenberger 1979a, 1980a). According to this hypothesis, polygyny evolves because it represents the best reproductive option for some unmated females.

A graphical model best illustrates the hypothesis (Figure 11–5). The data on reproductive success show that the best option for an unmated female is to become the sole mate of a resident male in a colony. Once every male in a colony has one mate, the next best option is to pair with an already mated male in a colony. This is a better reproductive option than breeding in an isolated area away from the colony (Downhower and Armitage 1971). Since polygynous matings reduce the success of resident females, resident females should try to prevent unmated females from entering the territories of their mates. If some resident females can effectively exclude unmated females while others cannot, some males should acquire two mates while others acquire only one. Eventually, opportunities for pairing with monogamous males are exhausted, and any remaining females can be most successful by pairing with bigamous males rather than breeding in isolated areas (see Downhower and Armitage 1971). If the concerted aggression of two females prevents entry of unmated females on some territories but not on others, some males will acquire only two mates while others will acquire three. At some point the combined aggression of all resident females on each territory prevents further entry into the colony, and any remaining unmated females are forced to breed in isolated areas or defer breeding altogether. Thus polygyny could have evolved because relatively few males exclude the remaining males from all suitable colony sites, with the result that many females must accept polygynous status in order to breed within a colony.

All the evidence is consistent with the competitive model. Monogamous females are more successful than bigamous females, who are more successful than trigamous females, who in turn are more successful than isolated females (Downhower and Armitage 1971). Resident females are aggressive toward unmated females, though to varying degrees (Svendsen 1974). The most aggressive females remain monogamous, while less aggressive females are joined by others to form polygynous breeding groups.

Two important points of the model have not been tested. The first point concerns the assumption that habitats are limiting. The hypothesis presumes that some males do not establish territories because they cannot find suitable habitat. It remains possible that males remain nonterritorial for some other reason. However, even if they do, the number of territorial males in colonies is still limited relative to the number of females attempting to breed in colonies. Therefore, violation of this assumption does not negate the competitive model. The second point concerns the assumption that female aggression is responsible for limiting the number of females breeding on each territory. Experimental studies, perhaps involving hormone implants, could test this point.

Either the cooperative or competitive version of the polygyny threshold model may ultimately provide the best explanation of polygyny in

Figure 11–5

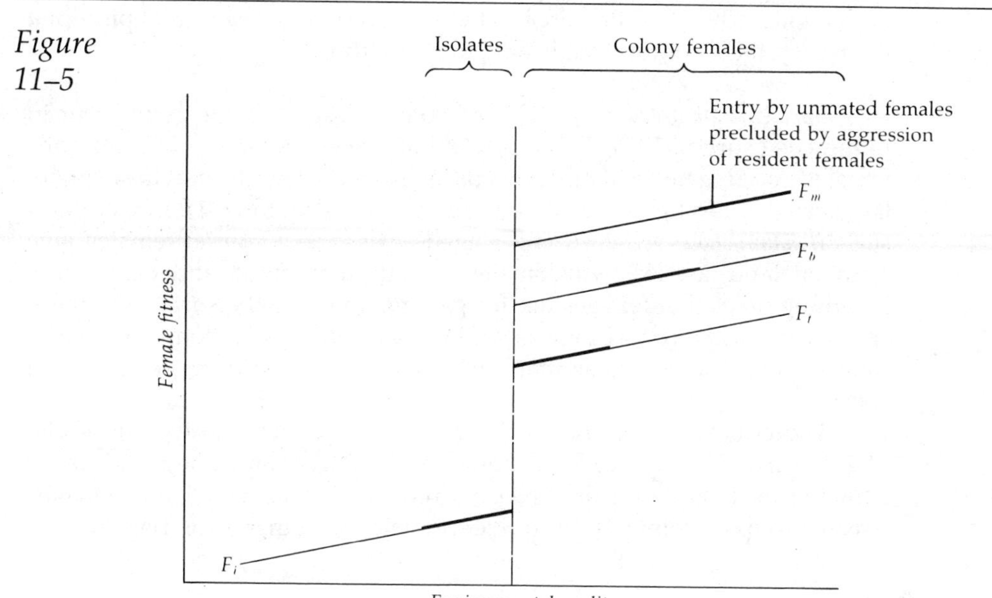

A model for explaining the evolution of polygyny in yellow-bellied marmots, based on the competitive version of the polygyny threshold model.

F_m = fitness of monogamous females; F_b = fitness of bigamous females; F_t = fitness of trigamous females; F_i = fitness of isolated females that breed outside colonies. Polygyny should occur once all males in a colony are mated. Females should not breed in isolated areas unless they are excluded from a colony by the aggression of colony members. The number of females attracted to each territory depends to a large extent on how aggressive resident females are on that territory.

Source: From J. F. Wittenberger, "The Evolution of Mating Systems in Birds and Mammals," in P. Marler and J. G. Vandenbergh, eds., *Handbook of Behavioral Neurobiology, Vol. 3: Social Behavior and Communication* (New York: Plenum Press, 1979), p. 290. Reprinted by permission.

marmots. Regardless of which model eventually proves correct, the polygyny threshold concept seems adequate for analyzing the mating system of yellow-bellied marmots.

Polygyny in pinnipeds

The extreme polygyny and marked sexual dimorphism evident in elephant seals, fur seals, and sea lions has long attracted the interest of biologists. The evolution of polygyny in pinnipeds is now reasonably well understood. The question here is whether current explanations of pinniped mating systems conform with the polygyny threshold model.

Species that breed on land. Two basic forms of polygyny occur in pinnipeds (Stirling 1975, 1977). In elephant seals, fur seals, and sea lions females congregate in dense groups or *pods* on remote beaches, where large-sized males fight to control access to them (Photo 11–1). In Weddell's seals (*Leptonychotes weddelli*) and ringed seals (*Phoca hispida*) females give birth on land-fast ice, while males remain submerged and defend the breathing holes that females use for gaining access to the surface. In these species breeding congregations are not as dense, and dominant males cannot control access to as many females as they can in more terrestrial species.

The critical characteristics underlying the extreme polygyny of elephant seals, fur seals, and sea lions are terrestrial copulation and dense crowding of females on the breeding rookeries. This combination makes female groups a defensible resource for males, and males compete fiercely

Photo 11–1

Two high-ranking male elephant seals compete for dominant status by challenging each other aggressively.

These two males are threatening each other by inflating special air sacs in their snouts and throwing back their heads, but physical fights are not uncommon. The winner of such disputes gains increased access to females and is therefore reproductively more successful.

Photo from B. J. Le Boeuf and R. S. Peterson, "Social Status and Mating Activity in Elephant Seals," *Science* 163 (3 January 1969):91–93.

for control of them. In species with aquatic copulation females are difficult to monopolize because they do not congregate together in stable assemblages during their period of receptivity. Males also cannot defend exclusive access to female groups underwater because they must surface periodically for air.

The origins of terrestrial copulation are unclear. Bartholomew (1970) argues that terrestrial copulation is an extension of terrestrial parturition that evolved because it allows females to select genetically superior mates. However, this hypothesis seems inadequate for at least two reasons. Harbor seals (*Phoca vitulina*) and monk seals (*Monachus schauinslandi*) give birth on land but copulate predominately in the water (Venables and Venables 1955, 1957, 1959; Kenyon and Rice 1959), implying that terrestrial parturition does not necessarily give rise to terrestrial copulation. Also, the hypothesis does not explain why improved opportunities for mating with genetically superior males are advantageous enough to favor terrestrial copulation in some species but not in others. Stirling (1975) argues that terrestrial copulation is a legacy from nonaquatic ancestral forms that has been retained in some phylogenetic lines while being lost in others. However, the legacy hypothesis is less than satisfying, as it implies that terrestrial and aquatic copulations are selectively neutral. At present, the evolution of terrestrial copulation has not been satisfactorily explained.

Female elephant seals, fur seals, and sea lions congregate into dense pods on breeding rookeries for two reasons (see pp. 322–323). First, breeding sites are restricted to relatively few islands as a defense against terrestrial predators (Bartholomew 1970; Stirling 1975), and, second, females crowd together to reduce risks of injury for themselves and their pups. Female fur seals and sea lions leave the breeding rookery periodically to forage, returning about once a week to suckle their pups (Bartholomew and Hoel 1953; Peterson and Bartholomew 1967; Miller 1975). To escape harassment from territorial males, they congregate as near to water as possible and avoid traversing territories of males while going off to sea. Female elephant seals do not leave their pups to forage, but they seek the center of pods to reduce the risk of being crushed by males and to minimize harassment from subdominant males (Le Boeuf 1972; Le Boeuf et al. 1972; Le Boeuf and Briggs 1977). The presence of dominant males near the center of pods reduces the frequency of fights because few males challenge dominant males and subdominant males are discouraged from approaching too closely. Females near the center of pods may also be inseminated more frequently by genetically superior males, which might improve the quality of their offspring (McLaren 1967; Cox and Le Boeuf 1977). The density of pods is ultimately limited by the aggression of central females, who defend a small personal space around themselves (Christianson 1974).

Polygyny occurs in terrestrial-breeding pinnipeds because a fraction of all males can control access to female pods, an argument identical to the polygyny threshold model for discontinuous breeding distributions. Male fur seals and sea lions defend territories where females congregate, while male elephant seals compete for high dominance status. This difference results from a difference in female settling patterns on the beach. Female fur seals and sea lions settle predictably along shorelines, allowing males to establish fixed territories in advance of female arrivals. In contrast, female elephant seals may settle anywhere on a beach, and males cannot predict in advance where females will be. Consequently, they establish dominance over large sections of beach so that they can later claim whichever spot where females eventually congregate.

Accepting polygynous status is not without cost for females. Males pay no attention to the presence of small pups, who were normally fathered by different males the preceding year, and often crush pups during fights over females (Bartholomew 1952; Bartholomew and Hoel 1953; Le Boeuf et al. 1972; Le Boeuf and Briggs 1977). Another major cause of mortality results from pups becoming separated from their mothers, often because the mother is displaced by an aggressive female or is distracted by other newborn pups shortly after giving birth (Le Boeuf and Briggs 1977). Lost pups are occasionally adopted, but they are more often bitten severely when they try to suckle from a strange female (Photo 11–2). These two factors account for most pup mortality in elephant seals, and they are both exacerbated by high female densities (Le Boeuf and Briggs 1977). They appear much less important in other terrestrial seals except grey seals (*Halichoerus grypus*). Another cause of offspring mortality in elephant seals is harassment of weaned pups, or weaners, by subordinate males (Le Boeuf et al. 1972). Such harassment causes weaners to congregate into separate pods away from the main breeding groups. In fur seals and sea lions high breeding densities may lead to local depletion of food resources, thus forcing females to spend longer periods at sea and hence increasing the risk of pups becoming lost or injured. Some evidence suggests that resources become depleted around Alaskan fur seal rookeries (Chapman 1961), but the evidence is not conclusive. The high costs of crowding into a dense rookery may explain why females sometimes mate with immature males in isolated areas. However, isolated females run the risk of not being found by a male, and this risk probably explains why most females breed in established rookeries.

Species that breed on land-fast ice. Polygyny is less extreme among seals that breed on land-fast ice. Weddell's seals breed on circumpolar fast ice surrounding the Antarctic continent and its associated islands, where terrestrial predators are absent (Lindsey 1937; Bertram 1940; Stirling 1969, 1977). Ringed seals breed on fast ice in the Arctic Ocean and are vulnerable to polar bears (*Ursus maritimus*) and, to a lesser extent, Arctic foxes and wolves (McLaren 1958; T. G. Smith 1973; Stirling 1977). The best information is for Weddell's seals because ringed seals give birth in lairs beneath the snow to escape terrestrial predators (Smith and Stirling 1975; Stirling 1977).

The reason polygyny is less extreme among Weddell's and ringed seals is that breeding densities are lower than for more terrestrial seals

Photo 11–2

A female elephant seal biting a lost pup that has strayed too near.
Female aggression toward orphaned and lost pups is one important source of pup mortality in elephant seals. Pups that die from becoming lost represent one cost of breeding in crowded rookeries instead of on dispersed beaches.
Photo by B. J. Le Boeuf.

(Stirling 1975, 1977). Breeding densities of female Weddell's seals are lower for at least two reasons. First, females are tied to access holes through the ice, which are the only connections with the surface, and only a limited number of females can utilize each hole. Second, females and pups are not susceptible to trampling because males do not emerge onto the ice, and the deleterious effects of crowding tend to space them apart. In ringed seals a third factor is also important. Because females are highly vulnerable to polar bears, they are spaced more widely apart to escape detection and hence often do not share breathing holes.

Male Weddell's seals defend a three-dimensional space below their breathing holes and can control access to the several females using each hole (Stirling 1975, 1977). The extent to which males can monopolize females is unknown and may be limited. The mating system may therefore be more nearly promiscuous than polygynous. Since the number of females at each hole is limited, the number of females a male can monopolize is also limited. Male ringed seals probably also defend a three-dimensional space below their breathing holes. However, only one or at most two females use each hole, so males can control only one or two females. Ringed seals are therefore nearly monogamous. Thus polygyny in Weddell's and ringed seals occurs because males can control access to a discrete set of resources (i.e., access holes) required by females. A comparison of all polygynous seals shows a clear relationship between the extent of female clumping and the extent of male monopolization of females.

Polygyny in passerine birds

Polygyny in passerine birds has been correlated with variations in both food abundance and vegetation structure, but the correlations must be interpreted cautiously. Variables correlated with male pairing success may be secondary correlates of other factors that are actually responsible for polygyny, and their importance as selective factors cannot be accepted unless they have a direct effect on female fitness. Correlations with vegetation structure are especially difficult to interpret because in most polygynous birds vegetation structure has not been shown to affect nestling survival, and vegetation structure may well be a secondary correlate of food availability. Since many insects depend on vegetation for food and cover, their population densities are bound to vary with species composition and density of vegetation. Correlations between polygyny and food abundance are more likely to reflect an important role of food abundance in selecting for polygyny, but, even so, evidence showing that changes in food abundance actually affect female fitness is still required. Since evidence of this sort is usually lacking, a theoretical basis for predicting which factors should be most important would be helpful. Such a theory would also clarify how specific environmental conditions shape avian mating systems.

Polygynous passerine birds fall neatly into two well-defined categories (Table 11–2) (Wittenberger 1976a). In the first category territorial males nearly always attract at least one mate, a majority are polygynous, and some attract as many as five or more mates. Yearling males usually do not establish territories and consequently remain unmated. Sexual dimorphism in plumage characteristics or size generally occurs, and yearling males are usually more similar in plumage to females than to adult males. These species all exhibit clumped nesting distributions and hence fit the discontinuous habitat distribution model. In the second category territorial males often fail to attract even one mate, only a minority are polygynous, and none attract as many as five females. Yearling males usually establish territories and often acquire one mate. Sexual dimorphism in plumage is often absent, sexual dimorphism in size is usually minimal or absent, and when sexual plumage dimorphism is present, yearling males resemble adult males rather than females. These species all exhibit dispersed nesting and hence fit the continuous habitat distribution model.

Males should establish territories only where they have some chance of attracting mates. The regular occurrence of unmated territorial males therefore implies that males cannot accurately predict female habitat preferences at the time they establish territories (Wittenberger 1976a). Hence a reasonable conclusion is that males cannot assess the quality of marginal

Table 11–2 *Effect of nesting dispersion on male behavior, morphology, and pairing success in polygynous passerine birds.*

Nesting dispersion	Number of genera	Territorial males often remain unmated	Age dimorphism among males in size or plumage	Sexual dimorphism in size or plumage	High degree of polygyny
Noncolonial (dispersed)	12	92–100%	0%	50%	0%
Colonial (clumped)	11	0%	82%	91%	82%

Source: From J. F. Wittenberger, "The Evolution of Mating Systems in Birds and Mammals," in P. Marler and J. G. Vandenbergh, eds., *Handbook of Behavioral Neurobiology, Vol. 3: Social Behavior and Communication* (New York: Plenum Press, 1979):296. Reprinted by permission.

habitats accurately in species with dispersed nesting distributions but can in species with clumped nesting distributions.

Referring to Figure 11–2, one can see a sharp discontinuity between marginal habitats that attract females and marginal habitats that do not. This discontinuity usually results from structural discontinuities in the environment. Colonial birds generally nest at relatively inaccessible sites such as isolated trees, marshes, thorny thickets, or cliff ledges to reduce vulnerability of nests to predation (Lack 1968). In addition, colony sites are often used traditionally year after year because females are more successful breeding in large established colonies than in small newly established ones (see Chapter 8). Thus males who can establish territories within colony sites are virtually assured of attracting mates, while males who are forced into other areas are virtually assured of remaining unmated. Males rarely establish territories outside suitable colony sites, and hence territorial males rarely remain unmated. Since habitat quality depends largely on physiognomic structure of the nesting substrate, physical or vegetational factors are responsible for enabling a fraction of the males to monopolize all the females.

The situation for species with dispersed nesting is quite different. Referring to Figure 11–1, one can see no discontinuity between marginal habitats that attract females and marginal habitats that do not. The exact location of the polygyny threshold is unpredictable for at least two reasons (Wittenberger 1976a, 1979a). First, the shape of the fitness curves varies between years due to annual variations in weather conditions, vegetational growth, and food availability. Second, males establish territories a week or more before females select mates and have less reliable cues available for assessing habitat quality. Males apparently select habitats according to physiognomic cues such as moisture conditions or vegetation structure because nestling food resources are generally absent or unpredictable at the time territories are established (Svärdson 1949; Hildèn 1965a; Klopfer 1969; Wittenberger 1976b). If females also base their choices on physiognomy, males should be able to predict female preferences fairly well unless unpredictable seasonal changes occur. However, if females base their choices on food availability, males would not be able to predict female preferences very well. The comparative evidence for species with dispersed nesting indicates that males cannot predict female preferences accurately, and hence it suggests that females base their choices on factors such as food availability which are unpredictable at the time males establish territories. Thus variations in food availability among territories may be a major reason why some males attract two or more females while others attract none.

One point about the analysis above should be kept in mind. The dichotomy between clumped and dispersed nesting is artificial, and intermediate conditions are possible. Classifying species into these two categories is useful because present evidence does not allow a quantitative measure of degree of clumping, but deviations from general trends, such as those indicated in Table 11–2, are likely to arise among species with intermediate nesting distributions.

Selective factors favoring polygyny in colonial passerine birds

The theory above predicts that polygyny should evolve in colonial birds when relatively few males can control all suitable breeding sites. The suitability of a site may depend on both structural characteristics of the habitat and the presence of an established colony. Polygyny should not occur if a single resident female can prevent other females from breeding on each male territory, if a female cannot reproduce successfully without undivided male parental assistance, or if a territorial male cannot control enough space in a colony to accommodate more than one female.

The evidence confirms that polygyny has evolved in colonial birds when relatively few males can control all suitable breeding sites (Wittenberger 1976a, 1979a) and has not evolved when every male can control a suitable breeding site (Wittenberger and Tilson 1980). The best evidence is for polygynous weaver finches of Africa and Asia, which nest in isolated trees on open savannas or in cultivated areas (Ali 1931; Crook 1958, 1960, 1964, 1965; Collias and Collias 1964; Ambedkar 1972). Weaver finches do not defend the abundant seed crops they exploit for food, but they do defend nest sites within colony trees. Both size and number of established breeding colonies are limited (e.g., Photo 11–3), and many males cannot acquire territories in colonies. Unmated females must therefore accept polygynous status in an existing colony or pair with an unmated male in a small, newly established colony. If female reproductive success is substantially lower in small colonies than in large ones, their best choice would

Photo 11–3

*A nesting colony of the village weaver (*Ploceus cucullatus*), a ploceine weaver finch of African savanna.*

Note that colony size is limited, creating an opportunity for males to control the nest sites used by several females.

Photo by Edward S. Ross.

be to accept polygynous status within an established large colony. Unfortunately, no data on nest success are available for colonies of differing size, so the hypothesis that females are making the best choice by accepting polygyny has not yet been validated.

Similar habitat conditions are apparently responsible for polygyny in colonial oropendolas, caciques, and grackles (Selander 1972; Wittenberger 1976a). Males of these species do not defend discrete territories within colonies, but they do exclude other males from mating with females at each colony site (Chapman 1928; Skutch 1954; Tashian 1957; Selander and Giller 1961; Drury 1962; Slud 1964). A single male may defend the colony if it is small, or several males may control access to it if the colony is large.

In marsh-breeding blackbirds the evolution of polygyny involves two interrelated questions. The first concerns why polygyny occurs at all, and the second concerns why some males within marshes acquire more mates than others (Wittenberger 1981). The distinction is important because different habitat variables may be involved at each level of analysis.

Polygyny occurs in marsh-breeding blackbirds because suitable marshes are in short supply and not all males can acquire territories within them. A shortage of suitable marshes is indicated by two lines of evidence. Experimental removal of territorial male red-winged blackbirds quickly leads to replacement by other adult males who were previously excluded

(Orians 1961a; Peek 1971), and nonterritorial males can often be seen in the vicinity of breeding marshes (Holm 1973).

Suitable marshes are limited in availability for several reasons. Blackbirds have generalized diets and concentrate on abundant seed resources during winter. As a result, winter mortality is relatively low compared to most birds and breeding populations increase to very high densities relative to the number of suitable marshes (Orians 1961a). Some marshes are not inhabitable by any given species for a variety of reasons. Yellow-headed blackbirds are restricted to the outer edges of reedbeds in the more productive marshes because their food requirements are high (Willson 1966; Orians 1980). Red-winged blackbirds are excluded from some marshes by the interspecific aggression of yellow-headed blackbirds and tricolored blackbirds (Orians 1961a; Orians and Collier 1963; Orians and Willson 1964). Red-winged blackbirds are also effectively excluded from some marsh areas by long-billed marsh wrens, which puncture blackbird eggs and kill young nestlings (Picman 1980).

Given a shortage of territorial males in marshes compared to the number of breeding females, many unmated females must choose between polygynous status in a marsh or monogamous status outside a marsh. Females should be able to reproduce as secondary or lower-ranking mates more successfully within marshes than as monogamous mates in uplands for at least two reasons. Marshes with high water nutrient levels produce much more food for feeding nestlings, and marshes provide greater protection from predation. Under current conditions of high breeding density and polygyny in marshes, intense competition for food offsets the advantages of breeding where food is more abundant, and the principal advantage of breeding in marshes is reduced predation. R. J. Robertson (1972) found that starvation rates of nestling red-winged blackbirds are similar in marshes and nearby uplands, while predation rates are substantially lower in marshes. This difference between marshes and uplands is not evident everywhere (e.g., Dolbeer 1976), but studies in agricultural districts may not be representative of past conditions if recent habitat changes or control programs strongly affect predator densities. The importance of predation pressure is substantiated by the fact that females faced with breeding in uplands near abundant food resources or in low-productivity marshes far from good food sources choose to breed in the marshes and travel long distances to forage (Holm 1973).

Not all marshes are equally suitable for breeding. Marshes vary tremendously in productivity of emergent aquatic insects (Orians 1980). Some types of vegetation and structural configurations offer better protection of eggs and young from predation or adverse weather (Fautin 1940; Holm 1973). Finally, marshes are less suitable for blackbirds if marsh wrens breed in them because these wrens frequently puncture blackbird eggs and kill young nestlings (Picman 1977, 1980; Picman and Picman 1980). Such factors result in higher average male pairing success in some marshes than in others, but they do not explain why some males within any given marsh acquire more females than others.

The reasons why some territories within a marsh attract more females than do others are poorly understood. One factor may be the amount of food on each territory. Food availability on territories may be especially important for yellow-headed blackbirds, at least in populations where they forage extensively on their territories (Willson 1966). Most food on a territory is concentrated along the outer edges of reedbeds, and the number of females attracted to each territory is correlated with edge length. However, in some populations female yellow-heads typically do not feed extensively on territories, and the number of females breeding on each territory is not correlated with direct measurements of food abundance (Wittenberger, unpublished data). Other factors that may affect female settling patterns in at least some marshes include food abundance in nearby undefended areas, composition and density of marsh vegetation, water depth, proximity to marsh wrens, and the aggression of resident females (Nero and Emlen 1951; Nero 1956; Holm 1973; Lenington 1977, 1980; Picman 1977, 1980; Wittenberger 1981). The importance of each factor probably varies between marshes and between years.

The determinants of territory quality within marshes are complex and difficult to unravel, but they are not directly responsible for the prevalence of polygyny in marsh-breeding blackbirds. They determine where females settle within each marsh and to some extent the reason why female densities vary among marshes, but polygyny is prevalent in the first place because a fraction of the total male population can control all suitable marsh habitats.

Selective factors favoring polygyny in noncolonial passerine birds

The theory discussed earlier predicts that food abundance or other unpredictable habitat characteristics should be important determinants of territory quality in polygynous birds with dispersed nesting distributions. Some data support this prediction, but some do not. One problem in evaluating this prediction has been that many studies failed to measure food abundance, with the result that its importance has often been discounted on the basis of inadequate evidence.

The best evidence that food availability is an important factor favoring polygyny comes from a study of bobolinks. The principal food fed to nestling bobolinks is caterpillars, and both the order in which females select territories and the number of females attracted to each territory are correlated with caterpillar densities (Figure 11–6) (Wittenberger 1976b, 1980b). Females forage both on and off their territories during the nestling period, but food abundance on their territories appears to affect nestling survival. Secondary females have lower success than primary females largely because more of their nestlings starve (Martin 1971, 1974; Wittenberger 1978b, 1980c), and in at least some populations secondary females rely on food resources near their nests to compensate for lost male parental assistance (Martin 1974). Males select territories before caterpillars emerge, and they

Figure 11–6

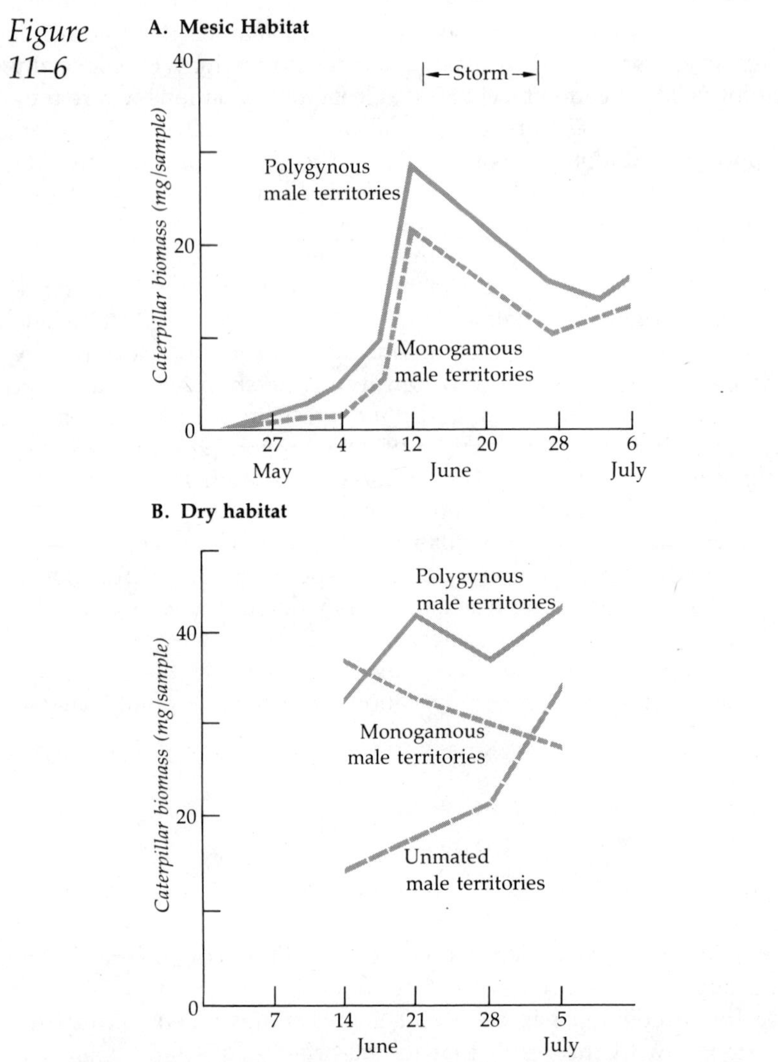

Relationship between caterpillar abundance and the occurrence of polygyny on bobolink territories.

A. A comparison of territories held by polygynous and monogamous males in mesic grassland habitat, which is preferred by both sexes over dry grassland habitat. The difference between territories of polygynous and monogamous males was statistically significant from 28 May to 6 June, which encompassed the time secondary females selected mates in mesic areas. B. A comparison of territories held by polygynous, monogamous, and unmated males in dry grassland habitat. The difference between polygynous and monogamous males was statistically significant only on 5 July. The difference between territories of polygynous and unmated males was statistically significant from 7 June to 28 June, which encompassed the period when secondary females selected mates in dry areas.

Source: From J. F. Wittenberger, "Vegetation Structure, Food Supply, and Polygyny in Bobolinks *(Dolichonyx oryzivorus),*" *Ecology* 61 (1980):147, 148. Reprinted by permission of Duke University Press. Copyright 1980 by the Ecological Society of America.

show no ability to discriminate between habitats that later differ in productivity within physiognomically similar habitats (Wittenberger 1976b). Also, the order males select habitats is only weakly correlated with the order females select habitats and the number of females attracted to them. Thus the inability of males to predict where caterpillar abundance will be high, along with a female preference for habitats with high caterpillar abundance, apparently explains why some males establish territories in habitats that do not attract any females.

Martin (1971) attributed polygyny in bobolinks to variations in nesting cover among territories rather than variations in food abundance. His evidence consists entirely of small differences in vegetation cover between territories that were correlated with male pairing success and the propensity for primary females to feed off territories. He did not measure food availability on territories, and secondary females did forage frequently on territories. The evidence does not show that nest success is directly affected by small differences in vegetation cover, and the observed differences could as easily have arisen as secondary correlates of food abundance.

Nesting failure in dickcissels, another grassland bird, results most often from predation or cowbird brood parasitism (Zimmerman 1966, 1971). Male pairing success is weakly correlated with vegetation structure, and Zimmerman attributed polygyny to variations in nesting cover among male territories. However, the relationship between cover and nest success is again unknown, and several inconsistencies in the data suggest that other factors are also important (see Wittenberger 1976a). Harmeson (1974) found a correlation between male pairing success and arthropod densities present on territories in June, suggesting that food availability may affect female mating choices.

Polygyny is apparently associated with food availability in at least three species of polygynous wrens and one species of Old World warbler. In European wrens polygyny is prevalent in productive forests but is absent on barren islands where food is scarce (Armstrong 1955; Armstrong and Whitehouse 1977). A male's pairing success is correlated with the number of nests he builds on his territory (Garson 1980). It seems likely that the number of nests a male can build depends on how much food is present on the territory, although the role of food abundance has not been studied. Male experience may also be a factor, since females seem to discriminate between good and poor nest sites and experienced males are more likely to build in good sites. In house wrens (*Troglodytes aedon*) polygyny arises when males can spend time seeking additional mates while still feeding nestlings of their first broods (Kendeigh 1941), and this is most likely to be possible on territories where food is most abundant. In long-billed marsh wrens male pairing success is correlated with edge length of emergent vegetation on their territories, which is likely to be correlated with abundance of emergent aquatic insects (Verner 1964). It is also correlated with the number of nests a male builds, which again may depend on food availability on the territory (Verner and Engelsen 1970). In the unrelated great reed warbler polygyny is most prevalent on territories located near trees and bushes adjacent to the reedbeds where nesting occurs, and much of the foraging by females is in those trees and bushes (Dyrcz 1977).

In lark buntings, another grassland bird, food abundance is apparently *not* important in determining territory quality (Pleszczynska 1978; Pleszczynska and Hansell 1980). Females forage mainly off territories, and the number of females attracted to each territory is not correlated with food abundance on that territory. The principal causes of nestling mortality are predation, heavy precipitation, and desiccation of nestlings during hot weather. Lark buntings nest in drier grasslands than bobolinks or dickcissels, and overhead cover for nests is much sparser. Since experimental shading of nests increases nestling survival, the amount of available cover should be an important determinant of territory quality. One way to assess vegetation cover is to measure light penetration to ground level, and Pleszczynska found that the number of females attracted to each territory is negatively correlated with mean light intensities at ground level. Vegetation cover is predictable in advance of nesting, and Pleszczynska could accurately predict male pairing success on the basis of light measurements alone. This evidence leaves unanswered the question of why territorial males regularly settle in habitats that later fail to attract any females.

Red-winged blackbirds are predominantly marsh-nesting birds, but they do regularly breed in upland habitats, especially in grasslands, pastures, and cultivated fields. Polygyny occurs in uplands as well as in marshes. The upland populations best fit a continuous habitat distribution, which creates an opportunity for testing the effect of nesting distribution on a single species. Several studies of upland red-wing populations have been conducted (e.g., Case and Hewitt 1963; Stowers et al. 1968; R. J. Robertson 1972; Blakley 1976; Dolbeer 1976; Monahan 1977), but none have clearly identified any environmental correlates of polygyny in upland habitats.

To summarize, food abundance is implicated as an important selective factor for many but not all polygynous birds with dispersed nesting distributions. A general pattern has not yet been established, but the evidence is still very incomplete. Even basic information pertaining to how vegetation cover affects nest success and the extent to which females forage on male territories during the nestling period is inadequate or lacking. Until such information becomes available, a clear understanding of why these species are polygynous will remain elusive.

Promiscuity

Promiscuity results from selection against prolonged pair bonding. When neither sex contributes parental care, prolonged pair bonding usually has little intrinsic value and promiscuity is the rule. The evolution of promiscuity is most interesting when parental contributions from at least one sex are necessary for offspring to survive. Once obligatory parental care evolves, females can potentially acquire protection or direct energetic contributions from the male while reproducing, and because they are time- and energy-limited, females should take advantage of every opportunity for exploiting male contributions. Promiscuity should therefore evolve only if opportunities for obtaining male assistance are absent or if the costs of taking advantage of them are too high.

Male emancipation hypothesis

The problem in trying to explain promiscuity in altricial birds is that females seem to suffer a high cost from loss of male parental assistance without receiving any compensatory benefits. Some authors have therefore argued that promiscuity evolves when females can raise just as many offspring without male help. Then loss of male parental assistance would involve minimal cost and females would gain little by forming pair bonds. According to this view, promiscuity allows females to mate with genetically superior males instead of often being restricted to inferior unmated males, while at the same time emancipating males from parental duties and freeing them to seek additional mates (Snow 1962a; Orians 1969; Snow 1970; Willis et al. 1978).

The male emancipation hypothesis has commonly been invoked to explain promiscuity among tropical frugivorous songbirds. The general argument is that clutch sizes are limited in frugivorous species by the low protein content of fruit rather than the ability of parents to feed nestlings. Low clutch sizes might in turn enable females to feed nestlings adequately without help from the male.

A major impetus for the male emancipation hypothesis has been the strong correlation between promiscuity and frugivory among tropical passerine birds (see Lack 1968). A majority of adequately studied tropical frugivores are promiscuous, and very few promiscuous passerines are nontropical or nonfrugivorous (for exceptions see Van Someren 1945; Murray 1969; Willis 1972b; Willis et al. 1978). Frugivory has not led to promiscuity in temperate birds, for reasons to be discussed below.

The male emancipation hypothesis implies that clutch sizes are limited by the ability of parents to feed nestlings in monogamous passerines but not in promiscuous ones. Otherwise, males should also be emancipated from parental duties in monogamous species. The hypothesis therefore hinges on the factors limiting clutch size in tropical birds.

Clutch sizes are consistently lower in tropical birds than in related high-latitude species (Cody 1966; Klomp 1970), but the underlying reasons are controversial. According to one hypothesis, clutch sizes are lower in the tropics because nests are more vulnerable to predation (Skutch 1949). Skutch argued that larger clutch sizes would necessitate more parental visits to the nest during the nestling period and hence make nests more conspicuous to predators. When fewer young are raised at a time, more nests would escape predation and average reproductive success would be higher. Since Skutch's hypothesis implies that males should be emancipated from parental duties in nearly all tropical birds and not just in frugivorous species, its validity would make the male emancipation hypothesis less plausible for explaining promiscuity.

Present evidence does not support Skutch's (1949) hypothesis. The hypothesis assumes that predators find nests by watching parent birds rather than searching in likely places for nests. Hence it predicts that predation rates should be higher during the nestling period than during incubation. The typical pattern, however, is for predation rates to remain the same or even decrease slightly during the nestling period (Snow 1962a; Lack 1966; Ricklefs 1969, 1977a). Moreover, clutch sizes are just as low for tropical birds that nest in holes or build pendant nests, even though their nests are much less vulnerable to predation than nests of other tropical species. If predation is responsible for reducing clutch sizes in tropical birds, it should not affect species that nest in relatively safe places.

An alternative hypothesis is that clutch sizes are limited by the rate at which parents can feed nestlings (Lack 1947–1948, 1954). According to this hypothesis, clutch size should be adjusted to maximize the number of offspring parents can rear to independence. Lack's hypothesis has gained convincing support from studies of temperate birds (see Lack 1968; Klomp 1970). If it applies equally well to all tropical birds except promiscuous species, the male emancipation hypothesis would be plausible.

Skutch (1949) has questioned the applicability of Lack's hypothesis to tropical birds on several grounds. Most tropical birds feed nestlings at much slower rates than do temperate birds; they often spend considerable periods of time in apparent idleness instead of feeding nestlings; and they are invariably capable of stepping up their feeding rates when broods are unusually large. Moreover, brood sizes are generally similar regardless of whether one or both parents feed young, even though two parents should be able to feed more young than one parent. Thus Lack's hypothesis does not conform with the available evidence.

The main inadequacy of Lack's hypothesis is that it is based on maximizing annual reproductive rate rather than lifetime reproductive output (Charnov and Krebs 1974; Ricklefs 1977b). As explained in Chapter 9, individuals should adjust annual reproductive rate according to a trade-off between immediate reproduction and future reproductive potential. Since increased reproductive effort is likely to decrease survival prospects (Trivers 1972; Goodman 1974; Pianka and Parker 1975), smaller clutch sizes may prevail in the tropics because they increase expected longevity enough to compensate for the reduction in immediate reproductive output (Snow and Lill 1974; Ricklefs 1977a).

Two factors make this hypothesis plausible. First, tropical birds are generally sedentary, making it difficult for young birds to find vacant habitats where they can breed. An increase in annual reproductive rate may therefore not greatly increase an individual's fitness because the additional offspring would have little chance of breeding. Second, tropical birds are more long-lived than temperate birds (Snow and Lill 1974; Fry 1977), and high expected longevity favors a shift from high immediate reproductive effort to deferred reproduction and enhancement of survival (see pp. 351–353) (MacArthur and Wilson 1967; Pianka 1970; Wittenberger 1979b).

If curtailments in immediate reproductive effort allow females to survive longer, any increase in parental care would be costly to females. Hence loss of male parental assistance would reduce female fitness in every species, regardless of dietary constraints. Since the male emancipation hypothesis assumes that loss of male parental assistance is not costly to females, it is incompatible with the life history explanation of why clutch sizes are low in the tropics. The male emancipation hypothesis also assumes that clutch sizes are limited to smaller sizes in tropical frugivores, which is not the case. Clutch sizes of promiscuous frugivores are no smaller than clutch sizes of related monogamous insectivores. Females of the former species should therefore not be affected any differently by loss of male parental assistance than are females of the latter species. Thus the male emancipation hypothesis is not supported by current evidence.

Male desertion hypothesis

It may be that females do not benefit from promiscuity. They may suffer costs that are not compensated by attendant benefits. Under certain conditions males may benefit by deserting their mates, and then females would have no choice but to accept a promiscuous mating system (Wittenberger 1979a).

Males should desert their mates when opportunities for new matings lead to greater reproductive success than would be possible were they to help rear current offspring. The costs of deserting may even be relatively low because the female should ordinarily remain with the clutch and rear her young unaided. In an ancestral population where most males are not deserters, a deserted female would have difficulty finding a new mate other than another desertion-prone male, since more faithful males would be busy caring for broods of their own mates. Also, starting over with a new mate involves a substantial time delay that may greatly reduce chances for offspring survival due to seasonal changes in food availability, predation pressure, or other environmental factors. Thus males who desert their mates often do not risk losing all the young in their first broods.

Promiscuity is most likely to arise from male desertion when food and other resources are not economically defensible. If resources are defensible, selection should favor male territory defense as a prerequisite to mating. For promiscuity to arise males would have to desert territories as well as mates, and deserting males would have to establish new territories elsewhere before they could mate again. The difficulty of establishing new territories effectively limits opportunities for mating again following desertion. Of course, a male could retain his territory and try to attract additional mates instead of providing parental assistance for his first mate, but then the system would be polygynous rather than promiscuous.

The fruit resources exploited by tropical frugivorous birds are often spatiotemporally clumped (see pp. 314–315) and hence economically indefensible. In addition, suitable nest sites used by most noncolonial frugivores appear unlimited in availability and hence are not worth defending. Males of many tropical frugivorous birds are not tied to resource-based territories, and males who ignore their mates following insemination have little difficulty in finding a place to seek additional mates. Moreover, males of tropical species have unusually favorable opportunities for mating again because breeding seasons are prolonged, nests frequently fail, and females often have time to rear several broods each year (e.g., Snow 1962a; Snow 1970). Male desertion probably reduces female fitness, but females would have no way of distinguishing between males prone to desert and males prone to feed nestlings. Thus the combination of indefensible resources and prolonged breeding seasons could favor male desertion and promiscuity among tropical frugivorous birds.

Several tropical frugivorous birds are monogamous even though they do not defend their food resources. If the male desertion hypothesis is valid, factors preventing male desertion should be evident among those species.

One species is the oilbird (*Steatornis caripensis*), an unusual colonial species that nests deep in caves and relies on echolocation to navigate through the darkness (Snow 1961–1962). Young oilbirds develop very slowly and cannot thermoregulate until they are three weeks old. Low ambient temperatures inside nesting caves make continuous brooding by a parent essential for nestling survival. A single parent is simply incapable of both brooding and feeding young nestlings. Male desertion is therefore precluded because both parents are necessary for rearing offspring.

Another group of monogamous frugivores are the hornbills of Africa and Asia (Photo 11–4). In many species females remain inside the nest cavity throughout the incubation and nestling periods, apparently as a defense against predation (Moreau 1937; Moreau and Moreau 1940; Kemp 1971, 1978). The male plasters the nest hole nearly shut with mud and single-handedly feeds both the female and her nestlings. Thus male parental assistance is again essential for nestlings to survive.

In the frugivorous toucans of South America, both sexes are required to maintain possession of a nesting hole (Van Tyne 1929; Skutch 1944b, 1958, 1971; Wagner 1944; Bourne 1974). Suitable nest holes are extremely scarce, and each pair uses the same hole every year. The pair begins defending a nest hole well in advance of breeding, and continuous defense is necessary to retain occupancy. A deserted female would soon be evicted from her hole by a competing pair, and a deserting male would have little opportunity to breed again with a new mate.

Frugivory has not led to promiscuity among temperate birds for at least two reasons. A major factor is that peak fruit abundance frequently occurs late in the nesting season, often after nestlings have fledged (see Morton 1973). Nestlings are therefore usually fed insects instead of fruit, and insects are more likely to be a defensible resource. A second factor is climatic seasonality, which reduces the length of breeding seasons and limits male opportunities for finding new mates following desertion. This factor may explain why promiscuity has not evolved in the few nonterri-

Photo 11–4

A male red-billed hornbill (Tockus erythrorhynchus) *delivering food to the female, who is inside the nest.*

The male seals the nest entrance shut at the time the female begins incubating, and he feeds both the female and the nestlings for most of the parental care period.

Photo from Alan C. Kemp, "A Review of the Hornbills: Biology and Radiation," *Living Bird* 17 (1978):122.

torial frugivores of temperate regions. For example, in waxwings (*Bombycilla*) pairing occurs in winter and is rapidly completed early in the season (Putnam 1949). Mated pairs often rear two broods each year, but individuals usually do not change mates between broods. As a result, few unmated females are available once breeding has begun, and deserting males would have little opportunity for mating again.

A few promiscuous passerines are not frugivorous. Jackson's whydahs (*Coliuspasser jacksoni*) breed on African savannas and feed their nestlings seeds and insects (Van Someren 1945), sharp-tailed sparrows (*Ammospiza caudacuta*) breed in North American marshes and feed their nestlings insects (Murray 1969), and McConnell's flycatchers (*Pipramorpha macconnelli*) breed in South American secondary forests and feed their nestlings insects (Willis et al. 1978). Males of these species do not defend resources or nest sites, but defensibility of their resources has never been studied. An analysis of how the food resources of these species are distributed in space and time would therefore be especially useful for testing the male desertion hypothesis.

Female desertion hypothesis

An alternative to the male desertion hypothesis is that females can reproduce more successfully alone than when paired with a male. Females should then desert their mates or avoid males except for brief visits to copulate with them.

The female desertion hypothesis takes three different forms. Females may avoid males because they rely on concealment to escape predation and the increased conspicuousness caused by a male's presence costs her more than loss of male parental assistance (Wittenberger 1978a). Females may avoid males because the distribution of food resources favors defense of individual territories over defense of shared territories (C. C. Smith 1968; Willis 1972; Stiles 1973; Wittenberger 1979a). In this event males should probably also exclude females from their territories following insemination, and hence promiscuity results from mutually exclusive territories defended by each sex. Finally, females may desert males and offspring to garner more energy for making eggs or recouping energy losses stemming from previous clutches (Jenni 1974; Graul et al. 1977; Maynard Smith 1977).

The female desertion hypothesis is one proposed explanation for promiscuity in grouse (Wittenberger 1978a). In most monogamous grouse males desert during the incubation period, and their primary contribution to female success is vigilance during the egg-laying period. Energy acquisition is an important factor limiting female success. Food intake rate affects the rate at which eggs can be laid, which in turn limits clutch size (Johnsgaard 1973), and the nutritional condition of laying females affects hatchability of eggs and viability of chicks (Siivonen 1957; Jenkins et al. 1963, 1967; Lack 1966). When food is scarce in spring, laying females can forage at a faster rate by relying on male vigilance for protection from predators. To do this, females must form prolonged pair bonds with males. In contrast, females probably cannot increase food intake rates by relying on male vigilance when food is abundant because their foraging rates are probably limited by digestive efficiency or the rate at which they can detoxify secondary plant toxins. Females might then be most successful by relying on their own vigilance to detect predators and avoiding males to minimize their conspicuousness. Thus the female desertion hypothesis predicts that females of monogamous grouse rely on male vigilance to escape predators because the additional time available for foraging increases food intake, while females of promiscuous grouse rely on their own vigilance to escape predators because the amount of time available for foraging does not limit food intake.

Males of monogamous grouse all provide vigilance while their mates are foraging (Choate 1963; Cheng 1964; Dement'ev and Gladkov 1967; MacDonald 1970), and in rock ptarmigan (*Lagopus mutus*) foraging females never look up unless alerted by their mates (MacDonald 1970). Males of promiscuous grouse do not provide vigilance except during the brief period while females are being courted. Reliance on male vigilance has not been shown to increase food intake rates for any female grouse, but it does increase food intake rates for females of some ducks (Ashcroft 1976). Also, the evidence for flocking birds shows that vigilance by other individuals should increase food intake rates (see pp. 122–123). Males may be more concerned with detecting unmated rival males rather than predators, but the benefits derived by females are the same regardless of the reason why males maintain vigilance.

According to the female desertion hypothesis, females of monogamous grouse should be limited by the abundance or nutritive quality of food available in spring, while females of promiscuous grouse should not. Relatively few data are available to test this prediction. In the monogamous red grouse, females forage mostly on heather, a plant that grows in nearly continuous stands on Scottish moors. Both quantity and quality of heather available for laying females are limited, and both affect female reproductive success (Moss 1969; Watson and Moss 1972; Moss et al. 1975). In comparison, several promiscuous grouse eat conifer needles, which appear superabundant in quantity, and the nutritive quality of conifer needles does not affect either food availability or female reproductive success (Zwickel and Bendell 1967, 1972; Ellison 1976). Since conifer needles contain large amounts of plant secondary compounds, the rate of food intake for species exploiting them is most likely limited by the rate at which these compounds can be detoxified. Thus the evidence for grouse with monophagous diets supports the female desertion hypothesis. However, many monogamous and promiscuous grouse have more diverse diets, and data are still needed for those species.

Sexual bimaturism hypothesis

An alternative hypothesis for explaining promiscuity in grouse is based on the assumption that males and females have different optimal life history schedules (Wiley 1974). The hypothesis is actually one form of the male desertion hypothesis discussed earlier.

Wiley (1974) argued that a female-biased breeding sex ratio would develop if delayed breeding is more advantageous for males than for females. A skewed sex ratio would create many mating opportunities for males and hence cause males to desert their mates. Wiley did not postulate the requirement that resources be indefensible, but this would be necessary to explain why promiscuity should evolve rather than polygyny.

For the hypothesis to be complete, the reason why males begin breeding at an older age than females must be explained. Delayed breeding is most likely to evolve in long-lived species because high longevity provides more opportunities for future reproduction (MacArthur and Wilson 1967; Pianka 1970; Trivers 1972; Stearns 1976; Wittenberger 1979b). Using a parallel argument, Wiley (1974) proposed that males of promiscuous grouse delay breeding longer than females because they have greater longevity than females. He attributed their greater longevity to larger body size. Larger size could reduce the vulnerability of males to predation because fewer predators are capable of attacking large-sized prey, and it could create a better energy balance in winter by reducing heat loss and lowering basal metabolic rates. Females would not be able to reap those benefits because larger size would conflict energetically with egg production. Males of monogamous grouse presumably did not evolve larger size than females because their size was constrained by counterselection stemming from some unidentified environmental conditions.

The main support for the sexual bimaturism hypothesis is that males of promiscuous grouse are usually larger than females and begin breeding at an older age, while males of monogamous grouse are more similar in size to females and begin breeding at about the same age as females. However, both sexual size dimorphism and delayed breeding by males can also be explained as consequences of promiscuity that arose from intrasexual competition for mates (see pp. 430–431), in which case they could not explain how promiscuity evolved in the first place. Therefore, the issue here concerns which came first.

The sexual bimaturism hypothesis is built on the assumption that all promiscuous grouse exhibit more marked sexual dimorphism than any monogamous grouse. However, some promiscuous grouse are no more dimorphic in size than monogamous grouse. Males are 8–17% larger than females in the promiscuous spruce grouse (*Canachites canadensis*), lesser prairie chicken (*Tympanuchus pallidicinctus*), ruffed grouse (*Bonasa umbellus*), and sharp-tailed grouse (*Pedioecetes phasianellus*), compared to 10–15% larger in all monogamous grouse (data from Johnsgaard 1973; Wiley 1974). Differences in sexual dimorphism therefore cannot explain why the first four species are promiscuous while the various monogamous species are not.

The sexual bimaturism hypothesis also depends on the assumption that increased sexual size dimorphism allows males to survive longer than females. In contrast, sexual selection theory predicts that increased sexual size dimorphism reduces male survival relative to female survival. Since adult sex ratios reflect the relative survival rates of males and females, these alternatives can be tested by correlating the degree of sexual size dimorphism to adult sex ratios (Wittenberger 1978a). The best available data show that the correlation is negative, indicating that male survival decreases as sexual size dimorphism increases (Figure 11–7). Hence the evidence on relative survival rates contradicts the sexual bimaturism hypothesis and supports sexual selection theory.

Finally, the sexual bimaturism hypothesis implies that young males defer breeding because the risks of predation and the energetic costs associated with immediate breeding are too high and not because young males are unable to compete for mates with older males. Alternatively, sexual selection theory implies that young males defer breeding because once promiscuity has evolved they can no longer compete for mates. One way to test these alternatives is to remove adult males from an area and see whether yearling males attempt to breed. Removal experiments have been performed for blue grouse (Bendell and Elliott 1966, 1967; Bendell et al. 1972; Zwickel 1972), ruffed grouse (Boag and Sumanik 1969; Rusch and Keith 1971; Fischer and Keith 1974), and sharp-tailed grouse (Rippin and

Figure 11–7

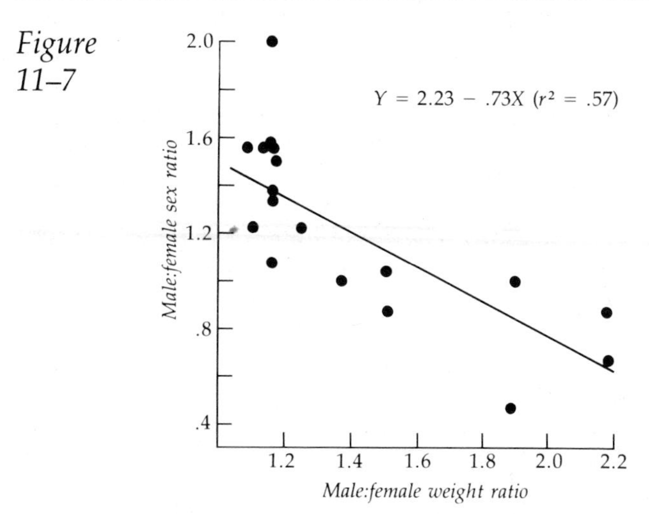

Relationship between adult sex ratio and sexual size dimorphism in grouse.

Each point represents a different species of grouse. The negative correlation implies that increased sexual size dimorphism reduces the survival of males relative to that of females.

Source: From J. F. Wittenberger, "The Evolution of Mating Systems in Grouse," *Condor* 80 (1978):128. Reprinted by permission.

Boag 1974). In every case yearling males entered the removal area, established territories, and began advertising for mates. Even in the absence of experimental removals, yearling males have been reported breeding in ruffed grouse, sharp-tailed grouse, and greater prairie chickens (*Tympanuchus cupido*) (W. H. Marshall 1965; Hjorth 1970; Hamerstrom 1972). Thus young males apparently defer breeding because they cannot compete with older males, not because predation risks or energetic costs associated with earlier breeding are too high.

Male reproductive strategies in social mammals

Overlap promiscuity. **Overlap promiscuity** (see Table 11–1) is almost an inevitable consequence of solitary habits in mammals. Males and females generally occupy overlapping home ranges or, occasionally, individual territories and come together solely to mate. In rodents, insectivores, antelope, primates, and various other mammals solitary habits are associated with densely vegetated habitats and often with nocturnal activity patterns (Eisenberg 1966; Jolly 1972; Jarman 1974). Females depend on secretive behavior and concealment for protection, and they can gain little benefit from male parental care or vigilance. In most cats solitary habits have evolved because cats are ambush predators and rely on concealment and stealth for capturing prey. Males could conceivably provide substantial help in capturing prey for cubs, but they can apparently reproduce more successfully by seeking new mates instead.

Harem polygyny. By joining social groups, females usually lose the option of forming prolonged monogamous pair bonds with a male (but see p. 508). They nevertheless retain the ability to select mates in many cases (see Ralls 1977). The interesting question in social mammals concerns how males compete for mates. In some species individual males control access to entire female groups, giving rise to **harem polygyny**, while in others males defend mating territories away from female groups or establish dominance hierarchies within them.

Harem polygyny evolves when females form small cohesive groups that can be defended by single males (Emlen and Oring 1977; Wittenberger 1979a, 1980a). Small groups are more defensible than large ones because fewer males compete for each group. The environmental determinants of group size and cohesiveness are critical in determining whether harem polygyny evolves, but discussion of these determinants will be deferred to Chapter 13.

Harem polygyny has evolved in a variety of animals, including some pheasants (Sharma 1969, 1970, 1972; Delacour 1977), coatimundis (Kaufmann 1962), peccaries and some swine (Frädrich 1965, 1974; Eisenberg and Lockhart 1972; Sowls 1974), red deer (Darling 1937), North American elk (M. Altmann 1956; Franklin et al. 1975), horses and most zebras (Joubert 1972; Tyler 1972; Klingel 1974a; Feist and McCullough 1975), some bats (Brosset 1976; Bradbury 1977a; Bradbury and Vehrencamp 1977), patas monkeys (*Patas erythrocebus*) (Hall 1968), and hamadryas baboons (Kummer 1968, 1971). In all these species females form small, cohesive social groups that are readily defendable by a single male.

Territorial harem polygyny. When female groups are not cohesive, males cannot defend exclusive mating opportunities with group members. Males may attempt to keep females from leaving the group by herding them together, but they rarely succeed when females persistently try to leave (Buechner and Schloeth 1965; Estes 1967; Jarman and Jarman 1974; Kitchen 1974; Leuthold 1977). Since defending groups is ineffective, males usually defend territories where female groups are found.

In lions female prides fragment into subgroups of varying composition, all of which share the same home range, and males defend the pride home range instead of the pride itself (Schaller 1972; Rudnai 1973; Bertram 1975a). If female prides are small, a single male defends each home range, but if prides are large, males form coalitions (Photo 11–5) and cooperate in defending the pride's home range. Since males are essentially guarding access to females by defending territories, the mating system of lions is harem polygyny.

Another form of territorial harem polygyny occurs in some antelopes. Females of these species move frequently from one social group to another and never form stable groups. Because groups are indefensible, males defend favorable habitats where females most often congregate. Female groups often do not remain on a male's territory for long, making the mating system nearly promiscuous, but in effect males control access to female groups by controlling access to the habitats that they visit.

The nature of the defended areas varies among species. Male pronghorn antelope defend areas containing water holes and high-quality grazing sites (Kitchen 1974). Male Grant's gazelles defend hilltops in open

Photo 11–5

A coalition of three nomadic male lions.

In certain parts of East Africa, competition for control of female prides is fierce, and single males are unable to maintain tenure on a pride home range. Males therefore form stable coalitions that may last for several years and cooperate to gain and maintain control of a pride home range. Such coalitions often, but not always, consist of brothers or other close relatives.

Photo by George B. Schaller.

savannas where females prefer to graze (Walther 1965; Estes 1967). Hilltops provide the best vantage points for spotting predators, and the forbs eaten by gazelles do not become overgrown with tall grass as much there as they do in wetter lowland swales. Male defassa waterbucks (*Kobus defassa*) and Coke's hartebeests (*Alcelaphus buselaphus*) defend riverine thickets or scrub edges adjacent to good grazing areas, where females can escape into dense cover when threatened by predators (Kiley-Worthington 1965; Spinage 1969; Gosling 1974). Male impala (*Aepyceros melampus*) defend habitats most preferred by females on a seasonal basis (Jarman 1979). During peak breeding in the wet season, males defend ridges where foraging is most favorable and females mainly congregate. In dry seasons when less breeding occurs and fewer males are territorial, the most successful males defend riverine lowlands where both water and food are most available. Thus males frequently defend habitat configurations where females can find better grazing and gain better protection from predators. Such habitat configurations do not exist or are not economically defensible for all antelopes. In those cases males adopt alternative strategies for gaining or controlling access to females.

Arena promiscuity. In many of the most gregarious African antelopes, males defend territories near female herds (Leuthold 1977; Owen-Smith 1977). These territories have evolved as mating stations where males can attract females. They are not centered in habitats preferred by females, and they do not provide males with better foraging. The reasons why males defend mating territories near female groups instead of establishing dominance hierarchies within female groups are not well understood.

One hypothesis is that males defend territories when female groups are sedentary and establish dominance hierarchies when female groups are nomadic (Geist 1974b). However, male wildebeest defend temporary territories while following nomadic herds (Estes 1969) and male Cape buffalo establish dominance hierarchies within sedentary herds (Sinclair 1974, 1977), so this hypothesis appears untenable.

Another hypothesis is based on the stability of group membership within female herds. When females continually move from one herd to another, they cannot be as familiar with the dominance status of individual males, and the best way for identifying genetically superior males might be to select a territorial male who has successfully fought off competing males (Wittenberger 1979a). This hypothesis is supported by the fact that Cape buffalos are unusual among antelopes in forming both stable herds and male dominance hierarchies. It is also supported by comparative evidence based on equid mating systems. Female horses and most zebras form closed social groups and each group is defended by a single dominant male stallion (Joubert 1972; Tyler 1972; Klingel 1974a; Feist and McCullough 1975). Male dominance hierarchies do not occur within groups because group sizes are small, but males defend females rather than territories. In contrast, female Grevy's zebras (*Equus greyvi*), feral donkeys, and wild asses (*E. africanus* and *E. hemionus*) do not form stable social groups, and males defend territories instead of female groups (Klingel 1974b, 1977; Moehlman 1974; Woodward 1979).

Hierarchical promiscuity. **Hierarchical promiscuity** based on male dominance hierarchies has evolved primarily in cohesive female groups that are too large for single males to control. It is especially prevalent in terrestrial or semiterrestrial primates (Crook 1970; Jolly 1972; Eisenberg et al. 1972). Male dominance hierarchies probably result from competition for access to receptive females and to limited resource items. However, the ability of dominant males to control access to females is debatable. Many studies show that dominant males copulate more often than subordinate males (e.g., Devore 1965, 1971; Hausfater 1975; Bernstein 1976), but this evidence must be viewed with a certain amount of skepticism. Males of high and low status are not equally visible to human observers, and many copulations of low-ranking males may be missed. For example, Drickamer (1974b) observed more copulations by high-ranking male rhesus monkeys, but after adjusting the data to correct for differences in observability, high-ranking males were no more successful than low-ranking males. Only one study has actually measured the number of offspring produced by males of differing status. Using serum proteins and other blood constituents to assess paternity, Duvall and co-workers (1976) found that the dominant male of one rhesus group did not father more offspring than several lower-ranking males.

In multi-male groups male mating strategies are complex and may involve a variety of tactics. Low-ranking male primates sometimes form coalitions to dominate an otherwise dominant male. They may also use furtive tactics to copulate with females out of view of high-ranking males. Female primates do exercise a choice when accepting mates, and they sometimes prefer subordinate males over dominant males (Dixson et al. 1973; Eaton 1973; Hausfater 1975; Lindburg 1975; Seyfarth 1978; Bachman and Kummer 1980). At the same time, dominant males use aggression to control access to receptive females, especially during peak estrus (e.g., Hausfater 1975). Thus males may use a variety of tactics, depending on their age, status, and perhaps other factors. The evolution of these tactics is not well understood.

Dominance hierarchies have evolved in mountain sheep (*Ovis canadensis*), American bison (*Bison bison*), and caribou (*Rangifer arcticus*) even though females form unstable herds (McHugh 1958; Lent 1965; Kelsall 1968; Geist 1971). In mountain sheep male dominance is signaled by horn development, and individual recognition is not necessary for identifying a male's dominance status (Photo 11–6) (Geist 1971). The same may be true in caribou. The use of such signals explains how dominance hierarchies can persist despite lack of familiarity among many herd members. Why males of these species compete for control of females within herds instead of defending mating stations is unclear.

Male reproductive strategies in frogs and toads

One important factor influencing the reproductive behavior of anuran amphibians (frogs and toads) is length of breeding season (Wells 1977a). Two general patterns are evident. *Explosive breeding* involves a short synchronous burst of breeding activity once or a few times each year. It characterizes many temperate anurans that breed in early spring and both temperate and tropical anurans that breed in temporary rain pools or other ephemeral habitats. *Prolonged breeding* lasts from two to three months in temperate anurans up to all year in some tropical anurans. It characterizes species that breed in permanent bodies of water or other permanent habitats. The duration of anuran breeding seasons is often limited by seasonal availability of breeding sites and climatic conditions, but ecological factors such as competition among larvae and vulnerability to predators may also be important.

Males of explosive breeders typically aggregate around suitable oviposition sites. These are usually temporary pools of water, but in some tree frogs they are trees overhanging temporary ponds (Duellman 1970; Pyburn 1970; Coe 1974; Scott and Starrett 1974; Howard 1980). The density of breeding aggregations varies considerably and affects how males compete for females (Wells 1977a).

At high densities males of explosive breeders usually obtain mates

Photo 11–6

Dominance in bighorn sheep is signaled by horn size.
The status of these rams increases from left to right in accordance with their increasing horn size. Because horn size is a clear indication of status, ewes can choose high-status mates despite their lack of familiarity with many individual rams.
Photo by Valerius Geist.

by actively searching for females rather than advertising for females from fixed locations (Wells 1977a). The searching strategy is more successful because females generally do not approach calling males and any who did so would be intercepted by a searching male before reaching the caller. Intense aggressive competition for females accompanies the searching strategy (Photo 11–7). Males frequently struggle with one another in attempts to dislodge the male who is most securely clasped to the female's back (Savage 1934, 1961; Wright and Wright 1949; M. Smith 1969; Calef 1973; Howard 1980). Contests usually consist of vigorous pushing and shoving among males, but in a few species they involve violent fights (Balinsky and Balinsky 1954; Wager 1965; Rivero and Esteves 1969).

The search patterns of males vary among species. Males of most species search the entire breeding area for females, but in a few species they remain in fixed locations for up to several hours and clasp any frog that approaches within a few centimeters of them (Wells 1977a). The latter behavior sometimes produces a regular spacing pattern similar to territoriality, and some evidence indicates that small individual territories may be defended (Whitford 1967; Lörcher 1969; Wahl 1969). The factors determining which searching strategy evolves are unknown.

Search patterns may also vary among males of a given species. In common European toads, for example, some males search for and pair with females on land who are still moving toward breeding ponds, while other males enter ponds unpaired and either clasp onto the backs of un-

Photo 11–7

Two male frogs fighting over a female.

Unmated male frogs commonly try to dislodge males who are already clasped to females when breeding occurs at high densities. This behavior often leads to intense wrestling contests, which on rare occasions result in displacement of the male originally clasped to the female.

Photo by Jane Burton/Bruce Coleman, Inc.

paired females in the water or attempt to dislodge males already on the backs of females (Davies and Halliday 1979). Larger males are better able to dislodge males in amplexus and avoid being displaced themselves. The proportion of males adopting each search pattern changes seasonally, with more males searching at the pond during later stages of the breeding period when a higher proportion of females are spawning. An ESS model indicates that the number of males adopting each search pattern is such that males who search on land and males who search ponds are about equally successful at breeding.

At low and intermediate densities males of explosive breeders often space themselves around the periphery of a pond and call from stationary positions (Wells 1977a). Active searching for females is rare or absent, and males obtain mates by attracting them with their calls. Males of some species call from concealed locations in vegetation, probably to reduce vulnerability to predation (Nelson 1973), and the wide spacing of males may have evolved to further enhance concealment.

The importance of density is shown by a shift in male reproductive behavior as density changes. Males of many explosive breeders adopt a calling strategy at low densities and a searching strategy at high densities (e.g., Bragg 1937, 1940; Thornton 1960; Karlstrom 1962; Nelson 1973; Howard 1980). Even within an aggregation males may switch from one strategy to the other. For example, male common European toads who arrive at a breeding pond early in the season call from isolated positions, but as density increases, calling ceases and those same males begin searching actively for females (Eibl-Eibesfeldt 1950; Heusser 1969). In several toads some males become silent at intermediate densities and gather around calling males to try and intercept females attracted by their calls (e.g., Axtell 1958; Brown and Pierce 1967). There is apparently some threshold density where searching for females becomes more efficient than attracting females, but no attempt has ever been made to predict where that threshold should lie.

The male reproductive strategy of prolonged breeders resembles the low-density strategy of explosive breeders. Males of virtually all prolonged breeders attract females by calling from fixed locations (Wells 1977a). They never search for females, and they rarely fight over individual females. Males of many species defend territories around oviposition sites such as burrows, tree holes, small pools of water, or pond edges (e.g., Clyne 1967, 1969b; Jenssen and Preston 1968; Schroeder 1968; Pengilley 1971; McDiarmid and Adler 1974; Wells 1977b, 1980; Howard 1978a, b). In other species males defend mating territories not associated with oviposition sites (Rabb and Rabb 1963a, b; Österdahl and Olsson 1963; Wiewandt 1971; Fellers 1979).

The significance of defending mating territories is not yet clear, but such defense may reduce interference by other males. Noncalling males occur in close association with calling males in some species of frogs and toads, and experimental releases of gravid females near these male associations show that noncalling males often achieve amplexus with females attracted by calling males (Perrill et al. 1978). Thus defending a mating station may reduce the probability of a noncalling male intercepting females attracted by the calling male.

The evolution of leks

In some, but by no means all, promiscuous animals, males congregate and display in groups to attract females (see Photo 7–1). The males on such **leks** typically defend individual territories, though in some species they do not. These territories do not contain any resources used by females, and the reason why males congregate in one place is not immediately obvious. Many hypotheses have been advanced to explain the evolution of lek behavior, but none are entirely satisfactory.

Hypotheses based on male reproductive strategies

One set of hypotheses considers lek behavior from a male perspective and evaluates the potential benefits derived by males.

Stimulation hypothesis. Lack (1939b) and Snow (1962a, 1963) suggested that males who display in groups have a stronger stimulating effect on females and hence obtain more mates than males displaying alone. This hypothesis is reminiscent of the Fraser Darling effect for colonial birds and, like the Fraser Darling effect, it only represents a proximate mechanism. The hypothesis is inadequate without some further explanation of why group displays should have a stimulating effect on females of lek species but not on females of nonlek species.

Conspicuousness hypothesis. A related hypothesis is that males are more conspicuous when displaying in groups and hence attract more females than males displaying alone (Chapman 1935; Lack 1939b; Hjorth 1970; Alexander 1975). According to this hypothesis, males should achieve higher average copulatory success when displaying in groups. However, average male success is not correlated with lek size in black grouse (*Lyrurus tetrix*), ruffs, or golden-headed manakins (*Pipra erythrocephala*) (Koivisto 1965; Hogan-Warburg 1966; Lill 1976). A positive correlation has been found in greater prairie chickens, but only for leks containing fewer than 15 males (Hamerstrom and Hamerstrom 1960). By itself, the hypothesis is again inadequate, because it does not explain why group displays should be an effective way to attract females in lek species but not in nonlek species. Finally, the reason why groups of calling males should be more conspicuous than single calling males is not at all clear (Bradbury 1980). Male groups can generate more calls per unit time, but the carrying power of their calls should not increase very much and the ease of detecting calls should not be appreciably improved at any given distance.

Mutual benefit hypothesis. A third hypothesis is that lek behavior allows males to derive mutual benefits (Vos 1979). Leks may act as information centers that enhance the ability of males to find good foraging areas, or they may provide protection from predators through mutual vigilance. In black grouse males forage in flocks away from the lek and often follow each other to unfamiliar feeding grounds. However, the extent to which this is true in other lek species of grouse is unclear, and comparable advantages are probably not obtainable in other birds, such as manakins, that display on leks (see Lill 1976).

Hypotheses based on female mating preferences

A second set of hypotheses considers lek behavior from the female perspective and evaluates the reasons why females might benefit from mating preferentially with males who display in groups.

Mate comparison hypothesis. One hypothesis is that females prefer males who display in groups because they can distinguish between superior and inferior males more easily when males display near one another (Alexander 1975; Borgia 1979; Vos 1979; Bradbury 1980). Present evidence indicates, however, that females of lek species often do not base their choices on male characteristics. Wiley (1973) conducted an intensive study of male courtship displays in sage grouse and found no correlation between display characteristics and mating success. Similarly, Lill (1974a, b) found no correlation between mating success and the morphology or display characteristics of male golden-headed manakins or white-bearded manakins

(*Manacus manacus*). He also experimentally removed the two most successful males from one lek of white-bearded manakins and found that the replacement males were just as successful in attracting females as the original males. Although the replacement males may by chance have been the two next best males on the lek, the experiment suggests that females choose their mates on the basis of territory characteristics rather than male characteristics. However, Lill could not identify any territory characteristics that correlated with a male's mating success either. On the other hand, Rhijn (1973) and Shepard (1975) found that the number of females attracted by resident male ruffs is correlated with the frequency of several male displays. Similarly, some evidence suggests that male behavior may also influence female mating preferences in sage grouse (Hartzler 1972; Donald Jenni, personal communication). Thus the role of male behavior is not yet well understood.

A modified version of the same hypothesis is that females select territories according to their geometric position on a lek, with male competition for those territories automatically separating superior males from inferior males. Female grouse, ruffs, and Uganda kob (*Kobus kob*) usually mate preferentially with males who hold central territories on a lek (Hogan-Warburg 1966; Hjorth 1970; Kruijt et al. 1972; Wiley 1973, 1978; Floody and Arnold 1975; but see Shepard 1975). Preferred males have won repeated contests for control of central territories and hence have demonstrated superior competitive ability, at least within aggressive contexts. However, since females do not prefer central males or males in any other consistent geometrical position on the leks of manakins, cocks-of-the-rock (*Rupicola rupicola*), or hammer-headed bats (*Hypsignathus monstrosus*) (Gilliard 1962; Lill 1974a, b, 1976; Bradbury 1977b), this hypothesis at best has only limited applicability.

The importance of male quality cannot be ruled out on the basis of present evidence, but female preferences based on easier identification of superior males cannot alone explain why leks have evolved. At the very least, the hypothesis must be expanded to explain why such preferences should lead to lek behavior in some species but not in others.

Predator defense hypothesis. Females may benefit from male lek behavior because mating within a group of males reduces their vulnerability to predation in certain types of habitats. In open habitats mutual vigilance may reduce vulnerability of females to predation during the mating process, while in closed habitats concealment may be the best way to escape predation.

Some of the comparative evidence supports this hypothesis. Promiscuous grouse that breed in open habitats all display on leks, while those breeding in forests all display on dispersed display sites (Koivisto 1965; Hjorth 1970; Wiley 1974; Wittenberger 1978a). Moreover, male blue grouse display in groups when breeding in the open and call from dispersed sites when breeding in dense underbrush (Hoffman 1956; Blackford 1958, 1963). Grouse leks are often, though not always, situated in elevated open terrain where visibility of surrounding areas is high (Hjorth 1970). Lek behavior is also associated with open grasslands in ruffs (Hogan-Warburg 1966), buff-breasted sandpipers (*Tryngites subruficollis*) (Pitelka et al. 1974), great bustards (*Otis tarda*), little bustards (*O. tetrax*) (Dement'ev and Gladkov 1967), Jackson's whydahs (Van Someren 1945), and Uganda kob (Buechner and Schloeth 1965; Leuthold 1966; Buechner and Roth 1974), although those of Jackson's whydah are on the ground in tall grass where visibility may be low. Leks of greater birds-of-paradise (*Paradisiaea apoda*) (Photo 11–8) are usually in isolated trees in forest clearings, while those of lesser birds-of-paradise (*P. minor*) are in isolated tree groves in open savanna or in the tops of tall trees emerging above the forest canopy (Gilliard 1969). Leks of male cocks-of-the-rock are on the ground in especially open parts of tropical rain forests where little understory vegetation exists (Gilliard 1962; Snow 1971a). On the other hand, manakin leks are in dense forest understory or thick canopy foliage where visibility of approaching predators is low (Snow 1962a, b, 1963; Sick 1967; Lill 1974a, b, 1976).

Females of many species behave in a manner that should reduce their vulnerability to predation while visiting a lek. Congregating on central territories should provide protection from terrestrial predators because more peripheral individuals are easier for a predator to surprise. Central territories also contain more alternative targets for an attacking predator because they are smaller and more tightly clustered. Females congregate in groups while on a lek (e.g., Buechner and Schloeth 1965; Hogan-Warburg 1966; Hjorth 1970; Wiley 1973), which allows them to benefit from the scattering and confusion effects described in Chapter 3. Finally, females

Photo 11–8

Two male greater birds-of-paradise displaying on a lek. *The evolutionary significance of lek displays in birds-of-paradise is still poorly understood.* Photo by G. Konrad.

usually visit leks for relatively brief periods each day, which minimizes the time they are exposed to attack while selecting a mate.

The predator defense hypothesis may offer the best explanation of lek behavior for many species, but it cannot apply to at least some species. The leks of Guy's hummingbirds (*Phaethornis guy*) are in thick underbrush and are associated with favorable foraging sites exploited by females (Snow 1968; Snow 1974). They probably evolved because males congregate near clumped food resources that attract many receptive females. Manakin leks are found in dense foliage and are not closely associated with local food concentrations. Why they have evolved is still a mystery. The leks of hammer-headed bats are found in both open and closed canopies of riparian forests along streams (Bradbury 1977b). It is difficult to believe that their leks could have evolved as a defense against predation, especially since many other bats in the same forests do not breed on leks. At the same time, their leks are not associated with local concentrations of food. Hence the adaptive significance of hammer-headed bat leks is still unclear.

Polyandry

Polyandry is an unusual mating system that occurs primarily in birds. It is always associated with sex role reversal (Jenni 1974). Polyandrous mating systems are very diverse, and few generalizations can be made about them. Two kinds of polyandry are discernable. In most "polyandrous" birds both males and females form relatively short-term pair bonds with several mates. Such mating systems are more accurately referred to as polygynous-polyandrous systems. True polyandry, in which each female forms exclusive pair bonds with several males, is much rarer but does occur in a few species.

Polygyny-polyandry in rheas

Rheas (*Rhea americana*) are large flightless birds of South American pampas. They are primitive ratite birds, distantly related to ostriches (*Struthio camelus*) and, as in all ratites except ostriches, males perform all parental duties unaided by the female. The basic social system is similar to harem polygyny, except each flock of females is controlled by a succession of dominant males.

During the nonbreeding season rheas form large flocks of over a hundred birds (Bruning 1973, 1974). As spring approaches, each flock splits apart into several small flocks of 2–15 females. At the same time, dominant males become more aggressive and expel subordinate males from the winter flocks, with the result that each small female flock is defended by a single dominant male. After a period of intense courtship, the females begin following the male around (Photo 11–9). The male chooses a nest site and constructs a large grass-lined nest. Each female lays one egg near the nest every other day for 7–10 days. The male remains on the nest almost continuously once laying begins, except for midday feeding trips

Photo 11–9

A male rhea followed by his small flock of females.

Each dominant male defends a small female flock, as occurs in many harem polygynous birds, but the female flocks eventually are controlled by a succession of males. Hence the rhea mating system is both polygynous and polyandrous.

Photo from D. F. Bruning, "Social Structure and Reproductive Behavior in the Greater Rhea," *Living Bird* 13 (1974):267.

of 10–60 minutes, and rolls each egg into the nest as it is laid. He becomes increasingly aggressive toward the females as laying proceeds, culminating by the tenth day in termination of laying (Photo 11–10). At about that time the females attract the notice of another male and leave to repeat the cycle with him. Each group of females mates with several males in succession, being chased away from each nest after the clutch is completed.

Polygynous matings in rheas occur because females form small flocks that are defensible by single males. Polyandrous matings occur because males chase their mates away from completed clutches and previously unmated males take control of expelled female flocks.

The most difficult problem is to explain why male rheas chase their mates away and assume all parental duties themselves. The females of each flock could conceivably lay their eggs in separate nests and care for offspring separately, a pattern that occurs in the harem polygynous red jungle fowl and peafowl (Sharma 1969, 1970, 1972; Delacour 1977). By not taking over parental duties, the male could retain control of the female flock and fertilize replacement clutches following nest failure. Alternatively, the male could abandon his first mates and seek new ones elsewhere. Opportunities for finding new mates seem high because many nests fail. Moreover, males run a high risk of complete failure by incubating all the eggs in a single nest because nests are often abandoned if even one egg rots and explodes (Bruning 1973, 1974). A male might therefore achieve better success by not putting all his eggs in one nest.

Photo 11–10

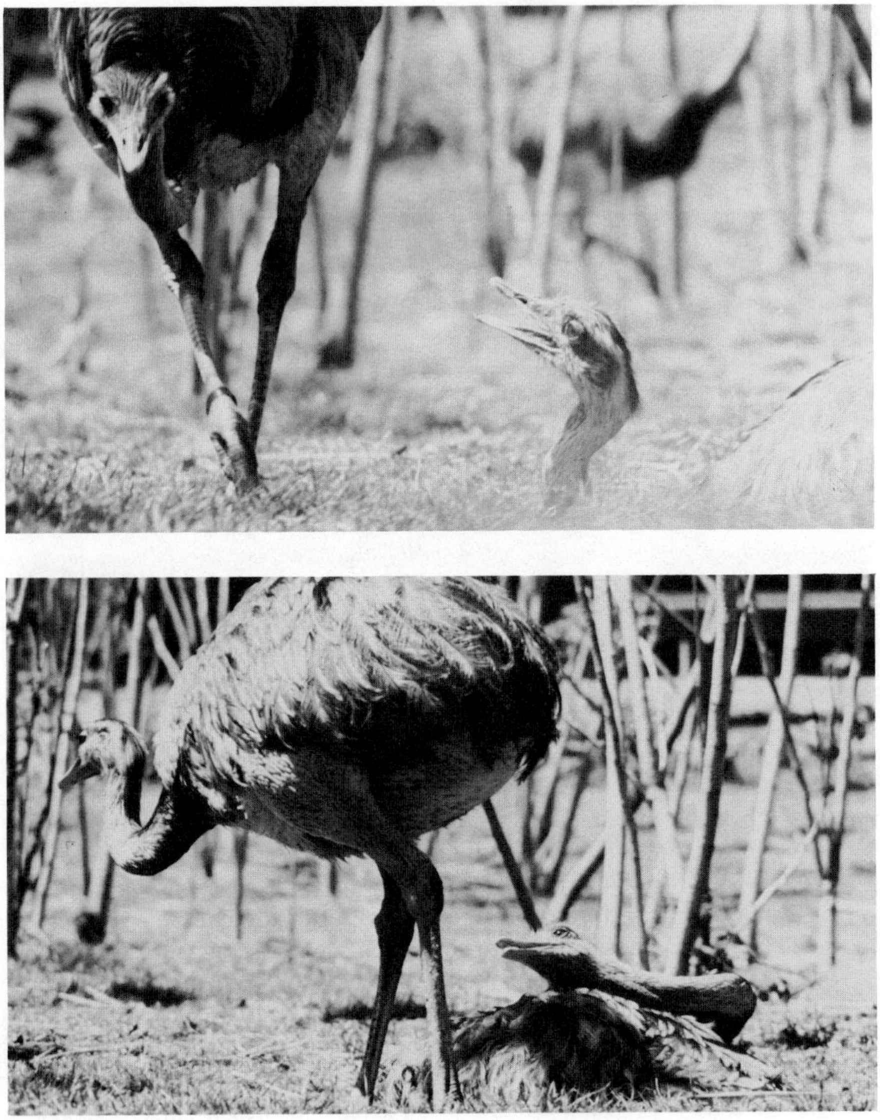

One possible explanation is that females would be unable to attend nests of their own adequately because they must recoup the energetic losses of egg laying (Wittenberger 1979a). This might be especially true during the laying period, when females must forage for more than short periods to obtain enough energy for producing eggs. Reduced nest attendance by each female at a separate nest may result in very high egg losses if nest attendance is crucial for protecting eggs from overheating. Males may therefore attend nests themselves because they can attend nests more continuously and thereby greatly increase egg survival. The result

The pair bonds between a male rhea and members of his female flock dissolve when the male becomes aggressive toward his mates.
The male remains continuously on the communal nest once laying begins, and after each female has contributed 4–5 eggs, he attacks the females and drives them away. An incubating male is shown here striking out at one of his mates just before his female group leaves and becomes associated with another male.
Photos from D. F. Bruning, "Social Structure and Reproductive Behavior in the Greater Rhea," *Living Bird* 13 (1974):272.

might be higher average success for males than would be possible were they to desert and seek additional mates elsewhere, especially since few late clutches survive the high daytime temperatures of midsummer (Bruning 1973).

Having been chased away by the male, females can either cease breeding or mate with a new male. They should choose the latter option if the benefits of continued laying are greater than any detrimental effects that continued breeding has on their survival. Few eggs laid late in the season result in surviving offspring but these few could favor continued laying if female survival is not appreciably reduced by the energetic costs involved.

Polygyny-polyandry in shorebirds

Females of several shorebird species form temporary pair bonds and lay complete clutches for their first mates and then desert to repeat the cycle with second and sometimes third mates (Jenni 1974). Males typically incubate the first clutch alone, though often not until after they spend several days seeking to copulate with additional females (Hildèn 1965b; Parmelee 1970; Hildèn and Vuolanto 1972; Raner 1972; Graul 1973; Nethersole-Thompson 1973; Parmelee and Payne 1973). Females typically incubate their last clutches alone, except in spotted sandpipers (*Actitis macularia*) where they receive help from their last mates (Hays 1972; Oring and Knudsen 1972).

Two questions require answers in these mating systems. Why do males care for clutches deserted by females, and why do females of some species desert while females of other species do not? The answers to these questions have not yet been fully worked out, but some reasonable hypotheses are now available.

Males should remain and care for clutches deserted by their mates when opportunities to mate again are limited (Parmelee and Payne 1973; Wittenberger 1979a). Most multiple-clutching shorebirds breed in the Arctic, where breeding seasons last only a couple of weeks. Deserted males can exploit most mating opportunities and still care for their first clutches simply by delaying incubation for a few days. Since opportunities for obtaining replacement clutches are limited, males can be most successful by caring for first clutches and letting second mates, if any, care for second clutches.

Short breeding seasons cannot explain why males remain with first clutches in spotted sandpipers and mountain plovers (*Eupoda montana*) because breeding seasons are more prolonged. In spotted sandpipers females defend large territories that contain the smaller territories of several males, and males are unable to mate with other females because their mates exclude other females from their territories (see Oring and Maxson 1978). The reason females can defend such large territories is unknown. In mountain plovers females do not defend territories that encompass the territories of several males. Emlen and Oring (1977) suggest that males incubate first clutches because females might then be more likely to remain with their original mates to lay second clutches. Females often do stay paired with the same male for both clutches, but no data are available for assessing how a male's parental behavior at the first nest affects the female's propensity to stay with him for the second clutch.

Given a male propensity to rear first clutches following desertion, females should be most likely to desert when they can garner energy for laying a second clutch or when recouping energy losses to enhance their survival prospects (Wittenberger 1979a). Females should not desert unless single parents have a good chance of rearing young successfully. In Arctic shorebirds adults must incubate eggs for most of the day to prevent chilling, especially during cold or rainy weather (Norton 1972). Hence adults must be able to find food rapidly if they are to succeed as single parents. Female desertion should therefore be most likely to evolve when food resources are locally abundant. Indirect evidence indicates that polygynous-polyandrous shorebirds do exploit spatially clumped and locally abundant food resources (Graul 1973, 1976; Parmelee and Payne 1973; Pitelka et al. 1974), but direct measurements of food abundance are not available.

Females need not change mates in order to desert their first clutch of eggs. They could remain paired with the same male and still lay a second clutch, as commonly occurs in quail and partridges (Goodwin 1953; Jenkins 1957; Schemnitz 1961; McMillan 1964; Francis 1965; Anthony 1970; Klimstra and Roseberry 1975). Changing mates should entail some cost because it requires searching for a new mate and forming a new pair bond. These costs could be avoided by not changing mates. Why then do double-clutching females regularly change mates in shorebirds? One possibility is that mate desertion allows females to seek males who do not yet have clutches to incubate (Pienkowski and Greenwood 1979). If such a male can be found, a female could lay a second clutch for him and then lay a third clutch which she could incubate herself. This hypothesis seems reasonable since some female shorebirds do occasionally lay more than two clutches in a season (e.g., Parmelee 1970; Parmelee and Payne 1973; Hildèn 1975).

Polyandry in jacanas

Jacanas or lily-trotters are tropical pond-dwelling birds that glean insects by walking on top of floating vegetation or searching in the grass of nearby lawns (Photo 11–11) (Jenni and Collier 1972). In American jacanas (*Jacana spinosa*) each male defends a small territory, while each female defends a larger territory that encompasses the territories of several males.

Photo 11–11

The jacana, or lily-trotter, one of the few birds known to be truly polyandrous.

A male (foreground) is shown here walking on lily pads in search of food, while his mate stands nearby. Note the very long toes that allow jacanas to walk on top of floating vegetation. Males are distinguishable from females by their substantially smaller body size and by the smaller-sized shield on their foreheads.

Photo by Terry Mace.

Sex role reversal is nearly complete with females being larger and more aggressive than males. Females lay clutches for every male on their territories, and they lay replacement clutches when nests of their mates fail. Males incubate clutches by themselves and care for chicks with little or no help from females (Jenni and Betts 1978).

Polyandry has apparently evolved in jacanas because breeding females can control access to several males. This is possible because breeding habitat is clumped and limited in availability. The situation parallels territorial polygyny in colonial songbirds, except that the sex roles are reversed (Wittenberger 1979a). Defending a large territory and still laying clutches for several males is energetically expensive and should not be possible unless food is very abundant. One might therefore expect jacanas to be monogamous in ponds with low food abundance, but data for testing this prediction are unavailable.

The difficult problem is to explain why sex roles are reversed in the first place. Short breeding seasons may allow females to desert clutches because males are predisposed to care for such clutches unaided. Then females could compete for additional mates and increase their reproductive output. Jacanas breed all year in the lowland rain forest habitat studied by Jenni and Collier (1972), but breeding is highly seasonal in other environments (Donald Jenni, personal communication). Nevertheless, short breeding seasons alone cannot explain polyandry because many birds with short breeding seasons are not polyandrous. Female energetics must also be an important factor.

Monogamy

Considerable confusion exists in the scientific literature regarding the evolution of monogamy, but a comprehensive set of hypotheses can be put together by using the theories available for explaining polygamous mating systems (Wittenberger 1979a; Wittenberger and Tilson 1980). The general rationale is that monogamy should evolve whenever it is advantageous for one sex, provided that environmental or social conditions do not predispose the other sex to desert. Females are more likely to benefit from monogamy than males because the reproductive success of females is ordinarily limited by time and energy constraints and can be increased by male parental assistance. Males are less likely to benefit because their reproductive success is ordinarily limited by the number of females they can inseminate.

Females incur several costs by forming monogamous pair bonds. Close association with a male may increase a female's conspicuousness and hence vulnerability to predation. It may also intensify competition for food resources unless males exploit different resources than those exploited by females. Maintaining an exclusive pair bond may require aggressive expulsion of unmated females from the male's territory, which would entail the costs of aggression discussed in Chapter 5. These costs must all be compensated by benefits before monogamy can evolve. The potential benefits gained by females include increased protection from predators, greater protection from the harassment of unmated males, reduced competition with other females, and acquisition of male parental assistance.

Monogamy can theoretically evolve when it is detrimental to females. If males benefit by controlling access to individual females, the costs of avoiding a male may be greater than the benefits. A female should then submit to a male's continued presence even though she would be better off without him. Also, a male may be less fit by having two mates rather than one. In that event males may aggressively exclude second females from their territories.

Monogamy does not necessarily evolve whenever it is advantageous to females. Environmental conditions may predispose males to desert their mates despite the best interests of females, as apparently occurs in tropical frugivorous birds. They may also promote social grouping behavior by females, in which case single males may be able to defend entire female groups or dominant males may be able to control access to receptive females. Thus monogamy cannot evolve unless conditions preclude males from deserting females or controlling access to more than one female at a time.

Hypothesis: Male help is essential to female success

Under certain conditions females are unable to breed without male assistance. Then males cannot benefit by deserting their mates, and monogamy is advantageous to both sexes. Lack (1968) has invoked this hypothesis as a general explanation of monogamy in birds, but the only convincing evidence for it comes from species breeding under harsh or intensely competitive conditions or far from their foraging grounds (Wittenberger and Tilson 1980). Most birds rear several offspring at a time, and it is probable that females of those species could rear at least one offspring without male assistance.

A variety of colonial birds lay one-egg clutches and can rear only one young at a time. One-egg clutches are typical of most colonial seabirds, including emperor and king penguins (*Aptenodytes forsteri* and *A. patagonica*), albatrosses, shearwaters, storm petrels, diving petrels, tropicbirds, frigatebirds, many alcids, and some tropical terns (Lack 1968). In addition, gannets and most boobies lay clutches of either one or two eggs but normally cannot rear more than one young to independence (Dorward 1962; Kepler 1969; Nelson 1970). The second egg merely acts as insurance in case the first egg fails to hatch or the first nestling dies shortly after hatching. Several experimental studies have shown that both parents are required to feed a single chick adequately in those species. No more than one chick ever survives when a second egg is added to the normal one-egg clutches of Laysan albatrosses (*Diomedia immutabilis*), manx shearwaters (*Puffinus puffinus*), Leach's petrels (*Oceanodroma leucorrhoa*), red-footed boobies (*Sula sula*), or common puffins (Rice and Kenyon 1962; Harris 1966; Lack 1968; Nelson 1969; Nettleship 1972). An additional experiment further shows that single parents cannot rear even one chick to independence without assistance in common puffins (Nettleship 1972).

The ability of seabirds to feed young is limited by intense competition for food around colony sites or long commute distances to foraging grounds (Lack 1968; Nelson 1970; Ashmole 1971). Long commute distances limit the rate at which food can be delivered to chicks, and they make attendance of chicks by the second parent vital during the long absences required for each foraging trip.

Several noncolonial birds also lay one-egg clutches (Lack 1968; Wittenberger and Tilson 1980). Of these, condors, certain vultures, and large eagles probably cannot rear more than one chick at a time despite the combined efforts of both parents (Photo 11–12). Many of the other species could probably rear additional young but do not because selection favors low annual reproductive effort in order to enhance survivorship (see Snow and Lill 1974). Experimental manipulations of brood size have not been conducted for any of these species, and without such experiments no definite conclusions can be drawn about the importance of male assistance in feeding young.

In some birds male parental assistance is essential because eggs or nestlings must be continuously attended if they are to survive. Continuous attendance of nests is necessary to prevent predation or interference from conspecific neighbors in herons, storks, pelicans, flamingos, some alcids, large gulls, avocets, and cranes, nearly all of which are colonial (Wittenberger and Tilson 1980). Continuous nest attendance is necessary to prevent chilling of eggs or young nestlings in penguins and oilbirds and to prevent overheating of eggs in desert or tropical species of flamingos, pratincoles, skimmers, doves, and terns. In some frugivorous toucans and toucanets both sexes must defend the nesting hole because suitable holes are extremely scarce and any undefended holes would quickly be taken over by competitors (see p. 482).

Hypothesis: The polygyny threshold is not reached

A more common situation in territorial animals is that females could raise some young without male assistance but can ordinarily raise substantially more young with male assistance. The costs of polygynous pair bonding may be too high, with monogamy prevailing because females are more successful pairing with unmated males in marginal habitats rather than with mated males in the best habitats.

Direct evidence that monogamous pair bonding in marginal habitats is more successful than polygynous pair bonding in the best habitats is difficult to obtain because the success females could achieve as second mates in high-quality habitats cannot be measured directly. An estimate of expected success for secondary females might be obtained by experimentally removing territorial males, provided that the males are not replaced by new arrivals and loss of male parental assistance is the main cost of polygyny for secondary females. Although many removal experiments

Photo 11–12

*A harpy eagle (*Harpia harpyja*) and her chick.*

The harpy eagle is a large endangered species that inhabits South American tropical rain forests and feeds primarily on monkeys and other large prey. It is probably monogamous because both parents are required to adequately feed their single chick.

Photo by Neil L. Rettig, Wolfgang A. Salb, and Alan R. Degen, courtesy of F.R.E.E. Ltd., Chicago, Illinois.

have been conducted to test whether territoriality limits local population density, none provide evidence concerning the success of unassisted females in the best habitats compared to that of assisted females in the most marginal habitats.

Demonstrations of an ideal free habitat distribution (see pp. 280–282) would show that the polygyny threshold has not been reached, but convincing evidence for ideal free distributions has not yet been obtained for any monogamous species. According to an ideal free distribution, females breeding in marginal habitats are just as successful as females breeding in the best habitats because higher density offsets the higher intrinsic quality of superior habitats. Monogamous females in every habitat would therefore have the same expected fitness, and secondary females would always be less fit no matter which mated male they chose. In the converse situation, an ideal dominance distribution would not imply that the polygyny threshold has been reached because the difference in female success between good and poor habitats may not be large enough to compensate females for the costs incurred by becoming second mates.

Polygyny has evolved in colonial animals because a fraction of all males can control all suitable breeding habitat. Hence monogamy should evolve when every male can acquire a territory within suitable habitat. Assessing the availability of suitable habitats is difficult, but when space remains unoccupied within established colonies, every male can presumably obtain an acceptable territory. Present evidence indicates that breeding habitat is not limiting for most monogamous colonial birds (Wittenberger and Tilson 1980). The principal exceptions are seabirds, herons, and other species where both parents are necessary for rearing young. In a few hole- and crevice-nesting parrots that breed in loose colonies, males apparently cannot defend more than one nest site at a time.

Present evidence is not conclusive, but many birds are probably monogamous because pairing with an unmated male in marginal habitat leads to higher reproductive success than pairing with a mated male in good habitat. The same may also be true of some monogamous mammals (see below). To the extent that the polygyny threshold model is valid for polygynous species, it should also be valid for many monogamous species.

Hypothesis: Female aggression prevents polygyny

Monogamy may evolve even though the polygyny threshold is exceeded if resident females are aggressive enough to prevent polygynous matings from occurring. Female aggression is commonplace in monogamous mammals and may be the reason why monogamy occurs, but it can evolve even though other factors are responsible for preventing polygyny from occurring (Wittenberger and Tilson 1980). For example, resident females may be aggressive to prevent encroachment by females who are seeking food on the territory, with monogamy prevailing because unmated females are never more successful by accepting polygynous status. Without evidence concerning whether unmated females would accept polygynous status in the absence of aggression by resident females, the female aggression hypothesis cannot be distinguished from the polygyny threshold hypothesis.

Monogamy has evolved in most mammals either because the polygyny threshold has not been reached or because female aggression prevents polygyny (Kleiman 1977a; Wittenberger and Tilson 1980). The polygyny threshold hypothesis appears most applicable when females lose considerable male assistance by becoming second mates or when resident females are not aggressive toward unmated females. The female aggression hypothesis is more plausible when resident females are aggressive and unmated females would not lose much male assistance by becoming second mates.

Males of most monogamous canids help feed pups, and female success would be substantially reduced by the loss of male assistance. This is true of wolves, coyotes, foxes, African wild dogs, and several other species of canids (Mech 1970; Kleiman and Brady 1978; Lamprecht 1979; Moehlman 1979). It is also true of dwarf mongooses (Rood 1974; Rasa 1975, 1977) and false vampire bats (*Vampyrum spectrum*) (Vehrencamp et al. 1977). Monogamy probably evolved in most of these species because the cost of losing male parental assistance is high and differences in territory quality are never sufficient to compensate for that cost. However, in social species female aggression appears more important. Dominant females prevent subordinate females from breeding in wolves (Mech 1970; Zimen 1975, 1976), coyotes (Ryden 1974; Camenzind 1978), African wild dogs (Lawick-Goodall and Lawick-Goodall 1970; Schaller 1972; Frame and Frame 1976), dwarf mongooses (Rasa 1972, 1975, 1977), and meerkats (Ewer 1963b; Wemmer and Flemming 1975). The same is also true of Asiatic clawless otters (*Amblonyx cinerea*) (Duplaix-Hall 1975).

The extent of male parental care varies considerably among monogamous primates. Males carry infants beginning within a few weeks of birth in tamarins, marmosets, night monkeys (*Aotus trivirgatus*), titi monkeys (*Callicebus*), sakis (*Pithecia*), and bearded sakis (*Chiropotes*) (Kleiman 1977a, b; Wittenberger and Tilson 1980). Infant carrying is an energetic burden and may increase risk of predation by hampering agility. Hence male carrying of infants may enable females to reproduce at a faster rate or survive longer than would otherwise be possible. It may be especially important in tamarins because females become pregnant again shortly after giving birth and in both tamarins and marmosets because females normally bear twins. The costs of losing male assistance may be high enough to prevent females from accepting polygynous status with an already mated male. However, female aggression also appears important. Monogamous tamarins and marmosets typically form social groups, many of which defend exclusive territories. Both sexes establish dominance hierarchies, and only the dominant male and dominant female breed. Subordinate individuals are unable to breed even when physiologically mature because like-sexed dominants prevent them from copulating or caring for young (Epple 1970, 1975, 1977; Christen 1974; Rothe 1975; Dawson 1976, 1977). Monogamous night monkey and titi monkey pairs occupy separate territories and do not breed in social groups, but female aggression may still be important because females defend territories against intruding females (Moynihan 1964; Mason 1968; Kinzey et al. 1977; P. C. Wright 1978).

Female aggression is largely absent in monogamous antelopes, and monogamy has probably evolved because the costs of competition make polygyny disadvantageous for unmated females (Wittenberger and Tilson 1980). Some monogamous antelopes, including klipspringers, Kirk's dikdiks (*Madoqua kirki*), oribis (*Ourebia ourebi*), and southern reedbucks (*Redunca arundinum*), live in relatively open habitats (Tinley 1969; Hendrichs and Hendrichs 1971; Jungius 1971; Viljoen 1975; Tilson 1979; Dunbar and Dunbar 1974, 1980). The male and female of each pair are always together and rely on each other for detecting predators. Mutual vigilance provided by larger social groups would probably be advantageous if food were sufficiently abundant, but food scarcity apparently prevents social groups and polygyny from evolving. The cooperative version of the polygyny threshold model appears applicable to these species, with monogamy prevailing because the polygyny threshold has not been reached.

The remaining monogamous antelope, namely, duikers and most dikdiks, live in dense cover and rely on concealment to escape predators (Dorst and Dandelot 1971; Heinichen 1972; Hendrichs 1975a; Dunbar and Dunbar 1979). Females spend much of their time apart from their mates and do not seem to benefit from male vigilance. Reliance on concealment keeps females apart from each other, and males can be polygynous only by defending territories large enough to encompass the home ranges of more than one female. The thickness of vegetation may prevent them from doing this effectively, although males of one species, the northern reedbuck (*Redunca redunca*), are able to defend several female home ranges at a time (Hendrichs 1975b).

Hypothesis: Males defend individual females

Males of some animals defend females instead of territories and can defend only one female at a time because females are dispersed. Defense of individual females should evolve when resources required by females are not economically defensible and when the sex ratio is skewed toward excess males. When sex ratios are skewed, the average success of breeding males would be less than one female per year, and males who control exclusive access to a single female would average higher success than males who take their chances competing for females in a promiscuous mating system (Parker 1974b; Wittenberger and Tilson 1980).

Monogamous pair associations have evolved in several arthropods because males sequester individual females. Starfish-eating shrimp (*Hymenocera picta*) live as stable pairs, with the members of each pair cooperating to kill their prey, various species of large starfish (Photo 11–13) (Wickler and Seibt 1970; Seibt and Wickler 1979). Males guard their mates by fighting off rival males. Copulations occur once each time the female molts, approximately at 18-day intervals, and males remain with their mates most of the time between molts instead of deserting to seek new mates as occurs in other crustaceans (see Ghiselin 1974) because they depend on the female's cooperation to catch prey. However, males do leave their mates temporarily during nonreceptive periods when they detect the pheromone of a receptive molting female (Seibt and Wickler 1979). Hence mated males do copulate with other females, though at an unknown frequency under natural conditions, making the species not strictly monogamous. Nevertheless, since males return to their original mates following such excursions, a true pair bond is maintained.

In wood roaches *(Cryptocercus punctulatus)* males defend the egg chambers of individual females within rotting logs (Cleveland et al. 1934; Ritter 1964). Chambers are not defended by females or unmated males, indicating that male defense evolved to control access to receptive females. Monogamy may have evolved for a similar reason in termites, which are closely related to wood roaches, in soil-burrowing roaches of the genus *Panesthia*, and in wood-boring beetles of the genus *Scolytus* (see Shaw 1925; Imms 1957; Roth and Willis 1960; Alexander 1974). In the horned beetle (*Typhoeus typhoeus*) males pair with individual females, cooperate in digging burrows, and defend their burrows by attacking intruders (Palmer 1978). Again, monogamy apparently results from male defense of individual females.

Males of many frogs seize females by climbing on their backs and defending them from rival males (Wells 1977a). This behavior is prevalent in explosive breeders that breed at high densities, and it often prevents males from copulating with more than one female per season (Savage 1961; Wells 1977a; Davies and Halliday 1979). In neotropical frogs of the genus

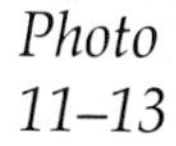

Photo 11–13

A starfish-eating shrimp feeding on a starfish.
These shrimp form relatively stable monogamous pair associations because they depend on cooperation to capture prey.

Photo from W. Wickler and U. Seibt, "Das Verhalten von *Hymenocera picta* Dana, einer Seesterne fressenden Garnde (Decapoda, Natantia, Gnathophyllidae)," *Zeitschrift für Tierpsychologie* 27 (1970):358.

Atelopus, monogamy is maintained by prolonged clasping of females, which lasts for many months (Dole and Durant 1974; Durant and Dole 1974). Males become extremely emaciated in the process because they have difficulty foraging while on the female's back. The adaptive value of such prolonged clasping may be related to an extremely low male-female encounter rate (Wells 1977a). If unmated females are rarely encountered, males might be more successful remaining with one female for the entire season rather than searching for additional mates.

Monogamy has evolved in migratory ducks because males defend access to individual females (Lack 1968). Many males cannot obtain mates because sex ratios are skewed to about 60 males per 40 females in every species (Johnsgaard and Buss 1956; Bellrose et al. 1961; Dzubin 1969; Aldrich 1973). As a result, males commonly harass and forcibly copulate with paired females (Johnsgaard 1975). The best tactic for males is to control access to one female, since otherwise their chances of mating are relatively low. Females also benefit from this behavior, as their mates protect them from the harassment of other males. The universal basis of pair formation in ducks is female "inciting" behavior, which induces males of well-established pairs to attack nearby unmated males (Johnsgaard 1965). Pair formation usually begins in late autumn on the wintering grounds, probably because competition for females is intense, and it lasts until females begin incubating eggs the following summer. Males do not provide any parental care, and the main benefit derived by females is protection from harassment and forced copulations.

Monogamy probably evolved for the same reason in quail and snowcocks. Males defend females rather than territories, and sex ratios are skewed toward excess males (Hickey 1955; Campbell and Lee 1956; Jenkins 1961a, b; and Schemnitz 1961; Anthony 1970; Campbell et al. 1973; Barclay and Bergerud 1975; Klimstra and Roseberry 1975). Mated males associate continuously with their mates and direct most of their aggression toward unmated males who follow the pair (Johnsgaard 1973). Males of most quail provide vigilance for females and their broods, and the potential loss of this assistance may make polygynous matings disadvantageous for females. However, snowcocks of the Himalayan Mountains are monogamous even though males do not provide any parental assistance (Dement'ev and Gladkov 1967), suggesting that monogamy results primarily from males defending access to females.

Males of some songbirds also defend females instead of territories. This behavior occurs, for example, in rosy finches, other cardueline finches, many Hawaiian honeycreepers, Smith's longspurs (*Calcarius pictus*), and some brood-parasitic cowbirds (references in Wittenberger and Tilson 1980). Males may defend females instead of territories in these species because resources are economically indefensible and sex ratios are male-biased. Cardueline finches exploit spatiotemporally clumped seed resources and Hawaiian honeycreepers exploit nectar, both of which are normally indefensible (see Newton 1972; Carpenter and MacMillen 1976). Sex ratios are skewed toward excess males in rosy finches, at least some other cardueline finches, probably all Hawaiian honeycreepers, and all cowbirds (Friedmann 1929; French 1959; Johnson 1965; Darley 1971; Berger 1972; Samson 1976; Shreeve 1980).

Conclusion

Theories of vertebrate mating systems clearly illustrate the value of analyzing the reproductive options open to each social class of individual separately. Male and female interests overlap, but they are never congruent. Both sexes are concerned with producing mature offspring, but the best way to maximize reproductive success differs for each sex. Males can usually produce more offspring by seeking additional mates, while females can usually produce more offspring by receiving assistance from their mates. Males may assist females in a variety of ways, all of which ease the energetic constraints placed on female reproductive output. Some common forms of male assistance are vigilance for predators, defense of food resources or nest sites, and care of offspring.

The degree of conflict between male and female interests intensifies as females become better able to utilize time and energy contributions provided by males. Since female fecundity is limited by energetics, females should exploit male contributions to the fullest extent possible. However, because male fecundity is limited by the number of females inseminated, the willingness of males to provide such contributions depends on the availability of alternative mating opportunities. The reproductive biology of a species determines how much females can potentially gain from male assistance, while the propensity for males to provide assistance determines how much females can tap that potential. The character of a mating system results from this interaction.

Several factors affect the ability of females to exploit male contributions. The nature of parental care is one important factor. If males can help feed or protect offspring, females can benefit considerably from male assistance. The nature of female defenses against predation is another important factor. If females rely on vigilance to evade predators, male vigilance may greatly enhance their survival and allow them to spend more time foraging or caring for offspring. On the other hand, when females rely on crypsis, male vigilance is not particularly helpful to female survival and may even be detrimental.

The extent to which males assist females depends on the relative value of alternative options. If additional mating opportunities are prevalent, males are likely to seek more mates rather than assist prior mates. A variety of factors create mating opportunities for males. Indefensible resources allow males to desert their mates following insemination to seek additional mates elsewhere. Female grouping behavior encourages males to ignore prior mates and court other females within social groups. Economically defensible but heterogeneously distributed resources may induce unmated females to choose already mated territorial males rather than any available unmated males.

Males are most likely to assist females when additional mating opportunities are absent. Several factors may restrict male mating opportunities. Aggression of dominant or resident females may prevent males from mating with subordinate females within social groups or attracting unmated females to their territories. Male-biased sex ratios may induce males to defend individual females, thereby limiting access of other males to receptive females. Economically defensible resources prevent males from deserting prior mates, while homogeneously distributed resources prevent them from attracting second mates to their territories. Short breeding seasons automatically limit mating opportunities. Any such factor or combination of factors that limits male mating opportunities predisposes males to assist their mates, although males may still desert mates to enhance their own survival prospects.

Thus both environmental and social constraints limit the reproductive options open to each sex. Environmental constraints shape the social fabric within which reproduction occurs. The reproductive options taken by each sex then limit even further the options left open to the other sex. The end result is the mating system typifying each species or population.

Insect sociality

The evolution of social behavior in insects has many interesting facets, but the one of greatest theoretical significance concerns the origin of sterile castes among ants, bees, wasps, and termites. The existence of nonreproductive worker castes poses a difficult problem for natural selection theory, as such individuals are clearly not maximizing their own reproductive potential. Instead, they work for the good of the colony in what appears to be an extreme form of altruism. Darwin (1859) regarded sterile insect castes, particularly the morphological castes of ants and termites, as the greatest challenge to his theory of natural selection, and with good reason.

Two theories for explaining sterile insect castes have been developed during the past two decades, and they have stirred lively debate. Because both theories have general significance, the problem of sterile insect castes has become a central issue in sociobiology. The most popular theory is that sterile insect castes evolved through indirect (kin) selection in combination with haplodiploidy (Hamilton 1964, 1972; Wilson 1971; Starr 1979). The principal competing theory is that sterile insect castes result from parental domination of offspring by means of chemical or other control mechanisms (Alexander 1974; Michener and Brothers 1974; Starr 1979). Both theories have gained some empirical support, and neither can yet be accepted to the exclusion of the other. This chapter explains why.

Degrees of sociality

The most advanced or **eusocial** insects can be distinguished as a group by their possession of three traits (Wilson 1971): (1) individuals of the same species cooperate in caring for young, (2) a division of labor exists in which more or less sterile individuals work on behalf of relatively few reproductive nest mates, and (3) at least two generations overlap enough for offspring to assist parents in performing colony labor during some part of their lives. All ants, most advanced wasps and bees, and all termites exhibit these traits and are considered eusocial.

Each of the three traits can occur independently of the others, and every combination of traits is exhibited in some species (Wilson 1971). A series of **presocial** levels of social organization can be recognized from combinations of two or fewer of the three traits. Two sequences reflecting

Chapter twelve

Table 12–1 *Degrees of sociality in social insects.*

Degrees of sociality	Qualities of sociality: Cooperative brood care	Reproductive castes	Overlap between generations
Parasocial sequence			
Solitary	–	–	–
Communal	–	–	–
Quasisocial	+	–	–
Semisocial	+	+	–
Eusocial	+	+	+
Subsocial sequence			
Solitary	–	–	–
Primitively subsocial	–	–	–
Intermediate subsocial I	–	–	+
Intermediate subsocial II	+	–	+
Eusocial	+	+	+

Source: From E. O. Wilson, *The Insect Societies* (Cambridge, Mass.: Belknap Press of Harvard University Press, 1971), p. 398. Reprinted by permission.

two probable courses of evolution from solitary to eusocial behavior have been reconstructed by Wheeler (1923, 1928), Evans (1958), Michener (1969), and Wilson (1971) (Table 12–1). In the **parasocial sequence** adults belonging to the same generation cooperate to various degrees during reproduction. In the **subsocial sequence** females show increasingly close associations with their offspring.

At the lowest or *communal* level of the parasocial sequence, adult females cooperate in constructing a shared nest but rear their broods separately. At the next or *quasisocial* level, adult females cooperate in attending broods, but every female lays eggs at some time during her life. At the *semisocial* level, some females attend eggs of other females without ever reproducing themselves. Hence they represent a true worker caste. Finally, at the *eusocial* level, reproductive females live long enough for generations to overlap, and their offspring help rear subsequent broods. The parasocial sequence probably reflects the evolutionary pathway to eusociality taken by most social bees (Michener 1969; Lin and Michener 1972).

At the *primitively subsocial* level of the subsocial sequence, females provide some parental care but depart before young eclose into adults. At the next or *intermediate subsocial I* level, females remain with their broods until young are fully mature, but their mature offspring never help rear subsequent broods. At the *intermediate subsocial II* level, a female's mature offspring remain with her and provide brood care for an additional brood.

At the eusocial level, mature offspring remain permanently with the female instead of eventually leaving to rear young of their own. The subsocial sequence reflects the evolutionary pathway to eusociality taken by ants, termites, social wasps, and some groups of social bees (Wilson 1971).

Caste differentiation in eusocial insects

Social bees and wasps

Some of the most elementary of all insect caste systems occur in semisocial halictine bees *(Augochloropsis* and *Pseudoaugochloropsis)* (Michener 1969, 1974). In *A. sparsilis* each colony contains several fertilized females and occasionally a few unfertilized females. Usually only one female possesses enlarged ovaries and lays eggs at any given time. The remaining fertilized females behave as workers. They do most of the foraging and do not lay eggs, though they may possibly become reproductives later in their lives. Unfertilized females are true workers and never reproduce. In one study only 1 of 86 females examined from the first generation of the year were unfertile, but 15%–20% of females from the second generation were unfertile (Michener and Lange 1958, 1959). Colonies of the semisocial *Pseudoaugochloropsis* may contain several egg-laying females and 0%–50% workers, many of which are unfertile (Michener 1969, 1974). The mean size of workers is smaller than that of egg layers in *Pseudoaugochloropsis* and perhaps also in *Augochloropsis.* The main distinction between workers and queens, other than behavior, is greater ovarian development in queens.

In eusocial halictine bees caste differentiation is generally somewhat greater and more variable (Wilson 1971; Michener 1969, 1974). Adult females are morphologically indistinguishable within perennial colonies of *Evylaeus marginatus,* but workers are unfertilized and do not live as long as queens (Plateaux-Quénu 1960, 1962). Each colony is started by a single female, and the founding female's daughters remain as unfertilized workers for four years. In the fifth and final year some of the daughters mate and start colonies of their own.

In other species of halictine bees workers are often considerably smaller than queens (Knerer and Atwood 1966; Knerer and Plateaux-Quénu 1966; Knerer 1980). Extreme differences in head size have been reported for a few species, and preliminary evidence suggests that large-headed females are probably the reproductives (Sakagami and Fukushima 1961). Castes are often produced at different seasons, with workers produced in early and mid summer and queens and males produced in late summer (Michener 1974).

In the highly evolved honeybee queens and workers are very different in appearance, and these differences are associated with very different behavioral roles in the colony. Queens are basically egg-laying machines. As such, they are larger than workers and have high metabolic rates, large abdomens, greatly enlarged ovaries, reduced mouth parts, no pollen-collecting apparatus, no wax-producing apparatus, and minute stingers (Wilson 1971; Michener 1974). Workers are unfertilized females who perform virtually all colony functions other than egg production, including colony defense, temperature regulation, brood care, and foraging, and they have morphology appropriate to those tasks.

Division of labor by workers is correlated with stage of development (Figure 12–1). During the first three weeks of adult life, the hypopharyngeal and mandibular glands, which are the principal sources of larval food, are fully developed and workers are primarily concerned with care of the larvae and construction of brood cells and honeycomb. Then those glands shrink in size, and workers become field bees principally engaged in food gathering and colony defense. However, this general pattern is not inviolate. Large workers become field bees about a week earlier than small ones (Kerr and Hebling 1964), and when a shortage of young workers occurs, some older workers rejuvenate their wax-producing glands and hypopharyngeal glands and revert back to constructing honeycomb and caring for broods (Nolan 1924; Rösch 1930; Free 1961, 1965).

Caste differentiation and the developmental sequence in division of labor are similar for the highly eusocial stingless meliponine bees (Wilson 1971). The main differences are that the period of wax construction begins earlier and lasts longer, life span is greater, and foraging begins later in life (Kerr and Santos Netos 1956; Hebling et al. 1964).

Caste differentiation in social wasps varies from primitive behavioral differentiation among female nest mates to well-defined morphological differentiation of queens and workers (Wilson 1971). Morphological differentiation never reaches the extremes that occur in highly eusocial bees. When morphological differentiation is well defined, division of labor mainly involves workers of different developmental stages. For example, workers of the paper wasp *Polistes fadwigae* do not forage during the first week after eclosion, and older workers are more prone to participate in nest work (Yoshikawa 1963). *Vespula* and *Vespa* workers assist in nest building the first few days and then begin feeding larvae (Ishay et al. 1967; Montagner 1967). Some but not all individuals forage as well as feed larvae.

Figure 12–1 *The time budget of a single worker honeybee during the first 24 days of adult life.*

Note the sequential change from cell cleaning to comb construction and maintenance to foraging as the worker becomes older.

Source: From E. O. Wilson, *The Insect Societies* (Cambridge, Mass.: Belknap Press of Harvard University Press, 1971), p. 176. (Redrawn from Ribbands 1953; based on data of Lindauer 1952). Reprinted by permission.

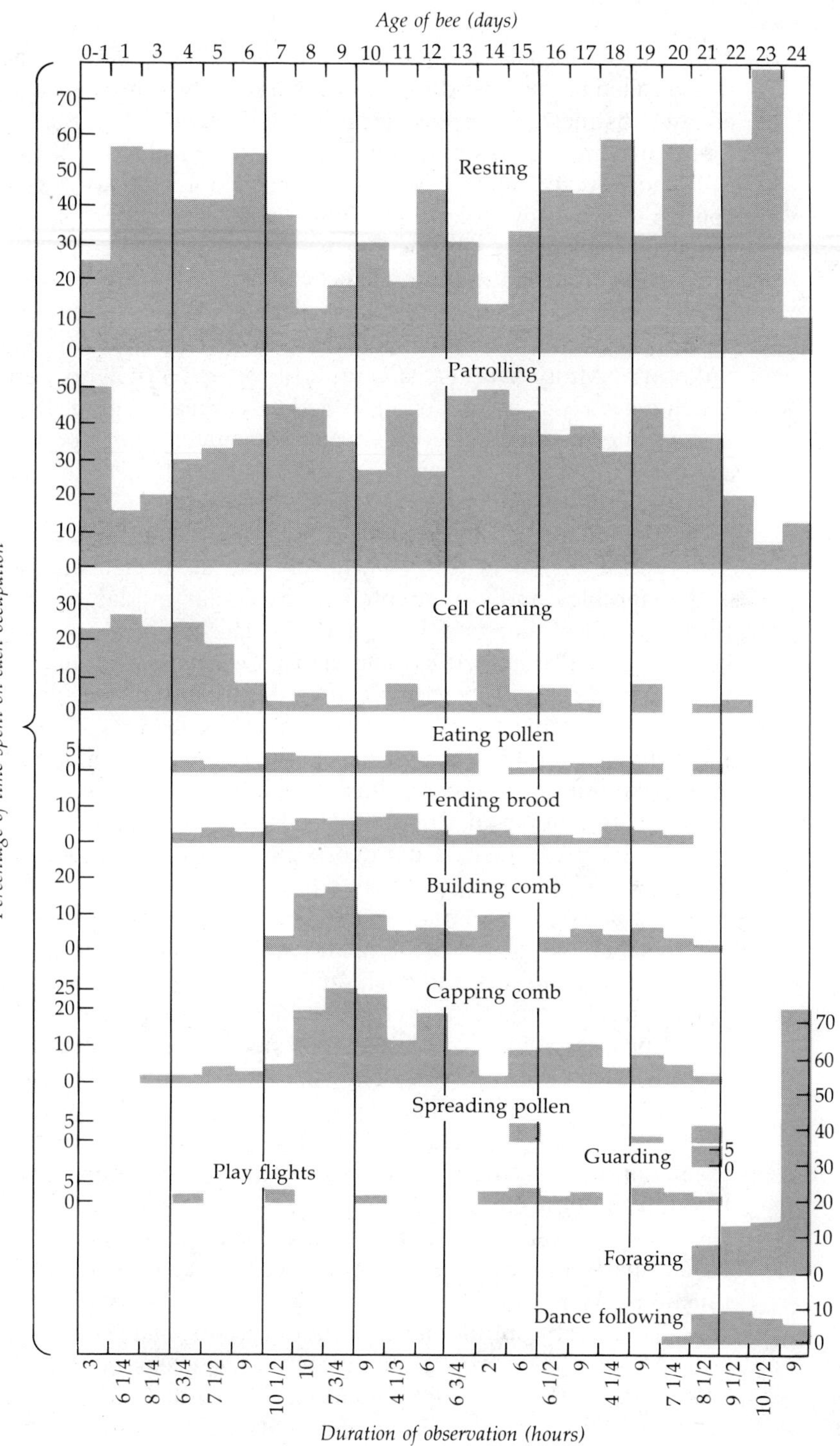

Age of bee (days)
0-1 1 3 4 5 6 7 8 9 10 11 12 13 14 15 16 17 18 19 20 21 22 23 24
Resting
Patrolling
Cell cleaning
Eating pollen
Tending brood
Building comb
Capping comb
Spreading pollen
Guarding
Play flights
Foraging
Dance following
Percentage of time spent on each occupation
3 6 1/4 8 1/4 6 3/4 7 1/2 9 10 1/2 10 7 3/4 9 4 1/3 6 6 3/4 2 6 6 1/2 9 4 1/4 9 7 1/4 8 1/2 9 1/2 10 1/2 9
Duration of observation (hours)

Ants

The three basic female castes in ants are the queen, the major worker (often called soldier), and the minor worker. These are usually, though not always, distinctive morphological forms (Wilson 1971). Males are not differentiated into castes and serve solely as reproductives. The queen is a fully reproductive female, whose main function is to lay eggs and in some species to start new colonies after mating. Major workers are unfertilized females. Their principal functions are colony defense and in some species food storage or seed milling (Glancey et al. 1973; Marikovsky 1974; Wilson 1974; Oster and Wilson 1978). Specialized castes occur among major workers in some ants but not in others. The reasons for this variability are unknown. Minor workers are unfertilized females who perform colony functions such as construction, brood care, and foraging. In some species a form intermediate between worker and queen castes occurs. Such individuals are called *ergatogynes.* They either complement the queen's reproductive activities or replace the queen entirely.

Several forms of major worker castes have evolved in the various genera of ants. These differ in morphological specializations of the head and mandibles and are adapted to particular modes of colony defense (Wilson 1971; Oster and Wilson 1978). The major workers of many ant genera (e.g., *Pheidole, Atta, Camponotus, Oligomyrmex, Zatapinoma*) possess mandibles designed for cutting integument and clipping off appendages of invading arthropods. Typical behavior of such workers is to wedge themselves several ranks deep in colony passageways, mandibles outward, and strike out at the most vulnerable body parts of invaders entering the colony (Creighton and Creighton 1959). The major workers of army ants *(Eciton, Anomma)* possess pointed, sickle- or hook-shaped mandibles designed for piercing an opponent's integument (Wilson 1971). During the nomadic phase, army ants move in columns, attacking any insects or vertebrates that cannot evade them, and major workers typically line up along the flanks of columns and rush any opponents threatening the columns. Finally, major workers of species that nest in cavities within living or dead plants possess shield- or plug-shaped heads used to block nest entrances (Figure 12–2) (Wilson 1971). They stand continuous guard, blocking the nest entrances at all times except when a minor worker returns to the nest (Creighton and Gregg 1954; Wilson 1974). Then, after touching antennae, the major worker crouches down and allows the minor worker to wriggle through the nest entrance.

Not all ants have a major worker caste, and in such species an undifferentiated or weakly polymorphic worker caste performs nest-guarding functions. Why specialized major worker castes have evolved in some species but not in others is still unknown (see Oster and Wilson 1978).

An unusual form of caste differentiation occurs in the leaf-cutting ant *Atta cephalotes* (Eibl-Eibesfeldt and Eibl-Eibesfeldt 1967). Minor workers accompany the larger workers, or *medias,* during leaf-cutting expeditions and often hitchhike on cut pieces of leaf carried back to the colony by medias.

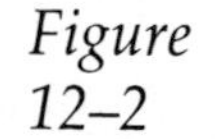

Figure 12–2

The major worker (soldier) caste of the European ant Camponotus truncatus *is specialized to block nest entrances from invaders.*

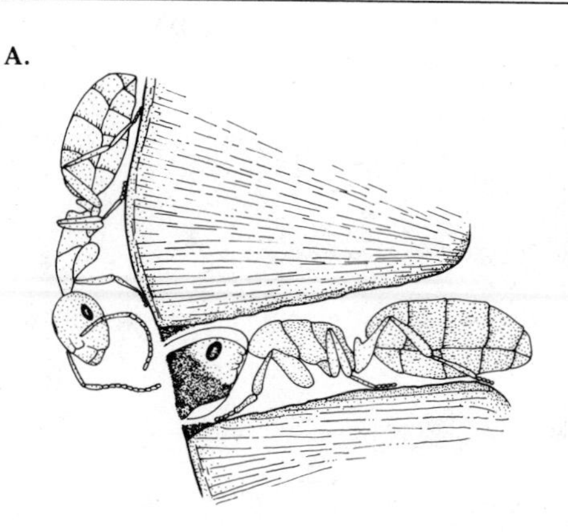

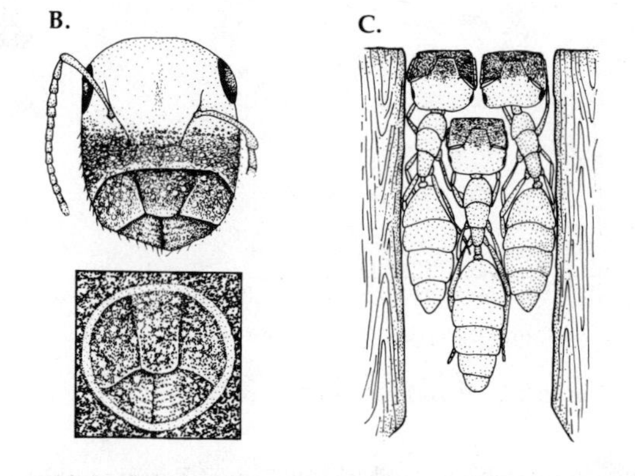

A. A minor worker returns from an excursion from the nest and is checked by the major worker before it is allowed to enter the nest. B. Two views of the major worker's head. The lower view is a frontal profile of a worker's head framed by the nest entrance, showing how tightly the entranceway is blocked. C. Larger nest entrances are plugged by a group of major workers acting in concert.

Source: From E. O. Wilson, *The Insect Societies* (Cambridge, Mass.: Belknap Press of Harvard University Press, 1971), p. 160. (From Szabó-Patay 1928.) Reprinted by permission.

They are not involved in leaf cutting and instead protect medias from parasitic phorid flies by snapping at the flies with their mandibles and repulsing them with their hind legs.

Within the minor worker caste of most ants, division of labor is based in part on age (Wilson 1971, 1976b; Brian 1979). Workers spend about the first 50 days inside the nest, where they care for the brood, the queen, and other workers, handle prey, and clean the nest (Figure 12–3). Then they shift to outside activities, particularly foraging and nest construction. As in social bees, this developmental pattern is not rigidly determined. For

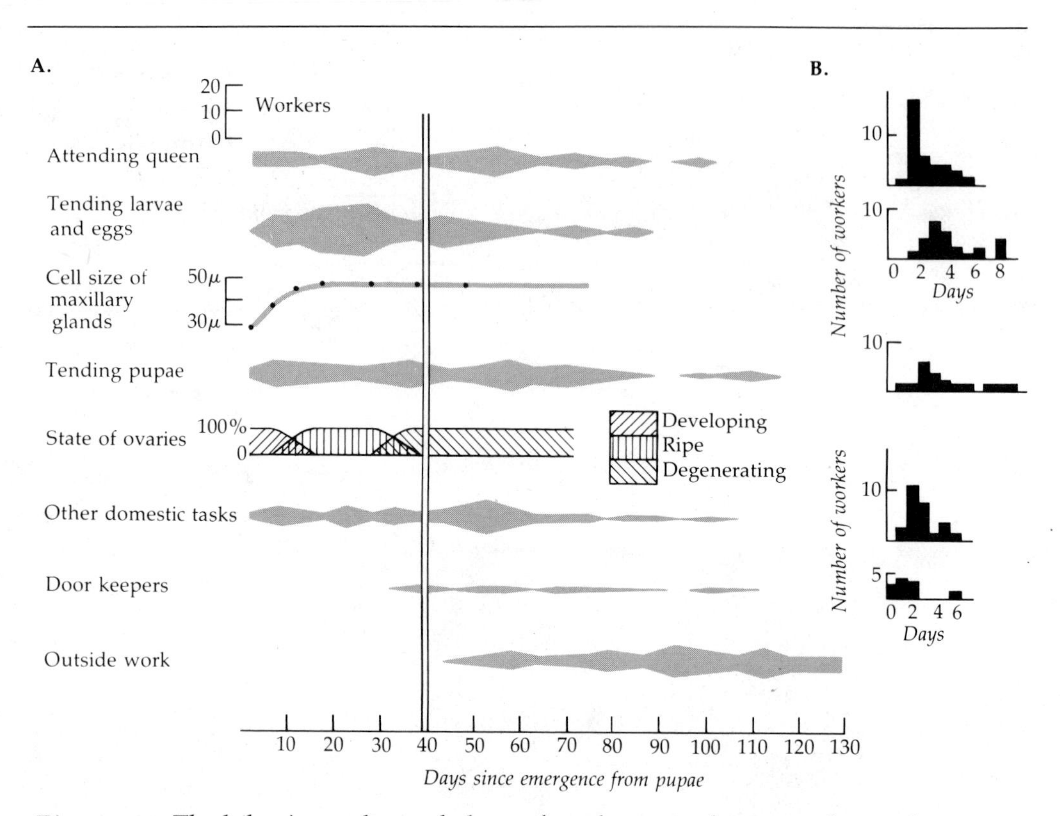

Figure 12–3 *The behavior and morphology of worker ants change as the workers grow older.*

A. Developmental changes in the behavior and morphology of Formica polyctena *workers as they grow older. Note the shift from indoor to outdoor activities, which occurs around days 40–50. B. Histograms are frequency distributions showing how many workers persist with a given task for the specified number of days.*

Source: From E. O. Wilson, *The Insect Societies* (Cambridge, Mass.: Belknap Press of Harvard University Press, 1971), p. 164. (Redrawn from Sudd 1967; based on data of Otto 1958.) Reprinted by permission.

example, when workers of *Myrmica rubra* who are strongly predisposed toward brood care are transferred from the colony nest to isolated cells containing only earth, they soon begin working on nest construction (Ehrhardt 1931). Similarly, when workers who are engaged in foraging and excavation are transferred to cells containing larvae but no earth, they soon become brood nurses. If both groups are subsequently returned to the home colony, all workers revert to their original roles. An exception to the typical pattern of development occurs in the primitive ponerine ant *Amblyopone pallipes* (Traniello 1978). This species is predaceous and shows no age-related differentiation of worker roles.

Termites

The caste system of termites is remarkably similar to that of social Hymenoptera (bees, wasps, and ants), even though termites are in an entirely different order of insects (Isoptera). Termite castes consist of primary reproductives, soldiers, workers, and sometimes *pseudergates,* or false workers (Miller 1969; Noirot 1969; Wilson 1971). Additional supplementary reproductive castes appear in colonies if one or both primary reproductives disappear. Termite castes are comprised of both females and males, a major difference between termites and social Hymenoptera. In social Hymenoptera only females are differentiated into castes.

The primary reproductives consist of a monogamously mated male (king) and female (queen), which mate on a nuptial flight following departure from their respective natal colonies (Wilson 1971). Once mated, the primary reproductives construct a nest and rear their first brood. After the workers and nymphs acquire their own digestive capability, they perform all nest construction, nursing, and foraging, and they begin feeding the primary reproductives with salivary secretions. The queen's abdomen soon swells greatly (Photo 12–1), and both queen and king then serve only a reproductive function.

A true worker caste does not exist in the lower termites, and colony labor is performed instead by nymphs and pseudergates (Wilson 1971). Nymphs are individuals capable of developing into functional reproductives by further molting but are normally inhibited chemically from doing so. Pseudergates are individuals who either have regressed from nymphs or have developed from larvae by nondifferentiating molts.

In higher termites a true worker caste of morphologically specialized workers develops. Division of labor among minor workers is not well studied but some examples are known (Wilson 1971). Division of labor based on the size of minor workers is absent in most termites, but it does occur among the Macrotermitinae. Among species that normally forage outside their nests, most foraging is done by large workers, all of whom are males (Noirot 1955, 1969). The smaller female workers work only within the nest. Present evidence indicates that no similar division of labor occurs among any species that forages in subterranean passageways.

A temporal division of labor has been documented in a few species. Pasteels (1965) found that the younger minor workers of *Nasutitermes lujae* are specialized as nurses, while older minor workers are responsible for laying odor trails and foraging. Similarly, Howse (1968) found that nymphs and older larvae of the primitive termite *Zootermopsis nevadensis* do most excavating and nest building, while older workers probably do most of the foraging.

Major workers are specialized for the single function of colony defense. Several kinds of morphological and behavioral specializations have evolved (Wilson 1971). Most commonly, major workers are *mandibulate,* with large armored heads, powerful muscles, and sharp, elongated mandibles. Mandibulate major workers are effective in defending the colony

Photo 12–1

*A termite queen (*Macrotermes natalensis*) with greatly enlarged abdomen next to the king.*

In the initial stage of colony growth, queens are not much larger than workers, but once the initial brood matures into a worker caste, the queen develops into an egg-laying machine.

Photo by John B. Free.

against other termites, but they are ineffective against smaller insect invaders (Stuart 1969). In some genera *(Termes, Acanthotermes, Macrotermes, Syntermes)* major workers possess large, extremely sharp, scimitar-like mandibles (Photo 12–2) (Deligne 1965; Wilson 1971). They are fully capable of defending colonies against large animals, including vertebrates. Their defense is not always effective, but the pain of their bites at least deters marauders. The snapping major workers of *Capritermes, Neocapritermes,* and *Pericapritermes* have asymmetrical mandibles that slide against each other to produce a snapping blow capable of stunning other insects and inflicting pain on vertebrates (Kaiser 1954; Deligne 1965; Wilson 1971). In *Cryptotermes* major workers have cylindrical heads for blocking passageways within the nest (Wilson 1971). Their mandibles are poorly developed and not well suited for defense. The major workers of many species also employ chemical defenses, particularly corrosive or sticky secretions and sprays (Moore 1968, 1969; Eisner et al. 1976).

Theories for the evolution of sterile castes

Haplodiploidy and indirect (kin) selection

In most species of Hymenoptera fertilized eggs develop into females while unfertilized eggs develop into males. This mode of sex determination

Photo 12–2

A mandibulate major worker termite that is specialized to bite any intruders who threaten the colony.

The bites inflicted by such workers can be extremely painful and are even an effective deterrent against many vertebrate marauders.

Photo by John B. Free.

is referred to as **haplodiploidy** because males are haploid and females are diploid.

An important consequence of haplodiploidy is that coefficients of relatedness among individuals become asymmetrical (Hamilton 1964). That is, the degree of relatedness between parents and offspring and among siblings depends on the sex of each individual.

In haplodiploid insects daughters are related to their parents by 1/2, just as they are in diploid organisms, because they receive half their genetic material from each parent. However, sons are related to their mothers by 1.0 and to their fathers by 0 because they receive all their genetic material from their mothers. Brothers are related to each other by 1/2 because for each chromosome pair of the mother they receive the same chromosome 50% of the time and different chromosomes 50% of the time. Sisters are related to each other by 3/4 because half their genetic material, that received from the father, is always identical while the other half is identical 50% of the time. Finally, sisters are related to brothers by 1/4 because half their genetic material, that received from the father, is never identical by common descent, while the other half is identical only 50% of the time. The coefficients of relatedness among siblings are of course averages, since some siblings share all their genes in common.

The most important effect is the high relatedness of sisters. Clearly, if all other factors are equal, females can propagate their genes faster by helping rear younger sisters rather than by producing daughters (Hamilton 1964). According to the indirect (kin) selection hypothesis, sterile worker castes evolved precisely for that reason.

Parental manipulation

An alternative to the indirect (kin) selection hypothesis is that sterile worker castes result from parental manipulation of offspring (Alexander 1974; Michener and Brothers 1974). If parental investments in offspring increase because they enhance a parent's reproductive success, it follows that parents may control the way those investments are allocated or used. For example, parents may limit the amount invested in each offspring to allow production of more offspring, each with poorer prospects of survival. They may restrict or withhold investments from the weakest or youngest offspring when resources become limited. They may kill some offspring or feed them to others. They may cause some offspring to be temporarily sterile helpers at the nest. Finally, they may cause offspring to be permanently sterile workers whose only reproductive option is to help rear siblings.

Clearly such controls create potential conflicts of interest between parents and offspring. Alexander (1974) argues that parents are most likely to win such conflicts for two reasons. First, parents are usually larger and stronger than offspring and hence are better able to impose their best interests. Second, parent-offspring interactions evolved because parental care is beneficial to the parents. If parental benefits are used by offspring to increase their own inclusive fitness at the expense of parents, selection should cause parents to invest less parental care. Moreover, offspring who reduce the fitness of their parents will in turn suffer the same costs once they produce offspring of their own (Williams 1966a). In the long run this consequence should eliminate any immediate benefits gained by winning conflicts to the detriment of parents. Nevertheless, present evidence and models suggest that compromise solutions to parent-offspring conflicts may sometimes be reached (see pp. 392–394).

An important issue here is the extent to which parents can control behavior of offspring among eusocial insects. In all insects mothers are in a position to control the development and behavior of offspring chemically. The extent to which they actually do is an empirical question for which there is considerable evidence, as will be shown below.

Predictions and evidence: The indirect (kin) selection hypothesis

The two hypotheses for explaining the sterile castes of eusocial insects are highly controversial. Many lines of evidence have been mustered to support or refute each hypothesis. For this reason, definitive conclusions are difficult to reach. Two guidelines are helpful in evaluating the evidence. First, each hypothesis can hold only if its underlying assumptions are valid. If the assumptions are invalid, the hypothesis must either be rejected or be restricted to cases where the assumptions do hold. Second, one hypothesis cannot be accepted over the other unless it uniquely explains some of the evidence. If each hypothesis uniquely explains some of the evidence, some combination of the two hypotheses may provide the best explanation.

Haplodiploidy and eusociality

The key underlying assumption of the indirect (kin) selection hypothesis is that eusocial insects are haplodiploid. While eusocial Hymenoptera are haplodiploid, eusocial termites are not. Therefore, indirect selection might explain sterile castes in the ants, bees, and wasps, but some other explanation is necessary for explaining such castes in termites. If the explanation for termites is not dependent on unique attributes of termite biology, it may be capable of explaining sterile castes in Hymenoptera as well. Thus the evolution of eusociality in termites will be a pivotal issue in evaluating the two hypotheses.

Eusocial Hymenoptera. A strong bias exists for eusociality to evolve in haplodiploid insects. Eusociality has evolved at least 12 separate times in the social Hymenoptera, compared to once in all other insects combined, and yet less than 20% of all insect species present since the Cenozoic era have belonged to the Hymenoptera (Wilson 1971, 1975a; Oster and Wilson 1978; see Evans 1977 for a contrary view). This bias led to formulation of the indirect (kin) selection hypothesis and has been a major factor leading to its widespread acceptance.

The close association between haplodiploidy and eusociality can be explained in at least three ways. First, it may exist because indirect (kin) selection, in conjunction with haplodiploidy, is responsible for the evolution of sterile castes in all but the termites (Hamilton 1964). Second, haplodiploidy may make colonial social organization easier to evolve because it gives females control over the sex of their offspring (Richards 1953). The ability of females to postpone production of male reproductives until late in the season or until the colony has reached an opportune stage of development may have facilitated the evolution of chemical controls for preventing female offspring from reproducing on their own. If mating opportunities are extremely limited, female offspring may be more successful by submitting to the control of their mother instead of dispersing. Third, female parental care is a prerequisite of eusociality, and it is more prevalent in Hymenoptera than in other insects. If haplodiploidy makes female parental care more advantageous or if early Hymenoptera were, for reasons independent of haplodiploidy, predisposed to exhibit female parental care, the correlation between haplodiploidy and eusociality would not imply that indirect (kin) selection has led to eusociality. Until alternative interpretations can be ruled out, the correlation between haplodiploidy and eusociality provides inconclusive support for the indirect (kin) selection hypothesis.

Not all haplodiploid insects are eusocial. Hence if indirect (kin) selection has led to the evolution of sterile castes, some additional factors must prevent the origin of such castes in some species. One possibility is that conditions prevent overlap of parental and offspring generations in parasocial species (Wilson 1971). If females cannot survive long enough to produce younger sisters for their daughters, sterile worker castes could not evolve. This hypothesis seems untenable, however, because sterile castes have not evolved in subsocial bees, wasps, and certain haplodiploid beetles (*Xyleborus*) despite overlapping generations and close parent-offspring associations (Wilson 1971).

Alternatively, changes in parasite or predation pressure may make colonial nesting disadvantageous in parasocial species. Coloniality may lead to increased parasite infestation or predation of adults or young, and this cost may offset the advantages that daughters might gain by remaining to help rear sisters (Lin and Michener 1972). Some evidence from halictine bees supports this hypothesis (Michener 1958; Michener and Lange 1958; Lin 1964).

The above considerations indicate that, at most, indirect (kin) selection represents one factor leading to eusociality. Ecological conditions favoring coloniality and prolonged survival of adult females are also important and should not be ignored.

Eusocial termites. The situation for termites is more complex than it is for ants, bees, and wasps. Termites are not haplodiploid, but some unusual properties of their genetic system may still make the indirect (kin) selection hypothesis viable (Lacy 1980).

Cytogenetic studies of several termite species indicate that a series of translocations effectively link about 10%–50% of the genome to the sex chromosomes (Syren and Luykx 1977; Fontana and Amorelli, cited by Lacy 1980). Males of the termites studied are always heterozygous for the translocations because cellular mechanisms during meiosis ensure that the translocated set of chromosomes is always transmitted to sons while the nontranslocated set is always transmitted to daughters. The result is a change in the genetic coefficients of relatedness among siblings.

Daughters share 1/2 of their non-sex-linked genes and 1/2 of their maternal sex-linked genes but share identical sets of paternal sex-linked genes (Lacy 1980). If 50% of the genome is functionally sex-linked, sisters will be related by .625 (the average of .50 for 3/4 of the genome and 1.0 for 1/4 of the genome). The same is true among brothers. Sons and daughters share 1/2 of the non-sex-linked and maternal sex-linked genes but do not share any of the paternal sex-linked genes, which are only transmitted to sons. Hence their average relatedness is .375 (the average of .50 for 3/4 of the genome and 0 for 1/4 of the genome). The bias is even stronger if the genes coding for worker behavior are contained in the sex-linked part of the genome. Then same-sexed siblings will be related by .75 with respect to the genes coding for worker behavior, while opposite-sexed siblings will be related by .25 with respect to those genes. Since parents remain half-related to both sons and daughters, offspring might increase their inclusive fitness by helping produce like-sexed siblings rather than direct progeny.

The argument for termites is similar to that proposed for eusocial Hymenoptera, except that here both sexes should become workers and each sex of workers should only rear larvae that are the same sex as the workers. Whether workers actually apportion their efforts on the basis of sex has not been ascertained. Data on that point would serve as an important test of the indirect (kin) selection hypothesis, at least insofar as it applies to termites.

Number of fathers

Theoretical considerations. An important assumption of the indirect (kin) selection hypothesis is that all workers in a eusocial insect colony have the same parents. If daughters are fathered by more than one male, their average relatedness ($\bar{r}$) is less than three-fourths. More precisely, if each male contributes equally to fertilizing eggs, the average relatedness of daughters is given by

$$\bar{r} = \frac{1}{2}\left(\frac{1}{2} + \frac{1}{n}\right)$$

where n = number of fathers (Wilson 1971). With two fathers daughters are related by only one-half, and the advantage of helping rear sisters rather than producing offspring of their own is lost.

If males do not contribute equally to fertilizing eggs, average relatedness of daughters is given by

$$\bar{r} = \frac{1}{2}\left(\frac{1}{2} + \sum_{i=1}^{n} f_i^2\right)$$

where f_i = the proportion of daughters fertilized by male i and $\Sigma_{i=1}^{n} f_i^2$ = the sum of the squared contributions of all n males (Wilson 1971). As long as one male fertilizes sufficiently more eggs than all other males combined, average relatedness of workers will exceed one-half, and the indirect (kin) selection hypothesis remains tenable. The threshold value for the contribution of the most successful male can be calculated as follows:

$$\bar{r} = \frac{1}{2}\left(\frac{1}{2} + f_1^2 + \sum_{i=2}^{n} f_i^2\right) > \frac{1}{2}$$

$$f_1^2 > \frac{1}{2} - \sum_{i=2}^{n} f_i^2$$

With two fathers average relatedness of daughters will be greater than one-half if one male fertilizes more than 50% of the queen's eggs. With more than two fathers one male would have to fertilize up to 70% of the queen's eggs before average workers would be more than 50% related to each other. The percentage of fertilizations by a single male must therefore be at least 50%–70% for the indirect (kin) selection hypothesis to hold in its original form, with the threshold percentage depending on the number of males involved and the proportion of eggs fertilized by each of the other males.

If one male does not predominate sufficiently in fertilizing a queen's eggs, the indirect (kin) selection hypothesis could still be salvaged with a small modification. Workers might still benefit by rearing the queen's progeny instead of their own if they selectively care for progeny who share the same paternity. This would require that workers be able to distinguish larvae on the basis of paternity. A recent study suggests that such discriminatory ability may be possible. Greenberg (1979) reared laboratory-bred bees *(Lasioglossum zephyrum,* a primitively eusocial species) in isolation and then placed them together to see if they showed any preferences based on relatedness when forming colonies. He found that individuals preferentially admitted close kin to newly formed colonies. Furthermore, the probability of admission to a colony was roughly proportional to the degree of relatedness. These results suggest that bees can identify degree of relatedness on the basis of genetically influenced traits. What those traits might be is unknown. Since Greenberg's study involved discriminations among adult workers, it remains to be seen whether workers can discriminate between developing larvae while providing them with brood care.

Evidence. The number of times females are inseminated varies considerably among eusocial Hymenoptera. In the honeybee matings occur during a nuptial flight and the queen is commonly pursued by several drones. Multiple inseminations are the rule, with the queen often being mated by as many as 7–12 males (Taber 1954; Alber et al. 1955; Alber 1956; Taber and Wendell 1958; Kerr et al. 1962; Gary 1963). Sperm is stored by the queen in specialized storage vesicles, called spermathecae, and the queen has physiological control of sperm release into the oviduct. Sperm from different males do not mix randomly within the spermathecae, and it is possible that a majority of sperm released into the oviduct at any one time are from a single male (Alber et al. 1955; Taber 1955). By artificially inseminating queens with sperm from one normal and one mutant drone, Taber (1955) showed that the normal drone fathered 63%–72% of the queen's daughters. It is possible that the preponderance of fertilizations by the normal male resulted from atypical pleiotropic effects of the mutant gene in the other male, but if a comparable preponderance of fertilizations by one male occurs in the wild, the indirect (kin) selection hypothesis remains tenable.

Multiple inseminations are apparently also the rule in most ants. Females are frequently receptive for relatively brief periods, but when mating occurs in swarms (Photo 12–3), several males may copulate with them. Proof of multiple inseminations is available for a few species, but most species have not been adequately studied. Multiple copulations have been observed in *Formica rufa* (Marikovsky 1961), *F. montana* and *F. subintegra* (Kannowski 1963), *Prenolepis imparis* (Talbot 1945), *Mycocepurus goeldii* (Kerr 1961), *Anomma wilwerthi* (Raigner et al. 1974), and four species of *Pogonomyrmex* (Nagel and Rettenmeyer 1973; Hölldobler 1976b). In the fungus-growing ant *Atta sexdens rubropilosa* virgin males contain 44–80 million spermatozoa, while newly mated queens contain 206–319 million spermatozoa, indicating that queens mate with four to nine males each (Kerr 1961). The proportion of eggs fertilized by any one male is not known for any of these species.

Single inseminations are the rule in a few eusocial Hymenoptera. Kerr and co-workers (1962) concluded on the basis of sperm counts that queens of the stingless bee *Melipona quadrifasciata* are inseminated by only one

Photo 12–3

An ant mating swarm with many males surrounding a receptive female.

The occurrence of multiple inseminations in such swarms casts doubt on the indirect (kin) selection hypothesis for explaining the evolution of sterile insect castes, but it does not disprove the hypothesis.

Photo by Jane Burton/Bruce Coleman, Inc.

male. In the myrmicine ant *Pheidole sitarches* queens drop quickly to the ground after copulating with a male, shed their wings, and hide (Wilson 1957). Such behavior ensures that they only copulate once. However, such cases are apparently exceptional. It is possible that multiple inseminations evolved following eusociality, possibly because queens need increased amounts of sperm to produce large numbers of workers (Wilson 1971; Hamilton 1972), but this seems unlikely because most presocial Hymenoptera females are also multiply inseminated (Alexander 1974). The most important questions are whether sperm from one male predominates sufficiently for workers to be related by more than 1/2 and whether workers can discriminate between larvae on the basis of relatedness. On all of these points present evidence is inadequate for reaching any definite conclusions.

Number of mothers

Mean relatedness among female progeny in a colony is reduced if more than one queen produces workers. With two queens mean relatedness becomes 3/8, assuming both queens produced equal numbers of progeny and each was inseminated by a single male. Assuming single inseminations, one queen would have to produce two-thirds of all the progeny for female progeny to have an average relatedness greater than 1/2. The proportion would have to be even larger if the queens are multiply inseminated. Thus the existence of multiple queens in eusocial insect colonies could pose a problem for the indirect (kin) selection hypothesis.

Species with subsocial ancestry. Most eusocial insect colonies contain only one queen, but there are exceptions. In a few genera of ants, some colonies contain two queens while others contain none or one (Buschinger 1974; Hölldobler and Wilson 1977). In the Argentine ant *Iridomyrmex humilis* separate colonies do not exist in densely occupied areas (Markin 1968). Colonies merge into large supercolonies containing millions of workers and hundreds, thousands, and sometimes millions of queens. The same is true of several other ant species (Lin and Michener 1972; Hölldobler and Wilson 1977; Herbers 1979).

The presence of multiple queens does not necessarily invalidate the indirect (kin) selection hypothesis. In small colonies that contain only a few queens, most or all of the progeny may possibly be produced by a single queen (Wilson 1971). Then most workers would still be related to daughters of the queen by more than one-half. No data are available to evaluate this possibility. In very large colonies workers may selectively rear progeny produced by their own mothers without contributing much assistance to progeny of other queens in the colony. This may be the case for *Pseudomyrmex,* as workers restrict their activities to small parts of the colony (Janzen, cited by Lin and Michener 1972), but it is probably not the case for *Iridomyrmex humilis* because workers move freely from one nest to another within the supercolonies (Markin 1968). Selective caring for relatives has not yet been demonstrated for any ant with multiple-queen colonies.

An important issue here is whether ancestral ants were monogynous (one queen) or polygynous (multiple queens). If polygyny arose secondarily from monogynous eusocial species, its existence today does not pose a problem for the indirect (kin) selection hypothesis. It would then be possible that polygyny evolved after eusociality had become an irreversible characteristic originally established by indirect (kin) selection.

Hölldobler and Wilson (1977) argue on theoretical grounds that ancestral ants were monogynous. They argue that queens should try to maintain monogyny because they are more closely related to daughters and sons than they are to nieces, nephews, and grandoffspring. Similarly, workers should try to maintain monogyny because they are more closely related to sisters and brothers than they are to half sisters and half brothers.The first argument is not convincing, however, because environmental circumstances can override the benefits of being closely related to progeny reared in the colony, as will be discussed below in reference to semisocial species.

Most polygynous ant colonies are founded by a single queen and become polygynous by incorporating queens produced in the colony, suggesting that polygyny is a secondarily derived trait (Hölldobler and Wilson 1977). However, colonies are initially founded by several queens in a sizeable minority of ant species, with monogyny eventually developing as a result of competition among the founding queens. Such observations suggest that polygyny may have been the primitive condition, with monogyny being secondarily derived. Furthermore, a microgyne or dwarf female caste regularly occurs among ants (Elmes 1976). Microgynes are much smaller than normal queens but nevertheless are apparently fully reproductive. Elmes argues on the basis of genetic characteristics that microgynes represent an ancestral form, implying that ancestral ants were not monogynous. Even so, present understanding of the microgyne caste does not allow for any definitive conclusions to be drawn.

Species with semisocial ancestry. Worker castes in semisocial insects are especially difficult to explain by indirect (kin) selection. Colonies of semisocial species are typically founded by a single reproductive queen who is later joined by one or more auxiliary females. These auxiliary females may reproduce, though to a lesser extent than the queen, but they also help rear progeny of the founding queen. Even if joining females are sisters of the founding queen, as they often are in *Polistes* wasps (West Eberhard 1969; Klahn 1979; Noonan 1980), their relatedness to daughters of the queen is only 3/8. The existence of semisocial colonies is therefore a major problem for the indirect (kin) selection hypothesis, as originally applied to eusocial insects, because auxiliary females are not more closely related to daughters of the queen than to their own daughters. Thus additional factors must be involved.

The origin of semisocial colonies can be explained by the spinster hypothesis (West 1967; West Eberhard 1975). Females vary greatly in ovarian development, probably due to environmental conditions prevailing during larval development, and females with the most fully developed ovaries become founding queens. Less developed females can either join a founding queen to form a semisocial colony or else nest solitarily. In *Polistes* wasps and probably other semisocial species, most newly founded nests fail. When nest success is higher for established colonies, as is likely when parasite or predation pressure is high (see Waloff 1957; Michener 1958; Michener and Lange 1958; Lin 1964; Wilson 1966), less developed females may achieve higher reproductive success by joining already established founding queens rather than starting new nests of their own. If founding females are sisters of auxiliary females, the auxiliary females may derive most of their fitness gain through progeny of the founding female. In *Polistes,* for example, the inclusive fitness of auxiliary females equals that of solitary foundresses largely for that reason (Noonan 1980). If founding females are unrelated to auxiliary females, the auxiliary female may still be more successful by joining a foundress as long as she produces some of the young. Why she should help rear progeny of an unrelated foundress is not clear, however.

Differential altruism by the two sexes

According to the indirect (kin) selection hypothesis, sterile castes should only evolve among females. Females are more closely related to sisters than to daughters and therefore benefit most by behaving altruistically toward sisters (Hamilton 1964). Males, on the other hand, are more closely related to daughters than to sisters and hence benefit most by behaving selfishly to increase their chances of surviving to mate. A striking characteristic of eusocial Hymenoptera is that workers *are* always females. In other words, females perform altruistic acts such as brood care, colony maintenance, and colony defense, while males do not. This evidence therefore supports the indirect (kin) selection hypothesis.

Although the fact that worker castes are restricted to females in social Hymenoptera has been taken as strong support for the indirect (kin) selection hypothesis (Wilson 1971; Oster and Wilson 1978), an alternative interpretation is available. Alexander (1974) points out, as does Hamilton (1964), that male parental care is rare in all Hymenoptera, including presocial species. This creates an initial genetic bias in ancestral forms for parental behavior to develop in females and not in males. According to the parental manipulation hypothesis, there has never been any selection to produce male worker castes because queens can control the proportion of workers produced simply by controlling the proportion of fertile and unfertile eggs they lay (see also West Eberhard 1975; Evans 1977).

A related prediction derived from the indirect (kin) selection hypothesis is that workers should be more altruistic toward sisters than toward brothers or nieces (Wilson 1971). Workers are 3/4 related to sisters, compared to 1/4 related to brothers and 3/8 related to nieces.

Workers are more altruistic toward sisters than toward brothers in some wasps and bees. When food is scarce, workers of *Polistes gallicus* readily regurgitate food to females but not to males, even though they regurgitate food to both sexes when food is abundant (Montagner 1964). Workers of *Tetramonium* feed reproductive females intensively after eclosion but do not feed reproductive males (Peakin 1972). Worker honeybees feed young drones, but as the drones grow older, they attack drones and drive them from the hive (Frisch 1955). These observations support the prediction drawn from the indirect (kin) selection hypothesis, but they are again open to alternative interpretations. For example, *Polistes* workers may restrict regurgitation to females during periods of scarcity because females become workers that can help maintain the colony, while males do not. *Tetramonium* may feed reproductive females but not reproductive males simply because the reproductive output of females is limited energetically while that of males is not. Such a difference would be expected if the reproductive payoff is higher for energy invested in females. Honeybee workers probably attack older drones because they are a burden on hive resources and are no longer worth supporting. The behavior is analogous to weaning in mammals. Thus the tendency for workers to behave more altruistically toward sisters is not uniquely explained by the indirect (kin) selection hypothesis.

The best way for workers to direct altruism toward sisters rather than toward nieces is to prevent production of nieces in the first place (Wilson 1971). One might therefore expect workers to prevent reproductive activities by a second queen. That is the case in several eusocial species. In many stingless bees only one virgin queen at a time can be supplied with an adequate supporting worker force, and surplus virgin queens are killed by workers shortly after one virgin queen makes a successful virgin flight (Wilson 1971; Michener 1974). When ant colonies with multiple queens are created artificially by combining laboratory colonies, workers attack and kill queens until only one remains (Baroni Urbani and Soulié 1962; Passera 1963; Soulié 1964; Wilson 1966). These observations also support the prediction drawn from the indirect (kin) selection hypothesis, but again an alternative interpretation is possible. Since the presence of extra queens in a colony is detrimental to the extant queen, she may produce chemical control agents that induce workers to eliminate competing queens. Queen killing by workers is therefore not uniquely explained by either hypothesis.

Production of male reproductives

A conflict of interest should exist between queens and workers with regard to who produces male reproductives. Since a queen is related to sons by 1/2 and to grandsons by 1/4, her fitness will be highest if she can produce all the males for the colony. However, since workers are related to brothers by 1/4 and to sons by 1/2, their fitness will be highest if they can produce all the male reproductives instead of the queen. According to the indirect (kin) selection hypothesis, workers apportion reproductive effort according to their own best interests, and hence workers should produce male reproductives. According to the parental manipulation hypothesis, queens control the behavior of workers, and hence workers should rear the queen's progeny instead of producing male progeny of their own.

In most social Hymenoptera workers are capable of laying unfertilized eggs that develop into males. However, the extent of laying by workers varies greatly. Workers produce most of the males in primitively eusocial bumblebees *(Bombus atratus)* (Zucchi 1966) and stingless bees *(Lasioglossum malachurum)* (Noll 1931), certain highly eusocial stingless bees *(Trigona postica* and others) (Beig 1972; Lin and Michener 1972), and possibly *Vespula* wasps (Montagner 1966; but see also Spradbery 1971). Laying by workers is widespread in both primitive and advanced ants (Freeland 1958; Passera 1965; Dejean and Passera 1974; Brian 1979). In the highly eusocial ant *Myrmica rubra* workers lay most of the eggs that develop into males (Brian 1968). Workers usually do not produce any males in honeybees and many other highly eusocial bees and ants, except in queenless colonies (Wilson 1971; Michener 1974). In these species workers begin laying unfertilized eggs shortly after the queen dies. Workers are not known to be the exclusive parents of males in any eusocial species (Oster and Wilson 1978).

At least three hypotheses can explain why both queens and workers produce males in some colonies (Oster and Wilson 1978). First, the cost of laying eggs may reduce the foraging efficiency of workers and hence the ability to rear female reproductives. Then workers would be most fit by producing some male reproductives themselves and also rearing males produced by the queen. Second, the ability of queens to prevent workers from laying unfertilized eggs may decrease as the colony becomes larger or as they approach senescence. Present evidence is not consistent with the second hypothesis (see Brian 1979), but more data are needed. Third, workers may altruistically rear some queen-produced males to increase average fitness of the colony in response to trait-group selection. In annually reproducing colonies trait-group selection could lead to a stable equilibrium between altruistic nonlaying workers and nonaltruistic laying workers provided that laying workers reduce average colony fitness by becoming less efficient foragers (see Mirmirani and Oster 1978).

The failure of workers to produce male progeny in many eusocial species lends support to the parental manipulation hypothesis for those species, as workers who rear brothers instead of their own progeny are apparently not behaving in their own best interests. However, workers may not produce sons because detrimental effects on foraging ability are too costly. In addition, workers may be unable to rear sons without help from other workers, and recruiting help might be difficult. If workers can distinguish males produced by the queen from males produced by other workers, they should help rear brothers and not nephews, since they are related to brothers by 1/4 and to nephews by only 1/8. Whether workers can make this discrimination is unknown.

Investment ratios

Predictions. In diploid organisms parents should invest equally in each sex of offspring, as was discussed in Chapter 10. One of the reasons for this is that parents are equally related to sons and daughters. In haplodiploid organisms the expected investment patterns are somewhat different (Trivers and Hare 1976; Oster et al. 1977; Charnov 1978; Craig 1980). Queens are still equally related to sons and daughters and should invest equally in each, but workers are three times more closely related to sisters than to brothers. Workers should therefore invest three times as much care in sisters as in brothers. The worker investment ratio is an important issue, because if workers invest equally in sisters and brothers, their average fitness gain would equal that obtained by producing direct progeny. Then workers would not benefit from rearing the queen's progeny.

According to the indirect (kin) selection hypothesis, workers should control investment ratios and invest three times more effort in female reproductives than in male reproductives. According to the parental manipulation hypothesis, queens should control worker investment ratios, and workers should invest equally in female and male reproductives. This difference therefore offers a good method to distinguish between the two hypotheses.

In computing investment ratios only the care invested in reproductive progeny should be compared. The theory of sex ratios and parental investment ratios is based on the relative reproductive value of male and female progeny. When workers of highly eusocial species ordinarily do not reproduce, they do not enter the breeding population and must be disregarded when computing investment ratios. When workers produce many of the male progeny, the situation is more complicated. The investment ratios predicted by each theory change, and the workers must be included in computations of investment ratios based on the extent to which they contribute direct progeny to future generations.

Evidence. For species without worker production of males, the investment ratio can be estimated from the cost of producing male and female reproductives multiplied by the sex ratio of reproductives (Trivers and Hare 1976). Since the main nonshareable investment in reproductives is food, the cost of producing male and female reproductives can be estimated from dry weights. Other forms of parental investment (e.g., colony maintenance and defense) are shareable and hence can be disregarded. Thus the investment ratio can be estimated from the dry-weight ratio of reproductives multiplied by their sex ratio.

Using this approach, Trivers and Hare (1976) calculated investment ratios for 21 monogynous ant species, in which colonies contain just one queen. Queens are known to produce most or all males in 3 of these species and were assumed to produce most males in the remaining 18 species. Although the estimated investment ratios vary considerably among the 21 species (range = 1.57 to 8.88), the average ratio is 3.90. A plot of weight ratio against sex ratio gives a regression coefficient of .33 (Figure 12–4), which is very similar to that expected for an investment ratio of 3.00. A potential error in the method used is that the caloric cost per dry weight may differ for male and female reproductives. In a detailed study of *Tetramorium caespitum*, for example, females cost only three-fourths as much per unit dry weight as males (Peakin 1972). When this correction is applied to the results, the average investment ratio is 2.92. Thus the estimated ratio is very close to that predicted by the indirect (kin) selection hypothesis and differs considerably from that predicted by the parental manipulation hypothesis.

For comparison, the investment ratio in slave-making ants should approximate 1.0 because the queen's brood is reared by workers of other species who were stolen from their natal nests before they became adults

Figure 12–4

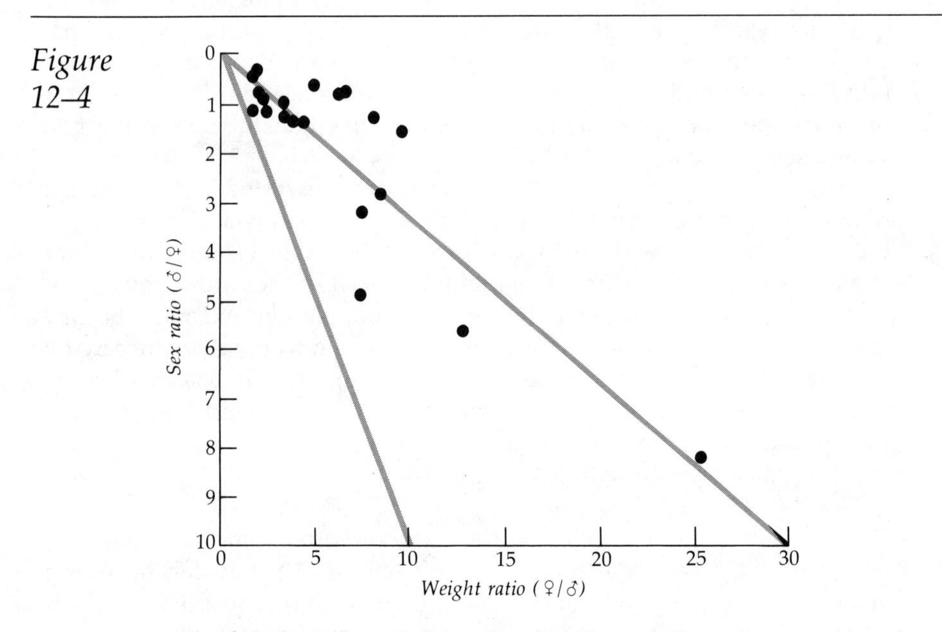

The female:male weight ratio of various ant species as a function of the male:female sex ratio.

According to the indirect (kin) selection hypothesis, the regression line should have a slope of .33. This prediction is indicated by the right-most line shown on the graph. According to the parental manipulation hypothesis, the regression line should have a slope of 1.0. That prediction is indicated by the left-most line shown on the graph. The data for 21 monogynous ant species (solid points) most closely fit the line with slope of .33, as predicted by the indirect (kin) selection hypothesis.

Source: From R. L. Trivers and H. Hare, "Haplodiploidy and the Evolution of the Social Insects," *Science* 191 (23 January 1976):255. Copyright 1976 by the American Association for the Advancement of Science. Reprinted by permission.

(Trivers and Hare 1976). Slaves have no stake in the queen's progeny, and hence the queen should control the investment ratio. Data are available for two slave-making species, and their respective investment ratios are 1.25 and .87.

The investment ratio of solitary bees and wasps should also approximate 1.00. Adult females of those species lay a single fertilized or unfertilized egg in each nest cell, which they provision themselves. There is no worker caste, and the laying female should control the investment ratio. For those species the caloric cost of male and female progeny can be estimated from either dry weight of mature adults or the volume of their respective nest cells (Trivers and Hare 1976). Nest cell volume is a measure of cost because nest cells are ordinarily filled completely by the provisions left for developing larvae. The estimated investment ratios average 1.92 for 20 species of wasps and 2.11 for 8 species of bees based on dry-weight ratios. They average 1.39 for wasps and 1.28 for bees based on cell volume ratios. If the caloric cost per unit of dry weight is higher for males than for females, as in ants, the estimates based on dry-weight ratios would approach more closely those based on cell volume ratios.

Few data are available for calculating investment ratios in eusocial bees and wasps. Moreover, the expected ratios are less clear because greater investment in males should occur when workers produce some of the male reproductives themselves. Trivers and Hare (1976) estimated the investment ratios for five eusocial species of bumblebees and found an average ratio of 2.08. They interpreted this lower ratio as a reflection of worker production of males. Data are also available for the paper wasp *Polistes fuscata* (Noonan 1978). The predicted investment ratio according to the indirect (kin) selection hypothesis is 3.00 because workers do not produce any of the males. The average investment ratio computed for 17 colonies is .99, which does not support the indirect (kin) selection hypothesis.

Since termites are diploid, the expected investment ratio is 1.00 for both queens and workers. This is because queens are equally related to sons and daughters and any bias in investment ratios of one worker sex should be offset by an opposite bias in investment ratios of the other sex (Lacy 1980). [While the average ratio for workers should be 1.00, the ratio for each sex of worker should be strongly skewed toward greater investment in reproductives of the same sex (see pp. 528–529).] Few data are available for calculating investment ratios in termites. The mean investment ratio for nine species is 1.91, which is "significantly closer to 1:1 than are the ratios for monogynous ants" (Trivers and Hare 1976). Nevertheless, the ratio is also considerably higher than the predicted ratio of 1.00.

Problems with interpreting the evidence. At face value the data on investment ratios mostly support the indirect (kin) selection hypothesis and contradict the parental manipulation hypothesis, with the exception of Noonan's (1978) study of paper wasps. Workers of eusocial ants invest about three times as much effort in female reproductives as they do in males, as predicted by the indirect (kin) selection hypothesis. Investment ratios are higher than predicted by either hypothesis in solitary bees and wasps, but they are closer to 1.00 than to 3.00. In eusocial bumblebees investment ratios are less than the predicted value of 3.00, possibly because some males are produced by workers instead of the queen, but they are higher than 1.00, the ratio predicted by the parental manipulation hypothesis. Unfortunately, several problems arise when interpreting the evidence.

One problem concerns the investment ratio predicted by the indirect (kin) selection hypothesis (Alexander and Sherman 1977). The predicted value of 3.00 assumes that all the queen's eggs are fertilized by a single male and all male reproductives are produced by the queen. When queens are inseminated by two males, each of whom fertilized half the eggs, the predicted investment ratio becomes 2.00 instead of 3.00 because workers are then related to female reproductives by 1/2 and to brothers by 1/4. If one male predominates in fertilizing eggs, the investment ratio will lie between 2.00 and 3.00, but it will always be less than 3.00.

When some or all males develop from eggs laid by workers, the average relatedness of workers to male reproductives becomes complicated. Workers are related to their sons by 1/2, to their brothers by 1/4, and to their nephews by 1/8. The average relatedness of workers to male reproductives will therefore vary, depending on the relative proportion of males who are sons, brothers, or nephews. In general, laying by workers will reduce relatedness between workers and male reproductives. If all males are produced by workers, with every worker contributing equally, most male reproductives would be related to each worker as nephews rather than sons. In this extreme case, relatedness of workers to male reproductives would be close to 1/8. When some males are produced by workers and some are produced by the queen, mean relatedness would lie somewhere between 1/8 and 1/4. Thus laying by workers should *increase* worker investment ratios unless workers can discriminate among sons, brothers, and nephews, in which case it would decrease investment ratios.

In the eusocial ants analyzed by Trivers and Hare (1976), worker production of males can probably be discounted. However, since multiple insemination is the rule in ants, the predicted ratio of 3.00 may be too high. This is not a serious problem because the results were certainly closer to the ratio predicted by the indirect (kin) selection hypothesis than to that predicted by the parental manipulation hypothesis. Problems do arise for eusocial bumblebees. Since workers produce many of the male reproductives, the investment ratio should be higher than 3.00. Nevertheless the observed ratio was only 2.08.

A second problem is that local mate competition among related males affects adult sex ratios and hence parental investment ratios (see pp. 407–410) (Hamilton 1967). Local mate competition is often associated with high inbreeding in a population, and at least some evidence indicates that inbreeding is widespread in eusocial Hymenoptera (Lin and Michener 1972; Alexander and Sherman 1977; Cowan 1979). [Note that high outbreeding does not negate the importance of local mate competition (contra Craig 1980) because local mate competition depends on the size of local breeding groups and not on the degree of inbreeding per se (see pp. 410–411).]

Alexander and Sherman (1977) argue that the female bias in worker investment ratios among ants results from local mate competition and not the higher relatedness of workers to female reproductives. To evaluate this possibility, they reanalyzed the data presented by Trivers and Hare (1976) for nonsocial Hymenoptera. Local mate competition is likely to be high when individuals nest solitarily but mature together, and it is likely to be low in group-nesting species or solitary-nesting species that lay their eggs singly. Four species in the first category averaged an investment ratio of 1.81, while 12 species in the second category averaged an investment ratio of .99. Hence the data suggest that local mate competition does bias investment ratios toward higher production of females. In addition, the investment ratios for eusocial ants are highly variable and are as high as 8.88 in one species. Similar variability also exists in the data for termites. Such variability would not be expected by the indirect (kin) selection hypothesis, since workers should invest in siblings in a ratio approximating 3.00 in every species, but it is readily explained by variability in degree of local mate competition. Trivers and Hare (1976) suggest that the higher-than-expected investment ratios found in some species represent sampling errors resulting from small sample sizes. Another possibility is that high investment ratios arose in some species because most males are derived from eggs laid by workers.

Another potentially serious problem is that the logic underlying the predictions about investment ratios may be erroneous. Herbers (1979) presents a game theory model suggesting that a stable, optimum investment ratio may not exist for eusocial Hymenoptera. According to the model, local optima may exist under specific conditions, but no general optimum that will apply across all species can be predicted from sex ratio theory. The extreme variability of investment ratios observed both within and between species is consistent with this model. If Herbers's model is correct, the rationale underlying the analyses of Trivers and Hare (1976) is erroneous, and data on investment ratios have no value in testing the two hypotheses for explaining sterile castes.

A general model of sex ratio and investment ratio developed by MacNair (1978) also raises doubts about the predicted investment ratio for workers. The 3.00 prediction is valid if workers control the primary sex ratio (i.e., male:female ratio among eggs), as shown by Oster and co-workers (1977), but it is no longer valid if queens control the primary sex ratio. If the queen enforces a 1:1 primary sex ratio, as she should according to sex ratio theory, the optimal investment ratio for workers is not necessarily 3.00 or any other particular value. However, it remains possible that workers control the primary sex ratio by manipulating the proportion of female larvae that develop into virgin queens. If so, the primary sex ratio should be 1:3 and the predicted investment ratio of 3.00 would be valid.

There are also statistical and procedural problems with the way Trivers and Hare (1976) analyzed their data (Alexander and Sherman 1977; MacNair 1978). For example, the slope of .33 shown in Figure 12–4 is due to data from only 6 of the 21 species studied, and the investment ratios for 13 of the species are near 1.0 (MacNair 1978). Hence it is possible that some species show one investment pattern, while other species show a different investment pattern.

In view of the above arguments, the data on investment ratios do not provide conclusive evidence for testing the indirect (kin) selection and parental manipulation hypotheses.

Queen control of caste differentiation: Evidence for parental manipulation

The central assumption of the parental manipulation hypothesis is that queens of eusocial insects control the activities of their progeny and cause them to develop into workers. Selection for such a control mechanism is not readily explainable by the indirect (kin) selection hypothesis because workers are supposed to be acting in their own best interests when they rear siblings rather than direct progeny. If the indirect (kin) selection hypothesis is valid, queen control of workers would seem superfluous.

Social bees and wasps

An elaborate chemical control system is well documented in honeybees (Wilson 1971; Michener 1974). Larvae develop into new queens when they are reared in larger-than-normal wax cells and are given a special secretion called royal jelly, which is produced by the salivary glands of workers. Deposition of royal jelly is normally inhibited by a chemical pheromone, trans-9 keto-2 decenoic acid, which is manufactured in the queen's mandibular glands and circulated constantly through the colony (Barbier and Lederer 1960; Butler et al. 1961). Not only does this acid inhibit production of new queens, but it also acts in conjunction with at least one other pheromone to inhibit ovarian development in workers.

Queens must dispense about .1 mg of 9-ketodecenoic acid per worker each day to maintain control over the colony because workers continuously metabolize it (Butler and Paton 1962; Wilson 1971). Such a high rate of production can be maintained because workers pass the inactive metabolites back to the queen, who then converts the metabolites back into 9-ketodecenoic acid at a great savings in energy (Johnston et al. 1965). When the queen dies, the inhibitory effect of the pheromone is lost within hours, and workers begin producing new queens by feeding royal jelly to developing larvae. In addition, they begin laying unfertilized eggs to produce males.

In stingless meliponine bees caste differentiation is at least partly determined by the amount of food fed to larvae (Kerr et al. 1966). All queens develop from pupae that weigh more than 72 g, though not all heavy pupae develop into queens (possibly because caste differentiation is also influenced genetically). Females with five ganglia always develop into workers, but females with four ganglia can develop into either workers or virgin queens, depending on how much food they receive as larvae. The segregation ratio of ganglion numbers averages three females with five ganglia to each female with four, a pattern that can be explained by a simple two-locus genetic model (Kerr 1950a, b). However, this ratio is not immutable. Some colonies of *Melipona quadrifasciata* produce a much higher proportion of females with five ganglia, suggesting that an additional gene locus may be involved (Kerr and Nielsen 1966; Kerr 1969). Alternatively, an environmental factor may determine the segregation ratio instead of genetic factors. The proportion of female reproductives can be increased to as high as 50% of all females by overfeeding larvae, which implies that the 3:1 ratio is not a result of simple Mendelian segregation (Darchen and Delage-Darchen 1974a, b, 1975). The results obtained by Darchen and Delage-Darchen have been criticized by Velthius (1976), but they suggest that environmental rather than genetic factors control caste determination in *Melipona* (see Brian 1979).

In other stingless bees two sizes of brood cells are constructed by workers (Figure 12–5) (Michener 1974). Larvae in large cells develop into virgin queens, while those in small cells develop into workers or males. The factor determining whether a queen develops is the quantity of food placed in the brood cell. Darchen and Delage (1970; Darchen and Delage-Darchen 1971) were able to induce *Trigona* larvae in worker cells to develop into queens merely by feeding them additional food from other worker cells. Genetic control over caste differentiation is apparently not an important factor. Queens evidently inhibit production of large brood cells and, in many species, laying of unfertilized eggs by workers, since death of the queen generally triggers production of new queens and laying by workers (Michener 1974). The mechanisms underlying the queen's control of worker behavior are unknown.

In primitively eusocial bees and wasps queens commonly control workers by physical domination. Bumblebee queens attack workers who attempt to eat their eggs, often severely injuring or killing them (Wilson 1971). After attempts to steal eggs have been rebuffed for a few hours in a new colony, workers begin caring for the queen's eggs instead of trying to eat them. Workers may also be inhibited chemically from producing virgin queens. The queen's presence normally inhibits workers from producing virgin queens, and in *Bombus terrestris* this inhibitory effect vanishes when workers are prevented from touching the queen (Röseler 1970). Röseler postulated that the queen deposits a nonvolatile pheromone on workers that inhibits them from producing virgin queens. Similarly, the queen's presence inhibits ovarian development in workers of the primitive halictine bee *Lasioglossum zephrum* (Michener and Brothers 1974; see also Breed et al. 1978). Queens nudge workers, though not in an overtly aggressive way. This behavior may be at least partially responsible for the inhibitory effect since the frequency of nudging by the queen is correlated

Figure 12–5

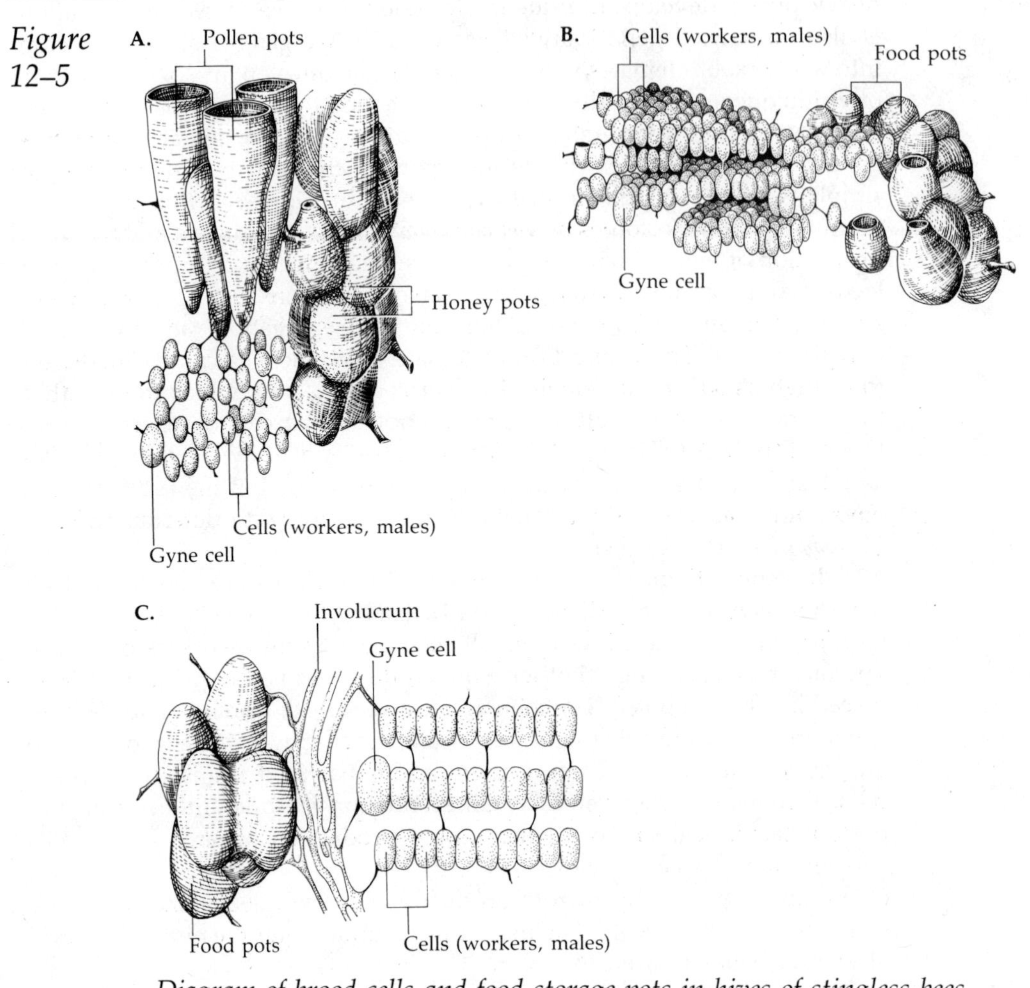

Diagram of brood cells and food storage pots in hives of stingless bees.

A. The cluster arrangement of Trigona flavicornis, *showing small brood cells that give rise to workers and males, a larger brood cell that gives rise to a virgin queen or gyne, and the two types of storage pots. B. The irregular comb arrangement of* T. ghilianii, *showing small brood cells, a single large gyne cell, and the one type of storage pot. C. The regular comb arrangement of most* Trigona *species, showing small brood cells, a single large gyne cell, the single type of storage pot, and an involucrum or sheath which surrounds the brood cells.*

Source: From C. D. Michener, *The Social Behavior of Bees* (Cambridge, Mass.: Belknap Press of Harvard University Press, 1974), p. 340. (Original drawings by J. M. F. de Camargo.) Reprinted by permission.

with the size of worker ovaries. Chemical control may also be important, but the role of chemical communication systems has not been studied.

In several species of *Evylaeus*, another group of halictine bees, aggression of the queen completely inhibits ovarian development of workers if colonies are small and partially inhibits ovarian development of workers if colonies are large (Knerer 1980).

In newly founded semisocial colonies of paper wasps *(Polistes),* aggressive interactions among founding females determine who is queen (West Eberhard 1969). A dominance order is established among the females, and only the most dominant females lay eggs (Table 12–2). Dominant females maintain their superior status in two ways. They receive more food than workers during periods of food scarcity, which allows them to remain larger in size, and they eat eggs laid by workers. Chemical control has not been demonstrated in paper wasps, but it has in vespine wasps. Workers of the oriental hornet are strongly attracted to the queen, frequently crowding around her and licking her body (Ishay et al. 1965). When a queen is removed, workers quickly become quarrelsome and no longer care for broods. This change in worker behavior can be prevented by allowing workers to lick an alcohol extract obtained from the queen, suggesting that a pheromone produced by the queen controls worker behavior. Similarly, removing the queen from a colony of *Paravespula germanica* causes workers to wander about, behave pugnaciously toward each other, and neglect or even eat larvae (Spradbery 1973; Ishay 1975). One pheromone, δ-η-hexadecalactone, has been identified in oriental hornets (Ikan et al. 1969). This substance induces queenless groups of workers to build queen cells in late summer, a behavior that normally occurs only in the presence of a queen.

Ants

Queen control over workers involves at least three forms of inhibition in ants (Wilson 1971). The presence of a queen can inhibit ovarian development in workers, affect the percentage of daughters that develop into queens, and suppress mating behavior of virgin queens and workers near the nest. Each type of inhibition has been documented in some ant genera, but the extent to which they occur among ants in general is unknown.

Ovarian development in workers is inhibited by the queen in *Leptothorax* (Bier 1954), *Formica* (Bier 1956), *Plagiolepis* (Passera 1965), and *Myrmica* (Mamsch and Bier 1966). These four genera are representative of three widely divergent subfamilies of ants, suggesting that inhibition of ovarian development in workers is widespread if not universal in ants (Wilson 1971). Worker ants do lay eggs in the presence of queens, but many such eggs are specialized trophic eggs incapable of development and used solely as a protein source for queens and larvae. Passera (1965) has shown, for example, that workers of *Plagiolepis pygmaea* lay both normal and trophic eggs when isolated from the queen but lay only trophic eggs when the queen is present.

Two physiological control mechanisms have been suggested to explain how queens inhibit ovarian development in workers. One postulated mechanism is that queens secrete a nonvolatile inhibitory pheromone that workers obtain by licking or touching the queen. Carr (1962) found that larval growth and hence chances of larvae developing into queens is sup-

Table 12–2 *Reproductive success and division of labor according to dominance rank within a group of colony-founding female paper wasps* (Polistes fuscatus).

Identification number	Dominance rank	Number of eggs laid	Number of eggs of others eaten	Number of new cells started	Foraging rate (loads per hour observed)	Number of loads received from associates
13	1	9	4	0	.08	25
34	2	5	2	1	.50	20
35	3	0	0	1	1.41	0
28	4	0	0	2	1.56	5
15	5	0	0	8	1.80	3
6	6	0	0	0	1.22	1
18	7	0	0	0	1.50	0

Note: Data taken between 18 May and 14 June 1965, based on 26 hours of observation.

Source: From M. J. West Eberhard, "The Social Biology of Polistine Wasps, *Miscellaneous Publications* 140 (1969), University of Michigan, Museum of Zoology. Reprinted by permission.

pressed in *Myrmica* if a dead queen is periodically placed in the colony. The effect does not occur if the dead queen is separated from the workers by a double-gauze barrier that prevents physical contact with the dead queen. Brian and Hibble (1963) then showed that larval growth can be inhibited by painting ethanol extracts of queen heads onto larvae or workers. The active portion is in the sterol fraction of those extracts (Brian and Blum 1969).

The second postulated inhibitory mechanism involves a shortage of "profertile substances" which are produced in the glands of workers (Bier 1958). These substances are believed to be necessary for full ovarian development. When the queen is present, she obtains most of the profertile substances produced in the colony, leaving too little for any workers or larvae to develop into queens. Indirect evidence for the existence of this mechanism was obtained by Mamsch (1967), who found that queens of *Myrmica ruginodis* are less able to suppress ovarian development in workers when no larvae are present.

Queens have been shown to be capable of reducing the proportion of larvae that develop into virgin queens in *Monomorium* (Peacock et al. 1954), *Myrmica* (Brian and Carr 1960; Carr 1962; Brian 1970, 1973), *Plagiolepis* (Passera 1965, 1969 a, b, 1974), *Aphaenogaster* (Ledoux and Dargagnan 1973), *Formica* (Plateaux 1971), *Crematogaster* (Delage-Darchen 1974), *Temnothorax* (Dejean and Passera 1974), and *Odontomachus* (Colombel 1978). The ability to suppress mating by virgin queens and workers near the colony has been shown only in *Formica* (Bier 1956). Thus queens clearly have some control over the primary sex ratio and worker behavior.

Termites

Caste differentiation in lower termites is controlled by pheromones produced by the queen and king. Termite eggs hatch into first-instar nymphs, which then molt five to seven times before becoming pseudergates (Lüscher 1953). Pseudergates can molt repeatedly without further differentiation, but at any molt they can develop into supplementary or primary reproductives or, through an additional instar, into the major worker caste. Two hormones, ecdysone and juvenile hormone, control caste differentiation and growth. Ecdysone is the general molting and differentiation hormone of insects. High ecdysone titers in pseudergates promote transformation into supplementary or primary reproductives (Lüscher 1969; Miller 1969). Low juvenile hormone titers inhibit caste differentiation. If ecdysone titers are high but juvenile hormone titers are low, pseudergates grow larger and molt repeatedly but do not differentiate into a reproductive caste. When juvenile hormone titers are high, pseudergates transform into major workers.

Pheromones produced by the queen and king apparently control ecdysone and juvenile hormone production in pseudergates, though the precise mechanisms are unknown. One pheromone, produced by the queen, prevents female pseudergates from developing into reproductives (Figure 12–6) (Lüscher 1961). Another pheromone, produced by the king, has the same effect on male pseudergates. A second pheromone produced by the king stimulates female pseudergates to develop into reproductives. Queens do not produce a pheromone with comparable effect on male pseudergates. Two additional pheromones allow supernumerary kings and queens to recognize other supernumeraries of the same sex and cause them to fight.

Developmental stages are somewhat different in higher termites (Termitidae) (Wilson 1971; Brian 1979). The process is exemplified by *Amitermes hastatus,* one of the more primitive species of higher termites (Skaife 1954). Individuals of both sexes pass through two undifferentiated instars and then enter one of five caste forms. New primary reproductives are produced only in flourishing colonies older than ten years of age. Secondary reproductives are produced in about 20% of the nests, even though the king and queen are present. They assist workers in foraging and serve as an instant reserve of reproductives capable of taking over from the primary reproductives. They do not normally reproduce while the primary reproductives are present. Tertiary reproductives are rare in *Amitermes,* and their significance is unknown. In more advanced higher termites the developmental process is similar but a trend toward sexual division of labor and increased morphological specialization of the worker caste is present (Noirot 1969, 1974).

The control of caste differentiation among higher termites is presently unknown. Removal of primary reproductives is known to stimulate production of replacement reproductives in many species, indicating that the primary reproductives have an inhibitory influence on developing instars

Figure 12–6

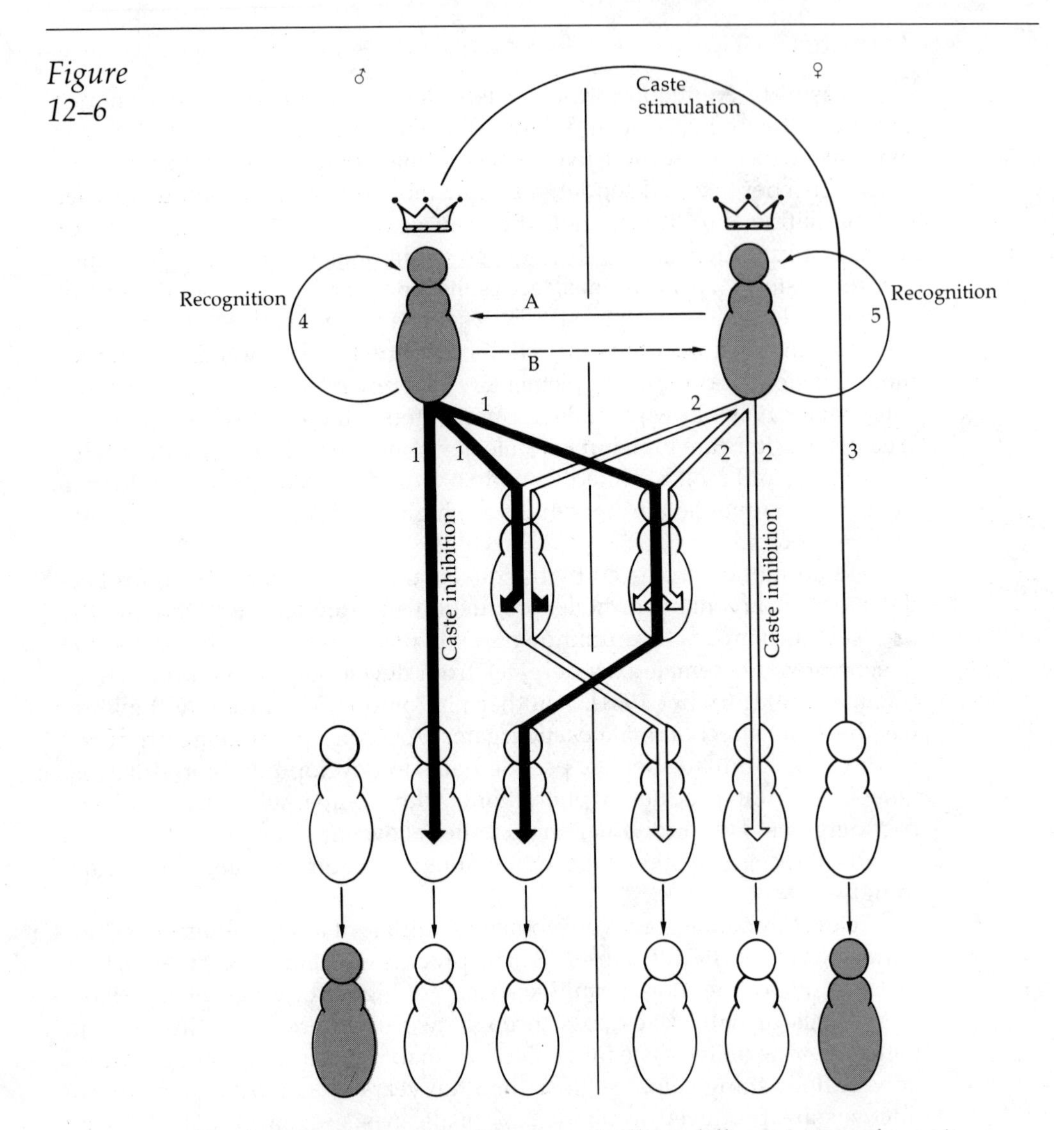

*The known chemical pathways controlling differentiation of termites (*Kalotermes flavicollis*) into reproductive castes.*

The functional king is the "crowned" figure to the upper left. The functional queen is the "crowned" figure to the upper right. The remaining figures all represent pseudergates. The king and queen produce chemical substances (labeled 1 and 2, respectively) that inhibit differentiation of same-sexed pseudergates into the royal caste. These substances are passed directly to the pseudergates and are also circulated through the digestive tracts of the pseudergates. Another male substance (labeled 3) stimulates differentiation of female pseudergates into functional queens. Supernumerary kings recognize each other by means of substance 4 and fight. Supernumerary queens recognize each other by means of substance 5 and fight. Finally, kings stimulate production of substance 2 by queens, and queens stimulate production of substance 1 by kings. The nature of the stimulus (labeled A and B, respectively) affecting each member of the royal caste is unknown.

(Wilson 1971; Brian 1979). However, little is known about the nature of this influence.

To summarize, present evidence shows that primary reproductives generally possess chemical control over gonadal development, caste differentiation, and behavior of workers in all groups of eusocial insects. Such control mechanisms have been thoroughly studied in relatively few species, but it seems probable that they will prove widespread or universal among eusocial and perhaps also semisocial and subsocial species.

Genetic models

Two genetic models have been devised to compare the parental manipulation and indirect (kin) selection hypotheses (Charlesworth 1978; Craig 1979). The mathematics involved are beyond the scope of the present discussion, but the results prove interesting. Both models compute a threshold value of k, defined as the ratio of beneficiary's gain to altruist's loss necessary for sterile castes to evolve according to each hypothesis. For eusocial insects beneficiary's gain is the number of the queen's offspring reared by workers, and altruist's loss is the number of worker offspring foregone as a result of rearing the queen's brood. The outcomes of both models predict that the threshold value of k is only half as high for the parental manipulation hypothesis as it is for the indirect (kin) selection hypothesis. The models are oversimplified representations of reality, but the results nevertheless suggest that worker altruism is more likely to evolve by parental manipulation than by indirect (kin) selection.

Conclusion

Two principal hypotheses have been advanced to explain sterile castes and worker altruism among eusocial insects. The indirect (kin) selection hypothesis proposes that asymmetrical coefficients of relatedness, generated primarily by haplodiploidy, allow workers to propagate genes faster through same-sexed siblings than through direct progeny. The parental manipulation hypothesis proposes that reproductive females (and males in termites) force their offspring to help them rear reproductive progeny. Many lines of evidence have been adduced to evaluate each hypothesis, but none are conclusive.

The main lines of evidence supporting the indirect (kin) selection hypothesis are the close association between haplodiploidy and eusociality, sex-linked worker altruism in all eusocial Hymenoptera, and higher worker

---

Source: From E. O. Wilson, *The Insect Societies* (Cambridge, Mass.: Belknap Press of Harvard University Press, 1971), p. 191. (Modified from Lüscher 1961). Reprinted by permission.

investment in reproductive females than in reproductive males. Each line of evidence is open to alternative interpretations or criticisms regarding its validity. Hence none of the evidence is conclusive.

Two lines of evidence contradict the indirect (kin) selection hypothesis, but neither is decisive enough to reject the hypothesis. First, the prevalence of multiple inseminations of the queen and the presence of multiple queens in some colonies reduce mean relatedness between workers and larvae reared by workers. However, workers may still be sufficiently related to larvae for indirect (kin) selection to work if most larvae are fathered by a single male and produced by a single queen. Second, the apparent origin of eusociality from semisocial species implies that indirect (kin) selection alone cannot explain eusociality. Even if founding females of semisocial colonies are relatives, additional factors must be postulated. Until more data on the actual relatedness between workers or auxiliary females and larvae become available, such evidence will remain inconclusive.

The main lines of evidence supporting the parental manipulation hypothesis are queen control over caste determination and worker brood behavior and the outcomes of two genetic models of the evolutionary process. The strongest evidence is the widespread or universal ability of queens to control caste determination, influence the primary sex ratio by controlling production of virgin queens, and suppress the ability of workers to produce male reproductives. Queens vary in their ability to suppress worker production of males, but this may result from limitations on the queen's ability to maintain control in large colonies or during senescence. The evidence implies that queens do manipulate offspring to further their own genetic interests, and the existence of queen control is not readily explained by the indirect (kin) selection hypothesis. Control over the primary sex ratio and suppression of worker production of males could have evolved secondarily once indirect (kin) selection led to eusociality, since worker interests conflict with those of the queen, but if workers benefit by rearing daughters of the queen, the origin of queen control over caste determination is not so easily explained. The evidence from genetic models is suggestive, as it implies that parental manipulation can lead to eusociality more easily than indirect (kin) selection can, but it does not show which process actually did lead to eusociality.

The weight of present evidence supports parental manipulation as the most likely explanation for the origin of eusociality, but the possible importance of indirect (kin) selection cannot be discounted. Asymmetrical coefficients of relatedness provide an a priori reason for expecting indirect (kin) selection to be important in shaping worker behavior, but its exact role requires further analysis. Quite possibly, parental manipulation and indirect (kin) selection interact in a complex manner to generate the present social organization of eusocial insects.

Mammalian sociality

Vertebrate social groups are organized quite differently from eusocial insect societies. The most profound differences stem from the higher proportion of reproductive individuals and the absence of sterile castes in vertebrate groups. All mature members of most vertebrate groups are actual or potential reproductives. Though group members are often related, average relatedness is lower and individual interests are more in conflict. While cooperation is a common thread running through both insect and vertebrate societies, the elements of competition and strife are generally much more evident within vertebrate societies.

Some of the simpler social groups occurring among vertebrates were discussed in Chapter 4. Many of these groups exhibit only rudimentary internal organization, often because they persist for relatively short periods of time. Such groups have generally evolved to enhance survival of group members, either because they provide better protection from predators or because they enable more efficient exploitation of resources. Mammalian social groups frequently exhibit more structured internal organization, adding a new level of complexity to the analysis, but they have evolved for basically the same reasons as simpler social groups. The major evolutionary questions involve the costs and benefits associated with forming social groups and the selective factors shaping the internal organization of those groups.

Costs and benefits of group living

Several costs of group living impede the evolution of sociality in mammals (see Alexander 1974). Competition for resources and mates is intensified whenever individuals join together as social groups. Diseases and parasites are transmitted more readily, and conspicuousness to predators or prey is often increased. Females may lose their prerogative of selecting mates, along with opportunities for acquiring male parental assistance. Subadults of both sexes may lose opportunities to mate because they are dominated by older group members. Not every cost is relevant to any given species, but grouping behavior always entails some of these costs.

The costs encountered by mammals that live in social groups parallel the costs discussed earlier for bird flocks, fish schools, and vertebrate colonies. This is more than just coincidence. Living in close proximity always

Chapter thirteen

intensifies competition, increases conspicuousness, and facilitates parasite transmission, no matter what species or type of social organization is involved.

Living in groups offers three general benefits, none of which are necessarily advantageous for any given species. Group members may be less vulnerable to predation, they may be better able to find or capture food, or they may be forced into groups by extreme localization of important resources such as safe sleeping sites or suitable breeding sites (Alexander 1974). These same benefits have been encountered previously, in discussions of cooperation and coloniality, and they remain important for social mammals. The precise benefits vary according to species. Much research has been devoted to clarifying these benefits and identifying the environmental conditions underlying them.

Social ungulates

Benefits of sociality. Social ungulates clearly gain protection from predators by living in social groups. Nearly all ungulates are vulnerable to predators, at least when young, and sociality provides several modes of defense not available to solitary individuals. As was discussed in Chapter 4, members of social groups are probably better able to detect stalking predators by depending on mutual vigilance, and characteristic alert postures or alarm signals provide the means for making mutual vigilance work. Sudden flight, stiff-legged bounding gaits (see Photo 3–2), and conspicuous rump patches or hind leg markings all communicate danger. Individuals may thwart attacks by bunching tightly together, exploding erratically in all directions, zigzagging through a herd, or even counterattacking. These defenses are all effective, depending on type of predator, group size, and environmental circumstances, but not all ungulates employ them to the same degree. Some species live as solitary individuals, relying on concealment or other defenses for protection, while others live in groups of various sizes. A particularly interesting problem is to understand why some species rely on social defenses while others do not.

Origin of variability among species. Individuals can adopt social defense tactics only if the costs of sociality are not too high. A major cost, and one that is likely to vary greatly among species, is intensity of competition for food. Thus one hypothesis for explaining why species adopt different defensive tactics is that variations in distribution and abundance of food lead to differing competitive costs.

Jarman (1974) has evaluated this hypothesis for African antelopes by relating their foraging ecology to group size. Antelopes eat basically two kinds of food: grasses and browse. Spatial distribution, seasonal availability, and nutrient content of grass and browse are quite different. Grass grows as continuous swards, while browse plants grow as more widely dispersed individual plants. Few grasses grow tall enough to escape grazing by ungulates, but many browse plants do. Most parts of grasses are edible, but many parts of browse plants are either inedible or very low in nutrient content. Grasses have higher fiber content and lower nutrient content than edible portions of browse plants, particularly during dry seasons when there is little new growth. Finally, grasses vary seasonally in abundance and grow mainly during relatively short wet seasons, while browse plants produce some new foliage all year. Thus if spatial distribution, seasonal availability, and nutritive quality of food resources are important, group sizes of social ungulates should vary with diet.

To test this prediction, Jarman (1974) classified African antelopes into five categories, based on principal foraging mode (i.e., grazing or browsing) and degree of dietary selectivity. He found that the species in each category differ with respect to typical group size and other attributes of social behavior (Table 13–1). These differences can be interpreted as consequences of the differing diets and habitat characteristics typical of species in each category.

Class A antelopes. Class A antelopes are small-sized browsers with highly selective diets (Photo 13–1). They eat primarily the most nutritive parts of a wide variety of browse plants. Their food appears scarce and rather evenly distributed. Food scarcity favors small body size to minimize daily food requirements and nongregarious behavior to minimize competition for food. The even distribution of browse resources makes them more defensible (see pp. 285–287), and most Class A antelopes are territorial. Territory defense is based primarily on scent-marking rather than overt visual or vocal displays, probably because overt displays would increase conspicuousness to predators. Many but not all Class A species rely on concealment to evade predators. Duikers (*Cephalophus, Sylvicapra*), steenbok (*Raphicerus campestris*), grysbok (*R. melanotis*), and suni (*Neotragus moschatus*) live as monogamous pairs in forest undergrowth and depend on concealment for protection, with members of pairs often remaining apart to enhance concealment (Heinichen 1972; Manson 1974; Dunbar and Dunbar 1979; Ronald L. Tilson, personal communication). Klipspringers and Kirk's dikdiks (*Madoqua kirki*) live as monogamous pairs in more open areas and depend on escape up rock escarpments to evade predators (Tinley 1969; Hendrichs and Hendrichs 1971; Dunbar and Dunbar 1974, 1980; Tilson 1979). Mutual vigilance is advantageous, but groups larger than mated pairs are apparently precluded by competition for food.

Class B antelopes. Class B antelopes forage exclusively on either grass or browse, and they are very selective of plant parts. Their foraging style does not differ greatly from Class A species, but they live in somewhat more open and seasonably variable habitats than most Class A species. Females commonly form small unstable social groups, and their home ranges often overlap several male territories (Verheyen 1955; Kiley-Worthington 1965; Leuthold 1970; Hendrichs 1972). However, southern reedbucks and oribis (*Ourebia ourebi*) form monogamous pairs in at least

Table 13–1 *Relationships among behavior, ecological variables, and body size for African ungulates.*

Category	Social organization	Habitat	Feeding style	Typical range in body size (kg)	Antipredator behavior	Examples
Class A	Solitary individuals or pairs, sometimes with offspring; group size 1–3; sedentary; territorial	Forest, bush, open rock escarpments	Diverse diet; selective of nutritious plant parts	3–20 (max 64)	Concealment, freeze or lie down when predator detected; flight up rock escarpments	Dikdiks *(Madoqua)*, duikers *(Cephalophus)*, klipspringer *(Oreotragus)*
Class B	Females form small groups, sometimes with offspring; group size 1–12, usually 3–6; sedentary; males territorial	Swamp, bush, grassland on hills or near water	Feed on either browse or grass (not both); selective of nutritious plant parts	20–100 (max 205)	Same as Class A	Reedbucks *(Redunca)*, lesser kudu, sitatunga *(Tragelaphus)*, gerenuk *(Lutocranius)*, oribi *(Ourebia)*
Class C	Females form larger herds; group size 3–100, usually 5–60; sedentary or migratory; breeding males territorial; others travel singly or in all-male groups	Woodland, bush, grassland near water, savanna, swamp	Feed on both grass and browse; moderately selective of plant parts	20–200 (max 320)	Diverse, depending on habitat; freeze to avoid detection and flee once detected; explode in all directions, stot, give alarm calls, counterattack if predator is small (usually mothers protecting young only)	Waterbucks, kob, puku *(Kobus)*, impala *(Aepyceros)*, gazelles *(Gazella)*, spring bok *(Antidorcas)*, greater kudu *(Tragelaphus)*

Class D	When sedentary, similar to Class C; group size 6–150; when migratory, aggregate into herds of several thousand	Grassland, dry savanna, open woodlands, open bush	Grass; somewhat selective of plant patches	100–200 (max 272)	Flee large predators; sometimes counterattack smaller predators	Hartebeests *(Alcelaphus)*, topi, tsessebe, blesbok, bontebok *(Damaliscus)*, wildebeests *(Connochaetes)*
Class E	Females, males, and offspring form large cohesive herds; group size 15–2000, usually hundreds; males form dominance hierarchies within female herds (at least in Cape buffalo); not territorial	Savanna, open woodland, dry bush, deserts	Grass and browse; unselective of parts eaten	80?–700 (max 900)	Often counterattack against even the largest predators	Cape buffalo *(Syncerus)*, probably elands *(Taurotragus)*, oryx and gemsbok *(Oryx)*

Source: Based on Jarman 1974.

some areas (Jungius 1971; Viljoen 1975). Greater food abundance, at least seasonally, probably facilitates formation of social groups. These species typically rely on concealment or freezing to avoid detection by predators, but mutual vigilance probably increases their ability to hide or freeze before a predator detects them.

Class C antelopes. Class C antelopes forage on both grass and browse but are still somewhat selective in choosing the most nutritious parts of plants. Their diets and habitat preferences are seasonally variable. They eat more grass during wet seasons and more browse during dry seasons. Habitat structure also varies seasonally. Class C species often occupy more open grassy woodlands during wet seasons and denser cover during dry seasons. In open habitats they depend on mutual vigilance and other social defenses to escape predation. In closed habitats they depend more on concealment. Greater abundance of food resources, especially during wet seasons, enables larger body sizes and larger group sizes. Social groups usually contain 6–60 individuals and may contain as many as several hundred (Photo 13–2). Group size varies seasonally and geographically, probably as a response to food availability and openness of habitat. Males usually defend mating territories, often in habitats where foraging condi-

Photo 13–1

A monogamous pair of steenbok.

The steenbok is a Class A African antelope that lives in open woodlands and on open plains where grass is tall. Note the small size of these antelopes.

Photo by Ron Tilson.

tions are especially favorable (see pp. 488–489). In two species, the Uganda kob and the blesbok, male territories coalesce into leks, where most mating occurs (Buechner and Schloeth 1965; Leuthold 1966; Lynch 1974). Nonterritorial males remain in female herds, presumably as a defense against predation.

Class D antelopes. Class D antelopes are primarily grazers. They are relatively unselective of plant parts, but they prefer fresh grass over old dried-out grass when they can get it. They live in open grasslands, where concealment is unfeasible, and form large herds to gain protection from predators. The large size of their herds is made possible by the abundance of grass. Their social organization is similar to that of Class C species, except herds are often larger and either nomadic or migratory (e.g., Estes 1966, 1969). Herds are constantly moving in search of new grass, and migration patterns are based on seasonal availability of food. Class D antelopes are typically larger in size than Class C species, probably as an adaptation for reducing vulnerability to small predators. Their large body size is made possible by abundant food, but it precludes selective foraging because large mouths take large bites and cannot weed out the less nutritious plant parts. Even if they could, nutritious plant parts are too scarce to supply the amount of food required by large animals.

Photo 13–2

A social group of greater kudu (Tragelaphus strepsiceros).

The kudu is a Class C African antelope of open woodlands and bush. Females form small social groups which visit male territories during the mating season.

Photo by Ron Tilson.

Class E antelopes. Class E antelopes eat both grass and browse, and they are very unselective of plant parts. The unselective nature of their diets makes food relatively abundant, which in turn makes large body size and large social groups possible. Class E species form large, sedentary herds in open savanna. Herds are closed social groups with stable membership (e.g., Sinclair 1977). Males establish dominance hierarchies within herds instead of defending territories, apparently because herds are highly cohesive (see p. 490). Class E species are very large, since food is abundant, and they depend on their size and strength to actively ward off predatory attacks.

Overview. Any classification scheme as simple as the one above greatly oversimplifies the diversity of social behaviors involved. It lumps together species with quite different feeding specializations and glosses over many behavioral differences present among species within each category (Leuthold 1977). Nevertheless, Jarman's (1974) analysis reveals the general relationships existing between foraging ecology, habitat structure, group size, and body size. It makes clear that sociality in ungulates is related to habitat structure as well as food distribution and abundance, a point made by other authors as well (Eisenberg 1966; Eisenberg and McKay 1974; Estes 1974; Geist 1974b). Habitat structure affects the abundance of food resources available to ungulates, the amount of cover available for concealment, and the range of visibility for detecting predators. These factors all covary, with food usually being least abundant where concealment is most effective and most abundant where concealment is least effective. Thus the combined influences of food availability and antipredator tactics make social grouping behavior most prevalent in open habitats and least prevalent in closed habitats.

Social primates

Benefits of sociality. In comparison with ungulates, the observed frequency of predation on primates is small. This has sometimes been construed as evidence that predation is not an important selective factor favoring sociality. However, predation on primates may be more difficult to observe, or it may be rare because sociality is an effective defense against predation.

The number of predators capable of attacking primates are numerous. Lions, leopards, eagles, and jackals are known to attack and kill baboons (Photo 13–3) (Kruuk and Turner 1967; Altmann and Altmann 1970; Schaller 1972), and hyenas, African wild dogs, and cheetahs are at least potential predators of baboons. At least 16 different species, including eagles, poisonous snakes, and carnivorous mammals, are potential or proven predators of the vervet monkey (Struhsaker 1967b). Tigers commonly hunt the semiterrestrial Hanuman langur of India (Schaller 1967). Arboreal primates fall prey to leopards, hawks, eagles, pythons, and boa constrictors, and they are the main diet of some large eagles (Haddow 1952; Brown and

Photo
13–3

A lion trees a group of savanna baboons.
Lions are one of many potential or actual predators that threaten baboon troops.
Photo by Irven DeVore/Anthro-Photo.

Amadon 1968; Jolly 1972; Tilson and Tenaza 1977; Dawson 1979). Primates are even killed occasionally by other primates. Chimpanzees regularly hunt red colobus monkeys (*Colobus badius*), baboons, vervets, and red-tailed monkeys (*Cercopithecus ascanius*) (Goodall 1963b; Teleki 1973a, b; Struhsaker 1975; Busse 1977; Nishida et al. 1979), while baboons occasionally capture vervets (Struhsaker 1967b; Altmann and Altmann 1970). In some areas human hunters have long been important predators on monkeys (e.g., Bourlière et al. 1970; Tenaza 1976; Fairbanks and Bird 1978). Clearly the threat of predation represents a potentially important selective factor favoring the formation of social groups.

That sociality is an important defense against predators is indicated by several lines of evidence. Social primates generally rely on mutual vigilance to detect predators, with many species employing specialized alarm calls to signal danger (e.g., Struhsaker 1967a; Gautier and Gautier 1977; Oppenheimer 1977; Vencl 1977). Early detection of predators is an important advantage, as it provides group members with more time to escape up trees or onto cliffs. Male baboons and patas monkeys often act as sentinels while other troop members forage or conduct other activities (Hall 1960, 1965, 1968; Stoltz and Saayman 1970). Male patas monkeys and guenons often distract predators by leaping against small bushes and then leading them away from the female group (Hall 1965; Struhsaker and Gartlan 1970; Kummer 1971). Finally, males and occasionally females of some species threaten predators with barks or abortive charges, and they sometimes counterattack small predators (see pp. 116–119). Thus primate sociality does appear important as a means of predator defense.

Origin of variability within and among species. Social defenses against predation generally become more effective as group size increases, and yet primate groups vary greatly in size, both within and among species. This implies that the net benefits derived from sociality vary from one group to another, probably because vulnerability to predation or costs of competition vary between groups. Most theorists have attributed variations in group size to the latter factor.

Abundance and distribution of food resources should affect competition intensity and hence group size (Denham 1971; Crook 1972; Altmann 1974). When food abundance is scarce throughout a habitat, social grouping behavior is not feasible because the competitive costs are greater than any benefits derived from predator protection. Hence sparse food resources should favor solitary behavior or at most monogamous pairs. If food is scarce and evenly distributed both spatially and temporally, it should be defensible and territoriality should evolve. If food is spatiotemporally variable, it should be indefensible and individuals or pairs should occupy overlapping home ranges. As food resources become more abundant, social groups become more advantageous. Once social groups evolve, their size should depend on how much food is available during periods of scarcity. If food is relatively abundant and evenly distributed in space and time, it should be defended by social groups. If food is relatively abundant but spatiotemporally variable, it should be exploited by social groups with overlapping home ranges.

The best way to evaluate these predictions is with comparative evidence based on as many species as possible. Comparisons based on quantitative measures of food abundance and distribution are difficult to achieve because primate food resources are so diverse that they are difficult to measure. The best available evidence is therefore based on subjective evaluations of food availability.

Solitary prosimians. Many prosimian primates, including lorises, galagos, and some lemurs, are nocturnal and either solitary or monogamous (Crook and Gartlan 1966; Jolly 1972; Pollock 1979). They are generally insectivorous, although the sportive lemur (*Lepilemur mustelinus*) eats leaves and flowers, and the mongoose lemur *(Lemur mongoz)* eats nectar (Petter 1962; Hladik and Charles-Dominique 1974; Tattersall and Sussman 1975; Sussman and Tattersall 1976; Hladik 1978a). Food resources generally appear evenly distributed. Food may be relatively scarce for some species, but for species that live at high population densities it appears to be relatively abundant. Females are solitary in most prosimians, and they often defend small territories that are probably just sufficient to meet their energetic requirements, while males defend larger territories that overlap those of several females (Charles-Dominique 1972, 1974, 1977, 1978; Fogden

1974; Hladik and Charles-Dominique 1974; Hladik 1978a; Kawamichi and Kawamichi 1979; Pollock 1979). The larger territories of males provide them with more than enough food but enable them to mate with additional females because matings occur mainly between individuals with overlapping home ranges. Mongoose lemurs may form monogamous pairs, with each pair sharing a common territory or home range (Tattersall and Sussman 1975; Sussman and Tattersall 1976). The reasons for this behavior are unknown.

At least some nongregarious prosimians exploit very abundant food resources, contrary to the prediction made earlier. The absence of grouping behavior in these species may be due to their nocturnal habits. Mutual vigilance and other social defenses against predation lose much of their effectiveness at night, when predators can stalk prey under cover of darkness. A better defense at night may be stealth and concealment, which in turn would favor solitary habits. Consistent with this hypothesis is the fact that the only nocturnal higher primate, the night monkey, forms monogamous pairs instead of social groups (Moynihan 1964; Wright 1978). However, contrary to this hypothesis, lesser mouse lemurs (*Microcebus murinus*) form social groups even though they are nocturnal (Martin 1972). Mouse lemurs are unusual in having an omnivorous rather than insectivorous diet. Their diet consists mainly of leaves, fruits, and insects, and fruits at least may be spatiotemporally clumped. A clumped distribution of food resources may therefore be responsible for the social grouping behavior of lesser mouse lemurs, but more research is needed before the selective factors involved will be understood.

Diurnal arboreal primates. The grouping behavior of diurnal primates is more diverse, with group sizes ranging from monogamous pairs to very large multi-male social groups. Several early studies attempted to explain this variability by relating broad categories of social organization to general habitat characteristics (Crook and Gartlan 1966; Crook 1970; Eisenberg et al. 1972). However, many species did not neatly fit the proposed schema because the categories were too insensitive to details of primate ecology. More recent comparative studies are now affording a somewhat clearer understanding of environmental conditions responsible for variability in group sizes.

In one study of arboreal primates, Clutton-Brock (1974, 1975) showed how the differing group sizes of red colobus and black-and-white colobus monkeys (*Colobus guereza*) may be related to differences in food availability. Red colobus monkeys live in the treetops of African tropical forests as large noisy social groups, and they range over relatively large areas in search of food during the course of a year. They eat flowers, shoots, buds, and young leaves from many tree species. Their food is spatiotemporally clumped since each tree species is spatially clumped and flowers during a different part of the year. Such resources could theoretically be exploited by many small social units (individuals or groups) that occupy greatly overlapping home ranges or by a single larger social group that roams over a common home range. The intensity of competition should be similar in both cases, as should the efficiency of exploiting the resource, but formation of a larger social group should provide greater protection from predators. Red colobus monkeys are highly vulnerable to chimpanzees and other predators, and they depend on sociality for protection (Busse 1977). Hence large social groups may have evolved in red colobus monkeys because the benefits of gaining protection from predators are relatively high and the costs arising from local competition for food are relatively low.

In comparison, black-and-white colobus monkeys live in much smaller social groups even though they inhabit the same forest canopies (Clutton-Brock 1974, 1975; Dunbar and Dunbar 1976; Oates 1977). During the dry season, when food is seasonally scarce, their diet consists almost entirely of mature leaves from two tree species. Since leaves are more evenly distributed and are always available, black-and-white colobus monkeys can remain in a relatively small area throughout the year. Moreover, most folivorous primates prefer the more nutritious new leaves and shoots when they are available, and during some seasons competition for new foliage is likely to be severe (Hladik 1978b). Thus while the potential benefits of social defenses against predation may be equally high for red colobus and black-and-white colobus monkeys, the costs of local competition for food appear much higher for black-and-white colobus monkeys. This difference probably explains the smaller group sizes prevailing in the latter species. As a consequence of their smaller group sizes, black-and-white colobus monkeys rely more on concealment and less on vigilance to escape predators than do red colobus monkeys.

Although the above interpretation appears plausible, it has been questioned by some authors. Struhsaker and Oates (1975) agree that a diet of mature leaves is probably responsible for the small home ranges and small group sizes of black-and-white colobus monkeys, but they disagree with the hypothesized relationship between omnivory and large group sizes of red colobus monkeys. They point out that the daily movements of red colobus groups are determined by the activities of neighboring groups and not by location of food resources (Struhsaker 1974), implying that food distribution is not directly responsible for their large home ranges. This evidence does not negate Clutton-Brock's hypothesis, however, because neighboring groups may avoid each other as a proximate mechanism for minimizing local competition for food even though large home ranges may have evolved in response to spatiotemporally variable food resources. Struhsaker and Oates (1975) also point out that the red colobus groups studied by Struhsaker (1975, 1980) have broadly overlapping home ranges, unlike those studied by Clutton-Brock (1974, 1975). Such groups should, according to Clutton-Brock's hypothesis, amalgamate into a single larger group, since protection from predators would thereby be improved without changing the number of individuals exploiting food resources in the area. Nevertheless, Clutton-Brock's hypothesis could still be valid if food abundance at each fruiting tree limits the number of individuals that can feed there at any given time.

Similar comparisons have been made for several other arboreal primates. The Hanuman langur forms relatively large groups, is nonterritorial, and exploits spatiotemporally clumped food resources such as fruits, flowers, and young leaves, while the gray langur forms small groups, defends territories, and exploits more evenly distributed mature leaves in addition to fruits (Hladik 1975, 1978a). The ring-tailed lemur (*Lemur catta*) forms large groups and eats fruits and new foliage, while the brown lemur (*L. fulvus*) forms small groups and eats mostly mature leaves (Jolly 1966; Sussman 1974; Sussman and Richard 1974; Sussman 1977). Two factors appear important in favoring larger group sizes in Hanuman langurs and ring-tailed lemurs. First, both species exploit spatiotemporally clumped food resources, which reduces local competition intensity and hence the costs of forming larger social groups. Second, both Hanuman langurs and ring-tailed lemurs leave the protection of trees more often than gray langurs or brown lemurs. The more terrestrial behavior of the former two species may make them more vulnerable to predation, thereby increasing the benefits derived from living in larger social groups. The importance of this factor is suggested by the fact that Hanuman langurs form larger social groups during the dry season when they become more terrestrial (Hladik 1975). However, gaining better protection from predators may not be important for ring-tailed lemurs, as they live in Madagascar where terrestrial predators are largely absent. Another species that exploits spatiotemporally clumped food resources is the mangabey (*Cercocebus albigena*), which eats figs. Like red colobus monkeys, both group sizes and home range sizes are relatively large (Waser 1976, 1977). In general, frugivorous primates occupy larger home ranges than leaf-eating primates, which lends some support to Clutton-Brock's hypothesis (Milton and May 1976; Clutton-Brock and Harvey 1977a).

Terrestrial primates. Terrestrial primates always stand to benefit from sociality because vulnerability to predation is particularly high. Nevertheless, group sizes vary considerably both between and within species. Comparative studies of hamadryas baboons, geladas (*Theropithecus gelada*), and savanna baboons suggest that the distribution and abundance of both food resources and sleeping sites are important factors affecting group size and spacing behavior. The typical patterns shown by each species are illustrated in Figure 13–1.

Hamadryas baboons have a hierarchical social organization consisting of three levels (Kummer 1968, 1971). At night several groups or bands congregate as large troops to share a sleeping cliff. Only certain bands are compatible with each other at the cliffs, and fights between strange bands are not uncommon. In the morning the various bands separate and travel to different foraging grounds, usually by following dry riverbeds (Photo 13–4). When a band reaches its foraging grounds, it divides into small one-male units consisting of an adult male and one or more females (Photo 13–5). Each one-male unit disperses and forages as a separate subgroup until early afternoon when it rejoins the other one-male units to form a

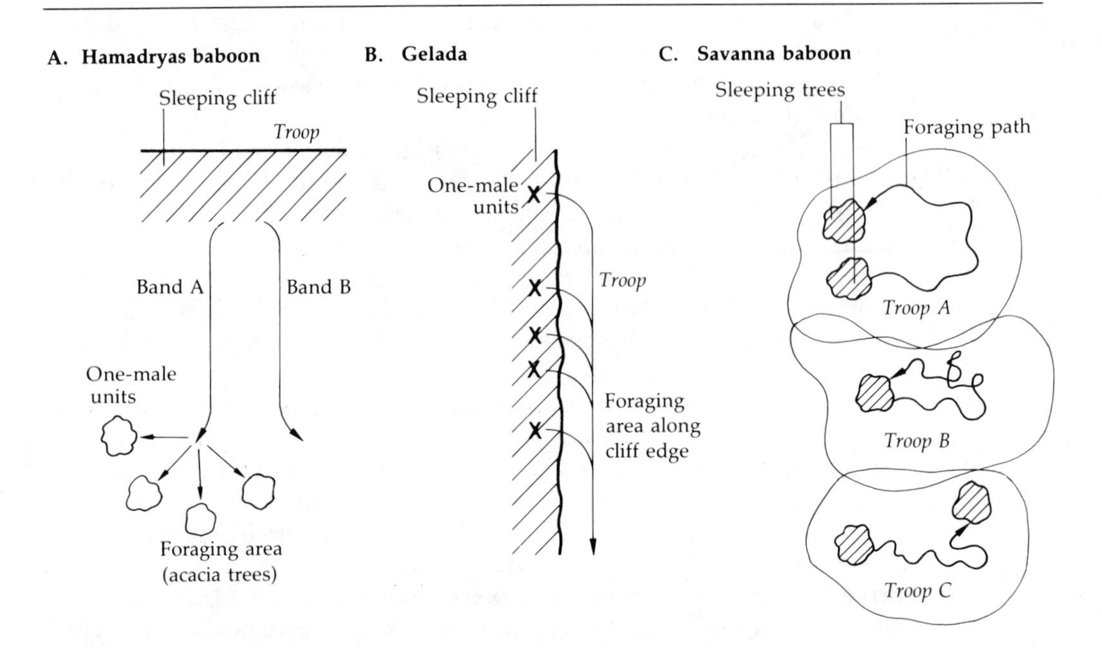

Figure 13–1 *The spatial organization of hamadryas baboons, geladas, and savanna baboons.*

A. Hamadryas baboons form large troops on sleeping cliffs. The troops separate into bands before traveling to foraging areas each morning. Bands fragment into one-male units while foraging and then reunite to travel back to the sleeping cliffs in mid afternoon. B. Geladas sleep on cliffs as one-male units. In the morning one-male units coalesce into a large troop, which remains together all day while foraging on the cliff top. C. Savanna baboons live as separate social groups with partially overlapping home ranges. Each group sleeps, travels, and forages as a separate entity. The size of savanna baboon groups is similar to that of hamadryas bands.

band again. The band then travels together to a water hole, where it remains for a period of social activity and rest for several hours before traveling back to a sleeping cliff.

Bands of hamadryas baboons congregate each evening because cliffs suitable for sleeping are relatively few in their Ethiopian desert habitat. Each sleeping cliff generally contains enough ledges to accommodate the one-male units from several bands, and cliffs are too few for each band to use a separate one. Bands separate during the day because food in any given place is not abundant enough to support more than one band. Food resources occur as widely scattered groves of acacia trees, whose pods, young leaves, flowers, and (during the dry season) mature leaves and roots are eaten by the baboons. Within these groves individual trees are rather widely dispersed, and each can support only a few baboons. Thus high local competition for food may be the reason bands fragment into small one-male units while foraging. During travel the costs of competition are

Photo 13–4

A band of hamadryas baboons traveling up a dry riverbed toward its foraging grounds.

The band consists of numerous one-male units and a few all-male units.

Photo from Hans Kummer, *The Primate Societies* (Chicago: Aldine-Altherton, 1971), p. 26.

Photo 13–5

A one-male unit of hamadryas baboons.

Each unit consists of an adult male (center), one or more adult females, offspring, and sometimes a subordinate follower male.

Photo by Hans Kummer.

no longer important, and individuals are vulnerable to ambush by lions and occasionally other predators. One-male units therefore coalesce and travel together in bands when moving between sleeping cliffs, foraging sites, and water holes.

Geladas are similar to hamadryas baboons in exhibiting a multiple-level social organization. However, the spacing behavior of geladas is quite different. Geladas live on arid plateaus of Ethiopian highlands, where the terrain is intersected by numerous impassable canyons (Crook 1966; Dunbar and Dunbar 1975; Kawai 1979). They disperse at night and sleep as either one-male units or all-male groups on widely scattered ledges along a cliff face. In the morning these units coalesce on top of the cliff to form a large herd (Photo 13–6). Each herd occupies its own home range, where it forages by progressing slowly along the plateau rim. Females with young and dominant males usually remain close to the cliff edge, where they are safest from predators, while subordinate males and all-male groups are usually farther from the cliffs (Crook 1966).

Herds are true social units in that herd membership and composition are comparatively stable (Ohsawa 1979a, b). They sometimes amalgamate into large multiherds, and they sometimes divide into two or more bands, but such changes are fairly temporary. Herds are not closed social units, as one-male units can move freely between them.

Each one-male unit within a herd concentrates its activities in a subsection of the herd's home range, though at one time or another every unit visits every part of the home range (Dunbar and Dunbar 1975). Particular one-male units tend to associate with one another as bands. Such bands are not cohesive or closed subgroups; they merely reflect affiliations among bands. New one-male units generally form by fissioning of large one-male units, and the resulting units tend to stay together following group fissioning. Hence females within each band may frequently be relatives.

Since geladas live mainly along plateau rims, sleeping cliffs are readily available. However, because sleeping ledges are small and widely scattered, troops must fragment into smaller units at night. The necessity of forming sleeping groups each night may help explain why one-male units persist as stable subgroups. Troops coalesce during the day because only a limited amount of space is available for foraging near the cliff edge. Foraging is concentrated near precipices where escape from predators is easiest, and one-male units apparently remain in troops to enhance their ability to detect predators. Large troops are possible because food availability is relatively high, at least during the rainy season (see Iwamoto 1979). During the dry season food becomes scarce in the highlands, and then troops disband into one-male or all-male units which sometimes range into lowland agricultural districts in search of food (Crook 1966; but see Ohsawa 1979b). Hence one-male units may also persist because they can disperse during the dry season when competition for food makes large herds disadvantageous (Crook 1966; Crook and Aldrich-Blake 1968). However, Iwamoto (1979) argues that recurrent food shortages during dry seasons are a result of recent human disturbances such as grazing cattle and have little evolutionary significance when explaining one-male units.

Savanna baboons do not exhibit a multiple-level social organization. They live as cohesive multi-male social groups, each with its own home

Photo 13–6

A.

B.

Social organization of geladas, a baboon-like terrestrial monkey of Ethiopian highlands.

A. A troop of geladas foraging near the precipice of a cliff. B. A one-male unit, consisting of an adult male and his group of females. One-male units are not spatially cohesive, unlike those of hamadryas baboons, and members often disperse through the troop while foraging.

Photo A from R. I. M. Dunbar and E. P. Dunbar, "Social Dynamics of the Gelada Baboons," *Contributions to Primatology* 6 (1975):7. Photo B from H. Ohsawa, "The Local Gelada Population and Environment of the Gich Area," in M. Kawa, ed., *Contributions to Primatology* 16 (1979):14.

range (Hall and Devore 1965; Altmann and Altmann 1970). Neighboring groups do not coalesce into larger troops or break into smaller one-male units. When groups do come together, they are usually antagonistic toward one another. Baboon groups typically occupy large, overlapping home ranges, but in desert canyons of the Namib Desert and in the Okavango Swamp of Botswana they defend smaller exclusive territories (Hamilton et al. 1975b).

Savanna baboons sleep in groves of trees or on the sides of rock stacks. Because suitable sleeping sites are relatively numerous and are dispersed throughout baboon habitat, neighboring troops ordinarily do not converge upon centrally located sleeping sites at night. Savanna baboons are highly omnivorous and eat a wide variety of plant and animal matter. The exact distribution of their food resources is unknown, but the opportunistic foraging habits of savanna baboons make food more abundant and less patchy than food exploited by hamadryas baboons. Greater food abundance reduces the costs of competition, making fragmentation into one-male units less advantageous, and the high density of predators in savanna environments makes larger groups more advantageous as a defense against predation.

The above patterns show how predation pressure and the distribution and abundance of safe sleeping sites and food resources affect the overall social structure of baboon groups. Many additional variations in both social structure and ecological conditions exist among geographically separated populations of each species, but comparative studies of such populations have only just begun to unravel the environmental factors responsible for such variations (Altmann 1974). Future studies promise to go far in explaining how differing environmental conditions lead to more subtle differences between baboon societies.

Social carnivores

Social carnivores can potentially benefit from cooperation in at least four ways (Eaton 1976). They might increase hunting success, improve their ability to defend or steal carcasses from competitors, reduce risks of being injured while attacking prey, or gain protection from other predators.

Benefits gained from cooperative hunting. All social carnivores studied to date hunt more effectively in groups than when alone, an advantage that is generally accepted as the most important reason favoring their sociality (Kleiman and Eisenberg 1973; Alexander 1974; Bekoff 1975). Cooperation improves hunting success in two ways. It enables capture of larger prey animals, and it increases the percentage of hunts that result in kills. The importance of each advantage varies, depending on the predator and prey species involved.

For lions the benefits gained from cooperatively hunting large prey are minimal during most times of year (Schaller 1972). Lions normally hunt small antelopes and other prey that can be killed by single individuals, except in certain areas or during certain seasons. In wet seasons lions of Serengeti National Park may depend on cooperative hunting of large prey because most smaller prey migrate out of pride home ranges and onto the open plains. In Lake Manyara National Park lions subsist primarily on Cape buffalo and young elephants all year because smaller prey are relatively scarce, and hence cooperative hunting of large prey is very important (Makacha and Schaller 1969). In the Kalahari Desert of Botswana cooperative hunting of large prey is unimportant because only small- and medium-sized prey are available (Eloff 1973).

Lions also benefit from cooperative hunting by capturing prey more efficiently. The percentage capture success of lions that hunt in groups of two or more is higher than that of solitary lions regardless of prey size (Schaller 1972; Caraco and Wolf 1975). Their higher success often results from cooperative ambush tactics. In many hunts several pride members encircle a herd of prey and lie in wait until the other pride members stampede the prey group toward them. In addition, more prey may be captured during the attack because erratically fleeing prey may successfully elude one lion only to blunder within reach of another. However, despite an increased percentage of successful hunts, the amount of food killed per lion actually decreases when more than two lions participate in a hunt (Figure 13–2). That is, greater capture success does not compensate for increased competition at the kill. Hence cooperative hunting by groups larger than two lions does not increase food intake per lion.

No other cat is as social as the lion, but tigers and cheetahs sometimes hunt in groups. The reason why tigers hunt in groups is unknown. In cheetahs females usually hunt alone or with immature cubs, but males and subadult litter mates often hunt together as stable groups (Eaton 1974, 1976). Solitary cheetahs primarily kill small prey such as gazelles and other small antelopes, and they depend on their great speed over short distances to outrun prey. Cheetah groups are more efficient than solitary individuals at capturing small prey, but for females, at least, the costs associated with group hunting (see below) are apparently too high for them to hunt cooperatively.

The size of spotted hyena hunting groups is attuned to the antipredator behavior of prey rather than to prey size (Kruuk 1972). Hyenas usually hunt alone or in groups of two or three individuals when pursuing small antelopes or medium-sized wildebeest, none of which ordinarily fight back when attacked. However, they depend strongly on cooperation when hunting zebras because zebras are likely to counterattack.

Group hunting increases capture success for hyenas when hunting wildebeest calves and zebras but not when hunting other kinds of prey (Kruuk 1972). Cooperative hunting is a more successful way of capturing wildebeest calves mainly because a wildebeest cow can fend off only one hyena at a time. A cow can often keep a solitary hyena away from her calf,

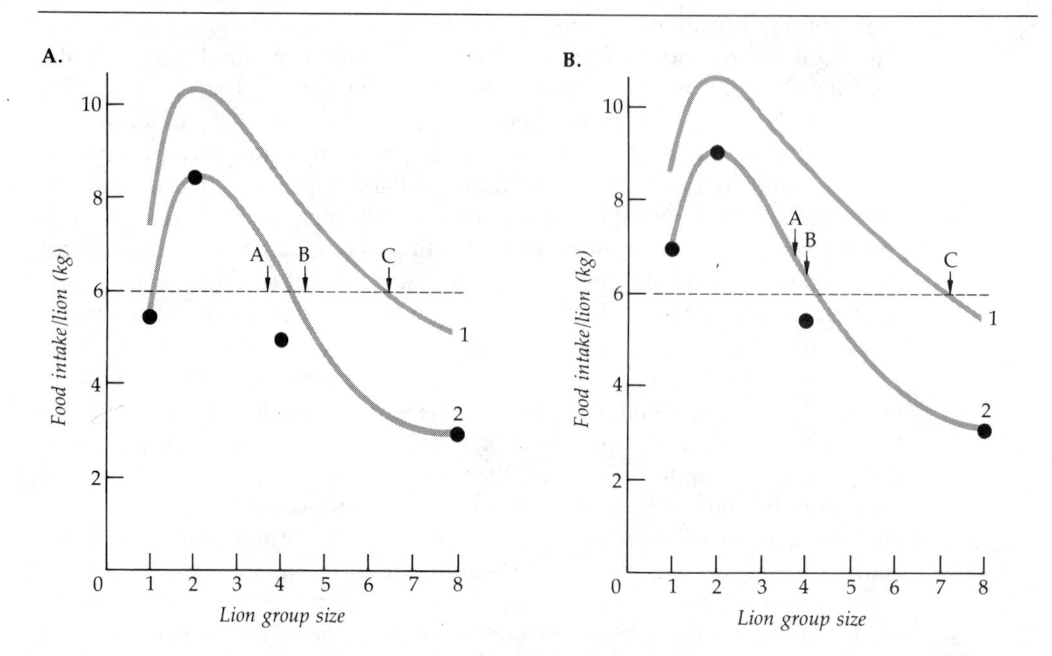

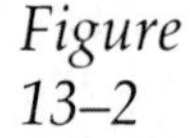

Figure 13–2 *The effect of lion hunting-group size on food intake rate per lion.*
Mean edible prey biomass (kg) captured per chase each day is plotted against the size of hunting groups. Curve 1 gives hypothesized intake rates during the wet season. Curve 2 gives hypothesized intake rates during the dry season. Dashed lines indicate minimum daily food requirements per lion. Observed mean lion group sizes are given for the eastern plains of Serengeti National Park (dry season) (A), the western woodlands (dry season) (B), and the border region (wet season) (C). A. Data for attacks on wildebeest. B. Data for attacks on zebra.

Source: Reprinted from T. Caraco and L. L. Wolf, "Ecological Determinants of Group Sizes of Foraging Lions," *American Naturalist* 109 (1975), p. 349, by permission of The University of Chicago Press. Copyright © 1975 by The University of Chicago Press.

but she can never ward off two or more hyenas. Cooperation also increases hunting success because hyenas lagging behind pack leaders in a chase can cut corners when the prey changes direction, and several individuals are more effective at bringing down a struggling prey.

Comparing the types of prey captured by social and nonsocial canids suggests that pack hunting is an important means of capturing large prey (Kleiman and Eisenberg 1973). Pack-hunting wolves and African wild dogs subsist primarily on large prey such as elk, moose, caribou, mountain sheep, or antelope, while foxes and other canids that ordinarily hunt alone subsist on smaller prey such as rodents and rabbits, along with various types of plant matter. In wild dogs a minimum of four to six individuals appear to be necessary for capturing most prey (Estes and Goddard 1967). Pack hunting may also increase hunting efficiency, as it does in lions and hyenas, but no data are available on this point.

Benefits gained from cooperatively protecting carcasses. In Africa cooperation improves the ability of carnivores to protect carcasses from scavengers or to steal carcasses from other predators (Kleiman and Eisenberg 1973; Eaton 1976; Lamprecht 1978). Lions and hyenas regularly compete for carcasses and often obtain significant amounts of food through thievery. A single lion usually cannot prevent a pack of hyenas or wild dogs from stealing a carcass, but two or more lions can (Schaller 1969; Rudnai 1973). Similarly, groups of lions can successfully steal food from hyena packs, while single lions cannot. Hyenas cooperate to steal food from cheetahs and wild dogs as well as from lions (Kruuk 1972), while wild dogs depend on cooperation to ward off hyenas (Estes and Goddard 1967).

Cooperation is not always an effective defense against scavengers. Grizzly bears sometimes usurp carcasses killed by wolf packs (Mech 1970), and groups of cheetahs cannot defend carcasses from other scavenging predators (Eaton 1976). In fact, grouping may even be deleterious for cheetahs because groups are more conspicuous and attract scavengers sooner than solitary individuals. The importance of this cost is suggested by a negative correlation between degree of grouping in cheetahs and number of mammalian predators coexisting with them in various parts of Africa (Eaton 1976).

Benefits gained from reduced risk of injury while hunting. Cooperation may reduce the risk of being injured or killed during a hunt, but the importance of this possible benefit is difficult to assess. Predators certainly risk injury when hunting large or dangerous prey. Lions, for example, are occasionally injured or killed by rhinoceros, crocodiles, porcupines, Cape buffalos, zebras, and large antelopes (Schaller 1972). Hyenas are sometimes injured or killed by zebra stallions defending their harems and zebra mares defending their foals (Cullen 1969). Wolves are regularly injured or killed by moose (Rausch 1967; Mech 1970). In each case cooperative hunting might reduce the risk of attacking large or dangerous prey by spreading the risk among more individuals (Eaton 1976). However, direct evidence to show that the risks are reduced is not available.

Benefits gained from cooperative defense against predators. A common benefit of sociality in vertebrates is better protection from predators, but adult carnivores gain little benefit in this regard because they are generally invulnerable to other predators. Most deaths of hyenas and wild dogs that can be attributed to other predators result from interspecific fighting over kills (Kruuk 1972). Lions rarely even eat carcasses of hyenas or wild dogs (Schaller 1972), suggesting that the flesh of those species may be distasteful. Wolves are not susceptible to predation in most parts of their range, since they are the top carnivore, but they may be vulnerable to tigers in Siberia. Cheetahs are too fast to be easily caught by other predators. They are most social in areas where other predators are absent, indicating that sociality has not evolved because it confers protection from predators (Eaton 1976).

Some social carnivores may rely on cooperation to protect offspring from predators. Young wolf and wild dog pups depend primarily on the safety of dens for protection, but in addition they are often attended by an adult while other pack members are out hunting (Kühme 1965a, b; Mech 1970; Lawick-Goodall and Lawick-Goodall 1970). Cooperation may be more important in lions, which do not use dens. An adult lioness often remains behind to baby-sit all the young cubs of a pride while the remaining lionesses are out hunting (Schaller 1972). Once a prey has been killed, the lioness and cubs rejoin the pride to share the kill. In contrast, hyenas do not cooperate in protecting cubs. Indeed, sociality poses a real risk for cubs because hyenas are cannibalistic (Kruuk 1972). For protection young hyenas rely on safe dens and an ability to burrow far back into narrow holes where predators and adult hyenas cannot reach them.

Origin of variability among species. Cooperative hunting has not evolved in all carnivores, and one reason for this may be related to habitat structure. Social carnivores all live in open habitats where cooperative pursuit of prey is most effective (Kleiman and Eisenberg 1973; Eaton 1976). Carnivores that hunt alone ordinarily live in forests or other habitats with dense cover, and they are usually more nocturnal. Such conditions make stealth and ambush tactics more effective than cooperative hunting. Moreover, food availability for large ungulates is generally lower in closed habitats. As a result, prey abundance is less in closed habitats, and costs of competition are greater than in open habitats. The combination of habitat structure and prey density is probably responsible for the difference in hunting tactics existing between social and nonsocial species.

Overview. Grouping behavior and cooperation are beneficial to social carnivores because they enable more efficient and safer hunting, better protection of carcasses, and enhanced capability of stealing carcasses. The relative importance of each benefit varies with species. Scavenging interactions are probably more important for lions, hyenas, and wild dogs than for wolves. Capturing large prey and reducing risks are probably most important for wild dogs, wolves, and, to a lesser extent, hyenas. Increasing capture efficiency is probably important in every species.

Determinants of group size

Previous sections have discussed in a qualitative way the benefits and costs associated with being social. The quantitative aspect of this same issue is group size. The magnitude of both costs and benefits should vary as a function of group size, and if it does, the net benefit derived from sociality will be maximal at some optimum size. This optimum size is very likely to differ for each individual in a group, since both costs and benefits should vary according to the individual's sex, age, and social status. Individuals are therefore likely to have conflicting interests regarding group size, and the group sizes actually observed are likely to be compromises between these various interests. An adequate theory of group size should predict optimal group size for each group member and show how the conflicting interests of group members should interact to produce the group sizes actually observed.

Models of optimal group size

One way of predicting optimal group size is to plot the separate effects of each selective factor affecting a particular individual's inclusive fitness as a function of group size (Figure 13–3) (Wilson 1975a). Optimal group size for any particular individual is given by the point where the combined effects of all selective factors maximize that individual's inclusive fitness. A different set of curves can be postulated for each class of individual found within a social group, and comparison of the separate graphs should reveal how individual interests conflict.

Wilson's (1975a) model is a useful way to visualize the complexities

Figure 13–3

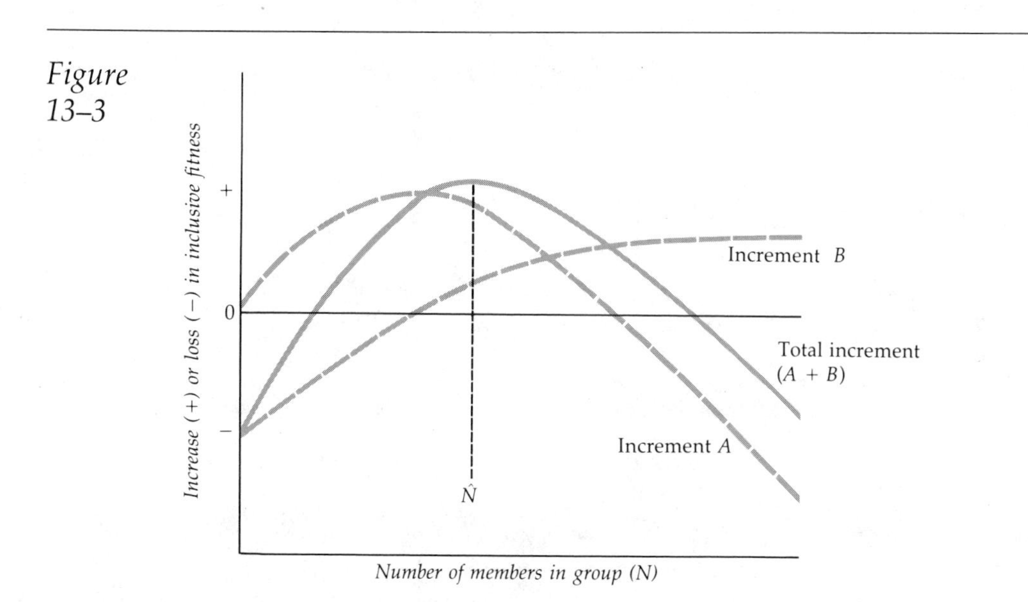

A model of optimum group size.

The vertical axis indicates the gain or loss of inclusive fitness arising as a result of a given group size. The horizontal axis gives group size. The curves labeled Increment A and Increment B (broken lines) indicate the fitness effects of two different selective factors (e.g., predator defense and cooperative hunting) that affect group size. The sum of increments A and B (solid line) represents the total effect of all selective factors affecting group size. Optimum group size ($\hat{N}$) occurs where the positive effect on inclusive fitness of all selective factors combined is maximal.

Source: From E. O. Wilson, *Sociobiology: The New Synthesis* (Cambridge, Mass.: Belknap Press of Harvard University Press, 1975), p. 136. Reprinted by permission.

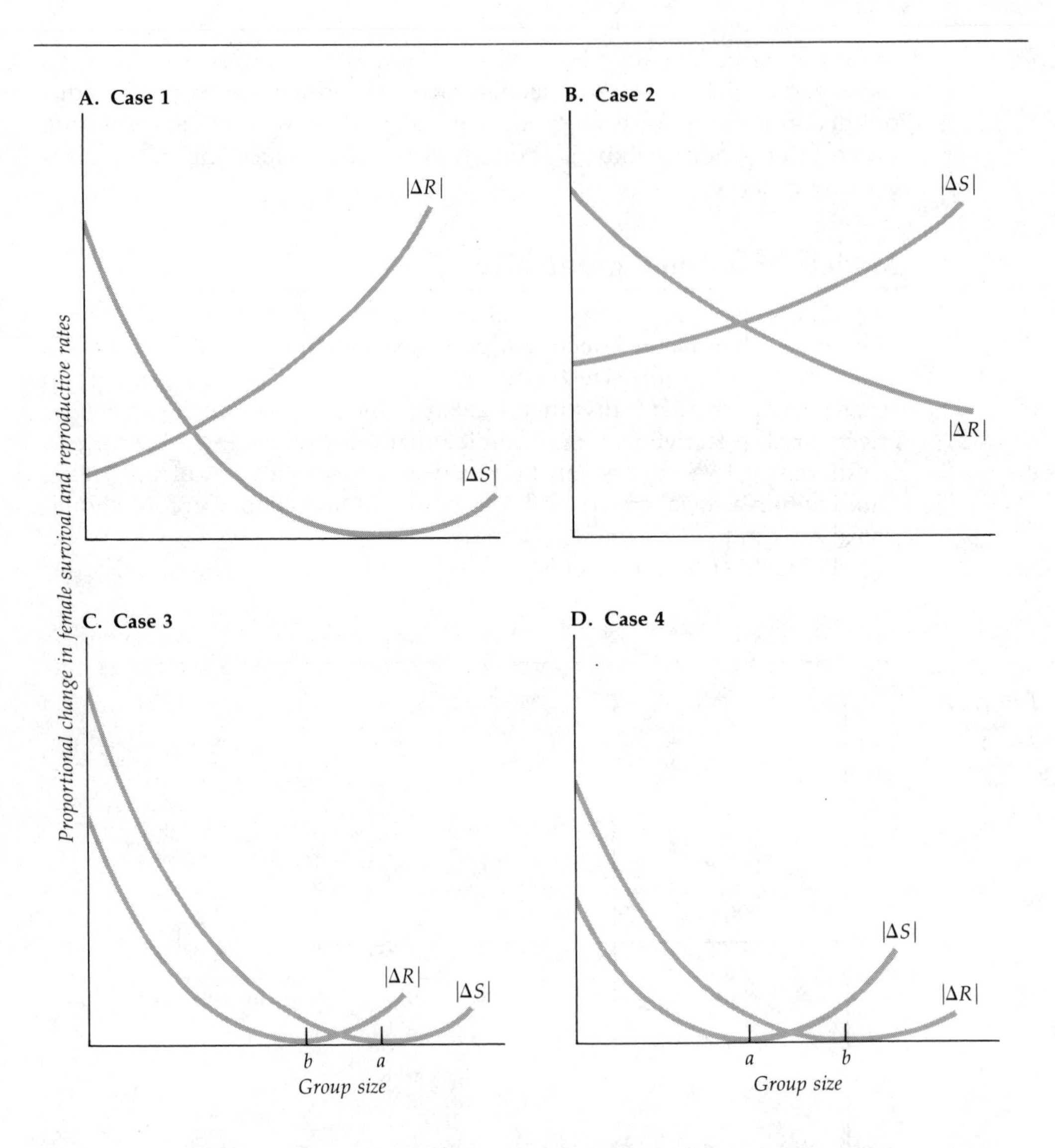

Figure 13–4 *The four cases of Wittenberger's model of optimum group size.*
The curve labeled |ΔR| represents the proportional change (either + or −) in female reproductive rate arising when a group containing a given number of females gains one additional adult female member. The curve labeled |ΔS| represents the proportional change (either + or −) in female survival rate arising when a group containing a given number of adult females gains one additional adult female member. The magnitude of change (ΔR and ΔS) varies with group size. Optimal group size occurs at the point where the |ΔR| and |ΔS| curves intersect. A. In Case 1 of the model, female reproductive rate decreases with increasing group size (ΔR < 0), while female survival rate increases (ΔS > 0). B. In Case 2 of the model, female reproductive rate increases with increasing group size (ΔR > 0), while female survival rate decreases (ΔS < 0). C. and D. In Cases 3 and 4 of the model, female reproductive rate increases until group size = b (ΔR > 0) and then

of the group size problem, but it is difficult to employ in practice. One problem with the model is its unrealistic assumptions. Each selective factor is assumed to affect inclusive fitness independently of all other selective factors, and all factors are assumed to ultimately have a negative effect once group size becomes sufficiently large. The latter assumption is probably reasonable, but the former is not. Another problem is the difficulty or impossibility of measuring the fitness effects of each selective factor independently of the others. Finally, one can never be certain that all important factors have been included in the graphs.

A second approach for predicting optimal group sizes is to compute lifetime reproductive output of group members as a function of group size (Wittenberger 1980a). The model developed from this approach calculates optimal group size for average females in a social group, but a better procedure would be to calculate separate optima for each individual or status level in the group.

The model is based on the Euler-Lotka life table equation (Wittenberger 1980a). It shows that optimal group size for an average female, defined as the point where her lifetime reproductive output is maximal, is reached when the *percentage change* in mortality rate caused by adding one more individual to the group equals the *percentage change* in reproductive rate. Costs and benefits enter the model by affecting mortality rate or reproductive rate. Mortality rate may decrease in a larger group because protection from predators is increased. It may increase because competition for resources is intensified or because disease is more prevalent. Reproductive rate may increase because offspring are safer from predators or better fed. It may decrease because increased competition for food leads to poorer nutritional condition of mothers or more prevalent malnutrition among offspring.

An important shortcoming of the model is that it does not include the costs of leaving an old group to join a new group of more nearly optimal size. Transferring to a new group may be hazardous if individuals must remain solitary during the transition period, and entering a new group may be difficult because members of the group may resist entry of new arrivals. These factors do not affect optimal group size, but they must certainly be taken into account when examining the group sizes actually occurring in a population.

The model predicts that sociality among vertebrates can evolve in four basic ways (Figure 13–4) (Wittenberger 1980b). In Case 1 adult female

---

decreases ($\Delta R < 0$). Female survival rate increases until group size = a ($\Delta S > 0$) and then decreases ($\Delta S < 0$). Case 3 occurs when reproductive rate begins decreasing before survival rate. Case 4 occurs when survival rate begins decreasing before reproductive rate.

Source: Reprinted from J. F. Wittenberger, "Group Size and Polygamy in Social Animals," *American Naturalist* 115 (1980), by permission of The University of Chicago Press.

survival rate *increases* with increasing group size, while adult female reproductive rate *decreases*. In Case 2 female survival rate *decreases* with increasing group size, while reproductive rate *increases*. In Case 3 female survival rate and reproductive rate *both increase initially* with increasing group size, but group size is limited by an eventual decline in *reproductive rate*. Case 4 is identical to Case 3 except that group size is limited by a decline in *survival rate* instead of reproductive rate.

Applying each case of the group size model

Evaluating the group size model requires quantitative data on how group size affects survival and reproductive rates of particular classes of individuals. Such data are generally unavailable, but the direction of these effects can be inferred from the costs and benefits discussed earlier. Such inferences allow predictions concerning which social mammals should fit each case of the model.

One hypothesis for sociality among colonial rodents is that females increase their survival rate at the expense of immediate reproductive success by breeding within social groups (see pp. 324–325). Colonial rodents depend on mutual vigilance for detecting predators and sometimes cooperative efforts to dig more elaborate burrow systems, but competition for food reduces reproductive output. They may therefore fit Case 1 of the group size model.

Case 2 of the model would be most applicable when adults are essentially immune to predation while offspring are not (Wittenberger 1980a). Then competition should reduce adult female survival while social defenses against predators should increase offspring survival. Some animals that may fit Case 2 conditions are elephants, gorillas, and whales.

The evidence presented earlier suggests that social ungulates and primates form groups to gain protection from predators and that group sizes are limited by competition for food (see also Wittenberger 1980a). Both adults and offspring probably survive better initially as groups increase in size, so either Case 3 or Case 4 should apply. No studies have examined the effects of group size on adult survival or reproductive rates, but food shortages are known to reduce birth rates and offspring survival in ungulates before they affect adult survival (Preobrazhenskii 1961; Klein 1970; Lamond 1970; Geist 1971; Sinclair 1977; Dunbar 1979a). Malnutrition probably has similar effects in primates, though less documentation is available. Hence Case 3 of the model appears most applicable.

In social carnivores cooperative hunting should enhance survival of both adults and offspring when group sizes are small, so either Case 3 or Case 4 of the model should apply. The size of lion prides may not be adjusted to maintain high adult survival, as group size does not correspond to the size that would maximize daily food intake rates (Caraco and Wolf 1975). Lion prides are typically as large as possible within the constraints set by minimum daily food requirements, suggesting that breeding females may sacrifice survival prospects to allow mature female offspring to remain with the pride. A major cause of adult mortality is fighting over kills (Schaller 1972), and the frequency of fights increases when competition at kills is intensified by prey scarcity (Schenkel 1966). Since competition is greater in large groups, increased group size is likely to increase adult female mortality. Cubs may not be affected by competition for food as early as adult females because males allow cubs to feed while keeping females away (Schaller 1972; Bertram 1976). Lionesses also suckle cubs other than their own, which counteracts detrimental effects of competition on subordinate females. Thus lions may best fit Case 4 of the model.

The situation is less clear for African wild dogs and wolves. Large group sizes may be advantageous for dominant females, who do most of the breeding, because it enhances pup survival even when competition reduces adult survival. Pups gain clear survival benefits from sociality. In wild dogs they are regularly allowed to feed while the adults keep hyenas at bay (Estes and Goddard 1967), and in both wild dogs and wolves they are fed communally by all pack members (Kühme 1965a, b; Mech 1970; Lawick-Goodall and Lawick-Goodall 1970). Pups therefore should receive more food when reared by larger packs. The reproductive benefits for subordinate females are less clear because the dominant female usually prevents them from breeding. However, since subordinate females are usually prior offspring of the dominant female, they can benefit by helping rear subsequent offspring of the dominant female. Thus wild dogs and wolves may also best fit Case 4 of the model.

The opposite situation prevails in hyenas. Hyena cubs are always in direct competition with adults and are never given prior access to food (Kruuk 1972). Very young cubs obtain food only when fed by their mothers, and older cubs are able to obtain food primarily because their mothers hunt alone instead of in groups. Hence group size in hyenas is probably limited by the effect of competition on female reproductive success rather than its effect on female survival.

Dynamics of group size

The underlying assumption of group size models is that individuals choose or influence group size in a way that increases their fitness. However, this assumption is not necessarily valid. Group size may result from demographic processes rather than adaptive choices of individuals.

Each social group can be viewed as a discrete population characterized by a birth rate, death rate, immigration rate, and emigration rate (Cohen 1969). If group size is determined primarily by birth and death rates, it should reach an equilibrium size reflecting the carrying capacity of the group's home range rather than a compromise size based on adaptive choices of group members. This outcome could arise if transferring from one group to another or ejecting subadult offspring cost more than living in a group whose size deviates from the optimum. If, on the other hand, group size is determined primarily by immigration and emigration, it should reflect the adaptive choices and conflicting interests of group members. Therefore, a first question concerns the extent to which individuals move from one group to another.

Group stability

Group sizes are most dependent on emigration-immigration rates when individuals form open social groups of unstable composition. In such systems individuals move continually from one group to another and normally do not associate with any particular individuals other than dependent offspring for prolonged periods. Why some species form open social groups while others form closed groups is poorly understood. One way to approach this problem is to analyze the opposing forces causing individuals to come together or move apart within social groups.

Internal spacing in stable social groups. The selective pressures contributing to internal spacing patterns within social groups generally involve the costs of competition for resources or mates and the benefits gained from parental care or cooperation. Two opposing forces of general importance are competition for food and defense against predation (Breder 1951; Emlen 1952b; Goss-Custard 1970). Competition for food creates a centrifugal impetus to spread out, while predation pressure creates a centripetal impetus to bunch together. Such effects can be seen in several social mammals. Gorillas, Cape buffalos, and Thomson's gazelles usually remain in tight social groups when resting or conducting social activities but spread out into looser groups when foraging (Schaller 1963; Sinclair 1977; Walther 1977). Nonforaging vervet monkeys stay closer together when cover is sparse (and hence vulnerability to predation is higher), while foraging individuals stay farther apart when food density is lower (Figure 13–5) (Fairbanks and Bird 1978). Red-tailed monkeys exploit sparser food resources than red colobus monkeys, and group members are more widely spaced while foraging than in red colobus groups (Struhsaker 1980).

Similar logic may explain why stable social groups are sometimes incohesive. In such cases group membership remains constant but groups often fragment into widely spaced subgroups. Lionesses, for example, form stable prides that share a group territory, but pride members rarely hunt together as a single group (Schaller 1972; Bertram 1975a). Individuals are most frequently alone or in the company of one or two other pride members. The size of these small hunting groups depends largely on size of the most available prey at the time. Lionesses frequently hunt alone when pursuing gazelles or other small prey, but they hunt in groups when pursuing wildebeest, zebra, or other large prey. Since lions can capture all sizes of prey more efficiently by hunting in groups, the size of hunting groups is apparently limited by competition for food at kills. Fragmentation of the main group is facilitated by invulnerability to predation, since mortality risk does not increase for smaller group sizes (at least for adults).

Hyena spacing behavior is similar to the spacing behavior of lions. In Ngorongoro Crater in Tanzania, hyenas live on permanent clan terri-

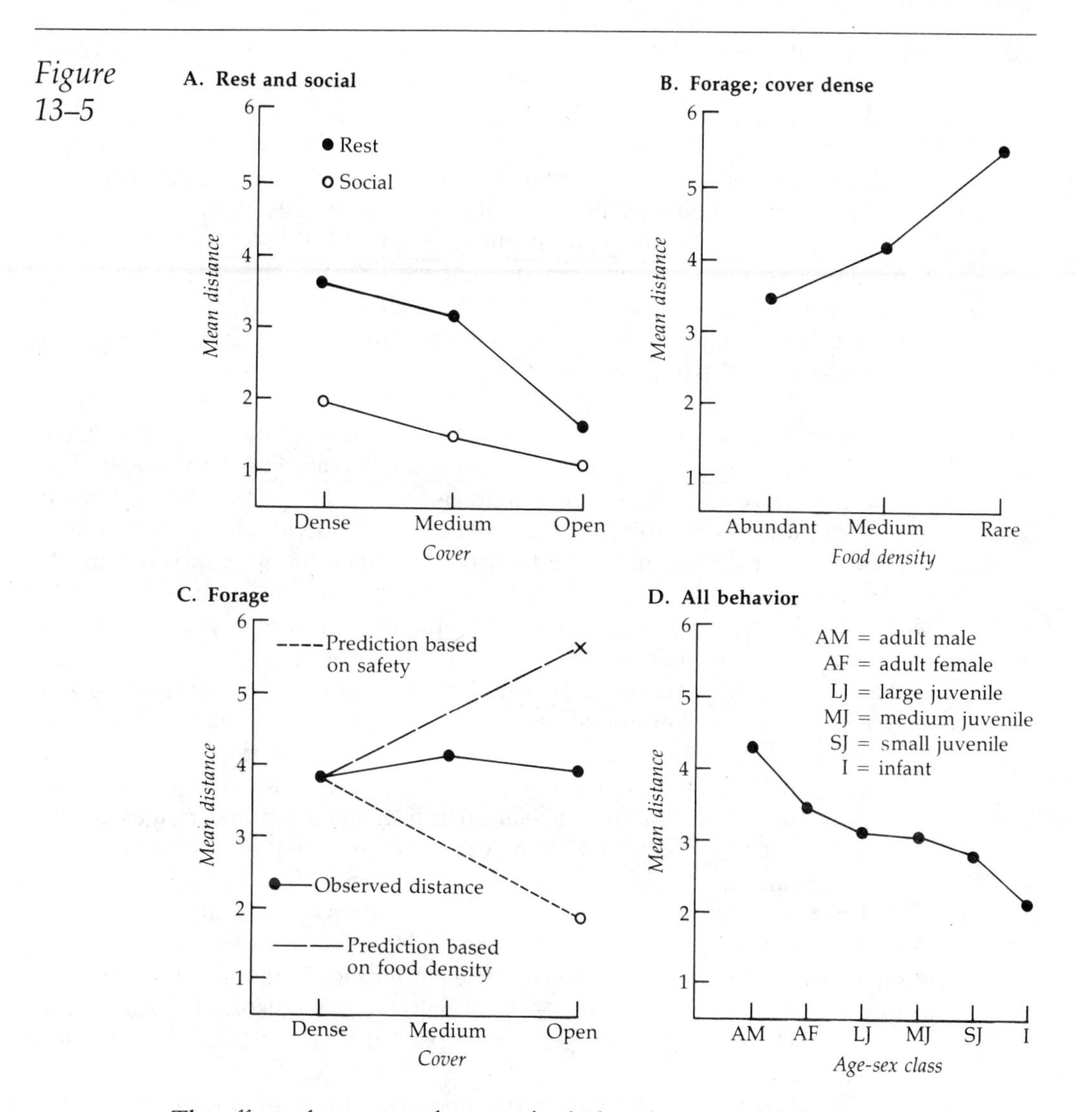

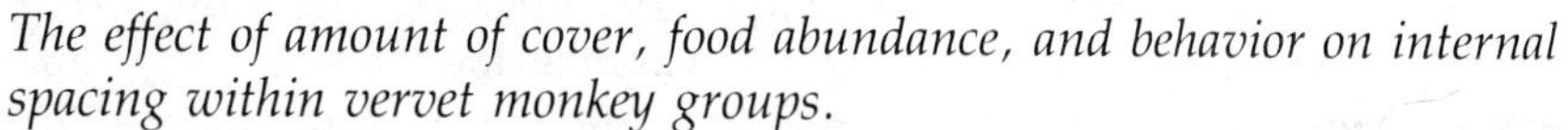
The effect of amount of cover, food abundance, and behavior on internal spacing within vervet monkey groups.

A. Mean distance to nearest neighbor within group while resting or performing social interactions as a function of cover density. B. Mean distance to nearest neighbor while foraging in dense cover as a function of food density. C. Predicted (broken lines) and observed (solid line) mean distance to nearest neighbor while foraging as a function of cover density. Predicted distances were computed by assuming that distance between individuals decreases as amount of cover decreases (prediction based on safety) or decreases as food density decreases (prediction based on food density). Note that cover does not affect spacing among foraging individuals, contrary to both predictions. D. Mean distance to nearest neighbor for all behaviors, broken down by age and sex classes of group members.

Source: From L. A. Fairbanks and J. Bird, "Ecological Correlates of Interindividual Distance in the St. Kitts Vervet (*Cercopithecus aethiops sabaeus*)," *Primates* 19 (1978):611. Reprinted by permission.

tories, but members of each clan hunt alone or in temporary subgroups (Kruuk 1972). The number of hyenas seen eating at carcasses increases with prey size, indicating that size of hunting groups is limited by competition for food at kills. On the nearby Serengeti Plains hyenas do not defend clan territories or maintain clan integrity. They still hunt in groups at times, but groups form on the spot when a large prey animal is attacked. As in lions, the size of hyena hunting groups is adjusted to competition intensity. Group sizes depend particularly on prey size, along with the defensive tactics employed by each kind of prey. Since hyenas are not vulnerable to attack themselves, they gain little by remaining together in stable groups of fixed size.

Chimpanzees exhibit a similar spacing pattern in that individuals join and depart groups frequently with little sign of antagonism (Goodall 1965; Reynolds and Reynolds 1965; Lawick-Goodall 1968; Sugiyama 1969). The chimpanzee social system can be characterized as a large group (often called a "community" or "unit group") comprised of individuals moving among many temporary subgroups. Individuals within each unit group are familiar with one another and may share a common home range (Itani and Suzuki 1967; Goodall 1973; Kawanaka and Nishida 1975). Internal spacing patterns within groups are not entirely clear, but they appear to be more complex than in lions or hyenas. Males roam throughout the group home range and actively defend boundaries from males of neighboring groups (Goodall et al. 1979; Nishida 1979). Nevertheless, boundaries overlap. Females may occupy smaller home ranges and not visit all parts of the unit group range (Wrangham 1979). The exact relationship of female home ranges to the group home range occupied by males is unclear. Females may be members of a particular unit group, or they may each occupy a separate home range that overlaps the home ranges of other females, with males simply defending as many female home ranges as they can (Wrangham 1979). In either event, females do enter and leave temporary subgroups within a unit group home range freely. Occasionally, females also visit neighboring groups on a temporary basis or transfer permanently to a new group (Pusey 1979; Nishida 1979).

The formation of temporary subgroups in chimpanzees may be a response to spatiotemporally variable food resources. Chimpanzees are predominantly frugivorous and prefer ripe fruits over alternative foods when they are available (Lawick-Goodall 1968; Jones and Sabater Pi 1971). The distribution of fruits is highly variable. In some seasons fruits are highly clumped, while in others they are more evenly dispersed. Chimpanzee group sizes are responsive to such changes. Groups are generally large when ripe fruits are localized in a few areas and small when they are more widely dispersed (Reynolds and Reynolds 1965; Lawick-Goodall 1968). Thus the incohesiveness of chimpanzee groups is apparently a response to seasonal variability in competition intensity at local food sources, and it is facilitated by the relative immunity of adult chimpanzees to predation.

The social structure of long-haired spider monkeys *(Ateles belzebuth)* also contains social groups whose members fragment into temporary subgroups (Klein and Klein 1975). Members of subgroups readily associate with one another and share a common home range. In contrast, individuals from neighboring home ranges are antagonistic when they encounter each other. Like chimpanzees, long-haired spider monkeys eat predominantly fruit. Klein and Klein analyzed the spatial distribution of fruiting trees and categorized them into five general types (Table 13–2). In seasons when spider monkeys forage largely on widely dispersed fruits, subgroups are small. Many isolated individuals can be seen, and the average group size (2.2) probably represents mother-daughter or other kin affiliations. In seasons when fruit occurs mainly in large localized clumps, subgroups are larger and relatively stable in size. During the remainder of the year, a large variety of trees with varied distribution patterns produce fruit, and average groups are intermediate in size but much less stable in composition. Thus group sizes are flexible and vary in response to seasonal variations in food availability.

Spider monkeys are potentially vulnerable to several predators, especially eagles and ocelots. Why they fragment into subgroups in the face of this threat is unclear. One possibility is that their extreme agility makes them more adept at eluding predators compared to most other arboreal

Table 13–2 *A classification system of tropical fruiting trees according to spatial and temporal dispersion of fruit availability.*

Classification of tree	Spatial dispersion	Temporal dispersion	Other characteristics
Ia	Widely dispersed throughout forest	Bears fruit for a few days to a week	Crown diameter greater than 60 ft
Ib	Same as Ia	Same as Ia	Crown diameter 25–60 ft
Ic	Same as Ia	Same as Ia	Crown diameter less than 25 ft
IIa	Clumped	Bears fruit for 1–2 weeks	Crown diameter very broad
IIb	Same as IIa	Same as IIa	Crown diameter moderately broad
IIc	Same as IIa	Same as IIa	Crown diameter narrow
IIIa	Widely dispersed throughout forest	Bears fruit for 2–4 weeks	Crown diameter greater than 60 ft
IIIb	Same as IIIa	Same as IIIa	Crown diameter 25–60 ft
IVa	Clumped	Bears fruit for 2–4 weeks	Crown diameter greater than 60 ft
IVb	Same as IVa	Same as IVa	Crown diameter 25–60 ft
V	Widely dispersed throughout forest	Bears fruit for 1–4 months	Crown diameter less than 25 ft

Source: Based on Klein and Klein 1975.

primates. Another possibility is that they rely on alarm calls of birds or other primates to detect predators, thereby reducing the value of remaining in larger social groups to maintain vigilance.

Comparable spacing behavior occurs among several other monkeys as well. Frugivorous black lemurs *(Lemur macaco)* forage separately in coherent bands but join together to sleep in the same tree (Petter 1962). Old World talapoins *(Cercopithecus talapoin)* and New World squirrel monkeys, both of which are frugivorous, forage in small subgroups but sleep together in large troops (Gautier-Hion 1970). Social groups of Syke's monkeys *(Cercopithecus mitis)* fragment into small subgroups during seasons when food is sparse and widely dispersed but coalesce into larger groups during seasons when food is highly clumped (Moreno-Black and Maples 1977). The congeneric vervet monkey remains in stable social groups all year because it has a more diverse diet and never concentrates on highly clumped foods. Why these species can afford to fragment into smaller subgroups despite an apparently high risk of predation is unclear.

Group instability. Some animals form unstable social groups that are not subgroups of larger, more permanent social groups on shared home ranges. Individuals occupy separate but overlapping home ranges and freely associate with other individuals or groups that happen to be on their home ranges at any given time. Many variations on this theme occur, particularly among ungulates (see Leuthold 1977). Selective pressures similar to those affecting the internal spacing behavior of carnivores and primates may be responsible for such spacing patterns.

The social groups formed by giraffes are among the least stable of all ungulates (Foster and Dagg 1976; Leuthold and Leuthold 1978; Leuthold 1979). Giraffes often remain alone, but they join and depart temporary groups at will. Individuals apparently show no preference for particular associates, and no higher-order social organization is evident. Giraffes are browsers that forage on a wide variety of green foliage from ground level to a height of about 5 m above ground level (Leuthold and Leuthold 1972). Competition for food is intense, as evidenced by the browse lines found in most parts of Africa where giraffes live (Photo 13–7) (Foster 1966; Leuthold and Leuthold 1972). Group sizes tend to be larger during dry seasons when giraffes are more dependent on the leaves of relatively clumped trees than during wet seasons when food is more widely dispersed as shrubs or herbs (Leuthold and Leuthold 1975). Since giraffes are relatively invulnerable to lions and other predators, due to their large size (see Schaller 1972; Moss 1975), their spacing behavior may be primarily a reflection of seasonal changes in food availability. However, giraffe grouping patterns are also influenced by the presence of young calves. Calves are vulnerable to predators, and females with calves often remain grouped together for prolonged periods of time (Leuthold 1979).

Open social groups with unstable membership occur in many other ungulates as well, but the more usual pattern is for small groups to move from one larger group to another (see Leuthold 1977). The environmental

Photo 13–7

A browse line at a height of about 5 m indicates intense competition for food in giraffes.

The browse line results from defoliation of all edible leaves and twigs within reach of adult giraffes.

Photo from B. M. Leuthold and W. Leuthold, "Food Habits of Giraffe in Tsavo National Park," *East African Wildlife Journal* 10 (1972):139.

bases of such grouping patterns have not been adequately studied. A particularly useful comparison might be made among the various species of the horse family (Equidae). Asiatic and African asses, Grevy's zebras, and feral donkeys form open social groups in which individuals constantly move from one group to another, while plains zebras, mountain zebras, and wild horses form highly cohesive stable social groups whose memberships rarely change through time (Klingel 1969, 1974a, b, 1977; Joubert 1972; Tyler 1972; Moehlman 1974). The reasons for this difference are presently unknown, but a comparative analysis of food distribution and abundance may prove enlightening.

Transferring from one group to another

Voluntarily leaving one group to join another entails both costs and benefits, some of which were already discussed above. The most important costs are increased risk of mortality, especially from predation, loss of social status, loss of kinship ties, and any costs suffered because group members resist entry of new individuals into their group. The most important benefits are improved survival or reproductive success gained from living in a group closer to optimal size, gains in social status, and avoidance of inbreeding. The magnitude of these costs and benefits differs for each individual. As a result, some individuals are more likely to change groups than others.

In open social groups, where group stability is low, individuals move freely and frequently from one group to another without meeting any resistance upon entering a new group. Individuals probably benefit from changing groups mainly because they can live in groups nearer to optimal size. Social status appears unimportant because dominance hierarchies are generally absent in such groups. The costs of transferring to a new group are often minimal in species with open social groups. Lions, hyenas, and chimpanzees are essentially invulnerable to predators, while ungulates often change groups by traveling in subgroups or by joining a new group when the transfer period is very short. Loss of kinship ties may be a significant cost in lions, chimpanzees, and spider monkeys, which may explain why subgroups remain together on a shared home range. It appears unimportant in ungulates, since individuals show little or no preference for associating with relatives.

In closed social groups, where group stability is high, the costs of transferring to a new group are likely to be high. Individuals often encounter considerable resistance when attempting to join a new group, and the transfer period may be prolonged if individuals must search at length for a group willing to accept them. Dominance hierarchies are more important in controlling access to resources, and loss of status may be a significant cost for dominant individuals. Loss of kinship ties may also be an important cost, especially if aid-giving behavior is directed preferentially at relatives.

Nevertheless, transfers from one group to another do regularly occur in species with closed social groups. Among primates such transfers usually involve subadult or sometimes adult males (Koford 1966; Nishida 1966; Struhsaker 1967b; Sade 1968; Boelkins and Wilson 1972; Drickamer and Vessey 1973; Packer 1975; Sugiyama 1976; Scott et al. 1980). Subadult males may be forced out of their natal groups by the aggression of dominant males, particularly in small groups where one adult male controls access to all females in the group. However, males voluntarily leave their natal groups in macaques and baboons, both of which live in multi-male troops. Several factors may explain why males rather than females transfer to new groups.

One important factor is that males stand to enhance reproductive success more than females. Males may eventually achieve higher social status and hence greater access to females by entering a new troop where other males lack information about their previous performance as adolescents. This advantage should be most important for males who were poor performers in juvenile play groups or were born of low-ranking females.

However, in rhesus monkeys such males are no more likely to change groups than males born of high-ranking females (Drickamer and Vessey 1973). Even so, the prospect of gaining higher social status cannot be discounted. Males often shift groups several times until they find one where they can penetrate the hierarchy, and then they are likely to remain indefinitely. Males may sometimes move from groups containing few females to groups containing more females. One study showed this to be true for rhesus monkeys (Drickamer and Vessey 1973), but earlier studies of the same population did not (Koford 1966; Boelkins and Wilson 1972). Finally, males may improve mating success by changing groups because females prefer mating with males who transferred into the group rather than males raised in the group (Sade 1967; Packer 1975). This preference may have evolved as a means of avoiding inbreeding (Itani 1972; Clutton-Brock and Harvey 1976), since inbreeding is known to reduce fitness of offspring (e.g., Schull and Neel 1965; Seemanova 1971; Hill 1974).

Additional factors may also be important, though they have not been adequately documented. Solitary males may be less vulnerable to predation than solitary females, making shifts to new groups less hazardous for them. Also, the cost of losing kinship ties may be higher for females than for males if aid-giving behavior follows kinship lines. In most primates cooperation appears more important for females, especially when accompanied by young infants.

Social dominance

What is dominance?

When animals live in stable groups, they repeatedly interact with the same individuals and soon learn the particular idiosyncrasies and capabilities of those individuals. This familiarity allows individuals to predict the outcome of social interactions before they occur, which in turn enables them to modify their behavior during the interactions. Such modifications of behavior are especially evident in competitive situations where aggression plays an important role.

When several animals are placed together in a cage for the first time, they are likely to engage in frequent and intense physical fights. As they gain familiarity with one another, they learn which individuals they can beat in a fight and which can beat them. Each individual then begins to give way to superior opponents without a fight, until finally a relatively stable dominance hierarchy emerges. **Social dominance,** then, is a priority system based on either overt or implied aggression. Dominant individuals gain prior access to resources or mates, while subordinate individuals must settle with the leavings.

Physical fights are frequently unnecessary for establishing dominance relationships. Morphological traits indicative of size, strength, or fighting ability usually suffice in dissuading weaker individuals from challenging stronger ones. It is only on the infrequent occasions when individuals are equally matched that overt fights are likely to occur, and then fighting may be intense and prolonged. As a result, dominance relationships often have the effect of maintaining an outward appearance of social tranquillity, even though that is not why they arise in the first place.

Advantages of being dominant

Dominance hierarchies evolve because aggressively superior competitors benefit by supplanting weaker rivals, while at the same time weaker individuals benefit by accepting inferior social status. The benefits gained by dominant individuals are relatively clear, although they vary among species.

An especially prevalent advantage of being dominant, at least among males, is gaining prior access to mates. Dominant males may succeed in totally excluding other mature males from social groups, as occurs in harem polygynous animals, or they may greatly hinder mating attempts of subordinate individuals within social groups. In an experimental study of laboratory mice, for example, DeFries and McClearn (1970) set up 22 groups consisting of three males and three females each. The males in each group quickly established stable dominance hierarchies. By using genetic markers, DeFries and McClearn could identify which males fathered each offspring. In 18 of the 22 groups the dominant male fathered all three litters, while in only one group did a subordinate male succeed in fathering as many as two litters. Overall, dominant males sired 92% of the offspring even though they constituted only one-third of all males in the experimental groups.

Equally extreme mating advantages are sometimes attained in nature (see p. 422). Ordinarily, dominant males are not that successful in monopolizing females, but positive correlations between high dominance status and mating success are the rule. Such correlations have been demonstrated in animals as diverse as domestic fowl (Guhl et al. 1945), Norway rats (Calhoun 1962), rabbits (Myers et al. 1971), elephant seals (Le Boeuf and Peterson 1969), ungulates (Schloeth 1961; Schaller 1967; Grimsdell 1969; Geist 1971), and primates (Bernstein 1976; Packer 1979b).

Dominant males often gain a mating advantage by hindering courtship or mounting attempts of subordinate males. However, part of their advantage also stems from a female preference for dominant males. Female elephant seals often protest copulation attempts by subordinate bulls, thereby inciting dominant bulls to attack the subordinates (Cox and Le Boeuf 1977). Likewise, recent evidence indicates that female preference may play a major role in determining male mating success in multi-male primate groups. In an experimental setting female pig-tailed macaques *(Macaca nemestrina)* were given the choice of releasing one of several males into their cages during estrus, and they showed a definite preference for particular males (Photo 13–8, Table 13–3) (Eaton 1973). The social status of preferred males was not determined in Eaton's study, but evidence from another study indicates that dominant males are at least sometimes pre-

Photo 13–8

The experimental setup used to study female mate preferences in pig-tailed macaques.

Females were exposed on regular occasions to all males used in the study for a period of one year prior to the tests. Each female was then given an opportunity to release each male into her cage during 5-minute tests. Females were only presented with one male at a time during the experimental tests.

Photo from G. G. Eaton, "Social and Endocrine Determinants of Sexual Behavior in Simian and Prosimian Females," in C. H. Phoenix, ed., *Primate Reproductive Physiology* (Basel: S. Karger, Symposium of the Fourth International Congress of Primatology, Vol. 2, 1973), p. 22.

Table 13–3

Percentage of tests in which female pig-tailed monkeys released particular males during preference tests.

Female number	Male number						
	2497	**2499**	**2500**	**2501**	**2502**	**2503**	**2504**
601	100.0	100.0	31.3	28.1	100.0	100.0	84.4
605	3.2	0	6.1	3.3	0	3.0	6.3
612	68.7	100.0	100.0	21.9	100.0	100.0	41.9
1275	23.5	96.7	12.5	6.5	96.9	100.0	16.7
2510	100.0	100.0	90.6	39.4	100.0	100.0	96.9
Mean	59.1	79.3	48.1	19.8	79.4	80.6	49.2

Note: Females were presented with one male at a time during all stages of the estrous cycle. Data represent percentages that each male was released during a 5-minute test period. Total number of tests = 1111; mean number of tests per pair = 31.7.

Source: From G. G. Eaton, "Social and Endocrine Determinants of Sexual Behavior in Simian and Prosimian Females," in C. H. Phoenix, ed., *Primate Reproductive Physiology* (Basel: S. Karger, Symposium of the Fourth International Congress of Primatology, Vol. 2, 1973), p. 25. Reprinted by permission.

ferred. Seyfarth (1978) found for one baboon troop containing only two adult males that consort preferences usually resulted from female choice and that most females strongly preferred the dominant male. Nevertheless, females sometimes prefer subordinate males or mate with both subordinate and dominant males during peak estrus (e.g., see Loy 1971; Rowell 1972). The reasons for such preferences have not been satisfactorily explained. One possibility is that females sometimes choose males according to qualities unrelated to aggressiveness or status, but the nature of such qualities, if they exist, is unknown.

A second benefit of dominant status is gaining prior access to food. This advantage is most important when survival is directly at stake. Gaining prior access to food is therefore an important advantage of being dominant in winter bird flocks (see Fretwell 1968, 1969b; Murton 1968; Murton et al. 1971). In social mammals survival is less often at stake, and individual food items may not be worth contesting. For example, the low value of individual mouthfuls of grass and foliage may explain why priority systems of food access have not evolved in most social herbivores or folivores (Geist 1974b; Clutton-Brock and Harvey 1976). In some cases rapid consumption rather than supplanting other group members is a more effective way to compete for food, as in most social carnivores (Bertram 1978b). Priority of access is important, however, when individual food items are worth contesting. For example, vertebrate prey that are captured on occasion by baboons and chimpanzees are highly desirable food items and are monopolized by dominant males (Altmann and Altmann 1970; Teleki 1973a). Females could potentially benefit by gaining prior access to food, since they are resource-limited, but in many social mammals they cannot dominate males (for exceptions see Ralls 1976).

A third advantage of high status is reduction of stress. Dominant individuals are relaxed and relatively unstressed because they need not concern themselves with potential aggression from other group members. Subordinate individuals, on the other hand, must remain constantly alert to the actions of dominants and either avoid or appease dominants whenever they approach too closely. A few examples illustrate the consequences of being subordinate. In wood pigeon flocks subordinate birds feed less rapidly than dominants because they spend time monitoring activities of dominants and forage on the periphery of flocks where dominants are less frequently encountered (Murton 1967; Murton et al. 1971). Furthermore, histological examinations show that subordinate birds have enlarged adrenal glands, one indicator of high chronic stress. Similarly, subordinate pumpkinseed fish initiate fewer aggressive interactions than dominants and develop larger interrenal glands, the source of corticosteroid hormones in fish (Erickson 1967). In rhesus monkeys the lowest frequencies of aggressive behavior occur among the most dominant and most subordinate individuals (Kaufmann 1967). Aggressive conflicts are most common among middle-ranking individuals, and those individuals probably suffer the greatest stress.

Why accept subordinate status?

Subordinate individuals often have reduced access to food and limited opportunities to mate. They must constantly defer to more dominant individuals, and they are continually stressed by the presence of dominants. Why then do individuals accept the costs of subordinate status?

The answer comes from considering the alternative options open to subordinate individuals. Animals with low status have basically three options. They can emigrate to join a new group, they can leave the group to live solitarily, or they can accept subordinate status within the group.

Emigrating to a new group is sometimes advantageous, as was discussed earlier. It usually does not lead to immediately increased status, but it may allow an individual to attain improved status sooner than would otherwise be possible. Of course, not all individuals can improve status by emigrating, since some individuals must necessarily be subordinate whenever others are dominant. The only real escape from subordinate status is solitary living.

Living solitarily may be advantageous if the risks of predation are not too high. Taking this option may improve an individual's survival prospects by increasing foraging efficiency, but for males it prevents breeding when females are all found in social groups. For many animals, living outside a social group is extremely hazardous because predation risks are high. The reasons why individuals form social groups in the first place are also the reasons why individuals are better off accepting subordinate status within a group instead of becoming solitary.

Remaining with a social group instead of living temporarily alone is also advantageous because it improves prospects for attaining high status at some future time. By remaining in social groups, subordinates can gain valuable experience that allows them to compete for higher status once the more dominant individuals have grown old. Having seniority in a group is often prerequisite to gaining high status, and seniority can be accumulated by temporarily accepting subordinate status.

Finally, subordinate animals are not completely excluded from resources and mates. They are usually able to obtain adequate amounts of food, except during periods of scarcity, even though they cannot gain access to the most desirable food items. In some social groups subordinate males can also successfully copulate with some females. Although they are not typically as successful as more dominant males, they are more successful than would be possible should they become solitary.

Criticisms of the dominance concept

According to the classical theory of social dominance, high status is attained by superior fighting ability and confers priority of access to desirable or contested resources, particularly food and mates. This view has been strongly criticized by many behavioral biologists, and some have abandoned the concept altogether.

A major criticism has been that different measures of status are often poorly correlated (Gartlan 1968; Richards 1974; Syme 1974; Bernstein 1976; Popp and Devore 1979). For example, the dominance rankings of male olive baboons *(Papio anubis)* when competing for food provided by an observer differ from their rankings when competing for estrous females (Hall and Devore 1965). Male Hanuman langurs who are dominant in agonistic interactions do not necessarily copulate with the most females (Jay 1965), although they may possibly copulate more often during peak estrus. The social rankings of male pig-tailed macaques differ depending on whether they are based on agonistic interactions, number of mountings, or grooming interactions (Bernstein 1970). A male's rank in one context is not predictive of his rank in another. The social status of male bonnet macaques *(Macaca radiata)* is not correlated with success in mating with females (Rahaman and Parthasarathy 1969). The dominance hierarchy of male ring-tailed lemurs during nonbreeding seasons breaks down when females begin estrus, and the previously most dominant males do not necessarily mate with the most females (Jolly 1966). Similar examples could be cited for cats, rats, mice, and other rodents, particularly in laboratory studies where dominance is measured in competitive tests for food, water, or shock avoidance (Syme 1974). The dominance concept loses its explanatory power when competitive orders change for each context. Little is gained by postulating a separate dominance order for each competitive situation, and a more parsimonius interpretation would be that individuals simply have different abilities to compete in each situation.

Dominance has been widely used as an interpretive concept because success in agonistic encounters is often correlated with increased access to food or mates. The fact that dominance rankings in different contexts are sometimes weakly correlated does not necessarily make the dominance concept useless, although it does point out deficiencies in overly simplistic notions of dominance.

One reason dominance rankings may vary in different competitive situations may be that the costs and benefits of gaining access to different requisites vary among individuals (Clutton-Brock and Harvey 1976). Popp and Devore (1979) have shown with their model of aggression (see pp. 179–180) that the outcome of an agonistic interaction should depend on both intrinsic competitive abilities of the individuals involved and the value of the resource to each individual. The value of a food item, for example, may be sufficiently higher for a competitively inferior individual that he or she will persist long enough to win a dispute for it despite his or her poorer competitive ability. Some empirical evidence supports this hypothesis. Rhesus monkeys kept on low-protein diets have higher dominance rank than those kept on high-protein diets when competing for food, even though they have lower dominance rank when competing to avoid shock treatments (Wise and Zimmerman 1973).

Low copulatory success of aggressively dominant males cannot be attributed to lower benefits derived by winning, since all males should benefit equally from high copulatory success. Discrepancies between aggressive superiority and copulatory success may arise for several reasons. Male copulatory success depends partly on female mating preferences, and females may select males partly on the basis of cues unrelated to a male's current dominance rank. Dominant males may tolerate copulations by subordinate males to encourage subordinates to remain in the group. Dominants may benefit from such encouragement because subordinates can detect predators or draw attacks away from dominants and they can reduce parasite infestations by grooming dominants. In some species dominant males may be better able to retain control of female groups when subordinates are present. This may explain why subadult male offspring are tolerated in some primate groups, giving rise to the age-graded male groups described by Eisenberg and his co-workers (1972). Finally, dominant males may father more offspring even though they copulate less often than subordinates, either because they ejaculate more often or because they restrict access to females only during peak estrus. In rhesus monkeys, for example, some of the most sexually active males have low social status, but these males ejaculate less often than dominant males (Hanby et al. 1971).

A second criticism of social dominance theory has been that dominance hierarchies are most evident in caged social groups, suggesting that they may be an artifact of captive situations that is caused by stress (Rowell 1967; Gartlan 1968). Subordinate individuals play a major role in maintaining dominance rankings, and since these individuals are subjected to the most stress during agonistic interactions, high levels of stress associated with captivity could plausibly explain how dominance hierarchies arise (Rowell 1974). This view is bolstered by evidence that artificial provisioning of primate groups in the wild generally leads to greatly increased frequencies of agonistic behavior (Hall and Devore 1965; Southwick et al. 1965; Rowell 1967; Sade 1967).

There is good reason to believe that social dominance is not just an artifact induced by artificial conditions. Some degree of hierarchical ranking is evident within natural groups of most social animals, and cases where hierarchical ranks have not been observed may reflect inadequate sampling rather than an actual absence of a dominance order (Clutton-Brock and Harvey 1976). In long-established social groups individuals are very familiar with one another, and deference to more dominant individuals may be expressed in extremely subtle ways without ever showing overt or even covert aggression. An observer could then easily fail to detect a dominance order by not recognizing the significance of subtle communications between group members. More frequent overt aggression in captive situations is not surprising, as individuals are less familiar with one another and hence less able to predict the outcome of competitive interactions without resorting to aggression. The fact that artificial provisioning in natural environments causes increased aggression is not surprising, since the sudden introduction of highly desirable food items greatly increases competition intensity in a local area. Similar increases in aggression also occur under natural conditions when a primate group suddenly encounters a locally clumped food source (Chalmers 1968). Stress is probably not responsible for social hierarchies in the wild, since subordinate individuals are generally not under sufficient stress to evoke the general adaptation syndrome (GAS) (Deag 1977). Although stress may make agonistic interactions more intense and more frequent in captivity, it probably is not the sole factor leading to social dominance rankings.

Perhaps the most cogent criticism is that social dominance theory overemphasizes aggression as an organizing factor in social groups (see Loy 1975; Bernstein 1976). While many competitive situations are resolved in light of dominance relationships, affiliative ties based on shared interests are of crucial importance for maintaining group cohesion and defining social relationships within groups. Social dominance theory should not be discarded, but dominance interactions should be viewed as just one of several ways that individuals pursue their own best interests within social groups.

Social roles

Criticisms of social dominance theory and observations showing that many behavioral differences among individuals do not result from competition for resources, mates, or other incentives led to a reinterpretation of behavior in terms of social roles (Bernstein 1964, 1966; Bernstein and Sharpe 1966; Gartlan 1968; Fedigan 1976). A **social role** is a set of behaviors that contributes to the organization or welfare of a social group. For example, the alpha (most dominant) male often plays the role of control animal, which consists of protecting the group from external disturbances such as predators or conspecific intruders, interfering with and terminating aggressive disputes among other group members, and sometimes determining group movements during daily ranging (Bernstein 1966).

Constructing "role profiles" such as the one presented by Gartlan (1968) for vervet monkeys (Figure 13–6) facilitates comparisons between social structures of different species or populations of the same species (Hinde 1974). Analyzing social roles also reveals how individuals of each age and sex class affect the internal dynamics of social groups.

A serious shortcoming is that the concept of social roles confuses theoretical issues regarding the evolution of social structure. The concept emphasizes the consequences of an individual's behavior on group welfare and diverts attention away from the consequences on each individual's inclusive fitness. The result is a tendency for researchers to interpret the adaptive significance of individual behaviors in terms of group welfare rather than inclusive fitness (e.g., Bernstein and Gordon 1974; Hinde 1974). Since all available evidence indicates that individuals behave in their own best interests, emphasis on an individual's contributions to group welfare is likely to produce misleading interpretations. Social roles are functional consequences of each individual's behavior on group dynamics and social organization, but these consequences do not necessarily reflect the adaptive significance of the behaviors involved. Thus the concept of social roles is

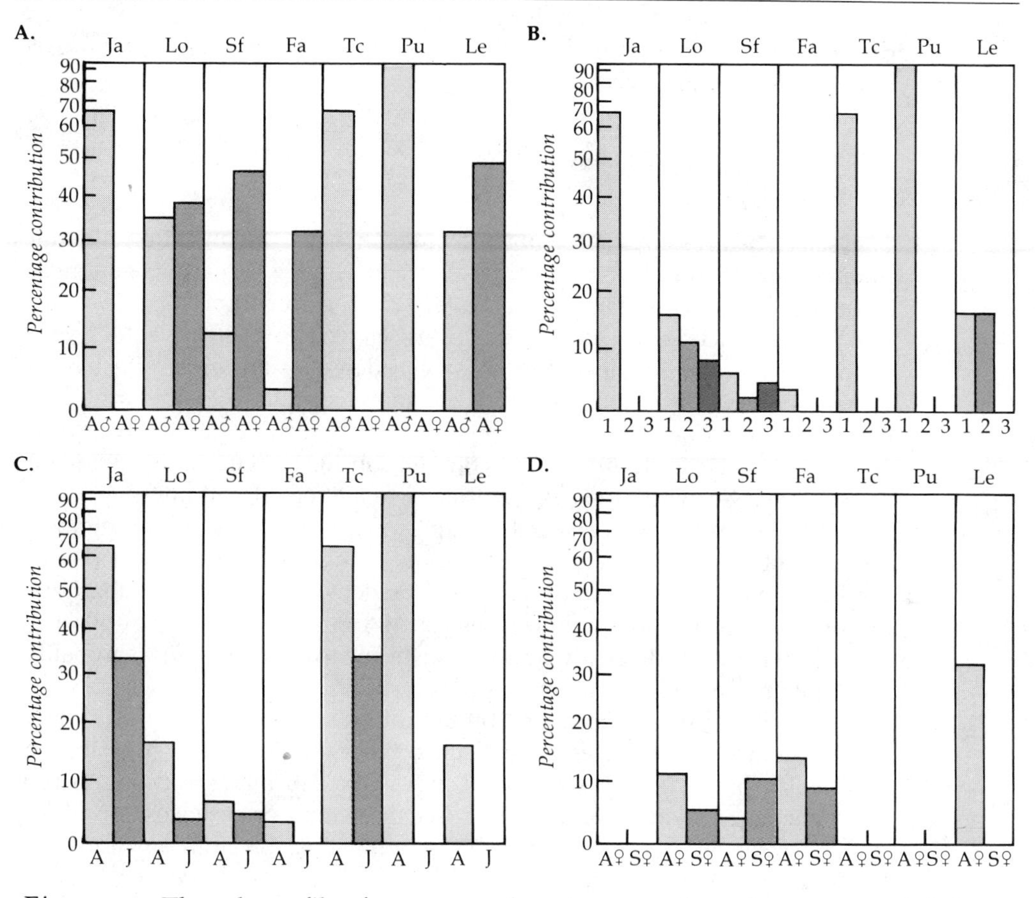

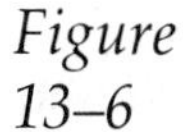

Figure 13–6 *The role profile of group members in a vervet monkey troop on Lolui Island in Lake Victoria, Uganda.*

Ja = jumping around (a territorial display); Lo = looking out (vigilance); Sf = social focus (object of friendly approaches); Fa = friendly approach; Tc = territorial chasing; Pu = punishing (interfering with intragroup aggressive disputes); Le = leading group movements. Note that the vertical axis is a logarithmic scale. A. The social roles of adult males (A♂) and adult females (A♀). B. The social roles of individual adult males (numbered 1–3). C. The social roles of adult (A) and juvenile (J) males. D. The social roles of adult (A♀) and subadult (S♀) females. Note that territorial behavior and interference with aggressive disputes were largely performed by a single adult male. Friendly approaches and the object of those approaches (social focus) primarily involved adult females. Leading behavior was exhibited only by adult males and females. Thus the various age and sex classes of group members show different behavioral repertoires which reflect their different individual interests within the group.

Source: From J. S. Gartlan, "Structure and Function in Primate Society," *Folia Primatologica* 8 (1968):108–9. Reprinted by permission.

useful for describing and analyzing the functional organization of group processes, but it is inappropriate for studying the adaptive significance of individual behaviors within social groups.

Conclusion

The selective pressures affecting mammalian social organization are not much different from those affecting other vertebrates. Mammals generally form social groups to gain better protection from predators or to capture prey more efficiently. Open habitats are especially conducive to group formation because in those habitats social defenses are most effective at deterring predators. Closed habitats are more conducive to solitary living because then concealment is often the best defense against predation. Group size is typically limited by competition for resources, and hence it is usually larger when food is clumped or relatively abundant.

The cohesiveness of social groups reflects intensity of competition and risk of predation. When competition changes seasonally and risk of predation is relatively low, groups often fragment into temporary subgroups of variable composition. When transferring from one group to another is not very hazardous, individuals are more prone to change groups, probably to maintain a more optimal group size.

Social organization often revolves around status relationships, which are frequently based on agonistic interactions. Aggressively superior individuals typically gain better access to resources or mates, although their mating success is often tempered by female mating preferences and alternative strategems employed by subordinate males. Social interactions are complex phenomena, reflecting the individual interests of group members and the compromises reached when those interests conflict. Perhaps the most important theoretical questions for future research will revolve around the ways that individuals pursue their own best interests within social contexts and how those pursuits are constrained by the activities of others. Social behavior is complex, but it can be understood by viewing it from the different perspectives of the various individual group members.

Human sociality

Human cultures are the most complex of all social systems found among animals, and the processes responsible for cultural change represent one of the major unsolved problems of science. Just as the theory of organic evolution is the central unifying principle of biology, a theory of cultural evolution could potentially be the central unifying principle of the social sciences.

In addressing this problem many sociobiologists maintain that natural selection provides a basis for understanding cultural change. They point out that human societies have evolved at least initially in biological contexts and argue that human behavior is not immune from the effects of natural selection. There is every reason to believe that human behavior is influenced genetically as well as environmentally, and hence it should be subject to natural selection processes (Alexander 1979). There is also every reason to believe that natural selection is not the only force shaping human behavior and that a theory of cultural selection will be necessary before the process of cultural evolution can be understood. The objectives of this chapter are to examine how cultural processes might lead to cultural change and to evaluate the role natural selection might play in those processes.

Processes of cultural change

The problem

Cultural change is a continual and often rapid process. New languages, new institutions, new political and economic systems arise, are modified, and sometimes disappear. Human societies diversify into different tribes and nationalities. New technologies and new belief systems foster new life styles and new ethical standards. Fashions change. Fads appear and disappear. People mourn after the good old days and yearn for progress. Cultures are crosscurrents of change, and yet they retain an identity based on relatively stable underlying attributes. History is a chronicle of cultural change as well as a progression of human events, and yet historical studies have not revealed why the changes took the directions they did. The challenge, therefore, is to explain and understand the processes responsible for the changing events of human history.

A general theory of cultural evolution must explain both long-term

Chapter fourteen

trends and short-term fads. It may not predict future directions of change, just as the theory of organic evolution cannot, but it should predict previously unidentified conditions that were responsible for changes that have already occurred.

No one theory of cultural evolution has gained widespread acceptance, but development of a unified theory should have high priority. The following sections will outline the shape a theory of cultural change might take, but no theory will be acceptable until it has proven its applicability across a wide range of differing cultures. Testing the various theories of cultural evolution will be formidable, and it is likely to preoccupy social and biological scientists for many years to come. Hence the ideas presented below should not be treated as proven doctrine but rather as building blocks toward further theoretical advances.

Cultural offspring and memes

The most promising approaches to cultural evolution are analogues to the theory of organic evolution (e.g., Campbell 1965; Ruyle 1973; Cloak 1975; Richerson 1977; Richerson and Boyd 1978). They postulate units of variation analogous to genes, competition, and a selective process. This kind of approach is useful, but only to the extent that dissimilarities as well as similarities between cultural and biological processes are kept in mind.

A first problem is to identify the units of variation upon which cultural selection acts. A fundamental unit should be one that is not itself changing in response to some internal selective process. Whole cultures, for example, would not qualify because cultures clearly evolve from within. The same can be said of institutions and other levels of organization that involve groups of individuals or combinations of innovations within cultures. The most likely units are ideas, innovations, norms, values, and similar sorts of information because such information forms the basis of what is selectively transmitted from one generation to the next (Cavalli-Sforza and Feldman 1973).

The nature of cultural units of variation is not easy to characterize. Richerson and Boyd (1978) coined the term *cultural offspring* to refer to them, based on an analogy with biological offspring. They argue that individuals compete to transmit items of culturally coded information to the next generation, with information that allows individuals to compete most successfully coming to predominate in a culture. Thus they envision cultural selection as a process that selects for phenotypic traits or behaviors that enhance an individual's ability to produce cultural offspring.

The argument presented by Richerson and Boyd implies that individuals compete to propagate cultural offspring to future generations. However, most people are not concerned about promoting particular ideas or innovations. They are primarily concerned with reaching particular goals or fulfilling particular needs. Ideas and innovations do not spread because they are effectively promoted; they spread because they have utilitarian or aesthetic value. The concept of cultural offspring is probably too close an analogy to biological offspring, and it is likely to facilitate erroneous thinking about how cultural processes work.

Dawkins (1976) has suggested an even more direct analogue to genes. He coined the term *meme* from a Greek root that conveys the idea of a unit of imitation and postulated that memes are the fundamental units of cultural variation. Competition among memes would then presumably determine which ideas and innovations would spread fastest, although Dawkins did not specify exactly what memes are competing for.

Here again there are problems with drawing too close an analogy with biological processes. No cultural characteristics can be identified that have the same properties as genes. Genes are discrete functional units on chromosomes. They can be inherited more or less independently of each other or as discrete linkage groups, and they are duplicated with high copying fidelity. They encode information by means of a universal genetic code, and they are transmitted in their original form to subsequent generations. They can be propagated only from parents to offspring and cannot spread in any other way.

No comparable units can be identified within cultures. Ideas and innovations are not discrete entities that are transmitted either wholly or not at all. They are continually modified and combined in new ways. Parts of one idea or innovation may be incorporated into one new belief system or technology, while other parts of the same idea or innovation are incorporated into another. There is no fundamental unit below which further subdivisions of an idea or innovation cannot be made; nor is there any unit that is transmitted in its original form to subsequent generations.

To make these points clearer, consider the scores for a symphony. One small portion of a symphony may be incorporated into a new musical score by one composer. Another portion, which contains part of that used by the first composer, may be incorporated into another new score by a second composer. Each portion represents an idea, but the ideas overlap and contain elements in common. No clear boundary exists between them. A nearly infinite number of ideas (sections of the original symphony) could be incorporated into new symphonies simply by taking out pieces that begin and end at different points in the original score. Moreover, the ideas from an original symphony may be somewhat modified to suit the new symphony, and yet the essence of the idea still exists. Clearly, ideas are not discrete entities that are replicated or mutated as indivisible functional units.

Thus an essential difference exists between genetic and cultural variation. Genetic variation involves variation among discrete units, each of which can be objectively separated from the others according to functional criteria, while cultural variation involves variation along a continuum that cannot be objectively partitioned into a set of discrete functional entities.

Assuming that fundamental units of cultural variation do not exist poses no theoretical problem. Cultural evolution involves continual modification of social organization, technology, belief systems, and modes of behavior. Every component of a culture can be modified in many ways, and some ways are more appropriate for performing a task or reaching a goal than others. Alternative ways of doing things need not be discretely different or mutually exclusive for a selection process to occur. They need only be different. Cultural evolution can occur in the absence of any fundamental units of variation, and cultural selection need not and, indeed, does not involve competition for survival among mutually exclusive fundamental units within a meme pool.

Levels of selection

A common view among anthropologists is that cultural evolution involves selection between whole cultures or social groups. Cohen (1968) argues, for example, that the essence of cultural adaptation is the facilitation of "the reproductive and survival capacity of the group." Cohen's argument reflects an especially prevalent view among ecologically oriented anthropologists, namely, that cultures evolve to regulate population density (e.g., Rappaport 1968; Gross 1975; Hayden 1975; Thomas 1975; Vayda and McCay 1975; Divale and Harris 1976; Harris 1977). The evidence from nonhuman animals consistently shows, however, that behavior does not evolve to that end, and the evidence from human cultures is all open to alternative interpretations (e.g., see Bates and Lees 1979). Indeed, the phenomenal growth of the human population during the past two centuries, which has been accompanied by equally dramatic cultural changes, argues against any built-in tendency for cultures to regulate population density.

Marvin Harris (1971, 1977) makes perhaps the strongest case for selection at the level of groups. He writes (1971), for example:

> *The most successful innovations are those that tend to increase population size, population density, and per capita energy production. The reason for this is that, in the long run, larger and more powerful sociocultural systems tend to replace or absorb smaller and less powerful sociocultural systems. The mechanism of innovation does not always require actual testing of one against another to determine which contributes most in the long run to sociocultural survival. Given a choice of a bow and arrow versus a high-powered rifle, the Eskimo adopts the rifle long before there is any change in the rate of population growth. In the short run, the rifle spreads among more and more people not because one group expands and engulfs the rest, but because individuals regularly accept innovations that seem to offer them more security, greater reproductive efficiency, and higher energy yields for lower energy inputs. Yet it cannot be denied that the ultimate test of any innovation is in the crunch of competing systems and differential survival and reproduction.*

While Harris accepts the importance of individuals choosing from among competing alternatives in the short run, he concludes that the ultimate survival of an innovation in the long run depends on its consequences for the entire culture. Let us examine this argument a little more closely.

Do cultures compete, with one culture eventually overrunning or engulfing another? In a sense, yes, they do. Conquering and subjugating foreign cultures has long played an important role in human evolution. This process certainly wreaks dramatic changes on both conquered and conquering cultures. However, there is a subtle distinction to be made here. Such changes could occur because they promote group interests (i.e., enhanced survival of the group), or they could occur because they promote the interests of individuals within the power structure of each culture. I will argue here that the latter interpretation is most appropriate.

Ruling individuals decide to wage war because they expect to reap benefits for themselves, and they gain followers because their followers also expect to reap benefits. Subjugated cultures change because their members choose to accept changes imposed by their conquerors as a better alternative than resistance. Innovations that enhance power or fighting ability spread because they give ruling members of one group the capability of pursuing their own interests at the expense of individuals in other (or their own) social groups. They do not spread because they enable one social group to replace another. Cultural evolution does not result from differential extinction or proliferation of cultures. It results from competition between individuals, either within the same culture or from different cultures. Individuals compete for desired goals that cannot be attained by everyone, and an understanding of those goals will allow an understanding of how cultural selection works.

A theory of cultural selection

Cultural selection occurs because individuals choose from among alternative ways of doing things to promote what they perceive as their own best interests. Ideas and innovations spread because they are imitated or learned and then are chosen for acceptance by individuals within a culture. Institutions and other higher levels of social organization arise because individuals organize them as useful tools for achieving their goals. When many individuals make the same choices, or when certain individuals gain the power to impose their choices on others, directional selection occurs.

Two elements are necessary for directional selection to occur. Competition between alternative ideas, innovations, or ways of doing things must be present, and certain selective criteria must have relatively general importance in determining how individuals choose.

Competition exists for several reasons. Technological choices involve economic investments and entail risks of economic loss. Alternative technologies therefore compete on the basis of cost effectiveness or economic feasibility. Behavioral choices have long-lasting effects on an individual's ability to acquire wealth, gain power, attain high social status, or achieve other goals, and acceptance of one option often precludes later acceptance of alternative options. Since time is limited, individuals cannot do or try everything. They must therefore choose how to use their time. Competition among ideas also exists. Books compete for readership. Music competes for listeners. Movies, plays, and sporting events compete for audiences. Since the outcome of such competition affects all individuals who have vested interests in the ideas, competition between ideas is actually competition between individuals behind the ideas. Every idea and innovation in a culture entails a cost for individuals who accept them, and hence not all can be accepted. This is the essence of competition among cultural attributes.

There are probably as many selective criteria as there are people within a culture, since each individual pursues a different set of goals and makes choices in a different way. Nevertheless, a few criteria have widespread importance. Some general goals pursued by many or most people within all cultures are (1) survival, (2) finding a mate or mates, (3) rearing children, (4) accumulating economic goods or wealth, (5) gaining influence or power, and (6) remaining or becoming socially accepted by particular subgroups within the society or by society as a whole. Probably not all cultural changes can be ascribed to selective pressures generated by people pursuing these goals, but certainly many can.

The role of the above criteria in shaping cultures has not been conclusively established, but their potential importance is suggested by the following assertions. Enhancing survival is an impetus for developing basic means of acquiring food, clothing, and shelter, for improved understanding of health and nutrition, for the proliferation of medical care and medical technology, for health fads, and for the invention and proliferation of more effective weaponry and methods of self-defense. The goal of marrying and raising children gives impetus to better ways of competing for or selecting mates, the existence of nuclear and extended families, and development of improved standards of living. The goal of accumulating wealth is an impetus for more cost-effective technology, increased worker productivity, entrepreneurism, the formation of corporations and armies, the development of offensive weapons, and the creation of various institutional mechanisms for retaining wealth once it has been accumulated. The goal of gaining power is an impetus for creating political systems, bureaucracies, armies, and various other social institutions. The goal of remaining or becoming socially accepted generates peer pressure, creates fads and fashions, makes blackmail and social ostracism important instruments for wielding power, gives rise to status symbols, and promotes charitable causes.

The origin or spread of any idea, innovation, or form of social organization should be examined in light of individual interests. Hypotheses can be constructed on the basis of possible goals that individuals pursue when choosing to accept or originate new ways of thinking or doing things. These hypotheses must be formulated with care. For example, one might argue that scientific ideas spread because they have explanatory power (e.g., Ruyle 1973), but one might argue instead that ideas with explanatory power spread because scientists who accept them gain or maintain prestige while those who do not lose prestige. The distinction is crucial. The first hypothesis implies that changes occur as a result of inherent qualities in the ideas themselves, while the second implies that changes occur as a result of social consequences for individuals who accept or reject them.

605 Role of natural selection

Two views. Two rather different views of how natural selection might influence human behavior have been advocated by sociobiologists and social scientists. According to one view, each specific human behavior or predisposition has some genetic basis and has been selectively retained in a culture because it has adaptive value in the biological sense (Wilson 1975a, 1978; Weinrich 1977; Barash 1979c; Freedman 1979). Some behaviors may have no adaptive value, but a legitimate approach from this viewpoint is to evaluate the adaptive significance of each specific behavior or cultural attribute as a separate entity. According to the other view, natural selection shapes the motivational makeup and underlying neurophysiological mechanisms of human behavior in such a way that humans tend to make choices in a biologically adaptive manner (Durham 1976, 1979). The specific behaviors or decision-making mechanisms involved need not have any genetic basis. Thus the first view is that natural selection affects cultures by promoting specific behaviors or decision-making mechanisms. The second view is that natural selection shapes general motivational systems and behavioral preferences without acting directly on specific behaviors or decision-making mechanisms.

Female infanticide: Biologically adaptive or culturally motivated? Sociobiologists who take the first viewpoint evaluate the importance of natural selection by determining whether specific behaviors lead to increased lifetime reproductive output. One example is an analysis of infanticide and reproductive strategies in socially stratified polygynous societies (Dickemann 1979). Dickemann's analysis was based on the prediction by Trivers and Willard (1973) that females could increase their fitness by varying the sex ratio of offspring in accordance with their ability to provide parental investment (see pp. 411–413). In polygynous animals females in good condition should, according to Trivers and Willard, produce more sons than daughters, because sons who receive high parental investment would later have a competitive advantage in attracting mates and hence would ultimately produce more descendants than could daughters. Females in poor condition should, by the same token, produce more daughters than sons, because sons who receive low parental investment would later be at a competitive disadvantage in attracting mates and hence would ultimately produce fewer descendants than could daughters. Using a parallel argument, Alexander (1974) predicted that in polygynous human societies "female-preferential infanticide is more likely among women married to high-ranking men and less likely among women married to low-ranking men or not legitimately married at all."

Dickemann (1979) tested Alexander's prediction with data from India, imperial China, and medieval and early modern western Europe. The best information is for several subcultures in India; only suggestive information is available for China and western Europe. In every instance the prediction was substantiated. From these results one might conclude that natural selection has led to the prevalent killing of young daughters among higher social strata of several human cultures.

Such a conclusion is not justified unless alternative hypotheses have been evaluated and rejected. This was not done in Dickemann's analysis, and a plausible alternative can be suggested. In the societies studied by Dickemann, men were permitted to marry women from the same or lower social strata but not women from higher social strata. The principal means by which individuals improved their social status was to marry their daughters to men from higher social strata. Economic wealth was not of itself sufficient to improve status; it merely conferred the means of providing a large enough dowry for daughters to marry into a higher social class. Daughters were therefore an asset to families seeking to improve social status because they were the means of cementing social alliances. However, daughters were a liability for families who already held high status because they could not directly contribute to the family's wealth or power and they required a large dowry for marrying a socially acceptable man of equally high status. Thus killing daughters may have been more prevalent in high social strata because economic considerations and status seeking led to different decisions for members of high- and low-ranking families. The general motivations to accumulate wealth and improve status can explain the observed results without attributing them directly to natural selection.

A similar explanation can be devised for virtually any human behavior that might affect biological fitness. Without evidence that the specific behavior in question has a genetic basis and cannot be explained by strictly cultural processes, an active role of natural selection in promoting that particular behavior cannot be demonstrated.

Biases that channel changes in adaptive directions. Invoking motivational mechanisms to explain specific behavioral choices may seem like a proximate rather than an ultimate explanation of human behavior, but this is not the case if motivational mechanisms are themselves the selective criteria responsible for a behavior's prevalence in a culture. As Durham (1976, 1979) has argued, natural selection may have favored the capacity for behavioral flexibility, along with some motivational biases for channeling that flexibility in adaptive directions, without promoting any of the specific behaviors actually adopted. People may behave adaptively because they are physiologically biased to do so and not because the behaviors themselves spread by the differential propagation of genes coding for them.

Durham (1979) lists four biases that could channel behavior in adaptive directions. One bias is the process of socialization, which teaches young children to make choices according to particular criteria or standards, as well as to adhere to social norms and accept traditional patterns of behavior (LeVine 1973; Pulliam and Dunford 1979). Children are taught to make decisions that are partly in the best interests of themselves and partly in the best interests of their parents and society as a whole. These best interests could be in terms of survival and reproductive success, but they need not be.

A second bias is what Ruyle (1973, 1977) calls the "bias of satisfaction." People experience positive reinforcement and feelings of satisfaction from some kinds of behavior, and they experience negative reinforcement or pain from others. This bias is the mode of action for motivational systems. Presumably, in early stages of hominid evolution, selection for behavioral tendencies that enhanced survival or reproductive success led to neurophysiological mechanisms that confer positive reinforcement, while selection against behavioral tendencies that detracted from survival or reproductive success led to mechanisms that confer negative reinforcement (see Pulliam and Dunford 1979). Thus our motivational makeup should reflect preferences that were at least initially adaptive.

A third bias is the structure and functioning of our sensory receptors and brain. We are more attentive to some kinds of stimuli than to others, and our brains filter and process information in particular patterns before it is stored or acted upon. Such biases in sensory input and response patterns may well channel our behavior in adaptive directions. Although direct evidence for this supposition is unavailable for humans, comparable mechanisms are well documented for other animals (see Marler and Hamilton 1966), and it seems probable that human behavior is channeled in a similar way.

A fourth bias is what Cloak (1977) refers to as "circumstantial bias." In cultures where parents ordinarily rear and enculturate their own children, Cloak argues, "a cultural instruction whose behavior helps its human carrier-enactor (or his/her relatives) to acquire more children thereby has more little heads to get copied into." As a result, instructions that enhance survival and reproductive success should be differentially propagated across generations until they come to prevail in the culture. The importance of this effect is debatable, however, because children do not necessarily follow cultural instructions instilled by their parents and much cultural information is learned from nonparents. The extent that particular cultural attributes are promoted because they favor large families remains to be shown.

Durham (1979) argues that these four biases induce individuals to choose biologically adaptive cultural attributes even though natural selection is not directly selecting for the particular choices involved. These biases do not automatically channel human behavior in adaptive directions, but they may produce a strong tendency in that direction. Although the adaptive value of many cultural attributes remains to be demonstrated, some recent studies suggest that a great variety of human behaviors do seem to enhance biological fitness (e.g., Weinrich 1977; Alexander and Noonan 1979; Kurland 1979).

Biological bases of the criteria underlying cultural selection

Six general goals that many people pursue during their lifetimes were proposed above as the principal criteria underlying cultural selection. People seek these goals to satisfy deep-seated psychological or motivational needs. Such motives might themselves have originated from the processes

Figure 14–1

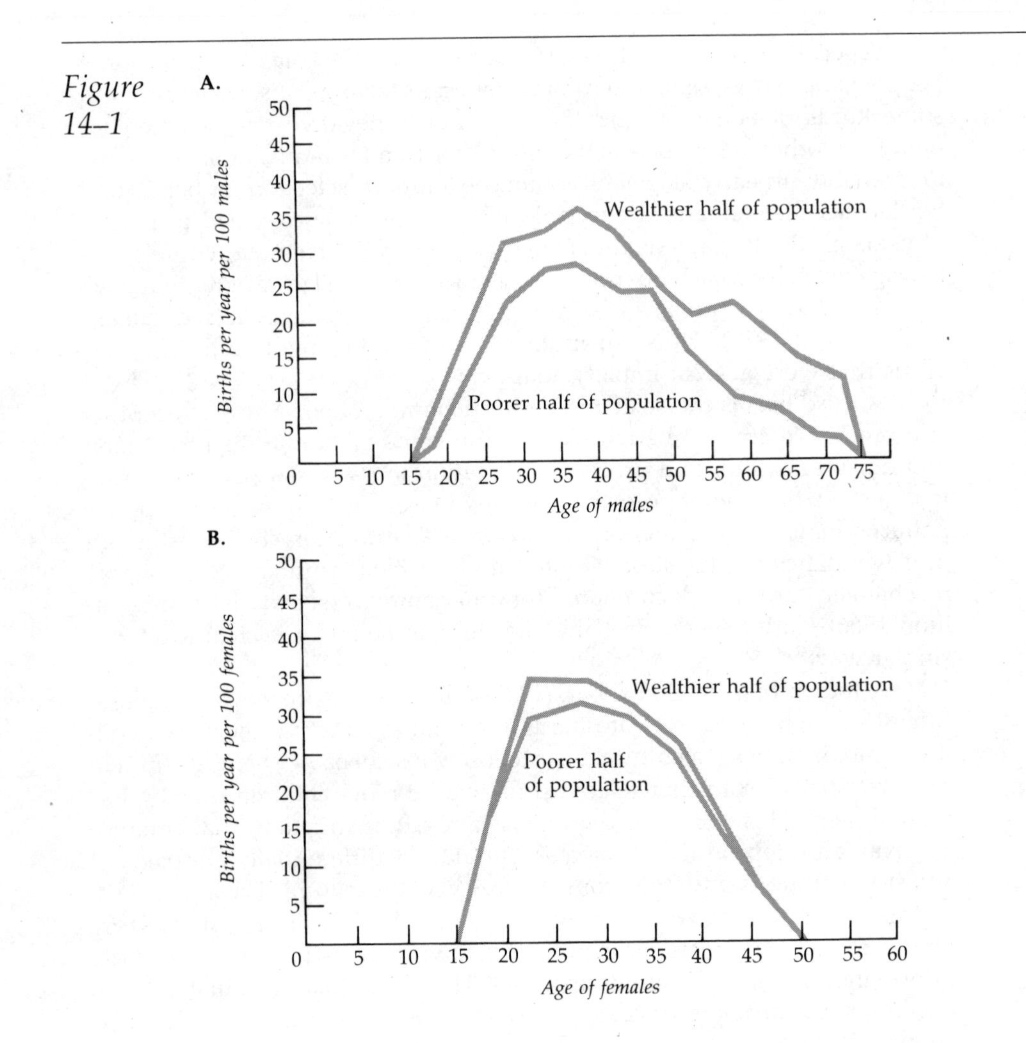

Age-specific reproductive and survival rates of male and female Turkman of Iran as a function of wealth.

A. Difference in male fertility between wealthier and poorer halves of the population. B. Difference in female fertility between wealthier and poorer halves of the population. The observed difference is not statistically significant. C. Difference in male survival between wealthier and poorer halves of the population. D. Difference in female survival between wealthier and poorer halves of the population. The observed difference is statistically significant.

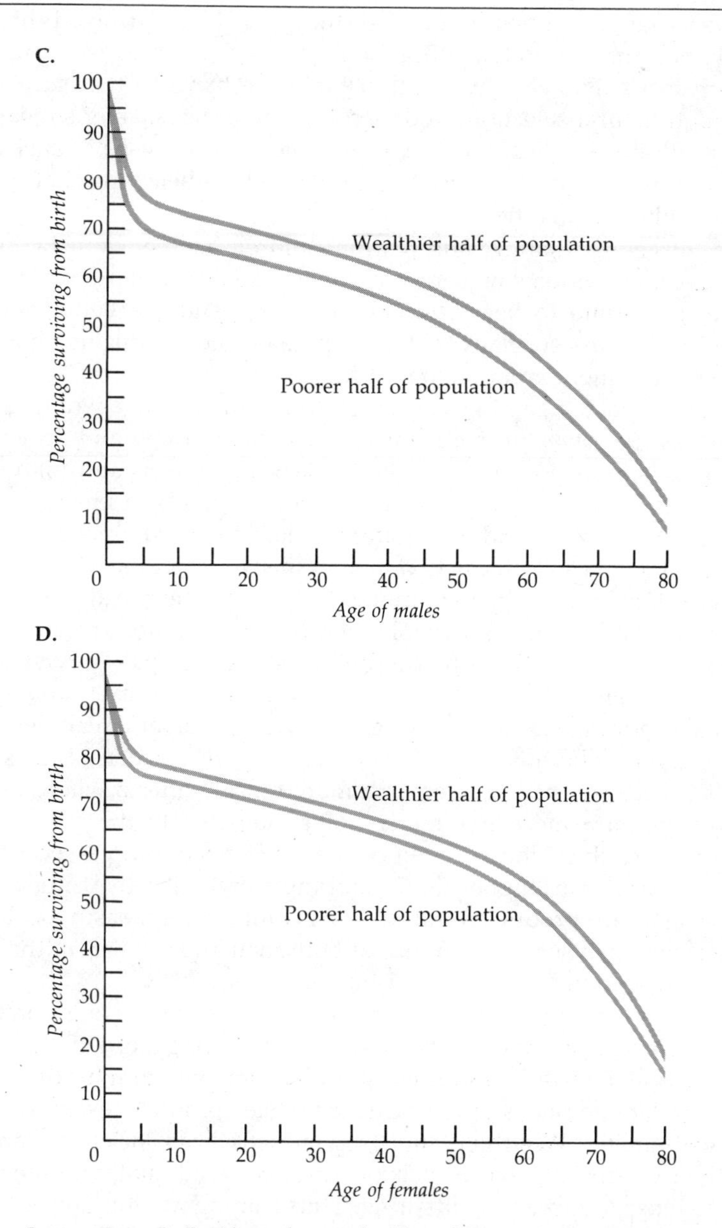

Source: From R. D. Alexander, et al., "Sexual Dimorphisms and Breeding Systems in Pinnipeds, Ungulates, Primates, and Humans," in N. A. Chagnon and W. Irons, eds., *Evolutionary Biology and Human Social Behavior: An Anthropological Perspective* (North Scituate, Mass.: Duxbury Press, 1979), p. 264–7. © 1979 Wadsworth, Inc., Belmont, Ca. Reprinted by permission of the publisher, Duxbury Press.

of socialization or enculturation, or they might have evolved through natural selection. A purely cultural origin for such motives seems unlikely for two reasons. First, motives reflect brain mechanisms that surely evolved through natural selection, and, second, the prevalence of similar motives in all cultures implies that they originated by a process not specific to any one culture. In addition, good reasons exist for believing that these goals have biological significance.

Surviving, mating, and rearing children have obvious biological significance. Individuals who survive longer can father more offspring (men), have more time to bear children (women), and are better able to rear children and assist relatives. Finding mates and producing children have direct consequences on biological fitness.

The goals of gaining wealth, power, and status generally go hand in hand, as great wealth is a means for wielding power and achieving high status. The question here is whether wealth, power, and high status increase an individual's reproductive output. Irons (1979) has recently evaluated this question with demographic data collected among the Turkmen of Persia. His results show that men in the wealthier half of the population survive longer and produce more children; women in the wealthier half of the population also survive longer, but they do not produce more children (Figure 14–1). Wealthier individuals of both sexes survive longer because they enjoy better diets, receive better medical care, and spend less time performing strenuous or high-risk types of labor. Wealthier men have more children because their families can afford to acquire brides for them earlier, because they can remarry more quickly after a wife dies, and because they are more frequently polygynous. Only the richest men can afford more than one wife, since the cost of acquiring a second or third wife is three times higher than acquiring a first wife. By using age-specific survival and reproductive rates, Irons computed a measure of Darwinian fitness for each sex and found that both men and women in the wealthier half of the population achieved higher fitness.

Irons's results may prove to be generally true among nonindustrialized societies, but they are probably not true among industrialized societies. R. A. Fisher (1930), for example, showed from demographic data for Britain and the United States that wealthier families produce fewer children than poorer families. Wealthier families in industrialized societies channel much of their wealth into self-protection and improved quality of life instead of using it to produce more offspring. Thus gaining wealth, power, and status probably increased an individual's inclusive fitness during most stages of human evolution, but it may no longer have that effect in technologically advanced societies.

Social animals that form closed social groups often show hostility toward outsiders and uneasiness or fear when separated from the group (Wilson 1975a). Outsiders pose a threat to the status and well-being of every group member because outsiders may be able to outcompete some or all group members for food or mates. Becoming separated from the group represents a real danger when group membership is important for avoiding predators or finding food. Thus gaining and maintaining acceptance within a social group has adaptive value for social mammals, and in early stages of hominid evolution, humans were probably no exception. The goal of being accepted requires responsiveness to social pressures and acceptance of conformity. That such behaviors were once biologically adaptive does not necessarily imply that they still are, but it does imply that a motivational bias favoring them was instilled by natural selection.

Conclusion

Human social organization is a complex combination of cultural attributes that originated and developed through a process of cultural evolution. In early stages of hominid evolution, behavioral flexibility was a key trait allowing humans to survive and reproduce. For most of human history the capacity to learn and innovate undoubtedly had great adaptive value. Natural selection channeled this capacity in adaptive directions by shaping the motivational makeup and neurophysiological mechanisms underlying human behavior, but once cultural attributes came to predominate, motivational systems originally instilled by natural selection gained a selective impetus of their own. This impetus may well lead to behaviors without biologically adaptive value in the greatly changed social and economic environments of modern societies. The more rapid rate of cultural selection allows it to take directions contrary to those favored by natural selection, and because the ability to learn and use cultural attributes makes many genotypes equally fit in the biological sense, counteracting effects of natural selection may be effectively neutralized.

Human behavior is not closely tied to underlying genetic mechanisms; nor is it entirely immune from genetic influences. Individuals make decisions based on their motivational biases and past experiences. Their biases and preferences are no doubt influenced genetically, but their immediate goals are as much social and economic as they are biological. It remains to be seen whether human behavior is by and large adaptive in the biological sense, but in all probability cultural processes generated by widespread motivational biases to pursue particular physiological, social, and economic goals are primarily responsible for the ways that cultures change.

Glossary

Adaptive landscape. A graphical "map" on which the altitude at any given location indicates Darwinian fitness for a specified combination of alleles or linkage groups.

Adaptive value. A measure of the tendency for a particular allele or linkage group to change in frequency within a gene pool. Equivalent to selective value or Darwinian fitness.

Adrenocorticotrophin (ACTH). A hormone produced by the anterior lobe of the pituitary gland in vertebrates that stimulates secretion of steroid hormones from the cortex of the adrenal glands.

Aggregation. A temporary assemblage of individuals, not necessarily of the same species, that is brought together by locally abundant resources or physical forces and is not internally organized nor engaged in cooperative activities.

Aggression. An overt behavioral act or threat directed at harming another individual with the intent of gaining some advantage.

Aggressive mimicry. A close resemblance of a predatory species to a nonpredatory species that enables the predator to capture prey more readily by deluding prey into a false sense of security.

Aggressive neglect. A reduction in parental care that results from parents spending time and energy on aggressive interactions and reduces the probability of offspring surviving to independence.

Aggressiveness. The predisposition or tendency to behave aggressively.

Agonistic behavior. The set of behavioral acts occurring in aggressive interactions, including all acts of aggression and submission.

Alarm signal. A visual, vocal, or chemical signal given during periods of potential danger that alerts other individuals to the danger.

Allele. A specific form of a gene distinguishable from other forms of the same gene on the basis of biochemical structure and functional effects.

Allelopathy. The release by plants of chemical substances into the soil that are toxic to other potentially competitive plants.

Allogrooming. Grooming of another conspecific individual's pelage or plumage.

Allomother. See *alloparent.*

Alloparent. An individual who assists genetic parents in caring for offspring.

Alloparental care. The assistance provided by individuals other than genetic parents who help care for offspring.

Altricial. Pertaining to young animals, especially birds, that are born naked and are essentially incapable of locomotion for a substantial period following birth or hatching. (Contrast with *precocial.*)

Altruism. Any behavior that benefits others while harming or imperiling the individual performing it. In animals benefit and harm are measured in terms of direct fitness.

Androgens. The male sex hormones of vertebrates.

Anisogamy. The characteristic of many sexually reproducing organisms where the female primary sex cell (ovum) is larger than the male primary sex cell (sperm). (Contrast with *isogamy.*)

Aposematism. The advertisement of dangerous or defensive mechanisms by venomous or distasteful prey animals. Generally involves bright coloration and/or conspicuous morphological structures or external markings.

Attendant species. In mixed-species bird flocks the various species that join and leave the flock periodically and do not remain with the flock for an entire season. (Contrast with *nucleus species.*)

Batesian mimicry. A close resemblance of vulnerable prey species to distasteful or venomous prey species that evolved as a means of dissuading predators from attacking otherwise poorly protected prey animals.

Beater effect. The flushing up of prey by the movements of animals or vehicles, with the consequence of making prey more vulnerable to capture by predators.

Benefit. In evolutionary terms any consequence of a phenotypic trait that increases Darwinian fitness. Sometimes, benefits are operationally defined in terms of direct and indirect fitness, especially during empirical studies or theoretical analyses pertaining to them. (Contrast with *cost.*)

Biodeterioration zone. In a central-place system the zone around a core area within which resources have been depleted but not exhausted.

Broadcast promiscuity. A form of sexual reproduction in which male and female gametes are released into the surrounding medium, where fertilization takes place. Normally, gamete release occurs in some coordinated manner and may involve large numbers of individuals of both sexes.

Brood parasitism. A reproductive pattern exhibited by certain birds in which eggs are laid in the nests of other species and all parental care is subsequently provided by the host species.

Carrying capacity. The largest density of a given species that can persist indefinitely in a particular habitat or locality without habitat deterioration. Usually symbolized by the parameter K.

Central-place system. A form of spatial organization in which one or more individuals or other social units focuses its activities around a single central location such as a den, nest site, breeding colony, or roosting site.

Codon. A sequence of three nucleotides on DNA or RNA that specifies either a particular amino acid sequence or the end of a functional unit (i.e., allele).

Colony. In vertebrates a group of organisms larger than a mated pair or family group whose activities are focused around a single fixed location. In invertebrates a group of organisms who maintain fixed spatial relationships with respect to each other or whose activities are focused around a single fixed location. Colonies are usually internally organized but may or may not involve cooperative interactions among colony members.

Commons. A resource or habitat whose depletion or deterioration affects all individuals using it to an equal extent regardless of imbalances in the rate that particular individuals deplete or harm it.

Communication. The imparting of information from one organism to another in a way that evokes a detectable response from the recipient at least some of the time.

Competition. Demand by two or more individuals, groups, or species for a common resource that is limited in availability.

Contest competition. A form of competition in which individuals rely on aggression to control access to a limited resource. Equivalent to interference competition. (Contrast with *scramble competition.*)

Cooperation. A type of behavioral interaction from which all participating individuals derive a net benefit. In animals benefit is measured in terms of direct fitness.

Coordination signal. A type of communication employed by individuals to coordinate their movements or activities during a cooperative or mutually beneficial interaction. (Contrast with *persuasion signal.*)

Core area. An area of especially intense use within a home range or the central location of a colony or other central-place system around which activities and movements are focused.

Cost. In evolutionary terms any consequence of a phenotypic trait that decreases Darwinian fitness. Sometimes, costs are operationally defined in terms of direct and indirect fitness, especially during empirical studies or theoretical analyses pertaining to them. (Contrast with *benefit.*)

Cost of anisogamy. The twofold reduction in Darwinian fitness that results from producing male progeny in anisogametic, sexually reproducing organisms.

Cost of meiosis. The alleged twofold reduction in Darwinian fitness that results from being only 50% related to each progeny in sexually reproducing organisms, as compared to the 100% relatedness present in asexually reproducing organisms.

Covalent bond. A strong chemical bond that involves the sharing of one or more electrons by two adjacent atoms.

Crossover. During meiosis the exchange of genetic material by two paired chromosomes as a result of both chromosomes breaking and subsequently fusing with the corresponding parts of their opposites.

Cultural selection. A nongenetic process based on transmission of learned information or concepts that leads to directional evolutionary change in the social structure of human and some animal societies.

Darwinian fitness. A measure of the rate at which allele or linkage-group frequencies change within a gene pool through time. Equivalent to adaptive value and selective value. (Contrast with *individual fitness, inclusive fitness, direct fitness,* and *indirect fitness.*)

Defensive aggression. A form of aggression used by some prey animals, often as a last resort, to defend themselves or their offspring from predators.

Density-dependent mortality. A form of mortality whose frequency or rate of occurrence increases as population density increases. (Contrast with *density-independent mortality.*)

Density-independent mortality. A form of mortality that remains constant under particular environmental or social conditions regardless of population density. (Contrast with *density-dependent mortality.*)

Determinate growth. A pattern of growth in which developing young grow only until they reach an approximately fixed adult size. (Contrast with *indeterminate growth.*)

Diploidy. The genetic condition characterized by chromosomes occurring as complementary pairs within the cell nucleus. One chromosome of each pair typically stems from the mother, while the other typically stems from the father. (Contrast with *haploidy* and *haplodiploidy.*)

Direct fitness. A measure of an individual's ability to leave direct descendants in future generations relative to the ability of other individuals in the same population or species. Equivalent to individual fitness. Direct fitness plus indirect fitness equals inclusive fitness. (Contrast with *Darwinian fitness.*)

Directional selection. A form of natural selection that leads to a change in the average phenotypic characteristics of a population. (Contrast with *disruptive selection* and *stabilizing selection.*)

Direct selection. A form of natural selection that affects gene frequencies as a consequence of individuals producing direct descendants at different rates. The combined effect of direct and indirect selection is referred to as kin selection (except when trait-group selection is involved). (Contrast with *indirect selection.*)

Displacement behavior. A seemingly irrelevant display behavior given in conflict situations where it has no direct functional significance. The behavior presumably does have communicatory significance.

Display. A behavior that has been modified or specialized by natural selection to communicate information or persuade a rival.

Disruptive selection. A form of natural selection that creates a bimodal pattern of phenotype frequencies in a population. (Contrast with *directional selection* and *stabilizing selection.*)

Dominance. See *social dominance.*

Dominion. A form of spatial organization in which individuals defend nonexclusive use of their home ranges. The degree to which intruders are excluded usually decreases with increasing distance from a central area of exclusive use.

Drone. A male in eusocial bees.

Ecological niche. See *fundamental niche* and *realized niche.*

Economic defensibility. The worthwhileness of aggressively controlling a particular resource, in terms of inclusive fitness. A resource is economically defensible if the benefits of defense exceed the costs.

Ectoparasite. A parasite that lives on the external surface of its host, where it obtains nutrients by feeding on host tissues or fluids.

Epideictic display. A postulated type of social display whose function is to convey information about local population density. The term is used solely in the context of interdemic selection for population-regulating mechanisms.

Epinephrine. A hormone produced by specialized cells in the adrenal medullae of vertebrates. Equivalent to adrenalin.

Ergatodyne. In eusocial insects any morphological form intermediate between worker and queen castes.

Eusocial. A form of insect society characterized by cooperative brood care, division of labor in performing reproductive tasks, and overlap of at least two generations that are concurrently involved in reproductive activities.

Evolutionarily stable strategy (ESS). In game theory a behavioral strategy that leads to higher inclusive fitness than any alternative strategy when that strategy predominates in the population.

Expected benefit. The amount of benefit, in fitness terms, that an individual can expect to receive at some future time by performing a particular behavior rather than some alternative behavior. Expected benefit equals benefit multiplied by the probability of realizing the benefit.

Expected cost. The amount of cost, in fitness terms, that an individual can expect to incur at some future time by performing a particular behavior rather than some alternative behavior. Expected cost equals cost multiplied by the probability of incurring the cost. (See also *risk.*)

Extrapair copulation. A copulation involving a pair-bonded individual and an individual outside the pair bond.

Fecundity. The number of eggs or offspring produced during a specified period of time.

Fitness. See *Darwinian fitness, direct fitness, inclusive fitness, indirect fitness, individual fitness, population fitness.*

Fixed-action pattern. A particular behavioral response that is always given in the same stereotyped manner whenever it is evoked by a stimulus situation. Sometimes equated with instinct.

Forced copulation. A copulation forced onto an unwilling female by an aggressive or overpowering male. Sometimes referred to as *rape,* a term which should be applied only to humans because of its emotional connotations.

Founder effect. The genetic differentiation of isolated populations arising because founders of those populations differ genetically from each other and average members of the parent population.

Fraser Darling effect. Stimulation or enhancement of reproductive activity that is generated by the presence and reproductive behavior of nearby conspecifics other than the mated pair.

Frequency-dependent selection. A form of natural selection in which the inclusive fitness of a phenotype depends on the frequency of alternative phenotypes in the population. Frequency dependence is typical of disruptive and sexual selection but sometimes characterizes other forms of natural selection as well.

Fundamental niche. The range of all environmental conditions, including microclimatic regimes and resource characteristics, within which a species can persist indefinitely. (Compare with *realized niche.*)

Gene. A functional unit in the genetic material (DNA or RNA) of an organism. Each gene generally contains the genetic information responsible for synthesizing one type of protein.

Gene flow. The spreading of genes through a gene pool as a consequence of dispersal and mating patterns occurring within or between populations.

Gene locus. The position on a chromosome where a particular gene resides.

Gene pool. The complete set of genes carried by all members of a population.

General adaptation syndrome (GAS). The combination of physiological responses arising in vertebrates as a result of external stress.

Genetic drift. A change in gene frequencies that arises due to chance processes alone.

Genetic load. The average reduction in Darwinian fitness of population members that arises because not all individuals are maximally fit.

Genic selection. The process whereby the genetic composition of a gene pool changes within or between generations. In modern organisms the genetic consequence of natural selection, which acts upon phenotypes.

Genome. The complete set of genetic materials carried by an organism in its primary sex cells.

Genotype. The particular combination of alleles carried by an individual at one or more gene loci.

Grain. The spatial pattern of a resource with respect to size and spacing of resource patches.

Group selection. A process whereby the relative frequency of genetically different groups (species, populations, or subunits within a population) changes through time.

Habitat. The place in a physical environment and biotic community where an organism lives at a particular time.

Habitat selection. The behavioral process of choosing a place to live from among several potentially suitable places that are available.

Handicap principle. The hypothesis that morphological or behavioral traits that hamper an individual's ability to survive can evolve because they are reliable indicators of the individual's competitive or genetic superiority.

Haplodiploidy. A mode of sex determination characterizing certain organisms, especially hymenopteran insects, in which unfertilized haploid eggs develop into males and fertilized diploid eggs develop into females.

Haploidy. The genetic condition characterized by chromosomes occurring as single copies within the cell nucleus. (Contrast with *diploidy* and *haplodiploidy*.)

Harem polygyny. A type of mating pattern where a single adult male controls exclusive sexual access to a social group of two or more breeding females, usually by means of aggressive behavior.

Hierarchical promiscuity. A type of mating pattern occurring in multi-male social groups where a male's sexual access to receptive females is affected by his status in a social hierarchy.

Hitchhiking effect. The process whereby an allele without positive selective value can increase in frequency because it is genetically linked to another allele with positive selective value.

Home range. The area within which an animal confines its activities during part or all of an annual cycle.

Homologous. Pertaining to similarities of morphology, physiology, or behavior that arose in different species because those species inherited them from a common ancestor.

Humanology. The scientific study of human behavior from an evolutionary (biological and cultural) perspective.

Hydrogen bond. A weak chemical bond that involves the sharing of a proton by two adjacent atoms.

Ideal dominance distribution. A theoretical dispersion pattern in which individuals occupying intrinsically higher-quality habitats achieve higher inclusive fitness than individuals occupying intrinsically poorer-quality habitats. (Contrast with *ideal free distribution*.)

Ideal free distribution. A theoretical dispersion pattern in which individuals occupying intrinsically higher-quality habitats achieve the same average fitness as individuals occupying intrinsically poorer-quality habitats. (Contrast with *ideal dominance distribution*.)

Inclusive fitness. A measure of an individual's ability to leave genetically related descendants in future generations relative to the ability of other individuals in the same population or species. Inclusive fitness equals direct fitness plus indirect fitness. (Contrast with *Darwinian fitness*.)

Indeterminate growth. A pattern of growth in which individuals continue to grow larger throughout their lives, with size at any given age dependent on how much energy was allocated to growth at earlier ages. (Contrast with *determinate growth*.)

Indirect fitness. A measure of the extent to which an individual increases the relative proportion of descendants left in future generations by a genetically similar conspecific or nondescendant relative. Indirect fitness plus direct fitness equals inclusive fitness.

Indirect selection. A form of natural selection that affects gene frequencies as a consequence of individuals helping relatives or other genetically similar individuals reproduce. Often referred to as kin selection. The combined effect of direct and indirect selection is referred to as kin selection (unless trait-group selection is involved). (Contrast with *direct selection.*)

Individual fitness. Equivalent to direct fitness.

Individual selection. Equivalent to direct selection.

Infanticide. The killing of infants by parents or, more typically, genetically unrelated conspecific adults.

Initiator. In the operon model of bacterial molecular genetics, a location on a gene where an activator substance binds and greatly increases the rate at which messenger RNA is transcribed.

Instinct. A highly stereotyped, species-specific behavior that is more complex than a simple reflex and is attributable largely to genetic influences. (Contrast with *learned behavior.*)

Intention movement. A preparatory motion that an animal exhibits prior to a complete behavioral response. Often given when a locomotory tendency and another type of behavioral tendency are in conflict.

Interdemic selection. A form of group selection involving differential proliferation or extinction of genetically different, genetically isolated local populations within a species.

Interference competition. Equivalent to contest competition.

Intersexual aggression. Aggression directed at opposite-sexed conspecifics of the same age class or reproductive class.

Intersexual selection. A selective process leading to the evolution of secondary sexual characteristics in one sex, usually the male, as a result of mating preferences expressed by the other sex. (Contrast with *intrasexual selection.*)

Interspecific competition. Competition between members of different species for a limited resource. (Contrast with *intraspecific competition.*)

Intrademic selection. Selection occurring within a local population or deme. Intrademic selection includes individual selection, kin selection, and trait-group selection. (Contrast with *interdemic selection.*)

Intrasexual aggression. Aggression directed at like-sexed conspecifics of the same age class or reproductive class.

Intrasexual selection. A selective process leading to the evolution of secondary sexual characteristics in one sex, usually the male, as a result of competition with other members of the same sex for access to mates. (Contrast with *intersexual selection.*)

Intraspecific competition. Competition between members of the same species for a limited resource. (Contrast with *interspecific competition.*)

Intrinsic competitive ability. The ability of an individual to inflict costs on an opponent during an aggressive interaction.

Inversion. A reversal of the typical sequence of genes along a chromosome. Inversions arise when a chromosome breaks in two places and reattaches with the orientation of the middle segment reversed.

Inversion polymorphism. A condition where the gene sequence along a chromosome occurs in both inverted and uninverted forms within a population at frequencies higher than could be maintained by chromosome breakage rates alone.

Isogamy. The characteristic of some sexually reproducing organisms in which gametes are all similar in size and not differentiated into sperm and ova. (Contrast with *anisogamy.*)

Iteroparity. A type of life history pattern in which adults normally reproduce at periodic intervals throughout their lives. (Contrast with *semelparity.*)

Kin selection. The process leading to genetic changes in a population due to differential success in producing genetically related descendants by the various phenotypes in a population. Kin selection equals the combined effects of direct selection and indirect selection (unless trait-group selection is involved). The term *kin selection* is often used as an equivalent to indirect selection operating through nondescendant relatives.

***K* selection.** With regard to life history patterns, a form of selection favoring high competitive ability and low fecundity in populations at or near carrying capacity. (Contrast with *r selection.*)

Learned behavior. A behavior arising as a result of experience that is usually not exhibited by all members of a species and is often not given in a stereotyped manner. (Contrast with *instinct*.)

Lek. A fixed location used repeatedly by a group of males who are advertising for mates. Sometimes used to refer to the group of males displaying at such a location.

Life history pattern. The way resources are allocated to survival, growth, and reproduction at each age throughout a typical individual's life.

Life table. A table showing the proportion of like-aged individuals surviving to each age and the rate surviving individuals reproduce at each age.

Linkage. The property of genes being located on the same chromosome. The closeness of linkage depends on how near to each other genes are on the chromosome.

Linkage group. A set of closely linked genes (see *linkage*).

Luteinizing hormone (LH). A hormone produced by the anterior lobe of the pituitary gland which stimulates production of testosterone in the testes of male vertebrates.

Major worker. A large nonreproductive caste in eusocial insects that performs defensive functions around the colony nest and sometimes also stores food in greatly enlarged abdomens. Formerly called soldiers. (Contrast with *minor worker*.)

Mate competition. Competition among members of one sex for sexual access to members of the opposite sex.

Meiotic drive. A property of male gametes in some species that makes some genetic types of sperm more likely to fertilize eggs than others.

Metapopulation. A set of genetically isolated populations belonging to the same species and existing at the same time.

Mimicry. The close resemblance of one species (the mimic) to another (the model). (See *aggressive mimicry, Batesian mimicry,* and *Müllerian mimicry*.)

Minor worker. A sterile but not necessarily nonreproductive caste in eusocial insects that performs most energy acquisition, colony maintenance, and brood care functions. Minor workers may be differentiated into several functionally different castes on the basis of age and/or ontogeny. (Contrast with *major worker*.)

Mobbing. A concerted aggressive assault on a predator undertaken by a social group or mated pair mounted with the purpose of driving the predator away.

Monogamy. A type of mating pattern where breeding adults of both sexes pair with only one mate at a time, with the duration of pair bonding lasting for a substantial fraction of the breeding season.

Monogyny. In eusocial insects the condition where a colony contains only one queen. (Contrast with *polygyny*.)

Müllerian mimicry. A close resemblance among two or more distasteful or toxic species that reduces predation for each species because predators need learn to avoid only one color pattern rather than several. (Contrast with *Batesian mimicry* and *aggressive mimicry*.)

Mutation. A change in the nucleotide sequence of a gene occurring during replication of DNA. The change may consist of loss of one or more nucleotides, insertion of one or more nucleotides, or substitution of one or more nucleotides for the original ones.

Natural selection. The process whereby the genetic composition of a population changes as a result of differential production or survival of offspring by genetically different members of the population.

Net expected benefit. The average net increase in fitness that an individual receives by choosing one of several possible behavioral options in a given context. Net expected benefit equals expected benefit minus expected cost.

Niche. The range of physical and biotic conditions within which a species is adapted to survive and reproduce. (See also *fundamental niche* and *realized niche*.)

Nonshareable parental investment. A form of parental investment that cannot be given to one offspring without reducing the ability of parents to rear other offspring. (Contrast with *shareable parental investment*.)

Norepinephrine. A hormone secreted mainly by neurons in the sympathetic nervous system to facilitate neural transmission across synapses.

Normalizing selection. Equivalent to stabilizing selection.

Nucleotide. A component of nucleic acids (DNA and RNA) that consists of a pentose sugar, a phosphate group, and a purine or pyrmidine base.

Nucleus species. In mixed-species bird flocks the focal species that is always present and serves as an attractant to other species that periodically join and leave the flock. (Contrast with *attendant species.*)

Ontogeny. The development of a fertilized (or unfertilized) egg into an adult organism.

Operator. In the operon model of bacterial molecular genetics, a location on a gene where a repressor substance binds and greatly decreases transcription of messenger RNA along a gene sequence. (Contrast with *initiator.*)

Operon. In bacterial genetics a functional unit of DNA that includes one to nine structural genes, a promoter region that facilitates transcription, and one or more regulatory sites that control the rate of transcription.

Overdispersion. A spatial dispersion pattern characterized by individuals being more evenly distributed than would be the case in a random dispersion pattern. (Contrast with *underdispersion.*)

Overlap promiscuity. A type of mating system in which individual males and females live on overlapping home ranges and come together only briefly for the purpose of mating.

Oviparity. A mode of reproduction involving the production and laying of eggs. (Contrast with *viviparity.*)

Parasocial. Pertaining to social insects that exhibit cooperative brood care and sometimes caste differentiation but not overlap of generations. (Contrast with *eusocial* and *subsocial.*)

Parental aggression. Aggression directed at progeny by a parent.

Parental care. All forms of assistance provided by parents to direct progeny following birth and prior to independence or maturity of the young. All parental investments other than a parent's investment in gametes.

Parental investment. Any investment by a parent in direct progeny that reduces the parent's ability to produce other progeny during its lifetime.

Patch. A resource clump or habitat island that is essentially homogeneous internally but differs in some important way from surrounding areas.

Patchiness. The extent that a resource or habitat is distributed in patches. Patchiness is measured by the difference in resource density within and outside resource patches.

Peer aggression. Aggression directed at conspecifics of similar age. Usually refers to aggression among immature individuals.

Pentose sugar. Any sugar that contains five carbon atoms.

Persuasion signal. A type of communication signal employed by individuals with conflicting interests to gain some advantage. Includes especially signals and displays given during aggressive interactions or courtship. (Contrast with *coordination signal.*)

Phenotype. The physical manifestation of an organism, including its morphology, physiology, biochemistry, and behavior.

Pheromone. A chemical substance used to communicate with a conspecific individual or group of individuals.

Phylogeny. The evolutionary history of a particular organism or group of organisms. Especially, the relationships of modern or fossil forms based on their common ancestry.

Plant secondary compounds. Any substance produced by a plant that is not directly involved in the biochemical processes underlying growth, tissue maintenance, or reproduction.

Pleiotropy. The phenomenon of one gene or linkage group affecting more than one phenotypic trait.

Polyandry. A mating pattern where females regularly maintain essentially exclusive, long-term pair associations with more than one male concurrently. (Contrast with *polygyny.*)

Polygamy. In the broad sense any nonmonogamous mating system. In the narrow sense any nonmonogamous mating system involving long-term pair associations (i.e., polygyny and polyandry). Used in this book in the broad sense.

Polygenic inheritance. The phenomenon of one phenotypic trait being influenced by or dependent upon the actions of more than one gene or linkage group.

Polygyny. In vertebrates a mating pattern where males regularly maintain essentially exclusive, long-term pair associations with more than one female concurrently. (Contrast with *polyandry.*) In eusocial insects the condition where a colony contains more than one actively reproductive queen. (Contrast with *monogyny.*)

Polygyny-polyandry. A type of vertebrate mating system in which males are regularly polygynous and females are regularly polyandrous within the same population.

Polygyny threshold. The difference in quality between high- and low-quality habitats that is theoretically necessary for territorial polygyny to evolve in vertebrates.

Polymorphism. The occurrence of more than one morphological or behavioral type of individual in a single population. In genetics the maintenance of two or more alternative alleles in a gene pool above the frequencies that can be maintained by mutation and gene flow alone.

Population fitness. The average inclusive fitness of all individuals in a population. Alternatively, the relative tendency for a given population to proliferate or become extinct within a metapopulation.

Preadaptation. An ancestral characteristic or trait that predisposes a directional evolutionary change to proceed in one direction rather than in another.

Precocial. Pertaining to young animals, especially birds, that are born fully covered (with feathers or fur) and are capable of locomotion shortly after hatching or birth.

Presocial. Pertaining to any social insect that lacks one or two of the three definitive traits that characterize eusocial species.

Primary sex ratio. The sex ratio of offspring at the time eggs are fertilized.

Primary sexual characteristic. Any trait that is directly involved in reproduction or parental care. (Contrast with *secondary sexual characteristic.*)

Principle of antithesis. The principle that a display with opposite function to another display involves postural characteristics and behavioral elements opposite to those of the other display.

Promiscuity. A mating pattern where males and females do not form long-term pair associations and individuals of at least one sex frequently mate with more than one individual of the opposite sex each reproductive season.

Protobiont. A hypothetical prebiotic complex of proteins and nucleic acids that eventually gave rise to self-replicating, and hence living, organisms.

Proximate causation. The immediate cause underlying expression of a phenotypic trait. (Contrast with *ultimate causation.*)

Pseudergate. A special undifferentiated caste in lower termites that performs worker functions but retains the capability of developing into more specialized castes by further molting.

Quality. In evolutionary biology the extent to which a particular mate, habitat, resource, or context can enhance an individual's inclusive fitness.

Quasisocial. Pertaining to social insects that use a composite nest and cooperate in brood care but do not exhibit caste differentiation or overlap of generations.

Realized niche. The range of all environmental conditions, including microclimatic regimes and resource characteristics, actually occupied or utilized by a species. Realized niche is a subset of fundamental niche.

Reciprocal altruism. An altruistic behavior performed with the expectation that the recipient will provide some future benefit in return.

Redirected aggression. Aggression directed at an inanimate object or a clearly inferior opponent rather than at the individual actually evoking the aggressive response.

Regulatory gene. A type of gene which codes for protein substances that act solely to regulate the transcription rate at particular structural genes or at other regulatory genes.

Regulon. In bacterial molecular genetics a functional unit comprised of two or more operons and a single regulatory gene.

Releaser. A specific stimulus responsible for evoking a particular fixed action pattern. A one-to-one correspondence normally exists between a releaser and its associated response.

Renewability. The ability of a resource to become replenished during a specified time interval following depletion.

Replicator. The simplest chemical entity capable of self-replication. In Dawkin's (1976, 1978) original usage, a naked gene or piece of DNA, not associated with any other molecules, that is capable of self-replication independently of other similar entities.

Resource. Any substance, object, or energy source utilized for body maintenance, growth, or reproduction.

Resource availability. The extent to which a given resource can be obtained or utilized by a given individual or species.

Risk. Essentially equivalent to expected cost. Risk usually refers to contexts where bodily harm can potentially occur, while expected cost can refer to any behavioral context.

Ritualization. The evolutionary process leading to elaboration, specialization, or increased stereotypy of a behavioral display.

***r* selection.** With regard to life history patterns, a form of selection favoring high colonizing ability and high fecundity in populations that exist far below environmental carrying capacity. (Contrast with *K selection*.)

Scramble competition. A form of competition in which individuals compete by exploiting a resource faster or more efficiently than their competitors without resort to aggression.

Search image. A mental image or temporary predisposition to search for particular prey types or to search for prey in particular locations.

Secondary sex ratio. Sex ratio of offspring at the time they become independent of their parents. (Contrast with *primary sex ratio*.)

Secondary sexual characteristic. Any trait involved in obtaining a mate but not otherwise directly involved in reproduction or parental care. (Contrast with *primary sexual characteristic*.)

Selection coefficient. In population genetics a measure of the rate a particular allele will change in frequency through time within a specified gene pool.

Selective value. Equivalent to adaptive value.

Selfish behavior. Any behavior that benefits the individual performing it while harming or imperiling others affected by it. In animals benefit and harm are measured in terms of direct fitness.

Semelparity. A type of life history pattern in which adults normally reproduce only once and then die. (Contrast with *iteroparity*.)

Semisocial. In social insects pertaining to species where females of the same generation cooperate in brood care and some of the females contribute little or nothing to actual egg production.

Sessile. Pertaining to organisms that are immobile and more or less permanently attached to a substrate as adults.

Sex reversal. A mode of reproduction where small individuals are functional females and the largest female undergoes a sex change to become a male when the current male of a group dies.

Sex role reversal. A reproductive pattern in vertebrates characterized by a reversal of the typical male and female roles. Females compete for mates and often exhibit secondary sexual characteristics associated with aggression, while males select mates and normally perform most or all of the parental care.

Sexual bimaturism. A life history pattern in which one sex, usually the male, first begins breeding at an older age than does the other sex.

Sexual dimorphism. A sexual difference in morphology or behavior, excluding primary sexual characteristics.

Sexual selection. A form of selection which leads to an increased frequency of alleles that enable individuals to either compete for mates more successfully or attract mates more successfully.

Shareable parental investment. A form of parental investment that can be given to one concurrent offspring without reducing the amount of parental care that can be given to other concurrent offspring. (Contrast with *nonshareable parental investment*.)

Signal. Any behavior that conveys information, regardless of whether it serves other functions as well.

Sign stimulus. Equivalent to releaser.

Social dominance. A hierarchical form of social organization based on overt or implied aggression. A status system based on agonistic interactions.

Social group. A group of conspecific individuals that persists because individuals gain mutual (cooperative) benefits by remaining together. Mated pairs and dependent offspring are not considered to be social groups. Groups with fixed internal spacing (i.e., colonies) are also not considered to be social groups, except for eusocial insects.

Sociobiology. The scientific study of the adaptive significance of animal behavior.

Spatiotemporally clumped resource. A resource that is clumped in space and whose availability at any given location varies temporally on a short-term basis.

Sperm competition. Competition among the sperm of more than one male to fertilize the eggs of one or more females. May occur in an external medium following gamete release or within a female's reproductive tract.

Spiteful behavior. A behavior that is harmful both to the recipient of the behavior and to the individual performing it. In animals harm is measured in terms of direct fitness.

Stabilizing selection. A form of natural selection that maintains the same mean and variance among phenotypes each generation. (Contrast with *directional selection* and *disruptive selection*.)

Stimulus filtering. The selective neural transmission of sensory input from sensory receptors to the brain or from lower-level brain centers to higher-level brain centers.

Strategy. A set of rules that determines which of several alternative behavioral tactics is employed to solve a particular problem or to achieve a particular goal.

Structural gene. A type of gene that codes for a particular protein that is used in metabolic processes.

Subsocial. Pertaining to social insects that exhibit overlap of generations and sometimes cooperative brood care but not differentiation into reproductive and worker castes. (Contrast with *parasocial* and *eusocial*.)

Symbiosis. A mutualistic interaction between two species, both of which benefit by associating with the other.

Sympathetic nervous system. A branch of the central nervous system in vertebrates, mediated by several cranial nerves, that conducts nerve impulses from the brain to internal organ systems preparatory to handling acutely stressful situations.

Tactic. A behavior pattern within a species' repertoire that can potentially be employed in a specified context to achieve a particular goal.

Teleology. A philosophical view that attributes conscious purposiveness to animal behavior.

Territorial polygyny. A type of vertebrate mating system in which males defend resource-based territories and commonly form pair bonds with more than one concurrent mate.

Territory. An area controlled more or less exclusively by an individual or group of individuals by means of implied or overt aggression.

Trait group. A recognizable entity comprised of more than one individual within a population that represents a unit upon which selection can act. For any given individual the trait group is the group of conspecifics with whom that individual interacts during a specified time interval.

Trampling zone. In a central-place system the area around the core area that has been damaged by the frequent movements of many individuals along or through it.

Transcription. The biochemical process whereby RNA is synthesized along one of the two chains in a DNA molecule.

Translation. The biochemical process whereby the genetic code of a messenger RNA molecule aligns amino acids into a specified sequence to produce a particular protein structure.

Translocation. A genetic process involving breakage of two nonpaired chromosomes and an exchange of parts between the two chromosomes.

Ultimate causation. The evolutionary cause or selective basis that gave rise to a phenotypic trait. (Contrast with *proximate causation*.)

Underdispersion. A spatial dispersion pattern characterized by individuals being closer together on average than would be the case in a random dispersion pattern. (Contrast with *overdispersion*.)

Ungulates. A general term for antelopes, deer, buffalo, elephants, and other members of the mammalian class Artiodactyla.

Viviparity. A mode of reproduction in which young are born alive. (Contrast with *oviparity*.)

You-first principle. A principle which states that an individual is safest by allowing other individuals to take the risk of entering a hazardous or unfamiliar area first.

References

Abel, E. F. 1960. Liàison facultative d'un poisson (*Gobius bucchichii* Steindachner) et d'une anémone (*Anemonia sculcata* Penn.) en Mediterranée. *Vie et Milieu* 11:517–531.

———. 1961. Freiwasserstudien über das Fortpflanzungverhalten des Monchfisches *Chromis chromis* Linne, einem Vertreter der Pomacentriden in Mittelmeer. *Zeitschrift für Tierpsychologie* 18:441–449.

Abele, L. G., and Gilchrist, S. 1977. Homosexual rape and sexual selection in acanthocephalan worms. *Science* 197:81–83.

Adams, D. B. 1971. Defence and territorial behavior dissociated by hypothalamic lesions in the rat. *Nature* 232:573–574.

Adams, L., and Davis, S. D. 1967. The internal anatomy of home range. *Journal of Mammalogy* 48:529–536.

Adams, M., and Neel, J. V. 1967. Children of incest. *Pediatrics* 40:55–61.

Adamson, R. H., and Davies, D. S. 1973. Comparative aspects of absorption, distribution, metabolism, and excretion of drugs. *Comparative Pharmacology* 2:851–911.

Ainsworth, M. D. 1962. The effects of maternal deprivation: A review of findings and controversy in the context of research strategy. In *Deprivation of maternal care: A reassessment of its effects*, pp. 97–165. Geneva: World Health Organization.

Alber, M. A. 1956. Multiple mating. *British Bee Journal* 83:134–135; 84:6–7, 18–19.

Alber, M., Jordan, R., Ruttner, F., and Ruttner, H. 1955. Von der Paarung der Hönigbiene. *Zeitschrift für Bienenforschung* 3:1–28.

Alberti, R. E., and Emmons, M. L. 1978. *Your perfect right*. 3rd ed. San Luis Obispo, Calif.: Impact.

Albrecht, H. 1969. Behaviour of four species of Atlantic damselfishes from Columbia, South America (*Abudefduf saxatilis, A. taurus, Chromis multilineata, C. cyanea:* Pisces, Pomacentridae). *Zeitschrift für Tierpsychologie* 26:662–676.

Alcock, J. 1975a. *Animal behavior. An evolutionary approach.* Sunderland, Mass.: Sinauer Associates.

———. 1975b. Territorial behaviour by males of *Philanthus multimaculatus* (Hymenoptera: Sphecidae) with a review of territoriality in male sphecids. *Animal Behaviour* 23:889–895.

Aldis, O. 1975. *Play fighting*. New York: Academic Press.

Aldrich, J. W. 1973. Disparate sex ratios in waterfowl. In *Breeding biology of birds*, ed. D. S. Farner, pp. 482–489. Washington, D.C.: National Academy of Sciences.

Alexander, M., and Perachio, A. A. 1973. The influence of target sex and dominance on evoked attack in rhesus monkeys. *American Journal of Physical Anthropology* 38:543–547.

Alexander, R. D. 1961. Aggression, territoriality, and sexual behavior in field crickets (Orthoptera: Gryllidae). *Behaviour* 17:130–223.

———. 1974. The evolution of social behavior. *Annual Review of Ecology and Systematics* 5:324–383.

———. 1975. Natural selection and specialized chorusing behavior in acoustical insects. In *Insects, science, and society*, ed. D. Pimentel, pp. 35–77. New York: Academic Press.

———. 1979. Evolution and culture. In *Evolutionary biology and human social behavior: An anthropological perspective,* ed. N. A. Chagnon and W. Irons, pp. 59–78. N. Scituate, Mass.: Duxbury Press.

Alexander, R. D., and Brown, W. L., Jr. 1963. Mating behavior and the origin of insect wings. *Miscellaneous Publications of the Museum of Zoology, University of Michigan* 133:1–62.

Alexander, R. D., Hoogland, J. L., Howard, R. D., Noonan, K. M., and Sherman, P. W. 1979. Sexual dimorphism and breeding systems in pinnipeds, ungulates, primates, and humans. In *Evolutionary biology and human social behavior: An anthropological perspective,* ed. N. A. Chagnon and W. Irons, pp. 402–435. N. Scituate, Mass.: Duxbury Press.

Alexander, R. D., and Noonan, K. M. 1979. Concealment of ovulation, parental care, and human social evolution. In *Evolutionary biology and human social behavior: An anthropological perspective,* ed. N. A. Chagnon and W. Irons, pp. 436–453. N. Scituate, Mass.: Duxbury Press.

Alexander, R. D., and Sherman, P. W. 1977. Local mate competition and parental investment in social insects. *Science* 196:494–500.

Ali, S. A. 1931. The nesting habits of the baya (*Ploceus philippinus*). *Journal of the Bombay Natural History Society* 34:947–964.

Allee, W. C., Collias, N. E., and Lutherman, C. Z. 1939. Modification of the social order in flocks of hens by the injection of testosterone propionate. *Physiological Zoology* 12:412–440.

Allee, W. C., Emerson, A. E., Park, O., Park, T., and Schmidt, K. P. 1949. *Principles of animal ecology.* Philadelphia: Saunders.

Allen, G. R. 1975. *The anemonefishes: Their classification and biology.* 2nd ed. Neptune City, N. J.: T.F.H. Pub.

Allen, M. R., and Kitts, W. D. 1961. The effect of yellow pine (*Pinus ponderosa* Laws) needles on the reproductivity of the laboratory female mouse. *Canadian Journal of Animal Science* 41:1–12.

Allport, G. W. 1954. *The nature of prejudice.* Reading, Mass.: Addison-Wesley.

Alper, J., Beckwith, J., and Miller, L. G. 1978. Sociobiology is a political issue. In *The sociobiology debate,* ed. A. L. Caplan, pp. 476–488. New York: Harper & Row.

Altmann, M. 1956. Patterns of herd behavior in free-ranging elk of Wyoming, *Cervus canadensis nelsoni. Zoologica* 41:65–71.

Altmann, S. A. 1956. Avian mobbing behavior and predator recognition. *Condor* 58:241–253.

———. 1974. Baboons, space, time, and energy. *American Zoologist* 14:221–248.

———. 1979. Baboon progressions: order or chaos? A study of one-dimensional group geometry. *Animal Behaviour* 27:46–80.

Altmann, S. A., and Altmann, J. 1970. *Baboon ecology: African field research.* Chicago: University of Chicago Press.

Altmann, S. A., Wagner, S. S., and Lenington, S. 1977. Two models for the evolution of polygyny. *Behavioral Ecology and Sociobiology* 2:397–410.

Altum, B. 1868. *Der Vogel und sein Leben.* Mūnster. Trans. E. Mayr, 1935, q.v.

Amadon, D. 1959. The significance of sexual difference in size among birds. *Proceedings of the American Philosophical Society* 103:531–536.

Ambedkar, V. C. 1972. On the breeding biology of the black-throated (*Ploceus benghalensis* Linnaeus) and the streaked (*Ploceus manyar flaviceps* Lesson) weaver-birds in the Kumaon Terai. *Journal of the Bombay Natural History Society* 69:268–282.

Anand, B. K., and Dua, S. 1955. Stimulation of limbic system of brain in waking animals. *Science* 122:1139.

———. 1956. Electrical stimulation of the limbic system of the brain ("visceral brain") in the waking animal. *Indian Journal of Medicinal Research* 44:107–119.

Andersen, D. C., Armitage, K. B., and Hoffman, R. S. 1976. Socioecology of marmots: Female reproductive strategies. *Ecology* 57:552–560.

Andrew, R. J. 1961. The displays given by passerines in courtship and reproductive fighting: A review. *Ibis* 103a:315–348.

Andrewartha, H. G., and Birch, L. C. 1954. *The distribution and abundance of animals.* Chicago: University of Chicago Press.

Andrews, R., and Rand, A. S. 1973. Reproductive effort in anoline lizards. *Ecology* 55:1317–1327.

Angst, W., and Thommen, D. 1977. New data and a discussion of infant killing in Old World monkeys and apes. *Folia Primatologica* 27:198–229.

Anthony, R. 1970. Ecology and reproduction of California quail in southeastern Washington. *Condor* 72:276–287.

Anzenberger, G. 1977. Ethological study of African carpenter bees of the genus *Xylocopa* (Hymenoptera, Anthophoridae). *Zeitschrift für Tierpsychologie* 44:337–374.
Archibald, E. E. A. 1948. Plant populations. I. A new application of Neyman's contagious distribution. *Annals of Botany, London. New Series* 12:221–235.
Ardrey, R. 1962. *African genesis.* London: Collins.
———. 1966. *The territorial imperative.* New York: Atheneum.
Arling, G. L., Ruppenthal, G. C., and Mitchell, G. D. 1969. Aggressive behaviour of the eight-year-old nulliparous isolate female monkey. *Animal Behaviour* 17:109–113.
Armitage, K. B. 1962. Social behaviour of a colony of the yellow-bellied marmot (*Marmota flaviventris*). *Animal Behaviour* 10:319–331.
———. 1974. Male behaviour and territoriality in the yellow-bellied marmot. *Journal of Zoology, London* 172:233–265.
Armitage, K. B., and Downhower, J. F. 1974. Demography of yellow-bellied marmot populations. *Ecology* 55:1233–1245.
Armstrong, E. A. 1947. *Bird display and behaviour.* 2nd ed. London: Lindsay Drummond.
———. 1955. *The wren.* London: Collins.
Armstrong, E. A., and Whitehouse, H. L. K. 1977. The behaviour of the wren. *Biological Reviews* 52:235–294.
Arnold, S. J. 1976. Sexual behaviour, sexual interference and sexual defense in the salamanders *Ambystoma maculatum, Ambystoma tigrinum* and *Plethodon jordani. Zeitschrift für Tierpsychologie* 42:247–300.
Arnold, S. J., and Wasserug, R. J. 1978. Differential predation on metamorphic anurans by garter snakes (*Thamnophis*): Social behavior as a possible defense. *Ecology* 59:1014–1022.
Ashcroft, R. E. 1976. A function of the pairbond in the common eider. *Wildfowl* 27:101–105.
Ashmole, N. P. 1963. The biology of the wideawake or sooty tern *Sterna fuscata* on Ascension Island. *Ibis* 103b:297–364.
———. 1968. Body size, prey size, and ecological segregation in five sympatric tropical terns (Aves: Laridae). *Systematic Zoology* 17:292–304.
———. 1971. Sea bird ecology and the marine environment. In *Avian biology*, vol. I, ed. D. S. Farner and J. R. King, pp. 223–286. New York: Academic Press.
Assem, J. van den. 1967. Territory in the three-spined stickleback, *Gasterosteus aculeatus* L. An experimental study in intra-specific competition. *Behaviour Supplement* 16:1–164.
———. 1971. Some experiments on sex ratio and sex regulation in the pteromalid *Lariophagus distinguendus. Netherlands Journal of Zoology* 21:373–402.
Auffenberg, W., and Weaver, W. G., Jr. 1969. *Gopherus berlandieri* in southeastern Texas. *Bulletin of the Florida State Museum, Biological Sciences* 13:141–203.
Axtell, R. W. 1958. Female reaction to the male call in two anurans (Amphibia). *Southwestern Naturalist* 3:70–76.
Bachmann, C., and Kummer, H. 1980. Male assessment of female choice in hamadryas baboons. *Behavioral Ecology and Sociobiology* 6:315–321.
Baenninger, R., Esters, R. D., and Baldwin, S. 1977. Anti-predator behaviour of baboons and impalas toward a cheetah. *East African Wildlife Journal* 15:327–329.
Bailey, J. E. 1952. Life history and ecology of the sculpin *Cottus bairdi punctulatus* in southwest Montana. *Copeia* 1952:243–255.
Bailey, T. N. 1974. Social organization in a bobcat population. *Journal of Wildlife Management* 38:435–446.
Baker, E. C. S. 1942. *Cuckoo problems.* London: H. F. & G. Witherby.
Baker, M. C., and Mewaldt, L. R. 1979. The use of space by white-crowned sparrows: Juvenile and adult ranging patterns and home range versus body size comparisons in an avian granivore community. *Behavioral Ecology and Sociobiology* 6:45–52.
Baker, R. R. 1972. Territorial behaviour in the nymphalid butterflies, *Aglaius urticae* (L.) and *Inachis io* (L.). *Journal of Animal Ecology* 41:453–470.
Baker, R. R., and Parker, G. A. 1979. The evolution of bird coloration. *Philosophical Transactions of the Royal Society of London, Series B* 287:63–120.
Bakken, A. 1952. Interrelationships of *Sciurus carolinensis* (Gmelin) and *Sciurus niger* (Linnaeus) in mixed populations. Ph.D. thesis, University of Wisconsin, Madison.
Balch, C. C., and Campling, R. W. 1965. Rate of passage of digestia through the ruminant digestive tract. In *Physiology of digestion in the ruminant*, ed. R. W. Dougherty. Washington, D.C.: Butterworth.
Balda, R. R., and Bateman, G. C. 1971. Flocking and annual cycle of the piñon jay, *Gymnorhinus cyanocephalus. Condor* 73:287–302.
———. 1972. The breeding biology of the piñon jay. *Living Bird* 11:5–42.

Baldwin, J. D. 1968. The social behavior of adult male squirrel monkeys (*Saimiri sciureus*) in a semi-natural environment. *Folia Primatologica* 9:281–314.

Baldwin, J. D., and Baldwin, J. I. 1974. Exploration and social play in squirrel monkeys (*Saimiri*). *American Zoologist* 14:303–315.

Balinsky, B. I., and Balinsky, J. B. 1954. On the breeding habits of the South African bullfrog, *Pyxicephalus adspersus*. *South African Journal of Science* 51:55–58.

Ballinger, R. E. 1973. Comparative demography of two viviparous iguanid lizards (*Sceloporus jarrovi* and *Sceloporus poinsetti*). *Ecology* 54:269–283.

Balon, E. K. 1975. Reproductive guilds of fishes: A proposal and definition. *Journal of the Fisheries Research Board of Canada* 32:821–864.

Balozet, L. 1971. Scorpionism in the Old World. In *Venomous animals and their venoms*. Vol. III. *Venomous invertebrates*, ed. W. Bücherl and E. E. Buckley, pp. 349–371. New York: Academic Press.

Banarjee, B. 1975. Growth mounds and foraging territories in *Odontotermes redemanni* (Wasmann) (Isoptera: Termitidae). *Insectes Sociaux* 22:207–212.

Bandura, A. 1973. *Aggression: A social learning analysis*. Englewood Cliffs, N.J.: Prentice-Hall.

Banfield, A. W. F. 1974. *The mammals of Canada*. Toronto: University of Toronto Press.

Barash, D. P. 1973. The social biology of the Olympic marmot. *Animal Behaviour Monographs* 6:172–245.

———. 1974. The social behavior of the hoary marmot, *Marmota caligata*. *Animal Behaviour* 22:256–261.

———. 1976. Social behaviour and individual differences in free-living alpine marmots (*Marmota marmota*). *Animal Behaviour* 24:27–35.

———. 1979a. Evolution as a paradigm for behavior. In *Sociobiology and human values*, ed. M. Gregory and A. Silvers, pp. 13–32. San Francisco: Jossey-Bass.

———. 1979b. Sociobiology: The underlying concept. *Science* 200:1106–1107.

———. 1979c. *The whisperings within: Explorations of human sociobiology*. New York: Harper & Row.

Barbier, M., and Lederer, E. 1960. Structure chimique de la 'substance royale' de la reina d' abeille (*Apis mellifica L.*). *Compte Rendu de l'Académie des Sciences, Paris, Séries D* 250:4467–4469.

Barclay, H. J., and Bergerud, A. T. 1975. Demography and behavioral ecology of California quail on Vancouver Island. *Condor* 77:315–323.

Bardach, J. E., and Todd, J. A. 1970. Chemical communication in fish. In *Advances in chemoreception*. Vol. 1. *Communication by chemical signals*, ed. J. W. Johnston, Jr., D. G. Moulton, and A. Turk, pp. 205–240. New York: Appleton-Century-Crofts.

Barlow, G. W. 1967. Social behavior of a South American leaf fish, *Polycentrus schomburgkii*, with an account of recurring pseudofemale behavior. *American Midland Naturalist* 78:215–234.

———. 1974. Contrasts in social behavior between Central American cichlid fishes and coral-reef surgeon fishes. *American Zoologist* 14:9–34.

Barnes, A. M., Ogden, L. J., and Campos, E. G. 1972. Control of the plague vector, *Opisocrostis hirsutis*, by treatment of prairie dog (*Cynonmys ludovicianus*) burrows with 2% carbaryl dust. *Journal of Medical Entomology* 9:330–333.

Barnett, S. A. 1964. Social stress. In *Viewpoints in biology*, ed. J. D. Carthy and C. L. Duddington, pp. 170–218. London: Butterworth.

Baron, R. A. 1977. *Human aggression*. New York: Plenum.

Baroni Urbani, C. 1979. Territoriality in social insects. In *Social insects*, vol. I, ed. H. R. Hermann, pp. 91–120. New York: Academic Press.

Baroni Urbani, C., and Soulié, J. 1962. Monogynie chez le fourmi *Cremastogaster suitellaris* [Hymenoptera Formicoidea]. *Bulletin de la Societé d'Histoire Naturelle, Toulouse* 97:29–34.

Barrette, C. 1977. Fighting behavior of muntjac and the evolution of antlers. *Evolution* 31:169–176.

Barrows, E. M. 1976. Mating behavior in halictine bees (Hymenoptera: Halictidae): II. Microterritorial and patrolling behavior in males of *Lasioglossum rohweri*. *Zeitschrift für Tierpsychologie* 40:377–389.

Bartholomew, G. A., Jr. 1942. The fishing activities of double-crested cormorants on San Francisco Bay. *Condor* 44:13–21.

———. 1952. Reproductive and social behavior in the northern elephant seal. *University of California Publications in Zoology* 47:369–472.

———. 1967. Discussion of paper by K. S. Norris. In *Aggression and defense: Neural mechanisms and social patterns*. Vol. 5: *Brain function*, ed. C. D. Clemente and D. E. Lindsley, pp. 232–241. Los Angeles: University of California Press.

———. 1970. A model for the evolution of pinniped polygyny. *Evolution* 24:546–559.
Bartholomew, G. A., Jr., and Hoel, P. G. 1953. Reproductive behavior of the Alaska fur seal, *Callorhinus ursinus. Journal of Mammalogy* 34:417–436.
Bastock, M. 1956. A gene mutation which changes a behavior pattern. *Evolution* 10:421–439.
———. 1967. *Courtship: An ethological study*. Chicago: Aldine.
Bateman, A. J. 1948. Intrasexual selection in *Drosophila. Heredity* 2:349–368.
Bates, B. C. 1970. Territorial behavior in primates: A review of recent field studies. *Primates* 11:271–284.
Bates, D. G., and Lees, S. H. 1979. The myth of population regulation. In *Evolutionary biology and human social behavior: An anthropological perspective*, ed. N. A. Chagnon and W. Irons, pp. 273–289. N. Scituate, Mass.: Duxbury Press.
Bate-Smith, E. C., and Metcalfe, C. R. 1957. Leuco-anthocyanins. 3. The nature and systematic distribution of tannins in dicotyledonous plants. *Journal of the Linnaean Society, Botany* 55:669–705.
Bate-Smith, E. C., and Riberéau-Gayon, P. 1959. Leuco-anthocyanins in seeds. *Qualitas Plantarum* 8:189–198.
Beach, F. A. 1950. The snark was a boojum. *American Psychologist* 5:115–124.
———. 1971. Hormonal factors controlling the differentiation, development and display of copulatory behavior in the ramstergig and related species. In *The biopsychology of development*, ed. L. Aronson and E. Tobach, pp. 249–296. New York: Academic Press.
Becker, P. H. 1977. Verhalten auf Lautäusserungen der Zwillingsart, interspezifische Territorialität und Habitatansprüche von Winter- und Sommergoldhähnchen (*Regulus regulus, R. ignicapillus*). *Journal für Ornithologie* 118:233–260.
Bédard, J. 1969a. Adaptive radiation in Alcidae. *Ibis* 111:189–198.
———. 1969b. Feeding of the least, crested and parakeet auklets around St. Lawrence Island, Alaska. *Canadian Journal of Zoology* 47:1025–1050.
———. 1976. Coexistence, coevolution and convergent evolution in seabird communities: A comment. *Ecology* 57:177–184.
Beecher, M. D., and Beecher, I. M. 1979. Sociobiology of bank swallows: Reproductive strategy of the male. *Science* 205:1282–1285.
Behle, W. H. 1944. The pelican colony at Gunnison Island, Great Salt Lake, in 1943. *Condor* 46:198–200.
Beig, D. 1972. The production of males in queenright colonies of *Trigona (Scaptotrigona) postica. Journal of Apiculture Research* 11:33–39.
Bekoff, M. 1974. Social play and play-soliciting by infant canids. *American Zoologist* 14:323–340.
———. 1975. Social behavior and ecology of the African Canidae: A review. In *The wild canids. Their systematics, behavioral ecology and evolution*, ed. M. W. Fox, pp. 120–142. New York: Van Nostrand Reinhold.
Bell, E. A. 1972. Toxic amino acids in the Leguminosae. In *Phytochemical ecology*, ed. J. B. Harborne, pp. 163–174. New York: Academic Press.
Bell, G. 1978a. Group selection in structured populations. *American Naturalist* 112:389–399.
———. 1978b. The evolution of anisogamy. *Journal of Theoretical Biology* 73:247–270.
Bellrose, F. C., Jr., Scott, T. G., Hawkins, A. S., and Low, J. B. 1961. Sex ratios and age ratios in North American ducks. *Illinois Natural History Survey, Bulletin* 27:391–474.
Bendell, J. F., and Elliott, P. W. 1966. Habitat selection in blue grouse. *Condor* 68:431–446.
———. 1967. Behaviour and the regulation of numbers in blue grouse. *Canadian Wildlife Service Report, Series 4*, pp. 1–76.
Bendell, J. F., King, D. G., and Mossop, D. H. 1972. Removal and repopulation of blue grouse in a declining population. *Journal of Wildlife Management* 36:1153–1165.
Benfield, E. F. 1972. A defensive secretion of *Dineutes discolor* (Coleoptera: Gyrinidae). *Annals of the Entomological Society of America* 65:1324–1327.
Benoit, J. 1950. Reproduction—charactères sexuels et hormones determinisme du cycle sexuel saisonnier. In *Traité de Zoologie. Oiseaux.* Tome XV. *Anatomie, Systématique, Biologie*, ed. P. Grassé, pp. 384–403. Paris: Masson et Cie.
Benson, W. W., and Emmel, T. C. 1973. Demography of gregariously roosting populations of the nymphaline butterfly *Marpesia berania* in Costa Rica. *Ecology* 54:326–335.
Bent, A. C. 1946. Life histories of North American jays, crows, and titmice. *United States National Museum Bulletin* 191:1–495.
Berger, A. J. 1972. *Hawaii birdlife*. Honolulu: University of Hawaii Press.
Berger, J. 1979. Weaning conflict in desert and mountain bighorn sheep (*Ovis canadensis*): An ecological interpretation. *Zeitschrift für Tierpsychologie* 50:188–200.

Berkowitz, L. 1965. The concept of aggressive drive: Some additional considerations. In *Advances in experimental social psychology*, vol. 2, ed. L. Berkowitz, pp. 301–329. New York: Academic Press.

———. 1969. The frustration-aggression hypothesis revisited. In *Roots of aggression: A re-examination of the frustration-aggression hypothesis*, ed. L. Berkowitz, pp. 1–28. New York: Atherton.

———. 1974. Some determinants of impulsive aggression: The role of mediated associations with reinforcements for aggression. *Psychological Review* 81:165–176.

Berndt, R., and Sternberg, H. 1968. Terms, studies and experiments on the problems of bird dispersion. *Ibis* 110:256–269.

Bernstein, I. S. 1964. Role of the dominant male rhesus monkey in response to external challenges to the group. *Journal of Comparative and Physiological Psychology* 57:404–406.

———. 1966. Analysis of a key role in a capuchin (*Cebus albifrons*) group. *Tulane Studies in Zoology* 13:49–54.

———. 1970. Primate status hierarchies. In *Primate behavior. Developments in field and laboratory research*, vol. I, ed. L. A. Rosenblum, pp. 71–109. New York: Academic Press.

———. 1976. Dominance, aggression and reproduction in primate societies. *Journal of Theoretical Biology* 60:459–472.

Bernstein, I. S., and Gordon, T. P. 1974. The function of aggression in primate societies. *American Scientist* 62:304–311.

Bernstein, I. S., and Sharpe, L. G. 1966. Social roles in a rhesus monkey group. *Behaviour* 36:91–104.

Bernstein, R. A. 1975. Foraging strategies of ants in response to variable food density. *Ecology* 56:213–219.

Bertram, B. C. R. 1975a. The social system of lions. *Scientific American* 232(5):54–65.

———. 1975b. Social factors influencing reproduction in wild lions. *Journal of Zoology, London* 177:463–482.

———. 1976. Kin selection in lions and in evolution. In *Growing points in ethology*, ed. P. P. G. Bateson and R. A. Hinde, pp. 281–301. Cambridge, England: Cambridge University Press.

———. 1978a. *Pride of lions*. London: Dent.

———. 1978b. Living in groups: Predators and prey. In *Behavioural ecology: An evolutionary approach*, ed. J. R. Krebs and N. B. Davies, pp. 64–96. Sunderland, Mass.: Sinauer Associates.

Bertram, G. C. L. 1940. The biology of the Weddell and crabeater seals. *Scientific Report of the British Graham Land Expedition, 1934–1937* 1:1–139.

Betts, M. 1955. The food of titmice in oak woodland. *Journal of Animal Ecology* 24:282–323.

Bevan, W., Jr., Bloom, W. L., and Lewis, G. T. 1951. Levels of aggressiveness in normal and amino-acid deficient albino rats. *Physiological Zoology* 24:231–237.

Bick, G. H., and Bick, J. C. 1965. Demography and behavior of the damselfly, *Argia apicalis* (Say), (Odonata: Coenagriidae). *Ecology* 46:461–472.

Bier, K. 1954. Über den Einfluss der Königin auf die Arbeiterinnenfertilität im Ameisenstaat. *Insectes Sociaux* 1:7–19.

———. 1956. Arbeiterinnenfertilität und Aufzucht von Geschlechtstieren als Regulationsleistung des Ameisenstaaten. *Insectes Sociaux* 3:177–184.

———. 1958. Die Regulation der Sexualität in den Insektenstaaten. *Erbegnisse Biologie* 20:97–126.

Birch, H. G., and Clark, G. 1946. Hormonal modification of social behavior: II. The effects of sex-hormone administration on the social dominance status of the female-castrate chimpanzee. *Psychosomatic Medicine* 8:320–321.

Bishop, S. C. 1941. The salamanders of New York. *New York State Museum Bulletin* 32:1–365.

Bisset, G. W., Frazer, J. F. D., Rothschild, M., and Schachter, M. 1960. A pharmacologically active choline ester and other substances in the garden tiger moth, *Arctica caja* (L.). *Proceedings of the Royal Society of London, Series B* 152:255–262.

Blackford, J. L. 1958. Territoriality and breeding behavior of a population of blue grouse in Montana *Condor* 60:145–158.

———. 1963. Further observations on the breeding behavior of a blue grouse population in Montana. *Condor* 60:145–158.

Blackman, G. E. 1942. Statistical and ecological studies in the distribution of species in plant communities. I. Dispersion as a factor in the study of changes in plant populations. *Annals of Botany, new series* 6:351–370.

Blaffer Hrdy, S. 1974. Male-male competition and infanticide among the langurs (*Presbytis entellus*) of Abu, Rajasthan. *Folia Primatologica* 22:19–58.

———. 1976. Care and exploitation of nonhuman primate infants by conspecifics other than the mother. *Advances in the Study of Behavior* 6:101–158.

———. 1977a. *The langurs of Abu. Female and male strategies of reproduction*. Cambridge, Mass.: Harvard University Press.

———. 1977b. Infanticide as a primate reproductive strategy. *American Scientist* 65:40–49.

———. 1977c. Reply to P. Dolhinow. *American Scientist* 65:266–268.

Blair, W. F. 1960. *The rusty lizard. A population study*. Austin: University of Texas Press.

Blakley, N. R. 1976. Successive polygyny in upland nesting redwinged blackbirds. *Condor* 78:129–133.

Blest, A. D. 1957. The function of eyespot patterns in the Lepidoptera. *Behaviour* 11:209–256.

Blum, M. S., Roberts, J. E., Jr., and Novak, A. F. 1961. Chemical and biological characteristics of venom of the ant *Solenopsis xyloni* McCook. *Psyche, Cambridge* 68:73–74.

Blum, M. S., Walker, J. R., Callahan, P. S., and Novak, A. F. 1958. Chemical, insecticidal, and antibiotic properties of fire ant venom. *Science* 128:306–307.

Blumer, L. S. 1979. Male parental care in bony fishes. *Quarterly Review of Biology* 54:149–161.

Boag, D. A., and Sumanik, K. M. 1969. Characteristics of drumming sites selected by ruffed grouse in Alberta. *Journal of Wildlife Management* 33:621–628.

Boardman, R. S., Cheetham, A. H., Oliver, W. A., Jr., Coates, A. G., and Bayer, F. M., eds. 1973. *Animal colonies. Development and function through time*. Stroudsburg, Pa.: Dowden, Hutchinson, & Ross.

Boch, R. 1956. Die Tanze der Bienen bei nahen und fernen Trachquellen. *Zeitschrift für vergleichende Physiologie* 38:136–167.

Boelkins, R. C., and Heiser, J. F. 1970. Biological bases of aggression. In *Violence and the struggle for existence*, ed. D. N. Daniels, M. F. Gilula, and M. Ochberg, pp. 15–52. Boston: Little, Brown.

Boelkins, R. C., and Wilson, A. P. 1972. Intergroup social dynamics of the Cayo Santiago rhesus (*Macaca mulatta*) with special reference to changes in group membership by males. *Primates* 13:125–140.

Boersma, P. D. 1976. An ecological and behavioral study of the Galapagos penguin. *Living Bird* 15:43–93.

Boersma, P. D., and Wheelwright, N. T. 1979. Egg neglect in the Procellariiformes: Reproductive adaptations in the fork-tailed storm-petrel. *Condor* 81:157–165.

Boggs, C. L., and Gilbert, L. E. 1979. Male contribution to egg production in butterflies: Evidence for transfer of nutrients at mating. *Science* 206:83–84.

Bonner, J. T. 1974. *On development. The biology of form*. Cambridge, Mass.: Harvard University Press.

Boorman, E., and Parker, G. A. 1976. Sperm (ejaculate) competition in *Drosophila melanogaster*, and the reproductive value of females to males in relation to female age and mating status. *Ecological Entomology* 1:145–155.

Boorman, S. A., and Levitt, P. R. 1972. Group selection on the boundary of a stable population. *Proceedings of the National Academy of Sciences* 69:2711–2713.

———. 1973. Group selection on the boundary of a stable population. *Theoretical Population Biology* 4:85–128.

Borgia, G. 1979. Sexual selection and the evolution of mating systems. In *Sexual selection and reproductive competition in insects*, ed. M. S. Blum and N. A. Blum, pp. 19–80. New York: Academic Press.

Boss, W. R. 1943. Hormonal determination of adult characters and sex behavior in herring gulls (*Larus argentatus*). *Journal of Experimental Zoology* 94:181–208.

Boucher, D. H. 1977. On wasting parental investment. *American Naturalist* 111:786–788.

Bourlière, F., Hunkeler, C., and Bertrand, M. 1970. Ecology and behavior of Lowe's guenon (*Cercopithecus campbelli lowei*) in the Ivory Coast. In *Old World monkeys. Evolution, ecology, and systematics*, ed. J. R. Napier and P. H. Napier, pp. 297–350. New York: Academic Press.

Bourne, G. R. 1974. The red-billed toucan in Guyana. *Living Bird* 13:99–126.

Bovbjerg, R. V. 1953. Dominance order in the crayfish, *Orconectes virilis* (Hagen). *Physiological Zoology* 26:173–178.

Bowers, W. S., Ohata, T., Cleere, J. S., and Marsella, P. A. 1976. Discovery of insect antijuvenile hormones in plants. *Science* 193:542–547.

Boycott, B. B. 1958. The cuttlefish—*Sepia*. *New Biology* 25:98–118.

Brach, V. 1977. *Anelosima studiosus* (Araneae: Theridiidae) and the evolution of quasisociality in theridiid spiders. *Evolution* 31:154–161.

Bradbury, J. W. 1977a. Social organization and communication. In *Biology of bats*, vol. 3, ed. W. Wimsatt, pp. 1–72. New York: Academic Press.

———. 1977b. Lek mating behavior in the hammer-headed bat. *Zeitschrift für Tierpsychologie* 45:225–255.

———. 1980. The evolution of leks. In *Natural selection and social behavior*, ed. R. D. Alexander and D. Tinkle. Ann Arbor: University of Michigan Press (in press).

Bradbury, J. W., and Emmons, L. 1974. Social organization of some Trinidad bats. I. Emballonuridae. *Zeitschrift für Tierpsychologie* 36:137–183.

Bradbury, J. W., and Vehrencamp, S. L. 1976. Social organization and foraging in emballonurid bats. II. A model for the determination of group size. *Behavioral Ecology and Sociobiology* 1:383–404.

———. 1977. Social organization and foraging in emballonurid bats. III. Mating systems. *Behavioral Ecology and Sociobiology* 2:19–29.

Braden, A. W. H., and McDonald, I. W. 1970. Disorders of grazing animals due to plant constituents. In *Australian grasslands*, ed. R. M. Moore, pp. 381–392. Canberra: Australian Natural History Press.

Bradt, G. W. 1938. A study of beavers in Michigan. *Journal of Mammalogy* 19:139–162.

Bragg, A. N. 1937. Observations on *Bufo cognatus* with special reference to breeding habits and eggs. *American Midland Naturalist* 18:273–281.

———. 1940. Habits, habitats, and breeding of *Bufo woodhousii woodhousii* in Oklahoma. *American Midland Naturalist* 24:306–321.

Brain, P. F. 1979a. Effects of the hormones of the pituitary-gonadal axis on behaviour. In *Chemical influences on behaviour*, ed. K. Brown and S. J. Cooper, pp. 255–329. New York: Academic Press.

———. 1979b. Effects of the hormones of the pituitary-adrenal axis on behaviour. In *Chemical influences on behaviour*, ed. K. Brown and S. J. Cooper, pp. 331–372. New York: Academic Press.

Brain, P. F., and Bowden, N. J. 1978. Sex steroid control of intermale fighting in albino laboratory mice. In *Current developments in psychopharmacology*, vol. 5, ed. W. B. Essman and L. Valzelli, pp. 403–465. New York: Spectrum.

Brain, P. F., and Nowell, N. W. 1969. Some behavioral and endocrine changes in the development of the albino laboratory mouse. *Communications in Behavioral Biology* 4:203–220.

Brain, P. F., Nowell, N. W., and Wouters, A. 1971. Some relationships between adrenal function and the effectiveness of a period of isolation in inducing intermale aggression in albino mice. *Physiology and Behavior* 6:27–29.

Brain, P. F., and Poole, A. E. 1976. The role of endocrines in isolation-induced intermale fighting in laboratory mice. I. Pituitary-adrenocortical influences. *Aggressive Behavior* 1:39–69.

Branch, G. M. 1975. Mechanisms reducing intraspecific competition in *Patella* spp: Migration, differentiation and territorial behaviour. *Journal of Animal Ecology* 44:575–600.

Brattstrom, B. H. 1974. The evolution of lizard social behavior. *American Zoologist* 14:35–49.

Bray, O. E., Kennelly, J. J., and Guarino, J. L. 1975. Fertility of eggs produced on territories of vasectomized red-winged blackbirds. *Wilson Bulletin* 87:187–195.

Breder, C. M., Jr. 1936. The reproductive habits of the North American sunfishes (family Centrarchidae). *Zoologica* 21:1–48.

———. 1939. On the life history and development of the sponge blenny, *Paraclinus marmoratus*. *Zoologica* 24:487–496.

———. 1941. On the reproductive behavior of the sponge blenny, *Paraclinus marmoratus* (Steindachner). *Zoologica* 26:233–236.

———. 1951. Studies on the structure of the fish school. *Bulletin of the American Museum of Natural History* 98:1–29.

Breder, C. M., Jr., and Rosen, D. E. 1966. *Modes of reproduction in fishes*. Garden City, N.Y.: Doubleday.

Breed, M. D., Silverman, J. M., and Bell, W. J. 1978. Agonistic behavior, social interactions, and behavioral specialization in a primitively eusocial bee. *Insectes Sociaux* 25:351–364.

Breitenbach, R. P., and Meyer, R. K. 1959. Effect of incubation and brooding on fat, visceral weights and body weight of the hen pheasant (*Phasianus colchicus*). *Poultry Science* 38:1014–1026.

Brian, M. V. 1955. Food collection by a Scottish ant community. *Journal of Animal Ecology* 24:336–351.

———. 1965. *Social insect populations*. London: Academic Press.

———. 1968. Regulation of sexual production in an ant society. *Colloques Internationaux du Centre National de la Rechereche Scientifique, Paris,* no. 173:61–76.

———. 1970. Communication between queens and larvae in the ant *Myrmica*. *Animal Behaviour* 18:467–472.

———. 1973. Queen recognition by brood-rearing workers of the ant *Myrmica rubra* L. *Animal Behaviour* 21:691–698.

———. 1979. Caste differentiation and division of labor. In *Social insects,* vol. I, ed. H. R. Hermann, pp. 121–222. New York: Academic Press.

Brian, M. V., and Blum, M. S. 1969. The influence of *Myrmica* queen head extracts on larval growth. *Journal of Insect Physiology* 15:2213–2223.

Brian, M. V., and Carr, C. A. H. 1960. The influence of the queen on brood rearing in ants of the genus *Myrmica. Journal of Insect Physiology* 5:81–94.

Brian, M. V., and Hibble, J. 1963. Larval size and the influence of the queen on growth in *Myrmica. Insectes Sociaux* 10:71–81.

Brian, M. V., Hibble, J., and Kelly, A. F. 1966. The dispersion of ant species in a southern English heath. *Journal of Animal Ecology* 35:281–290.

Brian, M. V., Hibble, J., and Stradling, D. J. 1965. Ant pattern and density in a southern English heath. *Journal of Animal Ecology* 34:545–555.

Brinckmann, M. 1954. Vom weissen Storch in Berzirk Osnabrück. *Die Vogelwarte* 75:194–200.

Bristowe, W. S. 1958. *The world of spiders.* London: Collins.

Britten, R. J., and Davidson, H. E. 1969. Gene regulation for higher cells: A theory. *Science* 165:349–357.

———. 1971. Repetitive and non-repetitive DNA sequences and a speculation on the origins of evolutionary novelty. *Quarterly Review of Biology* 46:111–138.

Brock, V. E., and Riffenburgh, R. H. 1960. Fish schooling: A possible factor in reducing predation. *Journal du Conseil, Conseil Permanent International pour l'Exploration de la Mer* 25:307–317.

Brockmann, H. J. 1973. The function of poster-coloration in the beaugregory, *Eupomacentrus leucostictus* (Pomacentridae, Pisces). *Zeitschrift für Tierpsychologie* 33:13–34.

Brockmann, H. J., and Hailman, J. P. 1976. Fish cleaning symbiosis: Notes on juvenile angelfishes (*Pomacanthus,* Chaetodontidae) and comparisons with other species. *Zeitschrift für Tierpsychologie* 42:129–138.

Brockway, B. F. 1964. Social influences on reproductive physiology and ethology of budgerigars *(Melopsittacus undulatus). Animal Behaviour* 12:493–501.

Bronson, F. H. 1967. Effects of social stimulation on adrenal and reproductive physiology of rodents. In *Husbandry of laboratory animals,* ed. M. L. Conalty, pp. 513–542. New York: Academic Press.

———. 1968. Pheromonal influences on mammalian reproduction. In *Perspectives in reproduction and sexual behavior,* ed. M. Diamond, pp. 341–361. Bloomington: Indiana University Press.

———. 1971. Rodent pheromones. *Biology of Reproduction* 4:344–357.

Bronson, F. H., and Desjardins, C. 1968. Aggresssion in adult mice: Modification by neonatal injections of gonadal hormones. *Science* 161:705–706.

———. 1969. Aggressive behavior and seminal vesicle function in mice: Differential sensitivity to an androgen given neonatally. *Endocrinology* 85:971–974.

———. 1971. Steroid hormones and aggressive behavior in mammals. In *The physiology of aggression and defeat,* ed. B. E. Eleftheriou and J. P. Scott, pp. 43–63. New York: Plenum.

Bronson, F. H., and Eleftheriou, B. E. 1963. Influence of strange males on implantation in the deermouse. *General and Comparative Endocrinology* 3:515–518.

Brosset, A. 1976. Social organization in the African bat, *Myotis boccagei. Zeitschrift für Tierpsychologie* 42:50–56.

Brower, J. V. Z. 1963. Experimental studies and new evidence on the evolution of mimicry in butterflies. *16th International Congress of Zoology* 4:156–161.

Brower, L. P. 1969. Ecological chemistry. *Scientific American* 220(2):22–29.

———. 1971. Prey coloration and predator behavior. In *Topics in the study of life. The BIO source book, section 6. Animal behavior,* ed. V. G. Dethier, pp. 360–370. New York: Harper & Row.

Brower, L. P., Alcock, J., and Brower, J. V. Z. 1971. Avian feeding behaviour and the selective advantage of incipient mimicry. In *Ecological genetics and evolution,* ed. R. Creed, pp. 261–274. Oxford: Blackwell Scientific.

Brower, L. P., and Brower, J. V. Z. 1964. Birds, butterflies, and plant poisons: A study in ecological chemistry. *Zoologica* 49:137–159.

Brower, L. P., Brower, J. V. Z., and Corvino, J. M. 1967. Plant poisons in a terrestrial food chain. *Proceedings of the National Academy of Sciences* 57:893–898.

Brown, J. L. 1963. Aggressiveness, dominance and social organization in the Steller's jay. *Condor* 65:460–484.

———. 1964. The evolution of diversity in avian territorial systems. *Wilson Bulletin* 76:160–169.

———. 1966. Types of group selection. *Nature* 211:870.
———. 1969a. Territorial behavior and population regulation in birds: A review and re-evaluation. *Wilson Bulletin* 81:293–329.
———. 1969b. The buffer effect and productivity in tit populations. *American Naturalist* 103:347–354.
———. 1970. Cooperative breeding and altruistic behaviour in the Mexican jay, *Aphelocoma ultramarina*. *Animal Behaviour* 18:366–378.
———. 1972. Communal feeding of nestlings in the Mexican jay (*Aphelocoma ultramarina*): Interflock comparisons. *Animal Behaviour* 20:395–403.
———. 1974. Alternate routes to sociality in jays with a theory for the evolution of altruism and communal breeding. *American Zoologist* 14:63–80.
———. 1975a. *The evolution of behavior.* New York: Norton.
———. 1975b. Helpers among Arabian babblers, *Turdoides squamiceps*. *Ibis* 117:243–244.
———. 1978. Avian communal breeding systems. *Annual Review of Ecology and Systematics* 9:123–155.
———. 1981. Fitness in complex avian social systems. In *Evolution of social behavior: Hypotheses and empirical tests,* ed. H. Markl. Basel: Verlag Chemie (in press).
Brown, J. L., and Balda, R. P. 1977. The relationship of habitat quality to group size in Hall's babbler, *Pomatostomus halli*. *Condor* 79:312–320.
Brown, J. L., and Brown, E. R. 1980. Kin selection and individual selection in babblers. In *Natural selection and social behavior: Recent research and new theory,* ed. R. D. Alexander and D. W. Tinkle. Portland, Oreg.: Chiron Press (in press).
Brown, J. L., Dow, D. D., Brown, E. R., and Brown, S. D. 1978. Effects of helpers on feeding of nestlings in the grey-crowned babbler (*Pomatostomus temporalis*). *Behavioral Ecology and Sociobiology* 4:43–59.
Brown, J. L., Hunsperger, R. W., and Rosvold, H. E. 1969. Defense, attack, and flight elicited by electrical stimulation of the hypothalamus of the cat. *Experimental Brain Research* 8:113–129.
Brown, J. L., and Orians, G. H. 1970. Spacing patterns in mobile animals. *Annual Review of Ecology and Systematics* 1:239–262.
Brown, L. 1978. Polygamy, female choice and the mottled sculpin (*Cottus bairdi*). Ph.D. thesis, Ohio State University, Columbus.
Brown, L., and Downhower, J. F. 1977. Resources and the mating system of the mottled sculpin, *Cottus bairdi*. *American Zoologist* 17:936 (abstract).
Brown, L. E., and Pierce, J. R. 1967. Male-male interactions and chorusing intensities of the Great Plains toad, *Bufo cognatus*. *Copeia* 1967:149–154.
Brown, L. H., and Amadon, D. 1968. *Eagles, hawks, and falcons of the world,* vols. 1 and 2. New York: McGraw-Hill.
Brown, L. H., and Urban, E. K. 1969. The breeding biology of the great white pelican *Pelecanus onocrotalus roseus* at Lake Shala, Ethiopia. *Ibis* 111:199–237.
Brown, R. G. B. 1967a. Breeding success and population growth in a colony of herring and lesser black-backed gulls *Larus argentatus* and *L. fuscus*. *Ibis* 109:502–515.
———. 1967b. Courtship behaviour in the lesser black-backed gull (*Larus fuscus*). *Behaviour* 29:122–153.
Brown, R. G. B., and Baird, D. E. 1965. Social factors as possible regulators of *Puffinus gravis* numbers. *Ibis* 107:249–251.
Brown, R. N. 1977. Character convergence in bird song. *Canadian Journal of Zoology* 55:1523–1544.
Bruce, H. M. 1961. Time relations in the pregnancy block induced in mice by strange males. *Journal of Reproduction and Fertility* 2:138–142.
———. 1963. Olfactory block to pregnancy among grouped mice. *Journal of Reproduction and Fertility* 6:451–460.
———. 1965. Effect of castration on the reproductive pheromones of male mice. *Journal of Reproduction and Fertility* 10:141–143.
———. 1966. Smell as an exteroceptive factor. *Journal of Animal Science, Supplement* 25:83–89.
Bruning, D. F. 1973. The greater rhea chick and egg delivery route. *Natural History* 82(3):68–75.
———. 1974. Social structure and reproductive behavior in the greater rhea. *Living Bird* 13:251–294.
Bryant, D. M. 1975. Breeding biology of the house martin *Delichon urbica* in relation to aerial insect abundance. *Ibis* 117:180–216.
———. 1979. Reproductive costs in the house martin. *Journal of Animal Ecology* 48:655–675.
Bücherl, W. 1971a. Venomous chilopods or centipedes. In *Venomous animals and their venoms.*

Vol. III. *Venomous invertebrates,* ed. W. Bücherl and E. E. Buckley, pp. 169–196. New York: Academic Press.

———. 1971b. Spiders. In *Venomous animals and their venoms.* Vol. III. *Venomous invertebrates,* ed. W. Bücherl and E. E. Buckley, pp. 197–277. New York: Academic Press.

———. 1971c. Classification, biology, and venom extraction of scorpions. In *Venomous animals and their venoms.* Vol. III. *Venomous invertebrates,* ed. W. Bücherl and E. E. Buckley, pp. 317–347. New York: Academic Press.

Bücherl, W., and Buckley, E. E., eds. 1971a. *Venomous animals and their venoms.* Vol. II. *Venomous vertebrates.* New York: Academic Press.

———, eds. 1971b. *Venomous animals and their venoms.* Vol. III. *Venomous invertebrates.* New York: Academic Press.

Bücherl, W., Buckley, E. E., and Deulofeu, V., eds. 1968. *Venomous animals and their venoms.* Vol. I. *Venomous vertebrates.* New York: Academic Press.

Buchli, H. H. R. 1969. Hunting behavior in the Ctenizidae. *American Zoologist* 9:175–193.

Buckley, E. E., and Porges, N., eds. 1956. *Venoms.* Publication no. 44. Washington, D.C.: American Association for the Advancement of Science.

Buechner, H. K., and Roth, H. D. 1974. The lek system in Uganda kob antelope. *American Zoologist* 14:145–162.

Buechner, H. K., and Schloeth, R. 1965. Ceremonial mating behavior in Uganda kob *(Adenota kob thomasi* Neumann). *Zeitschrift für Tierpsychologie* 22:209–225.

Bull, J. J., and Shine, R. 1979. Iteroparous animals that skip opportunities for reproduction. *American Naturalist* 114:296–303.

Bulmer, M. G. 1977. Periodical insects. *American Naturalist* 111:1099–1117.

Bumpus, H. C. 1899. The elimination of the unfit as illustrated by the introduced sparrow, *Passer domesticus.* (A fourth contribution to the study of variation.) *Biology Lectures at Wood's Hole for 1898,* pp. 209–226.

Burger, J. 1974a. Breeding biology and ecology of the brown-hooded gull in Argentina. *Auk* 91:601–613.

———. 1974b. Determinants of colony and nest-site selection in the silver grebe *(Podiceps occidentalis)* and the Rolland's grebe *(Rollandia rolland). Condor* 76:301–306.

———. 1979. Colony size: A test for breeding synchrony in herring gull *(Larus argentatus)* colonies. *Auk* 96:694–703.

Burger, J., and Shisler, J. 1980. The process of colony formation among herring gulls *Larus argentatus* nesting in New Jersey. *Ibis* 122:15–26.

Burgess, J. W. 1976. Social spiders. *Scientific American* 234(3):101–106.

Burley, N., and Moran, N. 1979. The significance of age and reproductive experience in the mate preferences of feral pigeons, *Columbia livia. Animal Behaviour* 27:686–698.

Burnett, T. 1971. Prey consumption in acarine predator-prey populations reared in the greenhouse. *Canadian Journal of Zoology* 49:903–913.

Burt, W. H. 1943. Territoriality and home range concepts as applied to mammals. *Journal of Mammalogy* 24:346–352.

Burton, R. 1976. *The mating game.* New York: Crown.

Buschinger, A. 1974. Monogynie und Polygynie in Insektensozietäten. In *Sozialpolymorphismus bei Insekten,* ed. G. H. Schmidt, pp. 862–896. Stuttgart: Wissenschaftliche Verlagsgesellschaft.

Buskirk, R. E. 1975. Coloniality, activity patterns and feeding in a tropical orb-weaving spider. *Ecology* 56:1314–1328.

———. 1979. Sociality in the Arachnida. In *Social insects,* vol. II, ed. H. R. Hermann. New York: Academic Press.

Buskirk, W. H. 1972. Foraging ecology of bird flocks in a tropical forest. Ph.D. thesis, University of California, Davis.

———. 1976. Non-breeding season social systems in a tropical avifauna. *American Naturalist* 110:293–310.

Buskirk, W. H., Powell, G. V. N., Wittenberger, J. F., Buskirk, R. E., and Powell, T. U. 1972. Interspecific bird flocks in tropical highland Panama. *Auk* 89:612–624.

Buss, A. 1961. *The psychology of aggression.* New York: Wiley.

Busse, C. D. 1977. Chimpanzee predation as a possible factor in the evolution of red colobus monkey social organization. *Evolution* 31:907–911.

Butler, C. G., Callow, R. K., and Johnston, N. C. 1961. The isolation and synthesis of queen substance, 9-oxodec-*trans*-2-enoic acid, a honeybee pheromone. *Proceedings of the Royal Society of London, Series B* 155:417–432.

Butler, C. G., and Paton, P. N. 1962. Inhibition of queen rearing by queen honey-bees (*Apis mellifera* L.) of different ages. *Proceedings of the Royal Entomological Society of London, Series A* 37:114–116.

Bygott, J. D., Bertram, B. C. R., and Hanby, J. P. 1979. Male lions in large coalitions gain reproductive advantages. *Nature* 282:839–841.

Cairns, J. 1963. The chromosome of *Escherichia coli. Cold Spring Harbor Symposia of Quantitative Biology* 28:43–46.

Calaby, J. H. 1951. Note on the little eagle; with particular reference to rabbit predation. *Emu* 51:33–56.

———. 1968. The platypus *(Ornithorhynchus anatinus)* and its venomous characteristics. In *Venomous animals and their venoms.* Vol. I. *Venomous vertebrates,* ed. W. Bücherl, E. E. Buckley, and V. Deulofeu, pp. 15–29. New York: Academic Press.

Caldwell, R. L. 1979. Cavity occupation and defensive behaviour in the stomatopod *Gonodactylus festai:* Evidence for chemically mediated individual recognition. *Animal Behaviour* 27:194–201.

Caldwell, R. L., and Dingle, H. 1976. Stomatopods. *Scientific American* 234(1):84–89.

Calef, G. W. 1973. Spatial distribution and "effective" breeding population of red-legged frogs *(Rana aurora)* in Marion Lake, British Columbia. *Canadian Field Naturalist* 87:279–284.

Calhoun, J. B. 1962. Population density and social pathology. *Scientific American* 206(2):139–148.

Calvert, W. H., Hedrick, L. E., and Brower, L. P. 1979. Mortality of the monarch butterfly *(Danaus plexippus* L.): Avian predation at five overwintering sites in Mexico. *Science* 204:847–851.

Camenzind, F. J. 1978. Behavioral ecology of coyotes on the National Elk Refuge, Jackson, Wyoming. In *Coyotes. Biology, behavior, and management.* ed. M. Bekoff, pp. 267–294. New York: Academic Press.

Campanella, P. J. 1975. The evolution of mating systems in temperate zone dragonflies (Odonata: Anisoptera). II: *Libellula luctuosa* (Burmeister). *Behaviour* 54:278–310.

Campbell, A. 1979. Structure of complex operons. In *Biological regulation and development.* Vol. I. *Gene expression,* ed. R. F. Goldberger, pp. 19–55. New York: Plenum.

Campbell, D. T. 1965. Variation and selective retention in sociocultural evolution. In *Social change in developing areas: A reinterpretation of evolutionary theory,* ed. H. R. Barringer, G. L. Blanksten, and R. W. Mack, pp. 19–49. Cambridge, Mass.: Schenkman.

Campbell, H., and Lee, L. 1956. Notes on the sex ratio of Gambel's and scaled quail in New Mexico. *Journal of Wildlife Management* 20:93–94.

Campbell, H., Martin, D. K., Ferkovich, P. E., and Harris, B. K. 1973. Effects of hunting and some other environmental factors on scaled quail in New Mexico. *Wildlife Monographs* 34:1–49.

Candland, D. K., and Leshner, A. I. 1974. A model of agonistic behavior: Endocrine and autonomic correlates. In *Limbic and autonomic nervous systems research,* ed. L. V. DiCara, pp. 137–163. New York: Plenum.

Caplan, A. L. 1976. Ethics, evolution, and the milk of human kindness. Hastings Center Report, reprinted in *The sociobiology debate,* ed. A. L. Caplan, pp. 304–314. New York: Harper & Row, 1978.

———, ed. 1978. *The sociobiology debate.* New York: Harper & Row.

Caraco, T., Martindale, S., and Pulliam, H. R. 1980. Avian flocking in the presence of a predator. *Nature* 285:400–401.

Caraco, T., and Wolf, L. L. 1975. Ecological determinants of group sizes of foraging lions. *American Naturalist* 109:343–352.

Carey, M. 1977. Aspects of the population dynamics of indigo buntings *(Passerina cyanea)* in two habitats: A test of the Verner-Willson-Orians model for the evolution of avian polygyny. Ph.D. thesis, Indiana University, Bloomington.

Carey, M., and Nolan, V., Jr. 1975. Polygyny in indigo buntings: A hypothesis tested. *Science* 190:1296–1297.

———. 1979. Population dynamics of indigo buntings and the evolution of avian polygyny. *Evolution* 33:1180–1192.

Carl, E. A. 1971. Population control in Arctic ground squirrels. *Ecology* 52:395–413.

Carpenter, C. C. 1965. The marine iguana of the Galapagos Islands, its behaviour and ecology. *Proceedings of the California Academy of Sciences* 34:329–376.

———. 1977. Variation and evolution of stereotyped behavior in reptiles. In *Biology of the Reptilia.* Vol. 7. *Ecology and behaviour, A,* ed. C. Gans and D.W. Tinkle, pp. 335–403. New York: Academic Press.

Carpenter, C. R. 1940. A field study in Siam of the behavior and social relations of the gibbon *(Hylobates lar). Comparative Psychology Monographs* 16:1–212.

———. 1942. Sexual behavior of free ranging rhesus monkeys *(Macaca mulatta). Journal of Comparative Psychology* 33:113–142.

Carpenter, F. L., and MacMillen, R. E. 1976. Threshold model of feeding territoriality and test with a Hawaiian honeycreeper. *Science* 194:639–642.

Carpenter, G. D. H. 1921. Experiments on the relative edibility of insects, with special reference to their coloration. *Transactions of the Royal Entomological Society of London* 69:1–105.

Carr, A. 1973. *So excellent a fishe.* New York: Doubleday (Anchor Books).

Carr, C. A. H. 1962. Further studies on the influence of the queen in ants of the genus *Myrmica. Insectes Sociaux* 9:127–211.

Carr, W. J., Roth, P., and Amore, M. 1971. Responses of male mice to odors from stressed vs. nonstressed males and females. *Psychonomic Science* 25:275–276.

Carrick, R. 1963. Ecological significance of territory in the Australian magpie, *Gymnorhina tibicen. Proceedings of the XIII International Ornithological Congress,* pp. 740–753.

———. 1972. Population ecology of the Australian black-backed magpie, royal penguin, and silver gull. In *Population ecology of migratory birds: A symposium.* United States Department of the Interior Wildlife Research Report no. 2, pp. 41–99.

Carroll, C. R., and Janzen, D. M. 1973. Ecology of foraging by ants. *Annual Review of Ecology and Systematics* 4:231–257.

Carthy, J. D., and Ebling, F. J. 1964. Prologue and epilogue. In *The natural history of aggression,* ed. J. D. Carthy and F. J. Ebling, pp. 1–5. London: Academic Press.

Caryl, P. G. 1979. Communication by agonistic displays: What can games theory contribute to ethology? *Behaviour* 68:136–169.

Case, N. A., and Hewitt, O. H. 1963. Nesting and productivity of the red-winged blackbird in relation to habitat. *Living Bird* 2:7–20.

Catchpole, C. K. 1978. Interspecific territorialism and competition in *Acrocephalus* warblers as revealed by playback experiments in areas of sympatry and allopatry. *Animal Behaviour* 26:1072–1080.

Cates, R. G., and Orians, G. H. 1975. Successional status and the palatability of plants to generalized herbivores. *Ecology* 56:410–418.

Cavalli-Sforza, L. L., and Bodmer, W. F. 1971. *The genetics of human populations.* San Francisco: Freeman.

Cavalli-Sforza, L., and Feldman, M. W. 1973. Models for cultural inheritance. I. Group mean and within group variation. *Journal of Theoretical Biology* 4:42–55.

Cavill, G. W. K., and Robertson, P. L. 1965. Ant venoms, attractants, and repellents. *Science* 149:1337–1345.

Cazier, M. A., and Linsley, E. G. 1963. Territorial behavior among males of *Protoxaea gloriosa* (Fox). *Canadian Entomologist* 94:547–556.

Chagnon, N. A., and Irons, W., eds. 1979. *Evolutionary biology and human social behavior: An anthropological perspective.* N. Scituate, Mass.: Duxbury Press.

Chai, C. K. 1976. *Genetic evolution.* Chicago: University of Chicago Press.

Chalmers, N. R. 1968. The social behaviour of free living mangabeys in Uganda. *Folia Primatologica* 8:263–281.

Chapin, J. P. 1954. The calendar wideawake fair. *Auk* 71:1–15.

Chapman, D. G. 1961. Population dynamics of the Alaska fur seal herd. *Transactions of the North American Wildlife and Natural Resources Conference* 26:356–369.

Chapman, F. M. 1928. The nesting habits of Wagler's oropendola *Zarhynchus wagleri* on Barro Colorado Island. *Bulletin of the American Museum of Natural History* 58:123–166.

———. 1935. The courtship of Gould's manakin *(Manacus vitellinus vitellinus)* on Barro Colorado Island, Canal Zone. *Bulletin of the American Museum of Natural History* 68:471–525.

Chapman, M., and Hausfater, G. 1979. The reproductive consequences of infanticide in langurs. A mathematical model. *Behavioral Ecology and Sociobiology* 5:227–240.

Chapman, R. F. 1966. *The insects: Structure and function.* London: English Universities Press.

Chapman, V. M., Desjardins, C., and Whitten, W. K. 1970. Pregnancy block in mice: Changes in pituitary LH, LTH, and plasma progesterone levels. *Journal of Reproduction and Fertility* 21:333–337.

Charles-Dominique, P. 1972. Ecologie et vie sociale de *Galago demidovii* (Fischer 1808; Prosimii). *Zeitschrift für Tierpsychologie, Supplement* 9:7–41.

———. 1974. Aggression and territoriality in nocturnal prosimians. In *Primate aggression, territoriality, and xenophobia: A comparative perspective,* ed. R. Holloway, pp. 31–48. New York: Academic Press.

———. 1977. Urine marking and territoriality in *Galago alleni* (Waterhouse, 1837—Lorisoidea, Primates) — A field study by radio-telemetry. *Zeitschrift für Tierpsychologie* 43:113–138.

———. 1978. Solitary and gregarious prosimians: Evolution of social structures in primates. In *Recent advances in primatology.* Vol. 3. *Evolution,* ed. D. J. Chivers and K. A. Joysey, pp. 139–149. New York: Academic Press.
Charlesworth, B. 1976. Recombination modification in a fluctuating environment. *Genetics* 83:181–195.
———. 1978. Some models of the evolution of altruistic behavior between siblings. *Journal of Theoretical Biology* 72:297–319.
———. 1979. A note on the evolution of altruism in structured demes. *American Naturalist* 113:601–605.
Charnov, E. L. 1973. Optimal foraging: Some theoretical expectations. Ph.D. thesis, University of Washington, Seattle.
———. 1975a. Optimal foraging: Attack strategy of a mantid. *American Naturalist* 110:141–151.
———. 1975b. Sex ratio selection in an age-structured population. *Evolution* 29:366–368.
———. 1977. An elementary treatment of the genetical theory of kin-selection. *Journal of Theoretical Biology* 66:541–550.
———. 1978. Sex-ratio selection in eusocial Hymenoptera. *American Naturalist* 112:317–326.
———. 1979. The gentical evolution of patterns of sexuality: Darwinian fitness. *American Naturalist* 113:465–480.
Charnov, E. L., and Krebs, J. R. 1974. On clutch size and fitness. *Ibis* 116:217–219.
———. 1975. The evolution of alarm calls: Altruism or manipulation? *American Naturalist* 109:107–112.
Charnov, E. L., Orians, G. H., and Hyatt, K. 1976. The ecological implications of resource depression. *American Naturalist* 110:247–259.
Charnov, E. L., Pyke, G. H., and Pulliam, H. R. 1977. Optimal foraging: A selective review of theory and tests. *Quarterly Review of Biology* 52:137–154.
Cheng Tso-hsin. 1964. *China's economic fauna: Birds.* Washington, D.C.: United States Department of Commerce Office of Technical Services, Joint Publication Research Service.
Choate, T. S. 1963. Habitat and population dynamics of white-tailed ptarmigan in Montana. *Journal of Wildlife Management* 27:684–699.
Christen, A. 1974. Fortpflanzungsbiologie und Verhalten bei *Cebuella pygmaea* und *Tamarin tamarin. Zeitschrift für Tierpsychologie, Supplement* 14:1–79.
Christian, J. J. 1956. Adrenal and reproductive responses to population size in mice from freely growing populations. *Ecology* 37:258–273.
———. 1961. Phenomena associated with population density. *Proceedings of the National Academy of Sciences* 47:428–449.
———. 1963. Endocrine adaptive mechanisms and the physiologic regulation of population growth. In *Physiological mammalogy.* Vol. 1. *Mammalian populations,* ed. W. V. Mayer and R. G. van Gelder, pp. 189–353. New York: Academic Press.
———. 1968. Endocrine-behavioral negative feed-back responses to increased population density. *Colloques Internationaux du Centre National de la Recherche, Paris,* no. 173:289–322.
Christian, J. J., and LeMunyan, C. D. 1958. Adverse effects of crowding on lactation and reproduction of mice and two generations of their offspring. *Endocrinology* 63:517–529.
Christianson, T. E. 1974. Aggressive behavior of the female northern elephant seal. Ph.D. thesis, University of California, Berkeley.
Clapham, A. R. 1936. Over-dispersion in grassland communities and the use of statistical methods in plant ecology. *Journal of Ecology* 24:232–251.
Clark, E. 1972. The Red Sea's garden of eels. *National Geographic* 142:724–735.
Clark, G., and Birch, H. 1945. Hormonal modifications of social behavior. I. The effect of sex-hormone administration on the social behavior of a male castrate chimpanzee. *Psychosomatic Medicine* 7:321–329.
Clark, T. W. 1977. Ecology and ethology of the white-tailed prairie dog *(Cynomys leucurus). Milwaukee Public Museum, Biology and Geology* 3:1–97.
Clarke, G. L. 1954. *Elements of ecology.* New York: Wiley.
Clarke, T. A. 1970. Territorial behavior and population dynamics of a pomacentrid fish, the garibaldi, *Hypsypops rubicunda. Ecological Monographs* 40:189–212.
Clemente, C. D., and Lindsley, D. B., eds. 1967. *Aggression and defense: Neural mechanisms and social patterns.* Vol. 5. *Brain function.* Los Angeles: University of California Press.
Cleveland, L. R., Hall, S. R., Sanders, E. P., and Collier, J. 1934. The wood-feeding roach *Cryptocercus,* its Protozoa, and the symbiosis between Protozoa and roach. *Memoirs of the American Academy of Arts and Sciences* 17:185–342.
Cloak, F. T., Jr. 1975. Is a cultural ethology possible? *Human Ecology* 3:161–182.
———. 1977. Comment on "The adaptive significance of cultural behavior." *Human Ecology* 5:49–52.

Cloudsley-Thompson, J. L. 1965. *Animal conflict and adaptation.* Chester-Springs, Pa.: Dufour.
Clutton-Brock, T. H., Albon, S. D., Gibson, R. M., and Guinness, F. E. 1979. The logical stag:
———. 1975. Feeding behaviour of red colobus and black-and-white colobus in East Africa. *Folia Primatologica* 23:165–207.
Clutton-Brock, T. H., Albon, S. D., Gibson, R. M., and Guinness, F. E. 1979. The logical stag: Adaptive aspects of fighting in red deer (*Cervus elaphus* L.). *Animal Behaviour* 27:211–225.
Clutton-Brock, T. H., and Harvey, P. H. 1976. Evolutionary rules and primate societies. In *Growing points in ethology,* ed. P. P. G. Bateson and R. A. Hinde, pp. 195–237. Cambridge, England: Cambridge University Press.
———. 1977a. Species differences in feeding and ranging behaviour of primates. In *Primate ecology: Studies of feeding and ranging behaviour in lemurs, monkeys and apes,* ed. T. H. Clutton-Brock, pp. 557–584. New York: Academic Press.
———. 1977b. Primate ecology and social organization. *Journal of Zoology, London* 183:1–39.
Clyne, D. 1967. Prothalamion in a garden pool. *Victorian Naturalist* 84:232–235.
———. 1969a. A spider that mimics the green tree ant. *Northern Queensland Naturalist* 36:3–4.
———. 1969b. *Australian frogs.* Melbourne: Lansdowne Press.
Coblentz, B. E. 1980. On the improbability of pursuit invitation signals in mammals. *American Naturalist* 115:438–442.
Cody, M. L. 1966. A general theory of clutch size. *Evolution* 20:174–184.
———. 1969. Convergent characteristics in sympatric species: A possible relation to interspecific competition and aggression. *Condor* 71:223–239.
———. 1971a. Finch flocks in the Mohave Desert. *Theoretical Population Biology* 2:142–158.
———. 1971b. Ecological aspects of reproduction. In *Avian biology,* Vol. I, ed. D. S. Farner and J. R. King, pp. 461–512. New York: Academic Press.
———. 1973a. Character convergence. *Annual Review of Ecology and Systematics* 4·189–211.
———. 1973b. Coexistence, coevolution and convergent evolution in seabird communities. *Ecology* 54:31–44.
———. 1974. Optimization in ecology. *Science* 183:1156–1164.
———. 1978. Habitat selection and interspecific territoriality among the sylviid warblers of England and Sweden. *Ecological Monographs* 48:351–396.
Cody, M. L., and Cody, C. B. J. 1972. Areal versus lineal territories in the wren, *Troglodytes troglodytes. Condor* 74:477–478.
Coe, M. 1974. Observations on the ecology and breeding biology of the genus *Chiromantis* (Amphibia: Rhacophoridae). *Journal of Zoology, London* 172:13–34.
Cofer, C. N., and Appley, M. H. 1964. *Motivation: Theory and research.* New York: Wiley.
Cohen, D., and Eshel, I. 1976. On the founder effect and the evolution of altruistic traits. *Theoretical Population Biology* 10:276–302.
Cohen, J. E. 1969. Natural primate troops and a stochastic population model. *American Naturalist* 103:455–477.
Cohen, Y. A. 1968. Culture as adaptation. In *Man in adaptation. The cultural present,* ed. Y. A. Cohen, pp. 40–60. Chicago: Aldine.
Coker, R. E. 1919. Habits and economic relations of the guano birds of Peru. *Proceedings of the United States National Museum* 56:449–511.
Cole, A. C., Jr. 1968. Pogonomyrmex *harvester ants. A study of the genus in North America.* Knoxville: University of Tennessee Press.
Cole, L. C. 1946. A study of the Cryptozoa of an Illinois woodland. *Ecological Monographs* 16:49–86.
Colin, P. L. 1973. Daily activity patterns and effects of environmental conditions on the behavior of the yellow-head jawfish, *Opisthognathus aurifrons,* with notes on its ecology. *Zoologica* 57:137–169.
———. 1975. *The neon gobies: The comparative biology of the gobies of the genus* Gobiosoma, *subgenus* elacantinus (Pisces: Gobiidae) *in the tropical western North Atlantic Ocean,* Neptune City, N. J.: T. F. H. Pub.
Collias, N. E., and Collias, E. C. 1964. The evolution of nest-building in weaverbirds (Ploceidae). *University of California Publications in Zoology* 73:1–162.
Colombel, P. 1978. Biologie d'*Odontomachus haematodes* L. (Hym. Form.). Déterminisme de la caste femelle. *Insectes Sociaux* 25:141–151.
Colwell, R. K., and Futuyama, D. J. 1971. On the measurement of niche breadth and overlap. *Ecology* 52:567–576.
Colwell, R. K., and Wilson, D. S. 1981. Female-biased sex ratios and local mate competition: Group selection is required for n > 1. *Nature* (in press).
Connell, J. H. 1955. Spatial distribution of two species of clams, *Mya arenaria* L. and *Petricola*

pholadiformes Lamarck, in an intertidal area. *Report on Fish and Invertebrates of Massachusetts* 8:15–25.

———. 1963. Territorial behavior and dispersion in some marine invertebrates. *Research in Population Ecology* 5:87–101.

Conner, R. L., and Levine, S. 1969. Hormonal influences on aggressive behaviour. In *Aggressive behaviour*, ed. S. Garattini and E. B. Sigg, pp. 150–163. Amsterdam: Excerpta Medica.

Cook, S. F., and Scott, K. G. 1933. The nutritional requirements of *Zootermopsis (Termopsis) angusticollis*. *Journal of Cellular and Comparative Physiology* 4:95–110.

Coppinger, R. P. 1969. The effect of experience and novelty on avian feeding behaviour with reference to the evolution of warning coloration in butterflies. I. Reactions of wild caught adult blue jays to novel insects. *Behaviour* 35:45–60.

———. 1970. The effect of experience and novelty on avian feeding behaviour with reference to the evolution of warning coloration in butterflies. II. Reactions of naíve birds to novel insects. *American Naturalist* 104:323–335.

Corbet, P. S. 1962. *A biology of dragonflies*. Chicago: Quadrangle.

Cott, H. B. 1940. *Adaptive coloration in animals*. London: Methuen.

———. 1947. The edibility of birds: Illustrated by five years experiments and observations (1941–1946) on the food preferences of the hornet, cat and man; and considered with special reference to the theories of adaptive coloration. *Proceedings of the Zoological Society of London* 116:371–524.

Cottam, C., Williams, C. S., and Sooter, C. A. 1942. Cooperative feeding of white pelicans. *Auk* 59:444–445.

Cottrille, W. P., and Cottrille, B. D. 1958. Great blue heron: Behavior at the nest. *Miscellaneous Publications of the Museum of Zoology*, University of Michigan 102:1–15.

Coulson, J. C. 1966. The influence of the pair-bond and age on the breeding biology of the kittiwake gull *Rissa tridactyla*. *Journal of Animal Ecology* 35:269–279.

———. 1968. Differences in the quality of birds nesting in the centre and on the edges of a colony. *Nature* 217:478–479.

Coulson, J. C., and Hickling, G. 1964. The breeding biology of the grey seal, *Halichoerus grypus* (Fab.), on the Farne Islands, Northumberland. *Journal of Animal Ecology* 33:485–512.

Coulson, J. C., and White, E. 1956. A study of colonies of the kittiwake *Rissa tridactyla* (L.). *Ibis* 98:63–79.

———. 1958. The effect of age on the breeding biology of the kittiwake *Rissa tridactyla*. *Ibis* 100:40–51.

———. 1960. The effect of age and density of breeding birds on the time of breeding of the kittiwake *Rissa tridactyla*. *Ibis* 102:71–86.

———. 1961. An analysis of the factors influencing the clutch size of the kittiwake. *Proceedings of the Zoological Society of London* 136:207–217.

Court-Brown, W. M. 1967. *Human population cytogenetics*. New York: Wiley.

Covich, A. P. 1976. Analyzing shapes of foraging areas: Some ecological and economic theories. *Annual Review of Ecology and Systematics* 7:235–257.

Cowan, D. P. 1979. Sibling matings in a hunting wasp: Adaptive inbreeding? *Science* 205:1403–1405.

Cox, C. R., and Le Boeuf, B. J. 1977. Female incitation of male competition: A mechanism in sexual selection. *American Naturalist* 111:317–335.

Cox, E. C., and Gibson, T. C. 1974. Selection for high mutation rates in chemostats. *Genetics* 77:169–184.

Craig, J. L. 1979. Habitat variation in the social organization of a communal gallinule, the pukeko, *Porphyrio porphyrio melanotis*. *Behavioral Ecology and Sociobiology* 5:331–358.

Craig, R. 1979. Parental manipulation, kin selection, and the evolution of altruism. *Evolution* 33:319–334.

———. 1980. Sex investment ratios in social Hymenoptera. *American Naturalist* 116:311–323.

Craighead, J. J., and Craighead, F. C. 1956. *Hawks, owls, and wildlife*. Harrisburg, Pa.: Stackpole.

Crane, J. 1967. Combat and its ritualization in fiddler crabs (Ocypodidae) with special reference to *Uca rapax* (Smith). *Zoologica* 52:49–76.

———. 1975. *Fiddler crabs of the world. Ocypodidae: Genus* Uca. Princeton: Princeton University Press.

Creighton, W. S., and Creighton, M. P. 1959. The habits of *Pheidole militicida* Wheeler (Hymenoptera: Formicidae). *Psyche, Cambridge* 66:1–12.

Creighton, W. S., and Gregg, R. E. 1954. Studies on the habits and distribution of *Cryptocercus texanus* Santschi (Hymenoptera: Formicidae). *Psyche, Cambridge* 61:41–57.

Crick, F. H. C. 1968. The origin of the genetic code. *Journal of Molecular Biology* 38:367–379.

———. 1971. General model for the chromosomes of higher organisms. *Nature* 234:25–27.

Cronin, E. W., Jr., and Sherman, P. W. 1976. A resource-based mating system: The orange-rumped honeyguide. *Living Bird* 15:5–32.

Crook, J. H. 1958. Études sur le comportement social de *Bubalornis a. albirostris* (Vieillot). *Alauda* 26:161–195.

———. 1960. Studies on the reproductive behavior of the baya weaver *(Ploceus phillipinus* (L.). *Journal of the Bombay Natural History Society* 57:1–44.

———. 1964. The evolution of social organization and visual communication in the weaver birds (Ploceinae). *Behaviour Supplement* 10:1–178.

———. 1965. The adaptive significance of avian social organizations. *Symposia of the Zoological Society of London* 14:181–218.

———. 1966. Gelada baboon herd structure and movement: A comparative report. *Symposia of the Zoological Society of London* 18:237–258.

———. 1970. The socio-ecology of primates. In *Social behaviour in birds and mammals: Essays on the social ethology of animals and man,* ed. J. H. Crook, pp. 103–166. New York: Academic Press.

———. 1972. Sexual selection, dimorphism, and social organization in the primates. In *Sexual selection and the descent of man, 1871–1971,* ed. B. G. Campbell, pp. 231–281. Chicago: Aldine.

Crook, J. H., and Aldrich-Blake, P. 1968. Ecological and behavioural contrasts between sympatric ground dwelling primates in Ethiopia. *Folia Primatologica* 8:192–227.

Crook, J. H., and Butterfield, P. A. 1970. Gender role in the social system of quelea. In *Social behaviour in birds and mammals: Essays on the social ethology of animals and man,* ed. J. H. Crook, pp. 211–248. New York: Academic Press.

Crook, J. H., and Gartlan, J. S. 1966. Evolution of primate societies. *Nature* 210:1200–1203.

Crow, J. F., and Kimura, M. 1965. Evolution in sexual and asexual populations. *American Naturalist* 99:439–450.

———. 1970. *An introduction to population genetics theory.* New York: Harper & Row.

Croze, H. 1970. Searching image in carrion crows. *Zeitschrift für Tierpsychologie, Supplement* 5:1–86.

Cruden, R. W. 1966. Observations on the behaviour of *Xylocopa c. californica* and *X. tabaniformis orpifex* (Hymenoptera: Apoidea). *Pan-Pacific Entomology* 42:111–119.

Cullen, A. 1969. *Window onto wilderness.* Nairobi: East African Pub.

Cullen, J. M., and Ashmole, N. P. 1963. The black noddy *Anous tenuirostris* on Ascension Island. Part 2. Behaviour. *Ibis* 103b:423–446.

Curio, E. 1959. Beiträge zur Populationsökologie des Trauerschnäppers *(Ficedula h. hypoleuca* Pallas). *Zoologische Jahrbücher Abteilung für Systematik, Ökologie und Geographie der Tierre* 87:185–230.

———. 1975. The functional organization of anti-predator behaviour in the pied flycatcher: A study of avian visual perception. *Animal Behaviour* 23:1–115.

———. 1976. The ethology of predation. *Zoophysiology and Ecology,* vol. 7. Berlin: Springer-Verlag.

Curio, E., Ernst, U., and Vieth, W. 1978. Cultural transmission of enemy recognition: One function of mobbing. *Science* 202:899–900.

Cushing, D. H., and Harden-Jones, F. R. 1968. Why do fish school? *Nature* 218:918–920.

Custer, T. W., and Osborn, R. G. 1978. Feeding habitat use by colonially-breeding herons, egrets, and ibises in North Carolina. *Auk* 95:733–743.

Daan, S., and Slopsema, S. 1978. Short-term rhythms in foraging behaviour of the common vole, *Microtus arvalis. Journal of Comparative Physiology* 127:215–227.

Daan, S., and Wickers, H. J. 1968. Habitat selection of bats hibernating in a limestone cave. *Zeitschrift für Säugetierkunde* 33:262–287.

Daanje, A. 1950. On locomotory movements in birds and the intention movements derived from them. *Behaviour* 3:48–98.

Dalquest, W. W., and Walton, D. W. 1970. Diurnal retreats of bats. In *About bats. A chiropteran biology symposium,* ed. B. H. Slaughter and D. W. Walton, pp. 162–187. Dallas: Southern Methodist University Press.

Dalton, K. 1964. *The premenstrual syndrome.* Springfield, Ill.: Thomas.

Daly, M. 1978. The cost of mating. *American Naturalist.* 112:771–774.

Daly, R. F. 1969. Mental illness and patterns of behaviour in 10 XYY males. *Journal of Nervous and Mental Disorders* 149:318–327.

Darchen, R., and Delage, B. 1970. Facteur déterminant les castes chez les Trigones. *Compte Rendu de l'Académie des Sciences, Paris, Séries D* 270:1372–1373.

Darchen, R., and Delage-Darchen, B. 1971. Le determinisme des castes chez les Triones *(Hyménoptères Apidés). Insectes Sociaux* 18:121–134.

———. 1974a. Nouvelles expériences concernant le déterminisme des castes chez les Mélipones (Hyménoptères apidés). *Compte Rendu de l'Académie des Sciences, Paris, Series D* 278:907–910.

———. 1974b. Les stades larvaires de *Melipona beecheii* (Hyménoptère, Apidé). *Compte Rendu de l'Académie des Sciences, Paris, Series D* 278:3115–3118.

———. 1975. Contribution a l'étude d'une abeille du Mexique *Melipona beecheii* B. (Hyménoptère: Apidé). *Apidologie* 6:295–339.

Darley, J. A. 1971. Sex ratio and mortality in the brown-headed cowbird. *Auk* 88:560–566.

Darling, F. F. 1937. *A herd of red deer.* London: Oxford University Press.

———. 1938. *Bird flocks and the breeding cycle. A contribution to the study of avian sociality.* Cambridge, England: Cambridge University Press.

Darwin, C. 1859. *On the origin of species by means of natural selection, or, the preservation of favoured races in the struggle for life.* London: John Murray.

———. 1871. *The descent of man and selection in relation to sex.* Vols. I, II. New York: Appleton.

Davenport, D. 1962. Physiological notes on actinians and their associated commensals. *Bulletin of the Institute of Oceanography, Monaco,* no. 1237:1–15.

———. 1966. Cnidarian symbiosis and the experimental analysis of behaviour. *Symposia of the Zoological Society of London* 16:361–372.

Davidson, E. H., and Britten, R. J. 1973. Organization, transcription, and regulation in the animal genome. *Quarterly Review of Biology* 48:565–613.

Davies, N. B. 1976a. Food, flocking and territorial behaviour in the pied wagtail *(Motacilla alba yarrelli* Gould) in winter. *Journal of Animal Ecology* 45:235–253.

———. 1976b. Parental care and the transition to independent feeding in the young spotted flycatcher *(Muscicapa striata). Behaviour* 59:280–295.

———. 1978a. Territorial defence in the speckled wood butterfly *(Pararge aegeria),* the resident always wins. *Animal Behaviour* 26:138–147.

———. 1978b. Ecological questions about territorial behaviour. In *Behavioural ecology: An evolutionary approach,* ed. J. R. Krebs and N. B. Davies, pp. 317–350. Sunderland, Mass.: Sinauer Associates.

Davies, N. B., and Halliday, T. R. 1979. Competitive mate searching in male common toads, *Bufo bufo. Animal Behaviour* 27:1235–1267.

Davis, D. E. 1951. The relation between level of population and pregnancy of Norway rats. *Ecology* 32:459–461.

———. 1963. The hormonal control of aggressive behavior. *Proceedings of the XIII International Ornithological Congress,* pp. 994–1003.

———. 1964. The physiological analysis of aggressive behavior. In *Social behavior and organization among vertebrates,* ed. W. Etkin, pp. 53–74. Chicago: University of Chicago Press.

Davis, J. W. F., and O'Donald, P. 1976. Sexual selection for a handicap: A critical analysis of Zahavi's model. *Journal of Theoretical Biology* 57:345–354.

Davis, R. B., Herreid, C. F., III, and Short, H. L. 1962. Mexican free-tailed bats in Texas. *Ecological Monographs* 32:311–346.

Dawkins, R. 1976. *The selfish gene.* London: Oxford University Press.

———. 1978. Replicator selection and the extended phenotype. *Zeitschrift für Tierpsychologie* 47:61–76.

———. 1979. Twelve misunderstandings of kin selection. *Zeitschrift für Tierpsychologie* 51:184–200.

Dawkins, R., and Carlisle, T. R. 1976. Parental investment, male desertion and a fallacy. *Nature* 262:131–133.

Dawkins, R., and Krebs, J. R. 1978. Animal signals: Information or manipulation? In *Behavioural ecology: An evolutionary approach,* ed. J. R. Krebs and N. B. Davies, pp. 282–309. Sunderland, Mass.: Sinauer Associates.

Dawson, G. A. 1976. Behavioral ecology of the Panamanian tamarin, *Saguinus oedipus* (Callitrichidae, Primates). Ph.D. thesis, Michigan State University, East Lansing.

———. 1977. Composition and stability of social groups of the tamarin, *Saguinus oedipus geoffroyi,* in Panama. In *The biology and conservation of the Callitrichidae,* ed. D. G. Kleiman, pp. 23–37. Washington, D.C.: Smithsonian Institution Press.

———. 1979. The use of time and space by the Panamanian tamarin. *Folia Primatologica* 31:253–284.

Deag, J. M. 1977. Aggression and submission in monkey societies. *Animal Behaviour* 25:465–474.

De Boer, J. N., and Heuts, B. A. 1973. Prior exposure to visual cues affecting dominance in the jewel fish, *Hemichromis bimaculatus* Gill 1862 (Pisces, Cichlidae). *Behaviour* 44:299–321.

Deevey, E. S., Jr. 1947. Life tables for natural populations of animals. *Quarterly Review of Biology* 22:283–314.

De Fries, J. C., and McClearn, G. E. 1970. Social dominance and Darwinian fitness in the laboratory mouse. *American Naturalist* 104:408–411.

Dejean, A., and Passera, L. 1974. Ponte des ouvrières et inhibition royale chez la fourmi *Temnothorax recedens* (Nyl.) (Formicidae, Myrmicinae). *Insectes Sociaux* 21:343–356.

Delacour, J. 1977. *The pheasants of the world.* 2nd ed. Surrey: Saiga.

Delage-Darchen, B. 1974. Écologie et biologie de *Crematogaster impressa* Émery, fourmi savanicole d'Afrique. *Insectes Sociaux* 21:13–34.

Delgado, J. M. R. 1960. Emotional behavior in animals and humans. *Psychiatric Research Reports* 12:259–271.

———. 1966. Aggressive behavior evoked by radio stimulation in monkey colonies. *American Zoologist* 6:669–681.

Deligne, J. 1965. Morphologie et fonctionnement des mandibules chez les soldats des Termites. *Biologia Gabonica* 1:179–186.

Demarest, W. J. 1977. Incest avoidance among human and non-human primates. In *Primate bio-social development,* ed. S. Chevalier-Skolnikoff and F. E. Poirier, pp. 323–342. New York: Garland.

De Martini, E. E. 1976. The adaptive significance of territoriality and egg cannibalism in the painted greenling, *Oxylebius pictus* Gill, a northeastern Pacific marine fish. Ph.D. thesis, University of Washington, Seattle.

Dement'ev, G. P., and Gladkov, N. A., eds. 1967. *Birds of the Soviet Union.* Vol. IV. Jerusalem: Israel Program for Scientific Translations.

———. 1968. *Birds of the Soviet Union.* Vol. II. Jerusalem: Israel Program for Scientific Translations.

Denham, W. W. 1971. Energy relations and some basic properties of primate social organization. *American Anthropologist* 73:77–95.

Devore, I. 1965. Male dominance and mating behavior in baboons. In *Sex and behavior,* ed. F. A. Beach, pp. 266–289. New York: Wiley.

———. 1971. The evolution of human society. In *Man and beast: Comparative social behavior,* ed. J. F. Eisenberg and W. S. Dillon, pp. 297–311. Washington D.C.: Smithsonian Institution Press.

———. 1972. Quest for the roots of society. In *The marvels of animal behavior,* ed. P. R. Marler, pp. 393–408. Washington D.C.: National Geographic Society.

———, ed. 1980. *Sociobiology and the social sciences.* Chicago: Aldine (in press).

Devore, I., and Hall, K. R. L. 1965. Baboon ecology. In *Primate behavior. Field studies of monkeys and apes,* ed. I. Devore, pp. 20–52. New York: Holt, Rinehart and Winston.

Devore, I., and Washburn, S. L. 1963. Baboon ecology and human evolution. In *African ecology and human evolution,* ed. F. C. Howell and F. Bourlière, pp. 335–367. New York: Wenner-Gren Foundation.

Dice, L. R. 1952. *Natural communities.* Ann Arbor: University of Michigan Press.

Dickemann, M. 1979. Female infanticide, reproductive strategies, and social stratification: A preliminary model. In *Evolutionary biology and human social behavior: An anthropological perspective,* ed. N. A. Chagnon and W. Irons, pp. 321–367. N. Scituate, Mass.: Duxbury Press.

Dingle, H., and Caldwell, R. L. 1969. The aggressive and territorial behaviour of the mantis shrimp *Gonodactylus bredini* Manning (Crustacea: Stomatopoda). *Behaviour* 33:115–136.

———. 1975. Distribution, abundance, and interspecific agonistic behavior of two mudflat stomatopods. *Oecologia* 20:167–178.

Dingle, H., Highsmith, R., Evans, K. E., and Caldwell, R. L. 1973. Interspecific aggressive behavior in tropical reef stomatopods and its possible ecological significance. *Oecologia* 13:55–64.

Divale, W. T., and Harris, M. 1976. Population, warfare, and the male supremacist complex. *American Anthropologist* 78:521–538.

Dixon, A. F. G., and Wratten, S. D. 1971. Laboratory studies on aggregation size and fecundity in the black bean aphid, *Aphis fabae,* Scop. *Bulletin of Entomological Research* 61:97–111.

Dixson, A. F., Everitt, G. F., Herbert, J., Rugman, S. M., and Scruton, D. M. 1973. Hormonal and other determinants of sexual attractiveness and receptivity in rhesus and talapoin monkeys. In *Primate reproductive behavior. Symposia of the 4th International Congress of Primatology,* vol. 2, ed. C. H. Phoenix, pp. 36–63. Basel: Karger.

Dmi'el, R. 1970. Growth and metabolism in snake embryos. *Journal of Embryology and Experimental Morphology* 23:761–772.

Dmi'el, R., Prevulotsky, A., and Shkolnik, A. 1980. Is a black coat in the desert a means of saving metabolic energy? *Nature* 283:761–762.

Dobrzańska, J. 1958. Partition of foraging grounds and modes of conveying information among ants. *Acta Biologica Experimentalis, Warsaw* 18:55–67.

Dobzhansky, T. H. 1950. Evolution in the tropics. *American Scientist* 38:209–221.

———. 1970. *Genetics of the evolutionary process.* New York: Columbia University Press.

Dobzhansky, T., Ayala, F. J., Stebbins, G. L., and Valentine, J. W. 1977. *Evolution.* San Francisco: Freeman.

Dolbeer, R. A. 1976. Reproductive rate and temporal spacing of nesting of red-winged blackbirds in upland habitat. *Auk* 93:343–355.

Dole, J. W., and Durant, P. 1974. Movements and seasonal activity of *Atelopus oxyrhynchus* (Anura: Atelopodidae) in a Venezuelan cloud forest. *Copeia* 1974:230–235.

Dolhinow, P. C. 1977. Letter to the editor. *American Scientist* 65:266.

Dolhinow, P. C., and Bishop, N. 1970. The development of motor skills and social relationships among primates through play. *Minnesota Symposium of Child Psychology* 4:141–198.

Dollard, J., Doob, L., Miller, N., Mowrer, O. H., and Sears, R. R. 1939. *Frustration and aggression.* New Haven: Yale University Press.

Done, B. S., and Heatwole, H. 1977. Effects of hormones on the aggressive behaviour and social organization of the scincid lizard, *Sphenomorphus kosciuskoi. Zeitschrift für Tierpsychologie* 44:1–12.

Dorst, J., and Dandelot, P. 1970. *A field guide to the larger mammals of Africa.* Boston: Houghton Mifflin.

Dorward, D. F. 1962. Comparative biology of the white booby and brown booby *Sula* spp. at Ascension. *Ibis* 103b:174–220.

Douglas-Hamilton, I. 1972. On the ecology and behaviour of the African elephant: The elephants of Lake Manyara. Ph.D. thesis, Oxford University, Oxford.

Douglas-Hamilton, I. and Douglas-Hamilton, O. 1975. *Among the elephants.* New York: Viking Press.

Downhower, J. F., and Armitage, K. B. 1971. The yellow-bellied marmot and the evolution of polygamy. *American Naturalist* 105:355–370.

Downhower, J. F., and Brown, L. 1979. Seasonal changes in the social structure of a mottled sculpin *(Cottus bairdi)* population. *Animal Behaviour* 27:451–458.

———. 1980. Mate preferences of female mottled sculpins, *Cottus bairdi. Animal Behaviour* 28:728–734.

Drent, R. 1975. Incubation. In *Avian biology,* vol. V, ed. D. S. Farner and J. R. King, pp. 333–420. New York: Academic Press.

Drickamer, L. C. 1974a. A ten-year summary of reproductive data for free-ranging *Macaca mulatta. Folia Primatologica* 21:61–80.

———. 1974b. Social rank, observability, and sexual behavior of rhesus monkeys *(Macaca mulatta). Journal of Reproduction and Fertility* 37:117–120.

Drickamer, L. C., and Vessey, S. H. 1973. Group changing in free-ranging male rhesus monkeys. *Primates* 14:359–368.

Driver, P. H., and Humphries, D. A. 1970. Protean displays as inducers of conflict. *Nature* 226:968–969.

Drury, W. H., Jr. 1962. Breeding activities, especially nest building, of the yellowtail *(Ostinops decumanus)* in Trinidad, West Indies. *Zoologica* 47:39–58.

Duellman, W. E. 1970. *Hylid frogs of Middle America.* Vols. 1 and 2. Lawrence: University of Kansas Press.

Du Mond, F. V., and Hutchinson, T. C. 1967. Squirrel monkey reproduction. The "fatted" male phenomenon and seasonal spermatogenesis. *Science* 158:1067–1070.

Dunbar, M. J. 1960. The evolution of stability in marine environments: Natural selection at the level of the ecosystem. *American Naturalist* 94:129–136.

———. 1972. The ecosystem as a unit of natural selection. *Transactions of the Connecticut Academy of Sciences* 44:114–130.

Dunbar, R. I. M. 1979a. Structure of gelada baboon reproductive units. I. Stability of social relationships. *Behaviour* 69:72–87.

———. 1979b. Energetics, thermoregulation and the behavioural ecology of klipspringer. *African Journal of Ecology* 17:217–230.

Dunbar, R. I. M., and Dunbar, E. P. 1974. Social organization and ecology of the klipspringer *(Oreotragus oreotragus)* in Ethiopia. *Zeitschrift für Tierpsychologie* 35:481–493.

———. 1975. Social dynamics of gelada baboons. *Contributions to Primatology* 6:1–157.

———. 1976. Contrasts in social structure among black-and-white colobus monkey groups. *Animal Behaviour* 24:84–92.

———. 1979. Observations on the social organization of common duiker in Ethiopia. *African Journal of Ecology* 17:249–252.
———. 1980. The pairbond in klipspringer. *Animal Behaviour* 28:219–229.
Dunford, C. 1977a. Kin selection for ground squirrel alarm calls. *American Naturalist* 111:782–785.
———. 1977b. Social system of round-tailed ground squirrels. *Animal Behaviour* 25:885–906.
Duplaix-Hall, N., 1975. River otters in captivity: A review. In *Breeding endangered species in captivity,* ed. R. D. Martin, pp. 315–327. New York: Academic Press.
Durant, P., and Dole J. W. 1974. Food of *Atelopus oxyrhynchus* (Anura: Atelopodidae) in a Venezuelan cloud forest. *Herpetologica* 30:183–187.
Durham, W. H., 1976. The adaptive significance of cultural behavior. *Human Ecology* 4:89–121.
———. 1979. Toward a coevolutionary theory of human biology. In *Evolutionary biology and human social behavior: An anthropological perspective,* ed. N. A. Chagnon and W. Irons, pp. 39–59. Scituate, Mass.: Duxbury Press.
Duvall, S. W., Bernstein, I. S., and Gordon, T. P. 1976. Paternity and status in a rhesus monkey group. *Journal of Reproduction and Fertility* 47:25–31.
Dyrcz, A. 1977. Polygamy and breeding success among great reed warblers *Acrocephalus arundinaceus* at Milicz, Poland. *Ibis* 119:73–77.
Dzubin, A. 1969. Assessing breeding populations of ducks by census counts. *Canadian Wildlife Service Report,* Series no. 6, pp. 178–230.
Eastwood, E. 1967. *Radar ornithology.* London: Methuen.
Eastwood, E., Isted, G. A., and Rider, G. C. 1962. Radar ring angels and the roosting behaviour of starlings. *Proceedings of the Royal Society of London, Series B* 156:242–267.
Eaton, G. G. 1973. Social and endocrine determinants of sexual behavior in simian and prosimian females. In *Primate reproductive behavior. Symposia of the 4th International Primatological Congress,* vol. 2, ed. C. H. Phoenix, pp. 20–35. Basel: Karger.
Eaton, G. G., and Resko, J. A. 1974. Plasma testosterone and male dominance in a Japanese macaque *(Macaca fuscata)* troop compared with repeated measures of testosterone in laboratory males. *Hormones and Behavior* 5:251–259.
Eaton, R. L. 1974. *The cheetah.* New York: Van Nostrand Reinhold.
———. 1976. The evolution of sociality in the Felidae. In *The world's cats,* vol. 3, ed. R. L. Eaton, pp. 95–142. Seattle: Carnivore Research Institute, Burke Museum, University of Washington.
Eaton, R. L., and Craig, S. J. 1973. Captive management and mating behaviour of the cheetah. In *The world's cats,* vol. 1, ed. R. L. Eaton, pp. 217–254. Winston, Oreg.: World Wildlife Safari.
Eberhard, W. G. 1979. The function of horns in *Podischnus agenor* (Dynastinae) and other beetles. In *Sexual selection and reproductive competition in insects,* ed. M. S. Blum and N. A. Blum, pp. 231–258. New York: Academic Press.
Ebersole, J. P. 1977. The adaptive significance of interspecific territoriality in the reef fish *Eupomacentrus leucostictus. Ecology* 58:914–920.
Edmunds, M. 1968. Acid secretion in some species of Doridacea (Mollusca, Nudibranchia). *Proceedings of the Malacological Society of London* 38:121–133.
———. 1974. *Defence in animals. A survey of anti-predator defences.* New York: Longman Group.
Edwards, A. W. F. 1970. The search for genetic variability of the sex ratio. In *The biosocial aspects of sex,* ed. G. A. Harrison and J. Peel, pp. 55–60. Oxford: Blackwell Scientific.
Edwards, D. A. 1968. Mice: Fighting by neonatally androgenized females. *Science* 161:1027–1028.
———. 1969. Early androgen stimulation and aggressive behavior in male and female mice. *Physiology and Behavior* 4:333–338.
Edwards, D. A., and Burge, K. G. 1971. Estrogenic arousal of aggressive behavior and masculine sexual behavior in male and female mice. *Hormones and Behavior* 2:239–245.
Edwards, G., Hosking, E., and Smith, S. 1949–1950. Reactions of some passerine birds to a stuffed cuckoo. *British Birds* 42:13–19; 43:144–150.
Edwards, J. S. 1960. Spitting as a defensive mechanism in a predatory reduviid. *11th International Kongress der Entomologie, Verhandlungen* 3:259–263.
———. 1961. The action and composition of the saliva of an assassin bug *Platymeris rhadamanthus* Gaerst. (Hemiptera, Reduviidae). *Journal of Experimental Biology* 38:61–77.
Ehrenfeld, D. W. 1979. Behavior associated with nesting. In *Turtles. Perspectives and research,* ed. M. Harless and H. Morlock, pp. 417–434. New York: Wiley.
Ehrhardt, S. 1931. Uber Arbeitsteilung bei *Myrmica-* und *Messor*-Arten. *Zeitschrift für Morphologie und Ökologie der Tiere* 20:755–812.
Ehrlich, P. R. 1975. The population biology of coral reef fishes. *Annual Review of Ecology and Systematics* 6:211–248.

Ehrlich, P. R., Talbot, F. H., Russell, B. C., and Anderson, G. R. V. 1977. The behaviour of chaetodontid fishes with special reference to Lorenz's "poster colouration" hypothesis. *Journal of Zoology, London* 183:213–228.

Ehrman, L., and Parsons, P. A. 1976. *The genetics of behavior.* Sunderland, Mass.: Sinauer Associates.

Eibl-Eibesfeldt, I. 1950. Ein Beitrag zur Paarungsbiologie der Erdkrote *(Bufo bufo* L.). *Behaviour* 2:217–236.

———.1959. Der Fisch *Aspidontus taeniatus* als Nachahmer des Putzers *Labroides dimidiatus. Zeitschrift für Tierpsychologie* 16:19–25.

———. 1961. The fighting behavior of animals. *Scientific American* 205(6):112–122.

———. 1970. *Ethology, the biology of behavior.* 2nd ed. New York: Holt, Rinehart and Winston.

———. 1974. Phylogenetic adaptation as determinants of aggressive behavior in man. In *Determinants and origins of aggressive behavior,* ed. J. de Wit and W. W. Hartup, pp. 29–57. The Hague: Mouton.

Eibl-Eibesfeldt, I., and Eibl-Eibesfeldt, E. 1967. Das Parasitenabwehren der Minima-Arbeiterinnnen der Blattschneider-Ameise *(Atta cephalotes). Zeitschrift für Tierpsychologie* 24:278–281.

Eichelman, B. S. 1971. Effect of subcortical lesions on shock-induced aggression in the rat. *Journal of Comparative and Physiological Psychology* 74:331–339.

Eickwort, G. C. 1977. Male territorial behavior in the mason bee *Hoplitis anthocopoides* (Hymenoptera: Megacilidae). *Animal Behaviour* 25:542–554.

Eisenberg, J. F. 1966. The social organization of mammals. *Handbuch der Zoologie* 10:1–92.

Eisenberg, J. F., and Lockhart, M. 1972. An ecological reconnaissance of Wilpattu National Park, Ceylon. *Smithsonian Contributions to Zoology* 101:1–118.

Eisenberg, J. F., and McKay, G. M. 1974. Comparison of ungulate adaptations in the New World and Old World tropical forests with special reference to Ceylon and the rainforests of Central America. In *The behaviour of ungulates and its relation to management,* ed. V. Geist and F. Walther, pp. 585–602. Moreges: International Union for Conservation of Nature and Natural Resources.

Eisenberg, J. F., Muckenhirn, N. A., and Rudran, R. 1972. The relation between ecology and social structure in primates. *Science* 176:863–874.

Eisner, T. 1968. Mongoose and millipedes. *Science* 160:1367.

———. 1970. Chemical defense against predation in arthropods. In *Chemical ecology,* ed. E. Sondheimer and J. B. Simeone, pp. 157–217. New York: Academic Press.

Eisner, T., and Davis, J. A. 1967. Mongoose throwing and smashing millipedes. *Science* 155:577–579.

Eisner, T., Kluge, A. F., Carrell, J. E., and Meinwald, J. 1971. Defense of phalangid: Liquid repellent administered by leg dabbing. *Science* 173:650–652.

Eisner, T., Kriston, I., and Aneshausley, D. J. 1976. Defensive behavior of a termite *(Nasutitermes exitiosus). Behavioral Ecology and Sociobiology* 1:83–125.

Eisner, T., Meinwald, J., Monro, A., and Ghent, R. 1961. Defence mechanisms of arthropods — I. The composition and function of the spray of the whipscorpion, *Mastigoproctus giganteus* (Lucas) (Aarachnida, Pedipalpida). *Journal of Insect Physiology* 6:272–298.

Eklund, C. R. 1961. Distribution and life history studies of the south-polar skua. *Bird-Banding* 32:187–223.

Eleftheriou, B. E., and Scott, J. P., eds. 1971. *The physiology of aggression and defeat.* New York: Plenum.

Elliott, J. P., Cowan, I. M., and Holling, C. S. 1977. Prey capture by the African lion. *Canadian Journal of Zoology* 55:1811–1828.

Elliott, P. F. 1975. Longevity and the evolution of polygamy. *American Naturalist* 109:281–287.

Ellison, L. N. 1971. Territoriality in Alaskan spruce grouse. *Auk* 88:652–664.

———. 1973. Seasonal social organization and movements of spruce grouse. *Condor* 75:375–385.

———. 1976. Winter food selection by Alaskan spruce grouse. *Journal of Wildlife Management* 40:205–213.

Elmes, G. W. 1976. Some observations on the microgyne form of *Myrmica rubra* L. (Hymenoptera, Formicidae). *Insectes Sociaux* 23:3–22.

Eloff, F. C. 1973. Ecology and behaviour of the Kalahari lion, *Panthera leo vernayi* (Roberts, 1929). In *The world's cats,* vol. 1, ed. R. L. Eaton, pp. 90–126. Winston, Oreg.: World Wildlife Safari.

Elton, C. S. 1927. *Animal ecology.* London: Sidgwick & Jackson.

———. 1932. Territory among wood ants *(Formica rufa* L.) at Picket Hill. *Journal of Animal Ecology* 1:69–76.

Emden, H. F., van. 1972. Aphids as phytochemists. In *Phytochemical ecology,* ed. J. B. Harborne, pp. 25–43. New York: Academic Prss.
Emlen, J. M. 1966. The role of time and energy in food preference. *American Naturalist* 100:611–617.
———. 1968. Optimal choice in animals. *American Naturalist* 102:385–390.
———. 1973. *Ecology: An evolutionary approach.* Reading, Mass.: Addison-Wesley.
———. 1978. Territoriality: A fitness set–adaptive function approach. *American Naturalist* 112:234–241.
Emlen, J. T. 1952a. Social behavior in nesting cliff swallows. *Condor* 54:177–199.
———. 1952b. Flocking behavior in birds. *Auk* 69:160–170.
———. 1957. Display and mate selection in the whydahs and bishop birds. *Ostrich* 28:202–213.
Emlen, J. T., and Lorenz, F. W. 1942. Pairing responses of free-living valley quail to sex-hormone pellet implants. *Auk* 59:369–378.
Emlen, S. T. 1971. Adaptive aspects of coloniality in the bank swallow. *American Zoologist* 11:47.
———. 1978. The evolution of cooperative breeding in birds. In *Behavioural ecology: An evolutionary approach,* ed. J. R. Krebs and N. B. Davies, pp. 245–281. Sunderland, Mass.: Sinauer Associates.
Emlen, S. T., and Ambrose, H. W., III. 1970. Feeding interactions of snowy egrets and red-breasted mergansers. *Auk* 87:164–165.
Emlen, S. T., and Demong, N. J. 1975. Adaptive significance of synchronized breeding in a colonial bird: A new hypothesis. *Science* 188:1029–1031.
Emlen, S. T., and Oring, L. W. 1977. Ecology, sexual selection, and the evolution of mating systems. *Science* 197:215–223.
Endler, J. A. 1978. A predator's view of animal color patterns. In *Evolutionary biology,* vol. 11 ed. M. K. Hecht, W. C. Steere, and B. Wallace, pp. 319–364. New York: Plenum.
Epple, G. 1970. Maintenance, breeding, and development of marmoset monkeys (Callitrichidae) in captivity. *Folia Primatologica* 12:56–76.
———. 1975. The behavior of marmoset monkeys (Callitrichidae). In *Primate behavior. Developments in field and laboratory research,* vol. 4, ed. L. A. Rosenblum, pp. 195–239. New York: Academic Press.
———. 1977. Notes on the establishment and maintenance of the pair bond in *Saguinus fuscicollis.* In *The biology and conservation of the Callitrichidae,* ed. D. G. Kleiman, pp. 231–237. Washington, D.C.: Smithsonian Institution Press.
Erickson, C. J., and Zenone, P. G. 1976. Courtship differences in male ring doves: Avoidance of cuckoldry? *Science* 192:1353–1354.
———. 1978. Aggressive courtship as a means of avoiding cuckoldry. *Animal Behaviour* 26:307–308.
Erickson, J. G. 1967. Social hierarchy, territoriality, and stress reactions in sunfish. *Physiological Zoology* 40:40–48.
Eshel, I. 1975. Selection of sex-ratio and the evolution of sex-determination. *Heredity* 34:351–361.
———. 1978. On the handicap principle — A critical defence. *Journal of Theoretical Biology* 70:245–250.
Estes, R. D. 1966. Behaviour and life history of the wildebeest. *Nature* 212:999–1000.
———. 1967. The comparative behavior of Grant's and Thomson's gazelles. *Journal of Mammalogy* 48:189–209.
———. 1969. Territorial behaviour of the wildebeest (*Connochaetes taurinus* Burchell 1823). *Zeitschrift für Tierpsychologie* 26:284–370.
———. 1974. Social organization of the African bovidae. In *The behaviour of ungulates and its relation to management,* ed. V. Geist and F. Walther, pp. 166–205. Moreges: International Union for Conservation of Nature and Natural Resources.
———. 1976. The significance of breeding synchrony in the wildebeest. *East African Wildlife Journal* 14:135–152.
Estes, R. D. and Estes, R. K. 1979. The birth and survival of wildebeest calves. *Zeitschrift für Tierpsychologie* 50:45–95.
Estes, R., and Goddard, J. 1967. Prey selection and hunting behavior of the African wild dog. *Journal of Wildlife Management* 31:52–70.
Etkin, W. 1964. *Social behavior from fish to man.* Chicago: University of Chicago Press.
Evans, F. C., and Holdenreid, R. 1943. A population study of the beechey ground squirrel in central California. *Journal of Mammalogy* 24:231–260.
Evans, H. E. 1958. The evolution of social life in wasps. *Proceedings of the 10th International Congress of Entomology* 2:449–457.

———. 1977. Extrinsic versus intrinsic factors in the evolution of insect sociality. *BioScience* 27:613–617.

Evans, W. F. 1968. *Communication in the animal world.* New York: Crowell.

Ewald, P. W. 1980. Energetics of resource defense: An experimental approach. *Acta XVII Congressus Internationalis Ornithologica.*

Ewald, P. W., Hunt, G. L., Jr., and Warner, M. 1980. Territory size in western gulls: Importance of intrusion pressure, defense investments and vegetation structure. *Ecology* (in press).

Ewer, R. F. 1959. Suckling behaviour in kittens. *Behaviour* 15:146–162.

———. 1963a. A note on the suckling behaviour of the viverrid, *Suricata suricatta* (Schreber). *Animal Behaviour* 11:599–601.

———. 1963b. The behaviour of the meerkat, *Suricata suricatta* (Schreber). *Zeitschrift für Tierpsychologie* 20:570–607.

———. 1967. The behaviour of the African giant rat (*Cricetomys gambianus* Waterhouse). *Zeitschrift für Tierpsychologie* 24:6–79.

———. 1968. *Ethology of mammals.* London: Logos Press.

———. 1973. *The carnivores.* London: Weidenfeld & Nicholson.

Fagan, R. 1974. Selective and evolutionary aspects of animal play. *American Naturalist* 108:850–858.

Fairbanks, L. A., and Bird, J. 1978. Ecological correlates of interindividual distance in the St. Kitts vervet (*Cercopithecus aethiops sabaeus*). *Primates* 19:605–614.

Falconer, D. S. 1960. *Introduction to quantitative genetics.* Edinburgh: Oliver & Boyd.

Falls, J. B. 1963. Properties of bird song eliciting responses from territorial males. *Proceedings of the XIII International Ornithological Congress,* pp. 259–271.

———. 1969. Functions of territorial song in the white-throated sparrow. In *Bird vocalizations. Their relations to current problems in biology and psychology. Essays presented to W. H. Thorp,* ed. R. A. Hinde, pp. 207–232. London: Cambridge University Press.

Farner, D. S. 1975. Photoperiodic controls in the secretion of gonadotropins in birds. *American Zoologist, Supplement* 15:117–135.

Farner, D. S., and Follett, B. K. 1979. Reproductive periodicity in birds. In *Hormones and evolution,* ed. E. J. W. Barrington, pp. 829–872. London: Academic Press.

Farr, J. A. 1976. Social facilitation of male sexual behavior, intrasexual competition, and sexual selection in the guppy, *Poecilia reticulata* (Pisces: Poeciliidae). *Evolution* 30:707–717.

———. 1977. Male rarity or novelty, female choice behavior, and sexual selection in the guppy, *Poecilia reticulata* Peters (Pisces: Poeciliidae). *Evolution* 31:162–168.

Fautin, R. W. 1940. The establishment and maintenance of territories by the yellow-headed blackbird in Utah. *Great Basin Naturalist* 1:75–91.

Feare, C. J. 1976. The breeding of the sooty tern *Sterna fuscata* in the Seychelles and the effects of experimental removal of eggs. *Journal of Zoology, London* 179:317–360.

Feddern, H. A. 1968. Hybridization between the western Atlantic angelfishes, *Holacanthus isabelita* and *H. ciliaris. Bulletin of Marine Science* 18:351–382.

Feder, H. M. 1966. Cleaning symbioses in the marine environment. In *Symbiosis,* vol. 1, ed. S. M. Henry, pp. 327–380. New York: Academic Press.

———. 1972. Escape responses in marine invertebrates. *Scientific American* 227(1):92–100.

Fedigan, L. M. 1976. A study of roles in the Arashiyama West troop of Japanese monkeys (*Macaca fuscata*). *Contributions to Primatology* 9:1–95.

Feeny, P. P. 1969. Inhibitory effect of oak leaf tannins on hydrolysis of proteins by trypsin. *Phytochemistry* 8:2119–2126.

———. 1970. Seasonal changes in oak leaf tannins and nutrients as a cause of spring feeding by winter-moth caterpillars. *Ecology* 51:656–681.

———. 1976. Plant apparency and chemical defense. In *Biochemical interaction between plants and insects. Recent Advances in Phytochemistry,* vol. 10, ed. J. W. Wallace and R. L. Mansell, pp. 1–40. New York: Plenum.

Feist, J. D., and McCullough, D. R. 1975. Reproduction in feral horses. *Journal of Reproduction and Fertility, Supplement* 23:13–18.

Feldman, M. W., and Lewontin, R. C. 1975. The heritability hang-up. The role of variance analysis in human genetics is discussed. *Science* 190:1163–1168.

Fellers, G. M. 1979. Aggression, territoriality, and mating behaviour in North American tree frogs. *Animal Behaviour* 27:107–119.

Felsenstein, J., and Yokoyama, S. 1976. The evolutionary advantage of recombination. II. Individual selection for recombination. *Genetics* 83:845–859.

Fentress, J. C., and Ryon, C. J. 1981. A long term study of distributed pup feeding and associated behavior in wolves. In *Portland wolf symposium,* ed. P. C. Paquet and J. O. Sullivan, New York: Garland Press, in press.

Feshbach, S. 1970. Aggression. In *Carmichael's manual of child psychology,* vol. II, ed. P. H. Mussen, pp. 159–259. New York: Wiley.

ffrench, R. P. 1967. The dickcissel on its wintering grounds in Trinidad. *Living Bird* 6:123–140.

Ficken, R. W., Ficken, M. S., and Hailman, J. P. 1978. Differential aggression in genetically different morphs of the white-throated sparrow (*Zonotrichia albicollis*). *Zeitschrift für Tierpsychologie* 46:43–57.

Fiedler, K. 1955. Vergleichende Verhaltenstudien an Seeadeln, Schlangennalde und Seepferdchen. *Zeitschrift für Tierpsychologie* 11:358–416.

Fielder, D. R. 1965. A dominance order for shelter in the spiny lobster *Jasus lalandei* (H. Milne-Edwards). *Behaviour* 24:236–245.

Fields, W. S. and Sweet, W. H., eds. 1975. *Neural bases of violence and aggression.* St. Louis: Warren H. Green.

Fink, B. 1959. Observations of porpoise predation on a school of Pacific sardines. *California Fish and Game Bulletin* 45:216–217.

Fischer, C. A., and Keith, L. B. 1974. Population responses of central Alberta ruffed grouse to hunting. *Journal of Wildlife Management* 38:585–600.

Fishelson, L. 1970. Behavior and ecology of a population of *Abudefduf saxatilis* (Pomacentridae, Teleostei) at Eilat (Red Sea). *Animal Behaviour* 18:225–237.

Fisher, A. E. 1955. The effects of differential early treatment on the social and exploratory behavior of puppies. Ph.D. thesis, Pennsylvania State University, University Park.

Fisher, J., and Hinde, R. A. 1949. The opening of milk bottles by birds. *British Birds* 42:347–357.

Fisher, J., and Lockley, R. M. 1954. *Sea-birds. An introduction to the natural history of the sea-birds of the North Atlantic.* Boston: Houghton Mifflin.

Fisher, R. A. 1930. *The genetical theory of natural selection.* Oxford: Clarendon Press.

Fitch, H. S. 1970. Reproductive cycles in lizards and snakes. *University of Kansas Museum of Natural History, Miscellaneous Publications* 52:1–247.

Fitch, H. S., Swenson, F., and Tillotson, D. F. 1946. Behaviour and food habits of the red-tailed hawk. *Condor* 48:205–237.

Flanders, S. E. 1956. The mechanisms of sex-ratio regulation in the (parasitic) Hymenoptera. *Insectes Sociaux* 3:325–334.

Floody, O. R., and Arnold, A. P. 1975. Uganda kob (*Adenota kob thomasi*): Territoriality and the spatial distributions of sexual and agonistic behaviours at a territorial ground. *Zeitschrift für Tierpsychologie* 37:192–212.

Fogden, M. 1974. A preliminary field-study of the western tarsier, *Tarsius bancanus* Horsefield. In *Prosimian biology,* ed. R. D. Martin, G. A. Doyle, and A. C. Walker, pp. 151–166. London: Duckworth.

Fontaine, R. 1980. Observations on the foraging association of double-toothed kites and white-faced capuchin monkeys. *Auk* 97:94–98.

Ford, E. B. 1964. *Ecological genetics.* London: Methuen.

Ford, H. A. 1971. The degree of mimetic protection gained by new partial mimics. *Heredity* 27:227–236.

Forrest, J. M. S. 1971. The growth of *Aphis fabae* as an indicator of the nutritional advantage of galling to the apple aphid *Dysaphis devecta. Entomologica experimentalis et applicata* 14:477–483.

Forrest, J. M. S., and Noordink, J. P. W. 1971. Translocation and subsequent uptake by aphids of ^{32}P introduced into plants by radioactive aphids. *Entomologica experimentalis et applicata* 14:133–134.

Forrester, J. W. 1971. Counterintuitive behavior of social systems. *Technology Review* 731:52–68.

Foster, J. B. 1966. The giraffe of Nairobi National Park: Home range, sex ratios, the herd, and food. *East African Wildlife Journal* 4:139–148.

Foster, J. B., and Dagg, A. I. 1976. *The giraffe: Its biology, behavior, and ecology.* New York: Van Nostrand Reinhold.

Fowden, L., Lewis, D., and Tristam, H. 1967. Toxic amino acids: Their action as antimetabolites. *Advances in Enzymology* 29:89–163.

Fox, M. W., and Andrews, R. V. 1973. Physiological and biochemical correlates of individual differences in behavior of wolf cubs. *Behaviour* 46:129–139.

Fox, S. W., and Dose, K. 1977. *Molecular evolution and the origin of life.* 2nd ed. San Francisco: Freeman.

Frädrich, H. 1965. Zur Biologie und Ethologie des Warzenschweines (*Phacohoeurus aethiopicus* Pallas), unter Berücksichtigung des Verhaltens anderer Suiden. *Zeitschrift für Tierpsychologie* 22:328–374, 375–393.

———. 1974. A comparison of behaviour in the Suidae. In *The behaviour of ungulates and its relation to management,* ed. V. Geist and F. Walther, pp. 133–143. Moreges: International

Union for Conservation of Nature and Natural Resources.
Fraenkel, G. 1969. Evaluation of our thoughts on secondary plant substances. *Entomologica experimentalis et applicata* 12:473–486.
Frame, L. H., and Frame, G. W. 1976. Female African wild dogs emigrate. *Nature* 263:227–229.
Frame, L. H., Malcolm, J. R., Frame, G. W., and Lawick, H., van. 1979. Social organization of African wild dogs (*Lycaon pictus*) on the Serengeti Plains, Tanzania 1967–1978. *Zeitschrift für Tierpsyhologie* 50:225–249.
Francis, W. 1965. Double broods in California quail. *Condor* 67:541–542.
Frankenhaeuser, M. 1971. Behavior and circulating catecholamines. *Brain Research* 31:241–262.
Frankie, G. W., Baker, H. G., and Opler, P. A. 1974. Comparative phenological studies of trees in tropical wet and dry forests in the lowlands of Costa Rica. *Journal of Ecology* 62:881–919.
Franklin, W. L. 1974. The social behaviour of the vicuña. In *The behaviour of ungulates and its relation to management*, ed. V. Geist and F. Walther, pp. 447–487. Moreges: International Union for Conservation of Nature and Natural Resources.
Franklin, W. L., Mossman, A. S., and Dole, M. 1975. Social organization and home range of Roosevelt elk. *Journal of Mammalogy* 56:102–118.
Franzen, E. A., and Myers, R. E. 1973. Neural control of social behavior: Prefrontal and anterior temporal cortex. *Neuropsychologia* 11:141–157.
Frazer, J. F. D., and Rothschild, M. 1960. Defense mechanisms in warningly colored moths and other insects. *International Kongress der Entomologie, Verhandlungen, 11th Kongress, Vienna* 3:249–257.
Free, J. B. 1961. Hypopharyngeal gland development and division of labour in honey-bee (*Apis mellifera* L.) colonies. *Proceedings of the Royal Entomological Society of London, Series A* 36:5–8.
———. 1965. The allocation of duties among worker honeybees. *Symposia of the Zoological Society of London* 14:39–59.
Freedman, D. G. 1979. *Human sociobiology: A holistic approach*. New York: Free Press.
Freeland, J. 1958. Biological and social patterns in the Australian bulldog ants of the genus *Myrmecia*. *Austrailian Journal of Zoology* 6:1–18.
Freeland, W. J., and Janzen, D. H. 1974. Strategies of herbivory by mammals: The role of plant secondary compounds. *American Naturalist* 108:269–289.
French, N. R. 1959. Life history of the black rosy finch. *Auk* 76:159–180.
Fretter, V., and Graham, A. 1962. *British prosobranch molluscs*. London: Ray Society.
Fretwell, S. 1968. Habitat distribution and survival in the field sparrow (*Spizella pusilla*). *Bird-banding* 34:293–306.
———. 1969a. On territorial behavior and other factors influencing habitat distribution in birds. III. Breeding success in a local population of field sparrows (*Spizella pusilla* Wils.). *Acta Biotheoretica* 19:45–52.
———. 1969b. Dominance behavior and winter habitat distribution in juncos (*Junco hyemalis*). *Bird-banding* 40:1–25.
Fretwell, S. D., and Calver, J. S. 1969. On territorial behavior and other factors influencing habitat distribution in birds. II. Sex ratio variation in the dickcissel (*Spiza americana* Gmel). *Acta Biotheoretica* 19:37–44.
Fretwell, S. D., and Lucas, H. L., Jr. 1969. On territorial behavior and other factors influencing habitat distribution in birds. I. Theoretical development. *Acta Biotheoretica* 19:16–36.
Fricke, H. 1974. Öko-Ethologie des monogamen Anemonenfisches *Amphiprion bicinctus* (Freiwasseruntersuchung aus dem Rotem Meer). *Zeitschrift für Tierpsychologie* 36:429–512.
Fricke, H., and Fricke, S. 1977. Monogamy and sex change by aggressive dominance in coral reef fish. *Nature* 266:830–832.
Friedmann, H. 1929. *The cowbirds*. Springfield, Ill.: Thomas.
———. 1948. The parasitic cuckoos of Africa. *Washington Academy of Sciences Monograph*, no. 1:1–204.
———. 1955. The honeyguides. *United States National Museum Bulletin* 208:1–292.
———. 1960. The parasitic weaverbirds. *United States National Museum Bulletin* 223:1–196.
———. 1963. Host relations of the parasitic cowbirds. *United States National Museum Bulletin* 233:1–276.
———. 1967. Avian symbiosis. In *Symbiosis*, vol. II, ed. S. M. Henry, pp. 291–316. New York: Academic Press.
Frisch, K. von. 1955. *The dancing bees: An accouont of the life and senses of the honey bee*. New York: Harcourt Brace Jovanovich.
Frisch, O. R. 1973. *Animal camouflage*. London: Collins.

Frith, H. J. 1962. *The mallee-fowl: The bird that builds an incubator*. Sydney: Angus & Robertson.
Fry, C. H. 1969. The recognition and treatment of venomous and non-venomous insects by small bee-eaters. *Ibis* 111:23–29.
———. 1972a. The social organization of bee-eaters (Meropidae) and co-operative breeding in hot-climate birds. *Ibis* 114:1–14.
———. 1972b. The biology of African bee-eaters. *Living Bird* 11:75–112.
———. 1975. Cooperative breeding in bee-eaters and longevity as an attribute of group-breeding birds. *Emu* 74:308–309.
———. 1977. The evolutionary significance of cooperative breeding in birds. In *Evolutionary ecology*, ed. B. Stonehouse and C. M. Perrins, pp. 127–136. London: MacMillan.
Fryer, G., and Iles, T. 1972. *The cichlid fishes of the Great Lakes of Africa: Their biology and evolution*. Neptune City, N.J.: T. F. H. Pub.
Fuller, J. L., and Clark, L. D. 1966. Genetic and treatment factors modifying the post-isolation syndrome in dogs. *Journal of Comparative and Physiological Psychology* 61:251–257.
Futuyama, D. J. 1979. *Evolutionary biology*. Sunderland, Mass.: Sinauer Associates.
Gadgil, M., and Bossert, W. H. 1970. Life historical consequences of natural selection. *American Naturalist* 104:1–24.
Gadow, H. 1901 (reprinted 1958). *Amphibia and reptiles*. Cambridge, England: Cambridge University Press.
Galton, F. 1871. Gregariousness in cattle and men. *MacMillan's Magazine, London* 23:353.
Gandelman, R., Paschke, R. E., Zarrow, M. X., and Deuenberg, V. H. 1970. Care of young under communal conditions in the mouse (*Mus musuclus*). *Developmental Psychobiology* 3:245–250.
Garattini, S., and Sigg, E. B., eds. 1969. *Aggressive behaviour*. Amsterdam: Excerpta Medica.
Garson, P. J. 1980. Male behaviour and female choice: Mate selection in the wren? *Animal Behaviour* 28:491–502.
Garson, P. J., Pleszczynska, W. K., and Holm, C. H. 1981. The "polygyny threshold" model: A reassessment. *Canadian Journal of Zoology* (in press).
Gartlan, J. S. 1968. Structure and function in primate society. *Folia Primatologica* 8:89–120.
———. 1969. Sexual and maternal behavior of the vervet monkey, *Cercopithecus aethiops*. *Journal of Reproduction and Fertility, Supplement* 6:137–150.
Gary, N. E. 1963. Observations of mating behaviour in the honeybee. *Journal of Apicultural Research* 2:3–13.
Gass, C. L., Angehr, G., and Centa, J. 1976. Regulation of food supply by feeding territoriality in the rufous hummingbird. *Canadian Journal of Zoology* 54:2046–2054.
Gaston, A. J. 1976. Brood parasitism by the pied crested cuckoo *Clamator jacobinus*. *Journal of Animal Ecology* 45:331–348.
———. 1977. Social behaviour within groups of jungle babblers (*Turdoides striatus*). *Animal Behaviour* 25:828–848.
———. 1978. Factors affecting the evolution of group territorial behaviour and co-operative breeding in birds. *American Naturalist* 112:1091–1100.
Gaston, A. and Perrins, C. 1975. The relationship of habitat to group size in the genus *Turdoides*, *Emu* 74 (supplement):1–309.
Gautier, J. P., and Gautier, A. 1977. Communication in old world monkeys. In *How animals communicate*, ed. T. A. Sebeok, pp. 890–964. Bloomington: Indiana University Press.
Gautier-Hion, A. 1970. L'organization sociale d'une bande de talapoins (*Miopithecus talapoin*) dans le Nord-Est du Gabon. *Folia Primatologica* 12:116–141.
Geist, V. 1966. The evolution of horn-like organs. *Behaviour* 27:175–214.
———. 1971. *Mountain sheep. A study in behavior and evolution*. Chicago: University of Chicago Press.
———. 1974a. On fighting strategies in animal combat. *Nature* 250:354.
———. 1974b. On the relationship of social evolution and ecology in ungulates. *American Zoologist* 14:205–220.
———. 1977. On weapons, combat, and ecology. In *Aggression, dominance, and individual spacing*, ed. L. Kramer, P. Pliner, and T. Alloway, pp. 1–30. New York: Plenum.
———. 1978. *Life strategies, human evolution, and environmental design: Toward a biological theory of health*. New York: Springer-Verlag.
Georgiev, G. P. 1969. On the structural organization of operon and the regulation of RNA synthesis in animal cells. *Journal of Theoretical Biology* 25:473–490.
———. 1972. The structure of transcriptional units in eukaryotic cells. *Current Topics in Developmental Biology* 7:1–60.
Gerling, D., and Hermann, H. R. 1978. Biology and mating behavior of *Xylocopa virginica* L.

(Hymenoptera, Anthophoridae). *Behavioral Ecology and Sociobiology* 3:99–111.

Ghent, A. 1960. A study of the group-feeding behaviour of larvae of the jack pine sawfly, *Neodiprion pratti bansianae* Roh. *Behaviour* 16:110–148.

Ghiselin, M. T. 1974. *The economy of nature and the evolution of sex.* Berkeley: University of California Press.

Gibb, J. A. 1956. Territory in the genus *Parus. Ibis* 98:420–429.

Gibson, F. 1971. The breeding biology of the American avocet (*Recurvirostra americana*) in central Oregon. *Condor* 73:444–454.

Gibson, R. M., and Guinness, F. E. 1980. Differential reproduction among red deer (*Cervus elaphus*) stags on Rhum. *Journal of Animal Ecology* 49:199–208.

Gilbert, W. M. 1975. Ecological energy relationships of honey bees, *Apis mellifera,* foraging at different distances from a common hive. Ph.D. thesis, University of California, Davis.

Giles, N. H. 1978. The organization, function, and evolution of gene clusters in eucaryotes. *American Naturalist* 112:641–657.

Gill, F. B., and Wolf, L. L. 1975. Economics of feeding territoriality in the golden-winged sunbird. *Ecology* 56:333–345.

Gill, J. C., and Thomson, W. 1956. Observations on the behaviour of suckling pigs. *British Journal of Animal Behaviour* 4:46–51.

Gilliard, E. T. 1962. On the breeding behavior of the cock-of-the-rock (Aves, *Rupicola rupicola*). *Bulletin of the American Museum of Natural History* 124:31–68.

———. 1969. *Birds of paradise and bower birds.* Garden City, N.Y.: Natural History Press.

Ginsburg, B., and Allee, W. C. 1942. Some effects of conditioning on social dominance and subordination in inbred strains of mice. *Physiological Zoology* 15:485–506.

Gladstone, D. E. 1979. Promiscuity in monogamous colonial birds. *American Naturalist* 114:545–557.

Glancey, B. M., Stringer, C. E., Jr., Craig, C. H., Bishop, P. M., and Martin, B. B. 1973. Evidence of a replete caste in the fire ant *Solenopsis invicta. Annals of the Entomological Society of America* 66:233–234.

Glesener, R. R. and Tilman, D. 1978. Sexuality and the components of environmental uncertainty. *American Naturalist* 112:659–673.

Gloor. 1975. In neural bases of violence and aggression, ed. W. S. Fields and W. H. Sweet. St. Louis: Warren H. Green.

Gochfeld, M. 1979a. Interspecific territoriality in red-breasted meadowlarks and a method for estimating the mutuality of their participation. *Behavioral Ecology and Sociobiology* 5:159–170.

———. 1979b. Breeding synchrony in the black skimmer: Colony versus subcolonies. *Proceedings of the Colonial Waterbird Group* 2:171–177.

Gol, A., Kellaway, P., Spaviro, M., and Hurst, C. M. 1963. Studies in hippocampectomy in the monkey, baboon and cat. *Neurology* 13:1031–1041.

Goldberger, R. F. 1979. Strategies of genetic regulation in prokaryotes. In *Biological regulation and development.* Vol. I. *Gene expression,* ed. R. F. Goldberger, pp. 1–18. New York: Plenum.

Goldstein, J. H. 1975. *Aggression and crimes of violence.* London: Oxford University Press.

Goldstein, J. L., and Swain, T. 1965. Changes in tannins in ripening fruits. *Phytochemistry* 4:185–193.

Goodall, D. W. 1970. Statistical plant ecology. *Annual Review of Ecology and Systematics* 1:99–124.

Goodall, J. 1963a. My life among wild chimpanzees. *National Geographic* 124(1):272–308.

———. 1963b. Feeding behavior of wild chimpanzees. (A preliminary report.) *Symposia of the Zoological Society of London* 10:39–47.

———. 1965. Chimpanzees of the Gombe Reserve. In *Primate behavior: Field studies of monkeys and apes,* ed. I. Devore, pp. 425–473. New York: Holt, Rinehart and Winston.

———. 1973. The behaviour of chimpanzees in their natural habitat. *American Journal of Psychiatry* 130:1–13.

———. 1977. Infant killing and cannibalism in free-living chimpanzees. *Folia Primatologica* 28:259–282.

Goodall, J., Bandora, A., Bergmann, E., Busse, C., Matama, H., Mpongo, E., Pierce, A., and Riss, D. 1979. Intercommunity interactions in the chimpanzee population of the Gombe National Park. In *The great apes,* ed. D. A. Hamburg and E. R. McCown, pp. 12–53. Menlo Park, Calif.: Benjamin/Cummings.

Goodman, D. 1974. Natural selection and a cost ceiling on reproductive effort. *American Naturalist* 108:247–268.

Goodman, L., and Gilman, A., eds. 1970. *The pharmacological basis of therapeutics.* 4th ed. New York: Macmillan.

Goodwin, D. 1953. Observations on voice and behaviour of the red-legged partridge *Alectoris rufa*. *Ibis* 95:581–614.

Gorlick, D. L., Atkins, P. D., and Losey, G. S., Jr. 1978. Cleaning stations as water holes, garbage dumps, and sites for the evolution of reciprocal altruism? *American Naturalist* 112:314–353.

Gosling, L. M. 1974. The social behaviour of Coke's hartebeest (*Alcelaphus buselaphus cokei*). In *The behaviour of ungulates and its relation to management*, ed. V. Geist and F. Walther, pp. 488–511. Moreges: International Union for Conservation of Nature and Natural Resources.

Goss-Custard, J. D. 1970. Feeding dispersion in some overwintering wading birds. In *Social behaviour in birds and mammals. Essays on the social ethology of animals and man*, ed. J. H. Crook, pp. 3–35. New York: Academic Press.

Gotto, R. V. 1969. *Marine animals. Partnerships and other associations*. London: English Universities Press.

Gottschalk, L. A., Kaplan, S. M., Gleser, G. C., and Winget, C. 1961. Variations in the magnitude of anxiety and hostility with phases of the menstrual cycle. *Psychosomatic Medicine* 23:448 (abstract).

Gould, S. J., and Lewontin, R. C. 1979. The spandrels of San Marco and the Panglossian paradigm: A critique of the adaptionist programme. *Proceedings of the Royal Society of London, Series B* 205:581–598.

Goy, R. W. 1968. Organizing effects of androgen on the behaviour of rhesus monkeys. In *Endocrinology and human behavior*, ed. R. P. Michael, pp. 12–31. London: Oxford University Press.

Graul, W. D. 1973. Adaptive aspects of the mountain plover social system. *Living Bird* 12:69–94.

———. 1976. Food fluctuations and multiple clutches in the mountain plover. *Auk* 93:166–167.

Graul, W. D., Derrickson, S. R., and Mock, D. W. 1977. The evolution of avian polyandry. *American Naturalist* 111:812–816.

Greaves, T. 1962. Studies of foraging galleries and the invasion of living trees by *Coptotermes acinaciformis* and *C. brunneus* (Isoptera). *Australian Journal of Zoology* 10:630–651.

Green, R. F. 1980. A note on K-selection. *American Naturalist* 116:291–296.

Greenberg, B. 1947. Some relations between territory, social hierarchy, and leadership in the green sunfish (*Lepomis cyanellus*). *Physiological Zoology* 20:267–299.

Greenberg, L. 1979. Genetic component of bee odor in kin recognition. *Science* 206: 1095–1097.

Greenberg, R. A., and White, C. 1967. The sexes of consecutive sibs in human sibships. *Human Biology* 39:374–404.

Greene, H. W. 1973. Defensive tail display by snakes and amphibians. *Journal of Herpetology* 7:143–161.

Greene, R., and Dalton, K. 1953. The premenstrual syndrome. *British Medical Journal* 1:1007–1014.

Greer, A. E. 1966. Viviparity and oviparity in the snake genera *Conopsis, Toluca, Gyalopion,* and *Ficimia* with comments on *Tomodon* and *Helicops*. *Copeia* 1966:371–373.

Gregory, M. S., Silvers, A., and Sutch, D., eds. 1978. *Sociobiology and human nature*. San Francisco: Jossey-Bass.

Greig-Smith, P. 1952. Ecological observations on degraded and secondary forest in Trinidad, British West Indies. II. Structure of the communities. *Journal of Ecology* 40:316–330.

———. 1964. *Quantitative plant ecology*. 2nd ed. London: Butterworth.

Grimes, L. G. 1976. The occurrence of cooperative breeding behaviour in African birds. *Ostrich* 47:1–15.

———. 1980. Observations of group behaviour and breeding biology of the yellow-billed shrike *Corvinella corvina*. *Ibis* 122:166–192.

Grimsdell, J. J. R. 1969. The ecology of the buffalo, *Syncerus caffer*, in western Uganda. Ph.D. thesis, Cambridge University, Cambridge, England.

Grinnell, J. 1903. Call notes of the bush-tit. *Condor* 5:85–87.

———. 1917. The niche relationships of the California thrasher. *Auk* 21:364–382.

———. 1924. Geography and evolution. *Ecology* 5:225–229.

Gross, D. 1975. Protein capture and cultural development in the Amazon basin. *American Anthropologist* 77:526–549.

Gross, M. R., and Shine, R. 1981. Parental care and mode of fertilization in fish. *Evolution* (in press).

Grossman, S. P. 1972. Aggression, avoidance, and reaction to novel environments in female rats with ventromedial hypothalamic lesions. *Journal of Comparative and Physiological Psychology* 78:274–283.

Guggisberg, C. A. W. 1972. *Crocodiles. Their natural history, folklore and conservation*. Harrisburg, Pa.: Stackpole.

Guhl, A. M. 1958. The development of social organization in the domestic chick. *Animal Behaviour* 6:92–111.

———. 1964. Psychophysiological interrelations in the social behavior of chickens. *Psychological Bulletin* 61:277–285.

———. 1968. Social inertia and social stability in chickens. *Animal Behaviour* 16:219–232.

Guhl, A. M., Collias, N. E., and Allee, W. C. 1945. Mating behaviour and the social hierarchy in small flocks of white leghorns. *Physiological Zoology* 18:365–390.

Gulger, H. D., Kaplan, W. D., and Kidd, W. D. 1965. The displacement of first-mating by second-mating sperm in the storage organs of the female. *Drosophila Information Service* 40:65.

Gullion, G. W., and Marshall, W. H. 1968. Survival of ruffed grouse in a boreal forest. *Living Bird* 7:117–167.

Gustafson, J. E., and Winokur, G. 1960. The effect of sexual satiation and female hormones upon aggressivity in an inbred mouse strain. *Journal of Neuropsychiatry* 1:182–184.

Guthrie, R. D. 1970. Evolution of human threat display organs. In *Evolutionary biology*, vol. 4, ed. M. K. Hecht, W. C. Steere, and B. Wallace, pp. 257–302. New York: Plenum.

———. 1971. A new theory of mammalian rump patch evolution. *Behaviour* 38:132–145.

Gwynne, D. T. 1978. Male territoriality in the bumblebee wolf, *Philanthus bicinctus* (Mickel) (Hymenoptera, Sphecidae): Observations on the behaviour of individual males. *Zeitschrift für Tierpsychologie* 47:89–103.

Haartman, L. von. 1951. Der Trauerfliegenschnäpper. II. Populationsprobleme. *Acta Zoologica Fennica* 67:1–60.

———. 1958. The incubation rhythm of the female pied flycatcher (*Ficedula hypoleuca*) in the presence and absence of the male. *Ornis Fennica* 35:71–76.

———. 1971. Population dynamics. In *Avian biology*, vol. I, ed. D. S. Farner and J. R. King, pp. 391–459. New York: Academic Press.

Haas, A. 1960. Vergleichende Verhaltensstudien zum Paarungsschwarm solitärer Apiden. *Zeitschrift für Tierpsychologie* 17:402–416.

Haas, R. 1976. Sexual selection in *Nothobranchius guentheri* (Pisces: Cyprinodontidae). *Evolution* 30:614–622.

Haddow, A. J. 1952. Field and laboratory studies on an African monkey, *Cercopithecus ascanius schmitti* Matschie. *Proceedings of the Zoological Society of London* 122:297–394.

Hafez, E. S. E., and Signoret, J. R. 1969. The behaviour of swine. In *The behaviour of domestic animals*, 2nd ed., ed. E. S. E. Hafez, pp. 349–360. London: Baillière, Tindall, & Cassell.

Hailman, J. P. 1964. Breeding synchrony in the equatorial swallow-tailed gull. *American Naturalist* 98:706–754.

Hairston, N. G. 1959. Species abundance and community organization. *Ecology* 40:404–416.

Hairston, N. G., Tinkle, D. W., and Wilbur, H. M. 1970. Natural selection and the parameters of population growth. *Journal of Wildlife Management* 34:681–690.

Hall, C. S. 1938. The inheritance of emotionality. *Sigma Xi Quarterly* 26:17–27.

Hall, C. S., and Klein, S. J. 1942. Individual differences in aggressiveness in rats. *Journal of Comparative and Physiological Psychology* 33:371–383.

Hall, J. R. 1970. Synchrony and social stimulation in colonies of the black-headed weaver *Ploceus cucullatus* and Vieillot's black weaver *Melanopteryx nigerrimus*. *Ibis* 112:93–104.

Hall, K. R. L. 1960. Social vigilance behaviour of the chacma baboon, *Papio ursinus*. *Behaviour* 16:261–294.

———. 1963. Variations in the ecology of the chacma baboon. *Symposia of the Zoological Society of London* 10:1–28.

———. 1965. Behaviour and ecology of the wild patas monkeys, *Erythrocebus patas*, in Uganda. *Journal of Zoology, London* 148:15–87.

———. 1968. Behaviour and ecology of the wild patas monkey, *Erythrocebus patas*, in Uganda. In *Primates: Studies in adaptation and variability*, ed. P. Jay, pp. 120–130. New York: Holt, Rinehart and Winston.

Hall, K. R. L., and Devore, I. 1965. Baboon social behavior. In *Primate behavior: Field studies of monkeys and apes*, ed. I. Devore, pp. 53–110. New York: Holt, Rinehart and Winston.

Halstead, B. W. 1971a. Venomous coelenterates: Hydroids, jellyfishes, corals, and sea anemones. In *Venomous animals and their venoms*. Vol. III. *Venomous invertebrates*, ed. W. Bücherl and E. E. Buckley, pp. 395–417. New York: Academic Press.

———. 1971b. Venomous echinoderms and annelids: Starfishes, sea urchins, sea cucumbers, and segmented worms. In *Venomous animals and their venoms*. Vol. III. *Venomous inver-*

tebrates, ed. W. Bücherl and E. E. Buckley, pp. 419–441. New York: Academic Press.
———. 1971c. Venomous fishes. In *Venomous animals and their venoms.* Vol. II. *Venomous vertebrates,* ed. W. Bücherl and E. E. Buckley, pp. 587–626. New York: Academic Press.
Hamburg, D. A. 1966. Effects of progesterone on behavior. In *Endocrines and the central nervous system.* ed. R. Levine, pp. 251–265. Baltimore: Williams & Wilkins.
Hamburg, D. A., Moos, R. H., and Yalom, I. D. 1968. Studies of distress in the menstrual cycle and the postpartum period. In *Endocrinology and human behaviour,* ed., R. P. Michael, pp. 94–116. London: Oxford University Press.
Hamerstrom, F. 1957. The influence of a hawk's appetite on mobbing. *Condor* 59:192–194.
———. 1972. Comments relating to the paper given by R. Robel. *Proceedings of the 15th International Ornithological Congress,* p. 184.
Hamerstrom, F. N., and Hamerstrom, F. 1960. Comparability of some social displays of grouse. *Proceedings of the 12th International Ornithological Congress,* pp. 274–293.
Hamilton, W. D. 1964. The genetical evolution of social behaviour. I, II. *Journal of Theoretical Biology* 7:1–52.
———. 1967. Extraordinary sex ratios. *Science* 156:477–488.
———. 1971. Geometry for the selfish herd. *Journal of Theoretical Biology* 31:295–311.
———. 1972. Altruism and related phenomena, mainly in social insects. *Annual Review of Ecology and Systematics* 3:193–232.
———. 1979. Wingless and fighting males in fig wasps and other insects. In *Sexual selection and reproductive competition in insects,* ed. M. S. Blum and N. A. Blum, pp. 167–220. New York: Academic Press.
Hamilton, W. J., Jr. 1963. Success story of the opossum. *Natural History* 72(2):17–25.
Hamilton, W. J., III. 1973. *Life's color code.* New York: McGraw-Hill.
Hamilton, W. J., III, Buskirk, R. E., and Buskirk, W. H. 1975a. Defensive stoning by baboons. *Nature* 256:488–489.
———. 1975b. Defense of space and resources by chacma *(Papio ursinus)* baboon troops in an African desert and swamp. *Ecology* 57:1264–1272.
———.1978a. Environmental determinants of object manipulation by chacma baboons *(Papio ursinus)* in two southern African environments. *Journal of Human Evolution* 7:205–216.
———. 1978b. Omnivory and utilization of food resources by chacma baboons, *Papio ursinus. American Naturalist* 112:911–924.
Hamilton, W. J., III, and Gilbert, W. M. 1969. Starling dispersal from a winter roost. *Ecology* 50:886–898.
Hamilton, W. J., III, Gilbert, W. M., Heppner, F. H., and Planck, R. J. 1967. Starling roost dispersal and a hypothetical mechanism regulating rhythmical movement to and from dispersal centers. *Ecology* 48:825–833.
Hamilton, W. J., III, and Hamilton, M. E. 1965. Breeding characteristics of yellow-billed cuckoos in Arizona. *Proceedings of the California Academy of Sciences* 32:405–432.
Hamilton, W. J., III, and Heppner, F. 1966. Radiant solar energy and the function of black homeotherm coloration: An hypothesis. *Science* 155:196–197.
Hamilton, W. J., III, and Orians, G. H. 1965. Evolution of brood parasitism in altricial birds. *Condor* 67:361–382.
Hamilton, W. J., III, and Watt, K. E. F. 1970. Refuging. *Annual Review of Ecology and Systematics* 1:263–286.
Hanby, J. P., Robertson, L. T., and Phoenix, C. H. 1971. The sexual behaviour of a confined troop of Japanese macaques *(Macaca fuscata). Folia Primatologica* 16:123–143.
Hansen, E. W. 1966. The development of maternal and infant behavior in the rhesus monkey. *Behaviour* 27:107–149.
Hardin, G. 1968. The tragedy of the commons. *Science* 162:1243–1248.
Harding, C. F., and Leshner, I. 1972. The effects of adrenalectomy on the aggressiveness of differently housed mice. *Physiology and Behavior* 8:437–440.
Hardy, J. W. 1961. Studies in behavior and phylogeny of certain New World jays (Garrulinae). *University of Kansas Science Bulletin* 42:13–149.
———. 1976. Comparative breeding behavior and ecology of the bushy-crested and Nelson San Blas jays. *Wilson Bulletin* 88:96–120.
Hargreaves, C. E. M., and Llewellyn, M. 1978. The ecological energetics of the willow aphid *Tuberolachnus salignus:* The influence of aphid aggregations. *Journal of Animal Ecology* 47:605–614.
Harlow, H. F., and Harlow, M. K. 1965. The affectional systems. In *Behavior of nonhuman primates,* vol. 2, ed. A. M. Schrier, H. F. Harlow, and F. Stollnitz, pp. 287–334. New York: Academic Press.

———. 1969. Effects of various mother-infant relationships on rhesus monkey behaviors. In *Determinants of infant behaviour,* vol. IV, ed. B. M. Foss, pp. 15–36. London: Methuen.

Harlow, H. F., Harlow, M. K., Dodsworth, R. O., and Arling, G. L. 1966. Maternal behavior of rhesus monkeys deprived of mothering and peer associations in infancy. *Proceedings of the American Philosophical Society* 110:58–66.

Harmeson, J. 1974. Breeding ecology of the dickcissel. *Auk* 91:348–359.

Harper, J. L. 1977. *Plant population biology.* New York: Academic Press.

Harper, L. V. 1970. Ontogenetic and phylogenetic functions of the parent-offspring relationship in mammals. *Advances in the study of Animal Behavior* 3:75–117.

Harris, G. W., and Levine, S. 1965. Sexual differentiation of the brain and its experimental control. *Journal of Physiology* 181:379–400.

Harris, M. 1971. *Culture, man, and nature: An introduction to general anthropology.* New York: Crowell.

———. 1977. *Cannibals and kings: The origins of cultures.* New York: Random House.

Harris, M. P. 1964. Aspects of the breeding biology of the gulls. *Larus argentatus, L. fuscus,* and *L. marinus. Ibis* 109:432–456.

———. 1966. Breeding biology of the manx shearwater *Puffinus puffinus. Ibis* 108:17–33.

———. 1969. The biology of storm petrels in the Galapagos Islands. *Proceedings of the California Academy of Sciences* 37:95–166.

Harrison, C. J. O. 1965. Allopreening as agonistic behaviour. *Behaviour* 24:161–209.

———. 1969. Helpers at the nest in Australian passerine birds. *Emu* 69:30–40.

Hartmann, F. 1979. Three spines on a stickleback. *Natural History* 88(10):32–35.

Hartzler, J. E. 1972. An analysis of sage grouse lek behavior. Ph.D. thesis, University of Montana, Missoula.

Harvey, P. H., and Greenwood, P. J. 1978. Anti-predator defence strategies: Some evolutionary problems. In *Behavioural ecology: An evolutionary approach,* ed. J. R. Krebs and N. B. Davies, pp. 129–151. Sunderland, Mass.: Sinauer Associates.

Haskins, C. P., and Haskins, E. F. 1949. The role of sexual selection as an isolating mechanism in three species of poeciliid fishes. *Evolution* 3:160–169.

Hausfater, G. 1975. Dominance and reproduction in baboons *(Papio cynocephalus). Contributions to Primatology* 7:1–150.

Haverschmidt, Fr. 1949. *The life of the white stork.* Leiden: E. J. Brill.

Haycock, K. A., and Threlfall, W. 1975. The breeding biology of the herring gull in Newfoundland. *Auk* 92:678–697.

Hayden, B. 1975. The carrying capacity dilemma. In *Population studies in archaeology and biological anthropology: A symposium. American Antiquity* 40, Part 2, Memoir 30, ed. A. C. Swedlund, pp. 11–21.

Hays, H. 1972. Polyandry in the spotted sandpiper. *Living bird* 11:43–57.

Hazlett, B. A., and Bossert, W. H. 1965. A statistical analysis of the aggressive communications systems of some hermit crabs. *Animal Behaviour* 11:357–373.

Heatwole, H. 1965. Some aspects of the association of cattle egrets with cattle. *Animal Behaviour* 13:79–83.

Hebb, D. O. 1953. Heredity and environment in mammalian behaviour. *British Journal of Animal Behaviour* 1:43–47.

Hebling, N. J., Kerr, W. E., and Kerr, F. S. 1964. Divisâo de trabalho entre operárias de *Trigona (Scaptotrigona) xanthotricha* Moure. *Papéis Avulsos do Departmento de Zoologica (Saô Paulo)* 16:115–127.

Hediger, H. 1949. Säugetier-Territorien und ihre Markierung. *Bijdragen tot de Dierkunde, Leiden* 28:172–184.

Heftman, E. 1970. Insect molting hormones in plants. *Recent Advances in Phytochemistry* 3:211–217.

Heiligenberg, W. 1965. A quantitative analysis of digging movements and their relationship to aggressive behaviour in cichlids. *Animal Behaviour* 13:163–170.

Heinichen, I. G. 1972. Preliminary notes on the suni, *Neotragus moschatus,* and the red duiker, *Cephalophus natalensis. Zoologica Africana* 7:157–165.

Heinlein, E. F. W., Holden, R. D., and Yoon, R. M. 1966. Comparative responses of horses and sheep to different forms of alfalfa hay. *Journal of Animal Science* 25:740–743.

Heisler, I. L. 1981. Offspring quality and the polygyny threshold: A new model for the "sexy son hypothesis." *American Naturalist* (in press).

Hendrichs, H. 1972. Beobachtungen und Untersuchungen zur Ökologie und Ethologie, insbesondere zur sozialen Organisation, ostafricanischer Säugetiere. *Zeitschrift für Tierpsychologie* 30:146–189.

———. 1975a. Changes in a population of dikdik, *Madoqua (Rhynchotragus) kirki* (Günther 1880). *Zeitschrift für Tierpsychologie* 30:55–69.
———. 1975b. Observations on a population of Bohor reedbuck, *Redunca redunca* (Pallas 1767). *Zeitschrift für Tierpsychologie* 38:44–54.
Hendrichs, H., and Hendrichs, U. 1971. Freilanduntersuchungen zur Ökologie und Ethologie der Zwergantilope *Madoqua (Rhynchotragus) kirki,* Günther 1880. In *Dikdik und elephanten,* ed. H. Hendrichs and U. Hendrichs, pp. 9–75. Munchen: Piper-Verlag.
Henry, C. S. 1972. Eggs and repagula of *Ululodes* and *Ascaloptynx* (Neuroptera: Ascalaphidae): A comparative study. *Psyche, Cambridge* 79:1–22.
Hensley, M. M., and Cope, J. B. 1951. Further data on removal and repopulation of the breeding birds in a spruce-fir forest community. *Auk* 68:483–493.
Henwood, K. 1975. Infrared transmittance as an alternative thermal strategy in the desert beetle *Onymacris plana. Science* 189:993–994.
Herald, E. S. 1959. From pipefish to seahorse—A study of phylogenetic relationships. *Proceedings of the California Academy of Sciences* 29:465–473.
Herbers, J. M. 1979. The evolution of sex-ratio strategies in hymenopteran societies. *American Naturalist* 114:818–834.
Heusser, H. 1969. Der rudimentäre Ruf der männlichen Erdkröte *(Bufo bufo). Salamandra* 5:46–56.
Hickey, J. J. 1955. Some American population research on gallinaceous birds. In *Recent studies in avian biology,* ed. A. Wolfson, pp. 326–396. Urbana: University of Illinois Press.
Hickling, R. A. O. 1959. The burrow excavation phase in the breeding cycle of the sand martin, *Riparia riparia. Ibis* 101:497–500.
Hildèn, O. 1965a. Habitat selection in birds. A review. *Annales Zoologici Fennici* 2:53–75.
———. 1965b. Zur Brutbiologie des Temminckstrandläufers, *Calidris temminckii* (Leisl.). *Ornis Fennica* 42:1–5.
———. 1975. Breeding system of Temminck's stint *Calidris temminckii. Ornis Fennica* 52:117–146.
Hildèn, O., and Vuolanto, S. 1972. Breeding biology of the red-necked phalarope *Phalaropus lobatus* in Finland. *Ornis Fennica* 49:57–85.
Hill, J. L. 1974. *Peromyscus:* Effect of early pairing on reproduction. *Science* 186:1042–1044.
Hinde, R. A. 1952. The behaviour of the great tit *(Parus major)* and some other related species. *Behaviour Supplement* 2:1–201.
———. 1956. The biological significance of the territories of birds. *Ibis* 98:340–369.
———. 1959. Behaviour and speciation in birds and lower vertebrates. *Biological Reviews* 34:85–128.
———. 1968. Dichotomies in the study of development. In *Genetic and environmental influences on behavior. Eugenics Society Symposium,* vol. 4, ed. J. M. Thoday and A. S. Parkes, pp. 3–14. Edinburgh: Oliver & Boyd.
———. 1970. *Animal behaviour. A synthesis of ethology and comparative psychology.* 2nd ed. New York: McGraw-Hill.
———. 1974. *Biological bases of human social behaviour.* New York: McGraw-Hill.
Hinde, R. A., and Fisher, J. 1951. Further observations on the opening of milk bottles by birds. *British Birds* 44:393–396.
Hingston, R. W. G. 1927. Field observations on spider mimics. *Proceedings of the Zoological Society of London* 1927(2):841–858.
———. 1933. *The meaning of animal colour and adornment.* London: E. Arnold.
Hinton, H. E., and Dunn, A. M. S. 1967. *Mongooses. Their natural history and behaviour.* Edinburgh: Oliver & Boyd.
Hirai, K., Shorey, H. H., and Gaston, L. K. 1978. Competition among courting male moths: Male-to-male inhibitory pheromone. *Science* 202:644–645.
Hirschfield, M. F., and Tinkle, D. W. 1975. Natural selection and the evolution of reproductive effort. *Proceedings of the National Academy of Sciences* 72:2227–2231.
Hirth, D. H., and McCullough, D. R. 1977. Evolution of alarm signals in ungulates with special reference to white-tailed deer. *American Naturalist* 111:31–42.
Hjorth, I. 1970. Reproductive behavior in Tetraonidae, with special reference to males. *Viltrevy* 7:183–596.
Hladik, C. M. 1975. Ecology, diet, and social patterning in Old and New World primates. In *Socio-ecology and psychology of primates,* ed. R. H. Tuttle, pp. 3–35. The Hague: Mouton.
———. 1978a. Adaptive strategies of primates in relation to leaf-eating. In *The ecology of arboreal folivores,* ed. G. G. Montgomery, pp. 373–395. Washington, D.C.: Smithsonian Institution Press.
———. 1978b. Phenology of leaf production in rain forest of Gabon: Distribution and com-

position of food for folivores. In *The ecology of arboreal folivores,* ed. G. G. Montgomery, pp. 51–71. Washington, D.C.: Smithsonian Institution Press.

Hladik, C. M., and Charles-Dominique, P. 1974. The behaviour and ecology of the sportive lemur *(Lepilemur mustelinus)* in relation to its dietary peculiarities. In *Prosimian biology,* ed. R. D. Martin, G. A. Doyle, and A. C. Walker, pp. 23–37. London: Duckworth.

Hobson, E. S. 1971. Cleaning symbiosis among California inshore fishes. *California Fisheries Bulletin* 69:491–523.

Hodgdon, H. E., and Larson, J. S. 1973. Some sexual differences in behaviour within a colony of marked beavers. *(Castor canadensis). Animal Behaviour* 21:147–52.

Hoedeman, J. J. 1974. *Naturalist's guide to fresh-water aquarium fish.* New York: Sterling.

Hoese, H. D. 1971. Dolphin feeding out of water in salt marsh. *Journal of Mammalogy* 52:222–223.

Hoffman, R. S. 1956. Observations on a sooty grouse population at Sage Hen Creek, California. *Condor* 58:321–337.

Hogan-Warburg, A. J. 1966. Social behavior of the ruff *Philomachus pugnax* (L.). *Ardea* 54:109–229.

Höhn, E. O. 1967. Observations on the breeding biology of Wilson's phalarope (Steganopus tricolor) in central Alberta. *Auk* 84:220–244.

Hölldobler, B. 1974. Home range orientation and territoriality in harvesting ants. *Proceedings of the National Academy of Sciences* 71:3274–3277.

———. 1976a. Recruitment behavior, home range orientation and territoriality in harvester ants, *Pogonomyrmex. Behavioral Ecology and Sociobiology* 1:3–44.

———. 1976b. The behavioral ecology of mating in harvester ants (Hymenoptera: Formicidae: *Pogonomyrmex*). *Behavioral Ecology and Sociobiology* 1:405–423.

———. 1976c. Tournaments and slavery in a desert ant. *Science* 192:912–914.

———. 1979a. Territoriality in ants. *Proceedings of the American Philosophical Society* 123:211–218.

———. 1979b. Territories of the African weaver ant *(Oecophylla longinoda* [Latreille]). A field study. *Zeitschrift für Tierpsychologie* 51:201–213.

Hölldobler, B., and Wilson, E. O. 1977. The number of queens: An important trait in ant evolution. *Die Naturwissenschaften* 64:8–15.

Holm, C. H. 1973. Breeding sex ratios, territoriality, and reproductive success in the red-winged blackbird *(Agelaius phoeniceus). Ecology* 49:682–694.

Holmes, R. T. 1970. Differences in population density, territoriality, and food supply of dunlin on arctic and subarctic tundra. In *Animal populations in relation to their food resources,* ed. A. Watson, pp. 303–317. Oxford: Blackwell Scientific.

Holmes, W. 1940. The colour changes and colour patterns of *Sepia officinalis* L. *Proceedings of the Zoological Society of London, Series A* 110:17–35.

Holmes, W. G. 1979. Social behavior and foraging strategies of hoary marmots *(Marmota caligata)* in Alaska. Ph.D. thesis, University of Washington, Seattle.

Hoogland, J. L. 1979a. Aggression, ectoparasitism, and other possible costs of prairie dog (Sciuridae, *Cynomys* spp.) coloniality. *Behaviour* 69:1–35.

———. 1979b. The effect of colony size on individual alertness of prairie dogs (Sciuridae: *Cynomys* spp.). *Animal Behaviour* 27:394–407.

———. 1980a. Nepotism and cooperative breeding in the black-tailed prairie dog (Sciuridae: *Cynomys ludovicianus).* In *Natural selection and social behavior: Recent research and new theory,* ed. R. D. Alexander and D. W. Tinkle. Portland, Oreg.: Chiron Press.

———. 1980b. The evolution of coloniality in white-tailed and black-tailed prairie dogs. (Sciuridae: *Cynomys leucurus* and *C. ludovicianus). Ecological Monographs.*

Hoogland, J. L., and Sherman, P. W. 1976. Advantages and disadvantages of bank swallow *(Riparia riparia)* coloniality. *Ecological Monographs* 46:33–58.

Hoogland, R., Morris, D., and Tinbergen, N. 1956–1957. The spines of sticklebacks *(Gasterosteus* and *Pygosteus)* as means of defence against predators *(Perca* and *Esox). Behaviour* 10:205–236.

Horn, H. S. 1968. The adaptive significance of colonial nesting in the Brewer's blackbird *(Euphagus cyanocephalus). Ecology* 49:682–694.

Horn, M. H. 1970. The swimbladder as a juvenile organ in stromateoid fishes. *Breviora,* no. 359:1–9.

Horner, B. E. 1947. Paternal care of young mice of the genus *Peromyscus. Journal of Mammalogy* 28:31–36.

Houck, L. D. 1977. Life history patterns and reproductive biology of neotropical salamanders. In *The reproductive biology of amphibians,* ed. D. H. Taylor and S. I. Guttman, pp. 43–72. New York: Plenum.

Howard, H. E. 1920. *Territory in bird life.* London: John Murray.
Howard, R. D. 1978a. The evolution of mating strategies in bullfrogs, *Rana catesbeiana. Evolution* 32:850–871.
———. 1978b. The influence of male-defended oviposition sites on early embryo mortality in bullfrogs. *Ecology* 59:789–798.
———. 1980. Mating behaviour and mating success in woodfrogs, *Rana sylvatica. Animal Behaviour* 28:705–716.
Howe, H. F. 1977. Sex-ratio adjustment in the common grackle. *Science* 198:744–746.
Howe, H. F., and Estabrook, G. F. 1977. On intraspecific competition for avian dispersers in tropical trees. *American Naturalist* 111:817–832.
Howse, P. E. 1968. On the division of labour in the primitive termite *Zootermopsis nevadensis* (Hagen). *Insectes Sociaux* 15:45–50.
Hudgens, G. A., Denenberg, V. H., and Zarrow, M. X. 1967. Mice reared with rats: Relations between mothers' activity level and offsprings' behavior. *Journal of Comparative and Physiological Psychology* 63:304–308.
Hudson, R. C. L. 1977. Preliminary observations on the behaviour of the gobiid fish *Signigobius biocellatus* Hoese and Allen, with particular reference to its burrowing behaviour. *Zeitschrift für Tierpsychologie* 43:214–220.
Huey, R. B., Pianka, E. R., Egan, M. E., and Coons, L. W. 1974. Ecological shifts in sympatry: Kalahari fossorial lizards *(Typhlosaurus). Ecology* 55:304–316.
Humphries, D. A., and Driver, P. M. 1967. Erratic displays as a device against predation. *Science* 156:1767–1768.
———. 1970. Protean defence by prey animals. *Oecologia* 5:285–302.
Hunkeler, C., Bourlière, F., and Bertrand, M. 1972. Le comportement social de la mone de Lowe *(Cercopithecus campbelli lowei). Folia Primatologica* 17:218–236.
Hunsperger, R. W., and Bucher, V. M. 1967. Affective behaviour produced by electrical stimulation in the forebrain and brain stem of the cat. *Progress in Brain Research* 27:125–127.
Hunt, G. L., Jr. 1980. Mate selection and mating systems in seabirds. In *Behavior of marine animals. Current perspectives in research: Marine birds,* ed. H. Winn, B. Olla, and J. Burger. New York: Plenum.
Hunt, G. L., Jr., and Hunt, M. W. 1976. Gull and chick survival: The significance of growth rates, timing of breeding, and territory size. *Ecology* 57:62–75.
Hunter-Jones, P. 1960. Fertilization of eggs of the desert locust by spermatozoa from successive copulations. *Nature* 185:336.
Hussell, D. J. T. 1972. Factors affecting clutch size in Arctic passerines. *Ecological Monographs* 42:317–364.
Hustings, E. L. 1965. Survival and breeding structure in a population of *Ambystoma maculatum. Copeia* 1965:352–362.
Hutchinson, G. E. 1957. Concluding remarks. *Cold Spring Harbor Symposia on Quantitative Biology* 22:415–427.
———. 1965. *The ecological theatre and the evolutionary play.* New Haven: Yale University Press.
———. 1967. *A treatise on limnology.* Vol. II. *Introduction to lake biology and the limnoplankton.* New York: Wiley.
Hutchinson, G. E., and MacArthur, R. H. 1959. On the theoretical significance of aggressive neglect in interspecific competition. *American Naturalist* 93:133–134.
Hutchinson, R. R., and Renfrew, J. W. 1966. Stalking attack and eating behaviors elicited from the same sites in the hypothalamus. *Journal of Comparative and Physiological Psychology* 61:360–367.
Hutchinson, R. R., Ulrich, R. E., and Azrin, N. H. 1965. Effects of age and related factors on the pain aggression reaction. *Journal of Comparative and Physiological Psychology* 59:365–369.
Hutchison, J. B., and Lovari, S. 1976. Effects of male aggressiveness on behavioural transitions in the reproductive cycle of the barbary dove. *Behaviour* 59:296–318.
Huxley, J. S. 1934. A natural experiment on the territorial instinct. *British Birds* 27:270–277.
———. 1938. Darwin's theory of sexual selection and the data subsumed by it, in the light of recent research. *American Naturalist* 72:416–433.
Hynes, H. B. N. 1950. The food of freshwater sticklebacks *(Gasterosteus aculeatus* and *Pygosteus pungitius)* with a review of methods used in studies of the food of fishes. *Journal of Animal Ecology* 19:36–58.
Idyll, C. P. 1969. Grunion. The fish that spawns on land. *National Geographic* 135:714–723.
Iersel, J. J. A. van. 1953. An analysis of parental behaviour of the male three-spined stickleback *(Gasterosteus aculeatus* L.). *Behaviour Supplement* 3:1–159.

Ikan, R., Gottlieb, R., Bergmann, E. D., and Ishay, J. 1969. The pheromone of the queen of the oriental hornet, *Vespa orientalis*. *Journal of Insect Physiology* 15:1709–1712.

Immelmann, K. 1973. Role of the environment in reproduction as source of 'predictive' information. In *Breeding biology of birds*, ed. D. S. Farner, pp. 121–147. Washington, D.C.: National Academy of Sciences.

Imms, A. D. 1957. *A general textbook of entomology*. 9th ed. London: Methuen.

Ingram, C. 1959. The importance of juvenile cannibalism in the breeding biology of certain birds of prey. *Auk* 76:218–226.

———. 1962. Cannibalism by nestling short-eared owls. *Auk* 79:714.

Irons, W. 1979. Cultural and biological success. In *Evolutionary biology and human social behavior: An anthropological perspective*, ed. N. A. Chagnon and W. Irons, pp. 257–272. N. Scituate, Mass.: Duxbury Press.

Ishay, J. 1975. Caste determination by social wasps: Cell size and building behaviour. *Animal Behaviour* 23:425–431.

Ishay, J., Bytinski-Salz, H., and Shulor, A. 1967. Contributions to the bionomics of the oriental hornet, *Vespa orientalis*. *Israel Journal of Entomology* 2:45–106.

Ishay, J., Ikan, R., and Bergmann, E. D. 1965. The presence of pheromones in the oriental hornet *(Vespa orientalis* F.). *Journal of Insect Physiology* 11:1307–1309.

Itani, J. 1959. Paternal care in the wild Japanese monkey, *Macaca fuscata fuscata*. *Primates* 2:61–93.

———. 1972. A preliminary essay on the relationship between social organization and incest avoidance in non-human primates. In *Primate socialization*, ed. F. E. Poirier, pp. 165–171. New York: Random House.

Itani, J., and Suzuki, A. 1967. The social unit of chimpanzees. *Primates* 8:355–381.

Ito, Y. 1960. Territorialism and residentiality in a dragonfly, *Orthetrum albistylum speciosum* Uhler (Odonata: Anisoptera). *Annals of the Entomological Society of America* 53:851–853.

Itzkowitz, M. 1974. A behavioural reconnaissance of some Jamaican reef fishes. *Zoological Journal of the Linnaean Society of London* 55:87–118.

———. 1977. Interrelationships of dominance and territorial behaviour in the pupfish *Cyprinodon variegatus*. *Behavioural Processes* 2:383–391.

Ivey, M. E., and Bardwick, J. M. 1968. Patterns of affective fluctuation in the menstrual cycle. *Psychosomatic Medicine* 30:336–345.

Iwamoto, T. 1979. Feeding ecology. In *Ecological and sociological studies of gelada baboons*, ed. M. Kawai, pp. 279–330. *Contributions to Primatology* 16:1–344.

Iwao, S. 1968. A new regression method for analyzing the aggregation pattern of animal populations. *Researches on Population Ecology* 10:1–20.

Jackson, J. B. C. 1968. Bivalves: Spatial and size-frequency distributions of two intertidal species. *Science* 161:479–480.

Jacob, F., and Monod, J. 1961. Genetic regulatory mechanisms in the synthesis of proteins. *Journal of Molecular Biology* 2:318–356.

Jacobs, M. E. 1955. Studies on territorialism and sexual selection in dragonflies. *Ecology* 36:566–586.

Jacobson, J., and Crosby, D. G., eds. 1971. *Naturally occurring insecticides*. New York: Dekker.

James, W. H. 1975. The distributions of the combinations of the sexes in mammalian litters. *Genetics Research* 26:45–53.

———. 1976. The combinations of the sexes in twin lambings. *Genetics Research* 28:277–280.

James, W. O. 1950. Alkaloids in the plant. In *The alkaloids: Chemistry and physiology*, vol. 1, ed. R. H. F. Manske and H. L. Holms, pp. 16–90. New York: Academic Press.

James, W. T. 1951. Social organization among dogs of different temperaments: Terriers and beagles reared together. *Journal of Comparative and Physiological Psychology* 44:71–77.

Janis, C. 1976. The evolutionary strategy of the Equidae and the origins of rumen and caecal digestion. *Evolution* 30:757–774.

Janzen, D. H. 1964. Notes on the behavior of four subspecies of the carpenter bee, *Xylocopa (Notoxylocopa) tabaniformis*, in Mexico. *Annals of the Entomological Society of America* 57:296–301.

———. 1966. Coevolution of mutualism between ants and acacias in Central America. *Evolution* 20:249–275.

———. 1967a. Interaction of the bull's horn acacia (*Acacia cornigera* L.) with an ant inhabitant (*Pseudomyrmex ferruginea* F. Smith) in eastern Mexico. *University of Kansas Science Bulletin* 47:315–558.

———. 1967b. Synchronization of sexual reproduction of trees within the dry season in Central America. *Evolution* 21:620–637.

———. 1970a. Altruism by coatis in the face of predation by *Boa constrictor*. *Journal of Mammalogy* 51:387–389.

———. 1970b. Herbivores and the number of tree species in tropical forests. *American Naturalist* 104:501–528.

———. 1971. Seed predation by animals. *Annual Review of Ecology and Systematics* 2:465–492.

———. 1972. Protection of *Barteria* (Passifloraceae) by *Pachysima* ants (Pseudomyrmecinae) in a Nigerian rain forest. *Ecology* 53:885–92.

———. 1974. Ephiphytic myrmecophytes: Mutualism by ants feeding plants. *Biotropica* 6:237–259.

———. 1977. A note on optimal mate selection by plants. *American Naturalist* 111:365–371.

Janzen, D. H., Juster, H. B., and Liener, I. E. 1976. Insecticidal action of the phytohemagglutinin in black beans on a bruchid beetle. *Science* 192:795–796.

Jarman, M. V. 1979. Impala social behaviour: Territory, hierarchy, mating, and the use of space. *Zeitschrift für Tierpsychologie, Supplement*. 21:1–93.

Jarman, P. J. 1974. The social organization of antelope in relation to their ecology. *Behaviour* 48:215–267.

Jarman, P. J., and Jarman, M. V. 1974. Impala behaviour and its relevance to management. In *The behaviour of ungulates and its relation to management*, ed. V. Geist and F. Walther, pp. 871–881. Moreges: International Union for Conservation of Nature and Natural Resources.

Jarvik, L. F., Klodin, V., and Matsuyama, S. S. 1973. Human aggression and the extra Y chromosome: Fact of fantasy? *American Psychologist* 28:674–682.

Jay, P. C. 1965. The common langur of North India. In *Primate behavior: Field studies of monkeys and apes*, ed. I. Devore, pp. 197–249. New York: Holt, Rinehart and Winston.

Jaycox, E. R. 1967. Territorial behaviour among males of *Anthidium banningense* (Hymenoptera: Megachilidae). *Journal of the Kansas Entomological Society* 40:565–570.

Jeanne, R. L. 1970. Chemical defense of brood by a social wasp. *Science* 168:1465–1466.

———. 1972. Social biology of the neotropical wasp *Mischocyttarus drewseni*. *Bulletin of the Museum of Comparative Zoology, Harvard University* 144:63–150.

Jenkins, D. 1957. The breeding of the red-legged partridge. *Bird Study* 4:97–100.

———. 1961a. Population control in protected partridges *(Perdix perdix)*. *Journal of Animal Ecology* 30:235–258.

———. 1961b. Social behaviour in the partridge *(Perdix perdix)*. *Ibis* 103a:155–188.

Jenkins, D., Watson, A., and Miller, G. R. 1963. Population studies on red grouse *(Lagopus lagopus scoticus* (Lath.)) in northeast Scotland. *Journal of Animal Ecology* 32:317–376.

———. 1967. Population fluctuations in the red grouse, *Lagopus lagopus scoticus* (Lath.), in northeast Scotland. *Journal of Animal Ecology* 36:97–122.

Jenkins, T. M. 1969. Social structure, position choice and microdistribution of two trout species *(Salmo trutta* and *Salmo gairdneri)* resident in mountain streams. *Animal Behaviour Monographs* 2:57–123.

Jenni, D. A. 1974. Evolution of polyandry in birds. *American Zoologist* 14:129–144.

Jenni, D. A., and Betts, B. J. 1978. Sex differences in nest construction, incubation, and parental behaviour in the polyandrous American jacana (*Jacana spinosa*). *Animal Behaviour* 26:207–218.

Jenni, D. A., and Collier, G. 1972. Polyandry in the American jacana (*Jacana spinosa*). *Auk* 89:743–765.

Jennings, J. B. 1965. *Feeding, digestion and assimilation in animals*. Oxford: Pergamon Press.

Jennrich, R. I., and Turner, F. B. 1969. Measurement of non-circular home range. *Journal of Theoretical Biology* 22:227–237.

Jenssen, T. A., and Preston, W. B. 1968. Behavioral responses of the male green frog, *Rana clamitans*, to its recorded call. *Herpetologica* 24:181–182.

Jewell, P. A. 1966. The concept of home range in mammals. *Symposia of the Zoological Society of London* 18:85–109.

Johns, J. E. 1969. Field studies of Wilson's phalarope. *Auk* 86:660–670.

Johnsgaard, P. A. 1965. *Handbook of waterfowl behavior*. Ithaca, N.Y.: Cornell University Press.

———. 1973. *Grouse and quails of North America*. Lincoln: University of Nebraska Press.

———. 1975. *Waterfowl of North America*. Bloomington: Indiana University Press.

Johnsgaard, P. A., and Buss, I. O. 1956. Waterfowl sex ratios during spring in Washington State and their interpretation. *Journal of Wildlife Management* 20:384–388.

Johnson, C. 1964. The evolution of territoriality in the Odonata. *Evolution* 18:89–92.

Johnson, R. E. 1965. Reproductive activities of rosy finches, with special reference to Montana. *Auk* 82:190–205.

Johnson, R. F., and Sloan, N. F. 1978. White pelican production and survival of young at Chase Lake National Wildlife Refuge, North Dakota. *Wilson Bulletin* 90:346–352.
Johnson, R. N. 1972. *Aggression in man and animals.* Philadelphia: Saunders.
Johnson, R. R., and West, G. C. 1973. Fat content, fatty acid composition and estimates of energy metabolism of adelie penguins (*Pygoscelis adeliae*) during the early breeding season fast. *Comparative Biochemistry and Physiology* 45:709–719.
Johnston, N. C., Law, J. H., and Weaver, N. 1965. Metabolism of 9-ketodec-2-enoic acid by worker honeybees (*Apis mellifera* L.). *Biochemistry* 4:1615–1621.
Jolly, A. 1966. *Lemur behavior: A Madagascar field study.* Chicago: University of Chicago Press.
———. 1972. *The evolution of primate behavior.* New York: Macmillan.
Jones, C., and Sabater Pi, J. 1971. Comparative ecology of *Gorilla gorilla* (Savage and Wyman) and *Pan troglodytes* (Blumenbach) in Rio Muni, West Africa. *Biblioteca Primatologica* 13:1–96.
Jones, D. A. 1979. Chemical defense: Primary or secondary function? *American Naturalist* 113:445–451.
Jones, D. A., Parsons, J., and Rothschild, M. 1962. Release of hydrocyanic acid from crushed tissues of all stages in the life-cycle of species of the Zygaeninae (Lepidoptera). *Nature* 193:52–53.
Jones, E. W. 1955–1956. Ecological studies on the rain forest of southern Nigeria. IV. The plateau forest of the Okomu Forest Reserve. *Journal of Ecology* 43:564–594; 44:83–117.
Jones, P. J., and Ward, P. 1979. A physiological basis for colony desertion by red-billed queleas (*Quelea quelea*). *Journal of Zoology, London* 189:1–19.
Joubert, E. 1972. The social organization and associated behaviour in the Hartmann zebra, *Equus zebra hartmannae*. *Madoqua* 6:17–56.
Joyner, D. E. 1976. Effects of interspecific nest parasitism by redheads and ruddy ducks. *Journal of Wildlife Management* 40:33–38.
Jungius, H. 1971. The biology and behaviour of the reedbuck (*Redunca arundinum* Boddaert 1785) in the Kruger National Park. *Mammalia Depicta*. Hamburg: Paul Parey.
Kaiser, P. 1954. Über die Funktion der Mandibeln bei den Soldaten von *Neocapritermes opacus* (Hagen). *Zoologischer Anzeiger* 152:228–234.
Kalela, O. 1954. Über den Revierbesitz bei Vögeln und Säugetieren als populationsökologisher Faktor. *Annales Zoologici Societatis Zoologicae Botanicae Fennicae "Vanamo" (Helsinki)* 16:1–48.
Källender, H. 1974. Advancement of laying of great tits by the provision of food. *Ibis* 116:365–367.
Kalmijn, A. J. 1971. The electric sense of sharks and rays. *Journal of Experimental Biology* 55:371–383.
Kannowski, P. B. 1963. The flight activities of formicine ants. *Symposia Genetica Biologia Italica* 12:74–102.
Karli, P. 1956. The Norway rat's killing response to the white mouse. *Behaviour* 10:81–103.
Karlstrom, E. L. 1962. The toad genus *Bufo* in the Sierra Nevada of California: Ecological and systematic relationships. *University of California Publications in Zoology* 62:1–104.
Karplus, I., Szlep, R., and Tsurnama, M. 1972. Associative behavior of the fish *Cryptocentrus cryptocentrus* (Gobiidae) and the pistol shrimp *Alpheus djiboutensis* (Apheidae) in artificial burrows. *Marine Biology* 15:95–104.
———. 1974. The burrows of alpheid shrimp associated with gobiid fish in the northern Red Sea. *Marine Biology* 24:259–268.
Karplus, I., and Tuvia, S. B. 1979. Warning signals of *Cryptocentrus steinitzi* (Pisces, Gobiidae) and predator models. *Zeitschrift für Tierpsychologie* 51:225–232.
Kaufman, D. W. 1974. Adaptive coloration in *Peromyscus polionotus*. Experimental selection by owls. *Journal of Mammalogy* 55:271–283.
Kaufmann, H. 1970. *Aggression and altruism.* New York: Holt, Rinehart and Winston.
Kaufmann, J. H. 1962. Ecology and social behavior of the coati, *Nasua narica*, on Barro Colorado Island, Panama. *University of California Publications in Zoology* 60:95–222.
———. 1967. Social relations of adult males in a free-ranging band of rhesus monkeys. In *Social communication among primates*, ed. S. A. Altmann, pp. 73–98. Chicago: University of Chicago Press.
Kawai, M., ed. 1979. Ecological and sociological studies of gelada baboons. *Contributions to Primatology* 15:1–344.
Kawamichi, T., and Kawamichi, M. 1979. Spatial organization and territory of tree shrews (*Tupaia glis*). *Animal Behaviour* 27:381–393.
Kawanaka, K., and Nishida, T. 1975. Recent advances in the study of interunit-group relationships and social structure of wild chimpanzees of the Mahali Mountains. *Proceedings*

of the 5th Congress of the International Primatological Society, pp. 173–186.
Keenleyside, M. H. A. 1972a. The behaviour of *Abudefduf zonatus* (Pisces, Pomacentridae). *Animal Behaviour* 20:763–774.
———. 1972b. Intraspecific intrusions into nests of spawning longear sunfish (Pisces: Centrarchidae). *Copeia* 1972:272–278.
———. 1978. Parental care behavior in fishes and birds. In *Contrasts in behavior*, ed. E. S. Reese and F. Lighter, pp. 1–19. New York: Wiley.
———. 1979. *Diversity and adaptation in fish behaviour. Zoophysiology and ecology*. Vol. 11. Berlin: Springer-Verlag.
Kelsall, J. P. 1968. *The migratory barren-ground caribou of Canada*. Ottawa: Department of Indian Affairs and Northern Development.
Kemp, A. C. 1971. Some observations on the sealed-in nesting method of hornbills (family: Bucerotidae). *Ostrich Supplement* 8:149–155.
———. 1978. A review of the hornbills: Biology and adaptive radiation. *Living Bird* 17:105–136.
Kemp, G. A., and Keith, L. B. 1970. Dynamics and regulation of red squirrel (*Tamiasciurus hudsonicus*) populations. *Ecology* 51:763–779.
Kendeigh, S. C. 1941. Territorial and mating behavior of the house wren. *Illinois Biological Monographs* 18:1–120.
———. 1972. Energy control on size limits in birds. *American Naturalist* 106:79–88.
Kendeigh, S. C., and Baldwin, S. P. 1937. Factors affecting yearly abundance of passerine birds. *Ecological Monographs* 7:91–124.
Kenward, R. E. 1978. Hawks and doves: Factors affecting success and selection in goshawk attacks on woodpigeons. *Journal of Animal Ecology* 47:449–460.
Kenyon, K. W., and Rice, D. W. 1959. Life history of the Hawaiian monk seal. *Pacific Science* 13:215–252.
Kepler, C. B. 1969. Breeding biology of the blue-faced booby on Green Island, Kure Atoll. *Publications of the Nuttall Ornithological Club* 8:1–97.
Kerbert, C. 1904. Zur Fortpflanzung von *Megalobatrachus maximus* Schlegel. *Zoologischer Anzeiger* 27:305–316.
Kerr, W. E. 1950a. Evolution of the mechanism of caste determination in the genus *Melipona*. *Evolution* 4:7–13.
———. 1950b. Genetic determination of castes in the genus *Melipona*. *Genetics* 35:143–152.
———. 1961. Acasalamento de raihnas com vários machos em duas espécies da Tribu Attini (Hymenoptera, Formicoidea). *Revista Brasileira de Biologia* 21:45–48.
———. 1969. Some aspects of the evolution of social bees (Apidae). In *Evolutionary biology*, vol. 4, ed. M. K. Hecht, W. C. Steere, and B. Wallace, pp. 119–175. New York: Plenum.
Kerr, W. E., and Hebling, N. J. 1964. Influence of the weight of worker bees on division of labor. *Evolution* 18:267–270.
Kerr, W. E., and Nielsen, R. A. 1966. Evidences that genetically determined *Melipona* queens can become workers. *Genetics* 54:859–866.
Kerr, W. E., and Santos Netos, G. R. dos 1956. Contribuicão para o conhecimento da bionomia dos Meliponini. 5. Divisão de trabalho entre operarias de *Melipona quadrifasciata quadrifasciata* Lep. *Insectes Sociaux* 3:423–430.
Kerr, W. E., Stort, A. C., and Montenegro, M. J. 1966. Importância de alguns fatôres ambientais na determinacão das castas do gênero *Melipona*. *Anais da Academia Brasileira de Ciências* 38:149–168.
Kerr, W. E., Zucchi, R., Nakadaira, J. T., and Botolo, J. E. 1962. Reproduction in the social bees. *Journal of the New York Entomological Society* 70:265–276.
Kershaw, K. D. 1964. *Quantitative and dynamic ecology*. London: Edward Arnold.
Kesner, R. P., and Keiser, G. 1973. Effects of midbrain reticular lesions upon aggression in the rat. *Journal of Comparative and Physiological Psychology* 84:194–206.
Kessel, E. L. 1955. The mating activities of balloon flies. *Systematic Zoology* 4:97–104.
Kettlewell, B. 1973. *The evolution of melanism: The study of a recurring necessity*. Oxford: Oxford University Press (Clarendon Press).
Kiley-Worthington, M. 1965. The waterbuck (*Kobus defassa* Ruppell 1835 and *K. ellipsyprymnus* Ogilby 1833) in East Africa: Spatial distribution: A study of the sexual behaviour. *Mammalia* 29:176–204.
Kimura, M. 1979. The neutral theory of molecular evolution. *Scientific American* 241(5):98–126.
Kimura, M., and Ohta, T. 1971. *Theoretical aspects of population genetics*. Princeton: Princeton University Press.
King, J. A. 1955. Social behavior, social organization, and population dynamics in a black-tailed prairiedog town in the Black Hills of South Dakota. *Contributions from the Laboratory of Vertebrate Biology, University of Michigan, Ann Arbor*, no. 67:1–123.

———. 1957. Relation between early social experience and adult aggressive behaviour in inbred mice. *Journal of Genetics and Psychology* 90:151–166.

———. 1959. The social behavior of prairie dogs. *Scientific American* 201(4):128–140.

King, J. A., and Gurney, N. L. 1954. Effect of early social experience on adult aggressive behavior in C_{57}BL/10 mice. *Journal of Comparative and Physiological Psychology* 47:326–330.

King, J. R. 1973. Energetics of reproduction in birds. In *Breeding biology of birds*, ed. D. S. Farner, pp. 78–107. Washington, D.C.: National Academy of Sciences.

Kinzey, W. G., Rosenberger, A. L., Heisler, P. S., Prowse, D. L., and Trilling, J. S. 1977. A preliminary field investigation of the yellow-handed titi monkey, *Callicebus torquatus torquatus*, in northern Peru. *Primates* 18:159–181.

Kirkpatrick, M. MS. Sexual selection and female choice. Submitted to *Evolution*.

Kislak, J. W., and Beach, F. A. 1955. Inhibition of aggressiveness by ovarian hormones. *Endocrinology* 56:684–692.

Kistchinski, A. A. 1975. Breeding biology and behaviour of the grey phalarope *Phalaropus fulicarus* in east Siberia. *Ibis* 117:285–301.

Kitchen, D. W. 1974. Social behavior and ecology of the pronghorn. *Wildlife Monographs* 38:1–96.

Klahn, J. E. 1979. Philopatric and nonphilopatric foundress associations in the social wasp *Polistes fuscatus*. *Behavioral Ecology and Sociobiology* 5:417–424.

Kleiman, D. G. 1977a. Monogamy in mammals. *Quarterly Review of Biology* 52:39–69.

———, ed. 1977b. *Biology and conservation of the Callitrichidae*. Washington, D.C.: Smithsonian Institution Press.

Kleiman, D. G., and Brady, C. A. 1978. Coyote behavior in the context of recent canid research: Problems and perspectives. In *Coyotes. Biology, behavior, and management*, ed. M. Bekoff, pp. 163–188. New York: Academic Press.

Kleiman, D. G., and Eisenberg, J. F. 1973. Comparisons of canid and felid social systems from an evolutionary perspective. *Animal Behaviour* 21:637–659.

Klein, D. R. 1970. Food selection by North American deer and their response to over-utilization of preferred plant species. In *Animal populations in relation to their food resources*, ed. A. Watson, pp. 25–44. Oxford: Blackwell Scientific.

Klein, L. L., and Klein, D. J. 1975. Social and ecological contrasts between four taxa of neotropical primates. In *Socio-ecology and psychology of primates*, ed. R. H. Tuttle, pp. 59–85. The Hague: Mouton.

Klimstra, W. D., and Roseberry, J. L. 1975. Nesting ecology of the bobwhite in southern Illinois. *Wildlife Monographs* 41:1–37.

Kling, A. 1972. Effects of amygdalectomy on social-affective behavior in nonhuman primates. In *The neurobiology of amygdala*, ed. B. E. Eleftheriou, pp. 511–536. New York: Plenum.

———. 1975. Brain lesions and aggressive behavior of monkeys in free living groups. In *Neural bases of violence and aggression*, ed. W. S. Fields and W. H. Sweet, pp. 146–158. St. Louis: Warren H. Green.

Kling, A., and Hutt, P. J. 1958. Effect of hypothalamic lesions on the amygdala syndrome in the cat. *American Medical Association Archives of Neurology and Psychiatry* 79:511–517.

Klingel, H. 1969. Reproduction in the plains zebra, *Equus burchelli boehmi:* Behaviour and ecological factors. *Journal of Reproduction and Fertility, Supplement* 6:339–345.

———. 1974a. A comparison of the social behaviour of the Equidae. In *The behaviour of ungulates and its relation to management*, ed. V. Geist and F. Walther, pp. 125–132. Moreges: International Union for Conservation of Nature and Natural Resources.

———. 1974b. Soziale Organisation und Verhalten des Grevy-Zebras (*Equus grevyi*). *Zeitschrift für Tierpsychologie* 36:37–70.

———. 1977. Observations on social organization and behaviour of African and Asiatic wild asses (*Equus africanus* and *E. hemiosus*). *Zeitschrift für Tierpsychologie* 44:323–331.

Klomp, H. 1970. The determination of clutch-size in birds: A review. *Ardea* 58:1–124.

Klopfer, P. H. 1969. *Habitats and territories. A study of the use of space by animals*. New York: Basic Books.

Klopfer, P. H., and Hatch, J. J. 1968. Experimental considerations. In *Animal communication: Techniques of study and results of research*, ed. T. A. Sebeok, pp. 31–43. Bloomington: Indiana University Press.

Kluijver, H. N. 1950. Daily routines of the great tit, *Parus major* L. *Ardea* 38:99–135.

———. 1951. The population ecology of the great tit. *Ardea* 9:1–135.

Kluijver, H. N., and Tinbergen, L. 1953. Territory and the regulation of density in titmice. *Archives Neerlandaises de Zoologie* 10:265–286.

Knapton, R. W., and Krebs, J. R. 1974. Settlement patterns, territory size, and breeding

density in the song sparrow (*Melospiza melodia*). *Canadian Journal of Zoology* 52:1413–1420.
Knerer, G. 1980. Evolution of halictine castes. *Die Naturwissensschaften* 67:133–135.
Knerer, G., and Atwood, C. E. 1966. Polymorphism in some Nearctic halictine bees. *Science* 152:1262–1263.
Knerer, G., and Plateaux-Quénu, C. 1966. Sur le polymorphisme des femelles chez quelques Halictinae (Insectes Hymĕnoptères) paléarctiques. *Compte Rendu de l'Académie des Sciences, Paris Séries D* 263:1759–1761.
Knopf, F.L. 1979. Spatial and temporal aspects of colonial nesting of white pelicans. *Condor* 81:353–363.
Kodric-Brown, A. 1977. Reproductive success and the evolution of breeding territories in pupfish (Cyprinodon). *Evolution* 31:750–766.
———. 1978. Establishment and defence of breeding territories in a pupfish (Cyprinodontidae: *Cyprinodon*). *Animal Behavior* 26:818–834.
Koenig, W. D., and Pitelka, F. A. 1979. Relatedness and inbreeding avoidance: Counterploys in the communally nesting acorn woodpecker. *Science* 206:1103–1105.
Koeppl, J. W., Slade, N. A., and Hoffman, R. S. 1975. A bivariate home range model with possible application to ethological data analysis. *Journal of Mammalogy* 56:81–90.
Koford, C. B. 1957. The vicũna and the puna. *Ecological Monographs* 27:153–219.
———. 1958. Prairie dogs, white faces, and blue grama. *Wildlife Monographs* 3:1–78.
———. 1966. Population changes in rhesus monkeys: Cayo Santiago, 1960–1964. *Tulane Studies in Zoology* 13:1–7.
Kogan, M. 1977. The role of chemical factors in insect–plant relationships. *Proceedings of the XV International Congress of Entomology*, pp. 211–227.
Koivisto, I. 1965. Behaviour of the black grouse, *Lyrurus tetrix* (L.), during the spring display. *Papers on Game Research, Helsinki* 26:1–60.
Kormondy, E. J. 1961. Territoriality and dispersal in dragonflies (Odonata). *Journal of the New York Entomological Society* 69:42–52.
Kortlandt, A. 1962. Chimpanzees in the wild. *Scientific American* 206(5):128–138.
———. 1965. How do chimpanzees use weapons when fighting leopards? *Yearbook of the American Philosophical Society*, pp. 327–332.
———. 1967. Experimentation with chimpanzees in the wild. In *Neue Ergebnisse der Primatologie. Progress in primatology*, ed. R. Schneider and H. J. Kuhn, pp. 208–224. Stuttgart: Gustav Fischer Verlag.
———. 1972. *New perspectives on ape and human evolution*. Amsterdam: Stichting voor Psychobiologie.
———. 1980. How might early hominids have defended themselves against large predators and food competitors? *Journal of Human Evolution* 9:79–112.
Kortlandt, A., and Kooij, M. 1963. Protohominid behaviour in primates. *Symposia of the Zoological Society of London* 10:61–88.
Kramer, D. L. 1973. Parental behavior in the blue gourami *Trichogaster trichopterus* (Pisces, Belontiidae) and its induction during exposure to varying numbers of eggs. *Behaviour* 47:14–32.
Krebs, C. J. 1978. *Ecology. The experimental analysis of distribution and abundance*. 2nd ed. New York: Harper & Row.
Krebs, J. R. 1970. The efficiency of courtship feeding in the blue tit *Parus caeruleus*. *Ibis* 112:108–110.
———. 1971. Territory and breeding density in the great tit, *Parus major* L. *Ecology* 52:2–22.
———. 1973. Social learning and the significance of mixed species flocks of chickadees (*Parus* spp.). *Canadian Journal of Zoology* 51:1275–1288.
———. 1974. Colonial nesting and social feeding as strategies for exploiting food resources in the great blue heron (*Ardea herodias*). *Behaviour* 51:99–131.
———. 1977. Song and territory in the great tit. In *Evolutionary ecology*, ed. B. Stonehouse and C. M. Perrins, pp. 47–62. London: Macmillan.
Krebs, J. R., MacRoberts, M. H., and Cullen, J. M. 1972. Flocking and feeding in the great tit *Parus major*—An experimental study. *Ibis* 114:507–530.
Krekorian, C. O. 1976. Field observations in Guyana on the reproductive biology of the spraying characid, *Copeina arnoldi* Regan. *American Midland Naturalist* 96:88–97.
Kruijt, J. P. 1964. Ontogeny of social behaviour in Burmese red junglefowl. *Behaviour Supplement* 12:1–20.
Kruijt, J. P., Vos, G. J. de, and Bossema, I. 1972. The arena system of black grouse [*Lyrurus tetrix tetrix* (L.)]. *Proceedings of the 15th International Ornithological Congress*, pp. 339–423.

Kruuk, H. 1964. Predators and anti-predator behaviour of the black-headed gull (*Larus ridibundus*). *Behaviour Supplement* 11:1–129.

———. 1972. *The spotted hyena: A study of predation and social behavior*. Chicago: University of Chicago Press.

———. 1975. *Hyaena*. London: Oxford University Press.

———. 1976. The biological function of gulls' attraction towards predators. *Animal Behaviour* 24:146–153.

Kruuk, H., and Turner, M. 1967. Comparative notes on predation by lion, leopard, cheetah, and wild dog in the Serengeti area, East Africa. *Mammalia* 31:1–27.

Kühme, W. 1965a. Freilandstudien zur Soziologie des Hyänenhundes (*Lycaon pictus lupinus* Thomas 1902). *Zeitschrift für Tierpsychologie* 22:495–541.

———. 1965b. Communal food distribution and division of labour in African hunting dogs. *Nature* 205:443–444.

Kullman, E. J. 1972. Evolution of social behavior in spiders (Araneae: Eresidae and Therididae). *American Zoologist* 12:419–426.

Kummer, H. 1968. *Social organization of hamadryas baboons: A field study*. Chicago: University of Chicago Press.

———. 1971. *Primate societies*. Chicago: Aldine-Atherton.

Kuo, Z. Y. 1967. *The dynamics of behavior development. An epigenetic view*. New York: Random House.

Kurland, J. A. 1977. Kin selection in the Japanese monkey. *Contributions to Primatology* 12:1–145.

———. 1979. Paternity, mother's brother, and human sociality. In *Evolutionary biology and human social behavior: An anthropological perspective*, ed. N. A. Chagnon and W. Irons pp. 145–180. N. Scituate, Mass.: Duxbury Press.

Kuroda, N. 1966. A note on the problem of hawk-mimicry in cuckoos. *Japanese Journal of Zoology* 1966:173–181.

Labitte, A. 1919. Observations sur *Rhodocera rhamni*. *Bulletin du Museum National Histoire Naturelle, Paris* 25:624–625.

Lack, D. 1933. Habitat selection in birds. *Journal of Animal Ecology* 2:239–262.

———. 1934. Habitat distribution in certain Icelandic birds. *Journal of Animal Ecology* 3:81–90.

———. 1939a. The behaviour of the robin. Part 1. The life history with special reference to aggressive behaviour, sexual behaviour and territory. *Proceedings of the Zoological Society of London, Series A* 109:169–178.

———. 1939b. The display of the black cock. *British Birds* 32:290–303.

———. 1940. Courtship feeding in birds. *Auk* 57:169–178.

———. 1943a. *The life of the robin*. London: H. F. & G. Witherby.

———. 1943b. The breeding of birds of Orkney. *Ibis* 85:1–27.

———. 1944. Ecological aspects of species-formation in passerine birds. *Ibis* 86:260–282.

———. 1947–1948. The significance of clutch size. *Ibis* 89:302–352; 90:25–45.

———. 1954. *The natural regulation of animal numbers*. Oxford: Clarendon Press.

———. 1956. *Swifts in a tower*. London: Methuen.

———. 1958. A quantitative breeding study of British tits. *Ardea* 46:91–124.

———. 1966. *Population studies of birds*. Oxford: Clarendon Press.

———. 1968. *Ecological adaptations for breeding in birds*. London: Methuen.

———. 1971. *Ecological isolation in birds*. Cambridge, Mass.: Harvard University Press.

Lack, D., and Emlen, J. T. 1939. Observations on breeding behavior in tricolored red-wings. *Condor* 41:225–230.

Lacy, R. C. 1980. The evolution of eusociality in termites: A haplodiploid analogy? *American Naturalist* 116:449–451.

Lagerspetz, K. Y. H. 1969. Aggression and aggressiveness in laboratory mice. In *Aggressive behaviour*, ed. S. Garattini and E. B. Sigg, pp. 77–85. Amsterdam: Excerpta Medica.

Lagerspetz, K. Y. H., Tivri, R., and Lagerspetz, K. M. J. 1968. Neurochemical and endocrinological studies of mice selectively bred for aggressiveness. *Scandanavian Journal of Psychology* 9:157–160.

Lagler, K. F., Bardach, J. E., and Miller, R. R. 1962. *Ichthyology*. New York: Wiley.

Lamb, K. P., Ehrhardt, P., and Moericke, V. 1967. Labelling of aphid saliva with Rubidium-86. *Nature* 214:602–603.

Lamba, B. S. 1963. The nidification of some common Indian birds. Part I. *Journal of the Bombay Natural History Society* 60:121–133.

Lamond, D. R. 1970. The influence of nutrition on reproduction in the cow. *Animal Breeding Abstracts* 38:359–372.

Lamprecht, J. 1978. The relationship between food competition and foraging group size in

some larger carnivores. A hypothesis. *Zeitschrift für Tierpsychologie* 46:337–343.
———. 1979. Field observations on the behaviour and social system of the bat-eared fox *Otocyon megalotis* Desmarest. *Zeitschrift für Tierpsychologie* 49:260–284.
Lancaster, J. 1971. Play-mothering: The relations between juvenile females and young infants among free-ranging vervet monkeys (*Cercopithecus aethiops*). *Folia Primatologica* 15:161–182.
Lance, A. N. 1970. Movements of blue grouse on the summer range. *Condor* 72:437–444.
———. 1978a. Survival and recruitment success of individual young cock red grouse *Lagopus 1,scoticus* tracked by radio-telemetry. *Ibis* 120:368–378.
———. 1978b. Territories and the food plants of individual red grouse. II. Territory size compared with an index of nutrient supply in heather. *Journal of Animal Ecology* 47:307–314.
Lande, R. 1976. The maintenance of genetic variability by mutation in a polygenic character with linked loci. *Genetics Research* 26:221–235.
———. 1980. Sexual dimorphism, sexual selection, and adaptation in polygenic characters. *Evolution* 34:292–305.
———. 1981. Rapid speciation by sexual selection on polygenic characters. *Proceedings of the National Academy of Sciences* (in press).
Langham, N. P. E. 1974. Comparative breeding biology of the sandwich tern. *Auk* 91:255–277.
Lawick-Goodall, H. van, and Lawick-Goodall, J. van. 1970. *Innocent killers*. Boston: Houghton Mifflin.
Lawick-Goodall, J. van. 1968. The behaviour of free-living chimpanzees in the Gombe Stream Reserve. *Animal Behaviour Monographs* 1:161–311.
Lawick-Goodall, J. van, and Lawick, H. van. 1966. Use of tools by Egyptian vultures. *Nature* 212:1468–1469.
Lawton, J. H. 1976. The structure of the arthropod community on bracken (*Pteridium aquilinium*) (L.) (Kuhn). *Botanical Journal of the Linnaean Society* 73:187–216.
Lazarus, J. 1979. The early warning function of flocking in birds: An experimental study with captive quelea. *Animal Behaviour* 27:855–865.
Lazarus, J., and Crook, J. H. 1973. The effects of luteinizing hormone, estrogen and ovariectomy on the agonistic behaviour of female *Quelea quelea*. *Animal Behaviour* 21:49–60.
Leakey, J. H. E. 1969. Observations made on king cobras in Thailand during May 1966. *Journal of the Natural Resources Council, Thailand* 5:1–10.
Le Boeuf, B. J. 1972. Sexual behavior in the northern elephant seal *Mirounga angustirostris*. *Behaviour* 41:1–26.
———. 1974. Male-male competition and reproductive success in elephant seals. *American Zoologist* 14:163–176.
Le Boeuf, B. J., and Briggs, K. T. 1977. The cost of living in a seal harem. *Mammalia* 41:167–195.
Le Boeuf, B. J., and Peterson, R. S. 1969. Social status and mating activity in elephant seals. *Science* 163:91–93.
Le Boeuf, B. J., Whiting, R. J., and Gantt, R. F. 1972. Perinatal behavior of northern elephant seal females and their young. *Behaviour* 43:121–156.
Ledoux, A., and Dargagnan, D. 1973. La formation des castes chez la fourmi *Aphaenogaster senilis* Mayr. *Compte Rendu de l'Académie des Sciences, Paris, Séries D* 276:551–553.
Lefevre, G., and Jonsson, U. B. 1962. Sperm transfer, storage, displacement, and utilization in Drosophila melanogaster. *Genetics* 47:1719–1736.
Leger, D. W., and Owings, D. H. 1978. Responses to alarm calls by California ground squirrels: Effects of call structure and maternal status. *Behavioral Ecology and Sociobiology* 3:177–186.
Lehrman, D. S. 1953. A critique of Lorenz's "objectivistic" theory of animal behavior. *Quarterly Review of Biology* 28:337–363.
Leigh, E. G. 1970. Sex ratio and differential mortality between the sexes. *American Naturalist* 104:205–210.
Lenington, S. 1977. The evolution of polygyny in red-winged blackbirds. Ph.D. thesis, University of Chicago, Chicago.
———. 1980. Female choice and polygyny in redwinged blackbirds. *Animal Behaviour* 28:347–361.
Lent, P. C. 1965. Rutting behavior in a barren-ground caribou population. *Animal Behaviour* 13:259–264.
———. 1974. Mother-infant relationships in ungulates. In *The behaviour of ungulates and its relation to management*, ed. V. Geist and F. Walther, pp. 14–55. Moreges: International Union for Conservation of Nature and Natural Resources.
Lerner, I. M. 1954. *Genetic homeostasis*. Edinburgh: Oliver & Boyd.
Leshner, A. I. 1978. *An introduction to behavioral endocrinology*. New York: Oxford University Press.

Leshner, A. I., Walker, W. A., Johnson, A. E., Kelling, J. S., Kreisler, S. J., and Svare, B. B. 1973. Pituitary adrenocortical activity and intermale aggressiveness in isolated mice. *Physiology and Behavior* 11:705–711.

Leutenegger, W., and Kelly, J. T. 1977. Relationship of sexual dimorphism in canine size and body size to social, behavioral, and ecological correlates in anthropoid primates. *Primates* 18:117–136.

Leuthold, B. M. 1979. Social organization and behaviour of giraffe in Tsavo East National Park. *African Journal of Ecology* 17:19–34.

Leuthold, B. M., and Leuthold, W. 1972. Food habits of giraffe in Tsavo National Park. *East African Wildlife Journal* 10:129–141.

———. 1978. Ecology of the giraffe in Tsavo East National Park, Kenya. *East African Wildlife Journal* 16:1–20.

Leuthold, W. 1966. Variations in territorial behavior of Uganda Kob *Adenota kob thomasi* (Neumann 1896). *Behaviour* 27:215–258.

———. 1970. Observations on the social organization of impala (*Aepyceros melampus*). *Zeitschrift für Tierpsychologie* 27:215–258.

———. 1977. *African ungulates: A comparative review of their ethology and behavioral ecology. Zoophysiology and Ecology.* Vol. 8. Berlin: Springer-Verlag.

Leuthold, W., and Leuthold, B. M. 1975. Patterns of social grouping in ungulates of Tsavo National Park, Kenya. *Journal of Zoology, London* 175:405–420.

Levin, B. R., and Kilmer, W. L. 1974. Interdemic selection and the evolution of altruism. *Evolution* 28:527–545.

Levin, D. A. 1975. Pest pressure and recombination systems in plants. *American Naturalist* 109:437–451.

———. 1976a. The chemical defenses of plants to pathogens and herbivores. *Annual Review of Ecology and Systematics* 7:121–159.

———. 1976b. Alkaloid-bearing plants: An ecogeographic perspective. *American Naturalist* 110:261–284.

Levine, M. D., Gordon, T. P., Peterson, R. H., and Rose, R. M. 1970. Urinary 17-OHCS response of high- and low-aggressive rhesus monkeys to shock avoidance. *Physiology and Behavior* 5:919–924.

LeVine, R. A. 1973. *Culture, behavior, and personality.* Chicago: Aldine.

Levins, R. 1968. *Evolution in changing environments. Some theoretical considerations. Monographs in population biology,* no. 2. Princeton: Princeton University Press.

———. 1970. Extinction. In *Some mathematical questions in biology,* ed. M. Gerstenhaber, pp. 77–107. Princeton: Princeton University Press.

Levinton, J. 1972. Spatial distribution of *Nucula proxima* Say (Protobranchia): An experimental approach. *Biological Bulletin* 143:175–183.

Leviton, A. E. 1971. *Reptiles and amphibians of North America.* Garden City, N.Y.: Doubleday.

Levy, J. V. 1954. The effects of testosterone propionate on fighting behavior in C57BL/10 young female mice. *Proceedings of the West Virginia Academy of Sciences* 26:14 (abstract).

Levy, J. V., and King, J. A. 1953. The effects of testosterone propionate on fighting behavior in young male C57BL/10 mice. *Anatomical Record* 117:562–563.

Lewontin, R. C. 1970. The units of selection. *Annual Review of Ecology and Systematics* 1:1–18.

———. 1974. *The genetic basis of evolutionary change.* New York: Columbia University Press.

Lewontin, R. C., and White, M. J. D. 1960. Interaction between inversion polymorphisms of two chromosome pairs in the grasshopper, *Moraba scurra. Evolution* 14:116–129.

Leyhausen, P. 1965. The communal organization of solitary mammals. *Symposia of the Zoological Society of London* 14:249–263.

Licht, P., and Gorman, G. C. 1970. Annual fat and reproductive patterns in West Indian *Anolis* lizards. *University of California Publications in Zoology* 95:1–52.

Liebert, R. M. 1974. Television violence and children's aggression: The weight of the evidence. In *Determinants and origins of aggressive behavior,* ed. J. de Wit and W. W. Hartup, pp. 525–531. The Hague: Mouton.

Ligon, J. D., and Ligon, S. H. 1978a. Communal breeding in green woodhoopoes as a case for reciprocity. *Nature* 276:496–498.

———. 1978b. The communal social system of the green woodhoopoe in Kenya. *Living Bird* 17:159–197.

Lill, A. 1974a. Social organization and space utilization in the lek-forming white-bearded manakin, *M. manacus trinitatis* Hartert. *Zeitschrift für Tierpsychologie* 36:513–530.

———. 1974b. Sexual behavior of the lek-forming white-bearded manakin (*Manacus manacus trinitatis* Hartert). *Zeitschrift für Tierpsychologie* 36:1–36.

———. 1976. Lek behavior in the golden-headed manakin, *Pipra erythrocephala,* in Trinidad (West Indies). *Zeitschrift für Tierpsychologie, Supplement* 18:1–83.

Limbaugh, C. 1961. Cleaning symbioses. *Scientific American* 205 (2):42–49.

Limbaugh, C., Pederson, H., and Chase, F. 1961. Shrimps that clean fishes. *Bulletin of Marine Science, Gulf and Caribbean* 11:237–257.

Lin, N. 1963. Territorial behaviour in the cicada killer wasp *Sphecius speciosus* (Drury) (Hymenoptera: Sphecidae, Larrinae). *Behaviour* 20:115–133.

———. 1964. Increased parasitic pressure as a major factor in the evolution of social behavior in halictine bees. *Insectes Sociaux* 11:187–192.

Lin, N., and Michener, C. D. 1972. Evolution and selection in social insects. *Quarterly Review of Biology* 47:131–159.

Lindauer, M. 1952. Ein Beitrag zur Frage der Arbeitsteilung im Bienenstaat. *Zeitschrift für vergleichende Physiologie* 34(4):299–345.

Lindburg, D. G. 1971. The rhesus monkey in North India. In *Primate behavior: Developments in field and laboratory research,* vol. 2, ed. L. A. Rosenblum, pp. 1–106. New York: Academic Press.

———. 1975. Mate selection in rhesus monkey *(Macaca mulatta). American Journal of Physical Anthropology* 42:315 (abstract).

Lindsey, A. A. 1937. The Weddell seal in the Bay of Whales, Antarctica. *Journal of Mammalogy* 18:127–144.

Linsley, E. G. 1965. Notes on male territorial behavior in the Galápagos carpenter bee (Hymenoptera; Apidae). *Pan-Pacific Entomology* 41:158–161.

Liversidge, R. 1971. The biology of the Jacobin cuckoo *Clamator jacobinus. Ostrich Supplement* 8:117–137.

Lloyd, C. W., and Weisz, J. 1975. Hormones and aggression. In *Neural bases of violence and aggression,* ed. W. S. Fields and W. H. Sweet, pp. 92–127.

Lloyd, M., and Dybas, H. S. 1966. The periodical cicada problem. I. Population ecology. II. Evolution. *Evolution* 20:133–149, 466–505.

Lobel, P. S. 1976. Predation on a cleaner fish *(Labroides)* by a hawkfish *(Cirrhites). Copeia* 1976:384.

Loefer, J. B., and Patten, J. A. 1941. Starlings at a blackbird roost. *Auk* 58:584–586.

Lofts, B., and Murton, R. K. 1973. Reproduction in birds. In *Avian biology,* vol. III, ed. D. S. Farner and J. R. King, pp. 1–107. New York: Academic Press.

Löhrl, H. 1955. Schlafgewohnheiten der Baumläufer *(Certhia brachydactyla, C. familiaris)* und andere Kleinvögel in kalten Winternächten. *Die Vogelwarte* 18:71–72.

Loper, G. M. 1968. Effect of aphid infestation on the coumestrol content of alfalfa varieties differing in aphid resistance. *Crop Science* 8:104–108.

Lörcher, K. 1969. Vergleichende bioakustische Untersuchungen an der Rotund Gelbbauchunke *Bombina bombina* (L.) und *Bombina variegata* (L.). *Oecologia* 3:84–124.

Lorenz, K. 1940. Durch Domestikation verursachte Störungen arteigenen Verhaltens. *Zeitschrift für angewandte Psychologie und Characterkunde* 59:56–75.

———. 1943. Die angeborenen Formen möglichen Erfahrung. *Zeitschrift für Tierpsychologie* 5:235–409.

———. 1950. The comparative method in studying innate behavior patterns. *Symposia of the Society of Experimental Biology* 4:221–268.

———. 1966. *On aggression.* New York: Harcourt Brace Jovanovich.

Lösch, A. 1954. *The economics of location.* 2nd ed. New Haven: Yale University Press.

Losey, G. S. 1972a. The ecological importance of cleaning symbiosis. *Copeia* 1972(4):820–833.

———. 1972b. Behavioral ecology of the cleaning fish. *Australian Natural History* 17:232–238.

———. 1972c. Predation protection in the poison-fang blenny, *Meiacanthus atrodorsalis* and its mimics, *Ecsenius bicolor* and *Runula laudandus* (Blenniidae). *Pacific Science* 26:129–139.

———. 1974. Cleaning symbiosis in Puerto Rico with comparison to the tropical Pacific. *Copeia* 1974:960–970.

———. 1978. The symbiotic behavior of fishes. In *The behavior of fish and other aquatic animals,* ed. D. I. Mostofsky, pp. 1–31. New York: Academic Press.

———. 1979. Fish cleaning symbiosis: Proximate causes of host behaviour. *Animal Behaviour* 27:669–685.

Losey, G. S., and Margules, L. 1974. Cleaning symbiosis provides a positive reinforcer for fish. *Science* 184:179–180.

Lott, D., Scholz, S. D., and Lehrman, D. S. 1967. Exteroceptive stimulation of the reproductive system of the female ring dove *(Streptopelia risoria)* by the mate and by the colony milieu. *Animal Behaviour* 15:433–438.

Louch, C. H., and Higginbotham, M. 1967. The relation between social rank and plasma corticosterone levels in mice. *General and Comparative Endocrinology* 8:441–444.

Low, R. M. 1971. Interspecific territoriality in a pomacentrid reef fish, *Pomacentrus flavicauda* Whitley. *Ecology* 52:648–654.

Lowe, F. A. 1954. *The heron.* London: Collins.

Lowe-McConnell, R. H. 1969. The cichlid fishes of Guyana, South America, with notes on their ecology and breeding behaviour. *Journal of the Linnaean Society of London* 48:255–302.

Lowery, G. H., Jr. 1974. *The mammals of Louisiana and its adjacent waters.* Baton Rouge: Louisiana State University Press.

Loy, J. 1971. Estrous behavior of free-ranging rhesus monkeys *(Macaca mulatta). Primates* 12:1–31.

———. 1975. The descent of dominance in *Macaca:* Insights into the structure of human societies. In *Socio-ecology and psychology of primates,* ed. R. H. Tuttle, pp. 153–180. The Hague: Mouton.

Ludwig, G. M., and Norden, R. M. 1969. Age, growth and reproduction of the northern mottled sculpin *(Cottus bairdi bairdi)* in Mt. Vernon Creek, Wisconsin. *Milwaukee Public Museum, Occasional Papers no. 2.*

Lumia, A. R. 1972. The relationships among testosterone, conditioned aggression, and dominance in male pigeons. *Hormones and Behavior* 3:277–286.

Lüscher, M. 1952. Die Produktion und Elimination von Ersatzgeschlectstieren bei der Termite *Kalotermes flavicollis* Fabr. *Zeitschrift für vergleichende Physiologie* 34:123–141.

———. 1953. The termite and the cell. *Scientific American* 188(5):74–78.

———. 1961. Social control of polymorphism in termites. *Symposium of the Royal Entomological Society of London* 1:57–67.

———. 1969. Die Bedeutung des Juvenilhormons für die Differenzierung der Soldaten bei der Termite *Kalotermes flavicollis. Proceedings of the 6th Congress of the International Union for the Study of Social Insects,* pp. 165–170.

Luther, W. 1971. Distribution, biology, and classification of salamanders. In *Venomous animals and their venoms.* Vol. II. *Venomous vertebrates,* ed. W. Bücherl and E. E. Buckley, pp. 557–568. New York: Academic Press.

Lutz, B. 1960. Fighting and an incipient notion of territory in male tree frogs. *Copeia* 1960:61–63.

———. 1971. Venomous toads and frogs. In *Venomous animals and their venoms.* Vol. II. *Venomous vertebrates,* ed. W. Bücherl and E. E. Buckley, pp. 423–473. New York: Academic Press.

———. 1973. *Brazilian species of Hyla.* Austin: University of Texas Press.

Lynch, C. D. 1974. A behavioural study of blesbok, *Damaliscus dorcas phillipsi,* with special reference to territoriality. Memoirs van die Nasionale Museum no. 8. Bloemfontein, South Africa.

Maas, W. K., Maas, R., Wiame, J. M., and Glansdorff, N. 1964. Studies on the mechanism of repression of arginine biosynthesis in *Escherichia coli.* I. Dominance of repressibility in zygotes. *Journal of Molecular Biology* 8:359–364.

MacArthur, R. H. 1958. Population ecology of some warblers of northeastern coniferous forests. *Ecology* 39:599–619.

———. 1965. Ecological consequences of natural selection. In *Theoretical and mathematical biology,* ed. T. Waterman and H. Morowitz, pp. 388–397. New York: Blaisdell.

———. 1972. *Geographical ecology.* New York: Harper & Row.

MacArthur, R. H., and Levins, R. 1964. Competition, habitat selection, and character displacement in a patchy environment. *Proceedings of the National Academy of Sciences* 51:1207–1210.

———. 1967. The limiting similarity, convergence, and divergence of coexisting species. *American Naturalist* 101:377–385.

MacArthur, R. H., and Pianka, E. R. 1966. On optimal use of a patchy environment. *American Naturalist* 100:603–609.

MacArthur, R. H., and Wilson, E. O. 1967. *The theory of island biogeography. Monographs in Population Biology,* no. 1. Princeton: Princeton University Press.

McBride, G. 1963. The "teat order" and communication in young pigs. *Animal Behaviour* 11:53–56.

McCord, W., McCord, J., and Howard, A. 1961. Familial correlates of aggression in nondelinquent male children. *Journal of Abnormal and Social Psychology* 63:493–503.

McCullough, D. R. 1969. The tule elk: Its history, behavior, and ecology. *University of California Publications in Zoology* 88:1–209.

McDiarmid, R. W., and Adler, K. 1974. Notes on territorial and vocal behavior of neotropical frogs of the genus *Centrolenella. Herpetologica* 29:188–191.

MacDonald, D. W. 1979. "Helpers" in fox society. *Nature* 282:69–71.
MacDonald, J. D. 1973. *Birds of Australia*. Sydney: A. H. & A. W. Reed.
MacDonald, S. D. 1970. The breeding behaviour of the rock ptarmigan. *Living Bird* 9:195–238.
MacGillivray, A. J., Paul, J., and Threlfall, G. 1972. Transcriptional regulation in eukaryotic cells. *Advances in Cancer Research* 15:93–162.
McHugh, T. 1958. Social behaviour of the American buffalo *(Bison bison bison)*. *Zoologica* 43:1–40.
McIlhenny, E. A. 1937. Life history of the boat-tailed grackle in Louisiana. *Auk* 54:274–295.
McKaye, K. R. 1977. Competition for breeding sites between the cichlid fishes of Lake Jiloa, Nicaragua. *Ecology* 58:291–302.
McKey, D. 1974. Adaptive patterns in alkaloid physiology. *American Naturalist* 108:305–320.
McKinney, T. D., and Desjardins, C. 1973. Postnatal development of the testis, fighting behavior and fertility in house mice. *Biology of Reproduction* 9:279–294.
McLaren, I. A. 1958. The biology of the ringed seal *(Phoca hispida* Schreber) in the eastern Canadian Arctic. *Bulletin of the Fisheries Research Board of Canada* 118:1–97.
———. 1967. Seals and group selection. *Ecology* 48:104–110.
———. 1972. Polygyny as the adaptive function of breeding territory in birds. *Transactions of the Connecticut Academy of Sciences* 44:189–210.
MacLean, P. D., and Delgado, J. M. R. 1953. Electrical and chemical stimulation of frontotemporal portion of limbic system in the waking animal. *Electroencephalography and Clinical Neurophysiology* 5:91–100.
McMichael, D. F. 1971. Mollusks—Classification, distribution, venom apparatus and venoms, symptomatology of stings. In *Venomous animals and their venoms*. Vol. III. *Venomous invertebrates*, ed. W. Bücherl and E. E. Buckley, pp. 373–393. New York: Academic Press.
McMillan, I. I. 1964. Annual population changes in California quail. *Journal of Wildlife Management* 28:702–711.
McNab, B. K. 1963. Bioenergetics and the determination of home range size. *American Naturalist* 97:133–140.
———. 1969. The economics of temperature regulation in neotropical bats. *Comparative Biochemistry and Physiology* 31:227–268.
MacNair, M. R. 1978. An ESS for the sex ratio in animals, with particular reference to the social Hymenoptera. *Journal of Theoretical Biology* 70:449–459.
MacNair, M. R., and Parker, G. A. 1979. Models of parent-offspring conflict. III. Intra-brood conflict. *Animal Behaviour* 27:1202–1209.
MacRoberts, B. R., and MacRoberts, M. H. 1972. Social stimulation of reproduction in herring and lesser black-backed gulls. *Ibis* 114:495–506.
MacRoberts, M. H., and MacRoberts, B. R. 1976. Social organization and behavior of the acorn woodpecker in central coastal California. *Ornithological Monographs* 21:1–115.
Madden, R., and Marcus, L. F. 1978. Use of the F distribution in calculating bivariate normal home ranges. *Journal of Mammalogy* 59:870–871.
Maiorana, V. C. 1976. Size and environmental predictability for salamanders. *Evolution* 30:599–613.
Major, P. F. 1978. Predator-prey interactions in two schooling fishes, *Caranx ignobilis* and *Stolephorus purpureus*. *Animal Behaviour* 26:760–777.
Makacha, S., and Schaller, G. 1969. Observations on lions in the Lake Manyara National Park, Tanzania. *East African Wildlife Journal* 7:99–103.
Makwana, S. C. 1979. Infanticide and social change in two groups of the Hanuman langur, *Presbytis entellus*, at Jodhpur. *Primates* 20:293–300.
Mallory, F. F., and Brooks, R. J. 1978. Infanticide and other reproductive strategies in the collared lemming, *Dicrostonyx groenlandicus*. *Nature* 273:144–146.
Mamsch, E. 1967. Quantitative Untersuchungen zur Regulation der Fertilität im Ameisenstaat durch Arbeiterinnern, Larven und Königin. *Zeitschrift für vergleichende Physiologie* 55:1–25.
Mamsch, E., and Bier, K. 1966. Das Verhalten von Ameisenarbeiterinnen gegenüber der Königin nach vorangeganger Weisellosigkeit. *Insectes Sociaux* 8:277–284.
Mandel, H. G. 1972. Pathways of drug biotransformation: Biochemical conjugations. In *Fundamentals of drug metabolism and drug distribution*, ed. B. N. La Du, H. G. Mandel, and E. L. Way, pp. 149–186. Baltimore: Williams & Wilkins.
Mangan, R. L. 1979. Reproductive behavior of the cactus fly, *Odontoloxozus longicornis*. Male territoriality and female guarding as adaptive strategies. *Behavioral Ecology and Sociobiology* 4:265–278.

Maniatis, T., and Ptashne, M. 1976. A DNA operator-repressor system. *Scientific American* 234(1):64–76.
Manson, J. 1974. Aspects of the biology and behaviour of the Cape grysbok, *Raphicerus melanotis,* Thunberg. M.Sc. thesis, University of Stellenbosch, South Africa.
Mansueti, R. 1963. Symbiotic behaviour between small fishes and jellyfishes, with new data on that between the stromateid, *Peprilus alepidotus,* and the scyphomedusa *Chrysaora quinquecirrha. Copeia* 1963:40–80.
Marikovskiy, P. I. 1962. On intraspecific relations of *Formica rufa* L. (Hymenoptera; Formicidae). *Entomological Reviews* 41:47–51.
Marikovsky, P. I. 1961. Material on sexual biology of the ant *Formica rufa* L. *Insectes Sociaux* 8:23–30.
———. 1974. The biology of the ant *Rossomyrmex proformicarum* K. W. Arnoldi (1928). *Insectes Sociaux* 21:301–308.
Mariscal, R. N. 1970a. The nature of the symbiosis between Indo-Pacific anemone fishes and sea anemones. *Marine Biology* 6:58–65.
———. 1970b. An experimental analysis of the protection of *Amphiprion xanthurus* Cuvier and Valenciennes and some other anemone fishes from sea anemones. *Journal of Experimental Marine Biology and Ecology* 4:134–149.
———. 1972. Behavior of symbiotic fishes and sea anemones. In *Behavior of marine animals,* vol. 2, ed. H. E. Winn and B. L. Olla, pp. 327–360. New York: Plenum.
Markert, C. L., and Ursprung, H. 1971. *Developmental genetics.* Englewood Cliffs, N.J.: Prentice-Hall.
Markin, G. P. 1968. Nest relationship of the Argentine ant, *Iridomyrmex humilis* (Hymenoptera: Formicidae). *Journal of the Kansas Entomological Society* 41:511–516.
Markow, T. A., Quaid, M., and Kerr, S. 1978. Male mating experience and competitive courtship success in *Drosophila melanogaster. Nature* 276:821–822.
Marler, P. 1955. Characteristics of some animal calls. *Nature* 176:6–7.
———. 1957. Specific distinctiveness in the communication signals of birds. *Behaviour* 11:13–39.
———. 1959. Developments in the study of animal communication. In *Darwin's biological work,* ed. P. R. Bell, pp. 150–206. Cambridge, England: Cambridge University Press.
———. 1961. The logical analysis of animal communication. *Journal of Theoretical Biology* 1:295–317.
———. 1976. On animal aggression. The roles of strangeness and familiarity. *American Psychologist* 31:239–246.
Marler, P. R., and Hamilton, W. J., III. 1966. *Mechanisms of animal behavior.* New York: Wiley.
Marmur, J., Rownd, R., and Schildkraut, C. L. 1963. Denaturation and renaturation of deoxyribonucleic acid. *Progress in Nucleic Acid Research* 1:231–300.
Marshall, A. J., and Roberts, J. D. 1959. The breeding biology of equatorial vertebrates: Reproduction of cormorants (Phalacrocoracidae) at latitude 0° 20′ N. *Proceedings of the Zoological Society of London* 132:617–625.
Marshall, N. B. 1965. *The life of fishes.* London: Weidenfeld & Nicolson.
Marshall, W. H. 1965. Ruffed grouse behavior. *BioScience* 15:92–94.
Martin, R. D. 1968. Reproduction and ontogeny in tree shrews *(Tupaia belangeri)* with reference to their general behavior and taxonomic relationships. *Zeitschrift für Tierpsychologie* 25:409–495, 505–532.
———. 1972. A preliminary field-study of the lesser mouse lemur *(Microcebus murinus* J. F. Miller 1977). *Zeitschrift für Tierpsychologie, Supplement* 9:43–89.
Martin, S. G. 1971. Polygyny in the bobolink: Habitat quality and the adaptive complex. Ph.D. thesis, Oregon State University, Corvallis.
———. 1974. Adaptations for polygynous breeding in the bobolink, *Dolichonyx oryzivorus. American Zoologist* 14:109–119.
Martinez, D. R., and Klinghammer, E. 1970. The behavior of the whale *Orcinus orca:* A review of the literature. *Zeitschrift für Tierpsychologie* 27:828–839.
Maschwitz, U. W. J., and Kloft, W. 1971. Morphology and function of the venom apparatus of insects—Bees, wasps, ants, and caterpillars. In *Venomous animals and their venoms.* Vol. III. *Venomous invertebrates,* ed. W. Bücherl and E. E. Buckley, pp. 1–60. New York: Academic Press.
Mason, R. R. 1970. Comparison of flight aggregation in two species of southern *Ips* (Coleoptera: Scolytidae). *Canadian Entomologist* 102:1036–1041.
Mason, W. A. 1968. Use of space in *Callicebus* groups. In *Primates: Studies in adaptation and variability,* ed. P. C. Jay, pp. 200–216. New York: Holt, Rinehart and Winston.
Massey, A. 1977. Agonistic aids and kinship in a group of pigtail macaques. *Behavioral Ecology and Sociobiology* 2:31–40.

Matessi, C., and Jayakar, S. D. 1976. Conditions for the evolution of altruism under Darwinian selection. *Theoretical Population Biology* 9:360–387.
Mathew, A. P. 1954. Observations on the habits of two spider mimics of the red ant, *Oecophylla smaragdina* (Fabr.). *Journal of the Bombay Natural History Society* 52:249–263.
Mathewson, S. F. 1961. Gonadotrophic control of aggressive behavior in starlings. *Science* 134:1522–1523.
Matthews, G. V. T. 1968. *Bird navigation.* 2nd ed. London: Cambridge University Press.
Matthews, L. H. 1937. The female sexual cycle in the British horseshoe bats *(Rhinolophus ferrum-equinum* and *R. hipposideros minutus). Transactions of the Royal Society of London* 23:224–266.
———. 1964. Overt fighting in mammals. In *The natural history of aggression,* ed. J. D. Carthy and F. J. Ebling, pp. 23–32. New York: Academic Press.
Mayfield, H. 1960. The Kirtland's warbler. *Bulletin of the Cranbrook Institute of Science* no. 40. Bloomfield Hills, Mich.
Maynard, E. C. L. 1968. Cleaning symbiosis and oral grooming on the coral reef. In *Biology of the month,* ed. P. Person, pp. 79–88. Washington, D.C.: American Association for the Advancement of Science.
Maynard Smith, J. 1956. Fertility, mating behaviour and sexual selection in *Drosophila subobscura. Journal of Genetics* 54:261–279.
———. 1958. *The theory of evolution.* Harmondsworth, England: Penguin.
———. 1964. Group selection and kin selection. *Nature* 201: 1145–1147.
———. 1965. The evolution of alarm calls. *American Naturalist* 99:59–63.
———. 1971. The origin and maintenance of sex. In *Group selection,* ed. G. C. Williams, pp. 163–175. Chicago: Aldine-Atherton.
———. 1974a. The theory of games and the evolution of animal conflicts. *Journal of Theoretical Biology* 47:209–221.
———. 1974b. Recombination and the rate of evolution. *Genetics* 78:299–304.
———. 1976a. Group selection. *Quarterly Review of Biology* 52:277–283.
———. 1976b. Evolution and the theory of games. *American Scientist* 64:41–45.
———. 1976c. A short-term advantage for sex and recombination through sib competition. *Journal of Theoretical Biology* 63:245–258.
———. 1976d. Sexual selection and the handicap principle. *Journal of Theoretical Biology* 57:239–242.
———. 1977. Parental investment: A prospective analysis. *Animal Behaviour* 25:1–9.
———. 1978. *The evolution of sex.* Cambridge, England: Cambridge University Press.
———. 1979. Game theory and the evolution of behavior. *Proceedings of the Royal Society of London, Series B* 205:475–488.
Maynard Smith, J., and Haigh, J. 1974. The hitch-hiking effect of a favorable gene. *Genetics Research* 23:23–25.
Maynard Smith, J., and Parker, G. A. 1976. The logic of asymmetric contests. *Animal Behaviour* 24:159–175.
Maynard Smith, J., and Price, G. R. 1973. The logic of animal conflict. *Nature* 246:15–18.
Mayr, E. 1935. Bernard Altum and the territory theory. *Proceedings of the Linnaean Society of New York,* nos. 45, 46:24–38.
———. 1941. Red-wing observations of 1940. *Proceedings of the Linnaean Society of New York,* nos. 52–53:75–83.
———. 1963. *Animal species and evolution.* Cambridge, Mass.:Harvard University Press (Belknap Press).
Meanley, B. 1954. Nesting of the water-turkey in eastern Arkansas. *Wilson Bulletin* 66:81–88.
Mech, L. D. 1970. *The wolf. The ecology and behavior of an endangered species.* Garden City, N.Y.: Natural History Press.
Medina, J. R., and Petit, C. 1979. L'effet hitch-hiking comme processus dispersif. *Journal of Theoretical Biology* 81:235–246.
Melchior, H. R. 1971. Characteristics of arctic ground squirrel alarm calls. *Oecologia* 7:184–190.
Menge, B. A. 1974. Effect of wave action and competition on brooding and reproductive effort in the sea star *Leptasterias hexactis. Ecology* 55:84–93.
Meyburg, B. 1974. Sibling aggression and mortality among nestling eagles. *Ibis* 116:224–228.
Michael, R. P. 1969. Effects of gonadal hormones on displaced and direct aggression in pairs of rhesus monkeys of opposite sex. In *Aggressive behaviour,* ed. S. Garattini and E. B. Sigg, pp. 172–178. Amsterdam: Excerpta Medica.
Michael, R. P., and Zumpe, D. 1970. Aggression and gonadal hormones in captive rhesus monkeys *(Macaca mulatta). Animal Behaviour* 18:1–10.
Michener, C. D. 1958. The evolution of social behavior in bees. *Proceedings of the 10th Inter-*

national Congress of Entomology, Montreal, pp. 441–447.
———. 1969. Comparative social behavior of bees. Annual Review of Entomology 14:299–342.
———. 1974. The social behavior of the bees. Cambridge, Mass.:Belknap Press.
Michener, C. D., and Brothers, D. J. 1974. Were workers of eusocial Hymenoptera initially altruistic or oppressed? Proceedings of the National Academy of Sciences 71:671–674.
Michener, C. D., and Lange, R. B. 1958. Distinctive type of primitive social behavior among bees. Science 127:1046–1047.
———. 1959. Observations on the behavior of Brazilian halictid bees (Hymenoptera, Apoidea). IV. Augochloropsis, with notes on extralimital forms. American Museum Novitates 1924:1–41.
Michener, J. R. 1951. Territorial behaviour and age composition in a population of mockingbirds at a feeding station. Condor 53:276–283.
Michod, R. 1979. Genetical aspects of kin selection: Effects of inbreeding. Journal of Theoretical Biology 81:223–233.
Michod, R. E., and Anderson, W. W. 1979. Measures of genetic relatedness and the concept of inclusive fitness. American Naturalist 114:637–647.
Miczek, K. A., Brykczynski, T., and Grossman, S. P. 1974. Differential effects of lesions in the amygdala, periamygdaloid cortex, or stria terminalis on aggressive behaviors in rats. Journal of Comparative and Physiological Psychology 79:37–45.
Miles, P. W. 1969. Interaction of plant phenols and salivary phenolases in the relationship between plants and hemiptera. Entomologia experimentalis et applicata 12:736–766.
Miley, W. M. 1973. Some effects of androgens in intermale fighting by laboratory mice in a spontaneous dominance situation. Ph.D. thesis, Temple University, Philadelphia.
Milinski, M. 1977. Do all members of a swarm suffer the same predation? Zeitschrift für Tierpsychologie 45:373–388.
Miller, A. H. 1942. Habitat selection among higher vertebrates and its relation to intraspecific variation. American Naturalist 76:25–35.
Miller, E. H. 1975. Social and evolutionary implications of territoriality in adult male New Zealand fur seals, Arctocephalus forsteri (Lesson, 1828), during the breeding season. Rapporte de Procès-Verbaux Réunions Conseil Internationale Exploration de Mer 169:170–187.
Miller, E. M. 1969. Caste differentiation in the lower termites. In Biology of termites, vol. 1, ed. K. Krishna and F. M. Weesner, pp. 283–310. New York: Academic Press.
Miller, G. R., and Watson, A. 1978. Territories and the food plant of individual red grouse. I. Territory size, number of mates, and brood size compared with the abundance, production and diversity of heather. Journal of Animal Ecology 47:293–306.
Miller, H. C. 1963. The behavior of the sunfish, Lepomis gibbosus (Linnaeus), with notes on the behavior of other species of Lepomis and the pigmy sunfish, Elassoma evergladei. Behaviour 22:88–151.
Miller, L. G. 1976. Fated genes—An essay review of E. O. Wilson, Sociobiology: The new synthesis. Journal of the History of the Behavioral Sciences 12:183–190.
Miller, N. E. 1941. The frustration-aggression hypothesis. Psychological Review 48:337–342.
———. 1948. Studies of fear as an acquirable drive. I. Fear as motivation and fear reduction as reinforcement in the learning of a new response. Journal of Experimental Psychology 1948:89–101.
Miller, R. C. 1922. The significance of the gregarious habit. Ecology 3:122–126.
Miller, R. S. 1973. The brood size of cranes. Wilson Bulletin 85:436–441.
Miller, S. L., and Orgel, L. 1974. The origin of life on the earth. Englewood Cliffs, N.J.: Prentice-Hall.
Milligan, S. R. 1976. Pregnancy blocking in the vole, Microtus agrestis. I. Effect of the social environment. Journal of Reproduction and Fertility 46:91–95.
Mills, J. A. 1973. The influence of age and pair-bond on the breeding biology of the red-billed gull, Larus novaehollandiae scopulinus. Journal of Animal Ecology 42:147–162.
Milton, K., and May, M. L. 1976. Body weight, diet and home range area in primates. Nature 259:459–462.
Minton, S. A., Jr., and Minton, M. R. 1969. Venomous reptiles. New York: Scribner.
Mirmirani, M., and Oster, G. F. 1978. Competition, kin selection, and evolutionarily stable strategies. Theoretical Population Biology 13:304–339.
Mirsky, A. 1955. The influence of sex hormones on social behavior in monkeys. Journal of Comparative and Physiological Psychology 48:327–335.
Mock, D. 1979. Repertoire shifts and "extramarital" courtship in herons. Behaviour 69:57–71.
Moehlman, P. 1974. Behavior and ecology of feral asses. Ph.D. thesis, University of Wisconsin, Madison.

———. 1979. Jackal helpers and pup survival. *Nature* 277:382–383.

Moffat, C. B. 1903. The spring rivalry of birds. Some views on the limit to multiplication. *Irish Naturalist* 12:152–166.

Mohnot, S. M. 1971. Some aspects of social changes and infant-killing in the Hanuman langur *(Presbytis entellus)* (Primates: Cercopithecidae) in western India. *Mammalia* 35:175–198.

———. 1974. Ecology and behavior of the common Indian langur, *Presbytis entellus* Dufresne. Ph.D. thesis, University of Jodhpur, India.

Mohr, H. 1960. Zum Erkennen von Raubvögeln, insbesondere von Sperber und Baumfalk, durch Kleinvögeln. *Zeitschrift für Tierpsychologie* 17:686–699.

Moir, R. J. 1968. Ruminant digestion and evolution. In *Handbook of physiology.* Section 6. Alimentary canal, vol. V, ed. C. F. Code, pp. 2673–2694. Washington, D.C.: American Physiological Society.

Moll, E. O., and Legler, J. M. 1971. The life history of a neotropical slider turtle, *Pseudemys scripta* (Schoepff), in Panama. *Bulletin of the Los Angeles City Museum of Natural History and Science,* no. 11.

Monahan, M. W. 1977. Determinants of male pairing success in the red-winged blackbird *(Agelaius phoeniceus):* A multivariate and experimental analysis. Ph.D. thesis, Indiana University, Bloomington.

Monro, J. 1967. The exploitation and conservation of resources by populations of insects. *Journal of Animal Ecology* 36:531–547.

Montagner, H. 1964. Étude du comportement alimentaire et des relations trophallactiques des mâles au sein de la société de guêpes, au moyen d'un ratio-isotope. *Insectes Sociaux* 11:301–316.

———. 1966. Sur l'origine des mâles dans les sociétés de guêpes du genre *Vespa. Compte Rendu de l'Académie des Sciences, Paris, Séries D* 263:785–787.

———. 1967. *Comportements trophallactiques chez les guêpes sociales.* Film produced by Service du Film de Rechereche Scientifique; 96, Boulevard Raspail, Paris. No. B2053.

Montevecchi, W. A. 1979. Predator-prey interactions between ravens and kittiwakes. *Zeitschrift für Tierpsychologie* 49:136–141.

Moore, B. P. 1968. Studies on the chemical composition and function of the cephalic gland secretion in Australian termites. *Journal of Insect Physiology* 14:33–39.

———. 1969. Biochemical studies in termites. In *Biology of termites,* vol. 1, ed. K. Krishna and F. M. Weesner, pp. 407–432. New York: Academic Press.

Moore, F. R. 1978. Interspecific aggression: Toward whom should a mockingbird be aggressive? *Behavioral Ecology and Sociobiology* 3:173–176.

Moore, N. W. 1960. The behaviour of the adult dragonfly. In *Dragonflies,* ed. P. Corbet, C. Longfield, and N. W. Moore, pp. 158–161. London: Collins.

———. 1964. Intra- and interspecific competition among dragonflies (Odonata). *Journal of Animal Ecology* 33:49–71.

Moore, W. G. 1948. Bat caves and bat bombs. *Turtox News* 26:262–265.

Moreau, R. E. 1937. The comparative breeding biology of the African hornbills (Bucerotidae). *Proceedings of the Zoological Society of London, Series A* 107:331–346.

Moreau, R. E., and Moreau, W. M. 1940. Hornbill studies. *Ibis* 1940:639–656.

Moreno-Black, G., and Maples, W. R. 1977. Differential habitat utilization of four Cercopithecidae in a Kenyan forest. *Folia Primatologica* 27:85–107.

Morrell, G. M., and Turner, J. R. G. 1970. Experiments on mimicry: I. The response of wild birds to artificial prey. *Behaviour* 36:116–130.

Morris, D. 1956. The function and causation of courtship ceremonies. In *L'instinct dans le comportement des animaux et de l'homme,* ed. P. P. Grass, pp. 261–287. Paris: Masson. (Reprinted in D. Morris, ed. 1967. *Primate ethology.* London: Weidenfeld & Nicolson.)

Morris, R. O., and Hunter, R. A. 1976. Factors influencing desertion of colony sites by common terns *(Sterna hirundo). Canadian Field Naturalist* 90:137–143.

Morton, E. S. 1973. On the evolutionary advantages of fruit eating in tropical birds. *American Naturalist* 107:8–22.

Morton, S. R., and Parry, G. D. 1974. The auxiliary social system in kookaburras: A reappraisal of its adaptive significance. *Emu* 74:196–198.

Moss, C. 1975. *Portraits in the wild. Behavior studies of East African mammals.* Boston: Houghton Mifflin.

Moss, R. 1969. A comparison of red grouse *(Lagopus l. scoticus)* stocks with the production and nutritive value of heather *(Calluna vulgaris). Journal of Animal Ecology* 38:103–122.

Moss, R., Watson, A., and Parr, R. 1975. Maternal nutrition and breeding success in red grouse *(Lagopus lagopus scoticus). Journal of Animal Ecology* 44:233–244.

Moyer, K. E. 1968. Kinds of aggression and their physiological basis. *Communications in Behavioral Biology* 2:65–87.
———. 1971. A preliminary model of aggression. In *The physiology of aggression and defeat,* ed. B. E. Eleftheriou and J. P. Scott, pp. 223–263. New York: Plenum.
———. 1975. A physiological model of aggression: Does it have different implications? In *Neural bases of violence and aggression,* ed. W. S. Fields and W. H. Sweet, pp. 161–195. St. Louis: Warren H. Green.
———. 1976. *The psychobiology of aggression.* New York: Harper & Row.
Moynihan, M. 1955. Some aspects of reproductive behavior in the black-headed gull *(Larus ridibundis ridibundis* L.) and related species. *Behaviour Supplement* 4:1–201.
———. 1962a. The organization and probable evolution of some mixed species flocks of neotropical birds. *Smithsonian Miscellaneous Collections* 143:1–140.
———. 1962b. Hostile and sexual behavior patterns of South American and Pacific Laridae. *Behaviour Supplement* 8:1–365.
———. 1964. Some behavior patterns of Platyrrhine monkeys. I. The night monkey *(Aotus trivirgatus). Smithsonian Miscellaneous Collections* 146:1–84.
Mueller, H. C. 1971. Oddity and specific search image more important than conspicuousness in prey selection. *Nature* 233:345–346.
———. 1975. Hawks select odd prey. *Science* 188:953–954.
Mühlethaler, F. 1952. Beobachtungen am Bergfinken—Schlafplatz bei Thun 1950/51. *Ornithologische Beobachter* 49:173–192.
Muller, H. J. 1932. Some genetic aspects of sex. *American Naturalist* 66:118–138.
Müller-Schwarze, D. 1971. Pheromones in black-tailed deer *(Odocoileus hemionus columbianus). Animal Behaviour* 19:141–152.
Müller-Velten, H. 1966. Über den Angstgeruch bei der Hausmaus. *Zeitschrift für vergleichende Physiologie* 52:401–429.
Mullins, D. E., and Keil, C. B. 1980. Paternal investment of urates in cockroaches. *Nature* 282:567–569.
Mulrow, P. J. 1973. The adrenals. In *Physiology and biophysics,* vol. III, 20th ed., ed. T. C. Ruch and H. D. Patton, pp. 224–247. Philadelphia: Saunders.
Munn, C. A., and Terborgh, J. W. 1979. Multi-species territoriality in neotropical foraging flocks. *Condor* 81:338–347.
Munro, I. S. R. 1967. *The fishes of New Guinea.* Port Moresby, New Guinea: Department of Agriculture and Stock Fish.
Munro, J. and Bédard, J. 1977. Gull predation and crèching behaviour in the common eider. *Journal of Animal Ecology* 46:799–810.
Murie, A. 1944. *The wolves of Mt. McKinley.* United States Fauna Series, no. 5, National Park Service. Washington, D.C.: Department of the Interior.
Murphy, G. I. 1968. Pattern in life history and the environment. *American Naturalist* 102:391–403.
Murphy, R. C. 1936. *Oceanic birds of South America.* Vols. I, II. New York: Macmillan.
Murray, B. G., Jr. 1969. A comparative study of the Le Conte's and sharp-tailed sparrows. *Auk* 86:199–231.
———. 1971. The ecological consequences of interspecific territorial behavior in birds. *Ecology* 52:414–423.
———. 1976. A critique of interspecific territoriality and character convergence. *Condor* 78:518–525.
Murray, R. D. 1980. The evolution and functional significance of incest avoidance. *Journal of Human Evolution* 9:173–178.
Murton, R. K. 1958. The breeding of woodpigeon populations. *Bird Study* 5:157–183.
———. 1967. The significance of endocrine stress in population control. *Ibis* 109:622–623.
———. 1968. Some predator-prey relationships in bird damage and population control. In *The problems of birds as pests. Symposia of the Institute of Biology,* no. 17, ed. R. K. Murton and E. N. Wright, pp. 157–169.
Murton, R. K., Isaacson, A. J., and Westwood, N. J. 1971. The significance of gregarious feeding behaviour and adrenal stress in a population of wood-pigeons *(Columba palumbus). Journal of Zoology, London* 165:53–84.
Murvosh, C. M., Fye, R. L., and LaBreque, G. C. 1964. Studies on the mating behavior of the house fly, *Musca domestica* L. *Ohio Journal of Science* 64:264–271.
Myers, G. R., and Waller, D. W. 1977. Helpers at the nest in barn swallows. *Auk* 94:596.
Myers, J. A. 1978. Sex ratio adjustment under food stress: Maximization of quality or numbers of offspring? *American Naturalist* 112:381–388.

Myers, J. P., Connors, P. G., and Pitelka, F. A. 1979a. Territoriality in non-breeding shorebirds. In *Shorebirds in marine environments. Studies in avian biology,* vol. 2, ed. F. A. Pitelka, pp. 231–246. Berkeley, Calif.: Cooper Ornithological Society.

———. 1979b. Territory size in wintering sanderlings: The effects of prey abundance and intruder density. *Auk* 96:551–561.

Myers, J. P., and Myers, L. P. 1980. The pampas shorebird community: Seasonal composition, habitat use, and spacing behaviors. *Ibis* (in press).

Myers, K., Hale, C. S., Mykytowycz, R., and Hughes, R. L. 1971. The effects of varying density and space on sociality and health in animals. In *Behavior and environment: The use of space by animals and man,* ed. A. H. Esser, pp. 148–187. New York: Plenum.

Myers, R. E. 1972. Role of prefrontal and anterior temporal cortex in social behavior and affect in monkeys. *Acta Neurobiologiae Experimentalis* 32:567–579.

Myrberg, A. A. 1972. Social dominance and territoriality in the bicolor damselfish, *Eupomacentrus partitus* (Poey) (Pisces: Pomacentridae). *Behaviour* 41:208–231.

Myrberg, A. A., Brahy, B. D., and Emery, A. R. 1967. Field observations on reproduction of the damselfish, *Chromis multilineata* (Pomacentridae), with additional notes on general behavior. *Copeia* 1967:819–827.

Myrberg, A. A., Jr., and Thresher, R. E. 1974. Interspecific aggression and its relevance to the concept of territoriality in reef fishes. *American Zoologist* 14:81–96.

Nadler, R., and Rosenblum, L. 1972. Hormone regulation of the "fatted" phenomenon in squirrel monkeys. *Anatomical Record* 173:181–187.

Nagel, H. G., and Rettenmeyer, C. W. 1973. Nuptial flights, reproductive behavior and colony founding of the western harvester ant, *Pogonomyrmex occidentalis* (Hymenoptera: Formicidae). *Journal of the Kansas Entomological Society* 46:82–101.

Nakanishi, K., Koreeda, M., and Imai, S. 1972. A catalogue of ecdysones. In *Some recent developments in the chemistry of natural products,* ed. S. Rangaswami and N. V. Subba Rae, pp. 194–213. New Delhi: Prentice-Hall.

Nauta, W. J. H. 1962. Neural associations of the amygdaloid complex in the monkey. *Brain* 85:505–520.

Neff, J. A., and Meanley, B. 1957. Status of Brewer's blackbird on the Grand Prairie of eastern Arkansas. *Wilson Bulletin* 69:102–105.

Neill, S. R. St. J., and Cullen, J. M. 1974. Experiments on whether schooling by their prey affects the hunting behaviour of cephalopod and fish predators. *Journal of Zoology, London* 172:549–569.

Neill, W. T. 1964. Viviparity in snakes: Some ecological and zoogeographical considerations. *American Naturalist* 98:35–55.

Nel, J. A. J. 1975. Aspects of the social ethology of some Kalahari rodents. *Zeitschrift für Tierpsychologie* 37:322–331.

Nel, J. J. C. 1968. Aggressive behaviour of the harvester termites, *Hodotermes mossambicus* (Hagen) and *Trinervitermes trinervoides* (Sjöstedt). *Insectes Sociaux* 15:145–156.

Nelson, C. E. 1973. Mating calls of the Microhylinae: Descriptions and phylogenetic and ecological considerations. *Herpetologica* 29:163–176.

Nelson, J. B. 1964. Factors influencing clutch-size and chick growth in the North Atlantic gannet, *Sula bassana. Ibis* 106:63–77.

———. 1965. The behaviour of the gannet. *British Birds* 58:233–288, 313–336.

———. 1966a. The breeding biology of the gannet *Sula bassana* on the Bass Rock, Scotland. *Ibis* 108:584–626.

———. 1966b. Population dynamics of the gannet *(Sula bassana)* at the Bass Rock, with comparative information from other Sulidae. *Journal of Animal Ecology* 35:443–470.

———. 1967a. The breeding behaviour of the white-footed booby *Sula dactylatra. Ibis* 109:194–232.

———. 1967b. Etho-ecological adaptations in the great frigate-bird. *Nature* 214:318.

———. 1968. Breeding behaviour of the swallow-tailed gull in the Galapagos. *Behaviour* 30:146–174.

———. 1969. The breeding ecology of the red-footed booby in the Galapagos. *Journal of Animal Ecology* 38:181–198.

———. 1970. The relationship between behaviour and ecology in the Sulidae with reference to other sea birds. *Oceanography and Marine Biology. An Annual Review* 8:501–574.

———. 1978. *The Sulidae: Gannets and boobies.* London: Oxford University Press.

Nelson, K. 1964. Behavior and morphology in the glandulocaudine fishes (Ostariophysi, Characidae). *University of California Publications in Zoology* 75:59–152.

Nero, R. W. 1956. A behavior study of the red-winged blackbird. *Wilson Bulletin* 68:5–37, 129–150.

Nero, R. W., and Emlen, J. T. 1951. An experimental study of territorial behavior in breeding red-winged blackbirds. *Condor* 53:105–116.

Nethersole-Thompson, D. 1973. *The dotterel.* London: Collins.

Nettleship, D. N. 1972. Breeding success of the common puffin (*Fratercula arctica* L.) on different habitats of Great Island, Newfoundland. *Ecological Monographs* 42:239–268.

Newton, I. 1966. Fluctuations in the weights of bullfinches. *British Birds* 59:393–419.

———. 1970. Irruptions of crossbills in Europe. In *Animal populations in relation to their food resources,* ed. A. Watson, pp. 337–353. Oxford: Blackwell Scientific.

———. 1972. *Finches.* New York: Taplinger.

Nice, M. M. 1933. The theory of territorialism and its development. In *Fifty Years of Progress in American Ornithology,* pp. 89–100. Lancaster, Pa.: American Ornithological Union.

———. 1937. Studies in the life history of the song sparrow. I. A population study of the song sparrow. *Transactions of the Linnaean Society of New York* 4:1–247.

———. 1941. The role of territory in bird life. *American Midland Naturalist* 26:441–487.

———. 1962. Development of behavior in precocial birds. *Transactions of the Linnaean Society of New York* 8:1–212.

Nichols, J. D., and Chabreck, R. H. 1980. On the variability of alligator sex ratios. *American Naturalist* 116:125–137.

Nicholson, A. J. 1957. The self-adjustment of populations to change. *Cold Spring Harbor Symposia on Quantitative Biology* 22:153–173.

Nicol, J. A. C. 1971. Physiological investigations of oceanic animals. In *Deep oceans,* ed. P. J. Herring and M. R. Clarke, pp. 225–246. London: Barker.

Nisbet, I. C. T. 1973. Courtship-feeding, egg-size and breeding success in common terns. *Nature* 241:141–142.

———. 1977. Courtship-feeding and clutch-size in common terns. In *Evolutionary ecology,* ed. B. Stonehouse and C. M. Perrins, pp. 101–109. London: Macmillan.

Nishida, T. 1966. A sociological study of solitary male monkeys. *Primates* 7:141–204.

———. 1979. The social structure of chimpanzees of the Mahale Mountains. In *The great apes,* ed. D. A. Hamburg and E. R. McCown, pp. 72–121. Menlo Park, Calif.: Benjamin/Cummings.

Nishida, T., Uehara, S., and Nyundo, R. 1979. Predatory behavior among wild chimpanzees of the Mahale Mountains. *Primates* 20:1–20.

Noble, G. K. 1939. The role of dominance in the social life of birds. *Auk* 56:263–273.

Noble, G. K., and Wurm, M. 1946. The effect of testosterone propionate on the black-crowned night heron. *Endocrinology* 26:837–850.

Noirot, C. 1955. Recherches sur le polymorphisme des termites supérieurs (Termitidae). *Annals des Sciences Naturelles* 17:399–595.

———. 1969. Formation of castes in the higher termites. In *Biology of termites,* vol. I, ed. K. Krishna and F. M. Weesner, pp. 311–350. New York: Academic Press.

———. 1974. Polymorphismus bei hoheren Termiten. In *Sozialpolymorphismus bei Insekten,* ed. G. H. Schmidt, pp. 740–765. Stuttgart: Wissenschaftliche Verlagsgesellschaft.

Noirot, E. 1969. Serial order of maternal responses in mice. *Animal Behaviour* 17:547–560.

Nolan, V., Jr. 1978. The ecology and behavior of the prairie warbler (*Dendroica discolor*). *Ornithological Monographs* 26:1–595.

Nolan, V., Jr., and Thompson, C. F. 1975. The occurrence and significance of anomalous reproductive activities in two North American non-parasitic cuckoos *Coccyzus* spp. *Ibis* 117:496–503.

Nolan, W. J. 1924. The division of labour in the honeybee. *North Carolina Beekeeper* 1924(10):10–15.

Noll, J. 1931. Untersuchungen über die Zeugnung und Staatenbildung des *Halictus malachurus. Zeitschrift für Morphologie und Ökologie der Tiere* 23:285–368.

Noonan, K. M. 1978. Sex ratio of parental investment in colonies of the social wasp *Polistes fuscatus. Science* 199:1354–1356.

———. 1980. Individual strategies of inclusive fitness maximizing in *Polistes fuscatus* foundresses. In *Natural selection and social behavior. Research and theory,* ed. R. D. Alexander and D. W. Tinkle. Portland, Oreg.: Chiron Press.

Norman, R. F., Taylor, P. D., and Robertson, R. J. 1977. Stable equilibrium strategies and penalty functions in a game of attrition. *Journal of Theoretical Biology* 65:571–578.

Norris, K. S. 1967. Aggressive behavior in Cetacea. In *Aggression and defense: Neural mechanisms and social patterns.* Vol. 5. *Brain function,* ed. C. D. Clemente and D. B. Lindsley, pp. 243–266. Los Angeles: University of California Press.

Norton, D. W. 1972. Incubation schedules of four species of calidridine sandpipers at Barrow, Alaska. *Condor* 74:164–176.

Novitski, C. 1947. Genetic analysis of an anomalous sex ratio condition in *Drosophila affinis*. *Genetics* 32:526–534.

Nursall, J. R. 1973. Territoriality in redlip blennies *(Ophioblennius atlanticus*—Pisces: Blenniidae). *Journal of Zoology, London* 182:205–223.

Oates, J. F. 1977. The social life of a black-and-white colobus monkey, *Colobus guereza*. *Zeitschrift für Tierpsychologie* 45:1–60.

O'Connor, R. J. 1978. Brood reduction in birds: Selection for fratricide, infanticide and suicide? *Animal Behaviour* 26:79–96.

O'Donald, P. 1962. The theory of sexual selection. *Heredity* 17:541–552.

———. 1963. Sexual selection and territorial behaviour. *Heredity* 18:361–364.

———.1972a. Natural selection of reproductive rates and breeding times and its effect on sexual selection. *American Naturalist* 106:368–379.

———. 1972b. Sexual selection by variations in fitness at breeding time. *Nature* 237:349–351.

———. 1973a. Models of sexual and natural selection in polygamous species. *Heredity* 31:145–156.

———. 1973b. Frequency-dependent sexual selection as a result of variations in fitness at breeding time. *Heredity* 30:351–368.

———. 1974. Polymorphisms maintained by sexual selection in monogamous species of birds. *Heredity* 32:1–10.

Odum, E. P. 1971. *Fundamentals of ecology*. 3rd ed. Philadelphia: Saunders.

Odum, E. P., and Kuenzler, E. J. 1955. Measurement of territory and home range size in birds. *Auk* 72:128–137.

Ohsawa, H. 1979a. The local gelada population and environment of the Gich area. In *Ecological and sociological studies of gelada baboons*, ed. M. Kawai, pp. 3–45. *Contributions to Primatology* 16:1–344.

———. 1979b. Herd dynamics. In *Ecological and sociological studies of gelada baboons*, ed. M. Kawai, pp. 47–80. *Contributions to Primatology* 16:1–344.

Oliver, J. A. 1956. Reproduction in the king cobra, *Ophiophagus hannah* Cantor. *Zoologica* 41:145–152.

Olsen, P. F. 1970. Sylvatic (wild rodent) plague. In *Infectious diseases of wild animals*, ed. J. W. Davis, L. H. Karstadt, and D. O. Trainer, pp. 200–213. Ames: Iowa State University Press.

Opler, P. A. 1974. Biology, ecology and host specificity of Microlepidoptera associated with *Quercus agrifolia* (Fagaceae). *University of California Publications in Entomology* 75:1–83.

Oppenheimer, J. R. 1970. Mouthbrooding in fishes. *Animal Behaviour* 18:493–503.

———. 1977. Communication in New World monkeys. In *How animals communicate*, ed. T. A. Sebeok, pp. 851–889. Bloomington: Indiana University Press.

Organ, J. A. 1961. Studies of the local distribution, life history, and population dynamics of the salamander genus *Desmognathus* in Virginia. *Ecological Monographs* 31:189–220.

Orians, G. H. 1960. Autumnal breeding in the tricolored blackbird. *Auk* 77:379–398.

———. 1961a. The ecology of blackbird *(Agelaius)* social systems. *Ecological Monographs* 31:285–312.

———. 1961b. Social stimulation within blackbird colonies. *Condor* 63:330–337.

———. 1969. On the evolution of mating systems in birds and mammals. *American Naturalist* 103:589–603.

———. 1980. *Adaptations of marsh-nesting blackbirds. Monographs in Population Biology*, no. 14. Princeton: Princeton University Press.

Orians, G. H., and Collier, G. 1963. Competition and blackbird social systems. *Evolution* 17:449–459.

Orians, G. H., and Horn, H. S. 1969. Overlap in foods and foraging of four species of blackbirds in the potholes of central Washington. *Ecology* 50:930–938.

Orians, G. H., and Janzen, D. H. 1974. Why are embryos so tasty? *American Naturalist* 108:581–592.

Orians, G. H., Orians, C. E., and Orians, K. J. 1977. Helpers at the nest in some Argentine blackbirds. In *Evolutionary ecology*, ed. B. Stonehouse and C. M. Perrins, pp. 137–151. London: Macmillan.

Orians, G. H., and Pearson, N. E. 1978. On the theory of central place foraging. In *Analysis of ecological systems*, ed. D. J. Horn, pp. 155–177. Columbus: Ohio State University Press.

Orians, G. H., and Willson, M. F. 1964. Interspecific territories of birds. *Ecology* 45:736–745.

Oring, L. W., and Knudsen, M. L. 1972. Monogamy and polyandry in the spotted sandpiper. *Living Bird* 11:59–73.

Oring, L. W., and Maxson, S. J. 1978. Instances of simultaneous polyandry by a spotted sandpiper *Actitis macularia*. *Ibis* 120:349–353.

Orr, R. T. 1965. Interspecific behavior among pinnipeds. *Zeitschrift für Säugetierkunde* 30:163–171.

Oster, G., Eshel, I., and Cohen, D. 1977. Worker-queen conflict and the evolution of social insects. *Theoretical Population Biology* 12:49–85.

Oster, G. F., and Wilson, E. O. 1978. *Caste and ecology in the social insects. Monographs in Population Biology*, no. 12. Princeton: Princeton University Press.

Österdahl, L., and Olsson, R. 1963. The sexual behaviour of *Hymenochirus boettgeri*. *Oikos* 14:35–43.

Otte, D., and Joern, A. 1975. Insect territoriality and its evolution: Population studies of desert grasshoppers on creosote bushes. *Journal of Animal Ecology* 44:29–54.

Otto, D. 1958. Über die Arbeitsteilung im Staate von *Formica rufa rufo-pratensis minor* Gössw. und irhe verhaltensphysiologischen Grundlagen: Ein Beitrag zur Biologie der Roten Waldameise. *Wissenschaftliche Abhandlungen der Deutschen Akademie der Landwirtschafts-wissenschaften zu Berlin* 30:1–169.

Owen, D. F. 1971. *Tropical butterflies*. Oxford: Clarendon Press.

Owen, D. F., and Chanter, D. O. 1969. Population biology of tropical African butterflies. Sex ratio and genetic variation in *Acraea encedon*. *Journal of Zoology, London* 157:345–374.

———. 1971. Polymorphism in West African populations of the butterfly, *Acraea encedon*. *Journal of Zoology, London* 163:481–488.

Owens, D. D., and Owens, M. J. 1979a. Communal denning and clan associations in brown hyenas *(Hyena brunnea*, Thunberg) of the central Kalahari Desert. *African Journal of Ecology* 17:35–44.

———. 1979b. Notes on social organization and behavior in brown hyenas *(Hyaena brunnea)*. *Journal of Mammalogy* 60:405–408.

Owens, N. W., and Goss-Custard, J. D. 1976. The adaptive significance of alarm calls given by shorebirds in their wintering grounds. *Evolution* 30:397–398.

Owen-Smith, N. 1977. On territoriality in ungulates and an evolutionary model. *Quarterly Review of Biology* 52:1–38.

Owings, D. H., and Coss, R. G. 1977. Snake mobbing by California ground squirrels: Adaptive variation and ontogeny. *Behaviour* 62:50–69.

Owings, D. H., and Virginia, R. A. 1978. Alarm calls of California ground squirrels *(Spermophilus beecheyi)*. *Zeitschrift für Tierpsychologie* 46:58–70.

Packard, G. C. 1966. The influence of ambient temperature and aridity on modes of reproduction and excretion of amniote vertebrates. *American Naturalist* 100:677–682.

Packard, G. C., Tracy, C. R., and Roth, J. J. 1977. The physiological ecology of reptilian eggs and embryos, and the evolution of viviparity within the class Reptilia. *Biological Reviews* 52:71–105.

Packer, C. 1975. Male transfer in olive baboons. *Nature* 255:219–220.

———. 1977. Reciprocal altruism in *Papio anubis*. *Nature* 265:441–443.

———. 1979a. Inter-troop transfer and inbreeding avoidance in *Papio anubis*. *Animal Behaviour* 27:1–36.

———. 1979b. Male dominance and reproductive activity in *Papio anubis*. *Animal Behaviour* 27:37–45.

Page, G., and Whitacre, D. F. 1975. Raptor predation on wintering shorebirds. *Condor* 77:73–83.

Paine, R. T. 1963. Food recognition and predation on opisthobranchs by *Navanax inermis* (Gastropoda: Opisthobranchia). *Veliger* 6:1–9.

Palmer, R. S., ed. 1962. *Handbook of North American birds*. Vol. 1. New Haven: Yale University Press.

Palmer, T. J. 1978. A horned beetle that fights. *Nature* 274:583–584.

Park, T. 1933. Studies in population physiology. II. Factors regulating initial growth of *Tribolium confusum* populations. *Journal of Experimental Zoology* 65:17–45.

Parke, D. V. 1968. *The biochemistry of foreign compounds*. Oxford: Pergamon Press.

Parker, G. A. 1970a. Sperm competition and its evolutionary consequences in the insects. *Biological Reviews* 45:525–567.

———. 1970b. Sperm competition and its evolutionary effect on copula duration in the fly *Scatophaga stercoraria*. *Journal of Insect Physiology* 16:1301–1328.

———. 1970c. The reproductive behavior and the nature of sexual selection in *Scatophaga stercoraria* L. IV. Epigamic recognition and competition between males for the possession of females. *Behaviour* 37:113–139.

———. 1974a. Assessment strategy and the evolution of fighting behaviour. *Journal of Theoretical Biology* 47:223–243.

———. 1974b. Courtship persistence and female-guarding as male time investment strategies. *Behaviour* 48:157–184.

———. 1978a. Selfish genes, evolutionary games, and the adaptiveness of behaviour. *Nature* 274:849–855.

———. 1978b. Selection on non-random fusion of gametes during the evolution of anisogamy. *Journal of Theoretical Biology* 73:1–28.

———. 1979. Sexual selection and sexual conflict. In *Sexual selection and reproductive competition in insects,* ed. M. S. Blum and N. A. Blum, pp. 123–166. New York: Academic Press.

Parker, G. A., Baker, R. R., and Smith, V. G. F. 1972. The origin and evolution of gamete dimorphism and the male-female phenomenon. *Journal of Theoretical Biology* 36:529–553.

Parker, G. A., and MacNair, M. R. 1978a. Models of parent offspring conflict. I. Monogamy. *Animal Behaviour* 26:97–110.

———. 1978b. Models of parent-offspring conflict. II. Promiscuity. *Animal Behaviour* 26:111–122.

———. 1979. Models of parent-offspring conflict. IV. Suppression: Evolutionary retaliation by the parent. *Animal Behaviour* 27:1210–1235.

Parker, S. 1976. The precultural basis of the incest taboo: Toward a biosocial theory. *American Anthropologist* 78:285–305.

Parkes, A. S., and Bruce, H. M. 1961. Olfactory stimuli in mammalian reproduction. *Science* 134:1049–1054.

Parmelee, D. F. 1970. Breeding behavior of the sanderling in the Canadian high Arctic. *Living Bird* 9:97–146.

Parmelee, D. F., and Payne, R. B. 1973. On multiple broods and the breeding strategy of arctic sandpipers. *Ibis* 115:218–226.

Parry, V. A. 1970. *Kookaburras.* Melbourne: Lansdowne Press.

———. 1973. The auxiliary social system and its effect on territory and breeding in kookaburras. *Emu* 73:81–100.

Parsons, J. 1971. Cannibalism in herring gulls. *British Birds* 64:528–537.

———. 1976. Nesting density and breeding success in the herring gull *Larus argentatus. Ibis* 118:537–546.

Partridge, L. 1978. Habitat selection. In *Behavioural ecology: An evolutionary approach,* ed. J. R. Krebs and N. B. Davies, pp. 351–376. Sunderland, Mass.: Sinauer Associates.

———. 1980. Mate choice increases a component of offspring fitness in fruit flies. *Nature* 283:290–291.

Passera, L. 1963. Les relations sociales chez la fourmi parasite *Plagiolepis xene* Stärcke. *Insectes Sociaux* 11:59–70.

———. 1965. Le cycle évolutif de la fourmi *Plagiolepis pygmaea* Latr. (Hyménoptères, Formicoidea, Formicidae). *Insectes Sociaux* 10:59–69.

———. 1969a. Biologie de la reproduction chez *Plagiolepis pygmaea* Latr. et ses deux parasites sociaux *Plagiolepis grassei* Le Masne et Passera et *Plagiolepis xene* St. (Hym. Formicidae). *Annales des Sciences Naturelles, Zoologie et Biologie Animale* 11:327–482.

———. 1969b. Interactions et fécondité des reines de *Plagiolepis pygmaea* Latr. et de ses parasites sociaux *P. grassei* Le Masne et Passera et *P. xene* St. (Hym. Formicidae). *Insectes Sociaux* 16:179–194.

———. 1974. Kastendetermination bei der Ameise *Plagiolepis pygmaea* Latr. In *Sozialpolymorphismus bei Insekten,* ed. G. H. Schmidt, pp. 513–532. Stuttgart: Wissenschaftliche Verlagsgesellschaft.

Pasteels, J. M. 1965. Polyéthisme chez les ouvriers de *Nasutitermes lujae* (Termitidae Isoptères). *Biologia Gabonica* 1:191–205.

Patterson, C. B. 1978. Male parental behavior in the redwinged blackbird *(Agelaius phoeniceus).* Ph.D. thesis, Indiana University, Bloomington.

Patterson, C. B., and Emlen, J. M. 1980. Variation in nestling sex ratios in the yellow-headed blackbird. *American Naturalist* 115:743–747.

Patterson, I. J. 1965. Timing and spacing of broods in the black-headed gull *Larus ridibundus. Ibis* 107:433–459.

Payne, A. P., and Swanson, H. H. 1971. Hormonal control of aggressive dominance in the female hamster. *Physiology and Behavior* 6:355–357.

Payne, R. B. 1969. The breeding seasons and reproductive physiology of tricolored blackbirds and redwinged blackbirds. *University of California Publications in Zoology* 90:1–137.

———. 1973a. The breeding season of a parasitic bird, the brown-headed cowbird, in central California. *Condor* 75:80–99.

———. 1973b. Individual laying histories and the clutch size and numbers of eggs of parasitic cuckoos. *Condor* 75:414–438.

———. 1977. The ecology of brood parasitism in birds. *Annual Review of Ecology and Systematics* 8:1–28.

———. 1979. Sexual selection and intersexual differences in variance of breeding success. *American Naturalist* 114:447–452.

Payne, R. B., and Payne, K. 1977. Social organization and mating success in local song populations of village indigobirds *Vidua chalybeata. Zeitschrift für Tierpsychologie* 45:113–173.

Paynter, R. A., Jr. 1949. Clutch-size and the egg and chick mortality of Kent Island herring gulls. *Ecology* 30:146–166.

Peacock, A. D., Smith, I. C., Hall, D. W., and Baxter, A. T. 1954. Studies in pharaoh's ant, *Monomorium pharaonis* (L.). 8. Male production by parthenogenesis. *Entomologist's Monthly Magazine* 90:154–158.

Peakin, G. J. 1972. Aspects of productivity in *Tetramorium caespitum* L. *Ekologia Polska* 20:55–63.

Pease, J. L., Vowles, R. H., and Keith, L. B. 1979. Interaction of snowshoe hares and woody vegetation. *Journal of Wildlife Management* 43:43–60.

Pechuman, L. L. 1967. Observations on the behavior of the bee *Anthidium manicatum* (L.). *Journal of the New York Entomological Society* 75:68–73.

Peek, F. W. 1971. Seasonal change in the breeding behavior of the male red-winged blackbird. *Wilson Bulletin* 83:383–395.

Pengilley, R. K. 1971. Calling and associated behavior of some species of *Pseudophryne* (Anura: Leptodactylidae). *Journal of Zoology, London* 163:73–92.

Penney, R. L. 1968. Territorial and social behavior in the adélie penguin. *Antarctic Research Series, Washington* 12:83–131.

Perrill, S. A., Gerhardt, H. C., and Daniel, R. 1978. Sexual parasitism in the green tree frog *(Hyla cinerea). Science* 200:1179–1182.

Perrins, C. M. 1965. Population fluctuations and clutch-size in the great tit, *Parus m. major* L. *Journal of Animal Ecology* 34:601–647.

———. 1968. The purpose of the high-intensity alarm calls in small passerines. *Ibis* 110:200–201.

Perrone, M., Jr. 1975. The relation between mate choice and parental investment patterns in fish who brood their young: Theory and case study. Ph.D. thesis, University of Washington, Seattle.

———. 1978. Mate size and breeding success in a monogamous cichlid fish. *Environmental Biology of Fish* 3:193–201.

Perrone, M., Jr., and Zaret, T. M. 1979. Parental care patterns in fish. *American Naturalist* 113:351–361.

Pesce, H., and Delgado, A. 1971. Poisoning from adult moths and caterpillars. In *Venomous animals and their venoms.* Vol. III. *Venomous invertebrates,* ed. W. Bücherl and E. E. Buckley, pp. 119–156. New York: Academic Press.

Peters, P. J., and Bronson, F. H. 1971. Neonatal androgen and the organization of aggression in mice. *American Zoologist* 11:621 (abstract).

Peters, P. J., Bronson, F. H., and Whitsett, J. M. 1972. Neonatal castration and intermale aggression in mice. *Physiology and Behavior* 8:265–268.

Peters, R. C., and Bretschneider, F. 1972. Electric phenomena in the habitat of the catfish *Ictalurus nebulosus* Les. *Journal of Comparative Physiology* 81:345–362.

Peterson, R. S., and Bartholomew, G. A., Jr. 1967. The natural history and behavior of the California sea lion. *Special publication no. 1,* American Society of Mammalogists, pp. 1–79.

Petter, J. J. 1962. Recherches sur l'ecologie et l'ethologie des Lémuriens malgaches. *Mémoires du Museum National d'Histoire Naturelle, Séries A* 27:1–146.

Pettingill, O. S. 1939. History of one hundred nests of arctic tern. *Auk* 56:420–428.

Pfeiffer, W. 1962. The fright reaction of fish. *Biological Reviews* 37:495–511.

———. 1963. The fright reaction of North American fish. *Canadian Journal of Zoology* 41:69–77.

———. 1967. Schreckreacktion und Schreckstoffzellen bei Ostariophysi und Gonorhynchiformes. *Zeitschrift für vergleichende Physiologie* 56:380–396.

———. 1974. Pheromones in fish and amphibia. In *Pheromones. Frontiers of Biology,* vol. 32, ed. M. C. Birch, pp. 269–296. Amsterdam: North Holland.

———. 1977. The distribution of fright reaction and alarm substance cells in fishes. *Copeia* 1977:653–665.

Phillips, R. R. 1971. The relationship between social behaviour and the use of space in the benthic fish *Chasmodes bosquianus* Lacépède (Teleostei, Blenniidae). I. Ethogram. *Zeitschrift für Tierpsychologie* 29:11–27.

———. 1977a. The relationship between social behaviour and the use of space in the benthic fish *Chasmodes bosquianus* Lacépède (Teleostei, Blenniidae). IV. Effects of topography on habitat selection and shelter choice. *Behaviour* 60:1–27.

———. 1977b. Behavioral field study of the Hawaiian rockskipper, *Istiblennius zebra* (Teleostei, Blenniidae). *Zeitschrift für Tierpsychologie* 43:1–22.

Phillipson, A. T., and McAnally, R. A. 1942. Studies on the fate of carbohydrates in the rumen of sheep. *Journal of Experimental Biology* 19:119–214.

Pianka, E. R. 1970. On r- and K-selection. *American Naturalist* 104:592–597.

———. 1972. r and k selection or b and d selection? *American Naturalist* 106:581–588.

———. 1974. *Evolutionary ecology*. New York: Harper & Row.

———. 1976. Competition and niche theory. In *Theoretical ecology. Principles and applications*, ed. R. M. May, pp. 114–141. Oxford: Blackwell Scientific.

Pianka, E. R., and Parker, W. S. 1975. Age-specific reproductive tactics. *American Naturalist* 109:453–464.

Pickens, A. L. 1935. Evening drill of chimney swifts during the late summer. *Auk* 52:149–153.

Pickles, W. 1935. Populations, territory and interrelations of the ants *Formica fusca*, *Acanthomyops niger* and *Myrmica scabrinodis* at Garforth (Yorkshire). *Journal of Animal Ecology* 4:22–31.

———. 1936. Populations and territories of the ants, *Formica fusca*, *Acanthomyops flavus*, and *Myrmica ruginodis*, at Thornhill (Yorks.). *Journal of Animal Ecology* 5:262–270.

Picman, J. 1977. Destruction of eggs by the long-billed marsh wren *(Telmatodytes palustris palustris)*. *Canadian Journal of Zoology* 55:1914–1920.

———. 1980. Impact of marsh wrens on reproductive success of red-winged blackbirds. *Canadian Journal of Zoology* 58:337–350.

Picman, J., and Picman, A. K. 1980. Destruction of nests by the short-billed marsh wren. *Condor* 82:176–179.

Pielou, E. C. 1977. *Mathematical ecology*. New York: Wiley-Interscience.

Pienkowski, M. W., and Greenwood, J. J. D. 1979. Why change mates? *Biological Journal of the Linnaean Society* 12:85–94.

Pietsch, T. W., and Grobecker, D. B. 1978. The compleat angler: Aggressive mimicry in an antennariid anglerfish. *Science* 201:369–370.

Pilecki, C., and O'Donald, P. 1971. The effects of predation on artificial mimetic polymorphisms with perfect and imperfect mimics at varying frequencies. *Evolution* 25:365–370.

Pitcher, T. J. 1973. The three-dimensional structure of schools in the minnow, *Phoxinus phoxinus* (L). *Animal Behaviour* 21:673–686.

Pitelka, F. A. 1959. Numbers, breeding schedule, and territoriality in pectoral sandpipers of northern Alaska. *Condor* 61:233–262.

Pitelka, F. A., Holmes, R. T., and MacLean, S. F., Jr. 1974. Ecology and evolution of social organization in arctic sandpipers. *American Zoologist* 14:185–204.

Plateaux, L. 1971. Sur le polymorphisme social de la fourmi *Leptothorax nylanderi* (Förster). II. Activité des ouvières et déterminisme des castes. *Annales des Sciences Naturelle, Zoologie et Biologie Animale* 13:1–90.

Plateaux-Quénu, C. 1960. Nouvelle preuve d'un déterminisme imaginal des castes chez Halictus marginatus Brullé. *Compte Rendu de l'Académie des Sciences, Paris, Séries D* 250:4465–4466.

———. 1962. Biology of *Halictus marginatus* Brullé. *Journal of Apicultural Research* 1:41–51.

Platt, J. R. 1964. Strong inference. *Science* 146:347–353.

Pleasants, J. M., and Pleasants, B. Y. 1979. The super-territory hypothesis: A critique, or why there are so few bullies. *American Naturalist* 114:609–614.

Pleszczynska, W. K. 1978. Microgeographic prediction of polygyny in the lark bunting. *Science* 201:935–937.

Pleszczynska, W. K., and Hansell, R. I. C. 1980. Polygyny and decision theory: Test of a model in lark buntings *(Calamospiza melanocorys)*. *American Naturalist* (in press).

Poignant, A. 1965. *The improbable kangaroo and other Australian animals*. Sydney: Angus & Robertson.

Poirier, F. E. 1968. The Nilgiri langur *(Presbytis johnii)* mother-infant dyad. *Primates* 9:45–68.

———. 1970. The Nilgiri langur *(Presbytis johnii)* of South India. In *Primate behavior: Developments in field and laboratory research*, vol. 1, ed. L. A. Rosenblum, pp. 251–383. New York: Plenum.

———, ed. 1972. *Primate socialization*. New York: Random House.

Poirier, F. E., Bellisari, A., and Haines, L. 1978. Functions of primate play behavior. In *Social play in primates*, ed. E. O. Smith, pp. 143–168. New York: Academic Press.

Poirier, F. E., and Smith, E. O. 1974. Socializing functions of primate play. *American Zoologist* 14:275–287.

Pollitzer, R. 1951. Plague studies. 1. A summary of the history and a survey of the present distribution of the disease. *World Health Organization Bulletin* 4:475–533.

Pollock, J. I. 1979. Spatial distribution and ranging behavior in lemurs. In *The study of prosimian behavior,* ed. G. A. Doyle and R. D. Martin, pp. 359–409. New York: Academic Press.

Pooley, A. C., and Gans, C. 1976. The Nile crocodile. *Scientific American* 234(4):114–124.

Popova, N. K., and Naumenko, E. V. 1972. Dominance relations and the pituitary-adrenal system in rats. *Animal Behaviour* 20:108–111.

Popp, J. L., and Devore, I. 1979. Aggressive competition and social dominance theory: Synopsis. In *The great apes,* ed. D. A. Hamburg and E. R. McCown, pp. 316–338. Menlo Park, Calif.: Benjamin/Cummings.

Popper, D., and Fishelson, L. 1973. Ecology and behavior of *Anthias squamipinnis* (Peters, 1855) (Anthiidae, Teleostei) in the coral habitat of Eilat (Red Sea). *Journal of Experimental Zoology* 184:409–424.

Popper, K. R. 1959. *The logic of scientific discovery.* New York: Basic Books.

Potts, G. W. 1973. The ethology of *Labroides dimidiatus* (Cuv. and Val.) (Labridae; Pisces) on Aldabra. *Animal Behaviour* 21:250–291.

Pournelle, G. H. 1968. Classification, biology, and description of the venom apparatus of insectivores of the genera *Solenodon, Neomys,* and *Blarina.* In *Venomous animals and their venoms.* Vol. 1. *Venomous vertebrates,* ed. W. Bücherl, E. E. Buckley, and V. Deulofeu, pp. 31–50. New York: Academic Press.

Powell, G. V. N. 1974. Experimental analysis of the social value of flocking by starlings (*Sturnus vularis*) in relation to predation and foraging. *Animal Behaviour* 22:501–505.

———. 1979. Structure and dynamics of interspecific flocks in a neotropical mid-elevation forest. *Auk* 96:375–390.

Preobrazhenskii, B. V. 1961. Management and breeding of reindeer. In *Reindeer husbandry,* ed. P. S. Zhigunov, pp. 78–128. Springfield, Va.: United States Department of Commerce.

Preston, J. L. 1978. Communication systems and social interactions in a goby-shrimp symbiosis. *Animal Behaviour* 26:791–802.

Price, E. O. 1978. Genotype versus experience effects on aggression in wild and domestic Norway rats. *Behaviour* 54:340–353.

Price, W. H., and Whatmore, P. B. 1967. Behavior disorders and patterns of crime among XYY males identified at a maximum security hospital. *British Medical Journal* 69:533–536.

Procter, D. L. C. 1975. The problem of chick loss in the south polar skua *Catharacta maccormicki. Ibis* 117:452–459.

Pryer, H. 1884. An account of a visit to the birds'-nest caves of British North Borneo. *Proceedings of the Zoological Society of London* 1884:532–538.

Pulliam, H. R. 1973. On the advantages of flocking. *Journal of Theoretical Biology* 38:419–422.

———. 1974. On the theory of optimal diets. *American Naturalist* 108:59–74.

Pulliam, H. R., and Dunford, C. 1979. *Programmed to learn. An essay on the evolution of culture.* New York: Columbia University Press.

Pusey, A. 1979. Intercommunity transfer of chimpanzees in Gombe National Park. In *The great apes,* ed. D. A. Hamburg and E. R. McCown, pp. 464–479. Menlo Park, Calif.: Benjamin/Cummings.

Putnam, L. S. 1949. The life history of the cedar waxwing. *Wilson Bulletin* 61:141–182.

Pyburn, W. F. 1970. Breeding behavior of the leaf-frogs *Phyllomedusa callidryas* and *Phyllomedusa dacnicolor* in Mexico. *Copeia* 1970:209–218.

Pycraft, W. P. 1910. *A history of birds.* London: Methuen.

Pyke, G. H. 1978. Are animals efficient harvesters? *Animal Behaviour* 26:241–250.

———. 1979. The economics of territory size and time budget in the golden-winged sunbird. *American Naturalist* 114:131–145.

Qasim, S. Z. 1957a. The biology of *Centronotus gunnellus* (L.) (Teleostei). *Journal of Animal Ecology* 26:389–401.

———. 1957b. The biology of *Blennius pholis* L. (Teleostei). *Proceedings of the Zoological Society of London* 128:161–208.

Quispel, A. 1968. Pre-biological evolution. *Acta Biotheoretica* 18:291–315.

Rabb, G., and Rabb, M. S. 1963a. On the behavior and breeding biology of the African pipid frog *Hymenochirus boettgeri. Zeitschrift für Tierpsychologie* 20:215–241.

———. 1963b. Additional observations on breeding behavior of the Surinam toad, *Pipa pipa. Copeia* 1963:636–642.

Ragge, D. R. 1965. *Grasshoppers, crickets and cockroaches of the British Isles*. London: Warne.

Rahaman, H., and Parthasarathy, M. D. 1969. Studies on the social behaviour of bonnet monkeys (*Macaca radiata*). *Primates* 10:149–160.

Raigner, A., Boven, J. van, and Ceusters, R. 1974. Der Polymorphismus der afrikanischen Wanderameisen unter biometrischen und biologischen Gesichtspunkten. In *Der Sozialpolymorphismus bei Insekten*, ed. G. H. Schmidt, pp. 668–693. Stuttgart: Wissenschaftliche Verlagsgesellschaft.

Raitt, R. J., and Hardy, J. W. 1979. Social behavior, habitat, and food of the Beechey jay. *Wilson Bulletin* 91:1–15.

Ralls, K. 1976. Mammals in which females are larger than males. *Quarterly Review of Biology* 51:245–276.

———. 1977. Sexual dimorphism in mammals: Avian models and unanswered questions. *American Naturalist* 111:917–938.

Rand, A. L. 1953. Factors affecting feeding rates of anis. *Auk* 70:26–30.

———. 1954. Social feeding behavior of birds. *Fieldiana Zoology* 36:1–71.

Rand, A. S. 1967. The adaptive significance of territoriality in iguanid lizards. In *Lizard ecology. A symposium*, ed. W. W. Milstead, pp. 106–115. Columbia: University of Missouri Press.

———. 1968. A nesting aggregation of iguanas. *Copeia* 1968:552–561.

Rand, W. M., and Rand, A. S. 1976. Agonistic behaviour in nesting iguanas: A stochastic analysis of dispute settlement dominated by the minimization of energy cost. *Zeitschrift für Tierpsychologie* 40:279–299.

Randall, J. E., and Randall, H. E. 1960. Examples of mimicry and protective resemblance in tropical marine fishes. *Bulletin of Marine Science, Gulf and Caribbean* 10:444–480.

Raner, L. 1972. Förekommer polyandri hos smalnäbbad simsnäppa (*Phalaropus lobatus*) och svartsnäppa (*Tringa erythropus*). *Fauna and Flora* 3:135–138.

Rappaport, R. A. 1968. *Pigs for the ancestors: Ritual in the ecology of a New Guinea people*. New Haven: Yale University Press.

Rasa, O. A. E. 1969. Territoriality and the establishment of dominance by means of visual cues in *Pomacentrus jenkinsi* (Pisces: Pomacentridae). *Zeitschrift für Tierpsychologie* 26:825–845.

———. 1972. Aspects of social organization in captive dwarf mongooses. *Journal of Mammalogy* 53:181–185.

———. 1975. Mongoose sociology and behavior as they relate to zoo exhibition. *International Zoo Yearbook* 15:65–73.

———. 1977. The ethology and sociology of the dwarf mongoose (*Helogale undulata rufula*). *Zeitschrift für Tierpsychologie* 43:337–406.

Rasweiler, J. J. 1973. Care and management of the long-tongued bat, *Glossophaga soricina*, in the laboratory with observations on estivation induced by food deprivation. *Journal of Mammalogy* 54:391–404.

Ratcliffe, F. N., Gay, F. J., and Greaves, T. 1952. *Australian termites, the biology, recognition, and economic importance of the common species*. Melbourne: Commonwealth Scientific and Industrial Research Organization.

Rausch, R. A. 1967. Some aspects of the population ecology of wolves, Alaska. *American Zoologist* 7:253–265.

Raven, P. H., Evert, R. F., and Curtis, H. 1976. *Biology of plants*. 2nd ed. New York: Worth.

Raw, A. 1975. Territoriality and scent marking by *Centris* males (Hymenoptera, Anthophoridae) in Jamaica. *Behaviour* 54:311–321.

Redfield, J. A. 1973. Variations in weight of blue grouse *(Dendragapus obscurus)*. *Condor* 75:312–321.

Rees, C. J. C. 1969. Chemoreceptor specificity associated with choice of feeding site by the beetle, *Chyrsolina brunsvicensis*, on its food plant, *Hypericum hirsutum*. *Entomologia experimentalis et applicata* 12:565–583.

Rees, H. H. 1971. Ecdysones. In *Aspects of terpenoid chemistry and biochemistry*, ed. T. W. Goodwin, pp. 181–222. New York: Academic Press.

Rees, W. J. 1966. *Cyanea lamarcki* Péron and Lesueur (Scyphozoa) and its association with young *Gadus merlangus* L. (Pisces). *Annual Magazine of Natural History* 9:285–287.

Reese, E. 1964. Ethology and marine zoology. *Oceanography and Marine Biology. An Annual Review* 1964:455–488.

———. 1975. A comparative field study of the social behavior and related ecology of reef fishes of the family Chaetodontidae. *Zeitschrift für Tierpsychologie* 37:37–61.

Rehr, S. S., Feeny, P. P., and Janzen, D. H. 1973. Chemical defense in Central American non-ant-acacias. *Journal of Animal Ecology* 42:405–416.

Reichstein, T., Euw, J. von, Parsons, J. A., and Rothschild, M. 1968. Heart poisons in the monarch butterfly. *Science* 161:861–866.

Reid, B. E. 1964. The Cape Hallett adelie penguin rookery—Its size, composition, and structure. *Records of the Dominican Museum* 5:11–37.

Rettenmeyer, C. W. 1970. Insect mimicry. *Annual Review of Entomology* 15:43–74.

Reynolds, G. S., Catania, A. C., and Skinner, B. F. 1963. Conditioned and unconditioned aggression in pigeons. *Journal of Experimental Analysis of Behavior* 1:73–75.

Reynolds, V., and Reynolds, F. 1965. Chimpanzees of the Budongo Forest. In *Primate behavior: Field studies of monkeys and apes,* ed. I. Devore, pp. 368–424. New York: Holt, Rinehart and Winston.

Rheingold, H. L. 1963. Maternal behavior in the dog. In *Maternal behavior in mammals,* ed. H. L. Rheingold, pp. 169–202. New York: Wiley.

Rhijn, J. G. van. 1973. Behavioural dimorphism in male ruffs, *Philomachus pugnax* (L.). *Behaviour* 47:153–229.

Rhine, R. J. 1975. The order of movement of yellow baboons (*Papio cynocephalus*). *Folia Primatologica* 23:72–104.

Rhine, R. J., and Owens, N. W. 1972. The order of movement of adult male and black infant baboons (*Papio anubis*) entering and leaving a potentially dangerous clearing. *Folia Primatologica* 18:276–283.

Rhoades, D. F. 1977. The anti-herbivore defenses of *Larrea.* In *Creosote bush,* ed. T. J. Mabry, J. Hunziker, and D. R. DiFeo, Jr., pp. 135–175. Stroudsburg, Pa.: Dowden, Hutchinson & Ross.

———. 1979. Evolution of plant chemical defense against herbivores. In *Herbivores: Their interaction with secondary plant metabolites,* ed. G. A. Rosenthal and D. H. Janzen, pp. 3–54. New York: Academic Press.

Rhoades, D. F., and Cates, R. G. 1976. Toward a general theory of plant antiherbivore chemistry. In *Biochemical interaction between plants and insects. Recent Advances in Phytochemistry,* vol. 10, ed. J. W. Wallace and R. L. Mansell, pp. 168–213. New York: Plenum.

Ribaut, J. 1964. Dynamique d'une population de Merles noires *Turdus merula* L. *Revue Suisse de Zoologie* 71:815–902.

Ribbands, C. Ronald. 1953. *Behavior and social life of honeybees.* New York: Dover.

Rice, D. W., and Kenyon, K. W. 1962. Breeding cycles and behavior of Laysan and black-footed albatrosses. *Auk* 79:517–567.

Rice, E. L. 1974. *Allelopathy.* New York: Academic Press.

Rice, J. C. 1978. Behavioural interactions of interspecifically territorial vireos. I. Song discrimination and natural interactions. *Animal Behaviour* 26:527–549.

Richards, O. W. 1927. Sexual selection and allied problems in the insects. *Biological Reviews* 2:298–364.

———. 1947. Observations on *Trypoxylon placidum* Cam. (Hym., Sphecoidea). *Entomologist's Monthly Magazine* 83:53.

———. 1953. *The social insects.* New York: Harper & Row.

Richards, S. M. 1974. The concept of dominance and methods of assessment. *Animal Behaviour* 22:914–930.

Richdale, L. E. 1957. *A population study of penguins.* Oxford: Clarendon Press.

Richerson, P. J. 1977. Ecology and human ecology: A comparison of theories in the biological and social sciences. *American Ethnologist* 4:1–26.

Richerson, P. J., and Boyd, R. 1978. A dual inheritance model of the human evolutionary process. I: Basic postulates and a simple model. *Journal of Social Biological Structure* 1:127–154.

Ricklefs, R. E. 1969. An analysis of nesting mortality in birds. *Smithsonian Contributions to Zoology* 9:1–48.

———. 1970. Clutch size in birds: Outcome of opposing predator-prey adaptations. *Science* 168:599–600.

———. 1977a. A note on the evolution of clutch size in altricial birds. In *Evolutionary ecology,* ed. B. Stonehouse and C. M. Perrins, pp. 193–214. London: Macmillan.

———. 1977b. On the evolution of reproductive strategies in birds: Reproductive effort. *American Naturalist* 111:453–461.

———. 1979. *Ecology.* 2nd ed. Portland, Oreg.: Chiron Press.

Ridley, M. W. 1978. Paternal care. *Animal Behaviour* 26:904–932.

———. 1980. The breeding behaviour and feeding ecology of grey phalaropes *Phalaropus fulicarius* in Svalbard. *Ibis* 122:210–226.

Riechert, S. E. 1978. Games spiders play: Behavioral variability in territorial disputes. *Behavioral Ecology and Sociobiology* 3:135–162.

Riemann, J. G., Moen, D. J., and Thorsen, B. J. 1967. Female monogamy and its control in the housefly, *Musca domestica. Journal of Insect Physiology* 13:407–418.

Ripley, S. D. 1959. Competition between sunbird and honeycreeper species in the Moluccan Islands. *American Naturalist* 93:127–132.

———. 1961. Aggressive neglect as a factor in interspecific competition in birds. *Auk* 78:366–371.

Rippin, A. B., and Boag, D. A. 1974. Recruitment to populations of male sharp-tailed grouse. *Journal of Wildlife Management* 38:616–621.

Ritter, J. 1964. Territoriality in *Cryptocercus punctulatus. Science* 143:1459–1460.

Rivero, J. A., and Esteves, A. E. 1969. Observations on the agonistic and breeding behavior of *Leptodactylus pentadactylus* and other amphibia species in Venezuela. *Breviora,* no. 321:1–14.

Roberts, T. A., and Kennelly, J. J. 1980. Variation in promiscuity among red-winged blackbirds. *Wilson Bulletin* 92:110–112.

Roberts, W. W., Steinberg, M. L., and Means, L. W. 1967. Hypothalamic mechanisms for sexual, aggressive, and other motivational behaviors in the opossum, *Didelphis virginiana. Journal of Comparative and Physiological Psychology* 64:1–15.

Robertson, D. R. 1972. Social control of sex reversal in a coral-reef fish. *Science* 177:1007–1009.

Robertson, R. J. 1972. Optimal niche space of the red-winged blackbird (*Agelaius phoeniceus*). I. Nesting success in marsh and upland habitat. *Canadian Journal of Zoology* 50:247–263.

———. 1973. Optimal niche space of the redwinged blackbird: Spatial and temporal patterns of nesting activity and success. *Ecology* 54:1085–1093.

Robertson, R. J., and Bierman, G. C. 1979. Parental investment strategies determined by expected benefits. *Zeitschrift für Tierpsychologie* 50:124–128.

Robertson, R. J., and Norman, R. F. 1976. Behavioral defenses to brood parasitism by potential hosts of the brown-headed cowbird. *Condor* 78:166–173.

———. 1977. The function and evolution of aggressive host behavior towards the brown-headed cowbird (*Molothrus ater*). *Canadian Journal of Zoology* 55:508–518.

Robinson, A. 1956. The annual reproductive cycle of the magpie, *Gymnorhina dorsalis* Campbell, in south-western Australia. *Emu* 56:233–336.

Robinson, B. W., Alexander, M., and Bourne, G. 1969. Dominance reversal resulting from aggressive responses evoked by brain telestimulation. *Physiology and Behavior* 4:749–752.

Robinson, M. H. 1968a. The defensive behaviour of the stick insect *Oncotophasma martini* (Griffini) (Orthoptera: Phasmatidae). *Proceedings of the Royal Entomological Society of London, Series A* 43:183–187.

———. 1968b. The defensive behaviour of the Javanese stick insect, *Orxines macklotti* De Haan, with a note on the startle display of *Metriotes dicocles* Westw. (Phasmatodea, Phasmidae). *Entomologist's Monthly Magazine* 104:46–54.

———. 1968c. The defensive behaviour of *Pterinoxylus spinulosus* Redtenbacher, a winged stick insect from Panama (Phasmatodea). *Psyche, Cambridge* 75:195–207.

Rodgers, R. J. 1979. Neurochemical correlates of aggressive behaviour: Some relations to emotion and pain sensitivity. In *Chemical influences on behaviour,* ed. K. Brown and S. J. Cooper, pp. 373–419. New York: Academic Press.

Roessler, C., and Post, J. 1972. Prophylactic services of the cleaning shrimp. *Natural History* 81:30–37.

Rogers, J. G., and Beauchamp, G. K. 1976. Some ecological implications of primer chemical stimuli in rodents. In *Mammalian olfaction, reproductive processes, and behavior,* ed. R. L. Doty, pp. 181–195. New York: Academic Press.

Rohwer, S. 1978. Parent cannibalism of offspring and egg raiding as a courtship strategy. *American Naturalist* 112:429–440.

Romaniuk, A. 1965. Representation of aggression and flight reactions in the hypothalamus of the cat. *Acta Biologiae Experimentalis (Warsaw)* 25:177–186.

Rood, J. P. 1974. Banded mongoose males guard young. *Nature* 248:176.

———. 1978. Dwarf mongoose helpers at the den. *Zeitschrift für Tierpsychologie* 48:277–278.

Root, R. B. 1967. The niche exploitation pattern of the blue-gray gnatcatcher. *Ecological Monographs* 37:317–350.

Rösch, G. A. 1930. Untersuchungen über die Arbeitsteilung im Bienenstaat. 2. Teil. Die Tätigkeit der Arbeitsbienen unter experimentelle veränderten Bedingungen. *Zeitschrift für vergleichende Physiologie* 12:1–71.

Rose, R. K., and Gaines, M. S. 1976. Levels of aggression in fluctuating populations of the prairie vole, *Microtus ochrogaster*, in eastern Kansas. *Journal of Mammalogy* 57:43–57.
Rose, R. M., Holaday, J. W., and Bernstein, I. S. 1971. Plasma testosterone, dominance rank and aggressive behaviour in male rhesus monkeys. *Nature* 231:366–368.
Röseler, P. F. 1970. Unterscheide in der Kastendetermination zwischen den Hummelarten *Bombus hypnorum* und *Bombus terrestris*. *Zeitschrift für Naturforschung* 25:543–548.
Ross, R. M. 1978. Territorial behavior and ecology of the anemonefish *Amphiprion melanopus* on Guam. *Zeitschrift für Tierpsychologie* 46:71–83.
Roth, L., and Willis, E. 1960. The biotic associations of cockroaches. *Smithsonian Miscellaneous Collections* 141:1–439.
Rothe, H. 1975. Some aspects of sexuality and reproduction in groups of captive marmosets (*Callitrix jacchus*). *Zeitschrift für Tierpsychologie* 37:255–273.
Rothschild, Lord. 1955. The spermatozoa of the honey-bee. *Transactions of the Royal Entomological Society of London* 107:289–294.
Rothschild, M., and Clay, T. 1952. *Fleas, flukes, and cuckoos: A study of bird parasites*. London: Collins.
Rothstein, S. I. 1975a. An experimental and teleonomic investigation of avian brood parasitism. *Condor* 77:250–271.
———. 1975b. Evolutionary rates and host defenses against avian brood parasitism. *American Naturalist* 109:161–176.
———. 1976. Experiments on defenses cedar waxwings use against cowbird parasitism. *Auk* 93:675–691.
———. 1979. Gene frequencies and selection for inhibitory traits, with special emphasis on the adaptiveness of territoriality. *American Naturalist* 113:317–331.
Rottman, S. J., and Snowdon, C. T. 1972. Demonstration and analysis of an alarm pheromone in mice. *Journal of Comparative and Physiological Psychology* 81:483–490.
Rovner, J. S. 1968. Territoriality in the sheet-web spider *Linyphia triangularis* (Clerck) (Araneae, Linyphiidae). *Zeitschrift für Tierpsychologie* 25:232–242.
Rowan, M. K. 1965. Regulation of sea-bird numbers. *Ibis* 107:54–59.
———. 1966. Territory as a density-regulating mechanism in some South African birds. *Ostrich Supplement* 6:397–408.
Rowell, T. E. 1963. Behaviour and female reproductive cycles of rhesus macaques. *Journal of Reproduction and Fertility* 6:193–203.
———. 1966. Forest living baboons in Uganda. *Journal of Zoology, London* 149:344–364.
———. 1967. A quantitative comparison of the behaviour of a wild and a caged baboon group. *Animal Behaviour* 15:499–509.
———. 1972. Female reproductive cycles and social behaviour in primates. *Advances in the Study of Behavior* 4:69–105.
———. 1974. The concept of social dominance. *Behavioral Biology* 11:131–154.
Rowell, T. E., Hinde, R. A., and Spencer-Booth, Y. 1964. "Aunt"-infant interactions in captive rhesus monkeys. *Animal Behaviour* 12:219–226.
Rowley, I. 1965. The life history of the superb blue wren, *Malurus cyanea*. *Emu* 64:251–297.
———. 1976. Co-operative breeding in Australian birds. *Proceedings of the 16th International Ornithological Congress*, pp. 657–666.
———. 1978. Communal activities among white-winged choughs *Corcorax melanorhampus*. *Ibis* 120:178–197.
Royama, T. 1966. A re-interpretation of courtship feeding. *Bird Study* 13:116–129.
Rudebeck, G. 1950. The choice of prey and modes of hunting of predatory birds with special reference to their selective effort. *Oikos* 2: 65–88, 200–231.
———. 1955. Some observations at a roost of European swallows and other birds in the south-eastern Transvaal. *Ibis* 97:572–580.
Rudnai, J. A. 1973. *The social life of the lion. A study of the behavior of wild lions* (Panthera leo [*Newman*]) *in the Nairobi National Park, Kenya*. Wallingford, Pa.: Washington Square East.
Rudran, R. 1973. Adult male replacement in one-male troops of purple-faced langurs (*Presbytis senex senex*) and its effect on population structure. *Folia Primatologica* 19:166–192.
Rusch, D. H., and Keith, L. B. 1971. Seasonal and annual trends in numbers of Alberta ruffed grouse. *Journal of Wildlife Management* 35:803–822.
Rutter, M. 1972. *Maternal deprivation*. Harmondsworth, England: Penguin.
Ruyle, E. E. 1973. Genetic and cultural pools: Some suggestions for a unified theory of biocultural evolution. *Human Ecology* 1:201–215.
———. 1977. Comment on "The adaptive significance of cultural behavior." *Human Ecology* 5:53–55.

Ryan, M. J. 1980. Female mate choice in a neotropical frog. *Science* 209:523–525.

Ryden, H. 1974. The "lone" coyote likes family life. *National Geographic* 146(2):279–294.

Rypstra, A. L. 1979. Foraging flocks of spiders. A study of aggregate behavior in *Cyrtophora citricola* Forskål (Araneae; Araneidae) in West Africa. *Behavioral Ecology and Sociobiology* 5:291–300.

Ryves, H. H., and Ryves, B. H. 1934. The breeding habits of the corn-bunting as observed in North Cornwall: With special reference to its polygamous habit. *British Birds* 28:2–26.

Sackett, G. P. 1967. Some persistent effects of differential rearing conditions on pre-adult social behavior of monkeys. *Journal of Comparative and Physiological Psychology* 64:363–365.

Sade, D. S. 1965. Some aspects of parent-offspring and sibling relations in a group of rhesus monkeys, with a discussion of grooming. *American Journal of Physical Anthropology* 23:1–17.

———. 1967. Determinants of dominance in a group of free-ranging rhesus monkeys. In *Social communication among primates*, ed. S. A. Altmann, pp. 99–114. Chicago: University of Chicago Press.

———. 1968. Inhibition of son-mother mating among free-ranging rhesus monkeys. *Science and Psychoanalysis* 12:18–38.

———. 1972. Sociometrics of *Macaca mulatta*. I. Linkages and cliques in grooming matrices. *Folia Primatologica* 18:196–223.

Sahlins, M. D. 1976. *The use and abuse of biology. An anthropological critique of sociobiology*. Ann Arbor: University of Michigan Press.

Sakagami, S. F., and Fukushima, K. 1961. Female dimorphism in a social halictine bee, *Halictus (Seladonia) aerarius* (Smith) (Hymenoptera, Apoidea). *Japanese Journal of Ecology* 11:118–124.

Sale, P. F. 1977. Maintenance of high diversity in coral reef communities. *American Naturalist* 111:337–359.

Salomonson, M. G., and Balda, R. P. 1977. Winter territoriality of Townsend's solitaires *(Myadestes townsendi)* in a piñon-juniper-ponderosa ecotone. *Condor* 79:148–161.

Salthe, S. N., and Mecham, J. S. 1974. Reproductive and courtship patterns. In *Physiology of the Amphibia*, vol. 2, ed. B. Lofts, pp. 309–521. New York: Academic Press.

Samson, F. B. 1976. Territory, breeding density and fall departure in the Cassin's finch. *Auk* 93:477–497.

Sassenrath, E. N. 1970. Increased adrenal responsiveness related to social stress in rhesus monkeys. *Hormones and Behavior* 1:283–290.

Saunders, F. J. 1968. Effects of sex steroids and related compounds on pregnancy and on development of the young. *Physiological Reviews* 48:601–643.

Savage, R. M. 1934. The breeding behaviour of the common frog, *Rana temporaria temporaria* Linn., and of the common toad, *Bufo bufo bufo* Linn. *Proceedings of the Zoological Society of London* 1934:55–74.

———. 1961. *The ecology and life history of the common frog*. New York: Hafner Press.

Schaffer, W. M. 1974. Optimal reproductive effort in fluctuating environments. *American Naturalist* 108:783–790.

Schäller, G. 1968. Biochemische Analysis des Aphidenspeichels und seine Bedeutung für die Gallenbildung. *Zoologische Jahrbucher. Abteilung für allgemeine Zoologie und Physiologie der Tiere*. 74:54–87.

Schaller, G. B. 1963. *The mountain gorilla: Ecology and behavior*. Chicago: University of Chicago Press.

———. 1964. Breeding behavior of the white pelican at Yellowstone Lake, Wyoming. *Condor* 66:3–23.

———. 1967. *The deer and the tiger: A study of wildlife in India*. Chicago: University of Chicago Press.

———. 1969. Life with the king of beasts. *National Geographic* 135:494–519.

———. 1972. *The Serengeti lion: A study of predator-prey relations*. Chicago: University of Chicago Press.

Schamel, D., and Tracy, D. 1977. Polyandry, replacement clutches, and site tenacity in the red phalarope (*Phalaropus fulicarius*) at Barrow, Alaska. *Bird Banding* 48:314–324.

Schemnitz, S. D. 1961. Ecology of the scaled quail in the Oklahoma panhandle. *Wildlife Monographs* 8:1–47.

Schenkel, R. 1966. Play, exploration, and territoriality in the wild lion. *Symposia of the Zoological Society of London* 18:11–22.

Schifferli, A. 1953. Der Bergfinken—Masseneinfall (*Fringilla montifringilla*) 1950/1951 in der Schweiz. *Ornithologische Beobachter* 50:65–89.

Schildtknecht, H. 1971. Evolutionary peaks in the defensive chemistry of insects. *Endeavour* 30:136–141.

Schildkraut, T. J., and Kety, S. S. 1967. Biogenic anines and emotion. *Science* 156:21–30.

Schlicter, D. 1970. *Thalassoma amblycephalus* ein neuer Anemonefisch—Typ. allgemeine Aspekte zur Beurteilung der Vergesellschaftung von Fiffanemone und ihren Partnern. *Marine Biology* 7:269–272.

Schloeth, R. 1961. Das Sozialleben des Camargue-Rindes. Qualitative und quantitative Untersuchungen über die sozialen Beziehungen—insbesondere die soziale Rangordnung—des halbwilden französischen Kampfrindes. *Zeitschrift für Tierpsychologie* 18:574–627.

Schmidt, J. O., and Blum, M. S. 1978. A harvester ant venom: Chemistry and pharmacology. *Science* 200:1064–1066.

Schmidt, R. S. 1958. Behavioural evidence on the evolution of Batesian mimicry. *Animal Behaviour* 6:129–138.

———. 1960. Predator behaviour and the perfection of incipient mimetic resemblances. *Behaviour* 16:149–158.

Schneirla, T. C. 1952. A consideration of some conceptual trends in comparative psychology. *Psychological Bulletin* 49:559–597.

———. 1971. *Army ants: A study in social organization.* San Francisco: Freeman.

Schoener, T. W. 1968. Sizes of feeding territories among birds. *Ecology* 49:123–141.

———. 1971. Theory of feeding strategies. *Annual Review of Ecology and Systematics* 2:369–404.

Schorger, A. W. 1937. The great Wisconsin passenger pigeon nesting of 1871. *Proceedings of the Linnaean Society of New York* 48:1–26.

———. 1955. *The passenger pigeon: Its natural history and extinction.* Madison: University of Wisconsin Press.

Schroeder, E. E. 1968. Aggressive behavior in *Rana clamitans. Journal of Herpetology* 1:95–96.

Schull, W. J., and Neel, J. V. 1965. *The effects of inbreeding in Japanese children.* New York: Harper & Row.

Schuster, L. 1964. Metabolism of drugs and toxic substances. *Annual Review of Biochemistry* 33:571–596.

Schüz, E. 1943. Über die Jungenaufzucht des weissen Storches. *Zeitschrift für Morphologie und Ökologie der Tiere* 40:181–237.

———. 1957. Das Verschlingen ergener Junger ("Kronismus") bei Vogeln und seine Bedeutung. *Die Vogelwarte* 19:1–15.

Schwagmeyer, P. 1979. The Bruce effect: An evaluation of male/female advantages. *American Naturalist* 114:932–939.

Scott, D. M., and Ankney, C. D. 1979. Evaluation of a method for estimating the laying rate of brown-headed cowbirds. *Auk* 96:483–488.

Scott, J. P. 1942. Genetic differences in the social behavior of inbred strains of mice. *Journal of Heredity* 33:11–15.

———. 1966. Agonistic behavior in mice and rats: A review. *American Zoologist* 6:683–701.

———. 1968. Evolution and domestication of the dog. In *Evolutionary biology*, vol. 2, ed. M. K. Hecht, W. D. Steere, and B. Wallace, pp. 244–275. New York: Plenum.

Scott, J. P., and Fredericson, E. 1951. The causes of fighting in mice and rats. *Physiological Zoology* 24:273–309.

Scott, J. P., and Fuller, J. L. 1965. *Genetics and the social behavior of the dog.* Chicago: University of Chicago Press.

Scott, N. J., Malmgren, L. A., and Glander, K. E. 1980. Grouping behaviour and sex ratio in mantled howling monkeys (*Alouatta palliata*). In *Proceedings of the 6th Congress of the International Primatological Society*, vol. 1, ed. D. J. Chivers and C. Harcourt, London: Academic Press.

Scott, N. J., and Starrett, A. 1974. An unusual breeding aggregation of frogs, with notes on the ecology of *Agalychnis spurrelli* (Anura: Hylidae). *Bulletin of the Southern California Academy of Sciences* 73:86–94.

Searcy, W. A. 1979. Male characteristics and pairing success in red-winged blackbirds. *Auk* 96:353–363.

———. 1980. Optimum body sizes at different temperatures: An energetics explanation of Bergmann's rule. *Journal of Theoretical Biology* 83:579–593.

Searcy, W. A., and Yasukawa, K. 1981. Does the sexy son hypothesis apply to mate choice in red-winged blackbirds? *American Naturalist* (in press).

Sears, R. R., Maccoby, E. E., and Levin, H. 1957. *Patterns of child rearing.* Evanston, Ill.: Row & Peterson.

Seay, B. 1966. Maternal behavior in primiparous and multiparous rhesus monkeys. *Folia Primatologica* 4:146–168.

Seemanova, E. 1971. A study of children of incestuous matings. *Human Heredity* 21:108–128.

Seibt, U., and Wickler, W. 1979. The biological significance of the pair bond in the shrimp *Hymenocera picta. Zeitschrift für Tierpsychologie* 50:166–179.

Seigler, D. S. 1977. Primary roles for secondary compounds. *Biochemical Systematics and Ecology* 5:195–199.

Seigler, D. S., and Price, P. H. 1976. Secondary compounds in plants: Primary functions. *American Naturalist* 110:101–105.

Seitz, P. E. D. 1954. The effects of infantile experiences upon adult behavior in animal subjects. I. Effects of litter size during infancy upon adult behavior in the rat. *American Journal of Psychiatry* 110:916–927.

Selander, R. K. 1964. Speciation in wrens of the genus *Campylorynchus. University of California Publications in Zoology* 74:1–305.

———. 1965. On mating systems and sexual selection. *American Naturalist* 99:129–141.

———. 1966. Sexual dimorphism and differential niche utilization in birds. *Condor* 68:113–151.

———. 1972. Sexual selection and dimorphism in birds. In *Sexual selection and the descent of man, 1871–1971*, ed. B. G. Campbell, pp. 180–230. Chicago: Aldine.

Selander, R. K., and Giller, D. R. 1961. Analysis of sympatry of great-tailed and boat-tailed grackles. *Condor* 63:29–86.

Selye, H. 1956. *The stress of life*. New York: McGraw-Hill.

Semler, D. F. 1971. Some aspects of adaptation in a polymorphism for breeding colours in the three-spined stickleback (*Gasterosteus aculeatus*). *Journal of Zoology, London* 165:291–302.

Sergeev, A. M. 1940. Researches on the viviparity of reptiles. *Moscow Society of Naturalists* 1940:1–34.

Sexton, O. J., Ortlet, E. P., Hathaway, L. M., Ballinger, L. M., Licht, R. E., and Licht, P. 1971. Reproductive cycles of three species of anoline lizards from the Isthmus of Panama. *Ecology* 52:201–215.

Seyfarth, R. M. 1976. Social relationships among adult female baboons. *Animal Behaviour* 24:917–938.

———. 1978. Social relationships among adult male and female baboons. I. Behaviour during sexual consortship. *Behaviour* 64:204–226.

Shah, S. A. 1970. *Report on the XYY chromosomal abnormality. National Institute of Mental Health Conference Report*. Washington, D.C.: United States Government Printing Office.

———. 1976. The 47,XYY chromosomal abnormality: A critical appraisal with respect to antisocial and violent behavior. In *Issues in brain/behavior control*, ed. W. L. Smith and A. Kling, pp. 49–67. New York: Spectrum.

Shalter, M. D. 1978. Localization of passerine seeet and mobbing calls by goshawks and pygmy owls. *Zeitschrift für Tierpsychologie* 46:260–267.

Shalter, M. D., and Schleidt, W. M. 1977. The ability of barn owls *Tyto alba* to discriminate and localize avian alarm calls. *Ibis* 119:22–27.

Shapiro, D. Y. 1979. Social behavior, group structure, and the control of sex reversal in hermaphroditic fish. *Advances in the Study of Behavior* 10:43–102.

Sharma, I. K. 1969. Habitat et comportemente du paon (*Pavo cristatus*). *Alauda* 37:219–223.

———. 1970. Analyse ècologique des parades du paon *(Pavo cristatus). Alauda* 38:290–294.

———. 1972. Ètude ècologique de la reproduction du paon, *Pavo cristatus. Alauda* 40:378–384.

Shaw, E. 1925. New genera and species (mostly Australian) of Blattidae, with notes, and some remarks on Tepper's types. *Proceedings of the Linnaean Society of New South Wales* 1:171–231.

Shaw, R. F. 1958. The theoretical genetics of the sex ratio. *Genetics* 43:149–163.

Shearer, J. W. 1976. Effect of aggregations of aphids *(Periphyllus* spp) on their size. *Entomologia experimentalis et applicata* 20:179–182.

Shepard, J. M. 1975. Factors influencing female choice in the lek mating system of the ruff. *Living Bird* 14:87–111.

Sherman, P. W. 1977. Nepotism and the evolution of alarm calls. *Science* 197:1246–1253.

———. 1980. The limits of ground squirrel nepotism. In *Sociobiology: Beyond nature/nurture?* ed. G. W. Barlow and J. Silverberg, pp. 505–544. Boulder, Colo.: Westview Press.

Shields, O. 1967. Hilltopping: An ecological study of summit congregation behavior of butterflies on a southern California hill. *Journal of Research on the Lepidoptera* 6:69–178.

Shields, W. 1980. Ground Squirrel alarm calls: Nepotism or parental care? *American Naturalist* 116:599–603.

Shine, R. 1978. Sexual size dimorphism and male combat in snakes. *Oecologia* 33:269–277.

Shine, R., and Bull, J. J. 1979. The evolution of live-bearing in lizards and snakes. *American Naturalist* 113:905–923.

Shreeve, D. F. 1980. Behaviour of the Aleutian grey-crowned and brown-capped rosy finches *Leucosticte tephrocotis*. *Ibis* 122:145–165.

Sick, H. 1967. Courtship behavior in the manakins (Pipridae): A review. *Living Bird* 6:5–22.

Siegfried, W. R. 1972. Breeding success and reproductive output of the cattle egret. *Ostrich* 43:43–55.

Sigg, E. B. 1969. Relationship of aggressive behavior to adrenal and gonadal function in male mice. In *Aggressive behaviour*, ed. S. Garattini and E. B. Sigg, pp. 143–149. Amsterdam: Excerpta Medica.

Siivonen, L. 1957. The problem of the short-term fluctuations in numbers of tetraonids in Europe. *Papers on Game Research, Helsinki* 19:1–44.

Sikes, S. K. 1971. *The natural history of the African elephant*. New York: Elsevier.

Simmons, K. E. L. 1951. Interspecific territorialism. *Ibis* 93:407–413.

———. 1952. The nature of the predator-reactions of breeding birds. *Behaviour* 4:161–172.

Simon, C. A. 1975. The influence of food abundance on territory size in the iguanid lizard *Sceloporus jarrovi*. *Ecology* 56:993–999.

Simonds, P. E. 1965. The bonnet macaque in South India. In *Primate behavior: Field studies of monkeys and apes*, ed. I. Devore, pp. 175–196. New York: Holt, Rinehart and Winston.

Sinclair, A. R. E. 1974. The social organization of the East African buffalo (*Syncerus caffer* Sparrman). In *The behaviour of ungulates and its relation to management*, ed. V. Geist and F. Walther, pp. 676–689. Moreges: International Union for Conservation of Nature and Natural Resources.

———. 1977. *The African buffalo: A study of resource limitation of populations*. Chicago: University of Chicago Press.

Singh, B. N., and Chalam, G. V. 1937. A quantitative analysis of the weed flora on arable land. *Journal of Ecology* 25:213–221.

Singh, B. N., and Das, K. 1938. Distribution of weed species on arable land. *Journal of Ecology* 26:455–466.

———. 1939. Percentage frequency and quadrat size in analytical studies of weed flora. *Journal of Ecology* 27:66–77.

Skaife, S. H. 1954. Caste differentiation among termites. *Transactions of the Royal Society of South Africa* 34:345–353.

Skellam, J. G. 1952. Studies in statistical ecology. I. Spatial pattern. *Biometrika* 39:346–362.

Skutch, A. F. 1935. Helpers at the nest. *Auk* 52:257–273.

———. 1944a. The life-history of the prong-billed barbet. *Auk* 61:61–88.

———. 1944b. Life history of the blue-throated toucanet. *Wilson Bulletin* 56:133–151.

———. 1949. Do tropical birds rear as many young as they can nourish? *Ibis* 91:430–455.

———. 1954. *Life histories of Central American birds*. Vol. 1. Pacific Coast Avifauna, no. 31. Berkeley, Calif.: Cooper Ornithological Society.

———. 1958. Roosting and nesting of aracari toucans. *Condor* 60:201–219.

———. 1961. Helpers among birds. *Condor* 63:198–226.

———. 1971. Life history of the keel-billed toucan. *Auk* 88:381–396.

Sláma, K. 1969. Plants as a source of materials with insect hormone activity. *Entomologia experimentalis et applicata* 12:721–728.

———. 1979. Insect hormones and antihormones in plants. In *Herbivores: Their interaction with secondary plant metabolites*, ed. G. A. Rosenthal and D. H. Janzen, pp. 683–700. New York: Academic Press.

Slatkin, M. 1975. Gene flow and selection in a two-locus system. *Genetics* 81:787–802.

Slud, P. 1964. The birds of Costa Rica. Distribution and ecology. *Bulletin of the American Museum of Natural History* 128:1–430.

Smith, A. J., and Robertson, B. I. 1978. Social organization of bell miners. *Emu* 78:169–178.

Smith, B. D. 1966. Effect of the plant alkaloid aparteine on the distribution of the aphid *Acrythosiphon spartii*. *Nature* 212:213–214.

Smith, B. G. 1907. The life history and habits of *Cryptobranchus alleganiensis*. *Biological Bulletin* 13:5–39.

Smith, C. C. 1968. The adaptive nature of social organization in the genus of tree squirrels *Tamiasciurus*. *Ecological Monographs* 38:31–63.

Smith, E. O. 1978. A historical view of the study of play: Statement of the problem. In *Social play in primates*, ed. E. O. Smith, pp. 1–32. New York: Academic Press.

Smith, H. M. 1943. Size of breeding populations in relation to egg-laying and reproductive success in the eastern red-wing (*Agelaius p. phoeniceus*). *Ecology* 24:183–207.

———. 1945. *The fresh-water fishes of Siam, or Thailand.* Smithsonian Institution Bulletin, no. 188. Washington, D.C.: Smithsonian Institution Press.

Smith, J. N. M. 1977. Feeding rates, search paths, and surveillance for predators in great-tailed grackle flocks. *Canadian Journal of Zoology* 55:891–898.

Smith, M. 1969. *The British amphibians and reptiles.* London: Collins.

Smith, N. 1973. Spectacular buteo migration over Panama Canal Zone, October 1972. *American Birds* 27:3–5.

Smith, S., and Hosking, E. 1955. *Birds fighting. Experimental studies of the aggressive displays of some birds.* London: Faber & Faber.

Smith, S. F. 1978. Alarm calls, their origin and use in *Eutamias sonomae. Journal of Mammalogy* 59:888–893.

Smith, S. M. 1978. The "underworld" in a territorial sparrow: Adaptive strategy for floaters. *American Naturalist* 112:571–582.

Smith, T. G. 1973. Population dynamics of the ringed seal in the Canadian eastern Arctic. *Bulletin of the Fisheries Research Board of Canada* 181:1–55.

Smith, T. G., and Stirling, I. 1975. The breeding habitat of the ringed seal *(Phoca hispida):* The birth lair and associated structures. *Canadian Journal of Zoology* 53:1297–1305.

Smith, W. J. 1968. Message-meaning analyses. In *Animal communication: Techniques of study and results of research,* ed. T. A. Sebeok, pp. 44–60. Bloomington: Indiana University Press.

———. 1969. Messages of vertebrate communication. *Science* 165:145–150.

———. 1977. *The behavior of communicating: An ethological approach.* Cambridge, Mass.: Harvard University Press.

Smith, W. J., Smith, S. L., Oppenheimer, C. C., Villa, J. G. de, and Ulmer, F. A. 1973. Behavior of a captive population of black-tailed prairie dogs: Annual cycle of social behavior. *Behaviour* 46:189–220.

Smith, W. L., and Kling, A., eds. 1976. *Issues in brain/behavior control.* New York: Spectrum.

Smythe, N. 1970a. On the existence of "pursuit invitation" signals in mammals. *American Naturalist* 104:491–494.

———. 1970b. The adaptive value of the social organization of the coati *(Nasua narica). Journal of Mammalogy* 51:818–820.

———. 1970c. Relationships between fruiting seasons and seed dispersal methods in a neotropical forest. *American Naturalist* 104:25–35.

———. 1977. The function of mammalian alarm advertising: Social signals or pursuit invitation? *American Naturalist* 111:191–194.

Snow, B. K. 1960. The breeding biology of the shag *Phalacrocorax aristotelis* on the island of Lundy, Bristol Channel. *Ibis* 102:554–575.

———. 1970. A field study of the bearded bellbird in Trinidad. *Ibis* 112:299–329.

———. 1974. Lek behaviour and breeding of Guy's hermit hummingbird *Phaethornis guy. Ibis* 116:278–297.

Snow, D. W. 1958. *A study of blackbirds.* London: Allen & Unwin.

———. 1961–1962. The natural history of the oil-bird *Steatornis caripensis. Zoologica* 46:27–47, 199–221.

———. 1962a. A field study of the black and white manakin, *Manacus manacus,* in Trinidad. Zoologica 47:65–104.

———. 1962b. A field study of the golden-headed manakin, *Pipra erythrocephala,* in Trinidad, W. I. *Zoologica* 47:183–198.

———. 1963. The evolution of manakin displays. *Proceedings of the XIII International Ornithological Congress,* pp. 553–561.

———. 1965a. The breeding of the red-billed tropic bird in the Galapagos Islands. *Condor* 67:210–214.

———. 1965b. A possible selection factor in the evolution of fruiting seasons in a tropical forest. *Oikos* 15:274–281.

———. 1968. The singing assemblies of little hermits. *Living Bird* 7:47–55.

———. 1971a. Evolutionary aspects of fruit-eating by birds. *Ibis* 113:194–202.

———. 1971b. Notes on the biology of the cock-of-the-rock *(Rupicola rupicola). Journal für Ornithologie* 112:322–333.

———. 1971c. Social organization of the blue-backed manakin. *Wilson Bulletin* 83:35–38.

Snow, D. W., and Lill, A. 1974. Longevity records for some neotropical birds. *Condor* 76:262–267.

Snow, D. W., and Snow, B. K. 1967. The breeding cycle of the swallow-tailed gull *Creagrus furcatus. Ibis* 109:14–24.

Snyder, N. F. R. 1967. An alarm reaction of aquatic gastropods to intraspecific extract. *Memoirs of the Cornell University Agriculture Experimental Station,* no. 403:1–122.

Snyder, N. F. R., and Snyder, H. 1970. Alarm response of *Diadema antillarum. Science* 168:276–278.

Snyder, N. F. R., and Wiley, J. W. 1976. Sexual size dimorphism in hawks and owls of North America. *Ornithological Monographs* 20:1–96.

Snyder, R. L. 1976. *The biology of population growth.* London: Croom Helm.

Sociobiology Study Group of Science for the People. 1975. Against sociobiology. *New York Review of Books,* 13 November.

———. 1976. Sociobiology—Another biological determinism. *BioScience* 26:182–186.

Sodetz, F. J., and Bunnell, B. N. 1967. Interactive effects of septal lesions and social experience in the hamster. Paper presented at the meeting of the Eastern Psychological Association. Cited in Moyer 1976.

Solomon, M. J., and Crane, F. A. 1970. Influences of heredity and environment on alkaloidal phenotypes in Solanaceae. *Journal of Pharmaceutical Science* 59:1670–1672.

Somanader, S. V. O. 1946. Koel chicks in crow's nests. *Zoo Life* 1:123–124.

Soulié, J. 1964. Le contrôle par les ouvrières de la monogynie des colonies chez *Sphaerocrema striatula* (Myrmicidae, Cremastogastrini). *Insectes Sociaux* 11:383–388.

Southwick, C. H. 1968. Effect of maternal environment on aggressive behavior of inbred mice. *Communications in Behavioral Biology* 1:129–132.

Southwick, C. H., Beg, M. A., and Siddiqi, M. R. 1965. Rhesus monkeys in North India. In *Primate behavior: Field studies of monkeys and apes,* ed. I. Devore, pp. 111–159. New York: Holt, Rinehart and Winston.

Southwick, C. H., and Bland, V. P. 1959. Effect of population density on adrenal glands and reproductive organs of CFW mice. *American Journal of Physiology* 197:111–114.

Sowls, L. K. 1974. Social behaviour of the collared peccary *Dicotyles tajacu* L. In *The behaviour of ungulates and its relation to management,* ed. V. Geist and F. Walther, pp. 144–165. Moreges: International Union for Conservation of Nature and Natural Resources.

Sparks, J. H. 1967. Allogrooming in primates: A review. In *Primate ethology: Essays on the socio-sexual behavior of apes and monkeys,* ed. D. Morris, pp. 190–225. Chicago, Aldine.

Spencer, H. 1851. *Social statics.* London: Chapman.

Spencer-Booth, Y. 1968. The behaviour of group companions towards rhesus monkey infants. *Animal Behaviour* 16:541–557.

Spinage, C. A. 1969. Territoriality and social organization of the Uganda defassa waterbuck *Kobus defassa ugandae. Journal of Zoology, London* 159:329–361.

Spradbery, J. P. 1971. Seasonal changes in the population structure of wasp colonies (Hymenoptera: Vespidae). *Journal of Animal Ecology* 40:501–523.

———. 1973. The European social wasp, *Paravespula germanica* (F.) (Hymenoptera: Vespidae) in Tasmania, Australia. *7th Congress of the IUSSI, London,* pp. 375–380.

Springer, V. G., and Smith-Vaniz, W. F. 1972. Mimetic relationships involving fishes of the family Blenniidae. *Smithsonian Contributions to Zoology* 112:1–36.

Sprunt, A., Jr. 1948. The tern colonies of the Dry Tortugas Keys. *Auk* 65:1–19.

Stacey, P. B. 1973. Kinship, promiscuity, and communal breeding in the acorn woodpecker. *Behavioral Ecology and Sociobiology* 6:53–66.

———. 1979. Habitat saturation and communal breeding in the acorn woodpecker. *Animal Behaviour* 27:1153–1166.

Stacey, P. B., and Bock, C. E. 1978. Social plasticity in the acorn woodpecker. *Science* 202:1298–1300.

Staedler, E. 1977. Sensory aspects of insect plant interactions. *Proceedings of the XV International Congress of Entomology,* pp. 228–248.

Stallcup, J. A., and Woolfenden, G. E. 1978. Family status and contributions to breeding by Florida scrub jays. *Animal Behaviour* 26:1144–1156.

Stammbach, E. 1978. On social differentiation in groups of captive female hamadryas baboons. *Behaviour* 67:322–338.

Stamps, J. A. 1977. Social behavior and spacing patterns in lizards. In *Biology of the Reptilia.* Vol. 7. *Ecology and behaviour,* A, ed. C. Gans and D. W. Tinkle, pp. 265–334. New York: Academic Press.

Stamps, J. A., Metcalf, R. A., and Krishna, V. V. 1979. A genetic analysis of parent-offspring conflict. *Behavioral Ecology and Sociobiology* 3:369–392.

Starr, C. K. 1979. Origin and evolution of insect sociality: A review of modern theory. In *Social insects,* vol. I, ed. H. R. Hermann, pp. 35–79. New York: Academic Press.

Stearns, S. C. 1976. Life-history tactics: A review of ideas. *Quarterly Review of Biology* 51:3–47.

———. 1977. The evolution of life history traits: A critique of the theory and a review of the data. *Annual Review of Ecology and Systematics* 8:145–171.
Stehn, R. A., and Richmond, M. E. 1975. Male-induced pregnancy termination in the prairie vole, *Microtus orchogaster. Science* 187:1211–1213.
Stenger, J. 1958. Food habits and available food of ovenbirds in relation to territory size. *Auk* 75:335–346.
Stenseth, N. C. 1978. Is the female biased sex ratio in wood lemming *Myopus schisticolor* maintained by cyclic inbreeding? *Oikos* 30:83–89.
Stephens, J. S., Jr., Hobson, E. S., and Johnson, R. K. 1966. Notes on distribution, behavior, and morphological variation in some chaenopsid fishes from the tropical eastern Pacific, with descriptions of two new species, *Acantemblemaria castroi* and *Coalliozetus springeri. Copeia* 1966:424–438.
Stephens, J. S., Jr., Johnson, R. K., Key, G. S., and McCosker, J. E. 1970. The comparative ecology of three sympatric species of California blennies of the genus *Hypsoblennius* Gill (Teleostomi, Blenniidae). *Ecological Monographs* 40:213–233.
Stewart, R. E., and Aldrich, J. W. 1951. Removal and repopulation of breeding birds in a spruce-fir forest community. *Auk* 68:471–482.
Stiles, F. G. 1973. Food supply and the annual cycle of the Anna hummingbird. *University of California Publications in Zoology* 97:1–109.
Stimson, J. 1970. Territorial behavior of the owl limpet, *Lottia gigantea. Ecology* 51:113–118.
———. 1973. The role of the territory in the ecology of the intertidal limpet *Lottia gigantea* (Gray). *Ecology* 54:1020–1030.
Stinson, C. H. 1979. On the selective advantage of fratricide in raptors. *Evolution* 33:1219–1225.
———. 1980. Weather-dependent foraging success and sibling aggression in red-tailed hawks in central Washington. *Condor* 82:76–80.
Stirling, I. 1969. Ecology of the Weddell seal in McMurdo Sound, Antarctica. *Ecology* 50:573–586.
———. 1975. Factors affecting the evolution of social behaviour in the Pinnipedia. *Rapporte de Procès-Verbaux Réunions Conseil Internationale Exploration de Mer* 165:205–212.
———. 1977. Adaptations of Weddell and ringed seals to exploit the polar fast ice habitat in the absence or presence of surface predators. In *Adaptations within Antarctic ecosystems,* ed. G. Llano, pp. 741–748. Houston: Gulf.
Stobo, W., and McLaren, I. A. 1975. *The Ipswich sparrow.* Halifax: Nova Scotian Institute of Science.
Stokes, A. W. 1974. Introduction. In *Territory. Benchmark Papers in Animal Behavior,* ed. A. W. Stokes, pp. 1–4. Stroudsburg, Pa.: Dowden, Hutchinson & Ross.
Stoltz, L. P., and Saayman, G. S. 1970. Ecology and behaviour of baboons in the northern Transvaal. *Annals of the Transvaal Museum, Pretoria* 26:99–143.
Stonehouse, B. S. 1960. The king penguin *Aptenodytes patagonica* of South Georgia. I. Breeding behaviour and development. *Scientific Report of the Falkland Islands Dependency Survey,* no. 23:1–81.
———. 1962. The tropic birds (genus *Phaethon*) of Ascension Island. *Ibis* 103b:409–422.
———. 1963. The frigate bird *Fregata aquila* of Ascension Island. *Ibis* 103b:409–422.
Storr, A. 1968. *Human aggression.* New York: Atheneum.
———. 1972. *Human destructiveness.* New York: Basic Books.
Stowers, J. F., Harke, D. T., and Stickley, A. R. 1968. Vegetation used for nesting by the red-winged blackbird in Florida. *Wilson Bulletin* 80:320–324.
Strawn, K. 1958. Life history of the pigmy seahorse, *Hippocampus zosterae* Jordan and Gilbert, at Cedar Key, Florida. *Copeia* 1958:16–22.
Streisinger, G. 1948. Experiments on sexual isolation in *Drosophila.* IX. Behavior of males with etherized females. *Evolution* 2:187–188.
Strobeck, C., Maynard Smith, J., and Charlesworth, B. 1976. The effects of hitch-hiking on a gene for genetic recombination. *Genetics* 82:547–558.
Struhsaker, T. T. 1967a. Auditory communication among vervet monkeys. In *Social communication among primates,* ed. S. A. Altmann, pp. 281–324. Chicago: University of Chicago Press.
———. 1967b. Ecology of vervet monkeys *(Cercopithecus aethiops)* in the Masai-Amboseli Game Reserve, Kenya. *Ecology* 48:891–904.
———. 1974. Correlates of ranging behavior in a group of red colobus monkeys *(Colobus badius tephrosceles). American Zoologist* 14:177–184.
———. 1975. *The red colobus monkey.* Chicago: University of Chicago Press.
———. 1977. Infanticide and social organization in the redtail monkey *(Cercopithecus ascanius*

schmidti) in the Kibale Forest, Uganda. *Zeitschrift für Tierpsychologie* 45:75–84.
———. 1980. Comparison of the behaviour and ecology of red colobus and redtail monkeys in the Kibale Forest, Uganda. *African Journal of Ecology* 18:33–51.
Struhsaker, T. T., and Gartlan, J. S. 1970. Observations on the behaviour and ecology of the patas monkey *(Erythrocebus patas)* in the Waza Reserve, Cameroon. *Journal of Zoology, London* 161:49–63.
Struhsaker, T. T., and Oates, J. F. 1975. Comparison of the behavior and ecology of red colobus and black-and-white colobus monkeys in Uganda: A summary. In *Socio-ecology and psychology of primates,* ed. R. H. Tuttle, pp. 103–123. The Hague: Mouton.
Stuart, A. M. 1969. Social behavior and communication. In *Biology of termites,* vol. I, ed. K. Krishna and F. M. Weesner, pp. 193–232. New York: Academic Press.
Sudd, J. H. 1967. *An introduction to the behaviour of ants.* London: Arnold, Ltd.
Sugiyama, Y. 1965. Behavioral development and social structure in two troops of Hanuman langurs *(Presbytis entellus). Primates* 6:213–247.
———. 1967. Social organization of Hanuman langurs. In *Social communication among primates,* ed. S. A. Altmann, pp. 221–236. Chicago: University of Chicago Press.
———. 1969. Social behavior of chimpanzees in the Budongo Forest, Uganda. *Primates* 10:197–225.
———. 1976. Life history of male Japanese monkeys. *Advances in the Science of Behaviour* 7:255–284.
Sumere, C. F. van, Albrecht, J., Dedonda, A., De Pooter, H., and Pé, I. 1975. Plant proteins and phenolics. In *The chemistry and biochemistry of plant proteins,* ed. J. B. Harborne and C. F. van Sumere, pp. 211–264. New York: Academic Press.
Summers-Smith, D. 1956. Mortality of the house sparrow. *Bird Study* 3:265–270.
Sussman, R. W. 1974. Ecological distinctions in sympatric species of *Lemur.* In *Prosimian biology,* ed. R. D. Martin, G. A. Doyle, and A. C. Walker, pp. 75–108. London: Duckworth.
———. 1977. Feeding behaviour of *Lemur catta* and *Lemur fulvus.* In *Primate ecology: Studies of feeding and ranging behaviour in lemurs, monkeys and apes,* ed. T. H. Clutton-Brock, pp. 1–36. New York: Academic Press.
Sussman, R. W., and Richard, A. 1974. The role of aggression among diurnal prosimians. In *Primate aggression, territoriality, and xenophobia: A comparative perspective,* ed. R. Holloway, pp. 49–76. New York: Academic Press.
Sussman, R. W., and Tattersall, I. 1976. Cycles of activity, group composition, and diet of *Lemur mongoz mongoz* Linnaeus 1766 in Madagascar. *Folia Primatologica* 26:270–283.
Svärdson, G. 1949. Competition and habitat selection in birds. *Oikos* 1:157–174.
Svare, B. B., and Gandelman, R. 1975. Aggressive behavior of juvenile mice: Influence of androgen and olfactory stimuli. *Developmental Psychobiology* 8:405–415.
Svendsen, G. E. 1974. Behavioral and environmental factors in the spatial distribution and population dynamics of a yellow-bellied marmot population. *Ecology* 55:760–771.
Swain, T. 1977. Secondary compounds as protective agents. *Annual Review of Plant Physiology* 28:479–501.
Swennen, C. 1968. Nest protection by eider ducks and shovelers by means of faeces. *Ardea* 56:248–258.
Swynnerton, C. F. M. 1915. Mixed bird-parties. *Ibis* 1915:346–354.
Syme, G. J. 1974. Competitive orders as measures of social dominance. *Animal Behaviour* 22:931–940.
Symons, D. 1974. Aggressive play and communication in rhesus monkeys *(Macaca mulatta). American Zoologist* 14:317–322.
———. 1978a. *Play and aggression. A study of rhesus monkeys.* New York: Columbia University Press.
———. 1978b. The question of function: Dominance and play. In *Social play in primates,* ed. E. O. Smith, pp. 193–230. New York: Academic Press.
Syren, R. M., and Luykx, P. 1977. Permanent segmental interchange complex in the termite *Incisitermes schwarzi. Nature* 266:167–168.
Szabó-Patay, J. 1928. A kapus-hangya. *Természettundományi Közlöny, Budapest* 1928:215–219.
Taber, S. 1954. The frequency of multiple mating of queen honey bees. *Journal of Economic Entomology* 47:995–998.
———. 1955. Sperm distribution in the spermathecae of multiple-mated queen honey bees. *Journal of Economic Entomology* 48:522–525.
Taber, S., and Wendell, J. 1958. Concerning the number of times queen bees mate. *Journal of Economic Entomology* 51:786–789.
Takai, T., and Mizokami, A. 1959. On the reproduction, eggs, and larvae of the pipefish,

Syngnathus schlegeli Kaup. *Journal of the Shimononseki College of Fisheries* 8:85–89.

Talbot, M. 1945. A comparison of flights of four species of ants. *American Midland Naturalist* 34:504–510.

Tashian, R. E. 1957. Nesting behavior of the crested oropendola *(Psarocolius decumanus)* in northern Trinidad, B. W. I. *Zoologica* 42:87–97.

Tattersall, I., and Sussman, R. W. 1975. Observations on the ecology and behavior of the mongoose lemur, *Lemur mongoz mongoz* Linnaeus, (Primates, Lemuriformes), at Ampijoroa, Madagascar. *Anthropological Papers of the American Museum of Natural History* 52:195–216.

Taylor, R. H. 1962. The adelie penguin *Pygoscelis adeliae* at Cape Royds. *Ibis* 104:176–204.

Taylor, R. J. 1976. Value of clumping to prey and the evolutionary response of ambush predators. *American Naturalist* 110:13–29.

———. 1977. The value of clumping to prey. Experiments with a mammalian predator. *Oecologia* 30:285–294.

Teleki, G. 1973a. Notes on chimpanzee interactions with small carnivores in Gombe National Park, Tanzania. *Primates* 14:407–411.

———. 1973b. *The predatory behaviour of wild chimpanzees.* Lewisburg, Pa.: Bucknell University Press.

———. 1973c. The omnivorous chimpanzee. *Scientific American* 228(1):32–42.

Tenaza, R. R. 1971. Behavior and nesting success relative to nest location in adélie penguins *(Pygoscelis adeliae). Condor* 73:81–92.

———. 1976. Songs, choruses and countersinging of Kloss' gibbons *(Hylobates klossii)* in Siberut Island, Indonesia. *Zeitschrift für Tierpsychologie* 40:37–52.

Tener, J. S. 1954. A preliminary study of the muskoxen of Fosheim Peninsula, Ellesmere Island, N. W. T. *Canadian Wildlife Service, Wildlife Management Bulletin,* 1st Series, no. 9:1–34.

———. 1965. *Muskoxen in Canada. A biological and taxonomic review.* Ottawa: Department of Northern Affairs and Natural Resources.

Tevis, L. 1950. Summer behavior of a family of beavers in New York State. *Journal of Mammalogy* 31:40–65.

Thibault, R. E., and Schultz, R. J. 1978. Reproductive adaptations among viviparous fishes (Cyprinodontiformes: Poeciliidae). *Evolution* 32:320–333.

Thomas, B. 1975. The ecology of work. In *Physiological anthropology,* ed. A. Damon, pp. 59–79. London: Oxford University Press.

Thompson, J. N., and Willson, M. F. 1979. Evolution of temperate fruit/bird interactions: Phenological strategies. *Evolution* 33:973–982.

Thompson, T. 1969. Aggressive behaviour of Siamese fighting fish. Analysis and synthesis of conditioned and unconditioned components. In *Aggressive Behaviour,* ed. S. Garattini and E. B. Sigg, pp. 15–31. Amsterdam: Excerpta Medica.

Thompson, T. E. 1960a. Defensive acid-secretion in marine gastropods. *Journal of the Marine Biology Association of the United Kingdom* 39:115–122.

———. 1960b. Defensive adaptations in opisthobranchs. *Journal of the Marine Biology Association of the United Kingdom* 39:123–134.

Thompson, T. E., and Bennett, I. 1969. *Physalia* nematocysts: Utilized by mollusks for defense. *Science* 166:1532–1533.

Thorn, R. 1962. Protection of the brood by a male salamander, *Hynobius nebulosus. Copeia* 1962:638–640.

Thornhill, R. 1976a. Sexual selection and nuptial feeding behavior in *Bittacus apicalis* (Insecta: Mecoptera). *American Naturalist* 110:529–548.

———. 1976b. Sexual selection and paternal investment in insects. *American Naturalist* 110:153–163.

———. 1980. Mate choice in *Hylobittacus apicalis* (Insecta: Mecoptera) and its relation to some models of female choice. *Evolution* 34:519–538.

Thornton, W. A. 1960. Population dynamics in *Bufo woodhousei* and *Bufo valliceps. Texas Journal of Science* 12:176–200.

Thorson, G. 1950. Reproduction and larval ecology of marine bottom invertebrates. *Biological Reviews* 25:1–45.

Thresher, R. E. 1976. Field analysis of the territoriality of the threespot damselfish *Eupomacentrus planiforms* (Pomacentridae). *Copeia* 1976:266–276.

Tiger, L., and Fox, R. 1971. *The imperial animal.* New York: Holt, Rinehart and Winston.

Tilley, S. G. 1968. Size-fecundity relationships and their evolutionary implications in five desmognathine salamanders. *Evolution* 22:806–816.

———. 1973. Life histories and natural selection in populations of the salamander *Desmognathus ochrophaeus*. *Ecology* 54:3–17.

———. 1977. Studies of life histories and reproduction in North American plethodontid salamanders. In *The reproductive biology of amphibians*, ed. D. H. Taylor and S. I. Guttman, pp. 1–41. New York: Plenum.

Tilson, R. L. 1977. Social organization of simakobu monkeys *(Nasalis concolor)* in Siberut Island, Indonesia. *Journal of Mammalogy* 58:202–212.

———. 1979. Klipspringer *(Oreotragus oreotragus)* social structure and predator avoidance in a desert canyon. *Madoqua* (in press).

Tilson, R. L., and Norton, P. M. MS. Alarm duetting and pursuit deterrence in an African antelope. Submitted to *American Naturalist*.

Tinbergen, L. 1960. The natural control of insects in pinewoods. 1. Factors influencing the intensity of predation by songbirds. *Archives Neerlandaises de Zoologie* 13:265–343.

Tinbergen, N. 1951. *The study of instinct*. Oxford: Oxford University Press.

———. 1952. *Social behaviour in animals with special reference to vertebrates*. London: Methuen.

———. 1956. On the functions of territories in gulls. *Ibis* 98:401–411.

———. 1959. Comparative studies of the behaviour of gulls (Laridae): A progress report. *Behaviour* 15:1–70.

———. 1960a. The evolution of behavior in gulls. *Scientific American* 203(6):118–130.

———. 1960b. *The herring gull's world*. Garden City, N.Y.: Doubleday.

———. 1963. On aims and methods of ethology. *Zeitschrift für Tierpsychologie* 20:410–433.

———. 1968. On war and peace in animals and man. *Science* 160:1411–1418.

Tinbergen, N., and Iersel, J. J. A. van. 1947. "Displacement reactions" in the three-spined stickleback. *Behaviour* 1:56–63.

Tinbergen, N., Impekoven, M., and Franck, D. 1967. An experiment on spacing-out as a defense against predation. *Behaviour* 28:307–321.

Tinkham, E. R. 1971a. The biology of the gila monster. In *Venomous animals and their venoms*. Vol. II. *Venomous vertebrates*, ed. W. Bücherl and E. E. Buckley, pp. 387–413. New York: Academic Press.

———. 1971b. The venom of the gila monster. In *Venomous animals and their venoms*. Vol. II. *Venomous vertebrates*, ed. W. Bücherl and E. E. Buckley, pp. 415–422. New York: Academic Press.

Tinkle, D. W. 1967. The life and demography of the side-blotched lizard. *Miscellaneous Publications of the Museum of Zoology, University of Michigan* 132:1–182.

———. 1969. The concept of reproductive effort and its relation to the evolution of life histories of lizards. *American Naturalist* 103:501–516.

———. 1973. A population analysis of the sagebrush lizard, *Sceloporus graciosus*, in southern Utah. *Copeia* 1973:284–295.

Tinkle, D. W., and Ballinger, R. E. 1972. *Sceloporus undulatus:* A study of the intraspecific comparative demography of a lizard. *Ecology* 53:570–584.

Tinkle, D. W., and Gibbons, J. W. 1977. The distribution and evolution of viviparity in reptiles. *Miscellaneous Publications of the Museum of Zoology, University of Michigan* 154:1–55.

Tinkle, D. W., and Hadley, N. F. 1975. Lizard reproductive effort: Caloric estimates and comments on its evolution. *Ecology* 56:427–434.

Tinkle, D. W., Wilbur, H. M., and Tilley, S. C. 1970. Evolutionary strategies in lizard reproduction. *Evolution* 24:55–74.

Tinley, K. L. 1969. Dik-dik *Madoqua kirki* in South-West Africa: Notes on distribution, ecology and behaviour. *Madoqua* 1:7–33.

Tollman, J., and King, J. A. 1956. The effects of testosterone propionate on aggression in male and female C57BL/10 mice. *British Journal of Animal Behaviour* 4:147–149.

Traniello, J. F. A. 1978. Caste in a primitive ant: Absence of age polyethism in *Amblyopone*. *Science* 202:770–772.

Treisman, M., and Dawkins, R. 1976. The "cost of meiosis": Is there any? *Journal of Theoretical Biology* 63:479–484.

Trivers, R. L. 1971. The evolution of reciprocal altruism. *Quarterly Review of Biology* 46:35–57.

———. 1972. Parental investment and sexual selection. In *Sexual selection and the descent of man, 1871–1971*, ed. B. G. Campbell, pp. 136–179. Chicago: Aldine.

———. 1974. Parent-offspring conflict. *American Zoologist* 14:249–264.

———. 1976. Foreword. In *The selfish gene*, R. Dawkins. London: Oxford University Press.

Trivers, R. L., and Hare, H. 1976. Haplodiploidy and the evolution of the social insects. *Science* 191:249–263.

Trivers, R. L., and Willard, D. E. 1973. Natural selection of parental ability to vary the sex ratio of offspring. *Science* 179:90–92.

Tschanz, B. 1979. Helfer-Beziehungen bei Trottellummen. *Zeitschrift für Tierpsychologie* 49:10–34.

Tsuneki, K., and Adachi, Y. 1957. The intra- and interspecific influence relations among nest populations of four species of ants. *Japanese Journal of Ecology* 7:166–171.

Tucker, B. W. 1943. In *The handbook of British birds,* ed. H. F. Witherby, F. C. R. Jourdain, N. Ticehurst, and B. W. Tucker, p. 270. London: H. F. & G. Witherby.

Turner, C. H., and Ebert, E. E. 1962. The nesting of *Chromis punctipinnis* (Cooper) and a description of their eggs and larvae. *California Fish and Game* 48:243–248.

Turner, E. A. 1954. Cerebral control of respiration. *Brain* 77:448–486.

Turner, E. R. A. 1965. Social feeding in birds. *Behaviour* 24:1–46.

Turner, F. B., Hoddenbach, G. A., Medica, P. A., and Lannom, J. R. 1970. The demography of the lizard, *Uta stansburiana* Baird and Girard, in southern Nevada. *Journal of Animal Ecology* 39:505–519.

Turner, F. B., Jennrich, R. I., and Weintraub, J. D. 1969. Home ranges and body size of lizards. *Ecology* 50:1076–1081.

Tyler, S. 1972. The behaviour and social organization of the New Forest ponies. *Animal Behaviour Monographs* 5:85–196.

Ulrich, R. E., Johnston, M., Richardson, J., and Wolff, P. C. 1963. The operant conditioning of fighting behavior in rats. *Psychological Record* 13:465–470.

Ulrich, R. E., and Symannek, B. 1969. Pain as a stimulus for aggression. In *Aggressive behaviour,* ed. S. Garattini and E. B. Sigg, pp. 59–69. Amsterdam: Excerpta Medica.

Urquhart, F. A. 1976. Found at last: The monarch's winter home. *National Geographic* 150:160–173.

Valzelli, L. 1969. Aggressive behaviour induced by isolation. In *Aggressive behavior,* ed. S. Garattini and E. B. Sigg, pp. 70–76. Amsterdam: Excerpta Medica.

———. 1974. Aggressiveness by isolation in rodents. In *Determinants and origins of aggressive behavior,* ed. J. de Wit and W. W. Hartup, pp. 299–308. The Hague: Mouton.

Valzelli, L., and Garattini, S. 1968. Behavioral changes and S-hydroxytryptamine turnover in animals. In *Proceedings of the International Congress on the Biological Role of Indolealkylamine Derivatives,* ed. S. Garattini and P. A. Shore. New York: Academic Press.

Vandenbergh, J. G. 1971. Effects of gonadal hormones on aggressive behavior of adult golden-hamsters *(Mesocricetus auratus). Animal Behaviour* 19:589–594.

Vandernoot, G. W., Symons, L., Lyman, R., and Fonnesbeck, P. 1967. Rate of passage of various food stuffs through the digestive tract of horses. *Journal of Animal Science* 26:1309–1311.

Van Someren, V. D. 1945. The dancing display and courtship of Jackson's whydah *(Coliuspasser jacksoni* Sharpe). *Journal of the East African Natural History Society* 18:131–141.

Van Tyne, J. 1929. Life history of the toucan *Ramphastos brevicarinatus. Miscellaneous Publications of the Museum of Zoology, University of Michigan* 19:5–43.

Van Valen, L. 1975. Group selection, sex, and fossils. *Evolution* 29:87–94.

Varley, M., and Symmes, D. 1966. The hierarchy of dominance in a group of macaques. *Behaviour* 27:54–75.

Vayda, A. P., and McCay, B. 1975. New directions in ecology and ecological anthropology. *Annual Review of Anthropology* 4:293–306.

Veen, J. 1977. Functional and casual aspects of nest distribution in colonies of the sandwich tern (*Sterna s. sandvicensis* Lath). *Behaviour Supplement* 20:1–193.

Vehrencamp, S., Stiles, F. G., and Bradbury, J. W. 1977. Observations on the foraging behavior and avian prey of the neotropical carnivorous bat, *Vampyrum spectrum. Journal of Mammalogy* 58:469–478.

Velthius, H. H. W. 1976. Environmental, genetic and endocrine influences in stingless bee caste determination. In *Phase and caste determination in insects: Endocrine aspects. 15th International Congress of Entomology,* ed. M. Lüscher, pp. 35–53. Oxford: Pergamon Press.

Velthius, H. H. W., and Camargo, J. M. F. de. 1975a. Observations on male territories in a carpenter bee, *Xylocopa (Neoxylocopa) hirsutissima* Maidl (Hymenoptera, Anthophoridae). *Zeitschrift für Tierpsychologie* 38:409–418.

———. 1975b. Further observations on the function of male territories in the carpenter bee *Xylocopa (Neoxylocopa) hirsutissima* Maidl (Anthophoridae, Hymenoptera). *Netherlands Journal of Zoology* 25:516–528.

Venables, U. M., and Venables, L. S. V. 1955. Observations of a breeding colony of the seal

Phoca vitulina in Shetland. *Proceedings of the Zoological Society of London* 125:521–532.
———. 1957. Mating behaviour of the seal *Phoca vitulina* in Shetland. *Proceedings of the Zoological Society of London* 128:387–396.
———. 1959. Vernal coition of the seal *Phoca vitulina* in Shetland. *Proceedings of the Zoological Society of London* 132:665–669.
Vencl, F. 1977. A case of convergence in vocal signals between marmosets and birds. *American Naturalist* 111:777–782.
Verheyen, R. 1955. Contribution a l'éthologie du waterbuck, *Kobus defassa*, et de l'antilope harnáchée, *Tragelaphus scriptus*. *Mammalia* 19:309–319.
Vermeer, K. 1963. The breeding ecology of the glaucous-winged gull (*Larus glaucescens*) on Mandarte Island, B. C. *Occasional Papers of the British Columbia Provincial Museum*, no. 13.
Verner, J. 1964. The evolution of polygamy in the long-billed marsh wren. *Evolution* 18:252–261.
———. 1965. Breeding biology of the long-billed marsh wren. *Condor* 67:6–30.
———. 1977. On the adaptive significance of territoriality. *American Naturalist* 111:769–775.
Verner, J., and Engelsen, G. H. 1970. Territories, multiple nest building, and polygyny in the long-billed marsh wren. *Auk* 87:557–567.
Verner, J., and Willson, M. F. 1966. The influence of habitats on mating systems of North American passerine birds. *Ecology* 47:143–147.
———. 1969. Mating systems, sexual dimorphism and the role of male North American passerine birds in the nesting cycle. *Ornithological Monographs* 9:1–76.
Viljoen, P. C. 1975. Oribis—The vanishing highvelders. *Fauna and Flora* 26:12–14.
Vine, I. 1971. Risk of visual detection and pursuit by a predator and the selective advantage of flocking behaviour. *Journal of Theoretical Biology* 30:405–422.
Vine, P. J. 1974. Effects of algal grazing and aggressive behaviour of the fishes *Pomacentrus lividus* and *Acanthurus sohal* on coral-reef ecology. *Marine Biology* 24:131–136.
Vos, G. J. de. 1979. Adaptedness of arena behaviour in black grouse (*Tetrao tetrix*) and other grouse species (Tetraonidae). *Behaviour* 68:277–314.
Waage, J. K. 1973. Reproductive behavior and its relation to territoriality in *Calopteryx maculata* (Beauvois) (Odonata: Calopterygidae). *Behaviour* 47:240–256.
———. 1979a. Dual function of the damselfly penis: Sperm removal and transfer. *Science* 203:916–918.
———. 1979b. Adaptive significance of postcopulatory guarding of mates and nonmates by male *Calopteryx maculata* (Odonata). *Behavioral Ecology and Sociobiology* 6:147–154.
Waddington, C. H. 1957. *The strategy of the genes: A discussion of some aspects of theoretical ecology.* London: Allen & Unwin.
Wade, M. J. 1977. An experimental study of group selection. *Evolution* 31:134–153.
———. 1978. A critical review of the models of group selection. *Quarterly Review of Biology* 53:101–114.
———. 1979. Sexual selection and variance in reproductive success. *American Naturalist* 114:742–746.
Wade, M. J., and Arnold, S. J. 1980. The intensity of sexual selection in relation to male sexual behaviour, female choice, and sperm precedence. *Animal Behaviour* 28:446–461.
Wade, T. D. 1979. Inbreeding, kin selection, and primate social evolution. *Primates* 20:355–370.
Wager, V. 1965. *The frogs of South Africa.* Capetown: Prunell & Sons.
Wagner, H. O. 1944. Notes on the life history of the emerald toucanet. *Wilson Bulletin* 56:65–76.
Wahl, M. 1969. Untersuchungen zur Bio-Akustik des Wasserfrosches *Rana esculenta* (L.). *Oecologia* 3:14–55.
Wahl, T. R., and Heinemann, D. 1979. Seabirds and fishing vessels: Co-occurrence and attraction. *Condor* 81:390–396.
Walker, B. W. 1952. A guide to the grunion. *California Fish and Game* 38: 409–420.
Wallace, B. 1968. *Topics in population genetics.* New York: Norton.
———. 1973. Misinformation, fitness, and selection. *American Naturalist* 107:1–7.
Waloff, N. 1957. The effect of the number of queens of the ant *Lasius flavus* (Fab.) (Hym., Formicidae) on their survival and on the rate of development of the first brood. *Insectes Sociaux* 4:391–408.
Waloff, N., and Blackith, R. E. 1962. The growth and distribution of the mounds of *Lasius flavus* (Fabricius) (Hym: Formicidae) in Silwood Park, Berkshire. *Journal of Animal Ecology* 31:421–437.
Walsberg, G. E. 1977. Ecology and energetics of contrasting social systems in *Phainopepla nitens* (Aves: Ptilogonatidae). *University of California Publications in Zoology* 108:1–63.
Walter, H. 1979. *Eleonora's falcon. Adaptations to prey and habitat in a social raptor.* Chicago: University of Chicago Press.

Walters, J. 1979. Interspecific aggressive behaviour by long-toed lapwings (*Vanellus crassirostris*). *Animal Behaviour* 27:969–981.

Walther, F. R. 1965. Verhaltensstudien an der Grantgazelle (*Gazella granti* Brooke, 1872) im Ngorongoro-Krater. *Zeitschrift für Tierpsychologie* 22:167–208.

———. 1969. Flight behaviour and avoidance of predators in Thomson's gazelle (*Gazellus thomsoni* Guenther 1884). *Behaviour* 34: 184–221.

———. 1977a. Artiodactyla. In *How animals communicate,* ed. T. A. Sebeok, pp. 655–714. Bloomington: Indiana University Press.

———. 1977b. Sex and activity dependency of distances between Thomson's gazelles (*Gazella thomsoni* Günther 1884). *Animal Behaviour* 25:713–719.

Ward, P. 1965a. Feeding behaviour of the black-faced dioch *Quelea quelea* in Nigeria. *Ibis* 107:173–214.

———. 1965b. The breeding biology of the black-faced dioch *Quelea quelea* in Nigeria. *Ibis* 107:326–349.

———. 1965c. Seasonal changes in the sex ratio of *Quelea quelea* (Ploceidae). *Ibis* 107:397–399.

Ward, P., and Zahavi, A. 1973. The importance of certain assemblages of birds as "information-centres" for food finding. *Ibis* 119:517–534.

Ward, R. W. 1967. Ethology of the paradise fish, *Macropodus opercularis*. I. Differences between domestic and wild fish. *Copeia* 1967:809–813.

Waring, G. H. 1970. Sound communications of black-tailed, white-tailed, and Gunnison's prairie dogs. *American Midland Naturalist* 83:167–185.

Warner, R. R. 1975. The adaptive significance of sequential hermaphroditism in animals. *American Naturalist* 109:61–82.

Warner, R. R., Robertson, D. R., and Leigh, E. G., Jr. 1975. Sex change and sexual selection. *Science* 190:633–638.

Waser, P. M. 1976. *Cercocebus albigena:* Site attachment, avoidance, and intergroup spacing. *American Naturalist* 110:911–935.

———. 1977. Mangabey feeding, movements, and group size. In *Primate ecology: Studies of feeding and ranging behavior in lemurs, monkeys, and apes,* ed. T. H. Clutton-Brock, pp. 183–222. New York: Academic Press.

Waser, P. M., and Wiley, R. H. 1979. Mechanisms and evolution of spacing in animals. In *Handbook of behavioral neurobiology.* Vol. 3. *Social behavior and communication,* ed. P. Marler and J. G. Vandenbergh, pp. 159–223. New York: Plenum.

Washburn, S. L. 1978. Human behavior and the behavior of other animals. *American Psychologist* 33:405–418.

Watson, A. 1970. Territorial and reproductive behaviour of red grouse. *Journal of Reproduction and Fertility, Supplement* 11:3–14.

———. 1977. Population limitation and the adaptive value of territorial behaviour in Scottish red grouse, *Lagopus l. scoticus*. In *Evolutionary ecology,* ed. B. Stonehouse and C. M. Perrins, pp. 19–26. London: Macmillan.

Watson, A., and Jenkins, D. 1964. Notes on the behaviour of the red grouse. *British Birds* 57:137–170.

———. 1968. Experiments on population control by territorial behaviour in red grouse. *Journal of Animal Ecology* 37:595–614.

Watson, A., and Miller, G. R. 1971. Territory size and aggression in a fluctuating red grouse population. *Journal of Animal Ecology* 40:367–383.

Watson, A., and Moss, R. 1971. Spacing as affected by territorial behavior, habitat and nutrition in red grouse (*Lagopus l. scoticus*). In *Behavior and environment. The use of space by animals and men,* ed. A. H. Esser, pp. 92–111. New York: Plenum.

———. 1972. A current model of population dynamics in red grouse. *Proceedings of the 15th International Ornithological Congress,* pp. 134–149.

Watson, J. D. 1965. *Molecular biology of the gene.* New York: W. A. Benjamin.

Wautier, V. 1971. Un phénomène social chez les coléopteres: le grégarisme des *Brachinus* (Caraboidea Brachinidae). *Insectes Sociaux* 18:1–84.

Way, M. J. 1954. Studies of the life history and ecology of the ant *Oecophylla longinoda* Latreille. *Bulletin of Entomological Research* 45:93–134.

Way, M. J., and Cammell, M. 1970. Aggregation behaviour in relation to food utilization by aphids. In *Animal populations in relation to their food resources,* ed. A. Watson, pp. 229–247. Oxford: Blackwell Scientific.

Weatherhead, P. J., and Robertson, R. J. 1977. Male behavior and female recruitment in the red-winged blackbird. *Wilson Bulletin* 89:583–592.

———. 1979. Offspring quality and the polygyny threshold: "The sexy son hypothesis." *American Naturalist* 113:201–208.

Weeden, J. S. 1965. Territorial behavior of the tree sparrow. *Condor* 67:193–209.

Weekes, H. C. 1935. A review of placentation among reptiles with particular regard to the function and evolution of the placenta. *Proceedings of the Zoological Society of London* 1935:625–645.

Weidmann, U. 1956. Observations and experiments on egg-laying in the black-headed gull (*Larus ridibundus* L.). *Animal Behaviour* 4:150–161.

Weinrich, J. D. 1977. Human sociobiology: Pair-bonding and resource predictability (effects of social class and race). *Behavioral Ecology and Sociobiology* 2:91–118.

Weiss, P. A. 1970. Life, order, and understanding: A theme in three variations. *Graduate Journal, University of Texas. Supplement* 8:1–157.

Welch, B. L., and Welch, A. S. 1969. Aggression and the biogenic amine neurohumors. In *Aggressive behaviour*, ed. S. Garattini and E. B. Sigg, pp. 188–202. Amsterdam: Excerpta Medica.

Welker, W. I. 1971. Ontogeny of play and exploration: A definition of problems and a search for conceptual solutions. In *The ontogeny of vertebrate behavior*, ed. H. Moltz, pp. 171–228. New York: Academic Press.

Weller, M. W. 1959. Parasitic egg-laying in the redhead (*Aythya americana*) and other North American Anatidae. *Ecological Monographs* 29:333–365.

Wells, K. D. 1977a. The social behaviour of anuran amphibians. *Animal Behaviour* 25:666–693.

———. 1977b. Territoriality and male mating success in the green frog (*Rana clamitans*). *Ecology* 58:750–762.

———. 1980. Behavioral ecology and social organization of a dendrobatid frog (*Colostethus inguinalis*). *Behavioral Ecology and Sociobiology* 6:199–209.

Welsh, D. A. 1975. Savannah sparrow breeding and territoriality on a Nova Scotia dune beach. *Auk* 92:235–251.

Welty, J. C. 1975. *The life of birds*. 2nd ed. Philadelphia: Saunders.

Wemmer, C., and Flemming, M. J. 1975. Management of meerkats, *Suricata suricatta*, in captivity. *International Zoo Yearbook* 15:73–77.

Werren, J. H. 1980. Sex ratio adaptations to local mate competition in a parasitic wasp. *Science* 208:1157–1159.

Werren, J. H., Gross, M. R., and Shine, R. 1980. Paternity and the evolution of male parental care. *Journal of Theoretical Biology* 82:619–631.

West, M. J. 1967. Foundress associations in polistine wasps: Dominance hierarchies and the evolution of social behavior. *Science* 157:1584–1585.

West Eberhard, M. J. 1969. The social biology of polistine wasps. *Miscellaneous Publications of the Museum of Zoology, University of Michigan* 140:1–101.

———. 1975. The evolution of social behavior by kin selection. *Quarterly Review of Biology* 50:1–34.

Wheatley, M. D. 1944. The hypothalamus and affective behavior in cats. *Archives of Neurology and Psychiatry* 52:296–316.

Wheeler, W. M. 1923. *Social life among the insects*. New York: Harcourt and Brace.

———. 1928. *The social insects: Their origin and evolution*. London: Kegan Paul, Trench, Trubner, & Co.

———. 1930. *Demons of the dust*. New York: Norton.

White, M. J. D. 1978. *Modes of speciation*. San Francisco: Freeman.

Whitford, W. G. 1967. Observations on territoriality and aggressive behavior in the western spadefoot toad, *Scaphiopus hammondii*. *Herpetologica* 23:318.

Whitney, C. L., and Krebs, J. R. 1975. Mate selection in Pacific tree frogs. *Nature* 255:325–326.

Whittaker, R. H., Levin, S. A., and Root, R. B. 1973. Niche, habitat, and ecotope. *American Naturalist* 107:321–338.

Wickler, W. 1968. *Mimicry in plants and animals*. New York: McGraw-Hill.

———. 1972. *The sexual code. The social behavior of animals and men*. Garden City, N.Y. Doubleday.

Wickler, W., and Seibt, U. 1970. Das Verhalten von *Hymenocera picta* Dana, einer Seesterne fressenden Garnde (Decapoda, Natantia, Gnathophyllidae). *Zeitschrift für Tierpsychologie* 27:352–368.

Wicklund, C. 1975. Pupal color polymorphism in *Papilio machaon* L. and the survival in the field of cryptic versus non-cryptic pupae. *Transactions of the Royal Entomological Society of London* 127:73–84.

Wiens, J. A. 1966. Group selection and Wynne-Edwards' hypothesis. *American Scientist* 54:273–287.

———. 1973. Interterritorial habitat variation in grasshopper and savannah sparrows. *Ecology* 54:873–884.

Wiewandt, T. A. 1971. Breeding biology of the Mexican leaf-frog. *Fauna* 2:29–34.
Wigglesworth, V. B. 1929. Digestion in the tsetse-fly: A study of structure and function. *Parasitology* 21:288–321.
———. 1931. Digestion in *Chrysops silacea* Aust. (Diptera: Tabanidae). *Parasitology* 23:73–76.
Wilbur, H. M. 1972. Competition, predation and the structure of the *Ambystoma-Rana sylvatica* community. *Ecology* 53:3–21.
Wilbur, H. M., and Collins, J. P. 1973. Ecological aspects of amphibian metamorphosis. *Science* 182:1305–1314.
Wilbur, H. M., Rubenstein, D. I., and Fairchild, L. 1978. Sexual selection in toads: The roles of female choice and male body size. *Evolution* 32:264–270.
Wilbur, H. M., Tinkle, D. W., and Collins, J. P. 1974. Environmental certainty, trophic level, and resource availability in life history evolution. *American Naturalist* 108:805–817.
Wiley, R. H. 1973. Territoriality and non-random mating in sage grouse, *Centrocercus urophasianus. Animal Behaviour Monographs* 6:85–169.
———. 1974. Evolution of social organization and life-history patterns among grouse. *Quarterly Review of Biology* 49:201–227.
———. 1978. The lek mating system of sage grouse. *Scientific American* 238(5):114–125.
Wiley, R. H., and Wiley, M. S. 1980. Spacing and timing in the nesting ecology of a tropical blackbird: Comparison of populations in different environments. *Ecological Monographs* 50:153–178.
Wilkinson, P. F., and Shank, C. C. 1977. Rutting-fight mortality among musk oxen on Banks Island, Northwest Territories, Canada. *Animal Behaviour* 24:756–758.
Williams, C. M. 1970. Hormonal interactions between plants and insects. In *Chemistry of phenolic compounds*, ed. J. B. Pridham, pp. 87–95. Oxford: Pergamon Press.
Williams, G. C. 1964. Measurement of consociation among fishes and comments on the evolution of schooling. *Papers of the Museum of Michigan State University, Biology Series* 2:351–383.
———. 1966a. *Adaptation and natural selection: A critique of some current evolutionary thought.* Princeton: Princeton University Press.
———. 1966b. Natural selection, the costs of reproduction, and a refinement of Lack's principle. *American Naturalist* 100:687–690.
———. 1975. *Sex and evolution. Monographs in Population Biology,* no. 8. Princeton: Princeton University Press.
———. 1979. The question of adaptive sex ratio in outcrossed vertebrates. *Proceedings of the Royal Society of London, Series B* 205:567–580.
Williams, L. 1952. Breeding behavior of the Brewer blackbird. *Condor* 54:3–47.
Willis, E. O. 1967. The behavior of bicolored antbirds. *University of California Publications in Zoology* 79:1–127.
———. 1972a. Do birds flock in Hawaii, a land without predators? *California Birds* 3:1–8.
———. 1972b. The behavior of plain-brown woodcreepers *Dendrocincla fuliginosa. Wilson Bulletin* 84:377–420.
Willis, E. O., Wechsler, D., and Oniki, Y. 1978. On behavior and nesting of McConnell's flycatcher (*Pipramorpha macconnelli*): Does female rejection lead to male promiscuity? *Auk* 95:1–8.
Willis, I. 1972. Adapting to a way of life. *Birds* 4:11–15.
Willson, M. F. 1966. The breeding ecology of the yellow-headed blackbird. *Ecological Monographs* 36:51–77.
Wilson, D. S. 1975. A theory of group selection. *Proceedings of the National Academy of Sciences* 72:143–146.
———. 1977. Structured demes and the evolution of group-advantageous traits. *American Naturalist* 111:157–185.
———. 1979. Structured demes and trait-group variation. *American Naturalist* 113:606–610.
———. 1980. *The natural selection of populations and communities.* Menlo Park, Calif.: Benjamin/Cummings.
Wilson, D. S., and Colwell, R. K. 1981. The evolution of sex ratio in structured demes. *Evolution* (in press).
Wilson, E. O. 1957. The organization of a nuptial flight of the ant *Pheidole sitarches* Wheeler. *Psyche, Cambridge* 64:46–50.
———. 1966. Behaviour of social insects. *Symposia of the Royal Entomological Society of London* 3:81–96.
———. 1971. *The insect societies.* Cambridge, Mass.: Belknap Press.
———. 1973. Group selection and its significance for ecology. *BioScience* 23:631–638.

———. 1974. The soldier of the ant, *Camponotus (Colobopsis) fraxinicola*, as a trophic caste. *Psyche, Cambridge* 81:182–188.

———. 1975a. *Sociobiology: The new synthesis*. Cambridge, Mass.: Belknap Press.

———. 1975b. For sociobiology. *New York Review of Books*, 11 December.

———. 1975c. Enemy specification in the alarm-recruitment system of an ant. *Science* 190:798–800.

———. 1976a. Academic vigilantism and the political significance of sociobiology. *BioScience* 26:183–190.

———. 1976b. The organization of colony defense in the ant *Pheidole dentata* Mayr (Hymenoptera: Formicidae). *Behavioral Ecology and Sociobiology* 1:63–81.

———. 1976c. Behavioral discretization and the number of castes in an ant species. *Behavioral Ecology and Sociobiology* 1:141–154.

———. 1978. *On human nature*. Cambridge, Mass.: Harvard University Press.

Wilson, E. O., and Bossert, W. H. 1971. *A primer of population biology*. Sunderland, Mass.: Sinauer Associates.

Wilson, H. 1946. The life history of the western magpie (*Gymnorhina dorsalis*). *Emu* 45:233–244, 271–286.

Wilsson, L. 1968. *My beaver colony*. Garden City, N.Y.: Doubleday.

———. 1971. Observations and experiments on the ethology of the European beaver (*Castor fiber* L.). *Viltrevy* 8:115–266.

Wiltbank, J. N., Rowden, W. W., Ingalls, J. E., Gregory, K. E., and Koch, R. M. 1962. Effect of energy level on reproductive phenomena of mature Hereford cows. *Journal of Animal Science* 21:219–225.

Wingfield, J. C., and Farner, D. S. 1978a. The annual cycle of plasma irLH and steroid hormones in feral populations of the white-crowned sparrow, *Zonotrichia leucophrys gambelii*. *Biology of Reproduction* 19:1046–1056.

———. 1978b. Reproductive endocrinology of the white-crowned sparrow (*Zonotrichia leucophrys pugatensis*). *Physiological Zoology* 51:188–205.

Wise, D. A. 1974. Aggression in the female golden hamster: Effects of reproductive state and social isolation. *Hormones and Behavior* 5:235–250.

Wise, L. A., and Zimmerman, R. R. 1973. The effect of protein deprivation on dominance measured by shock avoidance competition and food competition. *Behavioral Biology* 9:317–329.

Witkin, H. A., Mednick, S. A., Schulsinger, F., Bakkestrom, E., Christiansen, K. O., Goodenough, D. R., Hirschhorn, K., Lundsteen, C., Owen, D. R., Philip, J., Rubin, D. B., and Stocking, M. 1976. Criminality in XYY and XXY men. *Science* 196:547–555.

Wittenberger, J. F. 1976a. The ecological factors selecting for polygyny in altricial birds. *American Naturalist* 110:779–799.

———. 1976b. Habitat selection and the evolution of polygyny in bobolinks (*Dolichonyx oryzivorus*). Ph.D. thesis, University of California, Davis.

———. 1978a. The evolution of mating systems in grouse. *Condor* 80:126–137.

———. 1978b. The breeding biology of an isolated bobolink population in Oregon. *Condor* 80:355–371.

———. 1979a. The evolution of mating systems in birds and mammals. In *Handbook of behavioral neurobiology*, Vol. 3. *Social behavior and communication*, ed. P. Marler and J. Vandenbergh, pp. 271–349. New York: Plenum.

———. 1979b. A model for delayed reproduction in iteroparous animals. *American Naturalist* 114:439–446.

———. 1980a. Group size and polygamy in social mammals. *American Naturalist* 115:197–222.

———. 1980b. Vegetation structure, food supply, and polygyny in bobolinks (*Dolichonyx oryzivorus*). *Ecology* 61:140–150.

———. 1980c. Feeding of secondary nestlings by polygynous male bobolinks. *Wilson Bulletin* 92:330–340.

———. 1981. Male quality and polygyny: The "sexy son" hypothesis revisited. *American Naturalist* (in press).

Wittenberger, J. F., and Tilson, R. L. 1980. The evolution of monogamy: Hypotheses and evidence. *Annual Review of Ecology and Systematics* 11:197–232.

Wolf, A., and Haxthausen, E. F. von. 1960. Toward the analysis of the effects of some centrally acting sedative substances. *Arzneimittel-Forschung* 10:50–68.

Wolf, L. L. 1978. Aggressive social organization in nectivorous birds. *American Zoologist* 18:765–778.

Woodell, S. R. J., Mooney, H. A., and Hill, A. J. 1969. The behaviour of *Larrea divaricata* (creosote bush) in response to rainfall in California. *Journal of Ecology* 57:37–44.

Woodworth, S. L. 1979. The social system of feral asses (*Equus asinus*). *Zeitschrift für Tierpsychologie* 49:304–316.

Woodworth, C. H. 1971. Attack elicited in rats by electrical stimulation of the lateral hypothalamus. *Physiology and Behavior* 6:345–353.

Woodworth, S. L. 1979. The social system of feral asses *(Equus asinus). Zeitschrift für Tierpsychologie* 49:304–316.

Woolfenden, G. E. 1973. Nesting and survival in a population of Florida scrub jays. *Living Bird* 12:25–49.

———. 1975. The effect and source of Florida scrub jay helpers. *Auk* 92:1–15.

———. 1976. Cooperative breeding in American birds. *Proceedings of the 16th International Ornithological Congress*, pp. 674–684.

Woolfenden, G. E., and Fitzpatrick, J. W. 1978. The inheritance of territory in group-breeding birds. *BioScience* 28:104–108.

Wrangham, R. W. 1979. Sex differences in chimpanzee dispersion. In *The great apes*, ed. D. A. Hamburg and E. R. McCown, pp. 480–489. Menlo Park, Calif.: Benjamin/Cummings.

Wright, A. H., and Wright, A. A. 1949. *Handbook of frogs and toads*. Ithaca, N.Y.: Cornell University Press.

Wright, P. C. 1978. Home range, activity pattern, and agonistic encounters of a group of night monkeys (*Aotus trivirgatus*) in Peru. *Folia Primatologica* 29:43–55.

Wright, S. 1932. The roles of mutation, inbreeding, crossbreeding, and selection in evolution. *Proceedings of the 6th International Congress of Genetics*, pp. 356–366.

———. 1970. Random drift and the shifting balance theory of evolution. In *Mathematical topics in population genetics*, ed. K. Kojima, pp. 1–31. Berlin: Springer-Verlag.

———. 1977. *Evolution and the genetics of populations: A treatise*. Vol. 3. *Shifting balance theory*. Chicago: University of Chicago Press.

———. 1978. *Evolution and the genetics of populations: A treatise*. Vol. 4. *Variability within and among populations*. Chicago: University of Chicago Press.

Wynne-Edwards, V. C. 1962. *Animal dispersion in relation to social behaviour*. Edinburgh: Oliver & Boyd.

———. 1977. Society versus the individual in animal evolution. In *Evolutionary ecology*, ed. B. Stonehouse and C. M. Perrins, pp. 5–17. London: Macmillan.

Yaron, Z. 1972. Endocrine aspects of gestation in viviparous reptiles. *General and Comparative Endocrinology Supplement* 3:663–674.

Yasokochi, G. 1960. Emotional response elicited by electrical stimulation of the hypothalamus in cat. *Folia Psychiatrica et Neurologica Japonica* 14:260–267.

Yasukawa, K. 1979. Territory establishment in red-winged blackbirds: Importance of aggressive behavior and experience. *Condor* 81:258–264.

Yasuno, M. 1965. Territory of ants in the Kayano grassland at Mt. Hakkôda. *Scientific Report of Tôhoku University, Sendai, Japan. Series 4 (Biology)* 31:195–206.

Yeaton, R. I. 1972. Social behavior and social organization in Richardson's ground squirrel (*Spermophilus richardsonii*) in Saskatchewan. *Journal of Mammalogy* 53:139–147.

Yeaton, R. I., and Cody, M. L. 1974. Competitive release in island song sparrow populations. *Theoretical Population Biology* 5:42–58.

Yoshiba, K. 1968. Local and intertroop variability in ecology and social behavior of common Indian langurs. In *Primates: Studies in adaptation and variability*, ed. P. Jay, pp. 217–242. New York: Holt, Rinehart and Winston.

Yoshikawa, K. 1963. Introductory studies on the life economy of polistine wasps. II. Superindividual stage. 2. Division of labor among workers. *Japanese Journal of Ecology* 13:53–57.

Young, H. 1951. Territorial behavior in the eastern robin. *Proceedings of the Linnaean Society of New York* 58:1–37.

———. 1956. Territorial activities of the American robin. *Ibis* 98: 448–452.

Youngbluth, M. J. 1968. Aspects of the ecology and ethology of the cleaning fish *Labroides phthirophagus* Randall. *Zeitschrift für Tierpsychologie* 25:915–932.

Zahavi, A. 1971a. The social behaviour of the white wagtail *Motacilla alba alba* wintering in Israel. *Ibis* 113:203–211.

———. 1971b. The function of pre-roost gatherings and communal roosts. *Ibis* 113:106–109.

———. 1974. Communal nesting by the Arabian babbler: A case of individual selection. *Ibis* 116:84–87.

———. 1975. Mate selection—A selection for a handicap. *Journal of Theoretical Biology* 53:205–214.

———. 1976. Cooperative nesting in Eurasian birds. *Proceedings of the 16th International Ornithological Congress*, pp. 685–693.

———. 1977a. Reliability in communication systems and the evolution of altruism. In *Evolutionary ecology*, ed. B. Stonehouse and C. M. Perrins, pp. 253–259. London: Macmillan.

———. 1977b. The cost of honesty (further remarks on the handicap principle). *Journal of Theoretical Biology* 67:603–605.

Zenone, P. G., Sims, M. E., and Erickson, C. J. 1979. Male ring dove behavior and the defense of genetic paternity. *American Naturalist* 114:615–626.

Zillmann, D. 1978. *Hostility and aggression*. Hillsdale, N.J.: Earlbaum Associates.

Zimen, E. 1975. The social dynamics in a wolf pack. In *The wild canids. Their systematics, behavioral ecology and evolution*, ed. M. Fox, pp. 336–362. New York: Van Nostrand Reinhold.

———. 1976. On the regulation of pack size in wolves. *Zeitschrift für Tierpsychologie* 40:300–341.

Zimmerman, J. L. 1966. Polygyny in the dickcissel. *Auk* 83:534–546.

———. 1971. The territory and its density dependent effect in *Spiza americana*. Auk 88:591–612.

Zucchi, R. 1966. Aspectos evolutivas do compartamento social entre abelhas. *Reuniâo Anual do Sociedade Brasileiro Genetica, Programa e Resumos (Piracicaba)*, pp. 6–9.

Zwickel, F. C. 1972. Removal and repopulation of blue grouse in an increasing population. *Journal of Wildlife Management* 36:1141–1152.

Zwickel, F. C., and Bendell, J. F. 1967. Early mortality and the regulation of numbers in blue grouse. *Canadian Journal of Zoology* 45:817–851.

———. 1972. Blue grouse, habitat, and populations. *Proceedings of the 15th International Ornithological Congress*, pp. 150–169.

Index

Acanthasteridae, 235
Acanthephyra, 233
Acanthotermes, 524
Acarophenax, 411; *A. triboli*, 408
Accipiter cooperii, 336; *A. gentilis*, 87
Acinonyx jubatus, 118
Acrocephalus arcindinaceous, 460
Actitis macularia, 502
Adaptive landscapes, 49–53, 68
Adaptive value, 5, 9, 15, 46, 47–53
Aegithalos caudatus, 327
Aepyceros melampus, 82, 489, 556
Aethia pusilla, 350
Agelaius, 311; *A. phoeniceus*, 168, 382; *A. tricolor*, 123, 381
Aggregations, 303
Aggression, 135–192; adaptive significance of, 165–178; advertising competitive advantages, 182–183; aggressive neglect, 179; aggressive play, 176; attack strategies, 187–191; benefits of, 165–179; competitive ability and, 179–180; costs of, 178–180; in courtship, 165–170, 187, 420; defensive, 177–178; defined, 135–137, 138; developmental origins of, 142–147; and dominance, 590, 592, 593–595; endocrine controls of, 149–156; evolutionary model of, 163–165; evolutionarily stable strategies and, 187–190; exaggeration and deception in threat displays, 183–185; frustration-aggression model of, 140; hydraulic model of, 139–140; and mobbing, 115–120, 177, 387–389; monogamy and, 464, 508–512; motivational models of, 139–142, 162–163, 167; neurology of, 156–160; parental, 138, 170–175; peer aggression, 138, 175–176; persuasion vs. coordination signals, 181–182; physiological model of, 160–162; polygyny and, 452; predation and, 137–139; proximate causation of, 147–163; redirected, 138, 178, 184; self-restraint and, 185–186; sex and, 138, 165–170, 421, 429; stimuli, 148–149; and submission, 186–187; switchboard model of, 141; tactics, 180–191; and territoriality, 148–149, 151, 165, 184–185, 250, 252–253; threats and, 182, 184, 428–429; types of, 137–138
Aglaius urticae, 256
Aglaja inermis, 229
Agonistic behavior, 137
Agoutis, 119, 159
Alarm signals, 80–91
Albatrosses, 322, 505; Laysan, 505
Alcelaphus, 557; *A. buselaphus*, 489
Alcids, 99, 350, 362, 505, 506; coloniality in, 307, 311, 322; monogamy in, 505
Aldosterone, 155
Alectura lathami, 366
Alleles, 30, 34, 46, 47, 49–51, 60, 61, 68, 69, 70, 71
Allelopathy, 208
Alligator mississippiensis, 414
Alligator, 414
Allogrooming, 129–132
Altruism, 60, 61–62, 66, 68, 75–109; and alarm signals, 80, 87–91; and alloparental behavior, 105–108; and

cleaning symbioses, 77–80; definition of, 75; differential in worker castes, 534–535; and group selection, 62–70; and helping others reproduce, 95–100; reciprocal, 76–80, 90–91, 131; and reproductive curtailment, 92–95, 99–100; and self-restraint, 185–186; in social insect castes, 514–550
Alytes obstreticans, 367
Amblonyx cinerea, 509
Amblyopone pallipes, 522
Amblyrhynchus cristatus, 367
Amblyseius fallacis, 411
Ambystoma tigrinum, 427
Amia, 244
Amitermes hastatus, 547
Ammospiza caudacuta, 483
Amphibians, 368; anuran, 491; egg-laying behavior of, 367; fecundity of, 351; mating systems of, 491–494; parental behavior of, 262–263, 365, 367, 368; territoriality of, 262, 494
Amphibolurus barbatus, 242
Amphipods, 269
Amphiprion, 232; *A. akallopisos*, 415; *A. bicinctus*, 415
Amyciaea, 223
Anas clypeata, 238
Anchovies, 366; Hawaiian, 121, 125
Ancrya annamensis, 244, 245
Androgens, 149–152
Anemonefish, 232, 267–268, 377, 415
Anemones, sea, 232
Anhinga anhinga, 309
Anisogamy, cost of, 398; origin of, 404–405
Anomma, 520; *A. wilwerthi*, 531
Ant lions, 222
Antbirds, 124
Anteater, spiny, 229
Antelopes, 81, 111, 119, 571, 572, 573; alarm signals of, 81, 82; leks of, 259, 496–497; mating systems of, 487–490; monogamy in, 509; optimal group size in, 578, 580, 584; pronghorn, 82, 274; sable, 82; sociality of, 555–560; territoriality in, 274
Anthias squamipinnis, 257, 258
Anthidium manicatum, 255–256
Antidorcas, 556; *A. marsupialis*, 82
Antilocapra americana, 82, 274
Antilope cervicapra, 82
Antithesis, principle of, 187
Antlers, 428
Ants, 115, 175, 233, 393; acacia, 236; African weaver, 269–270; aggression in, 149; army, 124, 176, 520; caste differentiation of, 520–522; eusociality in, 520–522, 531–533, 537–542, 545–546; fire, 115, 236; harvester, 236, 270, 271; honeypot, 270; indirect (kin) selection in, 527–542; queen control in, 545–546; red wood, 115, 269; sex ratios in, 393, 537–542; stings of, 236; territoriality in, 269–272
Aotus trivirgatus, 509
Aphaenogaster, 546
Aphelocoma coerulescens, 96
Aphids, 233, 269, 270; nutrient enhancement by, 126–127
Apis mellifera, 235
Aposematism, 241
Aptenodytes patagonica, 350, 505; *A. forsteri*, 505
Ardea herodias, 315
Ardeola ibis, 328
Argia apicalis, 254
Armadillium, 228
Armadillos, 228
Ascaloptynx furcigei, 115, 116
Asio flammeus, 175
Aspidontus taeniatus, 223
Assassin bugs, 233
Assertiveness, 137
Asses, 490, 585
Ateles belzebuth, 582
Atepolus, 425, 511
Atherurus, 229
Atta, 520; *A. cephalotes*, 520; *A. sexdens rubropilosa*, 531
Augochloropsis, 517; *A. sparsilis*, 517
Auks, 262; least auklet, 350; little, 311
Aunting behavior, 105. *See also* Allogrooming
Avocet, 273, 506
Aythya americana, 383

Babblers: Arabian, 99, 104, 182, 435; grey-crowned, 96, 102–103; striated jungle, 101, 103, 104
Baboons, 111, 112, 159, 248, 274, 586, 590; chacma, 118; hamadryas, 184, 488, 565–568; mobbing by, 118–119; olive, 118, 132, 437; sociality in, 560, 561, 565–570
Bacteria, 40
Banteng, 82
Batfish, 223
Batrachoceps attenuatus, 359
Bats, 260, 337, 424, 488; coloniality in, 311, 314, 323; false vampire, 508; hammer-headed, 260, 496, 498; leks in,

496, 498; lesser horseshoe, 327–328; long-tongued, 328; Mexican free-tailed, 311; monogamy in, 508; roosts, 327–328; spear-nosed, 328
Bears, 177, 178; grizzly, 326, 573; polar, 468, 469
Beater effect, 123–124
Beaugregory, 261, 298–299
Beavers, 265
Bee-eaters, 240; red-throated, 103, 104; white-throated, 104
Bees, 115, 240, 385; carpenter, 256; caste differentiation of, 517–518; central place systems and, 340, 341; eusociality in, 517–519, 530–532, 534–536, 539, 542–545; halictine, 517, 528, 536, 543, 544; honeybee, 235, 340–341, 517–518, 531–532, 535, 542; indirect (kin) selection in, 527–542; queen control in, 542–545; sex ratios in, 393, 539; stingless, 531, 536, 542–543; stings of, 235; territoriality in, 255–256
Beetles, 237–238, 244, 428, 528; bark, 207–208; bombardier, 115, 233, 234; flour, 66; horned, 510; stag, 429; tenebrionid, 426; water, 237–238; wood-boring, 510
Betta splendens, 143
Birds, 64, 233; alarm signals of, 80, 81; allopreening in, 129–130; brood parasitism in, 383–390; camouflage, 231; clutch size in, 479–480; colonial, 173–175, 177, 227, 262, 305–343, 439, 471, 505, 506, 507; courtship of, 165–170, 443; distraction displays of, 244; egg-laying by, 365–366, 367; fecundity of, 351; flocking, 90, 119, 127–129; as helpers at nest, 95–104; mobbing by, 119, 389–391; monogamy in, 449, 482, 505–508, 511–512; mutual vigilance in, 112; parental care by, 365–366, 368, 381–383, 505–506; passerine, 80, 86–87, 284, 322, 470–478; polyandry in, 498–504; polygyny in, 381, 383, 453, 470–478; predatory, 222; promiscuity in, 479–487; sex role reversal in, 417–419; sexual selection in, 428, 431, 439–440; territoriality in, 257–260, 262–263, 264, 273, 274–285, 287–299
Bird-of-paradise, greater, 497; lesser, 497
Bison, American, 491
Bison bison, 491
Bittacus apicalus, 443
Blackbird, European, 350
Blackbirds, 104, 128, 227; Brewer's, 315, 316, 329, 336, 381; coloniality in, 305, 307, 309–310, 311, 315, 316, 319, 322; parental care in, 381–382; polygyny in, 473–475, 478; red-winged, 168, 309–310, 334, 382, 418, 473–474, 478; tricolored, 123, 330, 334, 342, 381; yellow-headed, 184, 298, 382, 475
Blackbuck, 82
Blarina, 237; *B. brevicauda*, 237
Blastophaga, 427
Blatella germanica, 444
Blenny: mussel, 268; poison fang, 237; redlip, 268; sabre-toothed, 223; striped, 261
Blesbok, 259, 557, 559
Bloodworms, 235
Boa constrictors, 560
Bobolink, 381, 460–461, 476, 477
Bombino, 239
Bombus atratus, 536; *B. terrestris*, 543
Bombycilla, 483; *B. cedrorum*, 389
Bonasa umbellus, 350, 486
Bontebok, 557
Boobies, 175, 505; blue-footed, 306; red-footed, 505; Peruvian, 306, 342
Boorman-Levitt model of intrademic selection, 67–68
Bos sondaicus, 82
Brachinus, 115; *B. crepitans*, 234
Brambling, 310, 311
Breeding behavior, 288; breeding synchrony, 329–335; coloniality and, 320–325; communal, 383; explosive, 491, 494, 510; quality of breeding situation, 456; of seals, 468–470; selective, 142–144; and territoriality, 260–263
Brevicoryne brassica, 126
Bristleworms, 235
Brood parasitism, 383–390; benefits of, 384–385; costs of, 386; cuckoo gentes and, 389; egg acceptors vs. rejectors, 389; egg mimicry and, 389; host defenses against, 386–389; mobbing and, 387–389; precursors of, 383–384
Brush turkey, 366
Bubalus bubabis, 116
Bubo virginianus, 337
Bubonic plague, 308
Bucorvus cafer, 123
Budgerigar, 335
Buffalo, Cape, 116, 490, 557, 571, 573, 580; water, 116
Bufo quericus, 440; *B. americana*, 440
Bug, assassin, 233
Bullfinch, 350
Bullfrog, 262, 263, 359

Bumblebees, 536, 540, 543
Bunocephalidae, 366
Buntings, corn, 382; indigo, 382, 460; lark, 381, 460, 478; snow, 350
Bushtit, common, 87
Bustard, great, 497; little, 497
Buteo albonotatus, 223; *B. jamaicensis*, 120, 273; *B. unicinctus*, 112
Butterfish, 232
Butterflies, 236, 238, 244, 425; brimstone, 425; hilltopping in, 256–257; monarch, 227–228, 240; peacock, 256–257; speckled wood, 256; territoriality in, 256–257
Butterflyfish, 273
Buzzard, bat-eating, 337; honey, 240

Caciques, 382, 473
Calamospiza melancorys, 381, 460
Calappa, 229
Calcarius lapponicus, 350; *C. pictus*, 512
Calidris alba, 275; *C. alpina*, 284
Callicebus, 509
Callichthyidae, 374
Calopteryx, 255; *C. maculata*, 424
Calorhinus ursinus, 307
Calypte anna, 291
Camouflage, 230–232, 237
Camponotus, 520; *C. truncatus*, 521
Canachites canadensis, 486
Canids, 124, 508, 572
Canis lupus, 104; *C. mesomelas*, 104
Cannibalism, 64, 269; and breeding synchrony, 330; egg, 441; infanticide and, 171, 173–174, 308, 574; and peer aggression, 175–176
Capreolus capreolus, 82
Capritermes, 524
Caranx ignobilis, 121
Caribou, 82, 491, 572
Carnivores, sociality in, 124–126, 224–225, 570–574. *See also* Predation
Carrying capacity, 354
Cassidix mexicanus, 122. *See also Quiscalus mexicanus*
Castor canadensis, 265; *C. fiber*, 265
Castration, 149–150, 151, 153
Caterpillars, 228, 236, 242, 243, 385
Catfish, 231, 237; armored, 374; banjo, 366
Catharacta maccormicki, 175
Cats: aggression in, 159–160, 186; ambush hunting in, 222; civit, 240; domestic, 159–160, 176; promiscuity in, 487; serval, 118; teat-order in, 176
Cell differentiation, 37–42
Centrocercus urophasianus, 259
Cephalophus, 555, 556
Ceratiidae, 223
Cercocebus albigena, 565
Cercopithecus aethiops, 80–81; *C. ascanius*, 561; *C. mitis*, 584; *C. talapoin*, 584
Certhia familiaris, 327
Cervus canadensis, 127–128; *C. elephas*, 82, 178; *C. elephas scoticus*, 82; *nippon*, 82
Chaetura pelagica, 336
Chaffinch, 127
Characid, spraying, 367
Chasmodes bosquianus, 261
Cheetah, 118, 560, 573; cooperative hunting in, 571; male coalitions in, 133
Chemical defenses: in animals, 233–241; in plants, 215–221
Chicken, domestic, 45, 151
Chimpanzees, 106, 151, 153, 590; grouping behavior in, 582; mobbing by, 116–118; predation by, 561–564
Chipmunk, Sonoma, 88
Chironex fleckeri, 235
Chiropotes, 509
Chondrichthyes, 237
Chough, white-winged, 96
Chromaffin cells, 154
Cicadas, 225–226
Cichlasoma maculicauda, 260, 440
Cichlid fish, 260, 366, 369, 377, 440
Ciconia ciconia, 171
Circaetus gallicus, 240
"Circumstantial bias," 607
Cistothorus palustris, 382
Civet, 240
Cladomylea, 223
Clamator jacobinus, 387
Clams, 207
Cleaner fish, 223, 414
Cleaner shrimp, 77–80; Pederson's, 78
Cleaner wrasse, 78
Cleaning symbioses, 77–80
Clicker, desert, 256
Cloudopsis scorpio, 264
Clustering, 327–328
Clutch size, 359, 361, 479–480
Coatimundi, 119, 488
Cobras, king, 367; spitting, 233
Coccyzus, 222
Cockroach, German, 444, 445
Cock-of-the-rock, 496, 497
Codons, 27, 28, 29
Coliuspasser jacksoni, 483
Colobus badius, 561; *C. guereza*, 563
Colobus, black-and-white, 563–564; red, 561, 563–564
Coloniality, 64, 303–343, 352, 402, 507;

benefits of, 311–328; central-place systems and, 337–338, 339, 340–341; colony sizes, extremes, 310–311; and competition, 337–342; costs of, 305–310; definition of, 304; disease and, 308; ectoparasitism and, 308; food distribution and, 312–318; Fraser Darling effect and, 329–335; geometrical model of, 312–313; habitat clumping and, 320–325; information center hypothesis and, 318–320; mass abandonment of, 340–342; massed flights and, 335–337; monogamy and, 507–508; mutual defense and, 325–326; polygyny and, 457, 463–465, 465–469, 470, 472–475; and predation, 307–308, 328–337; roosting and, 306, 307, 310–311, 319, 327–328, 335–337, 339, 342; and social groups, compared, 304

Color, 241, 230–232, 237, 428; morphs, 436

Columba livia, 152; *C. palumbus*, 112

Commons, 340–341, 342

Communication behavior, 181

Competition: and aggression, 163–164, 179–180; as cost of coloniality, 305–307; effects on coloniality, 337–342; in cultural evolution, 603–604; for mates, 415–420, 426–427; and niche space, 197–213; and optimal group size, 577, 580; and specialization, 197–199; sperm, 422–426; and territoriality, 252–254; types of, 197; persuasion vs. coordination signals, 181–183; threat displays, 184–185, 428–429

Complementarity, 27

Condors, 506

Connochaetes, 557; *C. taurinus*, 227

Cooperation, 75, 76, 95, 100–104, 110–133; definition of, 75, 76; allogrooming and, 129–132; and alarm signals, 84–87; to enhance foraging efficiency, 122–129; gaining information and, 127–128; and helping others reproduce, 100–105; hunting, 124–126, 224, 570–574; and male coalitions, 132–133; mobbing and active defense, 115–120; as defense against predation, 111–122; of predators, 124–126, 570–574; in reducing costs of sociality, 129–133

Copenia arnoldi, 367

Coptotermes acinaciformes, 272; *C. brunneus*, 272

Copulation, prolonged, 425–426

Coragyps atratus, 223

Corcorax melanorhamphus, 96

Cormorants, 123, 262; guanay, 310, 311, 312; shag, 330

Corvinella corvina, 96

Corvus corax, 125; *C. corone*, 231; *C. splendens*, 387

Cost-benefit analyses, 48, 57; of aggression, 163–165, 178–180; of group size, 577; of helping at nest, 96–98; of mammalian sociality, 553–574; of plant-herbivore interactions, 217; of territorial defense, 288–291

Cottus bairdi, 261, 440

Courtship behavior, and aggression, 165–170, 187, 420; courtship feeding, 442–443; and social stimulation, 329, 335, 494–495

Cowbirds, 383, 385, 388, 389, 512; and brood parasitism, 383–389; brown-headed, 384

Coyote, 508

Crabs, 229, 351; fiddler, 182; hermit, 228

Cranes, 175, 506

Crayfish, 267

Creeper, European tree, 327

Crematogaster, 546

Cricetomys gambianus, 176

Cricket, field, 264; wood, 226–227

Crocodiles, 239, 573; Nile, 123–124, 125, 367

Crocodylus niloticus, 123

Crocuta crocuta, 124

Crossbills, 315; red, 442

Crotophaginae, 383

Crowding, 154, 156

Crows, 321; carrion, 231; Indian house, 387

Crustaceans, 427, 428

Crypsis, 231, 232

Cryptobranchus alleghaniensis, 367

Cryptocercus punctulatus, 510

Cryptotermes, 524

Cuckoos, 222, 383; brood parasitism by, 383–389; drongo, 387; European, 388, 389; hawk, 387; jacobin, 387

Cuculus, 389; *C. canorus*, 389

Cultural influences, 12, 14–15; and cultural change, 599–611

Cuttlefish, 121, 233

Cyanocitta cristata, 240, 295; *C. stelleri*, 295

Cynomys ludovicianus, 88, 308; *C. leucurus*, 308

Cypriniformes, 81

Cyprinodon, 261

Dacelos gigas, 96

Dacus tryoni, 208

Daddy-long-legs, 233
Dama dama, 82
Damaliscus, 557; *D. dorcas*, 259
Damselflies, 227, 254–255, 315, 424
Danaus plexippus, 227
Daphnia magna, 121
Darwinian theories: antithesis principle, 187; natural selection, 8–9, 14, 75, 515; sexual selection, 420–421
Darwinism, social, 11
Dasyprocta punctata, 119
Data, role of, 16–17
Death, feigning of, 244
Deer, 488; black-tailed, 81, 82; fallow, 82; mule, 82, 178; red, 82, 178; roe, 82; sika, 82; white-tailed, 82, 84–85, 90
Defense, active, 115–120, 177–178
Defensibility, economic, 253–254, 285–295
Defensive circles, 115–116, 117
Delayed maturation, as life history trait, 352, 353, 358, 359–360; and polygyny, 470–471; and sexual selection, 431
Delichon urbica, 350
Demography, and group size, 576–577, 579–580; life tables and, 345–348
Dendragapus obscurus, 350
Dendroica discolor, 382, 460; *D. kirtlandii*, 277
Deprivation experiments, 45
Desmognathus ochrophaeus, 359
Detachable body parts, 243
Determinism, biological, 10
Developmental processes, 5, 13, 37–44
Devil, thorny, 229
Diadema antillarum, 84
Diadematidae, 235
Dickcissel, 280–281, 337, 382, 477
Dicrostichus, 223
Dicrostonyx groenlandicus, 172, 410
Didelphius marsupialis, 157
Dikdiks, 509, 555, 556
Dineutes discolor, 238
Diodonidae, 242
Diomedia immutabilis, 505
Diploidy, 27
Discus fish, 369
Disease, 129, 308
Displacement activities, 184–185
Displays, courtship, 165–167; distraction, 242–243; epideictic, 64; of hosts to cleaners, 79; intimidation, 242, 243; on leks, 53–54, 55, 259, 494–495; rump and tail, 80–83, 84–86; stotting, 83–85; submissive, 84, 187, 188; threat, 182, 184, 428–429
Distastefulness, 237–239
Distraction displays, 242–244
DNA, 27–42
Dogfish, spiny, 237
Dogs, 146, 176; African wild, 104–105, 119, 125, 508, 560, 572, 573, 574, 579; cooperative hunting by, 125, 572–574; helpers in, 104–105; selective breeding of, 142–143; sociality in, 572–574, 579; submissiveness in, 187, 188
Dolichonyx oryzivorous, 381, 460
Dolphins, 124
Dominance behavior, 132, 167, 176, 586, 587–596; and allopreening, 130–131; criticisms of, 591–594; hormones and, 151, 152–153; and mating success, 427, 439, 490–491, 588–590; and subordination, 156, 591
Dominions, 250, 294–295
Donkeys, feral, 490, 585
Doves, 506; ring, 170, 335; rock, 152, 440
Dragon, bearded, 243
Dragonflies, 227, 255, 315, 426, 428
Drongos, 124
Drosophilia, 408; *D. melanogaster*, 416, 435, 436, 437, 438
Ducks, 228, 238, 350, 383; common eider, 228, 238; monogamy in, 511–512; red-headed, 383; shoveler, 238
Duikers, 509, 555, 556
Dunlin, 284
Dytiscus marginalis, 237–238

Eagles, 560, 583; harpy, 507; monogamy in, 506; short-tailed, 240
Earthworms, 222, 233
Echidna, 229
Echinothuridae, 235
Eciton, 176, 520
Ecological niche, 200. *See also* Niche
Ectoparasites, 77–80, 129, 308
Ectopistes migratorius, 307, 311
Eel, garden, 264
Egg-laying behavior, 364–367; and brood parasitism, 383–390; and cannibalism, 441; and clutch size, 359, 361, 479–480; of fish, 372–373; and egg retention, 378–381
Egret, cattle, 328
Elands, 557
Elephants, defensive circles in, 116; helpers in, 104–105; optimal group size in, 578
Elk, 127–128, 572; fighting, costs of, 178; polygyny in, 488
Emberiza calandra, 382
Empididae, 443

Empimorpha comata, 443
Empis, 443, 444
Enhydra lutris, 229
Environmental influences, 15, 19, 20, 34, 35, 42, 44–45, 53, 57, 72, 195–245, 400–401; on aggression, 145–147
Enzymes, 42
Epinephrine, 153–154
Equidae, 585
Equus greyvi, 490; *E. africana*, 490; *E. hemionus*, 490
Erethizion, 229
Erichthonius braziliensis, 269
Erinaceus, 229
Erithacus rubecula, 148
Erythrocebus patas, 106
Escherichia coli, 403
Esox lucius, 121
Estrogen, 152–153
Eudynamis scolopacea, 389
Euphagus, 311; *E. cyanocephalus*, 315, 381
Eupoda montana, 502
Eupomacentrus leucostictus, 261
Eusociality, 516–550; indirect (kin) selection and, 524–526, 527–541, 549; parental manipulation and, 526, 542–549
Eutamias sonomae, 88
Evolutionarily stable strategies (ESS), 187–191, 393
Evolutionary theory, 5, 9; and adaptive value, 51; and aggression, 163–165; and cultural change, 599–600; and natural selection, 25–26; shifting balance theory, 68; and sociobiology, 6, 7, 14
Evylaeus species, 544
Eyespots, 242, 243

Falco columbarius, 112; *F. eleonorae*, 124; *F. peregrinus*, 337; *F. tinnunculus*, 227
Falcons, 122; Eleonora's, 124; peregrine, 337
Fecundity, 347–349, 360
Felis serval, 118
Ficedula hypoleuca, 119, 381
Filter-feeding, 222
Finches, 129, 512; bullfinch, 350; chaffinch, 127
Fish, 64, 182; alarm substances of, 81; angler, 223; benthic, 264; blennoid, 376; bowfin, 244; breeding sites of, 260–261; cichlid, 260, 366, 369, 377, 440; and cleaning symbioses, 77–80; egg-laying behavior of, 365, 366–367; fecundity of, 351; leaf, 441; man-of-war, 232; mating behavior of, 257, 258; morphological defenses of, 228–229; paradise, 144; parental care by, 368, 369, 372–378; pelagic, 366, 372; as prey, 244; reef, 257, 258, 260, 264, 272, 299; schooling of, 121; sex reversal in, 415; Siamese fighting, 143; territoriality of, 272–273; venomous, 237
Fitness, 45–47; Darwinian, 46–47, 49, 60; direct, 62; inclusive, 62–63; indirect, 62, 71; individual, 46
Flamingos, 305, 307, 506
Fleas, 308, 309
Flies, balloon, 443; cactus, 256; dung, 425, 426; fruit, 408, 416–417, 435, 436, 437, 438; hangflies, 443; house, 425; trypetid, 208
Flight, 232; erratic, 120–121, 232; "flight or fight" response, 154
Flocking behavior, 112, 121, 122, 124, 127–129
Flycatcher, McConnell's, 483; pied, 119, 381, 443
Food resources, 63–64, 556; availability of, 211–212; and coloniality, 312–320, 338–340, 342; cooperation and, 122–129; and courtship feeding, 442–443; defense of, 268–274; defensibility of, 287, 295, 298–299; distribution of, 206–210; and dominance, 590; and group size, 554–573, 578–579, 580–584; multipurpose territoriality and, 274–276; and niche space, 197–203; and parental care, 370; and plant-herbivore interactions, 213–221; and polygyny, 475–478, 488–489; and predator-prey interactions, 221–244; and promiscuity, 479–485; quality of, 212–213; storage of, 265, 270
Foraging, beater effect and, 123–124; central-place theory and, 338–340; and cooperation, 122–129; information, 127–129, 318–320; specialization vs. generalization, 197–199
Formica, 545, 546; *F. montana*, 531; *F. polyctena*, 522; *F. rufa*, 115, 269, 531; *F. subintegra*, 531
Founder effect, 66
Foxes, red, 104, 105, 326, 508, 572; arctic, 468; helpers in, 104–105
Fraser Darling effect, 329–335
Fratercula arctica, 311, 330
Fratricide, 64, 175
Fregata minor, 330
Frigatebirds, 505; great, 330
Fringilla coelebs, 127; *F. montifringilla*, 310, 311

Frogs, 359; female mating preference in, 439, 440; mating strategies in, 491–494; 510–511; predator defense in, 227, 238, 244; prolonged amplexus in, 425, 491; territoriality in, 262, 264–265
Frugivores, 314, 479, 481–482, 565

Gadus merlangus, 232
Galagos, 562
Gallus gallus, 146
Gannet, 175, 179, 329, 330, 332, 335, 505
Gasterosteus aculeatus, 121
Gastropods, 244
Gazella, 556; *G. granti*, 82, 85; *G. thomsoni*, 82, 85
Gazelles, 81, 111, 488–489, 556; Grant's, 82, 85; Thomson's, 82, 83, 85, 580
Gelada, 565, 566, 568, 569
Gemsbok, 557
General adaptation syndrome (GAS), 154–156, 594
Generalists, 197–199
Genes, 29; gene pool, 33, 34. *See also* Alleles
Genetic influences, 10, 12, 13, 601; on aggression, 142–144, 147; genetic drift, 66; genetic hitchhiking, 403–404; and instinct, 44–45; molecular genetics, 26–42
Genome, 5
Genotype, 31, 35–44
Gerenuk, 82, 556
Gibbons, 178
Gila monster, 236
Giraffe, 186, 584, 585
Glaucilla marginata, 235
Glaucus atlanticus, 235
Glossophaga soricinia, 328
Gnatcatcher, blue-gray, 202, 203
Goal-directedness, 48–49
Gobies, 260, 268; blind, 377
Gobius, 232
Gorgasia, 264
Gorillas, 578, 580
Goshawks, 87; chanting, 112
Gourami, blue, 374
Grackles, 473; common, 412; great-tailed, 122, 185, 382
Grain, resource, 209
Grasshoppers, 50, 228, 230, 244, 256
Grebe, silver, 334
Greenling, painted, 441
Grooming, 370; and allogrooming, 129–132
Ground squirrels, alarm calls in, 80–81, 88–90; arctic, 80, 324; Belding's, 88–90, 172, 173; California, 119; coloniality in, 305, 324–326; mobbing by, 119; round-tailed, 88
Group size, 574–587; dynamics of, 579–585; models of, 575–579
Grouse, 350–351, 431; black, 495; blue, 350, 486, 496; leks in, 257–259; promiscuity, 484–487; red, 92–95, 151, 275, 284; ruffed, 350–351, 486; sage, 259, 422, 495; sharp-tailed, 486; spruce, 486
Growth patterns: determinate, 350; indeterminate, 348, 361
Grunion, 367, 423
Gryllidae, 264
Grysbok, 555
Gulls, 119, 124, 443, 506; black-headed, 166, 328–329, 330; brown-hooded, 330; cannibalism in, 173–175, 308, 330; coloniality in, 305, 307, 328–329; courtship behavior of, 165–166, 169; Fraser Darling effect in, 329–335; glaucous-winged, 334; herring, 152, 174, 330, 334, 335; kittiwake, 125, 329, 330, 332, 335; lesser black-backed, 334; swallow-tailed, 330; western, 288
Guppies, 436
Gymnorhina, 382; *G. dorsalis*, 284; *G. tibicen*, 101, 284
Gymnorhinus cyanocephalus, 99

Habitat, 200; clumping and coloniality, 320–325; distribution model, 280–282; ephemeral, and plant or prey defenses, 213–214, 225; and life history, 352–353; limitation and polygyny, 465–470, 470–475; selection, 204–206; and viviparity, 379
Halichoerus grypus, 308, 468
Hamster, 153
Handicap principle, 100, 432–435
Hangflies, 443
Haplodiploidy, 27, 414, 524–526, 527–529
Haploidy, 27
Hare, snowshoe, 216
Harpia harpyja, 507
Hartebeests, 557; Coke's, 82, 489
Harvestfish, 232
Hawks, 314, 318, 321, 560; accipiter, 119; Cooper's, 336, 337; fratricide in, 175; goshawks, 87, 112; Harris, 112; red-tailed, 120, 273; territoriality in, 273, 298; tropical, 124; zone-tailed, 223
Hedgehog, 229
Heloderma suspectum, 236
Helogale parvula, 104

Helpers at the nest, 95–104
Hemachatus, 233
Hematopus, 229
Hemicentetes semispinosus, 229
Herding behavior, 121, 122
Herons, 19, 123, 128; black-crowned night, 151–152; coloniality in, 305, 307, 308, 309, 315, 318; great-blue, 315; monogamy in, 506; territoriality in, 265
Herrings, 366
Heterodon, 244
Heterodontidae, 237
Heteroeuthis, 233
Hilara species, 443, 444
Hilltopping behavior, 257
Hippocampus zosterae, 417
Hippotragus niger, 82
Hirundo rustica, 99
Hitchhiking, genetic, 68
Hodotermes mossambicus, 272
Home range, concept of, 247–249
Honeycreepers, 179; Hawaiian, 512
Honeyguides, 385, 386
Hormones: and aggression, 144, 149–156; female, 152–153; male, 149–152; in social insects, 547
Hornbills, 123, 124, 482; red-billed, 483
Hornet, oriental, 236, 545
Horns, 428
Horses, 488, 490, 585
Humans: aggression in, 144, 145–146, 159; brain of, 158; cultural evolution in, 599–611; and ecosystems, 3–4; endocrine system of, 150; and humanology, 6; sociobiology of, 6, 8, 9–12, 13
Hummingbirds, 298, 411; Anna's, 291; Guy's, 498; rufous, 275
Hyaena brunnea, 104
Hybridization, 435
Hyenas, 119, 560; brown, 104; sociality in, 571–574, 579, 580–582; spotted, 124, 214, 571–572
Hyla regilla, 439; *H. boans*, 264; *H. rosenbergi*, 264
Hylesia, 236
Hymenocera picta, 510
Hypnobius nebulosus, 367
Hypomesus pretiosus, 367
Hypotheses: evaluation of, 21–23; hypothetic-deductive logic, 17–18, 19
Hypsignathus monstrosus, 260, 496
Hypsoblennius gilberti, 268; *H. jenkinsi*, 268
Hypsopops rubicunda, 264

Ictalurus nebulosus, 231
Idarnes, 427
Iguana iguana, 180, 367
Iguanas, 180, 367; marine, 367
Iiwi, 290
Impala, 82, 489, 556
Inachis io, 257
Inbreeding, 50, 61, 410, 435–437, 586, 587
Incest, 436, 437
Infanticide, 171, 172–175, 468; human, 605–606
Insectivores, 487
Insects, cooperative foraging in, 125, 125–126; courtship feeding in, 443–444; cannibalism by, 269; colonial, 272; mating plugs in, 423–424; as prey, 314, 315; prolonged copulation in, 425–426; sex ratio deviations in, 407–411, sexual selection in, 423–424, 425–426, 428–429, 435–437, 443–444; social, 81, 115, 175–176, 269–272, 515–550; stick, 242; territoriality in, 254–257, 269–272. *See also* Social insects
Instinct, 44–45
Intimidation displays, 242
Ips grandicollis, 207
Iridomyrmex humilis, 532
Isogamy, 397
Istiblennius zebra, 268
Iteroparity, 347, 348

Jacana, American, 503–504
Jacana spinosa, 503
Jack, 121, 125
Jackal, black-backed, 104, 105
Jasus lalandei, 267
Jays, 329; blue, 240, 295; Florida scrub, 96, 101; piñon, 99, 315; Steller's, 295
Jellyfish, 232
Junco, 122; yellow-eyed, 112
Junco phaeonotus, 112
Jungle fowl, red, 146, 499

Kalotermes flavicollis, 548
Kestrel, 227
Kidnapping, 108
Kin selection, 60–63. *See also* Selection, indirect (kin)
Kingfisher, 96
Klipspringer, 86, 509, 555, 556
Kob, Uganda, 82, 259, 496, 497, 556, 559
Kobus, 556; *K. defassa*, 82; *K. ellipsiprymus*, 82; *K. kob*, 259, 496
Koel, Indian, 389
Kookaburra, 96, 100, 101
Kudu, greater, 82, 556, 559; lesser, 556
Kurtidae, 366

Labor, division of, 518
Labroides dimidiatus, 223, 414; *L. phthirophagus*, 78
Lagocephalus, 238
Lagopus lagopus scoticus, 92; *L. mutus*, 484
Langurs, Hanuman, 106, 107, 172, 173, 560, 565, 592; Nilgiri, 106; purple-faced or gray, 172, 565
Larus argentatus, 152; *L. furcatus*, 330; *L. fuscus*, 334; *L. glaucescens*, 334; *L. maculipennis*, 330; *occidentalis*, 288; *L. ridibundus*, 166, 328–329
Lasioglossum malachurum, 536; *L. rohweri*, 256; *L. zephyrum*, 530, 543
Leeches, 262
Leks, 53, 55, 259–260, 494–498, 559
Lemming, collared, 172, 173, 410; wood, 410
Lemur catta, 565; *L. fulvus*, 565; *L. macaco*, 584; *L. mongoz*, 562
Lemurs, 562; black, 584; brown, 565; lesser mouse, 563; mongoose, 562; ring-tailed, 565; sportive, 562
Leopards, 116–118, 560
Lepilemur mustelinus, 562
Lepomis cyanellus, 268; *L. gibbosus*, 261; *L. meglotis*, 441
Leptasterias hexactis, 358
Leptonychotes weddelli, 466
Leptothorax, 545
Lepus americanus, 216
Leuresthes tenuis, 367
Lice, 308, 309; wood, 228
Life, origin of, 31–34
Life history patterns, 345–362; bet-hedging theory, 356–359, 361; opportunistic strategy, 352–353, 356, 361; *r* and *K* selection theory, 354–356, 358–359, 361; reproductive trade-offs and, 348–351; sedentary strategy, 352–353, 356, 361
Ligurotettix coquilletti, 256
Limpets, 269; owl, 269
Lionfish, 238
Lions, 118, 119, 122, 170, 171, 178, 184, 560, 561; cooperative cub rearing in, 104–105, 574; cooperative hunting by, 124, 224, 570–571, 572, 573; infanticide in, 172–173; male coalitions in, 132–133, 488, 489; optimal group size in, 578–579, 580, 584; polygyny in, 488; sociality in, 570–571, 572, 573, 574, 580
Lizards, 151, 385; burrowing, 233; defenses against predation, 182, 228, 229, 233, 236, 242–243; distraction displays in, 242–243; egg-laying by, 365; iguanid, 276; life history patterns of, 359, 361, 362; monitor, 183; parental care in, 379; reproductive trade-offs in, 349; sagebrush, 359, 360, 362; scincid, 153; side-blotched, 359, 360; territoriality in, 274, 276; venomous, 236; viviparity in, 378–381, 417; Yarrow's spiny, 276, 359, 360
Lobipes lobatus, 419
Lobster, spiny, 267
Loligo vulgaris, 121
Longspur, Lapland, 350; Smith's, 512
Lophortyx californicus, 151
Lorises, 562
Lottia gigantea, 269
Loxia, 315; *L. curvirostra*, 442
Loxodonta africana, 104
Lucanus cervus, 429
Luscinia megarhynchos, 388
Luteinizing hormone (LH), 152
Lycaon pictus, 104
Lynx, 159
Lynx canadensis, 159
Lyrurus tetrix, 495

Macaca fuscata, 106; *M. mulatta*, 105; *M. nemestrina*, 588; *M. radiata*, 592
Macaques, 586; bonnett, 592; pig-tailed, 588–589. *See also* Monkeys
Machaeramphus alcinus, 337
Macrobdella decora, 262
Macroclemmys temmincki, 224
Macropodus operculus, 144
Macrotermes, 524; *M. natalensis*, 524
Madoqua, 556; *M. kirki*, 509, 555
Magicada, 225
Magpies, Australian, 101, 284, 382
Malurus cyaneus, 97
Mammals: aggression in, 182; coloniality in, 324–326; helpers in, 104–108; monogamy in, 508–510; parental care in, 104–108, 368, 509; polygamy in, 461–470, 487–491; sexual selection in, 428, 431, 437; sociality of, 553–596. *See also* Social mammals
Man-of-war, Portuguese, 235, 305
Manacus manacus, 496
Manakin, golden-headed, 495; white-bearded, 495–496
Mangabey, 565
Manis, 228
Mantids, 244
Mantises, praying, 222, 223
Marmosets, 509
Marmota flaviventris, 122, 324; *M. marmota*, 411

Marmots, 325; coloniality in, 324, 325–326; Olympic, 122, 325; polygyny in, 461–465; yellow-bellied, 122, 324, 325–326, 461–465
Marsupials, 424
Martin, house, 350
Mastigoproctus giganteus, 233
Mastophora, 223
Mating plugs, 423–424
Mating systems, 50, 449–513; of birds, 470–487, 498–508; 511–512; definitions, 449–452; of fish, 257, 260–261; of frogs and toads, 491–494; of insects, 64, 254–256, 529–534; and leks, 494–498; of mammals, 461–470, 487–491, 508–510; and monogamy, 504–512; and polyandry, 498–504; and polygyny, 452–479, 487–489; and promiscuity, 478–494; and territoriality, 254–261, 452–479, 488–489
Meerkat, grey, 176, 509
Megapodes, 365–366
Meiacanthus atrodorsalis, 237
Meiosis, cost of, 398
Meiotic drive, 408–409
Melanerpes formicivorous, 101
Melierax metabetes, 112
Melipona quadrifasciata, 531–532, 543
Melopstittacus undulatus, 335
Meme, 601
Mergansers, 123
Merlin, 112, 337
Merops bulocki, 103; *M. bulockoides*, 103
Methodology, 16–23; accepting hypotheses, 21–23; and advocacy, 18–19, 22; comparative method, 19–20; and crifical abilities, 21–23; experimental method, 20–21; hypothetic-deductive logic and, 17–18, 19
Mice, 145, 146, 153, 156; house, 81
Microcebus murinus, 563
Microtus arvalis, 227
Millipedes, 228
Mimicry, aggressive, 222–224; Batesian, 241–242; egg, 389
Mimus polyglottos, 295
Mirounga angustirostris, 308
Mischocyttarus drewseni, 233, 235
Mites, 408, 411
Mobbing, 115–120, 177, 389–391
Mockingbird, 295
Models, role of, 16
Molecular genetics, 26–42
Moles, 226, 232
Molluscs, 222, 228, 243, 351
Moloch horridus, 229
Molothrus ater, 384; *M. badius*, 383
Monachus schauinslandi, 467
Mongooses, 240; banded, 104, 229; dwarf, 104, 105, 508–509
Moniliformis dubius, 423
Monkeys, 124, 509; alloparental behavior of, 104–108; colobus, 561, 563–564; Japanese, 106, 108, 151; night, 509, 563; patas, 106, 488; red-tailed, 561, 580; rhesus, 105–106, 107, 132, 145, 146, 147, 156, 157, 159, 160, 437, 490–491, 587, 590, 592, 593; sociality in, 560–565; spider, 582–584, 586; squirrel, 151; Syke's, 584; titi, 509; vervet, 80–81, 106, 274, 580, 581, 594, 595
Monogamy, 20, 429–430, 504–512; and aggression, 464, 508–512; in arthropods, 510; in birds, 449, 482, 505–508, 511–512; definition of, 449–450; evolution of, 504–512; in frogs, 510–511; in mammals, 469, 508–510
Monogyny, in ants, 272, 533
Monomorium, 546
Moose, 572
Moraba scurra, 50
Morphological defenses, 228–232; weaponry, 428
Motacilla alba, 273
Moths, 223, 236, 238, 242, 425, 426
Mouse lemur, lesser, 563
Mungos mungo, 104
Murres, 262, 307; common, 99, 311; thick-billed, 311
Mus musculus, 81
Musca domestica, 425
Muskoxen, 115–116, 117, 178
Muskrat, 265
Mutations, 28, 30, 33–34, 43, 55, 403, 404, 407, 434
Myadestes townsendi, 275
Mycocepurus goeldii, 531
Myopus schisticolor, 410
Myrmecocystus mimicus, 270
Myrmeleon, 222
Myrmica, 545, 546; *M. rubra*, 521; *M. ruginodis*, 546
Myzomela, 179

Naja, 233
Nasonia vitripennis, 410
Nasua narica, 119
Nasutitermes lujae, 523
Nature-nurture dichotomy, 44–45
Nectarina, 179; *N. reichenowi*, 275
Nemobius sylvestris, 226

Neocapritermes, 524
Neodiprion pratti, 126
Neophron percnopterus, 230
Neotomys, 237; *N. fodiens*, 237
Neotragus moschatus, 555
Neurology, of aggression, 156–160
Niche, ecological, 197–205; defined, 199–203; fundamental, 201–202; hypervolume model, 200–202; realized, 202; response surfaces, 202–203; space, 199–203
Nightingales, 388
Nomeus gronowi, 232
Norepinephrine, 153–154
Nothobranchius guentheri, 438
Nucula proxima, 207
Nurseryfish, 366, 376
Nycticorax nycticorax, 151–152

Objectiveness, 18
Oceanodroma castro, 330; *O. leucorrha*, 505; *O. oceanicus*, 311
Ocelot, 583
Octopuses, 235
Odocoileus hemionus, 82; *O. hemionus columbianus*, 81, 82; *O. virginianus*, 82, 84–85
Odontoloxozus longicornis, 256
Odontomachus, 546
Odontotermes redemanni, 272
Oecophylla longinoda, 269
Ogcocephalus, 223
Oilbird, 320, 321, 482, 506
Olfactory signals, 81
Oligomyrmex, 520
Ondatra zibethicus, 265
Ontogeny, 5
Onychognathus morio, 179
Onymacris rugatipennis, 426
Operon, 37, 39
Ophioblennius atlanticus, 268
Ophiophagus hannah, 367
Opossum, North American, 157, 244
Opportunistic strategy, 352–353, 356, 361
Oratosquilla inorta, 264
Orcinus orca, 124
Oreotragus oreotragus, 86, 556
Oribi, 509, 555, 556
Ornithorhynchus paradoxus, 237
Oronectes virilis, 267
Oropendolas, 382, 473
Oryx, 557
Oryx, 557
Ostariophysi, 81
Ostrich, 498
Otis tarda, 497; *O. tetrax*, 497
Otter, Asiatic clawless, 509; sea, 229
Ourebia ourebi, 509, 555, 556
Ovenbird, 275
Ovibus moschatus, 115
Oviparity, 364–367
Ovis canadensis, 82, 393, 491; *O. dalli*, 345; *O. dalli stoni*, 82
Owlfly, social, 115, 116
Owls, 175, 273; barn, 86–87; great-horned, 337; short-eared, 175
Oxylebius pictus, 441
Oystercatchers, 229

Pachysima, 236
Panesthia, 510
Pangolins, 228
Panthera leo, 105; *P. pardus*, 116
Papio cynocephalus, 118; *P. hamadryas*, 184; *P. ursinus*, 118
Paradisiaea apoda, 497; *P. minor*, 497
Pararge aegeria, 257
Parasites, 3, 77, 197; brood parasitism, 383–390; ectoparasites, 77–80, 129, 308; parasitic wasps, 233, 410, 411, 414; parasitic worms, 423
Paravespula germanica, 545
Parental behavior, 75, 177, 362–394; aggression in, 170–175; alloparental behavior, 105–108; in amphibians, 367–368; in birds, 365–366, 368, 381–383, 505–506; brood parasitism and, 383–390; fecundity and, 351; in fish, 365, 366–367, 368, 372–378; by males, 20, 381, 383, 440–441, 454, 481, 482, 498, 499, 502, 505–506, 509; in mammals, 10–108, 368, 509; mouth-brooding, 377; parent-offspring conflict, 390–394; parental manipulation, 392–394, 526, 536, 540, 542–549; in polygynous birds, 381–383; reproductive trade-offs and, 345, 351, 353; in reptiles, 365, 367, 368; shareable vs. nonshareable, 369–371
Parental investment, 362–364; shareable vs. nonshareable, 369–371; and viviparity, 367–368, 378–381
Parrots, 508
Partridges, 503
Parus caeruleus, 442; *P. major*, 127, 282; *P. ater*, 282
Passer domesticus, 56
Passerculus sandwichensis, 382; *P. sandwichensis princeps*, 382, 460
Passerina cyanea, 382, 460
Patas erythrocebus, 106, 488
Patella, 269

Paternity, confidence of, 374–375, 419
Patchiness, 208–209
Pavo cristatus, 242
Peacock, 242; peafowl, 499
Peccaries, 488
Pecking order, 151
Pecking responses, 45
Pedioecetes phasianellus, 486
Pelecanus erythrorhynchus, 306; *P. onocrotalus*, 306
Pelicans, 123, 506; coloniality in, 305, 306, 307; cannibalism in, 173, 308; fratricide in, 175; white, 306, 328, 330, 334, 342, 506
Penguins, 506; adelie, 328, 333, 334; emperor, 505; king, 350, 505
Peprilus alepidotus, 232
Perenty, 183
Pericapritermes, 524
Periclemenes pedersoni, 78
Pernis apivora, 240
Petrels, 322; diving, 505; Leach's, 505; storm, 330, 505; Wilson's storm, 311
Petrochelidon pyrrhonota, 328
Phaethon aethereus, 262
Phaethornis guy, 498
Phainopepla, 314, 319
Phainopepla nitens, 314
Phalacrocorax aristotelis, 330; *P. bouganvillei*, 310, 311
Phalaropes, 418–419; northern, 419
Phasianus colchicus, 350
Phasmidae, 242
Pheasants, 488; ring-necked, 350
Pheidole, 520; *P. dentata*, 115; *P. sitarches*, 532
Phenotype, 31, 35–44
Pheromones, 81, 115
Philanthus bicinctus, 255
Philomachus pugnax, 53, 55
Phoca hispida, 466; *P. vitulina*, 467
Phoeniculus purpureus, 99
Phrynelox scaber, 223
Phyllocrania paradoxa, 223
Phyllostomus bastatus, 260; *P. discolor*, 260, 328
Physalaemus nattereri, 242, 243; *P. pustulosus*, 440
Physalia utriaulus, 235
Pigeon, passenger, 307, 310, 311, 315–316, 342; wood, 112, 590. *See also* Doves
Pig, 176
Pike, 121
Pinnipeds, 428; polygyny in, 465–470
Pipefish, 366, 376
Pipra erythrocephala, 495
Pipramorpha macconnelli, 483
Pison xanthopus, 241
Pithecia, 509
Plagiolepsis, 545, 546; *P. pygmaea*, 545
Plaice, 229
Plant-herbivore interactions, 213–221
Plant hopper, 244, 245
Platymeris rhadamantus, 233
Platypus, 237
Plautus alle, 311
Plectrophenax nivalis, 350
Pleiotrophy, 42–43
Pleuronectes, 229
Ploceinae, 382
Ploceus cucullatus, 330, 473
Plover, mountain, 502
Podiceps occidentalis, 334
Poecilia reticulata, 436
Pogonomyrmex, 270, 271, 531; *P. badius*, 236
Polioptila caerulea, 202
Polistes, 533–534, 535, 545; *P. fadwigae*, 518; *P. fuscatus*, 539, 546; *P. gallicus*, 534–535
Polyandry, 450, 451, 498–504; successive, 418
Polycentrus schomburgkii, 441
Polygamy, 429, 451
Polygenic inheritance, 43
Polygyny, 381, 420, 439, 449–450; in ants, 272; definition of, 449–450; evolution of, 452–478; harem, 487–489; in passerine birds, 470–478; in pinnipeds, 465–470; and polyandry, 498–503; polygyny threshold hypothesis, 454–460, 463–465; in social insects, 272, 533; territorial, 452–478, 488–489; in yellow-bellied marmots, 461–465
Polyps, coral, 222
Pomacentrids, 273, 299
Pomacentrus lividus, 273
Pomatostomus temporalis, 96
Population, 51, 64; ecology, 6, 7; fitness, 46; growth rates, 354–356; metapopulations, 66; regulation, 92, 93, 276–285; and sex ratio, 409; and territoriality, 93, 276–285
Porcupinefish, 242
Porcupines, 229, 573
Poronotus triacanthus, 232
Prairie chickens, greater, 487, 495; lesser, 486
Prairie dogs, 305; black-tailed, 88, 308, 309, 326; coloniality in, 324–325; white-tailed, 308, 309, 326
Pratincoles, 506
Predation, 3, 78, 195, 197, 211, 560, 561,

570–574; and aggression, 137–139; alarm signals and, 80–84; camouflage and, 230–232; chemical defenses and, 233–241; clumping and, 121–122, 227–228; and coloniality, 307–308, 325–326, 328–337; cooperation against, 111–122; cooperative hunting, 124–126, 224–225; deceiving predators, 241–244; defenses against, 80–84, 111–122, 225–244; of eggs, 262, 263, 366–367, 373, 375; flight from, 120–121, 232; hunting tactics, 124–126, 222–225; and inaccessibility of prey, 232–233; intimidating predators, 242; morphological defenses against, 228–232; and polygyny, 452, 453, 463; predator-prey interaction, 221–244; and sociality, 111–122, 124–126, 224–225, 553–574, 580–586; and spatiotemporal escape mechanisms, 225–228; synchronous breeding and, 225–227, 329–332
Pregnancy blockage, 424–425
Prenolepis imparis, 531
Preroosting aggregations, 64
Presbytis entellus, 106; *P. johnii*, 106; *P. senex*, 172
Primary sexual characteristics, 421
Primates, 159, 428; allogrooming in, 131–132; alloparental behavior in, 105–108; dominance in, 588–590, 592–594; group size in, 578, 580–584, 586; kin groups of, 90; male coalitions in, 132; mating systems of, 487, 491, 508; mobbing by, 116–119; monogamy of, 509; social roles in, 594–596; sociality of, 560–570; territoriality in, 274
Prolonged copulation, 425–426, 491
Progesterone, 152–153
Promiscuity, 450–451, 478–494; arena, 490; in birds, 479–487; definition of, 451; female desertion hypothesis, 484–485; evolution of, 478–494; hierarchical, 490; and leks, 494–498; male emancipation hypothesis, 479–481, 481–483; in mammals, 487, 490–491; sexual bimaturism hypothesis, 485–487; types of, 450–451
Protobionts, 33
Proximate causation, 4–5
Psaltriparus minimus, 87
Pseudergates, 523, 547, 548
Pseudoaugochloropsis, 517
Pseudaletia unipuncta, 426
Pseudomyrmex, 236, 532
Ptarmigan, rock, 484
Pterois volitans, 237
Pufferfish, 238, 242
Puffin, common, 311, 330, 505
Puffinus gravia, 311; *P. puffinus*, 505
Puku, 556
Pupfish, 261
Pygoscelis adeliae, 328
Pyrrhula pyrrhula, 350
Pythons, 560

Quail, 503, 512; California, 151
Quelea quelea, 112, 311
Quelea, red-billed, 112, 152, 305–306, 311, 342
Quiscalus mexicanus, 382; *Q. quiscula*, 412. *See also Cassidix mexicanus*

r and *K* selection theory, 354–356, 358–359, 361
Rabbits, 81, 146, 572, 588
Raia, 229
Rana catesbeiana, 262
Random segregation, Mendelian theory of, 413
Rangifer arcticus, 491; *R. tarandus*, 82
Raphicerus campestris, 555; *R. melanotis*, 555
Ratite birds, 498
Rats, 144, 146, 153, 156, 157, 158, 159, 588; brown Norway, 144; giant, 176
Rattus norvegicus, 144
Raven, 125
Rays, 229
Reciprocation, 76–80, 90–91, 100, 131
Recurvirostra americana, 273
Redunca, 556; *R. redunca*, 510
Reedbucks, 556; southern, 509, 510, 555
Regulon, 39
Releaser, 149
Removal experiments, 283–285, 486–487
Replication, DNA, 30–31, 34
Reproduction, sexual, 397–405
Reproductive curtailment, 92–95, 99–100
Reptiles, egg laying by, 367; fecundity in, 351; parental care in, 365, 367, 368; territoriality in, 274, 276; viviparity in, 378–381. *See also* Lizards, Snakes
Resources, avilability of, 211–212; characteristics, 206–213; distribution of, 206–210; quality of, 212–213
Rhamphomyia, 443
Rhea, 498–501
Rhea americana, 498
Rhinoceros, 573
Rhinolophus hipposideros, 327–328
Rhodocerca rhamni, 425
Rhynchops nigra, 330

Riparia riparia, 319
Rissa tridactyla, 125
Roach, soil-burrowing, 510; wood, 510
Robin, American, 295; European, 148
Rockskipper, Hawaiian, 268
Rodents, 424–425, 487, 572, 578
Rollandia rolland, 334
Royal jelly, 542
Ruff, 53–54, 55, 56, 57–60, 495, 497
Rump patches, 83, 84–86
Rupicola rupicola, 496

Saccopteryx bilineata, 260
Sagittarius serpentarius, 240
Saimuri sciureus, 151
Saki, 509; bearded, 509
Salamanders, 238, 262, 349, 359, 361, 362, 367; California slender, 359; mountain, 359; mountain, 427
Sample sizes, 22
Sanderling, 275, 288
Sandpipers, 222, 501–502; buff-breasted, 497; spotted, 502
"Satisfaction, bias of," 607
Sawfly, jack-pine, 126
Scatophaga stercoraria, 425
Scattering, explosive, 120–121
Scavenging, cooperative, 573
Sceloporus jarrovi, 276; 359, 360; *S. graciosus*, 359, 360; *S. undulatus*, 361
Schooling behavior, 64, 121
Scolytus, 510
Scorpionfish, 237
Sculpin, mottled, 261, 440
Sea lions, 305, 322–323, 466
Sea slugs, 229, 235
Sea stars, 358
Sea urchins, 84, 235, 351
Sea wasp, 235
Sea birds, 104, 130, 305, 507
Seahorses, 366, 376, 417; pygmy, 417
Seals, 178; coloniality in, 322–323; elephant, 305, 308–309, 328, 422, 439, 466–468, 588; fur, 305, 307, 322, 466–468; grey, 308, 468; harbor, 467; monk, 467; polygyny in, 465–470; pup-killing in, 308–309; ringed, 466, 468–470; Weddell's, 466, 468–470
Searciid fish, 233
Secondary sexual charcteristics, 420–421, 428–429
Secretary bird, 240
Sedentary strategy, 352–353, 356, 361
Seiurus avrocapillus, 275
Selaphorus rufus, 275
Selection, cultural 602–604, 607–611; direct, 61; directional, 53–54; disruptive, 57–60; frequency dependent, 57–60, 421; genic, 26, 35, 60; group 8–9, 26, 63–72; indirect (kin) 61, 87–90, 95–99, 108, 132–133, 515, 524–525, 527–542, 549–550; individual selection, 26, 35, 53–60, 400–403; interdemic selection hypothesis, 399–400; interdemic selection models, 64–72; intrademic group, 69–71; intersexual, 421, 431–444; intrasexual, 421, 422–431; kin, 60–63, 71, 390–391; molecular view of, 31–44; natural, 5, 8–9, 14, 15, 25–72, 75, 420–421, 515, 605–607; *r* and *K* selection theory, 354–356, 358–359, 361; runaway selection, 431–432, 433; sexual, 421–444; stabilizing, 54–57; trait group, 69–71, 410–411; types of individuals, 53–60
Selfish behavior, definition of, 75
Semelparity, 347, 348
Sepia officinalis, 121
Sex, evolution of, 397–405
Sex ratio, in alligators, 414; balanced, 405–407, 430, 453–454; in ducks, 511; and differential mortality, 306, 350; in eusocial insects, 393, 414, 537–542; evolution of, 405–415, 541; in grouse, 350, 485; local mate competition and, 407–410, 541; in mammals, 411, 412, 413; and monogamy, 510, 511, 512; and polygyny, 453–454; in quail and snowcocks, 350, 512; sexual selection and, 431; skewed, 306, 350, 407–415, 431, 485, 510, 511, 512, 537, 541; in songbirds, 306, 350, 412, 512; trait-group selection and, 410–411; Trivers-Willard hypothesis and, 411–414
Sex reversal, 415
Sex role reversal, 414–415, 417–419
Sexual bimaturism, 431, 485–487
Sexual dimorphism, 53, 428, 429, 431, 465, 470, 486
Sexual selection, 420–445; intersexual selection, 431–444; intrasexual selection, 422–431
Sexuality: and aggression, 165–170; anisogametic sex, 398, 404–405; and courtship, 442–444; evolution of, 397–405; female competition and male choice, 417–420; isogametic sex, 444, 445; male competition and female choice, 415–417; primary sexual characteristics, 421; secondary sexual characteristics, 420–421, 428–429; and selection, 420–445
Sharks, 237, 367

Shearwaters, 322, 505; manx, 505; great, 311
Sheep, bighorn or mountain, 82, 93, 491, 492, 572; Dall, 345–347; stone, 82
Shelter, 213; defense of, 265–268
Shorebirds, 501–502
Shrews, 237; short-tailed, 237; water, 237
Shrike, yellow-billed, 96
Shrimp, 233; alpheid, 268; cooperation in, 510; mantis, 264, 267; monogamy in, 510; Pederson's cleaner, 78; starfish-eating, 510, 511; territoriality in, 264, 267, 268
Sign stimulus, 149
Signigobius biocellatus, 260
Siluriformes, 81
Sitatunga, 556
Skimmer, 506; black, 330
Skua, 175, 443
Slugs, sea, 229
Snails, 84, 228, 229
Snakes, 119, 222, 240, 241, 560; cobras, 233, 367; hognose, 244; parental care in, 365; and sexual selection, 428; tail-twitching in, 242–243; venomous, 233, 236, 240; viviparity in, 379
Snowcocks, 512
Social insects: caste differentiation in, 542–549; degrees of sociality, 515–517; differential altruism of workers in, 534–535; haplodyploidy and, 527–529; indirect (kin) selection theory for eusociality, 524–526, 527–542; investment ratios in, 537–542; and male reproductives, 535–536; multiple insemination in, 529–532; number of mothers, 532–534; parental manipulation theory for eusociality, 526, 542–549; queen control in, 542–549; sterile caste evolution in, 524–526
Social mammals: dominance in, 586; 587–596; group size in, 574–587; carnivores, 124–126, 224–225, 570–574; mating systems in, 487–491; primates, 560–570; territoriality in, 274, 488–489; ungulates, 554–560. *See also* Mammals
Sociality, 304, 515–517; and caste differentiation, 517–524; human, 599–611; insect, 515–550; mammalian, 560–574
Socialization, 606
Sociobiology, 5–23; controversy in, 9–15; criticisms of, 9–15; historical background, 8–9; and human behavior, 6, 8, 9–12, 599–611; methodology of, 16–23; political criticisms of, 9–12; scientific criticisms of, 12–15; Study Group, 9–10, 11, 13; tenets of, 15
Solenodon, 237
Solenopsis geminata, 115; *S. saevissima*, 236; *S. xyloni*, 236
Solitaire, Townsend's, 275
Somateria mollisima, 228
Sparrows, field, 281; house, 56, 127, 350; Ipswich, 382, 460; rufous collared, 284; Savanna, 382; sharp-tailed, 483; tree, 278; white-crowned, 142, 143, 151
Specialists, 197–199
Specius speciosus, 255
Sperm competition, 422–426
Spermophilus beecheyi, 119; *S. beldingi*, 88–90; *S. tereticaudus*, 88; *S. undulatus*, 80
Sphenomorphus kosciuskoi, 151, 153
Spiders, 207, 222, 244; ant-mimicking, 223, 241; bola, 223; cooperation in, 126
Spiteful behavior, 75, 76
Spiza americana, 281, 382
Spizella arborea, 278; *S. pusilla*, 281
Sprays and secretions, as defense against predators, 233
Springbok, 82, 556
Squalidae, 237
Squid, 121, 233
Squirrels, free, 273–274, 276. *See also* Ground squirrels
Starfish, 229, 235, 244, 351, 511; sea stars, 358
Stargazers, 237
Starlings, 152; European, 112, 121–123, 152, 307, 311, 335, 339, 342; mutual vigilance in, 112, 122–123; predator defenses of, 112, 121, 335; red-winged, 179; roosting colonies of, 307, 311, 335, 339, 342
Steatornis caripensis, 482
Steenbok, 555, 558
Sterile castes, 515, 524–526
Sterna fuscata, 311, 329; *S. sandivicensis*, 330
Stickleback, three-spined, 121, 441
Stimulus filtering, 149
Stolephorus purpureus, 121
Stomatopoda, 267
Stonefish, 237, 239
Stork, 123, 128, 307; infanticide in, 171–172; monogamy in, 506; territoriality in, 265; white, 171–172, 265
Stotting, 83, 85
Streptopelia risoria, 170
Stress, 154–156
Struthio camelus, 498
Sturnus vulgaris, 112

Submission: and aggression, 186–187
Sula bassana, 179; *S. nebouxii*, 306; *S. sula*, 505; *S. variegata*, 306
Sunbirds, 179; golden-winged, 275, 288–289
Sunfish, green, 268; longear, 441; pumpkinseed, 268
Suni, 555
"Supermale" syndrome, 144
Surgeonfish, 228, 273, 366
Suricata suricata, 176
Surniculus lugubris, 387
Swallows, 336; bank, 319, 329, 330; barn, 99; cliff, 328, 329; coloniality in, 305, 314, 319
Swifts, 314, 337; chimney, 336
Swine, 488
Sylvatic plague, 308
Sylvicapra, 555
Symbioses, 77–80, 268
Symphysodon aequifasciata, 369
Synanceja horrida, 237
Syncerus caffer, 116
Synchrony, breeding, 225–227, 329–335
Syngnathidae, 366
Syntermes, 524

Tachyglossus, 229
Tadarida brasiliensis, 311
Tail flashing, 84–86
Talapoin, 584
Tamarins, 509
Tamiasciurus, 273
Taurotragus, 557
Teat order, 176
Temnothorax, 546
Tenrec, streaked, 229
Termes, 524
Termites, 510, 527, 539; caste differentiation in, 523–524, 547–549; cannibalism in, 175–176; chemical defense in, 115; eusociality, 528–529, 547–549; territoriality in, 269, 272
Terns, 173, 443, 506; coloniality in, 305, 311, 322, 329–331; monogamy in, 505; sandwich, 330; sooty or wide-awake, 311, 329, 330, 331
Terrapins, 228
Territoriality, 64, 167, 183, 208, 247–300; aggression and, 148–149, 151, 165, 184–185, 250, 252–253; breeding sites, defense of, 260–263; competition and, 252–254; definition of, 249–251; defensibility and, 253–254, 285–295; dominions and, 250, 294–295, 296; evolution of, 252–284; exclusiveness of defense, 293–295; food resources, defense of, 268–274; history of concept, 251–252; home range and, 247–249; interspecific, 295–299; of invertebrates, 254–257, 261, 267–272; mating stations, defense of, 254–260; multipurpose territories, 274–276; and parental care, 260–261, 297, 376; and polygyny, 452–478, 488–489; and population regulation, 276–285; shelter sites, defense of, 265–268; super-territory hypothesis, 291–293; territory size, 285–293; time and energy investments, defense of, 264–265; of vertebrates, 257–265, 266, 272–276
Testosterone, 94, 149–152
Tetramorium, 535
Tetraodontidae, 242
Theopompella westwoodi, 223
Theropithecus gelada, 565
Threat, 182, 184, 428–429
Thrush, song, 229
Tigers, 560, 571, 573
Titmice, 127, 282, 295; opening milk bottles, 127
Tits, blue, 442; coal, 282; great, 127, 128, 282–283, 350, 442; long-tailed, 327
Toadfish, 237
Toads, 81, 238–239, 244, 440, 491–494; Brazilian, 242, 243; European bell, 239; midwife, 367
Tockus erythrorhynchus, 483
Tolypeutes, 228
Topi, 557
Tortoises, 228
Toucans, 482, 506
Toxins, 218–219, 224, 233–241, 385
Tragelaphus, 556, 557; *T. strepsiceros*, 82
Trait-group selection, 69–71; and alarm calls, 69–71; 91; and distorted sex ratios, 410–411
Transcription, of RNA, 36–37
Translation, of RNA, 37
Tree shrew, 176
Tribolium confusum, 66
Trichogaster trichopterus, 374
Trigona postica, 536
Trinervitermes trinervoides, 272
Trionyx hurum, 242
Troglodytes aedon, 382, 477; *T. troglodytes*, 275, 382
Tropicbirds, 505; red-billed, 262
Trout, 268
Tryngites subruficollis, 497
Trypoxylon placidum, 241
Tsessebe, 557

Tupaia belangeri, 176
Turdoides squamiceps, 99; *T. striatus*, 101
Turdus merula, 350; *T. migratorius*, 295; *T. philomelos*, 229
Turtles, 228, 239; alligator snapping, 224; Burmese soft-shelled, 242
Tylonycteris pachypus, 260; *T. robustula*, 260
Tympanuchus pallidicinctus, 486; *T. cupido*, 487
Typhlogobius californensis, 377
Typhoeus typhoeus, 510
Tyto alba, 86–87

Uca, 182
Ultimate causation, 6
Ungulates, 119, 122, 428, 588; aggression in, 177, 186; alarms of, 81, 84; herds, 90; mating systems of, 274, 488–490, 497; mutual vigilance of, 112; optimal group size in 578, 580, 584–585; social, 554–560; territoriality in, 274
Urate, nodules, 444, 445
Uria aalge, 99, 311; *U. lomvia*, 311
Ursus chelan, 326; *U. maritimus*, 468
Uta stansburiana, 359, 360

Vampyrum spectrum, 508
Varanus giganteus, 183
Venoms, 3, 233–237, 239–241
Vespa orientalis, 236
Vespula, 518, 536
Vestiaria coccinea, 290
Vigilance, mutual, 111–120, 122–123, 325, 463
Vireo, Philadelphia, 297; red-eyed, 297
Vireo olivaceous, 297; *V. philadelphicus*, 297
Viverra civetta, 240
Viviparity, 367–368; evolution of, in reptiles, 378–381
Vole, common, 227
Vonomes sayi, 233
Vulpes fulva, 326; *V. vulpes*, 104
Vultures, 318, 506; black, 223; Egyptian, 229–230

Wagtail, white, 273, 384
Wapiti, 82; *See* Elk
Warblers, great reed, 477; Kirtland's, 277; prairie, 382, 460
Wasps, 115, 233, 235, 393; caste differentiation in, 518; cicada killer, 255; distorted sex ratios in, 408, 410–411, 414–539; eusociality in, 534–535, 536, fig, 408, 410, 411, 427; and mimicry, 241; paper, 518, 533–535, 539, 545, 546; parasitic, 233, 410, 411, 414; as prey, 240; queen control in, 543, 545, 546; semisociality and, 533–534; sphecid, 255; stings of, 235; territoriality in, 255
Water fleas, 121
Water resources, 212
Water turkey or anhinga, 309
Waterbucks, 489, 556; common, 82; defassa, 82
Waxwings, 483; cedar, 389
Weaning, 170, 391–392, 468
Weaver finches, 382, 385; black-headed or village, 330, 473; coloniality in, 305; polygyny in, 472–473; roosts of, 322, 330. *See also* Quelea
Weaverfish, 237
Whales, 178, 578; baleen, 222; killer, 124
Whipscorpion, 233
Whiting, 232
Whydah, Jackson's, 483, 497
Wildebeests, 177, 227, 557, 571, 580
Wolf, 115, 156, 177, 468; helpers in, 104–105; monogamy in, 508; sociality in, 572–574, 579
Wolf, bumblebee, 255
Woodchuck, 411
Woodhoopoe, green, 99, 100, 103, 104
Woodpeckers, 119, 298, 299; acorn, 101, 103, 265, 266
Worms, acanthocephalan, 423; bloodworms, 235; bristleworms, 235; earthworms, 222, 233; marine, 351; polychaete, 235, 243
Wrasses, 366
Wrens, European or winter, 382, 477; house, 382, 477; long-billed marsh, 382; marsh, 474; superb blue, 97–98; winter, 275

Xanthocephalus xanthocephalus, 184, 298, 382
Xyleborus, 528
Xylocopa, 256

Zabilius aridus, 230
Zatapinoma, 520
Zebrafish, 237, 238
Zebras, 111, 488, 490, 571, 573, 580, 585; Grevy's, 490, 585
Zonotrichia capensis, 284; *Z. leucophrys*, 142, 143
Zooplankton, 64
Zootermopsis nevadensis, 523

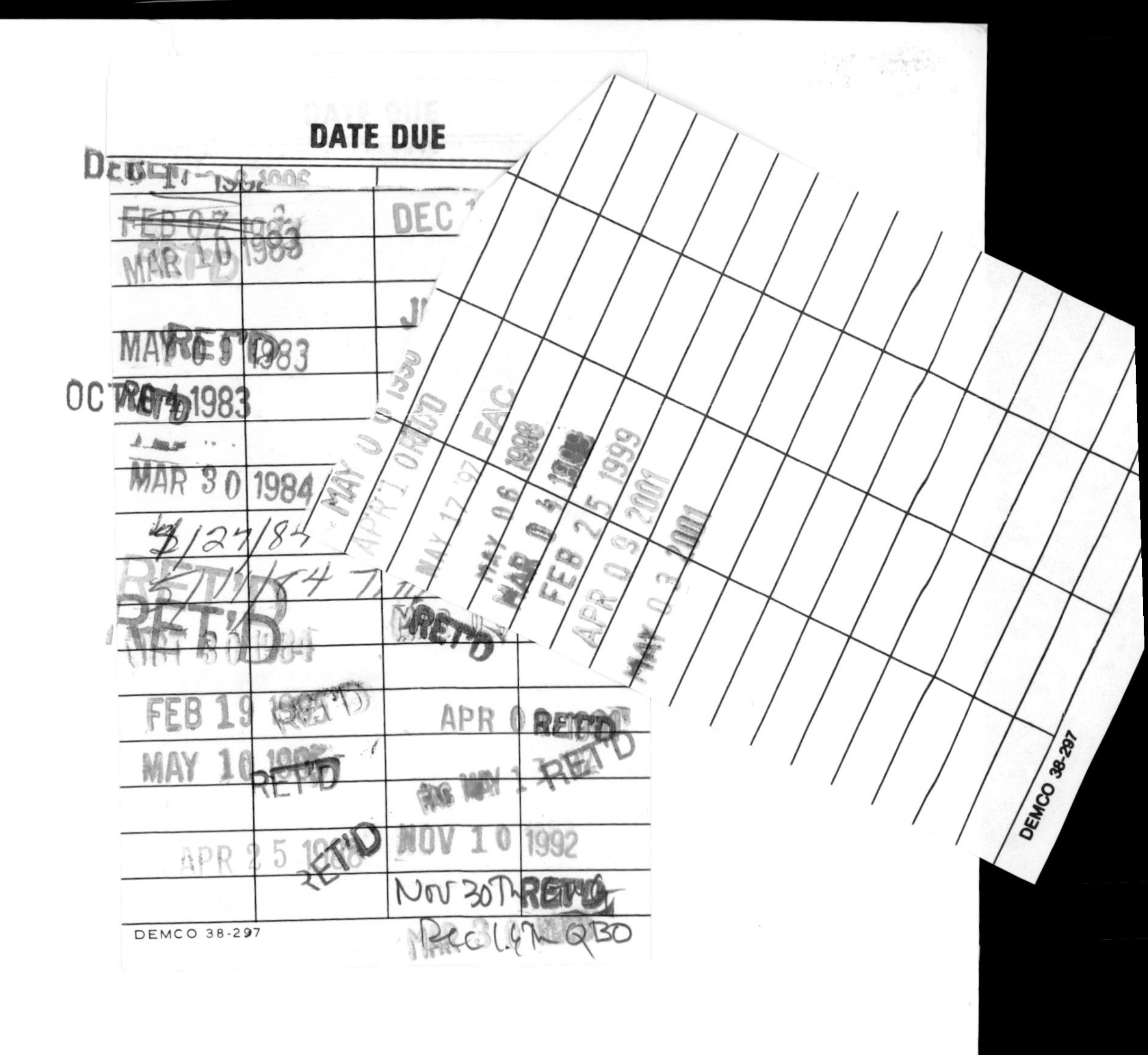
DATE DUE
DEMCO 38-297
DEMCO 38-297